Birds

o f C h i n a

LIU Yang and CHEN Shuihua

Translated and edited by

CAI Shangxiao, YANG Ziyou, LIANG Dan,

MU Tong, and Paul Holt

Princeton University Press

Princeton and Oxford

Published by Princeton University Press
41 William Street, Princeton, New Jersey 08540
99 Banbury Road, Oxford OX2 6JX

press.princeton.edu

Library of Congress Cataloging-in-Publication Data
Names: Liu, Yang (Ornithologist), author. | Chen, Shuihua, author. |
    Cai, Shangxiao, translator, editor. | Yang, Ziyou, translator, editor. |
    Liang, Dan, translator, editor. | Mu, Tong, translator, editor. | Holt,
    Paul I., translator, editor.
Title: Birds of China / Liu Yang and Chen Shuihua ; translated and edited
    by Cai Shangxiao, Yang Ziyou, Liang Dan, Mu Tong, and Paul Holt.
Other titles: Zhongguo niao lei guan cha shou ce. English
Description: Princeton : Princeton University Press, 2024. | "Originally
    published as The CNG Field Guide to the Birds of China by Liu Yang and
    Chen Shuihua; Text Copyright © Chinese National Geography Intellectual
    Property Co., Ltd." | Includes bibliographical references and index.
Identifiers: LCCN 2022056479 | ISBN 9780691237527 (paperback)
Subjects: LCSH: Birds—China—Identification. |
    Birds—China—Classification. | BISAC: NATURE / Birdwatching Guides |
    SCIENCE / Life Sciences / Zoology / Ornithology
Classification: LCC QL691.C5 .L589 2023 | DDC 598.0951—dc23/eng/20230223
LC record available at https://lccn.loc.gov/2022056479

British Library Cataloging-in-Publication Data is available

Editorial: Robert Kirk and Megan Mendonça
Production Editorial: Kathleen Cioffi
Cover Design: Benjamin Higgins
Production: Ruthie Rosenstock
Publicity: William Pagdatoon and Caitlyn Robson
Copyeditor: Charles J. Hagner
Cover image: ZHENG Qiuyang

This book has been composed in Myriad Pro and Song SC (headings)

Printed on acid-free paper. ∞

Typesetting of English language edition: D & N Publishing, Wiltshire, UK

Printed in Malaysia

10 9 8 7 6 5 4 3 2 1

Special Sponsors

Beijing Entrepreneur
Environmental
Protection Foundation

Mangrove Conservation
Foundation

Changjiang
Conservation
Foundation

Citizen Science Partner

Huatai Securities "Yixin Huatai"

Scientific Charity Partner

Tencent Foundation

Natural Education Partner

Naturewin

Foreword

Chinese Birders

From Emergence to Citizen Science

In 2019, I joined an expedition with an international team of researchers to visit the breeding ground of the Spoon-billed Sandpiper in Chukotka, in the Russian Far East, just south of the Arctic Circle. Motion sickness and storms at sea really wore me down, and I felt like I might lose my life after the 12-day voyage. But I couldn't help wondering during this "miserable" time how on earth Spoon-billed Sandpipers, tiny birds merely the size of a sparrow, managed their annual migrations between their wintering and breeding grounds, separated by more than 5000 km. How many storms and other obstacles did they face and conquer during their epic flights? When I eventually went ashore at Chukotka and saw for myself the breeding activities of the Spoon-billed Sandpiper, I burst into tears. "I am deeply touched and encouraged by the treacherous journeys of Spoon-billed Sandpipers. I will devote the rest of my whole life to protecting these birds!" was what I said when interviewed by a Russian journalist.

The COVID-19 pandemic that swept the globe in 2020 severely affected our daily lives. During the very first lockdown in Wuhan, Hubei, images of birders birding through apartment windows with binoculars and cameras went viral on Chinese social media. The activities of the birds brought not only hope and inner peace to people confined to their homes but also an opportunity for us to reconsider our relationships with, and our responsibilities toward, nature.

The vast range and diverse natural landscapes of China provide wildlife with a huge variety of habitats in which to live. China is home to more than 1400 bird species, around 14 percent of the global total! Three of the planet's nine major avian migratory flyways pass through China—the West Asian–East African, the Central Asian, and the East Asian–Australasian—and the country is visited annually by roughly 20–25 percent of the world's migratory bird populations. With more and more studies and surveys, the number of bird species recorded in China is continuously growing.

In recent years, birding has become increasingly popular as a hobby and leisure activity among the public. Clearly, this is linked to improving standards of living. The increase in the number of native birders in mainland China essentially started in the 1990s, and the growth has been exponential ever since. The establishment of bird-watching organizations in numerous provinces further catalyzed the increase of both birders and birding activities in China. However, in these early days, knowledge of birds and species identification was shared among birders mostly by word of mouth.

John MacKinnon's *A Field Guide to the Birds of China* was published in 2000, and it soon became *the guide* for birding in the country. It not only helped birders improve their bird-identification skills but also introduced many nonbirders to the world of birding. Publications covering birds of specific regions, habitats, and/or taxa soon emerged, and with the recent technological advances in new media, audio and video references have become increasingly widely available to assist bird identification. Those early field guides have proven extremely helpful in bird monitoring, research, and conservation, transforming birding into an enormous citizen science project. Whether they are bird researchers, conservation practitioners, or simply nature lovers, people who are into birding are collectively known as *niao ren* ("bird people" in Mandarin).

My career in wetland and bird conservation began in 2012, when I became the secretary-general of the Mangrove Conservation Foundation (MCF). When the Beijing Entrepreneur Environmental Protection Foundation and MCF initiated the Free Flying Wings Project four years later, I began to work more closely with bird researchers and experienced bird-watchers, birding and protecting birds with our joint forces.

I am delighted that *The CNG Field Guide to the Birds of China* (the Chinese version of *Birds of China*), produced by Chinese National Geography, came out in 2020. This guide, written and illustrated entirely by Chinese birders, incorporates the latest information on the taxonomy and distribution of China's avifauna, covering a staggering 1491 species in plates and text.

It has become a new *must-have* book on the shelves or in the backpacks of birders in China. I believe *Birds of China* will be a landmark for the transition of birding from a leisure activity into citizen science.

I see many birders devoting their lives as observers and practitioners to studying and protecting birds and their habitats. It is also what I hope to achieve.

SUN Lili

President (8th Session), Society of Entrepreneurs and Ecology

Co-founder and honorary council member, Mangrove Conservation Foundation

Key to the Plates and Accounts

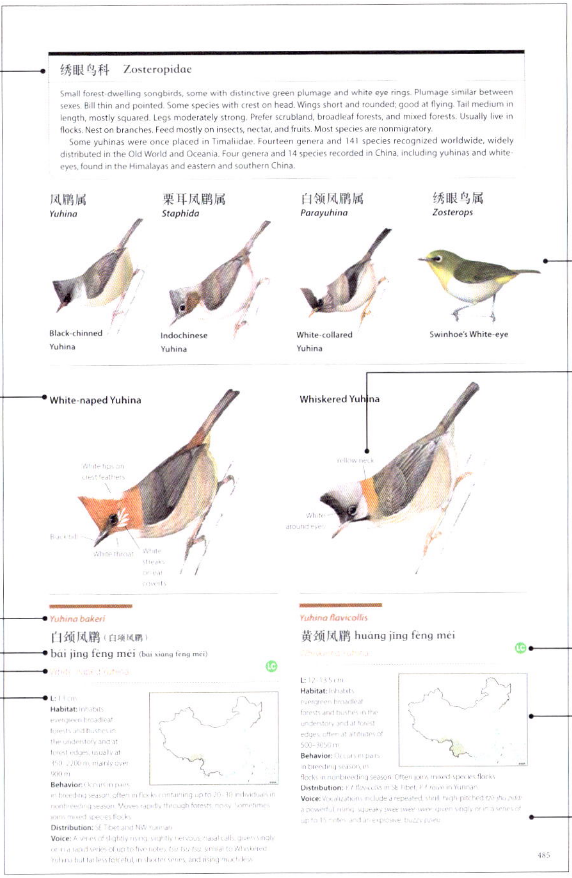

Avian order/family accounts

Plates
Illustrations showing the morphology and plumage of each species, drawing attention to the variations among species, subspecies, sexes, seasons, ages, and/or color morphs. Key features, behavioral vignettes, and different postures are also included when necessary.

Scientific names

Chinese common names and their Hanyu Pinyin
Widely used Chinese common names and alternative names (in parentheses, if any). Hanyu Pinyin is the romanization and phonetic transcription of Chinese characters for Standard Mandarin Chinese, listed here to assist pronouncing Chinese common names.

English common names

Body length (L)
Wingspan (WS) included for some species.

Representative species of families or genera in the group

Key field marks

Threatened status
Based on the status in the IUCN Red List of Threatened Species in 2020.

Distribution maps
Based on data from birdreport.cn, historical distribution, and recent published records.

Species accounts
Including habitat, behavior, distribution, and vocalizations.

绣眼鸟科　Zosteropidae

Small forest-dwelling songbirds, some with distinctive green plumage and white eye rings. Plumage similar between sexes. Bill thin and pointed. Some species with crest on head. Wings short and rounded; good at flying. Tail medium in length, mostly squared. Legs moderately strong. Prefer scrubland, broadleaf forests, and mixed forests. Usually live in flocks. Nest on branches. Feed mostly on insects, nectar, and fruits. Most species are nonmigratory.
Some yuhinas were once placed in Timaliidae. Fourteen genera and 141 species recognized worldwide, widely distributed in the Old World and Oceania. Four genera and 14 species recorded in China, including yuhinas and white-eyes, found in the Himalayas and eastern and southern China.

凤鹛属 *Yuhina*　栗耳凤鹛属 *Staphida*　白领凤鹛属 *Parayuhina*　绣眼鸟属 *Zosterops*

Black-chinned Yuhina　Indochinese Yuhina　White-collared Yuhina　Swinhoe's White-eye

White-naped Yuhina　Whiskered Yuhina

Yuhina bakeri　白领凤鹛 (白环凤鹛)
bái jǐng fèng méi (bai xiang feng mei)

Yuhina flavicollis　黄颈凤鹛 huáng jǐng fèng méi

Contents

An Overview of the Physical Geography and Bird Habitat Types in China

Physical Geography

The distribution of China's avifauna is, like anywhere, closely related to the country's physical geography. China has an immense and diverse landscape, encompassing a wealth of different landforms, including glaciers, lakes, rivers, ocean, loess, and karst, etc. With high elevation in the west and low elevation in the east, the land surface of China slopes down in three level steps. The first and highest level is the Qinghai-Tibet Plateau in Western China. Averaging more than 4000 m above sea level, it is the highest and largest plateau on Earth. The middle step consists of six major regions in the northern and central parts of China, including the Inner Mongolia Plateau, Loess Plateau, Yunnan-Guizhou Plateau, Junggar Basin, Sichuan Basin, and Tarim Basin. Most are below 3000 m above sea level. The third and lowest level comprises the low-lying plains and hills, generally lower than 1500 m, that lie between China's coast and a series of mountain ranges to the west. Aligned from northeast to southwest, these are the Greater Xing'an (Khingan), Taihang, Wu, Wuling, and Xuefeng Mountains. To the south and east of the coastline are the shallow seas covering the continental shelf. These are mostly less than 200 m in depth and are occasionally referred to as a fourth level.

The Qinghai-Tibet Plateau is bordered to the south by the Himalayas, the world's mightiest mountain range. The Himalayas encompass numerous peaks over 7000 m and include the world's tallest: Qomolangma (Mt. Everest), at 8848.86 m. On the northern edge of the plateau lie, from west to east, the Kunlun, Altun, and Qilian Mountains. Most peaks in the Qinghai-Tibet Plateau are covered by snow or glaciers year-round, and it is the snow- and icemelt that nourishes countless rivers and lakes, including the headwaters of many major rivers that run through East, South, and Southeast Asia. The largest lake in China, Qinghai Lake, lies in the northeastern part of the plateau, while the southeast is home to the largest expanse of primary forest found in China. The northern part of the Qinghai-Tibet Plateau is a huge area dominated by deserts and grasslands; cold, dry, and windy, with thin air and poor soil, the area is usually viewed as a dead zone, inhospitable to humans.

Habitats on the middle level include forests, grasslands, and deserts. The Tarim Basin in S Xinjiang is the largest desert area in China. Surrounded by mountains, the basin is a scorched dry land with intense sunlight and little rain. The Junggar Basin in N Xinjiang, however, sees the growth of varied types of desert vegetation, thanks to the precipitation resulting from the vapor-rich west winds blowing in through the gap left by mountains that border the basin. The steppe stretching from the Northeast China Plain to Qinghai Lake is a major component of the Great Steppe of Eurasia. In the southwest, the Yunnan-Guizhou Plateau, with its complex and diverse terrain, harbors the greatest biodiversity and the most diverse types of forest vegetation in the country.

The third level encompasses six major regions—the Northeast China Plain, North China Plain, Yangtze Plain, Liaodong Hills, Shandong Hills, and Southeast Hills. It is a vast lowland consisting of alluvial and coastal plains shaped by large rivers, such as the Yellow and Yangtze, as well as a few ancient and inactive mountain systems. Owing largely to the seasonal monsoon climate, this area hosts the majority of China's forests and wetlands. It is also the most populated area of the most populated nation on the planet and a region where most land has been converted to agricultural use. Although most of the forests in this area are not pristine, they still harbor a rich and diverse biome.

The fourth level is the shallow continental shelf and the open ocean. These areas span a vast three million square kilometers. They include the Bohai Sea, Yellow Sea, East China Sea, South China Sea, and Pacific Ocean off the east coast of Taiwan.

Zoogeography

The distribution of terrestrial animals is usually defined and separated by geographical and climatological barriers, leading to broad biogeographical patterns. The seven biogeographical realms of terrestrial animals are the Palearctic, Nearctic, Afrotropical, Indomalayan (Oriental), Neotropical, Oceanian, and Antarctic, among which China spans the Palearctic and Indomalayan realms. From west to east, these two realms are separated by the Himalayas, Hengduan Mountains, Qinling Mountains, and Huaihe River. Biogeographical realms can be further divided into regions and subregions. In 1999, ZHANG Rongzu (CHANG Yung-tsu) proposed a zoogeographical system for the terrestrial animals in China that consisted of 2 realms, 3 subrealms, 7 regions, and 19 subregions, each with its own representative fauna. (See table.)

Zoogeography of China (ZHANG 1999)

Realm	Subrealm	Region	Subregion	Representative fauna
Palearctic Realm	East Asia Bioregion	I Northeast China	IA Greater Xing'an Mountains	Fauna of cold temperate coniferous forest
			IB Changbai Mountains IC Northeast China Plain	Fauna of temperate forest, forest steppe, and cropland
		II North China	IIA North China Plain IIB Loess Plateau	
	Central Asia Bioregion	III Xinjiang-Inner Mongolia	IIIA Eastern steppe	Fauna of temperate steppe
			IIIB Western desert	Fauna of temperate desert and semidesert
			IIIC Tian Shan Mountains	Fauna of alpine forest steppe, meadow grasslands, and cold desert
		IV Qinghai-Tibet	IVA Changtang Plateau IVB Qinghai and S Tibet	
Indomalayan Realm	Indochina Bioregion	V Southwest China	VA Southwestern mountains	Fauna of subtropical forest, scrubland, grassland, and cropland
			VB Himalayas	
		VI Central China	VIA Eastern hills and plains VIB Western mountains and plateaus	
		VII South China	VIIA Coastal Fujian, Guangdong, and Guangxi VIIB Mountains of S Yunnan VIIC Hainan Island VIID Taiwan Island VIIE South China Sea Islands	Fauna of tropical forest, scrubland, grassland, and cropland

Bird Habitat Types

The considerable geographical range, the multitude of varied climatic types, and the ever-changing landscape of China provide a diverse array of habitats for birds. The varied collection of tundra, forests, steppe, deserts, rivers, lakes, coast, and islands in China jointly host some of the most diverse avifauna on Earth.

Plateau. A plateau may consist of glaciers, rivers, lakes, wetlands, forests, scrubland, grasslands, or deserts. Despite the variety of habitats, however, plateaus on their own can be viewed as a unique habitat type, given the suite of challenges they present to the birds living there. These include low temperature, low oxygen levels, and harsh weather. Wildlife must cope with these challenges through adaptations in behavior, physiology, morphology, and life history. Together with the historical influence from various glacial periods, the Qinghai-Tibet Plateau has become the home to a unique avifauna. The plateau's diverse habitats have led to dramatic distributions and radical avian diversity. The area encompassing the East Himalayas and the Hengduan Mountains hosts a great variety of vegetation spanning huge elevational ranges, including the country's densest and most pristine forests. It also hosts the most diverse avifauna anywhere in China. In contrast, the western part of the Qinghai-Tibet Plateau consists of mostly cold glacial terrain and arid deserts, and has the least diverse avifauna.

Grassland. Huge tracts of grasslands stretch across northern China from west to east. They consist of multiple different types, including typical steppe, meadow grassland, desert grassland, and alpine grassland. Among them, meadow grassland, which develops in a sub-humid climate, has the richest biodiversity and the most fertile soil. Typical steppe, as the name suggests, is the most typical and dominant type of grassland. It develops in a semiarid climate and has a lower biodiversity than meadow grassland. Desert grassland is the driest type, with the lowest productivity, and represents a transitional habitat between grassland and desert. Alpine grassland occupies the higher-altitude landscapes, is usually dry and cold, and supports only a short, sparse layer of vegetation. Grassland habitats support a unique avifauna that specializes in this relatively homogeneous landscape.

Desert. Deserts are widely distributed in the arid region of northwestern China. According

Fauna and scenery of plateau habitat.

Fauna and scenery of grassland habitat.

Fauna and scenery of forest habitat.

to their latitude, these deserts can be grouped as temperate, warm temperate, or plateau deserts. Despite their low plant diversity and meager biomass, the large expanse and low human population density of deserts leave the habitat relatively untouched, making them home to a number of birds that are well adapted to the harsh environment and flourish.

Forest. Forests in China are found mainly in the east. Ranging, roughly speaking, from north to south, they include cold temperate coniferous forest, temperate mixed broadleaf-coniferous forest, warm temperate deciduous forest, subtropical deciduous-evergreen broadleaf forest, subtropical evergreen broadleaf forest, subtropical monsoon forest, tropical monsoon forest, and tropical rain forest. The Hengduan

Mountains and the South Tibet Valley, with their expansive altitudinal ranges, host a broad collection of the forest types listed above within a restricted range. In the northwestern region of China, coniferous and deciduous forests are scattered among the Altai and Tian Shan Mountains and draw upon the limited humidity brought in from the Arctic and Atlantic Oceans. The diversity of forest ecosystems has nurtured a diverse community of forest birds, which serves as China's major bird habitats.

Wetland. There are five major types of wetlands in China: coastal wetlands, riverine wetlands, lakes, marshes, and artificial wetlands. Coastal wetlands can be further grouped into shallow seas, tidal flats, estuaries, shorelines, mangroves, reefs, and islands, etc. Some typical estuarine wetlands include the Yangtze River Delta, Yellow River Delta, and Liaohe River Delta. There are 2711 natural lakes in China with a water surface larger than one square kilometer, totaling more than 90,000 square kilometers in size. Waterbirds rely primarily on wetlands. Among the more important wetlands for waterbirds in China are Qinghai Lake, Poyang Lake, and Dongting Lake. Many of China's coastal wetlands provide critical stopover and wintering habitats for migratory waterbirds along the East Asian–Australasian Flyway.

Ocean and Sea. China has a coastline more than 18,000 km long. Beyond this are the shallow seas and open oceans, which altogether encompass about three million square kilometers. Most of China's coast and islands have been shaped by the interacting forces of rivers and seas and exhibit impressive degrees of diversity and complexity. This vast marine territory provides critical habitats for a suite of seabirds and migratory waterbirds, while the islands are used both as breeding sites by many seabirds and as stopover sites by migrants.

Fauna and scenery of wetland habitat.

A Brief History of Studies of the Systematics of the Birds of China

Studies of the taxonomy and distribution of the avifauna of China date back to 1863, when the British ornithologist Robert Swinhoe published *A Revised Catalogue of the Birds of China and Its Islands, with Descriptions of New Species, References to Former Notes, and Occasional Remarks*. They benefitted from the attention of ornithologists from China and abroad in the following 150 years. From the late 19th century to 1949, naturalists from Europe, Japan, and North America conducted extensive bird surveys and collected specimens in China. These efforts generated a wealth of new information about the then underexplored avifauna of the country and led to the discovery and description of many new species. In the 1930s, Chinese ornithologists, including CHONG Lin-Ting (CHANG Linding), FU Tung-Sheng (FU Tongsheng), YEN Kwok-Yung (REN Guorong), SHAW Tsen-Hwang (SHOU Zhenhuang), and CHENG Tso-Hsin (ZHENG Zuoxin), started to study the avifauna of China across the country. The Gold-fronted Fulvetta (*Schoeniparus variegaticeps*), named by YEN Kwok-Yung, was, proudly, the first bird species described by a Chinese ornithologist. After the establishment of the People's Republic of China, studies on the distribution, taxonomy, and evolution of the birds of China grew rapidly. A group of prominent and dedicated ornithologists, exemplified by CHENG Tso-Hsin, devoted their lives to studying the taxonomy of birds in China and mapping their distribution, gathering firsthand and fundamental information on the country's avifauna. *A Synopsis of the Avifauna of China*, the monumental work led by CHENG Tso-Hsin published in 1987, provided a comprehensive and systematic synopsis of such information, and it is still one of the most cited and widely used references about the distribution and taxonomy of the birds of China.

From the literature on the birds of China published in different periods, we can see how understanding of the avifauna of China has evolved. The number of bird species recorded in China increased from 1140 species in the *Keys to the Birds of China* (Cheng 1964) to 1244 species in *A Complete Checklist of Species and Subspecies of the Chinese Birds* (Cheng 1994) and to 1329 species in *A Field Guide to the Birds of China* (MacKinnon and Phillipps 2000). (The latter included some species that may occur in China.) The number continued to increase, to 1445 species, in *A Checklist on the Classification and Distribution of the Birds of China*, 3rd Edition (Zheng 2018).

This increase in the number of bird species recorded in China has been paralleled by an increase in the number of bird species recognized worldwide. Multiple factors have contributed to the growing number of species. First of all, with the theoretical advances in evolutionary biology, the definition and delineation of species has also evolved: While the classic biological species concept separates species based on morphology and distribution, emphasizing the reproductive isolation between different species, current taxonomic studies tend to adopt the phylogenetic species concept or the unified species concept, which focuses on the evolutionary history of species, guided by evolving theories in phylogenetics and evolutionary biology. In addition to the increasingly widespread use of DNA sequencing and systematics approaches in delineating species, the application of integrative taxonomy, which incorporates and consolidates multiple lines of evidence, including morphology, acoustics, behavior, ecology, and distribution, now generates more objective and consistent inferences on what constitutes a species. In light of these advances, the avifauna of China, being one of the most diverse of all Asian countries, has also been extensively studied and analyzed. The country's species-rich avifauna constitutes a critical component in the quest to reconstruct complete and comprehensive phylogenetic relationships for all major bird taxa. Being a hotspot for birds also makes China a hotspot for research in avian taxonomy and systematics, and as a result, many bird subspecies have been elevated to full species status following thorough investigations. In addition, new bird species are still being discovered in China, especially those "cryptic species" that share close phylogenetic relationships and morphological traits with their close relatives. One example is the recently described Sichuan Bush Warbler (*Locustella chengi*), named after CHENG Tso-Hsin. Another major factor contributing to the increasing number of bird species recorded in China is the growing community of bird-watchers and bird photographers. More and more people are observing and recording birds with binoculars and cameras, and some of these records have resulted in additions to China's checklist. Most of these additions have been recorded in the country's border regions, such as Xinjiang, Yunnan, and Tibet, as well as Taiwan and the east coast. In summary, although China is already home to one of the most diverse avifauna among all countries, it still holds high potential not only for exploring the occurrence and distribution of its birds, but also for studying their taxonomy and systematics.

Birds of China provides detailed accounts of 1491 bird species that have been recorded in China (up to November 2020), plus 74 more species that potentially occur or have become established in the wild. The taxonomy and organization of birds in the original Chinese version of the guide primarily followed the *IOC World Bird List* (version 10.2, published in July 2020, accessible at www.worldbirdnames.org). The list is maintained by the International Ornithologists' Union (IOU) and updated regularly to incorporate the latest changes in avian taxonomy worldwide. Both the scientific and English names in the current version have been updated to follow the eBird/Clements Checklist of Birds of the World v2022 as much as possible; in places where the *IOC World Bird List* and the eBird/Clements Checklist differ, the more relevant names were chosen based on our best judgment. The subspecies delineations have been kept as they are in the Chinese version, following the *IOC World Bird List*. The Chinese names are based primarily on *A Checklist on the Classification and Distribution of the Birds of China*, 3rd Edition (Zheng 2018), and, for species not included, on the *CBR Checklist of Birds of China*, version 8.0 (China Bird Report 2020). The Chinese names listed in the *CBR Checklist of Birds of China* are also included (after the main Chinese names, in parentheses) when they differ from those used in the primary reference.

A Brief Guide to Bird-Watching

Equipment

Optics. Most birds are shy and usually keep a distance from people, so binoculars (and sometimes spotting scopes) are the best tools to obtain "closer" and clearer views of birds. Choosing the right optics is essential for a good birding experience.

The definition of the "right optics" varies with the environment and conditions where the bird-watching takes place. Higher magnification usually means a smaller field of view, and a bigger objective-lens diameter usually means brighter views (especially important in low light conditions), but both lead to heavier pieces of equipment. Generally speaking, a pair of binoculars with 8–12x magnification and a 30–50 mm diameter objective lens gives the best experience for most bird-watching activities. For birds that are farther away—waterbirds on a tidal flat, for example—a spotting scope with 25–75x magnification and a 60–100 mm diameter objective lens may be needed. Comparing the two types of optics, binoculars are lighter in weight, more versatile, and deliver a larger field of view suitable for watching birds that move around actively, while spotting scopes have higher magnification power but are heavier, less stable, deliver a smaller field of view, and usually require a tripod, suitable for birds that move around less often. Binoculars with lower magnification and/or a smaller objective lens diameter are generally more portable and ideal for small kids and the elderly but also come with a smaller field of view and produce darker images.

Wild birds are usually very active and sometimes secretive, so becoming familiar with optics is also key to having a better birding experience. The sounds that birds make or the motion of vegetation are also helpful in finding birds.

Camera. Cameras are growing more and more popular and are widely used among bird-watchers, especially to obtain photos of species whose identification relies on minute morphological features, or to document new local or regional records. Taking bird photos usually requires a camera equipped with a telephoto lens, and both prime and zoom lenses would work. Recently, more and more smart phones are also equipped with a telephoto lens, making the documentation of birds much more convenient.

Audio Recorder. Noting the calls and songs of birds is also a critical part of birding. Bird sounds can be recorded easily by a smartphone or an audio recorder.

Location Device. The locations and tracks of birding activities can be recorded by smartphone apps or GPS devices.

Identifying Birds

Morphology is an important, but not the only, aid to identifying species. Clues for identifying birds usually come from the following three aspects. (Bird vocalization is also an important aspect in identification, and will be discussed in a separate section.)

Shape. Shape is usually the first thing you notice when a bird is spotted. Body length, wing span, and the length of different body parts all provide important information. However, it may not always be practical to get an accurate estimate of size and length in the field. This can be overcome by noting relative, rather than absolute, values by making references. For example, descriptions such as "beak relatively thick," "about twice the size of a sparrow," or "beak length similar to that of lore" can be equally as helpful. Shape, including the relative size and proportion of different body parts, provides more important clues to identification than does plumage. Sometimes, seemingly simple information, such as the shape and length of the beak or the general body contour, is the key to separating species.

Plumage. Plumage characters are also easy to pick up at a glance. The coloration of a bird contains a suite of information about the species, age, sex, molt, and health condition of the individual. Coloration also relates to the season and the bird's behavior.

Accurately describing a bird's plumage depends on familiarity with the bird's topography, or the name and location of its different body parts. Based on that, the distribution and arrangement of different colors on a bird can be correctly noted during an observation. It is worth mentioning that the same bird may have totally different plumages in different seasons, so familiarizing yourself with the differences between the breeding (or alternate) and nonbreeding (or basic) plumages will help you correctly identify a bird. Sometimes, noting and remembering the complex patterns of bird plumages will be very difficult; in such cases, shape will provide an even more important clue to a bird's identity.

Behavior. Behavior is another important clue to bird identification. Different groups or species of birds may exhibit totally different behavior in the wild. For example, laughingthrushes usually forage in the understory by searching among leaf litter, while swallows catch flying insects by means of aerobatic maneuvers, and spherical nests on tall trees or electrical poles are usually made by Oriental Magpies. The flying, perching, roosting, and foraging behaviors may all give away some specific features or "jizz" of a bird species

or group, and some of the typical behaviors can be particularly useful in species identification. Bird identification based on behavior usually requires some experience.

Habitat

What Is Habitat? Habitat is the environment in which a bird lives, particularly the areas where it rests, breeds, and forages.

Habitat Types. All kinds of habitat—forests, deserts, ocean, plateau, rain forests, or polar regions—host unique communities of birds. Different bird species usually have different habitat requirements, while the same bird species may also use different types of habitats under different conditions or states. Accordingly, understanding the relationships between birds and their habitat requirements can be particularly helpful for identification. For instance, birds that inhabit plateau habitat throughout the year normally won't appear in lowland, and those that usually hide in dense forests are less likely to visit very open habitat. The most efficient way to target a species is to look for it in its preferred habitat. Similarly, habitat types can also help to narrow down the range of likely species when other clues are not sufficient to identify the species.

Season, Tidal Cycle, and Time of Day

For birders, becoming familiar with birds' activity patterns is an important skill. The activity patterns of birds are affected by many factors, among which season, tidal cycle, and time of day are some of the most important.

Season. Birds that move seasonally between different habitats in a regular pattern are called migratory birds. Some birds—for example, many migratory shorebirds—conduct latitudinal migration in an approximately north-south direction, utilizing different breeding and wintering habitats. Other migratory species conduct altitudinal migration along an elevational gradient. This is commonly seen in forest birds that breed at higher elevation on mountain ranges.

Based on extensive observations and research, nine major flyways for migratory birds (primarily for latitudinal migrants) are recognized worldwide. Birds that occur in a specific region usually fall into four groups—resident (or more precisely, year-round) birds, winter visitors (wintering birds), summer visitors (mostly breeding birds), and passage migrants—indicating that the composition of a region's bird community also varies seasonally. As a result, season can also aid the identification of birds.

Time of Day. Different birds vary in their daily activity patterns and can be roughly grouped as diurnal and nocturnal species. Further observation may reveal more detailed patterns. For example, many diurnal birds are most active around dawn, while diurnal raptors usually soar and glide around midday, and some nocturnal birds become active around dusk.

Tidal Cycle. Coastal waters and tidal flats provide important foraging habitat for many waterbirds. Correspondingly, the activity patterns of these species are tied to tidal cycles. So it is vital to check the tide table before birding for waterbirds in coastal areas. Many coastal waterbirds forage on tidal flats after the tide recedes, when much of the foraging area is exposed. When the tide rises and covers the foraging ground, birds usually move to nearby ponds or salt pans. Arrange your visits to different habitats wisely, accounting for the tidal cycle and bird movements.

Recording Sightings

A complete checklist of bird sightings made during a birding visit usually includes the following information.

Species, Quantity, and Details. Generally speaking, the species and the number of individuals seen are the two most important pieces of information in a checklist, as they provide valuable data illustrating the distribution and activity patterns of the avifauna of a certain region in a specific period. Additional details about bird sightings, including the plumages, sexes, and special behaviors of birds, are also

worth recording, as they are helpful in understanding important ecological and behavioral traits of birds.

Location, Time, Habitat, and People. Apart from the above information about bird sightings, it's also important to record the location, time, habitat, and the person or persons who made the checklist. Observations of the species, quantity, and behavior of birds are meaningful only when associated with the location, time, and habitat of the birding activities.

Exploring and Sharing Checklists. Birders now explore other birders' checklists from a specific location and time and share their own checklists using platforms such as birdreport. cn and ebird.org. By sharing checklists online, birders and researchers can all benefit and get valuable information about birds around the world.

Observing and Reporting Banded Birds

Bird Banding. Banding is an important tool in bird research, especially for understanding migration dynamics. Metal bands made of alloy, and sometimes plastic color bands or flags, are put on birds, usually as their biometric measurements are also recorded and put on file. When the banded birds are sighted or recaptured, valuable scientific information about their movement, age, and population dynamics can be gathered.

Bands are usually put on birds' legs, but for some groups of birds, neck collars, wing tags, and nasal saddles are also used. Plastic color bands and flags are widely used in waterbirds; the color schemes and flag codes usually convey important information about the banding location and identity of the birds wearing them. The coordinated use of color flags along a flyway is especially helpful in understanding the movements of migratory birds, as the flag colors permit the banding locations of birds to be recognized.

Reporting Banded Bird. If you see a banded bird, report the information to the

Shorebird Color-Flagging Protocol along the East Asian–Australasian Flyway.
1. Queensland, Australia. 2. Gulf of Carpentaria, Australia. 3. Cambodia. 4. Jiangsu, China. 5. Yalu Estuary (Yalujiang), China. 6. Singapore. 7. Sri Lanka. 8. N Western Australia, Australia. 9. SW Western Australia, Australia. 10. Northern Territory, Australia. 11. Bangladesh. 12. Kamchatka, Russia. 13. Sakhalin Island, Russia. 14. Vietnam. 15. New South Wales, Australia. 16. Victoria, Australia. 17. Tasmania, Australia. 18. South Australia, Australia. 19. Sumatra, Indonesia. 20. West Papua, Indonesia. 21. E Yellow Sea, South Korea (old). 22. Java and Bali, Indonesia. 23. Myanmar. 24. Philippines. 25. Malaysia (used in Kamchatka, Russia, by mistake). 26. Thailand Peninsula, Thailand. 27. Gulf of Thailand, Thailand.

National Bird Banding Center of China or other relevant databases. By contributing such an observation, every birder becomes involved in the collection of valuable scientific data about bird migration.

Ethics

Be Honest. A basic rule in birding, as in life, is to be honest. The species, time, location, and people present should be correctly recorded. In the field, it is very common to be unable to identify, or even to see clearly, every bird encountered. This can be due to many factors—the influence of light, the angle of view, or the skill of the observers. When a bird cannot be identified to species, make a note of it and other useful information and leave it unidentified. When the number of individuals cannot to be counted accurately, providing an estimated number will also be helpful.

Avoid Disturbance. Avoiding or reducing disturbance to birds as much as possible is core to birding ethics. Normally, birders should not lure or feed birds, intentionally flush birds, come too close to birds, or destroy bird habitat.

When birds are resting, especially because of exhaustion or during inclement weather, they can be particularly vulnerable. Intentionally flushing birds under such conditions to get better views or photos may cause their body condition to deteriorate further and sometimes can be lethal.

Breeding is one of the most sensitive periods for birds. Excessive disturbance during incubation or when nestlings are being fed may result in nest abandonment and breeding failure. It is thus imperative to avoid disturbing breeding adults when nests, eggs, or chicks are encountered, and to shorten the observation period and keep a good distance. Particularly, trimming vegetation around a nest to get

28. S India. 29. Chongming Island, China. 30. Taiwan, China. 31. Hong Kong, China. 32. Chongming Island, China (old). 33. North Island, New Zealand. 34. South Island, New Zealand. 35. N India. 36. E Yellow Sea, South Korea. 37. Hainan and Guangxi, China. 38. Bohai Bay (Tangshan and Cangzhou), China. 39. Kyushu and Okinawa, Japan. 40. Lake Komuke, N Hokkaido, Japan. 41. Shunkunitai, E Hokkaido, Japan. 42. Tokyo Bay and Miyagi Prefecture, Japan. 43. Mongolia. 44. S Chukotka, Russia. 45. S Chukotka, Russia. 46. Wrangel Island, Russia. 47. N Chukotka, Russia. 48. N Alaska (Ikpikpuk and Prudhoe Bay), USA. 49. N Alaska (Canning River), USA. 50. NW Alaska (Cape Krusenstern), USA. 51. N Alaska (Utqiagvik), USA. 52. N Alaska (Utqiagvik), USA. 53. W Alaska (Nome), USA.

photos usually leads to the death of chicks and is strictly prohibited.

When you encounter unethical behaviors, stop them if possible. For an endangered or sensitive species, you may choose not to share the sighting or the precise location publicly to avoid excessive disturbance.

One thing to remember: whether it is watching or photographing birds, the welfare of the birds always comes first.

Bird Vocalizations and Identification

Vocalizations are an important means of communication among birds and include two major categories: calls and songs. Calls usually consist of simple notes used mainly to maintain contact with other individuals or to signal danger, fear, or excitement. In comparison, songs are often more complex and varied, are usually heard during breeding seasons, and are used for display, mate attraction, and territorial defense. Songs are not made by all species, while passerines (songbirds) are well known for their songs. In some species, the vocalizations are highly variable in different situations, while in other species, the vocalizations don't change very much.

Bird vocalizations are usually species specific, which is why they can be used to identify birds to species. If you are familiar with the vocalizations of the birds in a birding area, you will usually record more species than if you relied on visual identification alone, because many birds, especially those that inhabit densely vegetated areas, are more easily heard than seen. In many cases, a glance is not enough for accurate identification, so it is much easier to notice and identify secretive species by listening than by trying to get a clear and definitive view. Moreover, species in some taxonomic groups—for example, some cuckoos and warblers—differ markedly in their vocalizations but are similar in appearance. Using vocalizations to identify such species will be much easier.

Most differences in bird vocalizations can be readily recognized in the field by listening. Generally speaking, bird vocalizations can be grouped into the following five types.

1. **Attractive and complex songs**
   Many passerines (songbirds) have melodious and rhythmic songs. Chinese Hwamei (*Garrulax canorus*), Buffy Laughingthrush (*Pterorhinus berthemyi*), Red-billed Leiothrix (*Leiothrix lutea*), Oriental Magpie-Robin (*Copsychus saularis*), and larks (Alaudidae spp.) are highly accomplished vocalists. Some bird species, including several laughingthrushes and Chinese Blackbird (*Turdus mandarinus*), also mimic the songs of other birds; special attention should therefore be paid to these species.

2. **Mono- and multisyllabic repetitive notes**
   Most bird species give mono- or multisyllabic repetitive songs. Examples include Eastern Yellow Wagtail (*Motacilla tschutschensis*) and Plain Prinia (*Prinia inornata*), both of which make repetitive, monosyllabic songs. White-breasted Waterhen (*Amaurornis phoenicurus*), Common Cuckoo (*Cuculus canorus*), Black-streaked Scimitar-Babbler (*Erythrogenys gravivox*), and Zitting Cisticola (*Cisticola juncidis*) have repetitive, more disyllabic songs. Chinese Bamboo-Partridge (*Bambusicola thoracicus*), Large Hawk-Cuckoo (*Hierococcyx sparverioides*), Spotted Redshank (*Tringa erythropus*), and Streak-breasted Scimitar-Babbler (*Pomatorhinus ruficollis*) repeat trisyllabic vocalizations. Indian Cuckoo (*Cuculus Micropterus*), Common Redshank (*Tringa totanus*), and Rufous-capped Babbler (*Cyanoderma ruficeps*) give repeated four-

syllable vocalizations. Both Lesser Cuckoo (*Cuculus poliocephalus*) and Crested Bunting (*Emberiza lathami*) include repetitions of five or six syllables.

3. Loud whistles
The song of Plaintive Cuckoo (*Cacomantis merulinus*) starts with three consecutive whistles, followed by a series of short, muffled, gradually descending notes. Brown-chested Jungle Flycatcher (*Cyornis brunneatus*) gives two high-pitched notes, followed by a trisyllabic whistle. Brownish-flanked Bush Warbler (*Horornis fortipes*) starts with a long whistle and ends with two to three high-pitched whistles. Blue Whistling-Thrush (*Myophonus caeruleus*) gives a short whistled call reminiscent of a bicycle pump.

4. Sharp trills
Mostly given by smaller birds. For instance, Rufous-faced Warbler (*Abroscopus albogularis*) gives a long, high-pitched, bell-like trill, while Swinhoe's White-eye (*Zosterops simplex*), Common Kingfisher (*Alcedo atthis*), and Little Forktail (*Enicurus scouleri*) give a trill call in flight.

5. Simple and coarse notes
Noisy, monosyllabic, and coarse, given mostly by larger birds, such as Ring-necked Pheasant (*Phasianus colchicus*), ducks (Anatidae spp.), crows (*Corvus* spp.), Red-billed Blue-Magpie (*Urocissa erythroryncha*), and Long-tailed Shrike (*Lanius schach*).

If you have not heard it before, it is often challenging to imagine what a bird's vocalization may sound like just by reading the descriptions alone. Also, these vocalizations usually won't leave a clear impression in your mind unless they are heard multiple times. Fortunately, birds in the field are usually very vocal, providing ample chances for us to become familiar with their voices. When you hear an unfamiliar vocalization in the wild, it is a good idea to find and identify its origin species, and then to link the sound with this species. Alternatively, you can record the vocalization and find out which species made it afterward.

With the advances in sound recording and analyzing technologies, bird vocalization has become an important branch in ornithological research. Audio processing and analyzing software is growing fast, and many applications are freely available. Many new bird species are being discovered and described based on differences in vocalizations. In addition to species-specific characteristics, there are also subspecies- and population-level differences, or even individual-level variations in bird vocalizations.

While birding, some birders may use playback, anticipating that a target bird will respond by revealing itself. However, repeated and continuous use of playback will undoubtedly affect the normal behavior of birds. During the breeding season, the impact of tape playback is usually much stronger and may lead to a total failure of the breeding attempt. It is thus important to be cautious when using playback. Reduce the intensity and frequency of playback, and do not use playback intensively just to get a better photo while harming the welfare of birds.

Bird Conservation in China

Bird conservation is achieved mainly through the protection of bird habitats and individual species. In terms of habitat conservation, China has developed a comprehensive network of protected areas, including nature reserves, forest parks, wetland parks, ocean parks, and national parks. Among them, nature reserves and national parks are the two primary and most important types of protected areas. Nature reserves around the country, particularly, have provided effective protection for the avifauna of China. The first of its kind, the Dinghushan National Nature Reserve (NNR), was established in 1956. There are 2750 nature reserves in China today (in September 2019), covering 1.47 million square kilometers, or 15 percent of China's landmass. Depending on their significance, nature reserves in China are either NNRs or local nature reserves. In terms of the primary targets of protection, these nature reserves are focused on ecosystems, specific species, or natural relics. Some important and well-known NNRs focusing on bird protection are Zhalong NNR, Chongming Dongtan NNR, Yancheng NNR, Wuyanling NNR, Dongdongting Lake NNR, and Qinghai Lake NNR. Recently, China has begun to set up its national park system, aiming for a hierarchical and more comprehensive network of protected areas that are backboned by national parks, based primarily on nature reserves, and supplemented by various types of nature parks.

In terms of species protection, the International Union for Conservation of Nature (IUCN) has been building and maintaining the IUCN Red List of Threatened Species since 1964, through the dedicated work of researchers and IUCN's Species Survival Commission Specialist Groups. The IUCN Red List is updated regularly at present on its online platform, providing easier access and more timely information than in the early days, when the list was updated only once every several years. Based on criteria that consider important aspects of a species' population, including its population size, rate of population decline, geographical distribution, and the isolation or fragmentation of subpopulations, the IUCN Red List assigns each species to one of nine categories: Extinct (EX), Extinct in the Wild (EW), Critically Endangered (CR), Endangered (EN), Vulnerable (VU), Near Threatened (NT), Least Concern (LC), Data Deficient (DD), and Not Evaluated (NE). Species listed as CR, EN, and VU are collectively referred to as "threatened." The IUCN Red List has long provided the most authoritative and scientific guide to the status and protection of global biodiversity. In practice, China has made its own list of rare and threatened wildlife, with specific measures outlined in the Law of the People's Republic of China on the Protection of Wildlife (effective as of 1989, amended 2018). Wildlife under such special state protection are assigned either first- or second-class protection. An additional list of terrestrial species that are of important ecological, scientific, and social value also provides national-level protection to the listed species, while each provincial administrative division has also made its own list of locally protected species. In early 2020, the central government of China took a further step and put all the wild birds in China under protection.

In the 1980s, along with China's reform and opening-up policy, foreign birders started visiting the country for its rich and unique avifauna, introducing birding to China. In the 1990s, local birders appeared around China. As the country developed economically, the number of birders and bird-watching organizations in many provinces grew quickly, owing much to the publication of field guides and easier access to optics and cameras. The sheer amount of birding activity of the growing birding community is contributing valuable and firsthand information about the distribution, abundance, and threatened status of the birds of China. In this sense, birding is no longer merely a leisure activity; it is turning into a huge citizen science project. Many new bird species have been discovered for regions or the entire country since the very beginning of birding activities in China, and the distribution of species already recorded is also being updated continuously. Local bird-watching organizations have begun to organize and conduct surveys of bird diversity and distribution, gathering useful data that allows researchers to focus on more in-depth scientific questions. Undoubtedly, birding activities in China are also fueling the rapid advances in ornithology and bird conservation, as a part of the trend seen around the globe.

Bird Topography

Topography of a Passerine

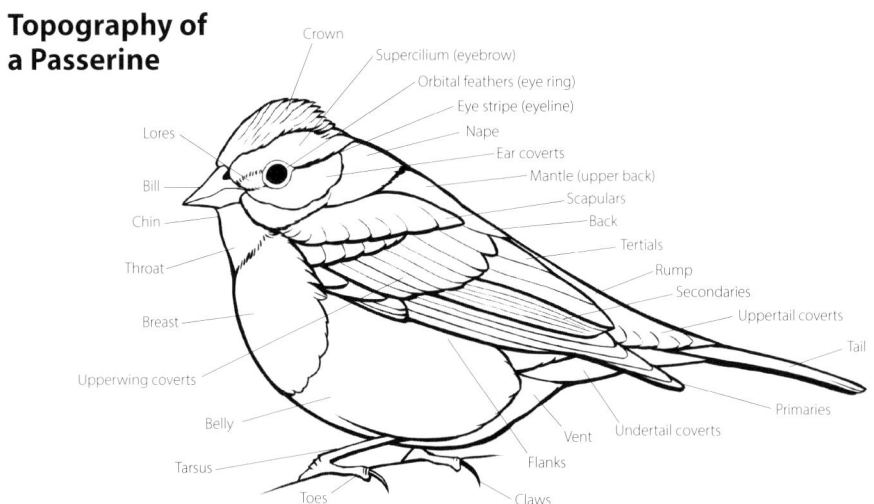

Crown
Supercilium (eyebrow)
Orbital feathers (eye ring)
Eye stripe (eyeline)
Nape
Ear coverts
Mantle (upper back)
Scapulars
Back
Tertials
Rump
Secondaries
Uppertail coverts
Tail
Primaries
Undertail coverts
Vent
Flanks
Claws
Toes
Tarsus
Belly
Upperwing coverts
Breast
Throat
Chin
Bill
Lores

Topography of a Passerine: Head Feathers

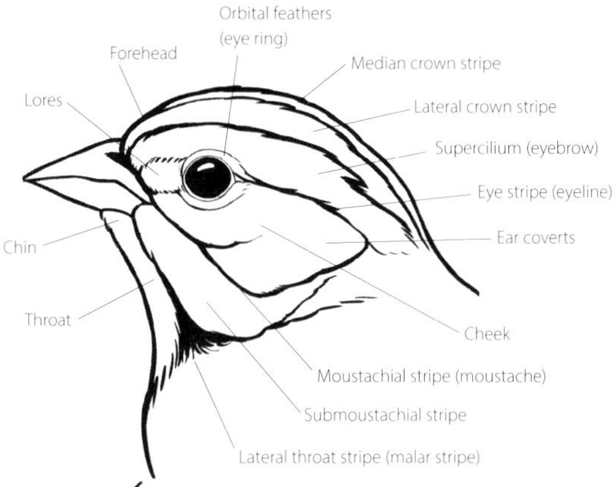

Orbital feathers (eye ring)
Forehead
Median crown stripe
Lores
Lateral crown stripe
Supercilium (eyebrow)
Eye stripe (eyeline)
Ear coverts
Chin
Throat
Cheek
Moustachial stripe (moustache)
Submoustachial stripe
Lateral throat stripe (malar stripe)

Topography of a Passerine: Wing Feathers

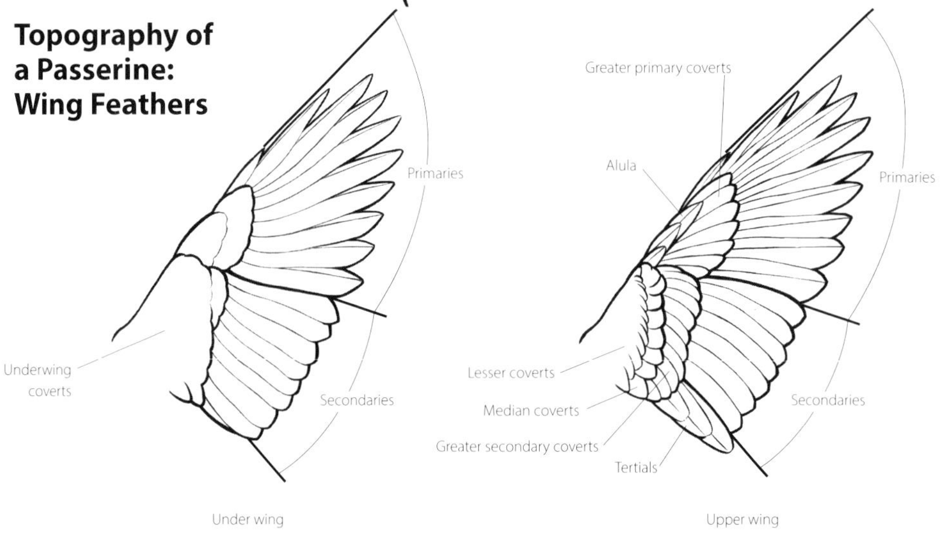

Primaries
Greater primary coverts
Alula
Primaries
Underwing coverts
Lesser coverts
Secondaries
Median coverts
Greater secondary coverts
Tertials
Secondaries

Under wing

Upper wing

Topography of a Duck

Crown

Eye stripe (eyeline)

Moustachial stripe
(moustache)

Neck

Mantle (upper back)

Scapulars

Tertials

Neck ring

Primaries

Breast

Belly

Undertail coverts

Tarsus

Flanks

Secondaries/Speculum

Vent

Topography of a Gull

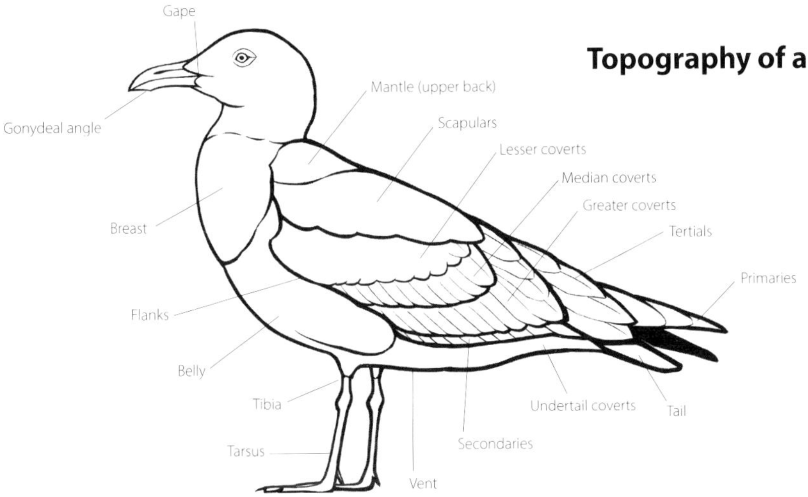

Gape

Mantle (upper back)

Scapulars

Gonydeal angle

Lesser coverts

Median coverts

Greater coverts

Breast

Tertials

Primaries

Flanks

Belly

Tibia

Undertail coverts

Tail

Tarsus

Secondaries

Vent

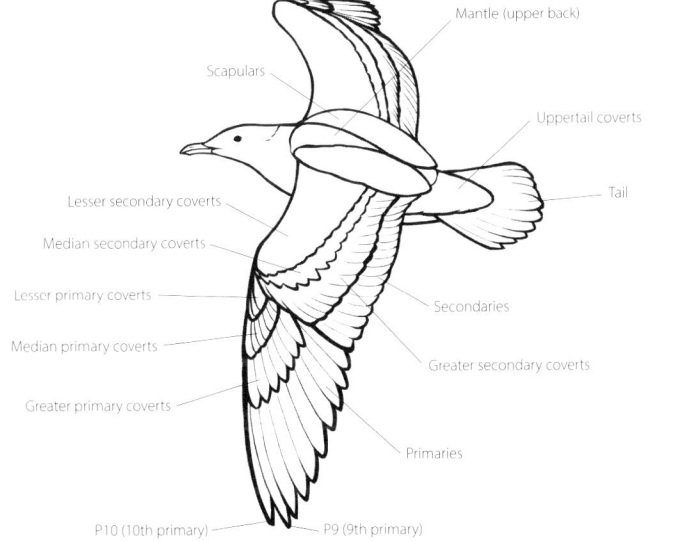

Mantle (upper back)

Scapulars

Uppertail coverts

Tail

Lesser secondary coverts

Median secondary coverts

Lesser primary coverts

Secondaries

Median primary coverts

Greater primary coverts

Greater secondary coverts

Primaries

P10 (10th primary)

P9 (9th primary)

27

Glossary

Bird Appearance and Structure

Albinism: An abnormality in bird plumage due to a deficiency in melanin, usually resulting in all-white plumage and red irises

Alula: Small and strong feathers that are attached to the first finger bone ("thumb") of a bird and allow the wing to achieve higher lift and avoid stalling

Bare part: Patches of skin not covered by feathers, including the lore, face, tarsus, etc.; bare parts in some species become brightly colored during breeding seasons

Breeding plumage: Boldly or brightly colored plumage that birds acquire for breeding seasons; males of some species may acquire beautiful ornamental feathers

Cere: Bare skin around the nostril at the base of the upper bill; seen in parrots, birds of prey, pigeons and doves, etc.

Ear coverts: Feathers surrounding the ear opening

Eclipse plumage: An intermediate stage of plumage that appears when birds molt from breeding to non-breeding plumage; commonly seen in waterfowl

Eyeline (eye stripe): The stripe that goes "through" the eye

Eye ring (orbital feathers): Feathers around the eye, usually light colored

Field mark: A key feature used to identify a species or separate similar species

Fingers: Fingerlike patterns on the wing tips visible in flying (usually soaring) birds; formed by the gaps between several outermost primaries; the number of fingers is useful in raptor identification

Flanks: Both sides of a bird, between the breast and vent

Leucistic: An abnormality in bird plumage due to a lack of melanin or other pigment, usually resulting in partially or entirely white plumage, with normal iris color

Lore: The bare part between the eye and bill

Mantle: Feathers on the upper back

Melanistic: Partially or entirely darker-than-normal plumage, resulting from an excessive amount of melanin

Molt: The process of shedding old feathers and growing new ones; the molt of flight feathers usually follows a set sequence; molt patterns and strategies differ among species

Morph: The (existence of) different plumage patterns exhibited by adult birds of the same species, due to genetic diversity

Nonbreeding plumage: Plumage during the nonbreeding seasons; normally duller or darker than breeding plumage but similar in some species

Primaries: Flight feathers attached to the manus ("hand") of a bird, usually 9–12 pieces, depending on the species

Rectrices (tail feathers): Flight feathers forming the tail of a bird, attached to or growing on the tail bones; usually 10–12 pieces, depending on the species

Scapulars: Feathers that cover the upper side of the wings when a bird is at rest

Secondaries: Flight feathers attached to the ulna ("forearm") of a bird, usually 10–20 pieces, depending on the species

Sexual dimorphism: Systematic and pronounced differences in size or plumage between adults of difference sexes in the same species

Speculum: A brightly (and usually iridescently) colored patch on the secondaries of some waterfowl, distinctly patterned and contrasting with adjacent coverts and flight feathers

Spur: A bony, clawlike structure, most commonly found on the tarsus of male Galliformes birds

Supercilium (eyebrow): A feather stripe right above the eye, differently colored from adjacent feathers

Tarsus (tarsometatarsus): The lower leg of birds, formed from the fusion of some tarsus and metatarsal bones; skin usually unfeathered, highly keratinized, and scalelike

Tertials: Several pieces of flight feathers arising in the brachial region, or the innermost part of the wings

Undertail coverts: Feathers that cover the base of the tail feathers from the ventral (under) side

Uppertail coverts: Feathers that cover the base of the tail feathers from the dorsal (upper) side

Vent: The lower belly of a bird between the legs (may also include the undertail coverts)

Wing bars: Stripe-like patterns on the wings, differently colored from the adjacent area

Wing coverts: Feathers that cover the base of the flight feathers on wings

Bird Ecology and Behavior

Accidental: Species that are uncommonly found in a region

Adult: A bird that is sexually mature and possesses definite plumage

Altricial: Chicks that are almost naked and whose eyes are closed upon hatching; altricial chicks rely on their parents for food and cannot leave the nest

Anisodactyl: Feet on which the second, third, and fourth toes face forward and the first toe faces backward, as on nearly all songbirds

Chick: The stage between hatching and gaining full

plumage when birds are naked or covered only by down feathers

Dispersal: Directional movement between a birthplace and a breeding area, or between two breeding areas

Endemic: Species that live exclusively in a single country or region

Heterodactyl: Feet on which the third and fourth toes face forward and the first and second toes face backward, as on trogons

Hybrid: The offspring of two different species

Immature: Usually referring to the stage between first full plumage and full adult plumage, including juvenile and subadult stages

Introduced species: Species that do not naturally occur in a region but are brought to it by human activities, including but not limited to escaped cage birds or birds that are purposely released; some introduced species may breed independently in the wild, with an established population

Irruption: Winter foraging movements of large groups of arctic or subarctic species (for example, tits and finches)

Juvenile: The stage between gaining the first set of full plumage (juvenile plumage) and the first molt (post-juvenile molt); sexually immature

Land birds: Species that are good at walking or running on the ground, with strong legs and feet (such as birds in Galliformes and Columbiformes)

Migration: The regular seasonal movement of birds, including latitudinal and altitudinal migration

Mixed-species flocks: Flocks of birds consisting of individuals of different species (typically passerines), usually observed in forest habitat during nonbreeding seasons

Passage migrants: Species that pass through a region in spring and/or fall during migration but neither winter nor breed in the region

Pelagic: Species that live mostly in open ocean or on oceanic islands; usually too far from shore to be observed on land frequently

Perching birds: Species that are good at climbing or perching on vertical surfaces, usually with zygodactyl, heterodactyl, or syndactyl feet

Precocial: Chicks that are covered with down upon hatching; once the down dries, they can move and forage, following adult birds, right away

Raptors: Predatory or scavenging carnivorous species, usually with sharp claws and bill (for example, hawks, eagles, falcons, owls)

Resident: Species that do not undertake long-distance migration but stay in the same region throughout the year

Songbirds: Birds with relatively small bodies that are good at singing with a well-developed vocal organ

Subadult: The stage between the first molt and gaining full adult plumage; an intermediate stage before sexual maturity that may last from several weeks to years, depending on the species; in some groups (such as raptors and gulls), the subadult stage lasts several years, with the plumage changing every year

Summer breeders: Species that visit a region to breed during summer only

Summer visitors: Species that visit a region during summer but do not breed

Syndactyl: Feet on which the second and third toes are wholly or partly united or webbed, as on rollers, bee-eaters, and kingfishers

Vagrant: A species that occurs outside its normal distribution range; usually thought to have "lost its way" due to inclement weather or a lack of experience

Waders: Species that prefer shallow waters, usually with a long neck, long legs, and a long bill (for example, cranes, herons, and egrets)

Waterfowl: Species that are good at swimming or diving, usually with webbed feet (such as geese, ducks, and swans)

Winter visitors: Species that visit a region during winter only

Zygodactyl: Feet on which the second and third toes face forward and the first and fourth toes face backward, as on parrots

Birding Activities

Big year: An activity in which a birder attempts to see as many species as possible within a year; usually competitive, requiring careful planning to cover the most target species in different regions and seasons

Digiscoping: Taking photos direct through optics (usually spotting scopes) with a camera or a mobile phone

Firsthand record: The initial finding of a species (usually a rare one); derived terms include *secondhand record* and *multihand record*

Life list: A list of species recorded by an individual birder in a specific country, province, or region in a lifetime: for example, a China life list or a Beijing life list

Lifer: A species seen or heard for the first time in a lifetime

Ornithology: A field of study focusing on bird research

Pishing: Making *pish*-like calls to attract birds, to get closer or clearer views

Twitcher: An enthusiastic birder who will make great efforts to see rare birds (usually involving long-distance travels) and usually keeps various types of bird lists

Twitching: The action of making a great effort to see birds (usually rare ones)

雁形目　ANSERIFORMES

Medium-sized to large waterbirds. Plumage usually different between sexes. Bill broad and flat, usually with nail at tip, ending with a small hook in some species. Wings narrow and pointed, enabling fast and long-distance flight. Feet typically webbed for efficient swimming. Tail short with well-developed preen (uropygial) gland. Nest in variable habitats from marshes to forests, on ground or in tree holes. Often gregarious during nonbreeding season. Omnivorous; diet ranges from plants and algae to aquatic insects, mollusks, and fish. Most species are migratory.

Three families, 56 genera, and 178 species recognized worldwide, with a cosmopolitan distribution. One family, 24 genera, and 61 species recorded in China, ubiquitously distributed nationwide.

鸭科
Anatidae

Mallard

Lesser Whistling-Duck

 Brown uppertail coverts

Brant

Small dark goose with broken white neck collar and bold white flash on flanks

---

Dendrocygna javanica

栗树鸭 lì shù yā

Lesser Whistling-Duck　(LC)

L: 38–40 cm
WS: 70–74 cm
Habitat: Often inhabits shallow waters surrounded by trees, such as small ponds and paddy fields; also gathers in flocks on large bodies of open water where there is little human disturbance and abundant floating plants.

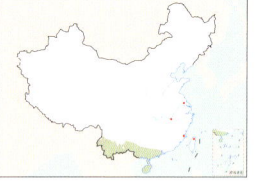

Behavior: Gregarious. Forages at night. Occurs in flocks on lakes and paddy fields, or rests in trees during the day; good at diving.
Distribution: Recorded primarily in Hainan and S and W Yunnan; also recorded in Guangdong, Guangxi, and Hong Kong, but uncommon; vagrant in Taiwan, Fujian, Jiangxi, Jiangsu, and Beijing.
Voice: High-pitched, shrill, whistled *wee* or often doubled, *tsee-tsee*, frequently given in flight.

Branta bernicla

黑雁 hēi yàn

Brant　(LC)

L: 55–66 cm
WS: 110–125 cm
Habitat: Generally winters along seashores; prefers large sandy or muddy gulfs, estuarine tidal flats, and saltwater marshes; feeds on seagrass. Also occurs inland on grass or farmland.

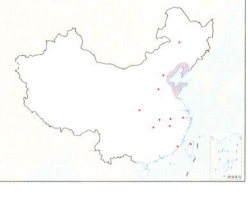

Behavior: Forages in grassy areas or shallow waters; also upends to feed in water. Associates more often with large ducks than geese; fast wingbeats in flight.
Distribution: Rare; may winter along the coasts of Yellow Sea and Bohai Sea. Vagrant in Beijing, Hebei, Inner Mongolia, Shaanxi, Liaoning, Taiwan, the Yangtze Plain, and the coastal areas of East and South China.
Voice: Single birds often silent; flocks make a noisy *raunk-raunk*, *ronk*, or similar.

Branta ruficollis
红胸黑雁 hóng xiōng hēi yàn
Red-breasted Goose **VU**

L: 53–56 cm
WS: 116–135 cm
Habitat: Inhabits lakes or wide river channels; forages on wet grasslands, lakes, or marshes.
Behavior: Often associates with other goose species, especially

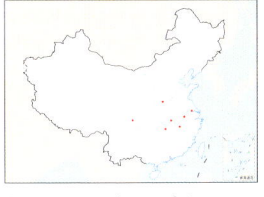

Greater and Lesser White-fronted Geese. Fast wingbeats in flight.
Distribution: Vagrant along the middle and lower reaches of the Yellow River, the coast of Yellow Sea and Bohai Sea, and wetlands along the Yangtze River from Sichuan to Jiangsu.
Voice: Unusual, sharp, high-pitched yapping, *kwii-kii*, similar to a rubbing noise from a poorly lubricated bicycle.

Branta canadensis
加拿大雁 jiā ná dà yàn
Canada Goose **LC**

L: 95–110 cm
WS: 122–183 cm
Habitat: No strict habitat requirement; occurs on lakes, slow-flowing streams, meadows, even in parks or other green spaces.
Behavior: Often forages on waters, also in grassy areas.
Distribution: Vagrant. Recorded in Tianjin, Beijing, and Liaoning.
Voice: Loud disyllabic honking, *h-ronk*, with the second note lower pitched.

Red-breasted Goose

Short small bill

Short neck

Canada Goose

Long slender neck

Long bill

Relatively large

Pale breast

Cackling Goose

Steep forehead

Short stubby bill

Relatively dark breast

Short stocky neck

B. h. minima

White collar sometimes broken

B. h. leucopareia

Branta hutchinsii
小美洲黑雁 xiǎo měi zhōu hēi yàn
Cackling Goose **LC**

L: 63–65 cm
WS: 108–111 cm
Habitat: Inhabits lakes and marshes; often forages on grass or farmland.
Behavior: Grazes on grass on land; also forages in water, employing

swanlike behavior where it submerges head underwater.
Distribution: Vagrant. *B. h. minima* recorded in Hubei; *B. h. leucopareia* in Hunan, Hubei, Jiangxi, Henan, and Jilin.
Voice: Similar to Canada Goose but higher pitched and squeakier.

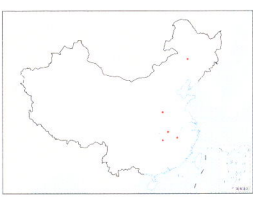

Branta leucopsis

白颊黑雁 bái jiá hēi yàn

Barnacle Goose　LC

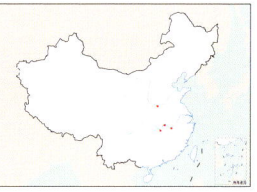

L: 58–71 cm
WS: 132–145 cm
Habitat: Nests on stony hillsides and lives on marshes in its normal range; found in wide river channels or on shallow lakes in China.
Behavior: Wingbeats relatively slow. Does not associate with other goose species in its normal wintering range, but vagrants in China tend to associate with large flocks of Taiga Bean-Goose, Tundra Bean-Goose, and Greater White-fronted Goose.
Distribution: Vagrant to Henan (Mengjin), Hunan (Dongting Lake), Hubei (Wuhan), and Jiangxi (Poyang Lake).
Voice: Short, shrill, barking *kah* or *kaw*, often repeated.

Anser indicus

斑头雁 bān tóu yàn

Bar-headed Goose　LC

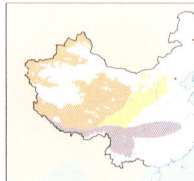

L: 62–85 cm
WS: 140–160 cm
Habitat: Breeds on plateau wetlands, usually at altitudes above 4000 m. Winters on lakes or marshes.
Behavior: Vigilant and gregarious. Migrates in flocks of dozens in V formation. Often calls in flight; circles once or twice before landing.
Distribution: Breeds on the Qinghai-Tibet Plateau and Tian Shan Mountains; winters on the Yunnan-Guizhou Plateau and the wetlands in S Tibet. Also occasionally seen on the wetlands of Yangtze Plain; rare in East China, vagrant in North China.
Voice: Rhythmic, low-pitched, slowly delivered honking: *hang-hang-hang*.

Anser canagicus

帝雁 dì yàn

Emperor Goose　NT

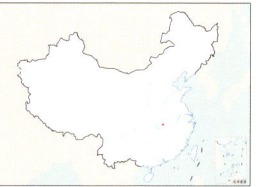

L: 66–72 cm
WS: 119–122 cm
Habitat: Generally seen on seashores; also occurs inland on wetlands or grass in winter.
Behavior: Rarely seen in large flocks. Often forages in intertidal zones, sometimes searches for shellfish by stamping its feet on muddy flats. Flies relatively low.
Distribution: Vagrant recorded in Hubei (Chenhu Lake), November 2019.
Voice: High-pitched *kla-ha* and relatively low-pitched and rapid *gagagaga*, reminiscent of Barnacle Goose but higher pitched.

Anser caerulescens

雪雁 xuě yàn

Snow Goose　LC

L: 66–84 cm
WS: 132–165 cm
Habitat: Winters on marshes, estuaries, and coastal flats; forages on grass or farmland.
Behavior: Vigilant. In China, often associates with Taiga Bean-Goose and Graylag Goose, which exhibit similar behaviors.
Distribution: Once considered a vagrant, but a few individuals are recorded almost annually in Jilin (Hunchun), Bohai Bay, coastal wetlands along S Yellow Sea, Poyang Lake, and Dongting Lake. Likely a rare winter visitor.
Voice: Sonorous, disyllabic, rising honking and cackling, *ga-luk* or *gaga*.

Anser anser

灰雁 huī yàn

Graylag Goose　LC

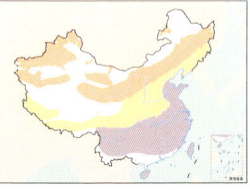

L: 76–89 cm
WS: 147–180 cm
Habitat: Prefers breeding on large marshes or lakes with expansive reeds; winters on wide rivers, lakes, and marshes.
Behavior: Gregarious and clumsy-looking. Generally forages during the day by standing in shallow water or upending in water; forages less frequently on land than Taiga Bean-Goose.
Distribution: Breeds in Northeast and northwestern China; migrates through most of China; winters in North China, Yangtze Plain, Yunnan-Guizhou Plateau, and South China. Commonly seen in suitable habitats.
Voice: Various loud honking notes similar to a domestic goose. Distinctive trisyllabic *ahng-ung-ung*.

Anser cygnoides

鸿雁 hóng yàn

Swan Goose　VU

L: 80–94 cm
WS: 165–185 cm
Habitat: Breeds on wetlands in grasslands and in the transition zone between grasslands and coniferous forests in Northeast China; winters at estuaries, coastal tidal flats, and lakes.
Behavior: Gregarious. Generally forages at night on farmland or in grassy areas; also digs plants in shallow waters or soft tidal flats.
Distribution: Breeds in Northeast China and E and C Inner Mongolia; migrates through much of eastern and central China. Locally common. Rare passage migrant in Xinjiang. Winters in the Yangtze Plain and the southeast coast.
Voice: Various loud honking and braying calls, most huskier than Graylag Goose: *gaa* or *gang-gang*.

Barnacle Goose

White forehead

Short bill

Short neck

Snow Goose

Base of bill curved, differing from bill of extralimital Ross's Goose

Short bill with black "grin patch"

Short neck

Black primaries

Bar-headed Goose

Yellow bill

Light gray nape

Graylag Goose

Emperor Goose

Head of some individuals tinged yellow

White head and hindneck

Short bill

White tail

Swan Goose

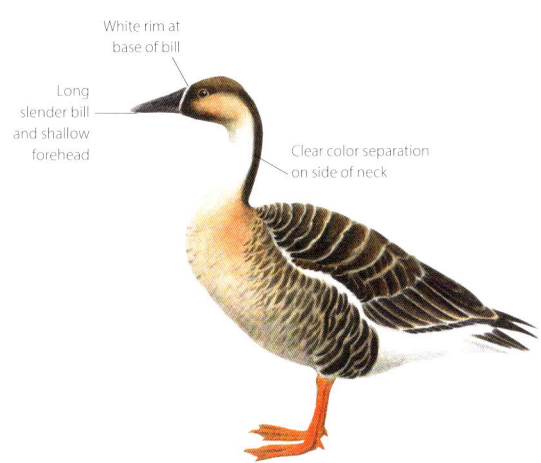

White rim at base of bill

Long slender bill and shallow forehead

Clear color separation on side of neck

Anser fabalis

豆雁 dòu yàn

Taiga Bean-Goose   LC

L: 70–89 cm
WS: 147–175 cm
Habitat: Breeds in marshes in taiga forests; winters in open wilderness, wetlands, and farmland.
Behavior: Gregarious. Often associates with other goose species; feeding habits similar to Swan Goose.
Distribution: *A. f. johanseni* in Western China; *A. f. middendorffii* throughout China except for southwestern China and the Qinghai-Tibet Plateau. Less abundant than Tundra Bean-Goose. The two bean-goose species are sometimes treated as a single species.
Voice: Typical goose-like honking: *kang-kung*.

Anser albifrons

白额雁 bái é yàn

Greater White-fronted Goose   LC

L: 70–86 cm
WS: 130–165 cm
Habitat: Forages on farmland, lakes, or muddy river bars.
Behavior: Gregarious, crepuscular. Forages on muddy river bars; takes off quickly from water.
Distribution: *A. a. flavrirostris* migrates throughout most provinces in eastern China and winters in the Yangtze Plain and the southeast coast; not uncommon, occasionally recorded in southwestern China and Taiwan. *A. a. albifrons* migrates through W Xinjiang and S Tibet, a rare passage migrant.
Voice: Honking call, *kyew-kyew*, higher pitched than calls of Graylag and bean-geese.

Sarkidiornis melanotos

瘤鸭 liú yā

Knob-billed Duck   LC

L: 64–79 cm
WS: 116–145 cm
Habitat: Inhabits freshwater marshes, rivers, lakes, or paddy fields.
Behavior: Spreads out in wet seasons, gregarious in dry seasons, flocks sometimes segregated by sex.
Distribution: Vagrant. Previously recorded in SE Tibet and S Yunnan. The most recent record in China was a male in Fuzhou, Fujian, in 1914.
Voice: Generally silent, rarely calls when disturbed. Utters high-pitched *ker* or *ker-ker* during breeding season.

Anser serrirostris

短嘴豆雁 duǎn zuǐ dòu yàn

Tundra Bean-Goose   NR

L: 66–89 cm
WS: 150–200 cm
Habitat: Breeds on tundra; winters on farmland, shallow lakes, and marshes. Prefers paddy fields and wet grasslands more than Taiga Bean-Goose.
Behavior: Gregarious; often forms large flocks containing up to tens of thousands of individuals; also associates with other geese, especially Graylag and Greater White-fronted. Vigilant and easily startled.
Distribution: The most common goose species in eastern China. *A. s. rossicus* winters in Xinjiang and occasionally Shaanxi; *A. s. serrirostris* occurs throughout central and eastern China; winters mainly in the Yangtze River Watershed and the southeast coast but regularly north to Beijing and Liaoning.
Voice: Very similar to Taiga Bean-Goose but perhaps lower pitched and sounding more "muffled."

Anser erythropus

小白额雁 xiǎo bái é yàn

Lesser White-fronted Goose   VU

L: 56–66 cm
WS: 115–135 cm
Habitat: Forages on farmland, river bars, grassy areas, etc.; spends the night on large lakes and wide river channels.
Behavior: Gregarious, but often does not flock with Greater White-fronted Goose. Forages in grassy areas, making fast motions.
Distribution: Vagrant in Xinjiang. Eastern population migrates through eastern China and winters in the Yangtze Plain and South China; generally rare and concentrated in a few wintering sites.
Voice: Honking *kya-kra-kra*, even higher and sharper than Greater White-fronted Goose.

Asarcornis scutulata

白翅栖鸭 bái chì qī yā

White-winged Duck   EN

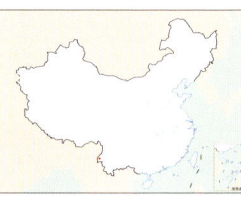

L: 66–81 cm
WS: 116–153 cm
Habitat: Inhabits secluded weedy ponds or marshes with slow-flowing water in forests. Occasionally in paddy fields.
Behavior: Often moves about solitarily, in pairs, or in small flocks; mostly sedentary or rests in trees, active only at dusk; not easy to see.
Distribution: Vagrant. First recorded in Yunnan (Yingjiang) in April 2019.
Voice: Often calls in flight—a high-pitched, honking, Ruddy Shelduck–like *ko-argh*, as well as the monosyllabic notes *ang* and nasal *hong*.

Taiga Bean-Goose

Yellow patch on bill

Long slender bill and shallow forehead, like Swan Goose

Long neck

Tundra Bean-Goose

Yellow patch on bill

Relatively rounded head

Extremes separable from Taiga Bean-Goose: often shorter bill with deeper base; lower mandible bulges at base

Short neck

Greater White-fronted Goose

Relatively rounded white patch on forehead

Bill longer than on Lesser White-fronted Goose

Wing tips align with tail at rest

Lesser White-fronted Goose

Relatively pointed white patch on forehead

Gold eye ring on smaller head

Relatively short stubby bill, often brighter pink

Shorter neck

Wings extend beyond tail at rest

Knob-billed Duck

Flecked head

White breast

White-winged Duck

Head less mottled in some individuals

Dark blue secondaries

White wing coverts

Dark yellow bill

Mottled head

Dark blue secondaries

Cygnus olor

疣鼻天鹅 yóu bí tiān é

Mute Swan ⓛⓒ

L: 125–160 cm
WS: 200–240 cm
Habitat: Generally breeds on lakes in grassland areas; migrates and winters on lakes, large reservoirs, and rivers and gulfs with slow-flowing water.
Behavior: Laborious

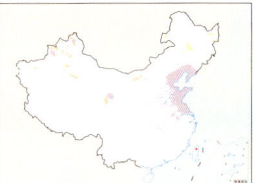

takeoff requiring a long run-up, vibrant wingbeats; when swimming, holds neck in S shape, with bill drooping and wings hunching, different from other swans. Carries chicks on its back similar to grebes.
Distribution: Breeds in the central and northern parts of northwestern China, the eastern edge of the Qinghai-Tibet Plateau, and W Inner Mongolia; migrates through Northeast and North China; winters along the Yellow Sea coast and (rarely) Qinghai Lake. Vagrant to Taiwan.
Voice: Various grunting, hissing, and snorting notes.

Cygnus columbianus

小天鹅 xiǎo tiān é

Tundra Swan ⓛⓒ

L: 115–150 cm
WS: 175–195 cm
Habitat: Inhabits wide, shallow lakes with abundant aquatic plants; also forages on wet grasslands or farmland.
Behavior: Moves in close family units, which at

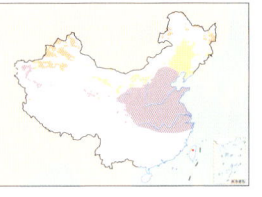

times gather in larger flocks; may associate with other swans and geese. Neck relatively short, usually held straight when swimming; takeoff and landing more nimble.
Distribution: Migrates through northwestern, Northeast, and North China; winters in the Yangtze Plain and southeastern China; occasionally recorded in southwestern and South China.
Voice: Often calls on water and in flight. Utters various pleasant, resounding, bugling notes similar to those of Whooper Swan.

Cygnus cygnus

大天鹅 dà tiān é

Whooper Swan ⓛⓒ

L: 140–160 cm
WS: 205–235 cm
Habitat: Similar to Tundra Swan.
Behavior: Similar to Tundra Swan. Often forages with head submerged in water; head and neck sometimes tinged dark yellow.

Distribution: Breeds in northern northwestern and northern Northeast China; migrates through northwestern and North China; winters farther north than Tundra Swan, from Beijing and the Yellow River Watershed to the Yangtze Plain; occasionally to the east coast.
Voice: Similar to Tundra Swan but often slightly lower pitched, louder, flatter in tone and given in longer sequences.

Tadorna tadorna

翘鼻麻鸭 qiào bí má yā

Common Shelduck ⓛⓒ

L: 55–65 cm
WS: 110–130 cm
Habitat: Prefers marshes and open lakes in breeding season; also appears on grasslands and semidesert areas. Winters in various types of open water but prefers estuaries and gulfs.

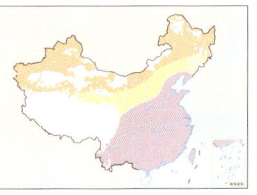

Behavior: Forages by standing in shallow water and sweeping bill from side to side; also digs in mud. Winters at sea in flocks. Slow wingbeats in flight.
Distribution: Throughout China except for the Qinghai-Tibet Plateau and Hainan. Breeds in Northeast, northwestern, and North China; winters in eastern China and Yangtze Plain. Relatively common.
Voice: Various soft, not unpleasant, whistling and whizzing calls in breeding season, while other calls include a rapid *gagagagagaga*, given by female in flight.

Tadorna ferruginea

赤麻鸭 chì má yā

Ruddy Shelduck ⓛⓒ

L: 58–70 cm
WS: 110–135 cm
Habitat: Adaptable to different water bodies, including rivers, lakes, and artificial wetlands; also occurs in vegetated areas in deserts. Distribution up to 4700 m, but rarely

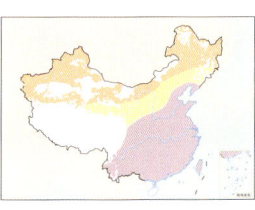

occurs at estuaries or marine habitats.
Behavior: Gregarious. Forages on shoals or in grassy areas; breeds in holes, on cliffs, even in trees, usually far from water.
Distribution: Throughout China, except for Hainan; generally very common.
Voice: Highly vocal; calls at rest and in flight. Like Common Shelduck, sexes differ. Frequent calls include a soft purring and a longer *aang* or *kakaka-kakaka*.

Mute Swan

Protruding forehead

Dark pinkish bill, black at base and around nostrils

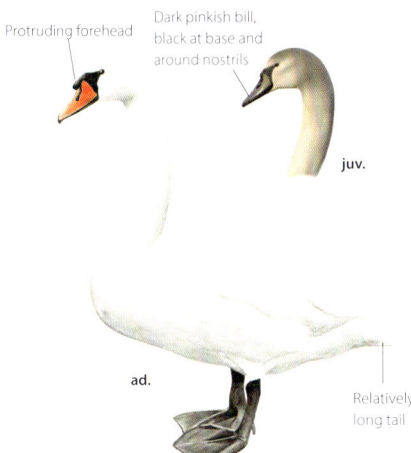

juv.

ad.

Relatively long tail

Tundra Swan

More rounded head

Less yellow on bill; color does not extend beyond nostril and is rarely wedge shaped

Whooper Swan

Triangular head

Long bill with extensive yellow that extends beyond nostril to sharp wedge

Common Shelduck

♂

Red knob on forehead

Black tail tip

♂ br.

Black belly

Lacks knob on forehead

♀

♂ non-br.

Ruddy Shelduck

♂

♂

Black neck ring

♀

37

Aix galericulata
鸳鸯 yuān yāng
Mandarin Duck LC

L: 41–51 cm
WS: 65–75 cm
Habitat: Inhabits relatively open waters in winter; also appears on streams. Often occurs on rivers in dense forests in breeding season; may also live and breed in city parks.

Behavior: Gregarious in winter; concentrates in one area to court in early spring, after which it spreads out to breed. Nests in tree cavities; also rests in trees. Often swims slowly on water, rarely dives.
Distribution: Breeds in most areas of central and eastern China north of the Yangtze River, also patchily in southwestern China; winters in North, South, and southwestern China and the Yangtze River Watershed; resident in Taiwan. Very common in water bodies of gardens and parks in cities such as Beijing and Hangzhou; common or occasional in other areas.
Voice: Relatively quiet. Call of male, made often in flight, is a high-pitched, nasal, drawn-out *hweee*. Also a short, soft, rising, nasal *quib*, lower pitched in female.

Nettapus coromandelianus
棉凫 mián fú
Cotton Pygmy-Goose LC

L: 31–38 cm
WS: 55–60 cm
Habitat: Inhabits quiet freshwater wetlands with floating or emergent vegetation.
Behavior: Forages by swimming among aquatic plants, rarely seen on land.

Solitary or in pairs in breeding season; nests in tree cavities; gregarious during migration. Shy, vigilant, takes off quickly from water in danger. Flies with buoyant wingbeats.
Distribution: Uncommon. Occurs mainly in C and SW Sichuan and the Yangtze Plain and regions to its south; rare in North China and Taiwan. Primarily visits there as summer breeder, and migrates south in winter.
Voice: Males give a peculiar short nasal yapping, *ka-ka-kalawa*; females' call is a weak quack.

Sibirionetta formosa
花脸鸭 huā liǎn yā
Baikal Teal LC

L: 36–43 cm
WS: 65–75 cm
Habitat: In nonbreeding season, inhabits fresh or brackish open waters, such as lakes, rivers, marshes, and reservoirs; also forages on paddy fields.

Behavior: Often gregarious in winter, forming flocks containing up to tens of thousands of individuals; also mixes with other dabbling ducks. Active feeding starts at dusk in winter.
Distribution: Uncommon and local. Wintering populations dropped precipitously in the last century in China. Winters in Central, East, and South China, including Taiwan and Hainan, and migrates through Northeast and North China; can be common during migration.
Voice: Typically silent in winter. Males occasionally give a chuckling laugh, *woh-ot-ot*; females a nondescript quack. Large flocks can be noisy.

Spatula querquedula
白眉鸭 bái méi yā
Garganey LC

L: 37–41 cm
WS: 59–67 cm
Habitat: Nests in grasslands, swampy meadows, and reed marshes. Forages on open freshwater lakes, ponds, estuaries, reservoirs, reed marshes, and wet grasslands; also forages on paddy fields.

Behavior: Gregarious during migration. Forages by dabbling in water.
Distribution: Common. Breeds in Xinjiang and Northeast China; winters in South China and along the southeast coast, including Taiwan and Hainan; migrates through most of China except for Tibet and Qinghai.
Voice: Song of male is a distinctive dry wooden rattle, varying in pitch. Various other unremarkable vocalizations.

Spatula clypeata
琵嘴鸭 pí zuǐ yā
Northern Shoveler LC

L: 44–52 cm
WS: 73–82 cm
Habitat: Prefers lakes, estuaries, coastal marshes, ponds, paddy fields, and other shallow wetlands rich in aquatic plants.
Behavior: Often gregarious and mixes

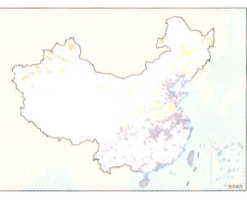

with other ducks. When foraging in water or on muddy areas, holds neck straight and sweeps bill from side to side to filter feed under water's surface.
Distribution: Common. Breeds in Northeast and northwestern China; winters in most of southern China, including Taiwan and Hainan; migrates through most of China.
Voice: Often silent, sometimes a soft nasal *took-took*.

Mandarin Duck

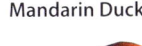

♂ br.

♀

Narrow white line behind eye

Gray bill

Pink bill

Narrow white line behind eye

Pale spots on flanks relatively rounded

Pale spots on flanks form stripes

♂ eclipse

Baikal Teal

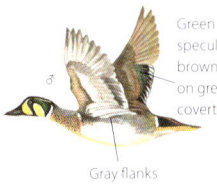

♂

Green speculum, brown tips on greater coverts

Gray flanks

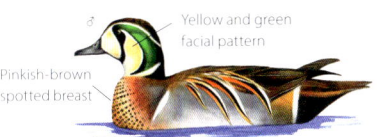

♂

Yellow and green facial pattern

Pinkish-brown spotted breast

♀

Conspicuous white Northern Pintail-like trailing edge on secondaries

♀

Bold head pattern with white loral spot

Cotton Pygmy-Goose

Black and white primaries ♂

Dark green back and coverts

Dark green neck ring

Crimson iris

White head and neck

Dark green back

♂

♀

Yellow bill

Brown iris

Dark eye stripe

♀

Garganey

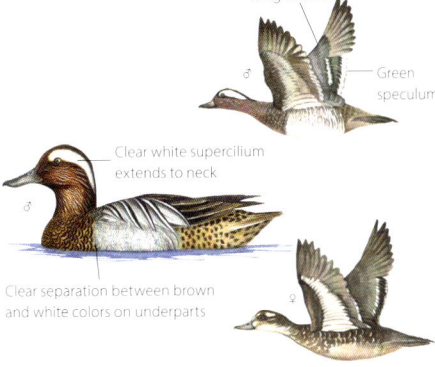

Bluish-gray wing coverts

♂

Green speculum

Clear white supercilium extends to neck

♂

Clear separation between brown and white colors on underparts

♀

Grayish-black bill

♀

Pale chin and throat

Northern Shoveler

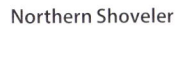

Bluish-gray wing coverts

♂

Green speculum

♀

Black bill, widened distally

♂

White breast contrasts strongly with chestnut flanks

Bill yellow to dark brown, widened distally

♀

♀

39

Mareca americana

绿眉鸭 lù méi yā

American Wigeon LC

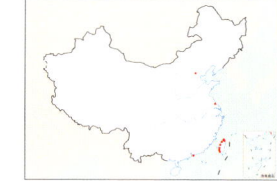

L: 45–56 cm
WS: 76–89 cm
Habitat: Inhabits freshwater wetlands, such as lakes, marshes, estuaries, and reservoirs; prefers freshwater marshes in coastal areas in winter. Also forages on paddy fields.
Behavior: Often occurs alone; also turns up with flocks of Eurasian Wigeon.
Distribution: Vagrant in Taiwan, Hong Kong, Jiangsu, and Beijing. Hybrids with Eurasian Wigeon occasionally found.
Voice: Male's whistle similar to that of Eurasian Wigeon but weaker.

Mareca falcata

罗纹鸭 luó wén yā

Falcated Duck NT

L: 46–54 cm
WS: 78–82 cm
Habitat: Prefers estuarine marshes and ponds with abundant aquatic vegetation in breeding season; inhabits large shallow water bodies, paddy fields, or flooded meadows in nonbreeding season.
Behavior: Gregarious in nonbreeding season, often mixes with other ducks. Shy and vigilant.
Distribution: Common in eastern China, but wintering populations have declined in the past decade. Breeds in Heilongjiang and Jilin; winters in eastern China, the Yangtze Plain, and the southeast coast, including Taiwan and Hainan.
Voice: Not very vocal. Males give short rising whistles, occasionally finished with an abrupt, fart-like buzz. Females make a hoarse quack.

Mareca strepera

赤膀鸭 chì bǎng yā

Gadwall LC

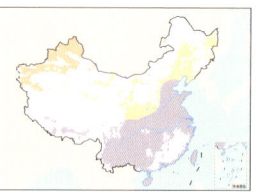

L: 45–57 cm
WS: 78–95 cm
Habitat: In breeding season, prefers estuarine marshes (both freshwater and brackish) with lush vegetation; tends to stay among tall and dense emergent plants. In nonbreeding season, often forages on paddy fields. Also occurs in coastal areas.
Behavior: Gregarious, often mixes with other ducks. Shy and vigilant.
Distribution: Common. Breeds in northern Northeast China and W Xinjiang; winters in S Tibet, Yunnan, Guizhou, Sichuan, the Yangtze Plain, and the southeast coast, including Taiwan and Hainan; migrates through Xinjiang, Qinghai, Inner Mongolia, and North China.
Voice: Short husky croaks and equally short rattles. Female makes a Mallard-like quack.

Mareca penelope

赤颈鸭 chì jǐng yā

Eurasian Wigeon LC

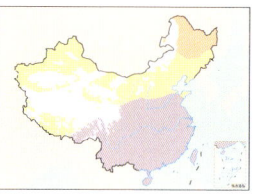

L: 42–51 cm
WS: 71–85 cm
Habitat: Inhabits freshwater marshes and lakes surrounded by woodlands in breeding season; moves to coastal lakes, marshes, estuaries, gulfs, fishponds, and reservoirs in nonbreeding season. Also forages on paddy fields.
Behavior: Gregarious, often mixes with other ducks. Shy and vigilant. In winter, also forages at night according to tides. Filter feeds phytoplankton and aquatic plants in water; also eats leaves, stems, roots, and grass seeds on land.
Distribution: Widely distributed across China. Breeds in Northeast China; winters along the Yellow River and in the regions to its south, including Taiwan and Hainan; migrates through Xinjiang, Inner Mongolia, southern Northeast China, and North China.
Voice: Male's call is a distinctive, loud, rising, then falling whistle: *whee-oh*. Both sexes make a gruff *ra-kaah-kar*.

Anas poecilorhyncha

印度斑嘴鸭 (印缅斑嘴鸭)

yìn dù bān zuǐ yā (yìn miǎn bān zuǐ yā)

Indian Spot-billed Duck LC

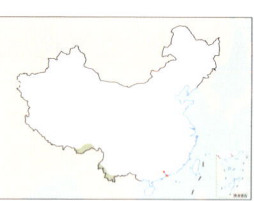

L: 58–63 cm
WS: 83–91 cm
Habitat: Inhabits various types of wetlands, including lakes, rivers, marshes, paddy fields, and estuaries.
Behavior: Often gregarious; also turns up in flocks of other ducks.
Distribution: Uncommon resident. *A. p. haringtoni* generally scarce but locally common in W & S Yunnan, vagrant to Guangdong and Hong Kong. *A. p. poecilorhyncha* recorded in Tibet (Shannan).
Voice: Hurried quacking, very similar to Mallard and Eastern Spot-billed Duck.

Anas zonorhyncha

斑嘴鸭 bān zuǐ yā

Eastern Spot-billed Duck LC

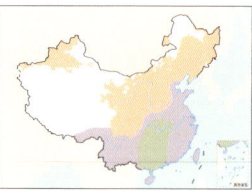

L: 58–63 cm
WS: 83–91 cm
Habitat: Similar to Indian Spot-billed Duck. Adaptable, highly tolerant of human disturbance.
Behavior: Often in small flocks; forms large flocks containing up to a few hundred individuals in winter. Male guards nesting female and helps protect chicks in breeding season.
Distribution: Common in eastern China. The northern breeding population migrates slightly south in winter. The southern resident population is expanding in range.
Voice: Mallard-like quacking.

American Wigeon

Relatively light-colored head with white crown

Dark iridescent green patch around eye

♂

Pinkish purple from breast to flanks

Whitish head and neck

Relatively dark around eye

♀

Reddish-brown flanks

Eurasian Wigeon

Large white patch on upperwing coverts

Narrow dark green speculum

♂

Yellowish crown

Chestnut head

♂

Black undertail coverts contrast strongly with white belly

Bluish-gray bill with black tip

♀

Heavily rufous-brown head, breast, and flanks

♀

Falcated Duck

♂

Dark green speculum; white tips on greater coverts

Glossy green on face and neck

♂

Yellowish-white patch on undertail coverts

Finely vermiculated from breast to flanks

Long scythe-shaped tertials

Dark bill

♀

♀

Relatively long tertials

Indian Spot-billed Duck

Orange legs

Bill black with yellow tip

Green speculum

A. p. haringtoni

Orange red at base of bill

A. p. poecilorhyncha

Gadwall

♂

White speculum; black and chestnut coverts

Black bill

♂

Black undertail coverts

Finely vermiculated breast and flanks

♀

Side of bill orange to brown

♀

White speculum visible at rest

Eastern Spot-billed Duck

Orange legs

Black bill with yellow tip

Bluish-purple speculum

Blackish-brown moustachial stripe

Anas luzonica

棕颈鸭 zōng jǐng yā

Philippine Duck **VU**

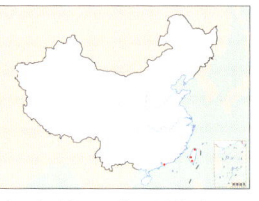

L: 48–58 cm
WS: 84 cm
Habitat: Inhabits freshwater lakes, marshes, and rivers.
Behavior: Generally in small flocks, often occurs solitarily or in pairs in China; mixes with flocks of other ducks. Forages on water and on land. Diet similar to Mallard.
Distribution: Rare. Vagrant in Taiwan and Hong Kong.
Voice: Soft nasal Common Shelduck–like note and others very similar to those of Mallard.

Anas platyrhynchos

绿头鸭 lǜ tóu yā

Mallard **LC**

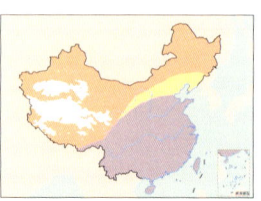

L: 55–70 cm
WS: 81–95 cm
Habitat: Inhabits wetlands, such as lakes, rivers, marshes, paddy fields, and estuaries. Adaptable, highly tolerant of human disturbance.
Behavior: Generally occurs in pairs or small flocks; occasionally seen in large flocks of hundreds, rarely thousands of individuals. Also mixes with flocks of other ducks. Omnivorous; eats seeds, stems, and leaves of aquatic and terrestrial plants, feeds on terrestrial or aquatic invertebrates, and occasionally amphibians and fish.
Distribution: Common, occurs in most of China. The northern breeding population migrates southward in winter. The southern resident population, perhaps of feral origin, is expanding in range.
Voice: Males give a soft *raeb*; females make a familiar quacking similar to many domestic ducks.

Anas acuta

针尾鸭 zhēn wěi yā

Northern Pintail **LC**

L: 51–76 cm (including the 10 cm long central tail feathers)
WS: 80–95 cm
Habitat: Inhabits freshwater marshes, lakes, rivers, estuaries, seashores, and other areas with shallow water. Also forages on paddy fields.
Behavior: Generally gregarious; often mixes with flocks of other ducks. Submerges the upper body in water when foraging; long neck adapted for foraging in water. Flies in straight line; wingbeats slower than other ducks.
Distribution: Common in China. Breeds in NW Xinjiang; winters across most areas south of the Yangtze River, including Taiwan and Hainan. Migrates through Northeast and North China and northern Yangtze Plain.
Voice: Male has a short descending note reminiscent of that of Eurasian Teal but shorter.

Anas crecca

绿翅鸭 lǜ chì yā

Eurasian Teal **LC**

L: 34–38 cm
WS: 53–59 cm
Habitat: Inhabits various types of wetlands, including lakes, rivers, reservoirs, paddy fields, and gulfs. Highly adaptable.
Behavior: Generally gregarious; often mixes with flocks of other ducks. Forages mainly by dabbling in water, upends and submerges head to feed. Also dives to feed in shallow water or filters food from mud. Individuals feeding in estuaries time their activities according to tides, foraging most actively during rising and falling tides. Flies with fast wingbeats.
Distribution: Common in China, but the wintering population has experienced a drastic decline in the past decade. Breeds in NW Xinjiang and northern and central Northeast China; winters from Beijing south, with large numbers in the Yangtze River Watershed and most of the southeast coastal region, including Taiwan and Hainan.
Voice: Male's call an attractive, penetrating whistle: *prip*. Females make a hoarse Mallard-like quack.

Anas carolinensis

美洲绿翅鸭 měi zhōu lǜ chì yā

Green-winged Teal **LC**

L: 34–38 cm
WS: 58 cm
Habitat: Same as Eurasian Teal.
Behavior: All records in China are of birds that turned up in flocks of Eurasian Teal.
Distribution: Rare. Vagrants in Hebei (Beidaihe), Jiangsu (Suzhou), Shanghai (Nanhui Dongtan), Zhejiang (Hangzhou), Guangdong (Shantou), and Hong Kong.
Voice: Similar to Eurasian Teal.

Marmaronetta angustirostris

云石斑鸭 yún shí bān yā

Marbled Teal **VU**

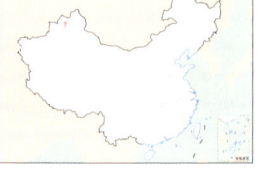

L: 39–48 cm
WS: 63–67 cm
Habitat: Inhabits dry areas; follows ephemeral ponds and rainfall. Also occurs at estuaries and deltas.
Behavior: Crepuscular; hides in trees near marshes in the daytime. Seldom flies; flies close to water's surface.
Distribution: On June 18, 1985, eight individuals were recorded in Xinjiang (Ailik Lake). The sighting was published by the Oriental Bird Club, but no other records thereafter. It was suspected by some that the record might be a misidentification of Red-crested Pochard.
Voice: Generally quiet. A high-pitched, asthmatic, wheezed *wee-uu* in breeding season.

Philippine Duck

Yellowish-brown legs

Bluish-green speculum

Bluish-gray bill without spots

Brown supercilium, cheek, and neck

Mallard

Dark iridescent green head and neck

Yellow bill

Chestnut-brown breast

Yellowish-brown bill with variably splotched upper mandible

Orange-red legs

Eurasian Teal

White bars on upperwing coverts

Green speculum

Broad dark green stripe runs through eye to neck

White stripe on side of body

Undertail coverts form creamy-yellow triangle

Inconspicuous supercilium

Green speculum often visible

Green-winged Teal

Whitish or buff bars on upperwing coverts

Inconspicuous pale fringe to green mask

Vertical white stripe on side of breast

Northern Pintail

Thin white stripe on side of neck

Black undertail coverts

Pin-shaped black tail feathers

Brown head and neck

Blue-gray bill

Relatively long neck

Pointed tail feathers

Marbled Teal

Relatively long and slender bill

Dark around eye

Short crest

No distinct crest

43

Netta rufina

赤嘴潜鸭 chì zuǐ qián yā

Red-crested Pochard

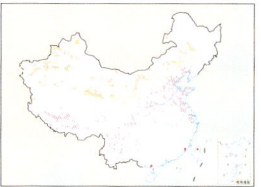

L: 53–57 cm
WS: 85–90 cm
Habitat: Inhabits open freshwater lakes or rivers with slow-flowing water; especially prefers deep water with emergent vegetation.
Behavior: Feeds mainly on aquatic plants; forages mainly by diving in water. Also adopts foraging techniques of dabbling ducks in shallow water.
Distribution: Breeds in N Xinjiang, W Inner Mongolia, and E Qinghai-Tibet Plateau; winters on freshwater lakes on plateaus in S Tibet and southwestern China; locally common. Expanding eastward. Occasionally stops over and winters in North and East China and the Yangtze River Watershed; vagrant in Fujian, Guangxi, and Taiwan.
Voice: Male gives a loud, wheezy, suppressed sneeze, as well as other quieter *seer* calls.

Aythya valisineria

帆背潜鸭 fān bèi qián yā

Canvasback

L: 48–63 cm
WS: 74–90 cm
Habitat: Inhabits lakes, ponds, shoals, marshes, estuaries, and streams with slow-flowing water.
Behavior: Often moves in tight flocks. Vigilant. Dives to feed; also extracts food by using feet to stir up sediment in water.
Distribution: Vagrant; recorded in Taiwan (Taipei, Changhua, and Hsinchu).
Voice: Vagrants are unlikely to be heard; quiet. In breeding season, *wee-woo-o-o.* Males sometimes make a slightly hoarse *errr* call.

Aythya americana

美洲潜鸭 měi zhōu qián yā

Redhead

L: 40–56 cm
WS: 74–85 cm
Habitat: Occurs on seashores and shallow coastal ponds during migration and winter; prefers alkaline waters.
Behavior: Similar to Common Pochard. Flight rapid, with strong wingbeats. In breeding season, sometimes lays eggs in nests of other duck species, which then incubate and care for its young along with their own.
Distribution: Vagrant in Jiangsu (Yancheng).
Voice: Male's call an *ahaa*; female's call an *err-err.*

Aythya ferina

红头潜鸭 hóng tóu qián yā

Common Pochard

L: 41–50 cm
WS: 67–75 cm
Habitat: Inhabits open slow-flowing waters; in breeding season often chooses lakes with lush fringing vegetation.
Behavior: Often gregarious, forms flocks of hundreds to thousands of individuals. Dives to feed; often on aquatic plants. Crepuscular, may feed all night.
Distribution: Breeds in N Xinjiang, western Northeast China; winters in C Sichuan, the Yangtze Plain, South China, and Taiwan; migrates through most of China. Fairly common, but the population is declining.
Voice: Trilling *errr.* Rarely calls after breeding season.

Red-crested Pochard

♂

♀

♀

♀

♂ eclipse

♂

Red iris

Large rounded head

Relatively thin red bill

♀ Relatively thin bill with light pink patch; differs from female Black Scoter

Canvasback

♂

♂

♂

♀

♀

♂ eclipse

Nearly trapezoidal head

Base of bill relatively thick

Nearly black forehead

Relatively large all-black bill, thick at the base ♂

Pale back

Relatively long neck

Redhead

♀

Bill with black tip and pale subterminal band

Steep forehead

Pale eye ring

Steep forehead

Gold iris

Dark gray back, darkest among all red-headed pochard species

Bill with black tip and leaden base ♂

♂ eclipse

Canvasback

Common Pochard

Redhead

Common Pochard

♀

♀

♂

♀

♀

Red iris; unlike yellow iris of closely related Redhead

Relatively rounded head

Bill smaller than female Canvasback

Dark gray bill without pale band

♀ sum.

Bill with pale gray subterminal band

Pale subterminal band

♀ win.

♂ eclipse

♂

45

Aythya marila
斑背潜鸭 bān bèi qián yā
Greater Scaup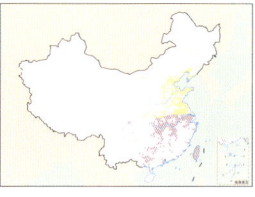

L: 42–49 cm
WS: 71–80 cm
Habitat: Often occurs on slow-flowing, shallow waters near coast. Inhabits both freshwater and saline wetlands. Also occurs on large inland lakes in winter.
Behavior: Often mixes with Tufted Duck, so extra attention should be given when observing large flocks of Tufted Duck.
Distribution: Uncommon in eastern China during migration; winters from the Yangtze River Watershed to South China (including Taiwan).
Voice: In flight, an unpleasant harsh growl, like Tufted Duck but lower pitched; rarely calls in nonbreeding season.

Aythya collaris
环颈潜鸭 huán jǐng qián yā
Ring-necked Duck

L: 37–46 cm
WS: 61–75 cm
Habitat: Prefers water bodies with abundant aquatic plants and sheltering places; winters in various shallow wetlands (less than 1.5 m deep), including marshes, flooded agricultural land, and impoundments.
Behavior: Relatively quiet in winter. Takes off directly from water, with no need for run-up.
Distribution: Vagrant in Taiwan and Shandong (Qingdao).
Voice: Unlikely to be heard in China; occasionally gives a brief *wow-wow-wow* call, like a barking dog.

Aythya affinis
小潜鸭 xiǎo qián yā
Lesser Scaup

L: 38–48 cm
WS: 64–74 cm
Habitat: Inhabits ponds and marshes with abundant aquatic plants.
Behavior: Gregarious, turns up with flocks of other diving ducks.
Distribution: Vagrant, Taiwan (Yilan).
Voice: Unlikely to be heard in China. Calls *arr* in breeding season, higher pitched than Greater Scaup.

Aythya fuligula
凤头潜鸭 fèng tóu qián yā
Tufted Duck

L: 34–49 cm
WS: 65–72 cm
Habitat: Prefers bodies of still or slow-flowing fresh water, such as lakes and ponds with aquatic plants; also occurs at estuaries. Rarely occurs offshore or in areas with strong wave action.
Behavior: Fast flier. Often dives for shellfish. Forms large flocks containing up to hundreds of individuals. Hybrids with other diving ducks commonly seen.
Distribution: Breeds in Western and Northeast China; winters in the Yangtze River Watershed and East and South China; migrates through most of China as one of the most common diving ducks, but less so in northwestern China.
Voice: "Song" an excitable sequence of bubbling notes. Calls include a mellow, coarse, churred *karrr*; rarely calls in nonbreeding season.

Greater Scaup

Bill wider and head broader than Lesser Scaup

Relatively rounded head; appears green in good light

Conspicuous large white patch with sharp border at base of bill

White secondaries; grayish-white areas on primaries increase in size and darken from P1 to P10

Pale crescent-shaped patch behind ear

♂ eclipse

♀ win.

♀ sum.

Finely vermiculated back appears pale gray from a distance; differs from Tufted Duck

Lesser Scaup

Bill smaller and narrower than Greater Scaup

Head tinged purplish in good light

Heavily vermiculated back

Lesser Scaup

Greater Scaup

Relatively peaked forehead

Relatively flat and straight nape

Grayish-brown primaries without white area

♂ eclipse

Ring-necked Duck

Metallic purple sheen on head

Yellow iris

Crown peaked, without prominent crest

Broad grayish-white band near bill tip

White line at base of bill

♂ win.

Brown iris with grayish-white eye ring and white line behind eye

Peaked crown

Base of bill pale

Relatively narrow grayish-white ring on bill

♀ win.

Conspicuous white flank spur

Tufted Duck

♂ ♀ ♂

Gold iris

Metallic purple or green sheen on head

Long crest

Black back

Gold iris

Diffuse white patch at base of bill in some individuals

Short crest

♀ win.

♂ eclipse

Aythya baeri

青头潜鸭 qīng tóu qián yā

Baer's Pochard

L: 42–47 cm
WS: 70–79 cm
Habitat: Often inhabits open and slow-flowing lakes, marshes, and ponds; breeds mainly in areas with ample floating plants and dense reedbeds.
Behavior: Often turns up in flocks of Ferruginous Duck. Crepuscular. Dives to feed. Spends most of the day swimming or resting on water.
Distribution: Breeds in Northeast, North, and Central China; winters in C Sichuan and the Yangtze Plain to South China (including Taiwan); migrates through eastern China. Once very abundant, but population numbers in sharp decline; nowadays only easily found in a few specific locations.
Voice: Poorly known. Apparently an *errr* similar to other diving ducks but higher pitched. Rarely calls in nonbreeding season.

Aythya nyroca

白眼潜鸭 bái yǎn qián yā

Ferruginous Duck

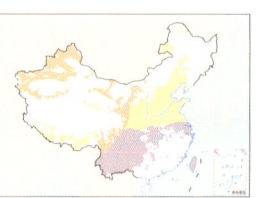

L: 33–43 cm
WS: 60–67 cm
Habitat: Inhabits slow-flowing lakes, ponds, and reservoirs; prefers closed water bodies more than other diving ducks. Occurs less frequently in saline waters.
Behavior: Moves solitarily, in pairs, or in flocks containing dozens to hundreds of individuals. Dives but for shorter time than other diving ducks. Fast wingbeats in flight.
Distribution: Breeds in Western China but expanding east; winters in C Sichuan and the Yangtze Plain to South China (including Taiwan). Seen in most of China during migration; relatively common.
Voice: Generally silent. Tufted Duck–like notes during breeding season.

Polysticta stelleri

小绒鸭 xiǎo róng yā

Steller's Eider

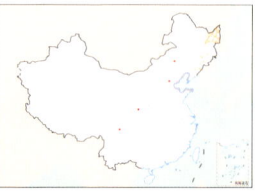

L: 42–48 cm
WS: 68–77 cm
Habitat: Breeds on tundra; winters mainly at estuaries. Prefers rocky shores; also occurs at harbors.
Behavior: Dives for mollusks and crustaceans. Moves in small flocks in winter. Fast wingbeats in flight.
Distribution: Winters in the sea north of Hokkaido. Rare vagrant in China; recorded in the mouth of Ussuri River, and, once, in Hebei.
Voice: Unlikely to be heard in China and generally very quiet even when breeding. Call is brief and hoarse in breeding season.

Histrionicus histrionicus

丑鸭 chǒu yā

Harlequin Duck

L: 38–45 cm
WS: 63–70 cm
Habitat: Inhabits mountain streams in breeding season; nests in tree cavities. Winters at sea. Often seen near coasts and prefers rocky shores.
Behavior: Gathers in small flocks in winter. Feeds by dipping head in water. Often rests on rocks in small tight gatherings.
Distribution: Rare breeder in Changbai Mountains, Jilin; rare passage migrant to Heilongjiang; rare winter visitor in coastal Hebei and Shandong; vagrant in Beijing, Sichuan, and Shaanxi.
Voice: Utters a high-pitched torrent-adapted *hig-hig* in breeding season, similar to the sound of a child's toy. Calls are typical ducklike *gaga* sounds the rest of year.

Baer's Pochard

Relatively rounded blackish head with dark green sheen; appears black in poor light

White iris

Brown lores

♀ win.

Brown iris

Relatively narrow chestnut area on flanks

♂ eclipse

Foreflanks mottled white

Ferruginous Duck

White iris

Brown iris

♀

Broad brown band between belly and undertail coverts

♂ eclipse

White undertail coverts

Steller's Eider

Relatively long bluish-gray tertials; conspicuous

Pale eye ring

Bluish speculum with broad white borders

Relatively long leaden bill

Harlequin Duck

Pale breast

♀ 1st win.

Pin-shaped tail feathers

♂

White spot behind ear

♀

Melanitta stejnegeri

斑脸海番鸭 bān liǎn hǎi fān yā

Stejneger's Scoter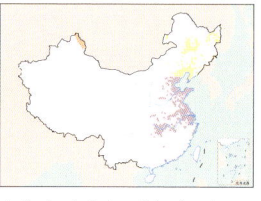

L: 51–58 cm
WS: 86–99 cm
Habitat: In China, breeds on freshwater lakes and large calm rivers in taiga forests; winters on sandy shores, seashores bordered by boulders, or large freshwater habitats.
Behavior: Good at diving, wings half-extended when diving in water. Head raised in flight.
Distribution: Breeds in the Altay region in N Xinjiang; likely also breeds in extreme northern Northeast China; migrates through Northeast China. Regular but uncommon passage migrant and winter visitor in the Bohai and Yellow Seas. Occasional wintering records along the East Sea coast and the Yangtze Plain.
Voice: Very quiet and poorly known. Occasionally calls a rapid, repetitive, and hoarse *ga*.

Melanitta fusca

绒海番鸭 róng hǎi fān yā

Velvet Scoter   VU

L: 51–58 cm
WS: 79–97 cm
Habitat: Similar to Stejneger's Scoter.
Behavior: Similar to Stejneger's Scoter; may turn up with Stejneger's Scoter in winter.
Distribution: Vagrant. Recorded in Shandong (Rizhao), Shanghai (Nanhui Dongtan), and Zhejiang (Ningbo) in winter. Extremely rare, but some records may be overlooked due to difficulty in identification.
Voice: Generally quiet. Calls similar to Stejneger's Scoter.

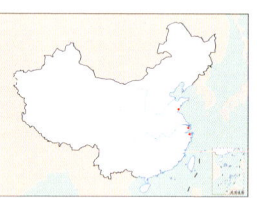

Melanitta americana

黑海番鸭 hēi hǎi fān yā

Black Scoter   NT

L: 43–54 cm
WS: 70–90 cm
Habitat: Winters on gulfs, harbors, and estuaries. Often occurs near coasts; prefers sandy shores.
Behavior: Often gregarious. Fast swimmer, tail often raised. Fast takeoff from water; flies close to water's surface.
Distribution: Vagrant in Chongqing, Fujian (Lianjiang), S Jiangsu, Shanghai, and Hong Kong.
Voice: Song is a melancholy series of slow, lengthy, low-pitched whistles rising or falling only slightly in pitch.

Clangula hyemalis

长尾鸭 cháng wěi yā

Long-tailed Duck   VU

L: 51–60 cm (♂)
    37–47 cm (♀)
WS: 65–82 cm
Habitat: Breeds on tundra crisscrossed by rivers; occurs at sea in winter, generally far offshore. Occasionally occurs on deep freshwater lakes.
Behavior: Gathers in large flocks and dives to great depths in winter. Male often raises tail when swimming.
Distribution: Locally common off the east coast of Liaodong Peninsula. Occasionally winters on the Bohai and Yellow Seas. Vagrant in northern China, Sichuan, Chongqing, Hunan, and East China Sea coast.
Voice: Noisy in breeding season, very quiet in winter. Distinctive, far-carrying, pleasant yodeling, *ow-ow-owdelee* or *kyar … kyar-w-owdelee*, rarely heard in China.

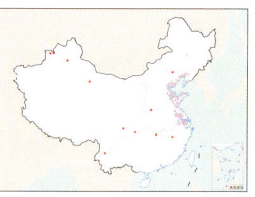

Bucephala clangula

鹊鸭 què yā

Common Goldeneye   LC

L: 40–48 cm
WS: 62–77 cm
Habitat: Inhabits forest-surrounded freshwater lakes in breeding season; winters in a variety of habitats, including rivers, lakes, and gulfs.
Behavior: Often gathers in small flocks; seldom turns up in flocks of other duck species. Frequently dives when foraging. Rapid wingbeats in flight. Throws head backward during display. Breeds in forests and nests in tree cavities.
Distribution: Common in China except for Hainan. Breeds in N Xinjiang and northern Northeast China; migrates through Northeast and northwestern China; winters in North China and areas to its south. Commonly seen in suitable habitats in northern China.
Voice: Hurried *gra-graaa*, most often heard in breeding season. Loud wing whir.

Stejneger's Scoter

♀

White secondaries

Knob at base of bill

Flat forehead

Relatively long upturned white mark behind eye

White speculum varies in width, invisible in some individuals

♂

Dark red bill with yellow lower edge

Base of forehead extends below nostrils

Pale patches at base of bill and behind eye

♀

Long-tailed Duck

♂ win.

Pink patch on short bill

Relatively short neck

Long tail

♂ win.

Long pale scapulars, curved downward

Pale band on side of neck

♀ sum.

Dark gray bill without pink patch

Central tail feathers not elongated

♀ win.

Velvet Scoter

Inconspicuous knob

♂

Short white mark behind eye, not conspicuously upturned

Relatively rounded forehead, more sloped from base of bill to crown

Yellow bill

♂

♀

Base of forehead relatively far from nostrils

Common Goldeneye

♂

Relatively peaked crown

Yellow iris

White spot on face ♂

Black bill

1st win.

Grayish-white breast

Two black bands on upper wings

♀

Yellow iris

Yellow spot at bill tip

♀

Black Scoter

P10 slightly shorter and narrower than P9

♂

Swollen yellow base of bill

♂

Red-crested Pochard–like head pattern

♀

♀

Mergellus albellus

斑头秋沙鸭（白秋沙鸭）
bān tóu qiū shā yā (bái qiū shā yā)

Smew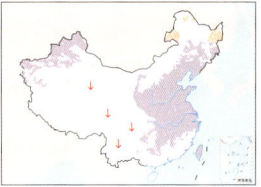

L: 38–44 cm
WS: 56–69 cm
Habitat: Occurs on freshwater lakes, rivers, or forest swamps in breeding season; winters on open waters but seldom occurs at sea.
Behavior: Dives to feed but submerges for shorter time than other mergansers; rapid takeoff from water.
Distribution: Locally common in China except for Hainan and W Qinghai-Tibet Plateau. Breeds in Northeast China. Commonly seen in North China and regions to its south in winter; occasionally winters in Taiwan.
Voice: Seldom calls; occasionally gives an unpleasant, deep, bullfrog-like croak, *kr-kr-kr-kr*.

Mergus serrator

红胸秋沙鸭 hóng xiōng qiū shā yā

Red-breasted Merganser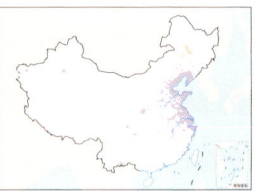

L: 52–60 cm
WS: 67–82 cm
Habitat: Generally winters at sea or on water bodies not far from coast; often occurs on the leeward side. Occasional records on inland waters.
Behavior: Similar to Common Merganser.
Distribution: Occurs in Northeast, North, East, and South China; also recorded in Xinjiang, Sichuan, Chongqing, and Yunnan. Commonly seen on the Bohai and Yellow Sea coasts in winter; occasionally elsewhere.
Voice: Seldom calls; occasionally calls *ko-ko*. Utters high-pitched and thin whistles in breeding season.

Mergus squamatus

中华秋沙鸭 zhōng huá qiū shā yā

Scaly-sided Merganser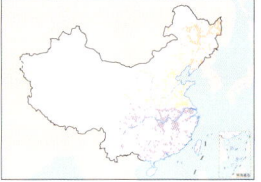

L: 49–64 cm
WS: 70–86 cm
Habitat: Occurs on wide fast-flowing rivers in breeding season. Breeds in tree cavities. Inhabits rivers and the upstream portion of reservoirs in nonbreeding season.
Behavior: Moves in pairs or small flocks. Other behaviors similar to Common Merganser.
Distribution: Breeds in Northeast China; migrates through North China; winters in the vast area stretching from the Yangtze River Watershed to South China, but very scattered. A limited number of wintering locations are known.
Voice: Rarely heard. Females give a loud, rapidly repeated, deep *grregrre* in breeding season.

Mergus merganser

普通秋沙鸭 pǔ tōng qiū shā yā

Common Merganser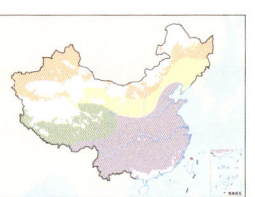

L: 54–68 cm
WS: 78–94 cm
Habitat: Occurs from open lakes to rivers; prefers deep water bodies but seldom occurs at sea.
Behavior: Moves in small to medium-sized flocks. Dives for fish; immerses head in water to look for prey before diving. Run-off required for takeoff from water; strong flier with fast speed and wingbeats.
Distribution: *M. m. orientalis* occurs in Western China; *M. m. merganser* ubiquitous in China except for Xinjiang and Hainan. Few records in South China but commonly seen elsewhere. Winters increasingly far north.
Voice: Relatively deep *kh-kh-krooh*; also *kre-kre-kre-kre*, similar to the sound of a tractor starting.

Oxyura leucocephala

白头硬尾鸭 bái tóu yìng wěi yā

White-headed Duck

L: 43–48 cm
WS: 62–77 cm
Habitat: Inhabits small lakes with lush reeds or reedy wetlands near the shore of medium- or large-sized lakes.
Behavior: Female secretive in breeding season, often crepuscular. Male inhabits mainly waters near reedbeds, tail sometimes erect; swims into reeds when disturbed. Occurs on large water bodies during migration. Often turns up in flocks of diving ducks and grebes. Dives to forage; prefers aquatic insects, but also small fish and other aquatic organisms.
Distribution: A few breeding populations occur on Irtysh River, Ulungur Lake, and north slope of the Tian Shan Mountains in N Xinjiang; found across the water bodies in N Xinjiang during migration. Uncommon passage migrant and winter visitor in S Xinjiang. Vagrants recorded in Inner Mongolia, Hubei, Tianjin, Sichuan, and Shaanxi in fall and winter.
Voice: Rarely calls. Males display with head high and tail cocked during courtship, then give low-pitched and piping sounds. Females utter deep hoarse calls.

Smew

Dark lores

Brown area on head extends to base of lower mandible, unlike Red-Crested Pochard or female Black Scoter

Short neck

♂

♀

♀

White median coverts

Red-breasted Merganser

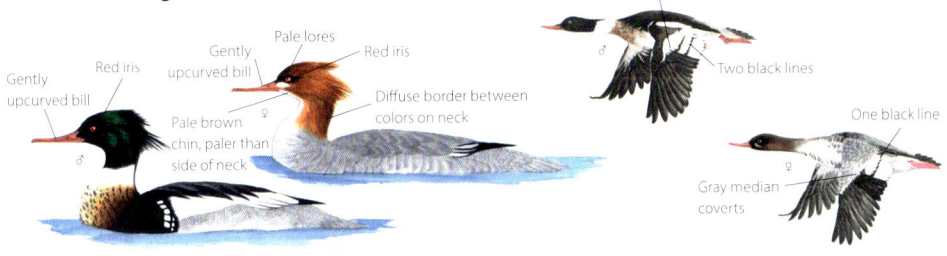

White median coverts

Pale lores

Gently upcurved bill

Red iris

Gently upcurved bill

Red iris

Pale brown chin, paler than side of neck

Diffuse border between colors on neck

♂

♂

Two black lines

One black line

♀

Gray median coverts

♂

Scaly-sided Merganser

Black nape glossed with green; extending to back

White median coverts

Two black lines

Prominent crest

Brownish-yellow cheeks, same color as side of neck

♀

Flat straight upper mandible

Prominent crest

Finely scaled flanks

♂

Gray median coverts

One black line

♂

♀

Finely scaled flanks

Common Merganser

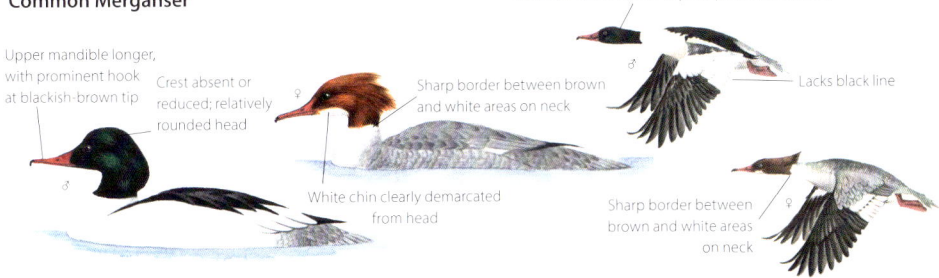

Greenish-black area on nape, separated from back

Upper mandible longer, with prominent hook at blackish-brown tip

Crest absent or reduced; relatively rounded head

Sharp border between brown and white areas on neck

Lacks black line

♂

♂

White chin clearly demarcated from head

Sharp border between brown and white areas on neck

♀

White-headed Duck

Blue bill with prominent swollen base

Stiff tail held erect or close to water surface

Black bill with slightly swollen base

Brown head with prominent white streak through cheek

White head with black crown

♂

♀

鸡形目 GALLIFORMES

Land birds that feed mostly on ground. Plumage different between sexes in some species, with males more brightly colored. Bill short but strong. Robust legs good for running and digging. Wings short and rounded. Tails typically long, with varying ornamentation. Inhabiting most terrestrial habitats, from deserts and tundra to tropical rain forests. Nest on ground or in trees. Omnivorous; feed mostly on plant seeds and fruits. Also take insects and other invertebrates. Reluctant fliers with characteristic flying style. Most species are residents; very few migratory.

Five families, 84 genera, and 299 species recognized worldwide, with a cosmopolitan distribution. One family, 27 genera, and 64 species recorded in China, ubiquitously distributed nationwide.

雉科
Phasianidae

Ring-necked Pheasant

Hazel Grouse

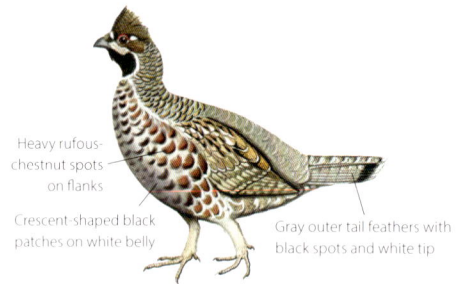

Heavy rufous-chestnut spots on flanks

Crescent-shaped black patches on white belly

Gray outer tail feathers with black spots and white tip

Severtzov's Grouse

Broad black bars on belly

Black outer tail feathers with fine white bars

Tetrastes bonasia

花尾榛鸡 huā wěi zhēn jī

Hazel Grouse   **LC**

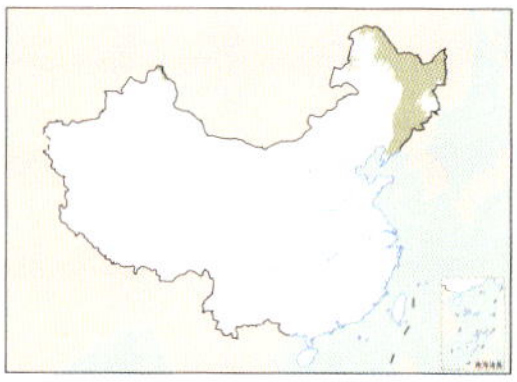

L: 33–40 cm
Habitat: Inhabits montane forests at altitudes of 400–1800 m. Prefers broadleaf forests and forest edges. Makes seasonal altitudinal movements.

Behavior: Often feeds on leaves and fruits in trees. Strong flier in forests. Moves in small flocks in fall and winter. Often spends the night in snow burrows. Nests in fallen trees or under tree roots.
Distribution: *T. b. sibiricus* seen in northern Northeast China and N Xinjiang; *T. b. amurensis* in NE Inner Mongolia, Heilongjiang, Jilin, and Liaoning; occasional and local.
Voice: Very high-pitched fine whistle, *tee-weewee titi*, faltering at the end. Wing noise as birds fly.

Tetrastes sewerzowi

斑尾榛鸡 bān wěi zhēn jī

Severtzov's Grouse   **NT**

L: 31–38 cm
Habitat: Inhabits alpine forests and scrubland at altitudes of 2500–4000 m; often migrates to low-altitude coniferous forests in winter.
Behavior: Similar to Hazel Grouse. Generally stays

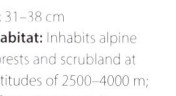

in trees in winter; spends more time on the ground in breeding season. Feeds mainly on plants.
Distribution: Chinese endemic. *T. s. sewerzowi* seen in S Gansu, N Sichuan, and NE Qinghai. *T. s. secundus* seen in E Tibet, E Qinghai, NW Yunnan, and W Sichuan.
Voice: Often quiet. A soft purring *gu, gu, gu* when eating or given as alarm. Wing fluttering or drumming in display.

Falcipennis falcipennis

镰翅鸡 lián chì jī

Siberian Grouse **NT**

L: 37–41 cm
Habitat: Inhabits taiga forests at altitudes of 200–1500 m.
Behavior: Exhibits lekking behavior. In breeding season, males' head and neck feathers turn fluffy, and individuals engage in fierce fights. Nests on the ground.
Distribution: No recent record in China; probably extirpated. Previously resident in northernmost Heilongjiang.
Voice: Poorly known. A rising, tremulous, low-pitched *coooo*, followed by one or two sharp clicks. Also a coarse *cha-cha*.

Tetrao urogalloides

黑嘴松鸡 hēi zuǐ sōng jī

Black-billed Capercaillie **LC**

L: 86–91 cm (♂)
   61–65 cm (♀)
Habitat: Inhabits coniferous forests in the cold temperate zone.
Behavior: Areas used for courtship display are fixed; males engage in fierce fights during courtship. Nests on the ground. Spends the night in snow burrows in winter.
Distribution: Rare local resident of Greater Xing'an (Khingan), Lesser Xing'an, and Changbai Mountains.
Voice: A series of loud and silvery *ga-kalada* during courtship.

Siberian Grouse

Black breast

Black and white bars on belly

White tail tip

Western Capercaillie

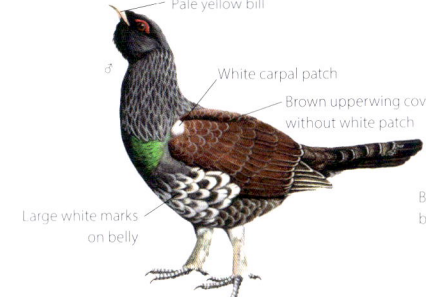

Pale yellow bill

White carpal patch

Brown upperwing coverts without white patch

Large white marks on belly

Black-billed Capercaillie

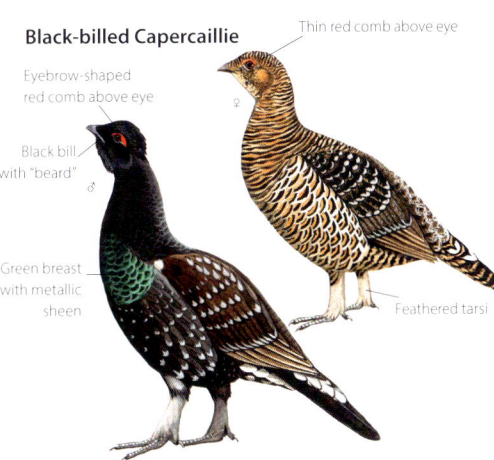

Thin red comb above eye

Eyebrow-shaped red comb above eye

Black bill with "beard"

Green breast with metallic sheen

Feathered tarsi

Brown breast lacks barring

Tetrao urogallus

松鸡 (西方松鸡) sōng jī (xī fāng sōng jī)

Western Capercaillie **LC**

L: 74–90 cm (♂)
   54–63 cm (♀)
Habitat: Inhabits secluded coniferous forests with little human disturbance at altitudes of 1500–2200 m.
Behavior: From April to May, males perform courtship displays in which they erect tail and throat feathers and are unafraid of people. Territorial in breeding season; polygynous. Generally spends the night on larch trees. Forages mainly for shoots and fruits of shrubs, such as those of *Lonicera caerulea* var. *altaica*; sometimes also eats shoots of larches.
Distribution: Rare resident in C and W Altai Mountains in N Xinjiang.
Voice: Song is an accelerating series of low-pitched popping notes terminated by a cork-popping note. Low-pitched, aggressive-sounding *ge-ge-ge* and other unpleasant belching notes.

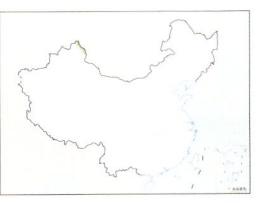

Lyrurus tetrix

黑琴鸡 hēi qín jī

Black Grouse LC

L: 54–61 cm (♂)
44–49 cm (♀)

Habitat: Inhabits low- and mid-altitude montane coniferous forests, mixed broadleaf-coniferous forests, and forest steppes.
Behavior: In breeding season, males aggregate at fixed lek sites, cock their tail up to show white undertail coverts, and perform courtship display by "running in circles." Hierarchy exists in males, where healthy adult males occupy the central position at lek sites. Nests on the ground.
Distribution: *L. t. ussuriensis* in Northeast China and N Hebei; *L. t. baikalensis* in northern Northeast China; and *L. t. mongolicus* in N Xinjiang. Occasional and local.
Voice: Lengthy series of hollow clicks, *gururu-gururu*, like young children fencing with wooden weapons, interspersed with a coarse disyllabic call.

Lagopus muta

岩雷鸟 yán léi niǎo

Rock Ptarmigan LC

L: 36–39 cm

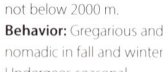

Habitat: Inhabits alpine meadows and scrubland above tree line at altitudes of 2000–3000 m. Generally not below 2000 m.
Behavior: Gregarious and nomadic in fall and winter. Undergoes seasonal changes of plumage where birds turn pure white in winter. Unafraid of people.
Distribution: Rare in Altai Mountains, Tarbagatai Mountains, and Saur Mountains in N Xinjiang.
Voice: Coarse fast rattle, almost Garganey-like in quality: *kuh, kuh, kwa-kwa-kwa*. Male's call a loud *arr, arr*.

Lagopus lagopus

柳雷鸟 liǔ léi niǎo

Willow Ptarmigan LC

L: 38–41 cm

Habitat: Inhabits low bushes of *Juniperus pseudosabina*, roses, birches, and willows at altitudes of 1500–2800 m.
Behavior: Similar to Rock Ptarmigan.
Distribution: Rare and local. *L. l. nadezdae* in Altai Mountains in N Xinjiang; *L. l. sserebrowsky* very rare along the Heilong (Amur) River in Northeast China.
Voice: A series of accelerating coarse barks with a terminal trill, *kek krrrrrrrrow … go-back*. Contact call of females is a harsh clucked *nyow*, sometimes in intense sequence.

Lerwa lerwa

雪鹑 xuě chún

Snow Partridge LC

L: 35 cm

Habitat: Inhabits alpine scrubland, meadows, and scree above tree line from 3000 to 5000 m; makes clear seasonal altitudinal movements.
Behavior: Often moves around in coveys. Bold; allows people to approach within a short range. Flushes immediately when startled.
Distribution: Seen in SE Tibet, NW Yunnan, W and N Sichuan, and S Gansu. Locally common.
Voice: Intensifying, staccato, jackass-like braying, *kyillu … kyillu … kyillu*, with each note repeated 8–10 times. Single calls like Red-billed Chough.

Tetraophasis obscurus

红喉雉鹑 (雉鹑)

hóng hóu zhì chún (zhi chún)

Verreaux's Partridge LC

L: 45–54 cm

Habitat: Inhabits coniferous forest edges and alpine scrubland above 3000 m.
Behavior: Good at walking and running on the ground. Gregarious. Often sings chorus in the morning. Nests on the ground or in bushes.
Distribution: Chinese endemic. Seen in E Qinghai, W and N Sichuan, and S Gansu. Occasional and local.
Voice: Discordant cacophony of loud chuckling, whistled squeals, and bubbling notes often rising to an intense climax. Groups often chorus together.

Tetraophasis szechenyii

黄喉雉鹑 huáng hóu zhì chún

Szechenyi's Partridge LC

L: 43–49 cm

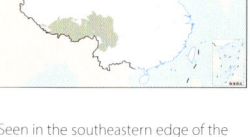

Habitat: Inhabits forest edges and scrubland from 3500 to 4500 m; makes seasonal altitudinal movements.
Behavior: Similar to Verreaux's Partridge. Often moves about in coveys. Adopts cooperative breeding.
Distribution: Chinese endemic. Seen in the southeastern edge of the Qinghai-Tibet Plateau (including Sichuan, Yunnan, and Tibet). Locally common.
Voice: Discordant cacophony of paired or tripled cackles and whistled squeals often rising to an intense climax. Groups often chorus together.

Black Grouse

Red comb

♀

Heavy
brown bars

Round tail

White wing bar

♂

Long outer tail feathers
curve outward

Snow Partridge

Red
bill

Dense black and white
bars on back

Lanceolate chestnut
bars on breast

Verreaux's Partridge

Rufous-chestnut
throat with white
outline

Rock Ptarmigan

Lacks black
lores

Red upper eyelid

Grayish-brown
upperparts

♀ sum.

Red supraorbital comb

Black lores

Grayish-brown
upperparts

♂ sum.

Black lores

All-white
body

♂ win.

Szechenyi's Partridge

Yellow throat without
white outline

Willow Ptarmigan

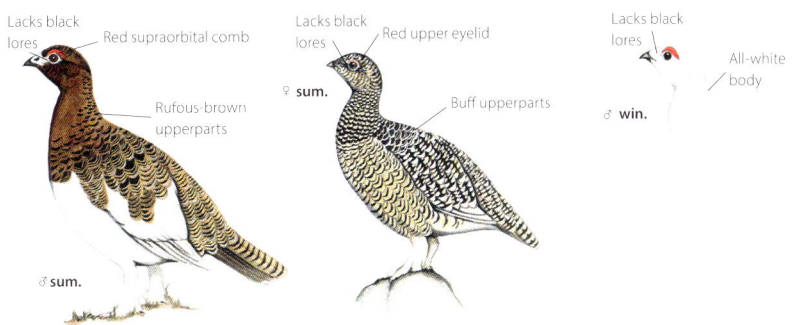

Lacks black
lores

Red supraorbital comb

Rufous-brown
upperparts

♂ sum.

Lacks black
lores

Red upper eyelid

Buff upperparts

♀ sum.

Lacks black
lores

All-white
body

♂ win.

57

Tetraogallus himalayensis
暗腹雪鸡 àn fù xuě jī
Himalayan Snowcock — LC

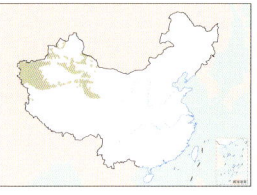

L: 52–60 cm
Habitat: Inhabits mainly alpine and subalpine rocky tundra meadows and bare rock zones at altitudes of 2500–5000 m, nearly approaching the snowline. In winter, moves down to scrubland above tree line or forest edges at altitudes of 2000 m or even 1500 m.
Behavior: Often in small flocks; generally occurs on cliffs, bare rock zones, or alpine tundra meadows and bare rocky hillsides littered with boulders. Often utters a loud and sonorous contact call. Shy and vigilant. Often glides downhill from mountaintops; moves uphill by running and walking.
Distribution: T. h. sauricus occurs on the Tarbagatai Mountains and Saur Mountains on the Xinjiang-Kazakhstan border. T. h. sewerzowi occurs on the Tian Shan Mountains in Xinjiang. T. h. himalayensis occurs on the Pamir Plateau and S Tian Shan Mountains of Xinjiang. T. h. grombszewskii occurs on the W Kunlun Mountains. T. h. koslowi occurs in W Inner Mongolia, Gansu, N Qinghai, and E Kunlun Mountains and Altun Mountains of Xinjiang.
Voice: Song a desolate, far-carrying, vaguely curlew-like, slightly stuttering, rising scream with a sudden drop in pitch at the end, pwiii-puer-e … shiii-pueer-e, and a relatively low wai-wain-guar-guar. Soft clucking contact calls, er-u, ger-u.

Tetraogallus tibetanus
藏雪鸡 zàng xuě jī
Tibetan Snowcock — LC

L: 50–64 cm
Habitat: Inhabits alpine meadows and scree above 3000 m; moves to lower altitudes in winter.
Behavior: Often in small flocks; vigilant. Forages on stony ridges and alpine meadows; almost never occurs in forests. Nimble on rocky slopes; moves downhill mostly by gliding.
Distribution: T. t. tibetanus in Xinjiang and W Tibet; T. t. tschimenensis in S Xinjiang and W Qinghai; T. t. aquilonifer in S Tibet; T. t. przewalskii in Qinghai, N Sichuan, and W Gansu; T. t. henrici in E Tibet and W Sichuan; and T. t. yunnanensis in NW Yunnan. Rare hybrids with Himalayan Snowcock when sympatric.
Voice: Rising curlew-like whistle, shorter and more faltering than that of Himalayan Snowcock. Also a series of loud staccato clucks that accelerate: chuck-chuck-chuck-chuck-chuck-aa-chuck-aa-chuck-chuck-chee-da-da-da.

Tetraogallus altaicus
阿尔泰雪鸡 ā ěr tài xuě jī
Altai Snowcock — LC

L: 58 cm
Habitat: Occurs at altitudes of 2000–3000 m on alpine and subalpine meadows and bare rock zones. In winter, descends to 2000 m in foothills.
Behavior: Nonbreeders often gather in flocks. Runs quickly uphill when disturbed, climbing to the mountaintop before taking off; flies fast but only short distances. Utters loud and sonorous calls from rocks or cliffs at dusk and dawn.
Distribution: Rare resident of N Xinjiang—found in the Altai Mountains and Baitag Bogd Mountain in NE Xinjiang.
Voice: Rising plaintive whistle, reminiscent of Himalayan Snowcock but finer and wavering in pitch, often combined with nasal clucking, geuk-geuk-geuk, or a short guk-guk-guk, followed by a series of repetitive rrruuuuuu notes, increasing in pace.

Alectoris chukar
石鸡 shí jī
Chukar — LC

L: 30–37 cm
Habitat: Inhabits low mountains, hills, the Loess Plateau, rocky hillsides, and forest-edge bushes.
Behavior: Gregarious. Fast runner; very vigilant. Well camouflaged; hard to see on rocky hillsides. Makes seasonal altitudinal movements.
Distribution: Locally common in North and northwestern China. A. c. falki seen in W and C Xinjiang; A. c. dzungarica in NW Xinjiang; A. c. pallida in W and S Xinjiang and N Qinghai; A. c. pallescens in Xinjiang and W Tibet; A. c. potanini in northwestern China; and A. c. pubescens in North China and eastern northwestern China.
Voice: Staccato, galloping ka-ka-ka-ka, occasionally developing to vaguely onomatopoeic chuk chuk chuk-chuk … chuckarr … chuckarr.

Alectoris magna
大石鸡 dà shí jī
Przevalski's Partridge — LC

L: 32–45 cm
Habitat: Similar to Chukar but prefers more desolate habitats.
Behavior: Similar to Chukar, often lives in small coveys.
Distribution: Chinese endemic. A. m. magna in Qinghai and A. m. lanzhouensis in Ningxia and Gansu. Some hybrids with Chukar. Locally common.
Voice: Very similar to Chukar—a series of brief ga-ga-ga notes, escalating in pitch, gradually turning into a disyllabic ga-ble.

Himalayan Snowcock

White area on side of neck bounded by two chestnut lines that form a chestnut band across upper breast

Grayish-black breast and belly

Altai Snowcock

Gray head and neck lack dark patches

Dirty-white lower breast and belly

Black patch on lower belly

Tibetan Snowcock

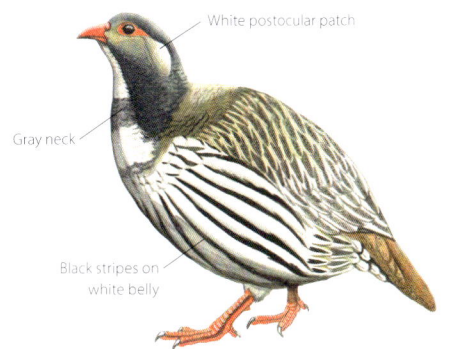

White postocular patch

Gray neck

Black stripes on white belly

Chukar

Thick black band across throat and upper breast

Discontinuous rufous-chestnut barring beside black bars on flanks

Przevalski's Partridge

Black band bordered below by a fuzzy rufous-chestnut band

Long and continuous rufous-chestnut barring beside black bars on flanks

Perdix perdix

灰山鹑 huī shān chún

Gray Partridge LC

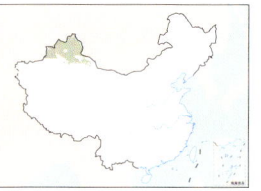

L: 29–31 cm
Habitat: Inhabits bushes and grassy areas on low mountains and hills at altitudes of 500–2000 m; in winter, descends to foothill plains, farmland, and meadows.
Behavior: Lives in family units in nonbreeding season. Good at running and hiding, taking off all together when being chased.
Distribution: Uncommon; very local resident in W Altai Mountains, Tarbagatai Mountains, Saur Mountains, Dzungarian Alatau, and the Tian Shan foothills in Ili, NW Xinjiang.
Voice: Song is a husky, grating *ki-errr-ik*; makes a low *grrrree-grrrree* when flushed.

Perdix dauurica

斑翅山鹑 bān chì shān chún

Daurian Partridge LC

L: 25–31 cm
Habitat: Inhabits forest steppes, shrubby meadows, and fallow agricultural land on low mountains and hills; prefers open habitats.
Behavior: Often in large flocks in fall and winter. Stands still when danger approaches; flies close to the ground, making fast and powerful wingbeats.
Distribution: *P. d. suschkini* often seen in Northeast and North China. *P. d. dauurica* seen throughout northwestern China.
Voice: Song and calls very similar to Gray Partridge.

Perdix hodgsoniae

高原山鹑 gāo yuán shān chún

Tibetan Partridge LC

L: 23–30 cm
Habitat: Inhabits alpine tundra and subalpine scrubland at altitudes of 2500–5000 m. Moves to lower altitudes in winter.
Behavior: Good at running; agile, occasionally makes rapid flights. Nests in shallow pits on ground; nests very crude and simple.
Distribution: *P. h. caraganae* in Xinjiang and W Tibet; *P. h. hodgsoniae* in SE Tibet; and *P. h. sifanica* in S Gansu, NW Yunnan, Qinghai, W Sichuan, and E Tibet. Locally common.
Voice: Series of usually five or six husky, grating, descending, and decelerating notes: *schrrrreeek-schrrrreeek*; also a clipped staccato *trik-trik-trik*, often repeated for long periods.

Coturnix coturnix

西鹌鹑 xī ān chún

Common Quail LC

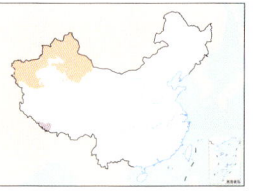

L: 16–22 cm
Habitat: Inhabits open plains, farmland, and grasslands.
Behavior: Generally lives on the ground. Fast runner; often hides in grass. Takes off suddenly and flies with fast wingbeats; returns to grass after flying a short distance.
Distribution: Breeds throughout Xinjiang; winters in S Tibet. Locally common.
Voice: Far-carrying, percussive, trisyllabic calls sounding like "wet my lips." At close range, a couple of muffled introductory *mau-waw* notes, sometimes audible.

Coturnix japonica

鹌鹑 ān chún

Japanese Quail NT

L: 15–20 cm
Habitat: Inhabits grassy plains, shrubby meadows, and farmland on low mountains and hills.
Behavior: Similar to Common Quail.
Distribution: Commonly seen throughout eastern China. Breeds in Northeast and North China; winters from Beijing south to southern China.
Voice: Abrupt, percussive, trisyllabic *chr-gruk-chrr*, very different from Common Quail. When flushed, a soft *wreee*.

Synoicus chinensis

蓝胸鹑 lán xiōng chún

Blue-breasted Quail LC

L: 12–14 cm
Habitat: Inhabits shrubby grasslands or thickets on open plains, hills, and low mountains.
Behavior: Generally lives in small coveys. Shy; often hides in grass during the day and forages at dusk and dawn.
Distribution: Rare in SE Yunnan, Guizhou, Guangxi, Guangdong, Hainan, Fujian, and Taiwan.
Voice: Resonant, two- or three-syllable, Red-wattled Lapwing–like territorial vocalization, *pu-pu-pe'er*, last note falling. Also a melancholy falling *pliuu*.

Gray Partridge

Lacks whiskers on throat

Gray upper breast

Horseshoe-shaped chestnut patch

♂

♀

Gray upper breast

Chestnut patch reduced or absent

Common Quail

Brown back

Black stripe on central throat

♂

Japanese Quail

Chestnut-brown throat and cheeks

Rufous-brown back

Buff throat

♂ br.

♀

Daurian Partridge

"Beard" of stiff feathers on side of throat

Orange-brown patches on throat and belly connect at breast

Large horseshoe-shaped black patch on belly

Blue-breasted Quail

Orange-brown throat

♀

Black barring on breast and flanks

chick

Triangular black patch on throat

White chest patch

Bluish-gray breast and flanks

Rufous-chestnut belly

♂

Tibetan Partridge

White supercilium

White throat

Rufous-chestnut neck

Black barring on white belly

Francolinus pintadeanus

中华鹧鸪 zhōng huá zhè gū

Chinese Francolin　LC

L: 29–35 cm

Habitat: Inhabits shrubby meadows and bamboo thickets near farmland on low mountains and hills; especially prefers dry grassy hillsides.

Behavior: Often makes rapid flights in a straight line. Calls loudly and frequently in breeding season.

Distribution: *F. p. pintadeanus* seen throughout southern China; *F. p phayrei* in Yunnan. Locally common.

Voice: Series of five or six loud, hoarse, crowing notes, *pwi-ta-tak … ta-kaa*, repeated at intervals.

Arborophila torqueola

环颈山鹧鸪 huán jǐng shān zhè gū

Hill Partridge　LC

L: 26–29 cm

Habitat: Inhabits evergreen broadleaf forests at altitudes of 1500–3800 m; often seen in open areas within forests and near ravines.

Behavior: Often lives in small coveys. Looks for food in thick layers of fallen leaves. Vigilant; fast runner.

Distribution: *A. t. torqueola* occasionally seen in SE Tibet; *A. t. batemani* seen in SW Yunnan.

Voice: Smoothly rising whistle lasting about 1.5 sec, *whooooo*, repeated every 3–4 sec. Female less commonly makes a repetitive disyllabic *kwi-kwi-kwi-kwi-kwi*.

Arborophila rufogularis

红喉山鹧鸪 hóng hóu shān zhè gū

Rufous-throated Partridge　LC

L: 25–29 cm

Habitat: Inhabits evergreen broadleaf forests and forest-edge bushes in low mountains and hills from 1000 to 2500 m; often seen near ravines.

Behavior: Prefers living in small coveys; often forages by digging on the ground. Rarely flies.

Distribution: *A. r. rufogularis* in SE Tibet; *A. r. intermedia* in W Yunnan; and *A. r. euroa* in SE Yunnan.

Voice: Individual notes at start of song are similar to those of Hill Partridge but flatter, slightly higher pitched, and not increasing in volume. They are often paired (with the second note being longer), and the speed of delivery increases, eventually becoming an intense rapid repetition of rising *gu-ger* whistles.

Arborophila atrogularis

白颊山鹧鸪 bái jiá shān zhè gū

White-cheeked Partridge　NT

L: 24–28 cm

Habitat: Inhabits moist forests and bamboo thickets on low mountains and hills below 1500 m.

Behavior: Similar to other *Arborophila* partridges.

Distribution: Rare and local in Yingjiang, SW Yunnan.

Voice: Song an intensifying series of repeated, mellow, paired whistles, *whi-huu … whi-huu … whi-huu*, the first note with a strong diphthong, the second flat and lower in pitch. Females sometimes join with downslurred purring *kew-kew-kew*.

Arborophila crudigularis

台湾山鹧鸪 tái wān shān zhè gū

Taiwan Partridge　LC

L: 27–30 cm

Habitat: Inhabits primary broadleaf forests from 300 to 2000 m; often occurs in areas with high canopy coverage and substantial leaf litter.

Behavior: Secretive and vigilant, often in small coveys. When foraging, walks and often makes contact calls. Omnivorous; forages by digging in leaf litter for seeds, fruits, and invertebrates. Highly territorial.

Distribution: Chinese endemic. Occurs in mountains in Taiwan; locally common.

Voice: Starts with a series of tremulous *guru* notes, then gradually increases slightly in pitch and accelerates into a repeated *weah-hu, weah-hu*. Like all *Arborophila*, calls mostly at dusk and dawn.

Arborophila mandellii

红胸山鹧鸪 hóng xiōng shān zhè gū

Chestnut-breasted Partridge　VU

L: 24–28 cm

Habitat: Inhabits dense evergreen broadleaf forests below 2500 m.

Behavior: Similar to other *Arborophila* partridges.

Distribution: Rare resident in SE Tibet.

Voice: Loud penetrating whistle similar to Hill Partridge but rising slightly in pitch and more tremulous, even occasionally stuttering or burry. Occasionally paired like Rufous-throated Partridge.

Chinese Francolin

White cheek

White throat

Prominent oval white spots on breast, belly, and upper back

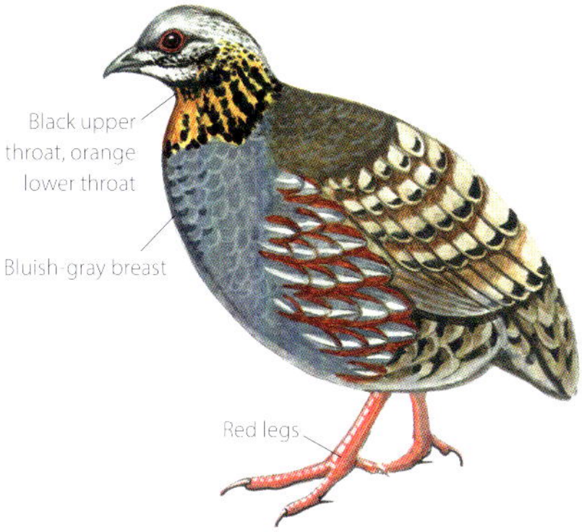

White-cheeked Partridge

Creamy-white cheeks

Black throat

Bluish-gray breast with short black streaks

Hill Partridge

♂

Rufous-chestnut crown and ear coverts

Black throat

Fine black bars on brownish-gray breast and upper back

Dark legs

♀

Taiwan Partridge

Broad black eye stripe

Black band across white throat

Slate-gray breast

Coral-red legs

Chestnut-breasted Partridge

Rufous-chestnut crown

Narrow black and white neck collar

Rufous-chestnut breast

White spots on flanks

Rufous-throated Partridge

Black upper throat, orange lower throat

Bluish-gray breast

Red legs

Arborophila brunneopectus

褐胸山鹧鸪 hè xiōng shān zhè gū

Bar-backed Partridge

L: 28–30 cm

Habitat: Inhabits evergreen broadleaf forests on low mountains below 1500 m.

Behavior: Similar to other *Arborophila* partridges.

Distribution: Uncommon resident in S Yunnan and SW Guangxi.

Voice: Song is series of intense paired whistles repeated at length, *we-hu, we-hu, we-hu*, to which the female often duets with a series of *diu-diu-diu*.

Arborophila rufipectus

四川山鹧鸪 sì chuān shān zhè gū

Sichuan Partridge

L: 28–30 cm

Habitat: Inhabits natural evergreen broadleaf and deciduous mixed forests from 1200 to 2000 m.

Behavior: Secretive; often forages on the ground in areas with substantial leaf litter and tree cover.

Omnivorous; when foraging, scratches among leaf litter with feet and pecks at invertebrates, seeds, and fruits. Highly territorial in breeding season; utters territorial calls to defend territories. Often roosts in woodlands with dense shrub cover at night.

Distribution: Chinese endemic. Occurs in C and S Sichuan, NE Yunnan. Local and uncommon resident.

Voice: Loud typical whistled *Arborophila* song, *whoo*, with notes similar to those of Hill Partridge but shorter and slightly rising. Pairs often duet.

Arborophila gingica

白眉山鹧鸪 bái méi shān zhè gū

White-necklaced Partridge

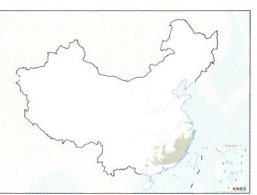

L: 25–30 cm

Habitat: Inhabits mountains at altitudes of 300–1800 m. Occurs in primary and secondary forests and bamboo thickets; prefers valleys and moist habitats by rivers.

Behavior: Secretive. Forages mainly on the ground. Often moves solitarily or in pairs; lives in small coveys in fall and winter.

Distribution: Chinese endemic. *A. g. gingica* widely distributed in mountains and hills in southeastern China, locally common. *A. g. guangxiensis* occurs in the Jiuwanda Mountain and Daming Mountain in Guangxi.

Voice: Very short, loud whistles given in an increasingly frenetic sequence, similar notes often paired: *whu-huuu … whu-huuu*. Can be heard almost year-round; especially prefers calling on rainy days.

Arborophila ardens

海南山鹧鸪 hǎi nán shān zhè gū

Hainan Partridge

L: 28–30 cm

Habitat: Inhabits tropical evergreen broadleaf and mixed broadleaf-coniferous forests at altitudes of 500–1200 m; prefers valleys and steeply sloping, moist forests by rivers. Often lives in forests with a high level of canopy cover and sparse undergrowth.

Behavior: Secretive. In breeding season, maintains territories by uttering repetitive calls at dusk and dawn.

Distribution: Chinese endemic. Occurs only in mountains in C and SW Hainan. Local and rare.

Voice: Male and female often duet in breeding season. Male starts with short, intense, fast-paced, slightly descending whistles, *wa-wa, wa-wa, wa-wa*; female replies with a longer, lower-pitched, slightly rising *phium phiu, phiu*.

Bar-backed Partridge

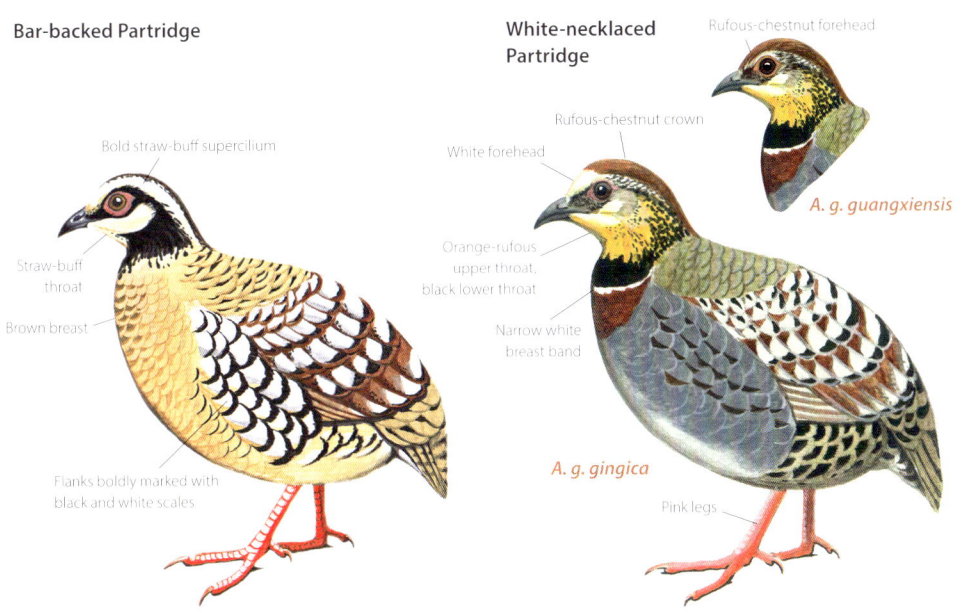

Bold straw-buff supercilium

Straw-buff throat

Brown breast

Flanks boldly marked with black and white scales

White-necklaced Partridge

Rufous-chestnut forehead

Rufous-chestnut crown

White forehead

Orange-rufous upper throat, black lower throat

Narrow white breast band

A. g. guangxiensis

A. g. gingica

Pink legs

Sichuan Partridge

Orange ear coverts

Black streaks on white throat

Rufous-chestnut breast band

Leaden legs

Hainan Partridge

Black on side of head

White ear coverts

Orange-red breast

Orange-yellow belly

Bambusicola fytchii

棕胸竹鸡 zōng xiōng zhú jī

Mountain Bamboo-Partridge

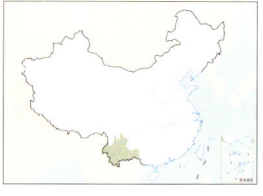

L: 30–36 cm
Habitat: Inhabits hillside shrubby meadows and bamboo thickets from low mountains up to 2500 m.
Behavior: Prefers foraging in tall grass or bushes near gully streams. Calls frequently. Nests on the ground.
Distribution: Seen in Yunnan, S Sichuan, SW Guizhou, and W Guangxi. Local and uncommon.
Voice: Intense song is a very fast repetition of scolding *kerche-kercher-kercher* notes that gradually decrease in volume. Calls include a rasping, rising *pweer* and a breathless, hissing *kssissh*.

Bambusicola thoracicus

灰胸竹鸡 huī xiōng zhú jī

Chinese Bamboo-Partridge

L: 27–35 cm
Habitat: Occurs from plains to mountains up to 1800 m. Uses various kinds of habitats; inhabits natural and artificial forests as well as farmland. Often walks through bushes.
Behavior: Omnivorous; often forages for fruits, seeds, and invertebrates on the ground. Often moves in pairs in breeding season and small coveys in nonbreeding season. Makes contact calls when walking, which can be noisy.
Distribution: Chinese endemic. Occurs throughout southern China (except for Hainan and Taiwan), north to S Shaanxi and west to the Sichuan Basin. Locally common.
Voice: Song resonant and loud; often starts with a series of single high-pitched notes, followed by a repetitive three-note "people pray, people pray, people pray," with gradually decreasing frequency and speed. Calls include a loud *kweear*.

Bambusicola sonorivox

台湾竹鸡 tái wān zhú jī

Taiwan Bamboo-Partridge

L: 30–32 cm
Habitat: Inhabits dense bushes and bamboo thickets at altitudes of 300–1200 m; makes short seasonal altitudinal movement.
Behavior: Moves about in small coveys. Scrapes for tender shoots, fruits, and invertebrates on the ground. Makes loud and resonant calls at dusk and dawn in breeding season.
Distribution: Chinese endemic. Restricted to Taiwan, where locally common.
Voice: Similar rhythm and tone to calls of Chinese Bamboo-Partridge but hoarser, a repetitive, trisyllabic *ji-go-gwai, ji-go-gwai, ji-go-gwai.* Alarm call is a repetitive monotone *jiu-jiu.*

Ithaginis cruentus

血雉 xuè zhì

Blood Pheasant

L: 37–46 cm
Habitat: Inhabits alpine coniferous and mixed forests and scrubland at altitudes of 1700–3500 m. Moves to lower altitudes in winter.
Behavior: Often forages in flocks on the ground, walking and pecking at food; especially prefers mosses but also eats leaves, fruits, seeds, and small invertebrates.
Distribution: All 12 subspecies occur in China. Green-winged group (males' greater upperwing coverts are green or tinged with green, head and breast have a tinge of red, feathers behind eyes do not extend into a conspicuous crest, including *I. c. cruentus, I. c. tibetanus, I. c. affinis, I. c. marionae, I. c. rocki, I. c. clarkei, I. c. kuseri,* and *I. c. geoffroyi*) occurs in Tibet, W and S Sichuan, Yunnan, and C and S Qinghai. Brown-winged group (males' greater upperwing coverts are brown, head and breast are not red, feathers behind eyes extend into a crest, including *I. c. sinensis, I. c. berezowskii, I. c. beicki,* and *I. c. michaelis*) occurs in NE and NW Qinghai, N Sichuan, Gansu, and S Shaanxi. Locally common.
Voice: Peculiar territorial calls include one or a series of strained, intense squeals, hisses, and occasional clucks, *kwee-kwee chiu-chiu.* Short sharp Eurasian Blackbird–like *chuck* in alarm. Complex geographic variation still unresolved.

Mountain Bamboo-Partridge

White supercilium, black stripe behind eye

White spots on brown breast

Bold black spots on belly

Taiwan Bamboo-Partridge

♀

Bluish-gray face and breast

Chestnut-brown chin and throat

♂

Large chestnut-brown spots on flanks

Chinese Bamboo-Partridge

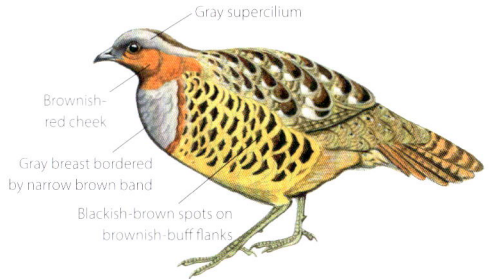

Gray supercilium

Brownish-red cheek

Gray breast bordered by narrow brown band

Blackish-brown spots on brownish-buff flanks

Blood Pheasant

Short white or gray crest

Crimson patch on head, throat, or breast

Green or grayish-green greater coverts

♂

I. c. tibetanus

Dark eye

I. c. geoffroyi

I. c. clarkei

Green-winged group

I. c. rocki

I. c. marionae

Short gray crest

Brown face

♀

Black crest

Lacks crimson patch on chin and breast

♂

Greater coverts rufous brown; some tinged green

Brown-winged group

I. c. berezowskii

Red legs

Tragopan melanocephalus
黑头角雉 hēi tóu jiǎo zhì
Western Tragopan **VU**

L: 55–60 cm (♂)
45–50 cm (♀)
Habitat: Inhabits
temperate broadleaf
forests and subalpine
coniferous forests at
altitudes of 1800–3600 m.
Behavior: Often eats
young leaves in trees; also
forages by digging on the ground. Often roosts solitarily or in pairs on
lateral branches of trees at night. Expands blue horns and lappet during
courtship display.
Distribution: Occurs in the northwestern part of Himalayas. In China,
restricted to extreme SW Tibet. Probably very rare.
Voice: Male's advertising call is like a mournful goat or short, childlike,
rising *waaaa*, similar to Satyr Tragopan but more nasal and fractionally
shorter. Female's calls are even shorter. Also a fast, repeated *pwe-pew-
pwe* when agitated.

Tragopan satyra
红胸角雉 hóng xiōng jiǎo zhì
Satyr Tragopan **NT**

L: 55–79 cm
Habitat: Inhabits dense
forests at altitudes of
2000–3800 m; moves to
broadleaf forests at lower
altitudes in winter.
Behavior: Omnivorous;
forages in trees and
on ground. Courtship
behavior typical of tragopans; nests in trees.
Distribution: Occurs in the central part of Himalayas. In China, seen in
Yadong and Chona Counties in Tibet. Local and uncommon.
Voice: Male's advertising call is typical of tragopans: a loud wailing
waaaa, rising slightly in pitch and repeated up to 15 times at intervals of
once every few seconds, as the volume increases.

Tragopan blythii
灰腹角雉 huī fù jiǎo zhì
Blyth's Tragopan **VU**

L: 65–70 cm (♂)
55–60 cm (♀)
Habitat: Inhabits
evergreen broadleaf
forests or rhododendron
forests between 1500 and
3000 m. Prefers areas with
abundant undergrowth
and close to rivers.
Behavior: Omnivorous; forages in trees and on ground. Courtship
display typical of tragopans; generally nests in trees.
Distribution: *T. b. molesworthi* in S Tibet; *T. b. blythii* in mountains west of
Nujiang (Salween) River, Yunnan. Local and very rare.
Voice: Advertising call is similar to that of Satyr Tragopan but very
slightly lower pitched: *waaaa*.

Tragopan temminckii
红腹角雉 hóng fù jiǎo zhì
Temminck's Tragopan **LC**

L: 65–70 cm (♂)
55–60 cm (♀)
Habitat: Inhabits moist
evergreen broadleaf
and mixed broadleaf-
coniferous forests at
altitudes of 1200–3000 m.
Behavior: Omnivorous;
forages in trees and on
ground. Courtship display typical of tragopans; generally nests in trees.
Distribution: Occurs in S Shaanxi, W Hubei, Chongqing, Sichuan,
Guizhou, W Hunan, NW Guangxi, Yunnan, and SE Tibet. Local but not
uncommon.
Voice: Typical tragopanlike bleating or crying, *waaa*, like a baby crying,
and a clipped, rapidly repeated *quip, qip, quip* when agitated.

Tragopan caboti
黄腹角雉 huáng fù jiǎo zhì
Cabot's Tragopan **VU**

L: 52–63 cm (♂)
45–50 cm (♀)
Habitat: Inhabits
subtropical montane
forests between 800 and
1800 m; often occurs in
natural broadleaf forests
or mixed broadleaf-
coniferous forests with
tall trees. Prefers moving about near river valleys with abundant supply
of water.
Behavior: Omnivorous; forages in trees and on ground. Often moves
solitarily or in pairs. Courtship displays typical of tragopans; nests on
lateral branches of tall trees.
Distribution: Chinese endemic. *T. c. caboti* seen in S Zhejiang, Jiangxi,
Fujian, N Guangdong, E and S Hunan; *T. c. guangxiensis* in NE Guangxi.
Local and occasional.
Voice: Both sexes give truncated, more nasal, and descending versions
of the typical tragopan call *wa*.

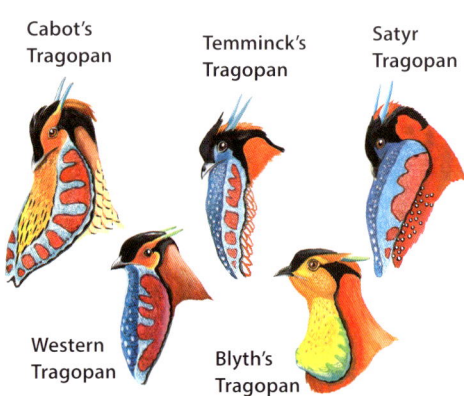

Inflated lappets and horns of five tragopan species

Western Tragopan

Black crown

Red face

Bright red hindneck and breast

Black belly

Dense white spots on body

Almost no white streaks on back

Temminck's Tragopan

Blue face

Crimson back

Red belly covered with oval grayish-white spots

Lanceolate white markings on breast and belly

Rufous-brown bands on tail

Satyr Tragopan

Black face

Bright red hindneck and breast

Dense white spots on underparts

Brown back

Relatively dark breast

Bands on tail finer than Temminck's Tragopan

Cabot's Tragopan

Orange face

Rufous-chestnut back with oval spots

Buff belly

Almost no white spots at center of back

Grayish-brown bands on tail

Blyth's Tragopan

Orange to yellow face

Rufous-brown back with rounded white spots

Red neck and breast

Dusky gray belly with oval-shaped spots

Back tinged rufous brown

Lanceolate white markings on belly

Pucrasia macrolopha

勺鸡 sháo jī

Koklass Pheasant

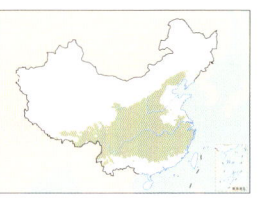

L: 40–63 cm
Habitat: Occurs mainly in shrubs and dense grass in mixed broadleaf-coniferous forests between 1000 and 3000 m.
Behavior: Forages on the ground for fruits and seeds. Secretive. Builds nests on the ground using dry leaves and grass.
Distribution: Distributed widely in the mountains spanning from North China to South and southwestern China and the Himalayas; local and occasional. *P. m. meyeri* seen in SE Tibet, W Sichuan, and NW Yunnan; *P. m. ruficollis* in S Shaanxi, N Ningxia, S Gansu, and Sichuan; *P. m. xanthospila* in the mountains of North China; *P. m. joretiana* in W Anhui; and *P. m. darwini* in Central and East China.
Voice: Unpleasant, hoarse, barked *ko-ko-ke-lass*, loud and far carrying.

Lophophorus impejanus

棕尾虹雉 zōng wěi hóng zhì

Himalayan Monal

L: 70–75 cm
Habitat: Inhabits montane forests and scrubland between 3000 and 4000 m.
Behavior: Prefers foraging on open forest edges and scrubby meadows. Male exhibits elaborate courtship behaviors by flapping wings and fanning tail, sometimes also accompanied by a flight display in which male swoops from sky. Roosts on high rocky cliffs at night; nests on ground or in large tree cavities.
Distribution: S and SE Tibet. Locally relatively common.
Voice: Loud and clear curlew-like whistles, with either a single *kuui* or multiple syllables, *klee-ih-wick*. Also high-pitched clucks. Calls from lofty perches in rocks or trees.

Lophophorus sclateri

白尾梢虹雉 bái wěi shāo hóng zhì

Sclater's Monal

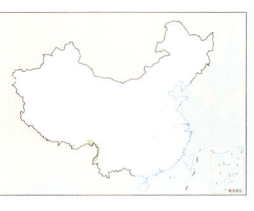

L: 58–68 cm
Habitat: Occurs above 3000 m; inhabits alpine meadows and bushes near forest edges.
Behavior: Often forages on rocky slopes of high mountains; nests on the ground or rocky platforms on steep cliffs.
Distribution: *L. s. sclateri* in SE Tibet; *L. s. orientalis* in NW Yunnan. Rare.
Voice: Usually calls from lofty rocky cliffs—loud far-carrying *wah-waheeee* whistles that resound in the valley.

Lophophorus lhuysii

绿尾虹雉 lǜ wěi hóng zhì

Chinese Monal

L: 76–81 cm
Habitat: Inhabits alpine meadows, scrubland, and rocky slopes above 3000 m.
Behavior: Moves about solitarily or in small flocks; often forages by digging or pecking with bill at bulbs, roots, and leaves on the ground. Nests on the ground or in cliff caves.
Distribution: Chinese endemic. Seen in W and N Sichuan, S Gansu, E Qinghai, and NW Yunnan. Uncommon.
Voice: Typical monal-like long, clear, far-carrying whistle, sometimes flat and plaintive, other times a disyllabic curlew-like *gu-li-*, *gu-li*. In alarm, repeats shorter, equally penetrating shrill piping notes.

Koklass Pheasant

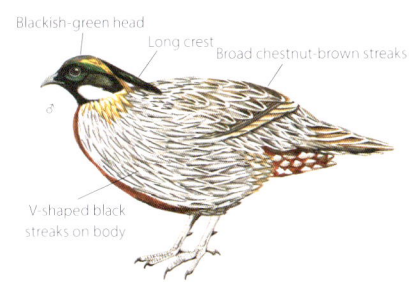

Blackish-green head
Long crest
Broad chestnut-brown streaks
♂
V-shaped black streaks on body

Pale brown patch below ear coverts
♀
Brown body

Sclater's Monal

No crest
Stout bill
♂
White uppertail coverts
White tail tip
Grayish-brown back and uppertail coverts with fine black barring
♀

Himalayan Monal

Green crest
♂
Green uppertail coverts
Cinnamon tail
♀
Brown back

Chinese Monal

♂
Purple crest
Iridescent uppertail coverts
Metallic green tail
♀
White back

Gallus gallus

红原鸡 hóng yuán jī

Red Junglefowl

L: 60–70 cm (♂)
    42–48 cm (♀)
Habitat: Inhabits shrubby plains or forests on low mountains and hills; sometimes also moves around cultivated lands and villages.
Behavior: Generally forages on ground like domestic chicken, but has a stronger ability to fly; roosts in trees at night.
Distribution: *G. g. spadiceus* occurs in W and S Yunnan; *G. g. jabouillei* occurs in SE Yunnan, SW Guangxi, Guangdong, and Hainan. Locally common.
Voice: Similar to the familiar *cock-a-doodle-do* of domestic chicken but hoarser.

Lophura leucomelanos

黑鹇 hēi xián

Kalij Pheasant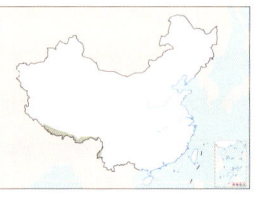

L: 63–70 cm (♂)
    50–60 cm (♀)
Habitat: Inhabits montane forests and scrubland below 1000 m (in far SW Yunnan) or above 2000 m (S Tibet).
Behavior: Often moves in pairs or small flocks; forages by digging in grass or under sparse bush cover. Scurries away or flies into trees when alarmed.
Distribution: *L. l. leucomelanos* seen in S Tibet; *L. l. lathami* in Yunnan (W Dehong). Known to hybridize with Silver Pheasant. Locally common.
Voice: Contact call is a low *gu-gu-gu*. Male's advertising call is a loud strained squeal, very similar to that of Silver Pheasant and sometimes accompanied by drumming or whirring wings.

Lophura nycthemera

白鹇 bái xián

Silver Pheasant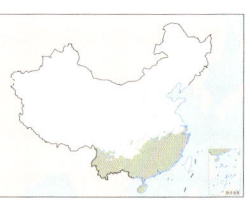

L: 90–130 cm (♂)
    70–90 cm (♀)
Habitat: Inhabits montane forests between 300 and 2000 m; especially prefers ravines close to water with dense forests and sparse undergrowth.
Behavior: Lives in flocks; flock size up to 30 or more individuals in winter. Vigilant; when in danger, scatters and produces a piercing alarm call. Nests on the ground; female highly protective of nests.
Distribution: Numerous subspecies, widespread south of the Yangtze River. Locally common. *L. n. omeiensis* in NE Yunnan, C Sichuan, and W Hubei; *L. n. rongjiangensis* in S and W Guizhou and Guangxi; *L. n. nycthemera* in E Yunnan, Guangxi, and Guangdong; *L. n. fokiensis* in Hunan, S Hubei, S Anhui, S Jiangsu, Zhejiang, Jiangxi, NW Fujian, and E Guangdong; *L. n. whiteheadi* in Hainan; *L. n. occidentalis* in W Yunnan; *L. n. rufipes* in SW Yunnan; *I. n. jonesi* in E Yunnan; and *L. n. beaulieui* in S Yunnan.
Voice: Usually very quiet; a low *gu-gu-gu-gu* for contact and Kalij Pheasant-like squeals.

Lophura swinhoii

蓝腹鹇 lán fù xián

Swinhoe's Pheasant

L: 60–80 cm (♂)
    50–60 cm (♀)
Habitat: Inhabits montane forests below 2700 m; often seen in primary broadleaf forests.
Behavior: Similar to Silver Pheasant. Omnivorous; vigilant. Male displays side-on during courtship.
Distribution: Chinese endemic and restricted to Taiwan, locally common.
Voice: Usually silent. Occasionally a series of repetitive and monotonous calls, *ge, ge, ge, ge*, and higher-pitched squeals when agitated.

Red Junglefowl

Fleshy red comb on crown

Black tail with metallic green

Brownish-yellow neck

♀

Silver Pheasant

Heavily black-barred upperparts

L. n. whiteheadi

Scaly pattern on breast and belly

♀

Kalij Pheasant

Scaly white pattern on rump

Lanceolate white feathers on belly

♂

L. l. leucomelanos

White-fringed body feathers

♀

Grayish-white legs

White back with V-shaped black streaks

L. n. jonesi

♂

Brown back with buff feather shafts

♀

Swinhoe's Pheasant

Short white crest

White upper back

Scaly bluish-green pattern on lower back

White central tail feathers

♂

Triangular buff patch on back

♀

♂

L. n. nycthemera

♀

Crossoptilon crossoptilon
白马鸡 bái mǎ jī
White Eared-Pheasant　NT

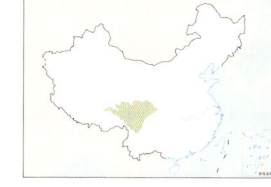

L: 80–100 cm
Habitat: Inhabits alpine or subalpine forests above 3000 m and scrubland above tree line. Moves to broadleaf forests at lower altitudes in winter.
Behavior: Generally lives in flocks; can form flocks of over 60 individuals. Powerfully built, with strong running ability. Often forages in open areas within forests or at forest edges. Diurnal, roosts in trees at night.
Distribution: Chinese endemic. *C. c. crossoptilon* seen in SE Tibet, SE Qinghai, and Sichuan; *C. c. dolani* in S Qinghai; *C. c. lichiangense* in NW Yunnan and SW Sichuan; and *C. c. drouynii* in SE Tibet, S Qinghai, and W Sichuan. Locally common.
Voice: Varied grunting, clucking, squealing, yelping, and nasal snorting, while advertising call is typical of an eared-pheasant: a far-carrying, raucous, barked sequence of three to six intensifying crowing notes, reminiscent of a braying donkey, *ge … ga-, geeer, geera, geera, geeraa.*

Crossoptilon harmani
藏马鸡 zàng mǎ jī
Tibetan Eared-Pheasant　NT

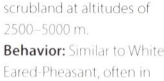

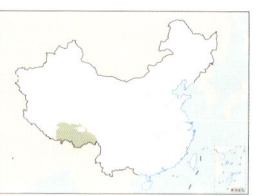

L: 81–86 cm
Habitat: Inhabits montane forests and scrubland at altitudes of 2500–5000 m.
Behavior: Similar to White Eared-Pheasant, often in small flocks.
Distribution: S Tibet, locally common.
Voice: Advertising call is a typical eared-pheasant sequence of three to six unpleasant barked notes that start with one, occasionally two, shorter notes before intensifying into a loud braying sequence, *ga, ga-gga-ge-ra, ga-ge-ra, ga-ge-ra, ga-ge-ra,* very similar to White Eared-Pheasant but with the first two notes uttered more slowly.

Crossoptilon mantchuricum
褐马鸡 hè mǎ jī
Brown Eared-Pheasant　VU

L: 90–100 cm
Habitat: Inhabits mixed broadleaf-coniferous forests between 1000 and 3000 m; moves to lower altitudes in winter.
Behavior: Mostly a ground dweller; forages for plants and small invertebrates. Moves in pairs in breeding season; may form large flocks of over 30 individuals in winter. Often nests under rocks or in ground depressions among bushes.
Distribution: Endemic to North China, local and rare.
Voice: Varied grunting, clucking, squealing, yelping, and nasal snorting, similar to other eared-pheasants. Advertising call also similar, but sequence of 8–14 (typically 13, rarely up to 24) braying notes considerably longer (up to 11 sec). Male also gives a *trip crrrr ah* call that commences softly but rapidly increases in volume.

Crossoptilon auritum
蓝马鸡 lán mǎ jī
Blue Eared-Pheasant　LC

L: 75–100 cm
Habitat: Inhabits subalpine forests at altitudes of 2000–4000 m. Moves to alpine meadows and scrubland at higher altitudes in summer.
Behavior: Similar to other eared-pheasants.
Distribution: Chinese endemic. Found in S Gansu, the Qilian Mountains, E Qinghai, N Sichuan, and the Helan Mountains in Ningxia. Locally common.
Voice: Very loud and harsh, similar to call of Brown Eared-Pheasant but clearer and with a narrower frequency range.

Chrysolophus pictus
红腹锦鸡 hóng fù jǐn jī
Golden Pheasant　LC

L: 86–100 cm (♂)
59–70 cm (♀)
Habitat: Inhabits montane broadleaf forests and forest-edge bushes at altitudes of 500–2000 m. Also forages on farmland in spring and winter.
Behavior: Elegant and agile medium-sized pheasant with strong running ability; flies swiftly. Vigilant; runs away immediately when in danger. Uses irregular lek sites in breeding season, where three to five males get together, surround females, and display side-on. During courtship, male displays neck feathers (cape) and brightly colored back. Lives in large flocks in fall and winter.
Distribution: Chinese endemic. Found in NW Sichuan, S Gansu, Qinling Mountains, Daba Mountains, and the eastern part of Yunnan-Guizhou Plateau. Locally common.
Voice: Male utters a mellow and rapid *gu-gu-gu*, the second note louder and higher pitched than in call of Lady Amherst's Pheasant. Female utters a hoarse *cha-cha.*

Chrysolophus amherstiae
白腹锦鸡 bái fù jǐn jī
Lady Amherst's Pheasant　LC

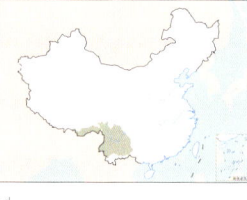

L: 110–150 cm (♂)
54–67 cm (♀)
Habitat: Inhabits montane forests and forest-edge bushes above 1500 m. Sometimes also forages on farmland.
Behavior: Good at running in montane forests; often nests on the ground.
Distribution: Found in the western portions of Yunnan-Guizhou Plateau, SE Tibet, and SW Sichuan. Local but often not uncommon.
Voice: Hoarse call similar to that of Golden Pheasant, *kuang*, often doubled. Gives a shrill, piercing alarm call.

White Eared-Pheasant

Short white ear tufts

Almost all-white body

Black wing feathers

C. c. lichiangense

White wing feathers

C. c. drouynii

Blue Eared-Pheasant

Long white ear tufts

All bluish-gray body

Tibetan Eared-Pheasant

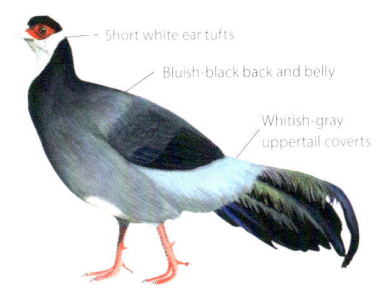

Short white ear tufts

Bluish-black back and belly

Whitish-gray uppertail coverts

Brown Eared-Pheasant

Black tips on mostly white tail feathers

Long white ear tufts

Grayish-brown body

Golden Pheasant

Orange cape with black edges

Red breast and belly

Black spots on brown tail

Slightly smaller than female Lady Amherst's Pheasant

Orange-yellow legs

Lady Amherst's Pheasant

White cape on neck

Black bars on white tail

White belly

Belly lacks barring

Leaden legs

Syrmaticus ellioti

白颈长尾雉 bái jǐng cháng wěi zhì

Elliot's Pheasant   NT

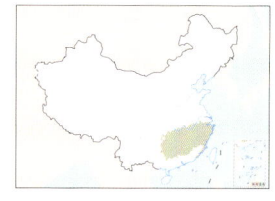

L: 81–90 cm (♂)
    45–50 cm (♀)

Habitat: Inhabits mountains and hills below 1000 m. Especially frequents broadleaf forests and mixed forests; moves to scrubland or farmland at lower altitudes in winter.

Behavior: Vigilant; very quiet when moving solitarily or in small flocks. Omnivorous; generally forages in the open understory of dense forests. Male displays side-on during courtship.

Distribution: Chinese endemic. Occasionally seen in Anhui, Zhejiang, Jiangxi, Hunan, Guangdong, Guangxi, and Fujian.

Voice: Rarely calls; occasionally makes a low-pitched *gu-gu-gu*.

Syrmaticus humiae

黑颈长尾雉 hēi jǐng cháng wěi zhì

Hume's Pheasant   NT

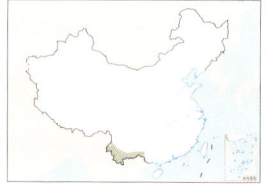

L: 96–104 cm (♂)
    47–50 cm (♀)

Habitat: Inhabits montane forests and forest-edge bushes at altitudes of 500–2500 m.

Behavior: Similar to Elliot's Pheasant. Vigilant and quiet. Nests on the ground close to trees, in forest understory with abundant bushes and herbaceous plants.

Distribution: S Yunnan and S Guangxi. Local and uncommon.

Voice: Makes a *ge-ge-ge* when foraging and a high-pitched whistle in alarm.

Syrmaticus mikado

黑长尾雉 hēi cháng wěi zhì

Mikado Pheasant   NT

L: 86–89 cm (♂)
    52–56 cm (♀)

Habitat: Inhabits primary broadleaf, coniferous, and mixed forests above 1500 m.

Behavior: Vigilant and quiet. Prefers foraging on the ground in areas with dense mature forests and sparse undergrowth. More active in the morning and afternoon; rests or takes dust bath around noon.

Distribution: Chinese endemic restricted to Taiwan; local and uncommon.

Voice: Muffled low-pitched clucking contact calls. Alarm call a nasal *wok*, repeated. Male also gives shrill whistles in breeding season and produces a wing whir.

Syrmaticus reevesii

白冠长尾雉 bái guān cháng wěi zhì

Reeves's Pheasant   VU

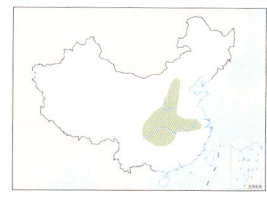

L: 140–190 cm (♂)
    56–70 cm (♀)

Habitat: Inhabits montane forests between 300 and 2500 m; especially common in mountainous areas with many gullies and steep slopes.

Behavior: Strong pheasant with good flying ability; can fly across valleys that are hundreds of meters wide. Good at running. Vigilant and quiet. Male highly territorial in breeding season; attracts female by flapping wings rapidly to produce drumming sounds.

Distribution: Chinese endemic. Found in Shanxi, Shaanxi, Henan, Guizhou, and Hunan. Range has shrunk rapidly in recent years.

Voice: Very rarely calls; occasionally makes a high-pitched cheeping and a mellow *gu-gu-gu* when foraging. Loud wing whir.

Phasianus colchicus

环颈雉 (雉鸡) huán jǐng zhì (zhì jī)

Ring-necked Pheasant   LC

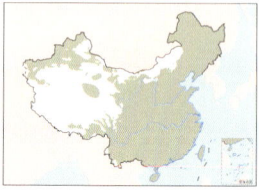

L: 80–100 cm (♂)
    57–65 cm (♀)

Habitat: Inhabits mainly low mountains, hills, and plains; also moves around farmland.

Behavior: Strong legs, good at running; can fly short distances. Gregarious in fall and winter. Often forages by scraping the ground for plants and invertebrates. Nests on the ground with withered grass and leaves.

Distribution: Common throughout most of China. Many subspecies; sometimes treated as three species: Chinese Pheasant (*P. vlangalii*) occurs in central and eastern China (including subspecies *P. c. vlangalii, P. c. satscheuensis, P. c. strauchi, P. c. sohokhotensis, P. c. alaschanicus, P. c. edzinensis, P. c. pallasi, P. c. kiangsuensis, P. c. karpowi, P. c. suehschanensis, P. c. takatsukasae, P. c. torquatus, P. c. decollatus,* and *P. c. formosanus*). Yunnan Pheasant (*P. elegans*) occurs in the Hengduan Mountains (including subspecies *P. c. elegans* and *P. c. rothschildi*). Turkestan Pheasant (*P. colchicus*) occurs in Xinjiang (including subspecies *P. c. mongolicus, P. c. shawii,* and *P. c. tarimensis*).

Voice: Male's advertising call is a familiar loud crowing, *ga-ka*, often accompanied by the sound of rapid wing whirring.

Elliot's Pheasant

White neck

Black throat

White belly

♂

Black throat

White belly

♀

Reeves's Pheasant

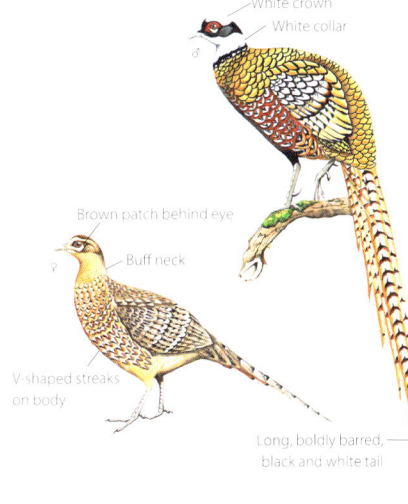

White crown

White collar

♂

Brown patch behind eye

Buff neck

V-shaped streaks on body

♀

Long, boldly barred, black and white tail

Hume's Pheasant

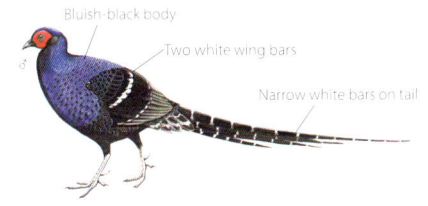

Bluish-black neck with metallic sheen

Chestnut body

Black bars on white tail

♂

Two narrow white wing bars

♀

Scaly, pale brown breast and belly

Mikado Pheasant

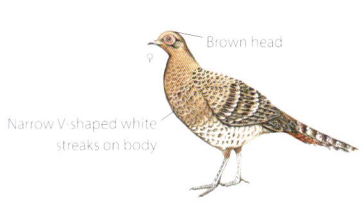

Bluish-black body

Two white wing bars

Narrow white bars on tail

♂

Brown head

Narrow V-shaped white streaks on body

Ring-necked Pheasant

Glossy blackish-green neck

White neck collar

Broad red orbital skin

Gray rump

Black bars on reddish-brown tail

♂

P. c. torquatus
"Chinese Pheasant"

White neck collar

Olive or chestnut rump

♂

Rufous-chestnut flanks

P. c. mongolicus
"Turkestan Pheasant"

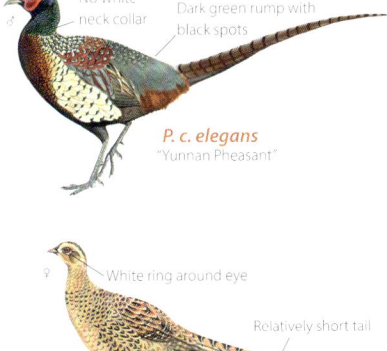

No white neck collar

Dark green rump with black spots

♂

P. c. elegans
"Yunnan Pheasant"

White ring around eye

♀

Relatively short tail

Tropicoperdix chloropus

绿脚树鹧鸪 lǜ jiǎo shù zhè gū

Scaly-breasted Partridge

L: 25–28 cm
Habitat: Inhabits montane evergreen forests and scrubland below 1500 m; occasionally moves to farmland.
Behavior: Often occurs under dense bushes in forest understory. Secretive; seldom flies. Omnivorous; behavior similar to that of *Arborophila* partridges.
Distribution: Scarce resident in S Yunnan.
Voice: Lengthy series of loud and gradually accelerating short, flat, intensifying whistles, sometimes paired, often ending with a flourish.

Polyplectron bicalcaratum

灰孔雀雉 huī kǒng què zhì

Gray Peacock-Pheasant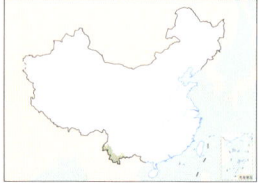

L: 57–76 cm (♂)
   46–51 cm (♀)
Habitat: Inhabits tropical monsoon forests or ravine rain forests below 1500 m.
Behavior: Secretive. Often forages in sparse understory of dense forests. Male poses front-on by displaying eyespots on its wings, back, and tail.
Distribution: Uncommon resident in S and W Yunnan.
Voice: A series of five to nine loud but somewhat muffled notes, *kok-kok-kok-kok-ko*, and a faster *ga-ga-ga-ga-ga*, used perhaps exclusively when alarmed.

Polyplectron katsumatae

海南孔雀雉 hǎi nán kǒng què zhì

Hainan Peacock-Pheasant

L: 53–65 cm (♂)
   40–45 cm (♀)
Habitat: Inhabits ravine rain forests, evergreen monsoon forests, and broadleaf forests between 800 and 1300 m.
Behavior: Extremely secretive; solitary almost all year-round. Forages by pecking on the ground. When in danger, either stays still or runs away through forest.
Distribution: Chinese endemic. Rare resident in the mountains of C and S Hainan.
Voice: Loud, rapid *ga-ga-ga*, reminiscent of Gray Peacock-Pheasant but faster.

Pavo muticus

绿孔雀 lǜ kǒng què

Green Peafowl

L: 230–250 cm (♂)
   100–110 cm (♀)
Habitat: Inhabits mountain valleys, monsoon forests with sparse undergrowth, or savannas below 2000 m; occasionally seen in conifer plantations.
Behavior: Huge, elegant. Often lives in a family unit of one male, three to five females, and several subadults. Male fans brightly colored tail during courtship display.
Distribution: Previously widespread in southern China but much depleted and now only rarely seen in dry-hot river valleys and monsoon rain forests in C and S Yunnan.
Voice: Loud *ke-aow*, rising then falling, sometimes more obviously disyllabic.

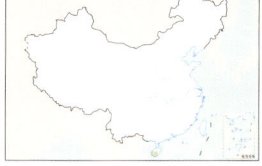

Scaly-breasted Partridge

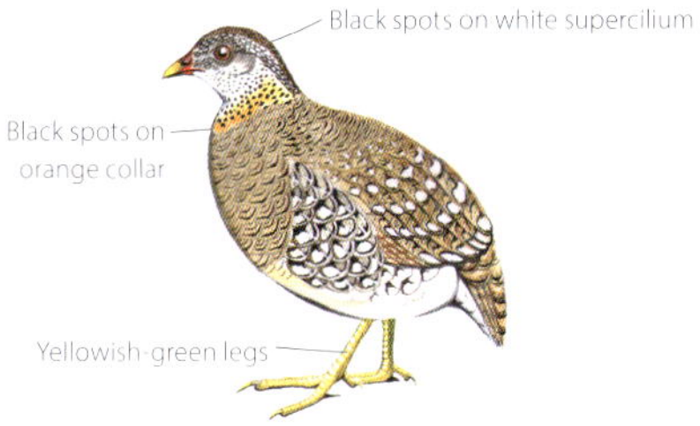

Black spots on white supercilium

Black spots on orange collar

Yellowish-green legs

Hainan Peacock-Pheasant

Relatively short crest

Black eyespots on back with inconspicuous bluish-purple sheen

♂

♀

Gray Peacock-Pheasant

White chin and throat

Bluish-purple or green eyespots on back, wings, and tail

Black spots on wings and back

♀

Bluish-green eyespots on tail

Green Peafowl

Vertical plume

♂

Iridescent green neck and back

Extremely elongated uppertail coverts become train

Vertical plume

Relatively small yellow patch on cheek

Iridescent green neck and back

Lacks train

潜鸟目　GAVIIFORMES

Strongly built, medium-sized to large seabirds. Body profile resembles ducks but with straight and pointed bill. Good at diving, similar to cormorants. Plumage similar between sexes, mostly black and white, some with purple or red. Neck long and thick. Tail short. Good at flying; rarely seen on land. Nest on or near ponds and lakes, and winter primarily in coastal waters. Feed mostly on fish. All species are migratory.

One family, one genus, and five species recognized worldwide, distributed in the Arctic and temperate waters in northern latitudes. Four species recorded in China, found wintering mostly in coastal waters of eastern China, and some may breed inland in northern China.

Gavia stellata
红喉潜鸟 hóng hóu qián niǎo
Red-throated Loon　LC

L: 53–69 cm
Habitat: Occurs on calm freshwater lakes, rivers, and ponds, rare in saltwater habitats.
Behavior: Often moves about solitarily or in pairs on water. Quiet. Seldom flies.

Distribution: Mainly a passage migrant and winter visitor to China; seen along the entire coast but regular only near Shanghai and Zhejiang. Rare migrant through E Heilongjiang, Beijing, SC Inner Mongolia, and SE Yunnan. May breed in Heilongjiang.
Voice: Unlikely to be heard in China but gives eerie catlike mewing and mournful wails in breeding season; rarely heard flight call is a ducklike *gaga*.

Gavia pacifica
太平洋潜鸟 tài píng yáng qián niǎo
Pacific Loon　LC

L: 60–68 cm
Habitat: Inhabits medium-sized to large lakes, rivers, and ponds on tundra or in coniferous forests; also seen along the coast in winter.
Behavior: Appears solitarily or in pairs on inland freshwater wetlands.

Distribution: Mainly a passage migrant or winter visitor in China. Migrates through Heilongjiang, E Liaoning, NE Hebei, and Shandong; winters in Jiangsu; vagrant in Beijing and Hong Kong.
Voice: Gives a long series of wails in breeding season; also gives a gull-like *qua-qua*.

Gavia arctica
黑喉潜鸟 hēi hóu qián niǎo
Arctic Loon　LC

L: 56–77 cm
Habitat: Occurs on medium-sized to large lakes, rivers, and ponds on tundra or in forests.
Behavior: Moves solitarily or in pairs on medium-sized to large freshwater wetlands; also occurs in saltwater habitats in winter.

Distribution: Rare passage migrant and scarce, local winter visitor in China. *G. a. viridigularis* on China's east coast, from Jilin in the north to Fujian and Taiwan in the south; occasional inland in SE Inner Mongolia; sporadic records in Shaanxi and Sichuan. *G. a. arctica* recorded in the Altai Mountains in N Xinjiang.
Voice: Unlikely to be heard in China but gives loud, disyllabic, upwardly inflected calls in breeding season; flight call is a monotone, monosyllabic, gull-like *qua-qua*.

Gavia adamsii
黄嘴潜鸟 huáng zuǐ qián niǎo
Yellow-billed Loon　NT

L: 75–100 cm
Habitat: Occurs on lakes and rivers on tundra and mountains; winters inshore.
Behavior: Appears in pairs or small flocks on large freshwater lakes and estuaries.

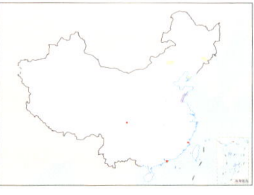

Distribution: Vagrants recorded in Jilin, Liaoning, Shandong, Inner Mongolia (Xilingol), Fujian (Lianjiang), Hong Kong, and Sichuan (Deyang). Small numbers winter offshore in Shandong and N Jiangsu.
Voice: Gives a series of lengthy, flat, catlike mewing calls, *heeaaoo*, and a shorter yapping *u-u-a*, rising in pitch, in breeding season.

Red-throated Loon

Streamlined head with no prominent peak

White lores and eye ring

juv.

More white on upper neck

non-br.

Bill more pointed and angled up than other loons

Triangular rufous-chestnut patch from throat to foreneck

br.

Neck often maneuvered in flight

chick

Arctic Loon

juv.

Dark area larger than white area on side of neck

Sharp straight black bill

Gray head and hindneck

Lower throat and foreneck black with green sheen

br.

non-br.

Conspicuous white thigh patch

chick

Pacific Loon

juv.

Thin black collar

non-br.

Lower throat and foreneck black with purple sheen

br.

No white patch on flanks

chick

Yellow-billed Loon

Conspicuous peak on forehead

Broad, long, pale yellow bill

br.

juv.

Large feet

non-br.

chick

鹱形目　PROCELLARIIFORMES

Pelagic birds of varying sizes. Body profile resembles gulls. Plumage similar between sexes, mostly black, brown, and white. Bill long and laterally flattened with hooked tip and tubular nostrils. Wings extremely long and narrow, equipped for dynamic soaring. Tail short to long, rounded, forked, or wedge shaped. Feet webbed. Roam the open tropical and temperate oceans almost their entire life, solitarily or in large groups, coming to land only when breeding. Feed mostly on fish, mollusks, and plankton. Most species are nomadic during nonbreeding seasons.

　Four families, 27 genera, and 147 species recognized worldwide, widely distributed across the world's oceans and seas. Four families, nine genera, and 17 species recorded in China, found mostly in the ocean and seas off the coast, occasionally recorded inland.

洋海燕科
Oceanitidae

Wilson's Storm-Petrel

海燕科
Hydrobatidae

Swinhoe's Storm-Petrel

信天翁科
Diomedeidae

Laysan Albatross

鹱科
Procellariidae

Streaked Shearwater

Oceanites oceanicus

黄蹼洋海燕 huáng pǔ yáng hǎi yàn
Wilson's Storm-Petrel   LC

L: 15–19 cm
WS: 38–42 cm
Habitat: Inhabits oceans in various climatic zones.
Behavior: Occurs solitarily or in flocks at sea; often forages in flocks.
Distribution: Vagrant to Jiangsu (Qiansan Islands) and Zhejiang (Hangzhou).
Voice: Unlikely to be heard in China but gives bizarre electronic buzzing and chatter calls when in flocks.

Wilson's Storm-Petrel

White rump

Toes project beyond tail tip in flight

Square, slightly notched tail

黑叉尾海燕 hēi chā wěi hǎi yàn

Swinhoe's Storm-Petrel   **NT**

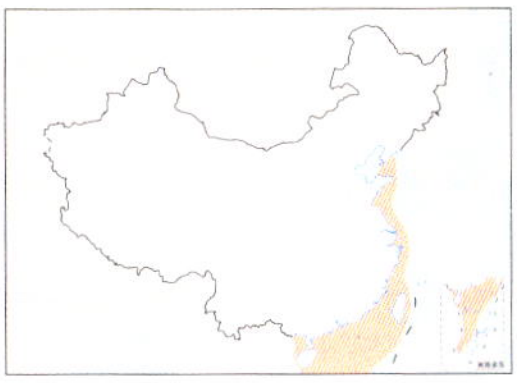

L: 18–22 cm
WS: 45–48 cm
Habitat: Seen on inshore and pelagic waters; also breeds on and forages around islands.
Behavior: Small coastal seabird. Often breeds on coasts and inshore islands. Can glide like tern and walk on the ground.
Distribution: Breeds on islands in the Yellow Sea, East China Sea, and South China Sea.
Voice: Hoarse Wilson's Storm-Petrel–like buzzing, *ga—ga ga*, and high-frequency whistles; rarely calls in flight.

白腰叉尾海燕 bái yāo chā wěi hǎi yàn

Leach's Storm-Petrel   **VU**

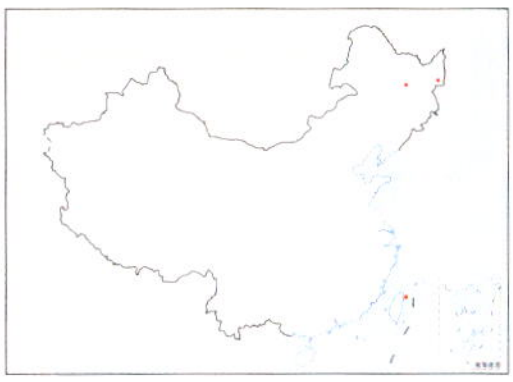

L: 19–25 cm
WS: 43–48 cm
Habitat: Occurs at sea from subtropical to polar regions; breeds on islands and coasts in the polar region.
Behavior: Pelagic seabird; prefers flying in flocks over the sea. Breeds on islands and coasts. Can walk fast on the ground.
Distribution: Vagrant in Heilongjiang (historic) and Taiwan.
Voice: Unlikely to be heard in China but gives an incessant *gi-gi-ga-ga*.

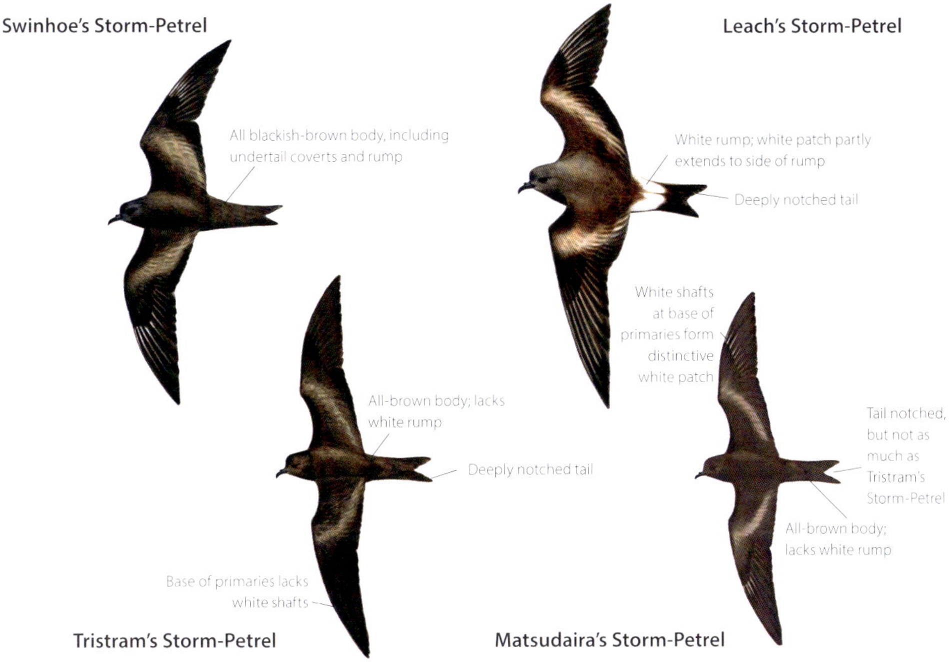

Swinhoe's Storm-Petrel

All blackish-brown body, including undertail coverts and rump

All-brown body; lacks white rump

Deeply notched tail

Base of primaries lacks white shafts

Tristram's Storm-Petrel

Leach's Storm-Petrel

White rump; white patch partly extends to side of rump

Deeply notched tail

White shafts at base of primaries form distinctive white patch

Tail notched, but not as much as Tristram's Storm-Petrel

All-brown body; lacks white rump

Matsudaira's Storm-Petrel

褐翅叉尾海燕 hè chì chā wěi hǎi yàn

Tristram's Storm-Petrel   **LC**

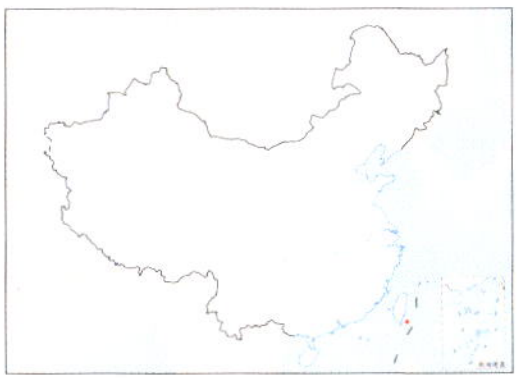

L: 24–27 cm
WS: 56 cm
Habitat: Occurs in subtropical and temperate oceans.
Behavior: Gregarious at sea; breeds on islands in the NW Pacific Ocean.
Distribution: Vagrant off Taiwan.
Voice: Unlikely to be heard in China but makes rapid and undulating *ga-go—gu—* calls.

日本叉尾海燕 rì běn chā wěi hǎi yàn

Matsudaira's Storm-Petrel   **VU**

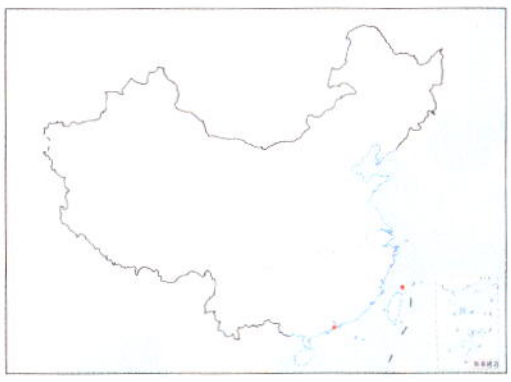

L: 24–25 cm
WS: 56 cm
Habitat: Occurs in subtropical and tropical oceans.
Behavior: Often flies over pelagic waters in small flocks; flight slow and somewhat cumbersome. Wanders far in nonbreeding season.
Distribution: Vagrant off N Taiwan and in the South China Sea.
Voice: Unknown.

Phoebastria immutabilis

黑背信天翁 hēi bèi xìn tiān wēng

Laysan Albatross · NT

L: 71–81 cm
WS: 195–203 cm
Habitat: Roams over the North Pacific Ocean year-round; breeds on islands in N North Pacific.
Behavior: Pelagic seabird; often flies solitarily or in pairs over the sea; rests on the sea's surface. Seldom comes ashore.
Distribution: Winter visitor to Fujian and Taiwan coasts.
Voice: Unlikely to be heard in China but utters *ta-ta* sounds by snapping the mandibles together; also makes a cowlike monosyllabic *moo-* and a thin *ying-ying-*.

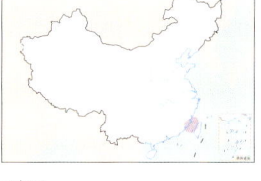

Phoebastria nigripes

黑脚信天翁 hēi jiǎo xìn tiān wēng

Black-footed Albatross · NT

L: 68–83 cm
WS: 193–213 cm
Habitat: Generally seen in open areas far out at sea; breeds in colonies on a few islands.
Behavior: Pelagic seabird; occurs at sea either solitarily or in pairs. Also follows ships. Moves to islands only in breeding season.
Distribution: Rare at sea off Shandong, Zhejiang, Fujian, Taiwan, and Hainan in summer.
Voice: Unlikely to be heard in China but makes a rapid *te-te-te-te* by snapping the mandibles together and soft, high-pitched *xu-xu-* whistles; can also produce a cowlike *moo-* and a ducklike *gwa-gwa*.

Phoebastria albatrus

短尾信天翁 duǎn wěi xìn tiān wēng

Short-tailed Albatross · VU

L: 84–100 cm
WS: 213–229 cm
Habitat: Generally inhabits marine waters; moves to land only during breeding season.
Behavior: Appears more often on inshore and coastal waters than other albatrosses; also less likely to follow ships than other albatrosses.
Distribution: Rare at sea off Shandong, Guangdong, and most regularly Taiwan.
Voice: Unlikely to be heard in China but gives *te-te* by snapping the mandibles together; also makes soft, high-pitched *xu-xu-* whistles.

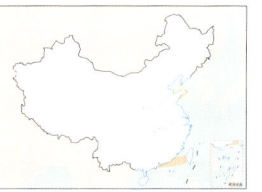

Fulmarus glacialis

暴风鹱 bào fēng hù

Northern Fulmar · LC

L: 43–52 cm
WS: 107 cm
Habitat: Occurs from inshore to pelagic waters in temperate and polar regions.
Behavior: Typical seabird; prefers foraging and resting in flocks at sea; almost never on land except during breeding season.
Distribution: Vagrant off E Liaoning (historic) and Taiwan.
Voice: Unlikely to be heard in China but makes a noisy *ga-ga*, like calls made by duck flocks.

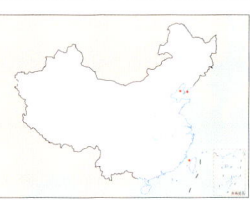

Pterodroma hypoleuca

白额圆尾鹱 bái é yuán wěi hù

Bonin Petrel · LC

L: 30–31 cm
WS: 63–71 cm
Habitat: Occurs from inshore to pelagic waters in temperate and tropical zones.
Behavior: Typical seabird; always out at sea except during breeding season. Gregarious.
Distribution: Passage migrant or vagrant to Fujian, Taiwan, Shanghai, and Zhejiang.
Voice: Unlikely to be heard in China but makes hoarse and dull *gu-gu-e* sounds; sometimes also gives a very low guttural *e—e*.

Pseudobulweria rostrata

钩嘴圆尾鹱 gōu zuǐ yuán wěi hù

Tahiti Petrel · NT

L: 38–40 cm
WS: 84 cm
Habitat: Occurs at sea near the Pacific Islands.
Behavior: Generally occurs at sea in flocks; breeds on islands.
Distribution: Vagrant to Taiwan.
Voice: Unlikely to be heard in China but makes prolonged *zi-ya—* whistles, like the sound of a door closing.

Laysan Albatross

Black smudge around eye, gray wash on cheeks and chin; appears to have "dark circles"

Yellow bill with dark tip

White overall, except for sooty-brown back and upper wing

Northern Fulmar

White head distinctive among petrels and shearwaters

Pale morph

Neck distinctly shorter and stouter than other petrels and shearwaters

Bill shorter than other petrels and shearwaters, with consistent thickness

Dark morph

Black-footed Albatross

White rump

Black bill tip

Black feet

Bonin Petrel

M-shaped pattern on back and upper wing

White forehead and cheek

Dark mask

Plain white underparts

Distinctive pattern on underwing coverts

Short-tailed Albatross

White back and rump

Pink bill

Feet project well beyond tail tip, compared to other albatrosses

Tahiti Petrel

Plain blackish-brown upperparts

Blackish-brown hood

Dark head and neck clearly demarcated from white underparts, differing from other petrels and shearwaters

Pale tips of underwing coverts form pale band across entire under wing

85

Calonectris leucomelas

白额鹱 bái é hù

Streaked Shearwater

L: 45–52 cm
WS: 122 cm
Habitat: Occurs from inshore to pelagic waters in temperate and tropical zones.
Behavior: The most common shearwater in W North Pacific; often
flocks on inshore and pelagic waters with other seabirds.
Distribution: Regular along the coasts of Yellow Sea, East China Sea, South China Sea, and the nearby islands; resident in Taiwan and Hainan; summer visitor or passage migrant in other regions; vagrants inland in Tianjin, Anhui, and Jiangxi.
Voice: Vocal in breeding season, relatively quiet in flight. At colony, gives various high-pitched loud whistles, growls, and grunts. Male's tremulous whistled *pee er-wee* at burrow is replied to with a gull-like *ke-argh* by female.

Ardenna pacifica

楔尾鹱 xiē wěi hù

Wedge-tailed Shearwater

L: 38–47 cm
WS: 97–105 cm
Habitat: Occurs from coastal to pelagic waters.
Behavior: Often occurs in flocks across tropical and subtropical oceans; breeds on islands and coasts. Very strong flier.
Distribution: Vagrant in Shanghai, Zhejiang, Hainan, and Taiwan.
Voice: Melodious moaning call, *wo-wu-u—e*, unlikely to be heard in China.

Ardenna grisea

灰鹱 huī hù

Sooty Shearwater

L: 40–51 cm
WS: 94–109 cm
Habitat: Occurs from inshore to pelagic waters.
Behavior: Breeds on coasts and offshore islands; flocks over open ocean with other shearwaters in nonbreeding season.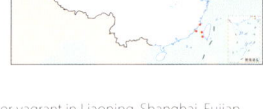
Distribution: Summer visitor or vagrant in Liaoning, Shanghai, Fujian, and Taiwan.
Voice: Unlikely to be heard in China but gives a childlike call, *a-a-ye—e*, on the ground.

Ardenna tenuirostris

短尾鹱 duǎn wěi hù

Short-tailed Shearwater

L: 35–45 cm
WS: 95–100 cm
Habitat: Occurs from inshore to pelagic waters.
Behavior: Occurs in flocks over open ocean in nonbreeding season; moves to land in breeding season. Strong flier.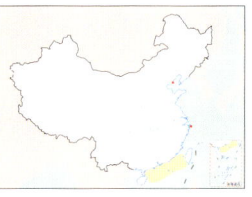
Distribution: Passage migrant or vagrant; accidental records in the Yellow Sea, East China Sea, and South China Sea; regular passage records in Taiwan and Hong Kong.
Voice: Unlikely to be heard in China but noisy at colonies, making *e-e-ge-ge-ga-ga* calls.

Ardenna carneipes

淡足鹱 dàn zú hù

Flesh-footed Shearwater

L: 40–48 cm
WS: 99–107 cm
Habitat: Occurs over oceans in tropical and subtropical zones.
Behavior: Plumage sootier than other shearwaters. Breeds on islands; wanders over
oceans in flocks during nonbreeding season.
Distribution: Breeds on the South China Sea Islands; vagrant in Taiwan.
Voice: At colony makes a sharp and hoarse *zi-zi-za-ya*.

Bulweria bulwerii

褐燕鹱 hè yàn hù

Bulwer's Petrel

L: 26–30 cm
WS: 68–73 cm
Habitat: Occurs over oceans in temperate and subtropical zones.
Behavior: Often gregarious in nonbreeding season. Appears more graceful and storm-petrel-like than other shearwaters.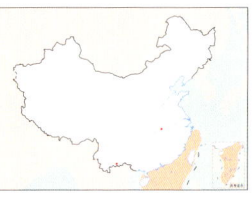
Distribution: Breeds on the East China Sea and South China Sea; vagrant in Hubei and Yunnan.
Voice: Relatively low-pitched disyllabic or trisyllabic *e—e* or *gu—gu-e* calls.

Streaked Shearwater

Head densely streaked and spotted; appears pale

Wedge-shaped tail

Distinctively streaked carpal patch on under wings

Short-tailed Shearwater

Flesh-footed Shearwater

Wedge-tailed Shearwater

Short-tailed Shearwater

Dark feet project well beyond tail tip in flight

Pale area on underside of wings relatively darker

Wedge-tailed Shearwater

Pale morph

Conspicuous wedge-shaped tail

Pale underparts and underwing coverts differ from other species in genus *Ardenna*

Feet do not project beyond tail tip

Dark morph

Pale mottled underwing coverts

Flesh-footed Shearwater

Pale bill with dark tip

Sooty-black body

Pale feet do not project beyond tail tip in flight

Bulwer's Petrel

Much smaller than other shearwaters, and slightly larger than storm-petrels

Blackish brown overall

Wedge-shaped tail appears long and pointed in flight, without notch

Sooty Shearwater

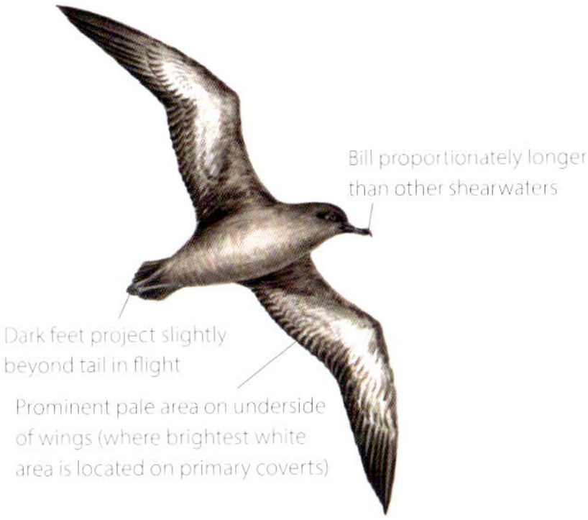

Bill proportionately longer than other shearwaters

Dark feet project slightly beyond tail in flight

Prominent pale area on underside of wings (where brightest white area is located on primary coverts)

鸊鷉目　PODICIPEDIFORMES

Small to medium-sized waterbirds. Profile resembles ducks, but with straight and pointed bill. Plumage similar between sexes, mostly black, chestnut, and grayish brown. Neck thin and straight. Tail short. Lobed feet well equipped for frequent diving. Inhabit freshwater habitats, such as rivers, lakes, marshes, and ponds, in pairs or small groups, rarely on land. Nest on water by building floating vegetative nests. Feed mostly on fish, crustaceans, and aquatic insects. Most species are migratory.

One family, six genera, and 23 species recognized worldwide, with a cosmopolitan distribution. Two genera and five species recorded in China, ubiquitously distributed nationwide.

Tachybaptus ruficollis
小鸊鷉 xiǎo pì tī
Little Grebe

L: 23–29 cm
Habitat: Inhabits freshwater lakes, marshes, ponds, and slow-flowing rivers; also seen on coastal waters.
Behavior: Usually seen solitarily or in pairs in freshwater areas; also gathers in small flocks in winter. Dives for aquatic animals. Requires long run-up before takeoff.
Distribution: Common throughout China. *T. r. capensis* in Xinjiang, S Tibet, and W Yunnan; *T. r. poggei* widespread in eastern China. *T. r. philippensis* occurs in Taiwan.
Voice: Varied. Most familiar "song" is a lengthy, descending, Ruddy-breasted Crake–like trill.

Podiceps grisegena
赤颈鸊鷉 chì jǐng pì tī
Red-necked Grebe

L: 40–57 cm
Habitat: Occurs on medium to large freshwater lakes, marshes, ponds, and backwaters; also seen on coasts and estuaries.
Behavior: Often moves around solitarily or in pairs on freshwater wetlands with abundant emergent or submerged plants. Seldom flies. More agile than Great Crested Grebe.
Distribution: Widespread but uncommon. *P. g. holbollii* breeds in Northeast and North China, occasionally south to Hebei; winters and migrates through North, northwestern, East, and South China and Sichuan, but almost no record in East and South China in the past 20 years and only sporadic wintering records in Sichuan. *P. g. grisegena* may once have bred in Xinjiang but is now a scarce migrant; a subadult was seen in Ili in fall 2019.
Voice: Gives varied harsh, far-carrying, chattering calls and strained Water Rail–like squeals, *ga—ku-ku-ku*, in breeding season and, rarely, a series of sharp *ke-ke-ke* trills in nonbreeding season.

Podiceps cristatus
凤头鸊鷉 fèng tóu pì tī
Great Crested Grebe

L: 45–51 cm
Habitat: Occurs on open lakes, marshes, reservoirs, ponds, rivers, and coasts.
Behavior: Often moves around on freshwater wetlands with abundant emergent plants; also gathers in large loose flocks in breeding and wintering seasons.
Distribution: Widespread, but absent in Hainan. Breeds in northern China and winters in southern China.
Voice: Variety of clicking or throaty moaning calls and a series of hoarse *e-e-e* trills.

Podiceps auritus
角鸊鷉 jiǎo pì tī
Horned Grebe

L: 31–39 cm
Habitat: Inhabits open lakes, ponds, and rivers with aquatic plants.
Behavior: Often moves around in pairs or solitarily; prefers open waters.
Distribution: Breeds in northwestern China; migrates through Northeast and North China; winters coastally and, rarely, in the Yangtze Plain and the regions to its south, including Taiwan.
Voice: Descending guttural wails, *geee-argh*, and other chatters, as well as vaguely Little Grebe–like trills.

Podiceps nigricollis
黑颈鸊鷉 hēi jǐng pì tī
Eared Grebe

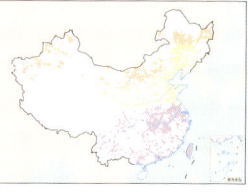

L: 25–35 cm
Habitat: Occurs on inland freshwater lakes, marshes, ponds, and rivers.
Behavior: Moves around in pairs or small flocks on open freshwater wetlands with aquatic plants; rarely seen on land. Generally winters in large flocks.
Distribution: Widespread but absent in Hainan and C Qinghai-Tibet Plateau. Breeds in Northeast and northwestern China; migrates through North China; winters mostly along the coast south of the Qinling-Huaihe Line. Locally common; a large wintering population containing up to a thousand individuals was recorded in Zhejiang (Changtan Reservoir, Taizhou) over several consecutive years. Flocks of several thousand occasionally seen on saltpans or other saline water bodies during migration.
Voice: Easily overlooked, shrill, off-key, rising, whistled *oor-kip* calls, sometimes accompanied by shorter *jiu-jiu-jiu* trills.

Little Grebe

Yellow iris

Chestnut chin and neck

br.

Dark stripes on face

chick

Bill shorter than other grebes, base of bill yellow

non-br.

juv.

Light brown neck

Red-necked Grebe

juv.

Yellow on base of both mandibles

Silver-white cheek

Rufous-chestnut neck

br.

Larger yellow area on lower mandible

non-br.

Gray neck, thicker and shorter than Great Crested Grebe

Great Crested Grebe

juv.

Black stripes on face and neck

Unique ornamental plumes extend from behind ear to neck

White foreneck

br.

White lores

Pink bill

non-br.

Horned Grebe

Flat forehead, not steep

Yellow ornamental plumes from behind eye to nape, shaped like a horn

br.

White-tipped bill straight, not upturned

Chestnut from foreneck to entire underparts

Crown peaks at rear

juv.

non-br.

Plain white cheeks sharply demarked from dark cap

Eared Grebe

Prominent ear tufts

br.

Black neck

chick

Crown peaks at center of head

Steep forehead

non-br.

Conspicuously upturned bill

Irregular division of dark cap and white cheeks

红鹳目　PHOENICOPTERIFORMES

Large waders with unique body profile. Long neck and legs superficially resemble those of storks, but readily told from all other birds by the thick and angled bill, specialized for filter feeding. Plumage similar between sexes, mostly white, pink, and black. Inhabit lagoons, marshes, and lakes from coast to inland in large aggregations; especially prefer saline water. Nest in large groups. Feed mostly on algae and other small aquatic organisms. Most species are nonmigratory but nomadic in nonbreeding seasons.

One family, three genera, and six species recognized worldwide, distributed in Africa, Europe, Asia, and the Americas. One genus and one species recorded in China, occasionally found in northern China.

红鹳科
Phoenicopteridae

Heavy, uniquely shaped, black-tipped bill bends downward from middle

White body; swims like swan

Greater Flamingo

Both upperwing and underwing coverts red

Extremely long pink legs

Phoenicopterus roseus

大红鹳 dà hóng guàn

Greater Flamingo

L: 120–145 cm

Habitat: Often seen on saltwater lakes, marshes, gulfs, and islands in the temperate zone; also seen on freshwater wetlands.

Behavior: inhabits and breeds in flocks in saltwater lakes and marshes.

Uses its unique bill to filter feed invertebrates, planktons, and algae in the water column. Requires a run-up for takeoff.

Distribution: Vagrants in Xinjiang. Also recorded in central and eastern China in recent years; some records could be escapes.

Voice: Short, goose-like, multisyllabic honking, *gu-gu-ga-ga*.

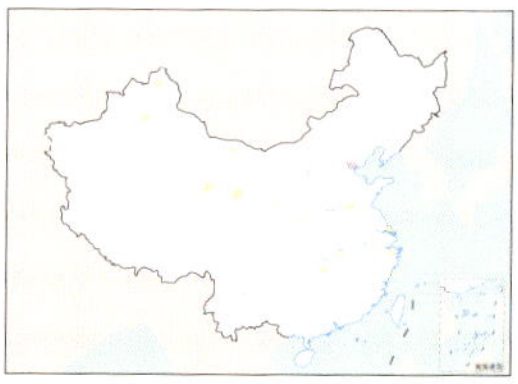

鹲形目
PHAETHONTIFORMES

Medium-sized seabirds of tropical oceans. Body profile resembles terns. Plumage similar between sexes, mostly white, black, and red. Bill straight and sharp. Tail rounded, with two extremely elongated central tail feathers (streamers). Excellent fliers usually seen over water solitarily or in pairs. Feed by diving, mostly on fish, squid, crustaceans, and other marine invertebrates. Nest on land, and nomadic in nonbreeding seasons.

One family, one genus, and three species recognized worldwide, mostly in tropical oceans. All three species recorded in China.

鹲科
Phaethontidae

Red-billed Tropicbird

Dense black bars on upper back

Bright red bill

Conspicuous black on wings

Red-tailed Tropicbird

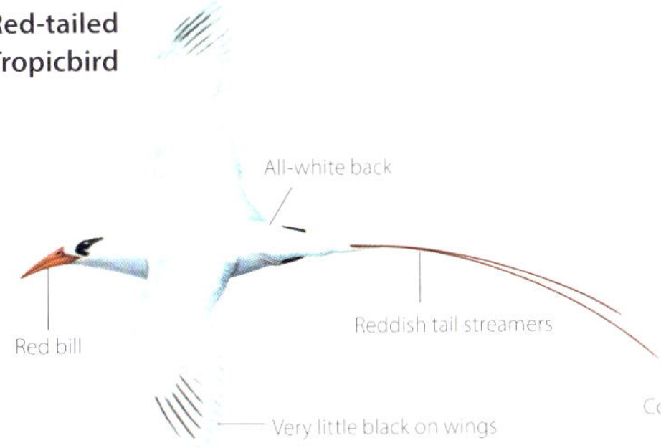

All-white back

Red bill

Reddish tail streamers

Very little black on wings

White-tailed Tropicbird

V-shaped pattern

Yellow bill

White tail streamers

Conspicuous black on wings

Phaethon aethereus
红嘴鹲 hóng zuǐ méng
Red-billed Tropicbird **LC**

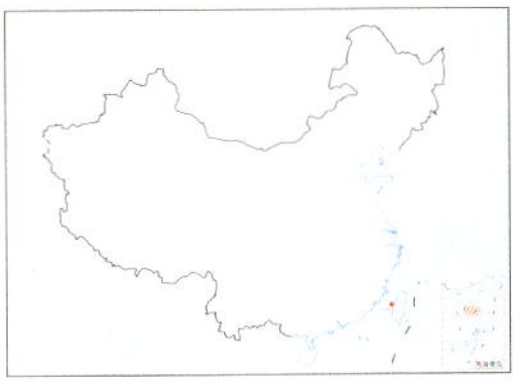

L: 45–107 cm (tail streamers 46–56 cm)
Habitat: Occurs on pelagic waters in tropical and subtropical zones.
Behavior: Tropical pelagic bird; generally wanders at sea solitarily for long periods. Has strong flying and swimming abilities.
Distribution: Vagrant in Taiwan.
Voice: Typically silent away from breeding colony, where it gives a series of shrill and ternlike *ki-ki-ki* trills, accompanied by sharp whistles.

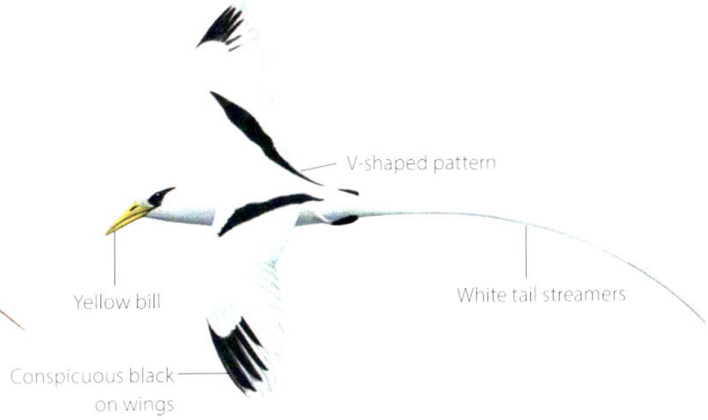

Phaethon rubricauda
红尾鹲 hóng wěi méng
Red-tailed Tropicbird **LC**

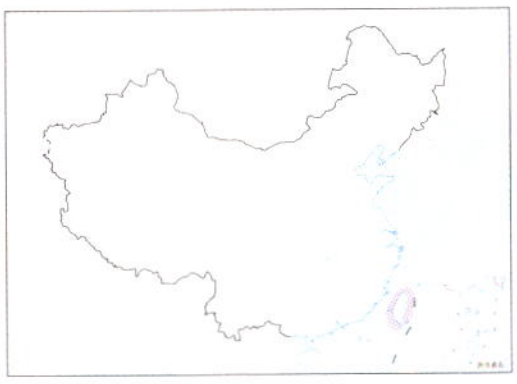

L: 45–102 cm (tail streamers 28–38 cm)
Habitat: Occurs on pelagic waters in subtropical and tropical zones.
Behavior: In nonbreeding season, often flies over oceans solitarily, looking for food; can dive to feed.
Distribution: Wanders off Taiwan.
Voice: Typically silent away from breeding colony, where it gives a series of hoarse calls, *ke-ke-ki-a—*.

Phaethon lepturus
白尾鹲 bái wěi méng
White-tailed Tropicbird **LC**

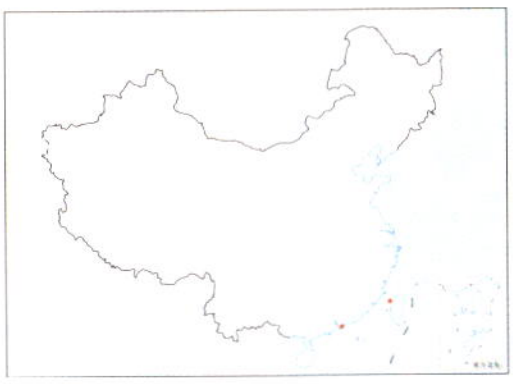

L: 37–99 cm (tail streamers 33–45 cm)
Habitat: Occurs over oceans in the tropical zone.
Behavior: Generally wanders at sea solitarily in nonbreeding season; a swift flier.
Distribution: Vagrants recorded off Hong Kong and Taiwan.
Voice: Typically silent away from breeding colony, where it gives a rapid and variable *ge-ki-ya*.

鹳形目　CICONIIFORMES

Large waders. Plumage similar between sexes, mostly black, white, red, and yellowish brown. Bill long, thick, and strong. Long neck and legs. Tail short. Inhabit primarily wetland habitats, such as rivers, lakes, and marshes, solitarily or in groups. Strong fliers. Nest on high places, such as trees, cliffs, and roofs. Feed mostly on fish, amphibians, reptiles, insects, and carrion. Species breeding in temperate zones are migratory.

Ciconiiformes was once a large order including multiple families but currently includes only storks and allies. One family, six genera, and 19 species recognized worldwide, with a cosmopolitan distribution. Four genera and seven species recorded in China, widely distributed except on the Qinghai-Tibet Plateau.

Mycteria leucocephala

彩鹳 cǎi guàn

Painted Stork   **NT**

L: 93–102 cm
Habitat: Occurs on lakes, rivers, marshes, and muddy areas with abundant vegetation; also seen on farmland, seashores, and saltmarshes.
Behavior: Generally occurs and breeds on freshwater wetlands in flocks; nests on large trees in wetlands.
Distribution: Relatively common historically. Very few recent records, with vagrants recorded in Yunnan, Guizhou, and Guangdong.
Voice: Generally silent except at breeding colonies, where it makes harsh screams and grunts, e—a-a-a, like the sounds of an infant.

鹳科
Ciconiidae

Painted Stork

Black Stork

Red bare skin on head

Black wing feathers and coverts

Drooping yellow bill

Black band across belly

Anastomus oscitans

钳嘴鹳 qián zuǐ guàn

Asian Openbill

L: 68–81 cm 
Habitat: Inhabits rivers, lakes, wet grass, paddy fields, and various kinds of wetlands. May move up to altitudes of 2000 m.
Behavior: In breeding season, builds nests and forages in large colonies; also moves about in flocks in nonbreeding season.
Often mixes with other storks and waterbirds. When flying, birds usually circle in flocks with their necks outstretched.
Distribution: Increasingly common in Yunnan, occasional north to Gansu, Sichuan, Chongqing, Guizhou, Guangxi, and Jiangxi (Poyang Lake).
Voice: Generally silent away from breeding colonies; occasionally gives deep moans and makes *ka-ka* sounds by snapping the mandibles.

Ciconia nigra

黑鹳 hēi guàn

Black Stork

L: 100–120 cm 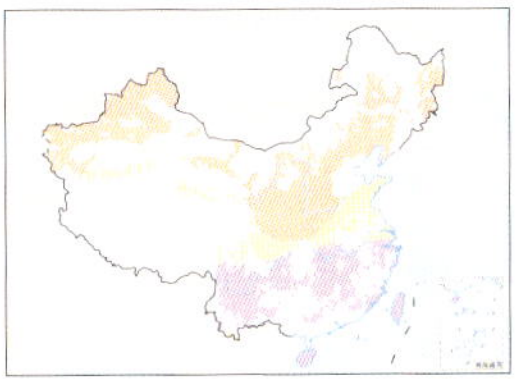
Habitat: Prefers foraging on rivers near cliffs in breeding season. Builds nests on cliffs; inhabits marshes and shallow lakes in winter.
Behavior: Generally gregarious. Forages in streams and lakes. Often rests in trees or on cliffs.
Distribution: Widespread except Tibet. Breeds in Northeast, North, and northwestern China; winters along the Yangtze River Watershed and on plateau lakes in southwestern China. Uncommon, but easier to see in breeding colonies and wintering sites.
Voice: Generally silent away from nest but sometimes gives a unique cooing call.

Asian Openbill

Large bill; gap remains between mandibles when closed

Black wing feathers

Black Stork

Brown head and neck, not glossy

imm.

Red bare skin around eye

ad.

Ciconia ciconia

白鹳 bái guàn

White Stork

L: 100–115 cm
Habitat: Inhabits open wetlands, such as lakes and shallow ponds on plains.
Behavior: Nests on trees, roofs, utility poles, and water towers.
Distribution: Once a common breeding bird in SW Tarim Basin but became locally extinct in Xinjiang around 1980.
Voice: Hisses and bill clattering at nest.

Ciconia episcopus

白颈鹳 bái jǐng guàn

Asian Woolly-necked Stork

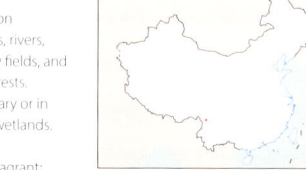

L: 86–95 cm
Habitat: Seen on freshwater lakes, rivers, marshes, paddy fields, and peat swamp forests.
Behavior: Solitary or in small flocks in wetlands. Shy.
Distribution: Vagrant; one record from Yunnan (Napa Lake).
Voice: Usually silent but makes various hisses and bill clattering near nest.

Ciconia boyciana

东方白鹳 dōng fāng bái guàn

Oriental Stork

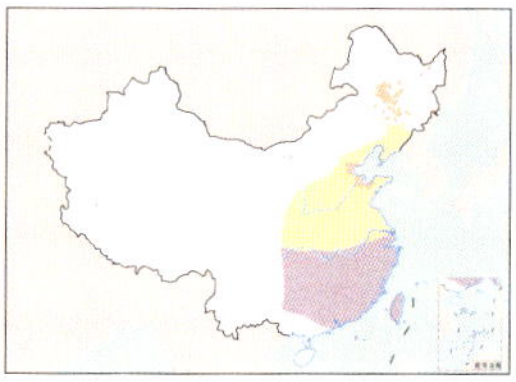

L: 110–115 cm
Habitat: Inhabits marshes and open fields; also forages in grassy areas.
Behavior: Nests on tall trees and transmission towers; gregarious in nonbreeding season. Flies with slow wingbeats, circling up on thermals during migration.
Distribution: Rare or locally common. Restricted to eastern China. Breeds in Northeast and North China as far south as the Yangtze River Watershed; winters in the Yangtze River Watershed and is occasionally recorded in southwestern and South China (including Taiwan).
Voice: Bill clattering heard year-round.

Leptoptilos javanicus

秃鹳 tū guàn

Lesser Adjutant

L: 110–135 cm
Habitat: Inhabits marshes, ponds, riverbeds, and paddy fields in tropical and subtropical zones; also seen on seashores and mudflats.
Behavior: Often on wetlands, farmland with plenty of insects, or even garbage dumps in flocks; nests on large trees.
Distribution: Vagrant to Yunnan, Sichuan, Chongqing, Jiangxi, and Hainan.
Voice: Essentially silent but makes a series of short, hoarse, and guttural *e-e-e* calls when breeding.

Oriental Stork

Pale yellow iris

Heavy black bill

White Stork

Red bill

Lesser Adjutant

Pinkish head, throat, and bill

Yellow bare skin on neck

Bill proportionately large and conspicuous

Asian Woolly-necked Stork

Black cap

White neck contrasts strongly with dark body

Red legs

鹈形目 PELECANIFORMES

Medium to large waterbirds. Plumage similar between sexes. Bill long, with prominent grooves on upper bill. Long neck and legs. Some species fly with neck folded. Wings wide. Tail short. In pelicans, bill developed large pouch, and feet are webbed. Inhabit wetland habitats, such as rivers, lakes, marshes, and coasts, solitarily or in groups. Nest on trees or ground, some in groups. Feed mostly on fish, amphibians, reptiles, insects, and small mammals. Most species are migratory.

Five families, 34 genera, and 118 species recognized worldwide, with a cosmopolitan distribution. Three families, 15 genera, and 35 species recorded in China, ubiquitously distributed nationwide.

鹈鹕科 Pelecanidae	鹮科 Threskiornithidae	鹭科 Ardeidae

Dalmatian Pelican

Crested Ibis

Little Egret

Pelecanus onocrotalus

白鹈鹕 bái tí hú

Great White Pelican   LC

L: 140–175 cm
Habitat: Inhabits large lakes, reservoirs, rivers, and marshes.
Behavior: Often lives in colonies; prefers foraging in flocks.
Distribution: Very rare passage migrant and winter visitor. In migration seasons and especially fall, occasional at large water bodies across Xinjiang. Summer visitors recorded in Tacheng and the Irtysh River Watershed in Xinjiang; historical breeding records in N Xinjiang (Ganjia Lake). Vagrant to Beijing, Henan, Qinghai, Gansu, Sichuan, Anhui, Jiangsu, and Fujian, some of which might be escapes.
Voice: Generally silent but can give guttural groans.

Pelecanus crispus

卷羽鹈鹕 juǎn yǔ tí hú

Dalmatian Pelican   NT

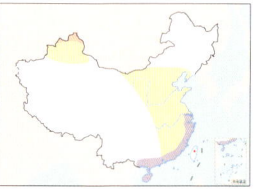

L: 160–183 cm
Habitat: Often seen on freshwater lakes, marshes, and estuaries; also seen at gulfs.
Behavior: Gregarious. Prefers open wetlands; breeds on the ground or in trees. Strong flier.
Distribution: Passage migrant or winter visitor to China. Migrates through most of North and East China; winters in water bodies along the east coast. Rare.
Voice: Cowlike, monotone mo-mo— and hoarse wo-wo.

Pelecanus philippensis

斑嘴鹈鹕 bān zuǐ tí hú

Spot-billed Pelican   NT

L: 127–156 cm
Habitat: Inhabits large lakes, marshes, seashores, and lagoons; also seen at estuaries.
Behavior: Moves around in flocks on open wetlands. Enjoys swimming;

good at exploiting thermals to glide in sky. Nests on trees.
Distribution: Historical records from Zhejiang, Fujian, Guangdong, Guangxi, and Yunnan, but no confirmed record in many years. Some historical northerly records might be misidentified Dalmatian Pelicans.
Voice: Cowlike, monosyllabic moo-moo-.

Great White Pelican

Black wing feathers

non-br.

non-br.

Extensive pink bare skin around dark eye

br.

Pink legs

Dalmatian Pelican

Black primaries with white shafts

non-br.

non-br.

Gray hue to neck and scruffy crest

Pale eye with limited facial skin

Leaden upper mandible

br.

Reddish-orange gular pouch

Dark legs

Spot-billed Pelican

Black from primaries to tertials

non-br.

non-br.

br.

Short straight crest on hindneck

Pinkish bill

Purplish gular pouch

Threskiornis melanocephalus

黑头白鹮 hēi tóu bái huán

Black-headed Ibis

L: 65–76 cm
Habitat: Inhabits marshes, lakes, riverbanks, and ponds with lush vegetation; also occurs on paddy fields.
Behavior: Often solitary or in small flocks on wetlands. Agile flier. Forages by probing bill into water or mud.
Distribution: The eastern population breeds in northern parts of Northeast China and migrates through the east coast but has much declined, with no record for many years. Occasional in Yunnan.
Voice: Usually quiet but makes a short nasal *eng-en-eng* near nest.

Pseudibis davisoni

白肩黑鹮 bái jiān hēi huán

White-shouldered Ibis

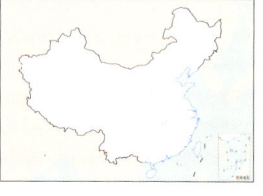

L: 60–85 cm
Habitat: Inhabits forested swamps, rivers, ponds, and paddy fields; also seen on cultivated lands.
Behavior: Often moves about solitarily or in small flocks on wetlands and cultivated fields. Solitary and quiet.
Distribution: Vagrant in SW Yunnan.
Voice: Loud, prolonged, and slightly trilling *er-er-er—*, similar to extralimital Red-naped Ibis (*P. papillosa*).

Nipponia nippon

朱鹮 zhū huán

Crested Ibis

L: 55–84 cm
Habitat: Inhabits wetlands on low mountains and hilly lowlands, including swamps, ponds, streams, and paddy fields.
Behavior: Occurs solitarily or in pairs on wetlands near forests; sometimes also gathers in small flocks. Flies slowly. Nests in large trees.
Distribution: Historically widespread across East Asia. Now restricted to S Shaanxi. Also reintroduced to S Henan, N Zhejiang, South Korea, and Japan.
Voice: Moaning and wailing notes near nest, *eee-ar-ar-arrh*.

Plegadis falcinellus

彩鹮 cǎi huán

Glossy Ibis

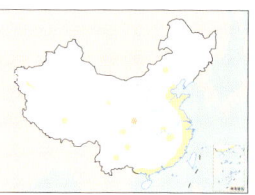

L: 49–66 cm
Habitat: Inhabits marshy lakes, rivers, ponds, paddy fields, and coastal wetlands.
Behavior: Generally moves around in wetlands in small or loose flocks. Nests in trees.
Distribution: Scattered records in Xinjiang, Qinghai, and eastern and southwestern China. Most frequent in Yunnan. A breeding population was recently discovered in Shaanxi (Hanzhong).
Voice: Short, hoarse, grunting *e-e—er*.

Platalea leucorodia

白琵鹭 bái pí lù

Eurasian Spoonbill

L: 80–95 cm
Habitat: Inhabits lakes, marshes, rivers, and reservoirs with abundant aquatic animals; also seen on coasts and estuaries.
Behavior: Prefers moving in flocks on open wetlands. Exhibits a unique foraging technique by sweeping bill from side to side. Nests in trees or on the ground in colonies.
Distribution: Widely distributed throughout China. Breeds in Northeast and northwestern China; migrates through North China and northwestern and southwestern China; winters in areas south from the Yangtze River Watershed.
Voice: Typically silent away from colony, where it gives nasal *un-ung-*.

Platalea minor

黑脸琵鹭 hēi liǎn pí lù

Black-faced Spoonbill

L: 60–79 cm
Habitat: Often seen on coastal tidal flats, lagoons, and shrimp ponds; also occurs on freshwater lakes, marshes, paddy fields, and ponds.
Behavior: Gregarious. Foraging technique similar to that of Eurasian Spoonbill.
Distribution: Rare and local breeder on coastal islands in Liaoning; uncommon migrant through the east coast and the Yangtze Plain; winters along the southeast coast, including Taiwan and Hainan.
Voice: Hoarse and low *un-un*.

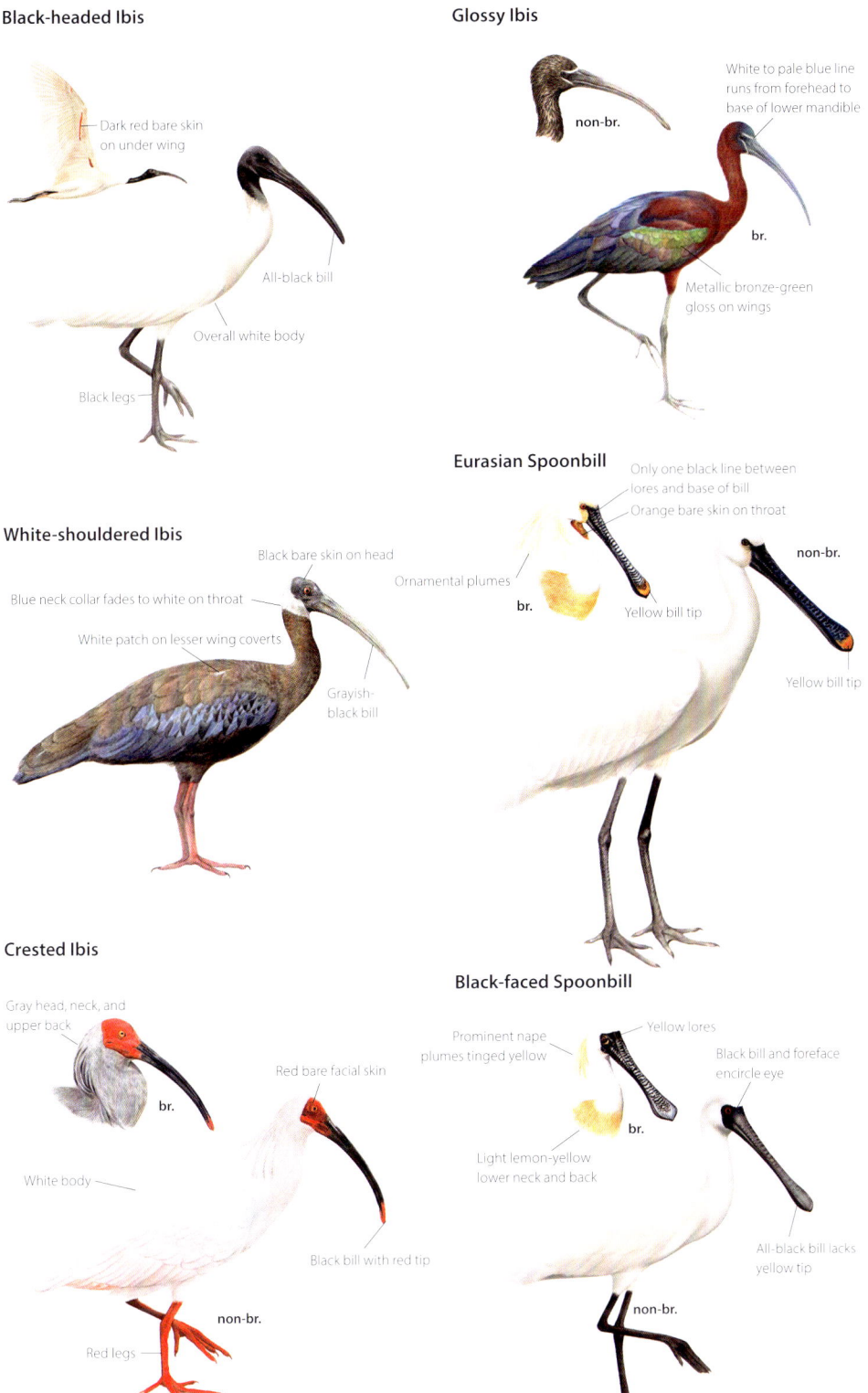

Black-headed Ibis

Dark red bare skin on under wing

All-black bill

Overall white body

Black legs

Glossy Ibis

non-br.

White to pale blue line runs from forehead to base of lower mandible

br.

Metallic bronze-green gloss on wings

White-shouldered Ibis

Black bare skin on head

Blue neck collar fades to white on throat

White patch on lesser wing coverts

Grayish-black bill

Eurasian Spoonbill

Only one black line between lores and base of bill

Orange bare skin on throat

Ornamental plumes

br.

Yellow bill tip

non-br.

Yellow bill tip

Crested Ibis

Gray head, neck, and upper back

br.

White body

Red bare facial skin

Black bill with red tip

non-br.

Red legs

Black-faced Spoonbill

Prominent nape plumes tinged yellow

Yellow lores

Black bill and foreface encircle eye

Light lemon-yellow lower neck and back

br.

All-black bill lacks yellow tip

non-br.

Botaurus stellaris

大麻鳽 dà má yān
Great Bittern

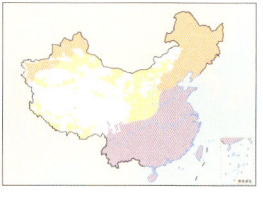

L: 64–78 cm
Habitat: Inhabits reedbeds near rivers and lakes.
Behavior: Plumage very well camouflaged. When alarmed, stands upright and still, resembling reeds; not frequently seen. Crepuscular; nests in reedbeds.
Distribution: Widespread in China except for C Qinghai-Tibet Plateau; locally common. Northern breeders winter in south.
Voice: Deep, very low-pitched series of two to seven booms made during breeding season. Flight calls include a Common Raven–like *argh*.

Ixobrychus minutus

小苇鳽 xiǎo wěi yān
Little Bittern

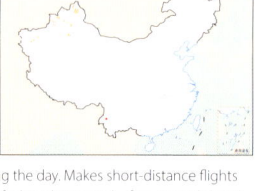

L: 31–38 cm
Habitat: Inhabits reedbeds or cattails in lakes, reservoirs, ponds, and rivers on plains and hills.
Behavior: Nocturnal; generally active at dusk and dawn. Hides in reeds or other dense vegetation during the day. Makes short-distance flights when alarmed. Secretive. When feeling threatened, often extends head and neck upward and stands still, becoming difficult to distinguish from surrounding reeds and branches.
Distribution: Often seen in reedbeds in S and N Xinjiang in breeding season; a few wintering records in S Xinjiang. Vagrant in C Yunnan.
Voice: Courtship call is a low grunting *gook*, repeated every 2–3 sec. Nasal flight call *qwer* or *kek-kek-kek*.

Ixobrychus sinensis

黄斑苇鳽 (黄苇鳽)
huáng bān wěi yān (huáng wěi yān)
Yellow Bittern

L: 30–40 cm
WS: 50–55 cm
Habitat: Inhabits dense reedbeds or paddy fields.
Behavior: Secretive; generally hides in emergent vegetation. When alarmed, stands still and points bill upward to mimic reeds. When flushed, tends not to fly high and returns to reedbeds after flying a short distance.
Distribution: Widespread in eastern China and absent from Xinjiang, Qinghai, and Tibet. Breeds in most regions; winter visitor or resident in South China. Generally common.
Voice: Makes a slow, low-pitched (but noticeably higher than call of Little Bittern) *woo-woo* in breeding season and a *kik-kik* in nonbreeding season.

Ixobrychus eurhythmus

紫背苇鳽 zǐ bèi wěi yān
Schrenck's Bittern

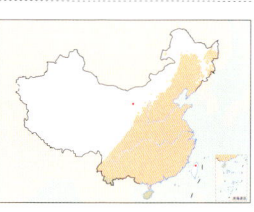

L: 33–42 cm
WS: 48–59 cm
Habitat: Similar to Yellow Bittern.
Behavior: Similar to Yellow Bittern.
Distribution: Summer breeder to the east of the Hu Line; resident or winter visitor to Hainan and S Yunnan; vagrant in Taiwan and Ningxia. Secretive and uncommon.
Voice: Calls at night in breeding season; gives *oo-oo-oo-oo* or *gup-gup-gup*, resembling the croaking sound of frogs and "song" of Greater Painted-Snipe; tempo faster than call of Yellow Bittern.

Ixobrychus cinnamomeus

栗苇鳽 lì wěi yān
Cinnamon Bittern

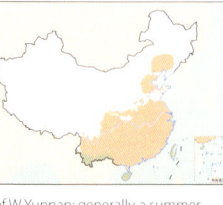

L: 40–41 cm
WS: 48–51 cm
Habitat: Similar to Yellow Bittern.
Behavior: Similar to Yellow Bittern.
Distribution: Occurs rarely as far north as Beijing and from there southwest to the vast regions east of W Yunnan; generally a summer breeder. Resident or winter visitor to S Yunnan, Guangdong, Guangxi, Hainan, and Taiwan; local and occasional.
Voice: Often calls at dusk and dawn. Gives a rhythmic, pulsed series of up to 12 short scops-owl–like *wup-up-up-up-up* calls that rise, then fall. Flight call similar to Little Bittern.

Ixobrychus flavicollis

黑鳽 hēi yān
Black Bittern

L: 54–66 cm
WS: 74–80 cm
Habitat: Occurs on riverbanks, lakes, ponds, reedbeds, marshes, flooded fields, mangroves, and bamboo thickets on low mountains and hills. Nests in riparian reedbeds, scrubland, willows, or bamboos.
Behavior: Generally moves about solitarily. Often forages at night, most actively at dusk and dawn and occasionally during the day.
Distribution: Breeding summer visitor to Central, East, and South China. Locally common.
Voice: Alarm call in breeding season a low-pitched resonant Collared Scops-Owl–like *whoo-*.

Great Bittern

Black crown

Black malar stripe

Brown body with dense black streaks

Long stout neck

Schrenck's Bittern

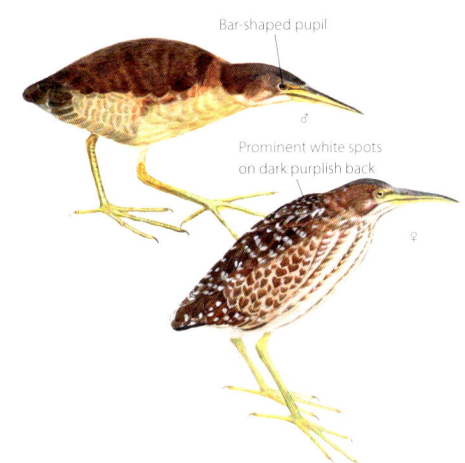

Bar-shaped pupil

♂

Prominent white spots on dark purplish back

♀

Little Bittern

Jet-black crown

♂

Black back

Blackish-brown streaks on brown back

Grayish-black crown

♀

Large buff-white patches on wing coverts

Large yellowish-brown patches on wing coverts

Cinnamon Bittern

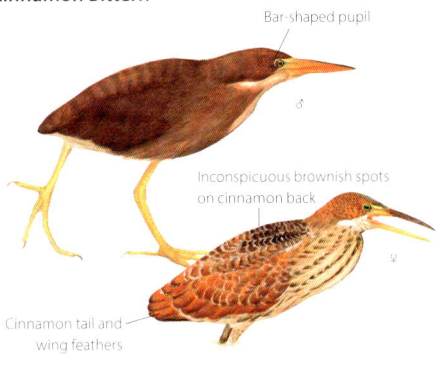

Bar-shaped pupil

♂

Inconspicuous brownish spots on cinnamon back

♀

Cinnamon tail and wing feathers

Yellow Bittern

Crown nearly black

Rounded pupil

♂

Light-colored crown

Dark patches on back

♀

Black Bittern

Male all black, female tinged dark brown

Yellowish side of neck

Blackish-brown breast and neck covered with long buff streaks

Gorsachius melanolophus

黑冠鳽 hēi guān yán

Malayan Night-Heron 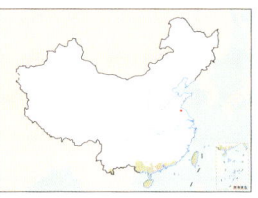

L: 47–51 cm
WS: 86–87 cm
Habitat: Inhabits densely
wooded areas near
marshes or streams. In
Taiwan, also occurs in
well-vegetated parks.
Behavior: Secretive; often
moves around solitarily at
dusk and dawn. Behavior similar to Japanese Night-Heron.
Distribution: Most frequent in Taiwan, occasionally in Guangxi,
Guangdong, Yunnan, Hainan, and Hong Kong. Vagrant in Jiangsu.
Voice: Deep, hollow, mournful-sounding series of *wu, wu, wu* notes,
very similar to Japanese Night-Heron and difficult to distinguish, but
individual notes slightly shorter.

Gorsachius goisagi

栗头鳽 (栗鳽) lì tóu yán (lì yán)

Japanese Night-Heron 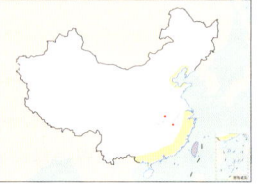

L: 48–49 cm
WS: 87–89 cm
Habitat: Inhabits low-
mountain broadleaf
forests near wetlands;
also appears in city parks
or other regions with
dense vegetation during
migration.
Behavior: Secretive and shy, mainly forages during daytime; male calls
at night during breeding season. Forages for worms on the ground.
When a prey item is spotted, the bird stares at it for a while before
quickly catching it. If alarmed, immediately flies away or into a tree.
Distribution: Rare passage migrant along the southeast coast, from
Shanghai to Guangxi; rare winter visitor in Hong Kong and Taiwan;
vagrant to Beijing, Bohai Bay, Jiangxi, and Hubei.
Voice: Slow and deep *bwoo-bwoo*, sounding almost like an owl.

Gorsachius magnificus

海南鳽 hǎi nán yán

White-eared Night-Heron 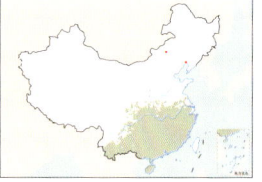

L: 54–56 cm
WS: 83–85 cm
Habitat: Occurs in
valleys, in streams, along
reservoirs, and on moist
farmland in or near dense
subtropical montane
forests.
Behavior: Nocturnal,
often hides in dense alpine forests during the day and comes out to
forage right before or after dark, then returns to rest in trees at dawn.
Forages all night and does not return to nests unless it is rearing chicks.
Solitary; adults never forage or fly in flocks, but recently fledged birds
forage in flocks on shoals near their nests.
Distribution: Uncommon resident in Central, East, South, and
southwestern China, including Hainan; some are migrants. Vagrants
reported in Liaoning and Inner Mongolia.
Voice: Seldom calls. Occasionally gives a deep, owl-like *whoaa* alarm call
in breeding season. Almost silent in nonbreeding season.

Nycticorax nycticorax

夜鹭 yè lù

Black-crowned Night-Heron 

L: 58–65 cm
WS: 90–100 cm
Habitat: Inhabits various
kinds of wetlands,
including streams,
marshes, shallow lakes,
artificial ponds, and
mangroves.
Behavior: Lives solitarily
or in flocks. Unafraid of people. Forages both day and night. Generally
waits near water and quickly catches prey, primarily fish and amphibians.
Retracts neck in flight. Breeds in colonies. Nests in trees; nesting area
often smelly.
Distribution: Widespread in China except for W Tibet. Summer breeder
in Northeast and northwestern China; summer breeder, resident, or
winter visitor south from North China. Generally common.
Voice: Noisy, especially when in flocks; gives hoarse croaks or *kwok*.

Nycticorax caledonicus

棕夜鹭 zōng yè lù

Nankeen Night-Heron 

L: 55–59 cm
WS: 95–110 cm
Habitat: Inhabits various
kinds of wetlands, such as
marshes, mangroves, and
rocky shores; especially
prefers areas with tall
trees.
Behavior: Similar to Black-
crowned Night-Heron.
Distribution: Vagrant in Taiwan (Kaohsiung and Taoyuan).
Voice: Throaty, harsh *gyaowk* or *waowk*, reminiscent of both Black-
crowned Night-Heron and Striated Heron.

Butorides striata

绿鹭 lù lù

Striated Heron 

L: 35–48 cm
WS: 52–60 cm
Habitat: Inhabits
mountain streams,
mangroves, and reservoirs
and lakes with abundant
vegetation.
Behavior: Wary, easily
startled. Often moves
solitarily at dusk and dawn; stands still near water for a long time to wait
for prey. Generally hides in trees or bushes near wetlands. When startled,
looks around and slowly flies away close to water's surface.
Distribution: *B. s. amurensis* breeds in Northeast, northwestern, and
North China; resident or winter visitor in East and South China (excluding
Taiwan and Hainan). *B. s. actophila* breeds in southwestern and East
China; resident or winter visitor to Central and South China (excluding
Taiwan and Hainan). *B. s. javanica* is a resident in Guangdong, Hong
Kong, Hainan, and Taiwan.
Voice: A relatively sharp *kyeow*.

Malayan Night-Heron

Grayish body conspicuously spotted and barred

juv.

Long black crest on crown, clearly demarcated from surrounding brown feathers

ad. White-tipped primaries

Black and white stripes on underwing coverts

Rufous-brown body

Black-crowned Night-Heron

Densely spotted

Red iris

imm.

ad.

Japanese Night-Heron

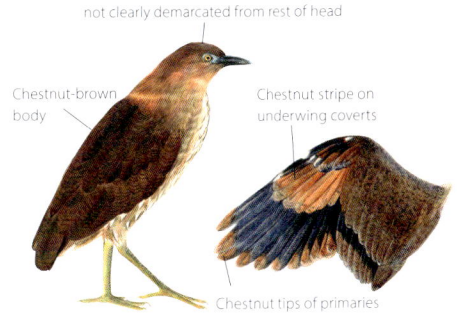

Chestnut crown, darker in some individuals, but not clearly demarcated from rest of head

Chestnut-brown body

Chestnut stripe on underwing coverts

Chestnut tips of primaries

Nankeen Night-Heron

Brown back

White-eared Night-Heron

Large protruded eyes

Yellow lores and eye ring

White postocular stripe extends above ear coverts

Light rufous-brown hindneck bordered black

Blackish-brown central foreneck

Medium-sized, very lanky

Yellowish-green legs

Striated Heron

Gray side of neck

Bright yellow lores and eye ring

Ardeola grayii
印度池鹭 yìn dù chí lù
Indian Pond-Heron

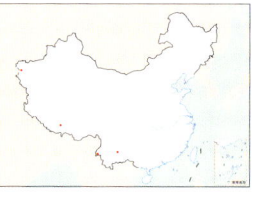

L: 39–46 cm
Habitat: Inhabits various wetlands, including rivers, lakes, wet grass, and paddy fields.
Behavior: Extralimital colonial nester. Moves around solitarily in nonbreeding season; often mixes with other herons and waterbirds. Secretive and quiet, can stay still for a long time to wait for prey.
Distribution: Vagrant in W Xinjiang, C Tibet, and W and C Yunnan.
Voice: Varied high-pitched squawks and croaks.

Ardeola bacchus
池鹭 chí lù
Chinese Pond-Heron

L: 42–52 cm
WS: 79–90 cm
Habitat: Adapted to various kinds of wetlands.
Behavior: Often solitary or in small flocks; sometimes also mixes with other herons and egrets. Unafraid of people. Stays still for a long time to wait for prey; catches prey immediately once detected.
Distribution: Occurs throughout China except for northern Northeast China and the Qinghai-Tibet Plateau; very common in eastern China, occasional or rare in Western China. Winters mostly south of the Yangtze River.
Voice: Gruff monosyllabic woa.

Ardeola speciosa
爪哇池鹭 zhǎo wā chí lù
Javan Pond-Heron

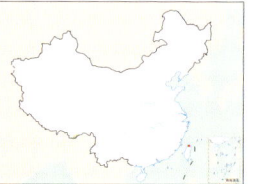

L: 45 cm
Habitat: Inhabits various wetlands, including rivers, lakes, wet grass, and paddy fields.
Behavior: In breeding season, generally forages and nests in pairs or colonies. Moves around solitarily in nonbreeding season; often mixes with other herons and waterbirds. Secretive and quiet; can stay still for a long time to wait for prey.
Distribution: Vagrant in Taiwan.
Voice: Relatively deep croaks, very similar to Chinese Pond-Heron.

Bubulcus ibis
牛背鹭 niú bèi lù
Cattle Egret

L: 46–53 cm
Habitat: Inhabits various wetlands on plains or low mountains; also occurs on pastures and farmland.
Behavior: Prefers pecking at invertebrates on newly plowed farmland. Bold, unafraid of people. Retracts neck in S shape in flight. Often nests on tall trees in colonies.
Distribution: Widespread throughout China. Mainly a resident in the regions south of the Yangtze River and a summer breeder to the north of the Yangtze River.
Voice: Rarely calls; occasionally gives a deep wak or rick-rack in flight.

Ardea cinerea
苍鹭 cāng lù
Gray Heron

L: 92–99 cm
Habitat: Inhabits various wetlands, including rivers, lakes, tidal flats, and paddy fields.
Behavior: Often stands in shallow waters for a long time, waiting patiently for foraging opportunities. Flies slowly, neck curved in S shape in flight. Nests in colonies, on cliffs, or in trees near wetlands.
Distribution: Occurs throughout China; common. A. c. cinerea is a resident in Xinjiang. A. c. jouyi is seen everywhere except for Xinjiang; summer breeder in northern China, passage migrant or resident in southwestern, East, and Central China, winter visitor in South China.
Voice: Hoarse fraaank in flight.

Ardea insignis
白腹鹭 bái fù lù
White-bellied Heron

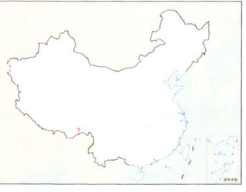

L: 127 cm
Habitat: Inhabits rivers and forested swamps at low altitudes in tropical and subtropical zones.
Behavior: Often lives in pairs in breeding season; solitarily in nonbreeding season. Shy, seldom mixes with other waterbirds. Often stands near water for a long time waiting for prey; also forages by flushing fish with legs in shallow water.
Distribution: Extremely rare in S Tibet and W Yunnan.
Voice: Loud deep croaks, like a person vomiting, have been documented.

Indian Pond-Heron

Ornamental plumes

Pale head and neck

br.

Cattle Egret

Orange ornamental plumes on head and neck

Pink base of bill

Yellow bill

br.

Pink legs

non-br.

Black legs

Chinese Pond-Heron

Reddish-brown head and neck

Relatively dark breast

br.

non-br.

Gray Heron

Black plumes

Orange bill

Two black streaks along median line of neck

Javan Pond-Heron

Pale crown

Gray back

Rufous to orange breast

br.

White-bellied Heron

Long thin white ornamental plumes on rear crown, neck, and breast

All-gray upperparts

Large body

White belly

Ardea purpurea
草鹭 cǎo lù
Purple Heron   (LC)

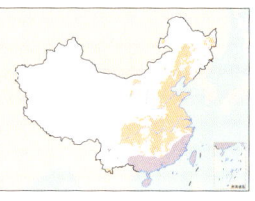

L: 84–97 cm
Habitat: Inhabits open wetlands on plains, low mountains, and hills. Generally occurs in shallow waters surrounded by abundant reeds or other aquatic vegetation.
Behavior: Generally forages solitarily; can stand on one leg and wait for a long time for fish to approach. Very slow flier. Nests in reedbeds or other dense vegetation.
Distribution: Summer breeder in Northeast, North, East, and Central China; resident or winter visitor in South China. Relatively common.
Voice: Harsh *krrk-*, shorter, drier, croakier, and more rasping than similar flight call of Gray Heron.

Ardea alba
大白鹭 dà bái lù
Great Egret   (LC)

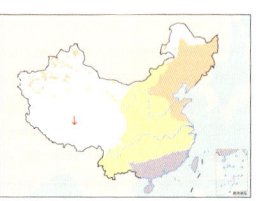

L: 90–98 cm
Habitat: Inhabits open wetlands on plains, low mountains, and hills.
Behavior: Often stands still in water to wait for prey or pecks at prey while walking in shallow water.
Distribution: *A. a. alba* breeds in Heilongjiang, Liaoning, and N Xinjiang and migrates through northwestern and southwestern China and the Qinghai-Tibet Plateau. *A. a. modesta* breeds in Jilin, Liaoning, and E Inner Mongolia, migrates through North, East, Central, and southwestern China, and winters in South China.
Voice: Lengthy dry *krrkkrkrkr* in flight.

Ardea intermedia
中白鹭 zhōng bái lù
Intermediate Egret   (LC)

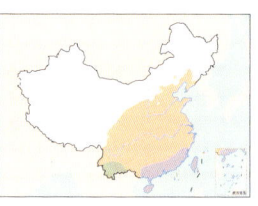

L: 62–70 cm
Habitat: Inhabits rivers, lakes, paddy fields, and seashores.
Behavior: Similar to Little Egret. Walks slowly, flies gracefully. Usually breeds in colonies, sometimes in colonies mixed with other egrets.
Distribution: Breeds mainly in regions south of the Yangtze River; winters in Guangdong, Hainan, and Taiwan; occasionally seen in North China.
Voice: Generally quiet; occasionally lengthy, almost hissing, harsh *krsssssshh* and a more cracked-sounding *krrkk*.

Egretta picata
斑鹭 bān lù
Pied Heron   (LC)

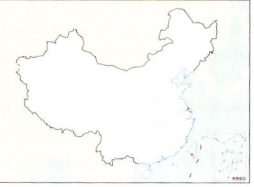

L: 43–55 cm
Habitat: Inhabits tropical lakes, ponds, and grassy areas in the vicinity of marshes; often occurs near residential areas.
Behavior: Generally moves around solitarily; bold, unafraid of people. Usually nests in mangroves or on other trees in colonies.
Distribution: Vagrant in Taiwan and Jiangsu (Rudong).
Voice: Deep and froglike *gu-gu-gua*.

Egretta novaehollandiae
白脸鹭 bái liǎn lù
White-faced Heron   (LC)

L: 65–70 cm
Habitat: Inhabits various kinds of lowland marshes, grassy areas, and seashores.
Behavior: Stands still in shallow waters or on tidal flats to forage for small aquatic animals. Flight slow and somewhat undulating.
Distribution: Vagrant in E Fujian and Taiwan.
Voice: Undistinguished hoarse *graak-*.

Egretta garzetta
白鹭 bái lù
Little Egret   (LC)

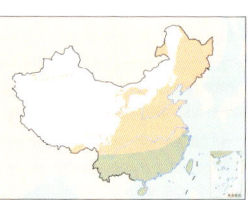

L: 55–68 cm
Habitat: Inhabits various kinds of wetlands, including lakes, marshes, and paddy fields.
Behavior: Flies relatively slowly; retracts neck into S shape in flight. Nests in tall trees.
Distribution: Widespread throughout much of China, very common.
Voice: Hoarse croaking *aaah-*.

Purple Heron

- Bluish-black crown
- Long, snakelike, chestnut-brown neck
- Long black stripe on side of neck

Pied Heron

- Slate-black crown and crest
- Bluish-gray upperparts
- White neck

Great Egret

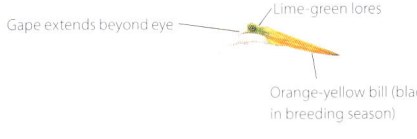

- Lime-green lores
- Gape extends beyond eye
- Orange-yellow bill (black in breeding season)

White-faced Heron

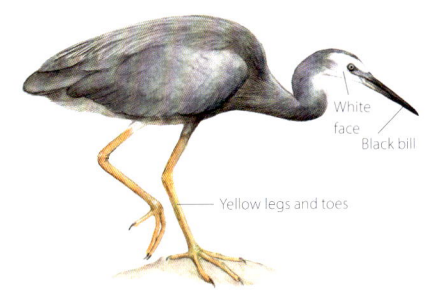

- White face
- Black bill
- Yellow legs and toes

Little Egret

- Yellowish-green lores
- Black bill (lower mandible yellow in breeding season)
- Black legs
- Yellow toes

non-br.

non-br.

- Black legs and toes

Intermediate Egret

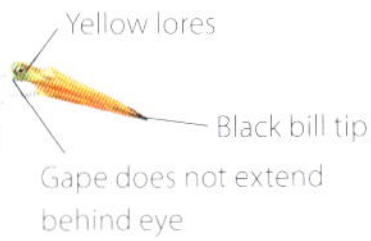

- Yellow lores
- Black bill tip
- Gape does not extend behind eye
- Black legs and toes

Egretta sacra

岩鹭 yán lù

Pacific Reef-Heron

L: 58–66 cm
WS: 90–100 cm
Habitat: A typical coastal bird. Prefers coastal habitats; almost never moves inland. Occurs mainly on tropical and subtropical offshore islands and coastal areas. Inhabits coastal mangroves, mudflats, seashores, and tidal rivers. Especially prefers foraging on reefs and coastal rocks.
Behavior: Highly territorial. In breeding season, moves about in pairs or family groups and occupies a small area of reef rock or coast. In nonbreeding season, often moves solitarily and occasionally wanders away from its regular territory. Usually stands on coastal rocks, walks and forages quietly along coastal reef rocks, or flies close to the sea's surface.
Distribution: Uncommon resident on the southeast and south coasts of China north to Zhejiang (Zhoushan), including Hainan and Taiwan.
Voice: Very quiet; rarely calls. Occasionally gives a nasal corvid-like *gyaaah* in flight.

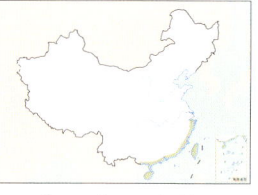

Egretta eulophotes

黄嘴白鹭 huáng zuǐ bái lù

Chinese Egret

L: 65–68 cm
WS: 99 cm
Habitat: Forages on gulfs, estuarine tidal flats, and saltpans; occasionally visits freshwater wetlands near coasts. Nests in clumps of trees or low bushes on cliffs on offshore islands, or in wooded areas in coastal areas.
Behavior: Often solitary, in pairs, or in small flocks. Sometimes gathers in flocks containing up to a few dozen individuals. Generally nests in colonies. In coastal areas, often breeds in or near colonies of Little Egret, Great Egret, Cattle Egret, and Black-crowned Night-Heron, whereas on coastal islands, breeds in or near colonies of Little Egret, Intermediate Egret, or Black-faced Spoonbill. Walks and forages elegantly in shallow waters or on tidal flats.
Distribution: Breeds on islands off coasts of Liaodong Peninsula, Shandong, Zhejiang, and Fujian. A few winter along coast of South China, including Hainan. Migrates along east coast of China.
Voice: Generally silent. When startled, gives a low, grating *aargh*, very similar to Little Egret.

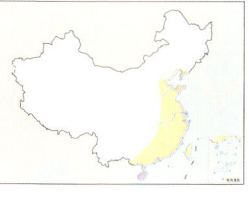

Pacific Reef-Heron

Dark morph

Bill grayish brown to dark yellow

Some white on chin and throat

Dusky slate-gray body

Relatively upright standing posture

Short stout yellowish-green legs

Yellow toes

Yellow-green lores

Relatively stout dirty-yellow bill

White morph

Legs appear short; tibia barely exposed when standing, and toes project slightly beyond tail in flight

Relatively stout, light yellow or yellow-green legs and toes

Chinese Egret

Light bluish-green bare skin at lores

Dusky black bill; yellow at base of lower mandible

non-br.

Lacks ornamental plumes

Light blue bare skin at lores, turns bright blue in breeding season

Greenish-yellow legs

Yellow toes

Dense long ornamental plumes on crown and neck

Scapulars align with tail tip

br.

Yellow bill turns orange yellow or orange red in breeding season

Long slender ornamental plumes on lower neck hang close to breast

Yellow or yellowish-green toes

Fulvous to blackish-brown legs

鲣鸟目　SULIFORMES

Medium to large waterbirds. Plumage similar between sexes, mostly black, white, red, and brown. Bill long and thick, with grooves and hooked tip on upper bill. Wings long and pointed, or short and rounded. Legs short, usually with webbed feet. Tail long, deeply forked or wedge shaped; long and straight in cormorants and darters. Strong fliers, inhabiting open ocean, lakes, rivers, and other wetland habitats. Nest on short trees, bushes, and cliffs. Most species good at diving or swimming. Feed mostly on fish and other aquatic animals. All species are migratory or nomadic.

Four families, eight genera, and 61 species recognized worldwide, with a cosmopolitan distribution. Three families, four genera, and 12 species recorded in China. Except for the widely distributed Great Cormorant, most species are found in the ocean, seas, and coastal areas of eastern and southern China.

军舰鸟科
Fregatidae

Great Frigatebird

鲣鸟科
Sulidae

Red-footed Booby

鸬鹚科
Phalacrocoracidae

br.

Great Cormorant

non-br.

Fregata andrewsi
白腹军舰鸟 bái fù jūn jiàn niǎo
Christmas Island Frigatebird

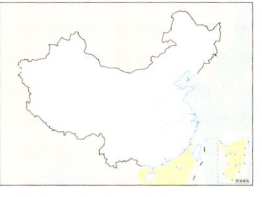

L: 89–100 cm
WS: 205–230 cm
Habitat: Inhabits oceans of the tropics.
Behavior: Generally occurs over tropical and subtropical oceans; rarely approaches coasts. Strong flier.
Distribution: Rare off the coasts of Fujian to the South China Sea.
Voice: Usually silent away from colonies, where it makes a humorous trumpeting *e-o-a-e-a*.

Fregata minor
黑腹军舰鸟 hēi fù jūn jiàn niǎo
Great Frigatebird

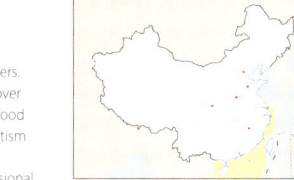

L: 80–105 cm
WS: 206–230 cm
Habitat: Occurs from inshore to pelagic waters in the tropics.
Behavior: Stays on oceanic islands in breeding season, ranges over oceans looking for food in nonbreeding season.
Distribution: Widespread but not commonly seen off the coast, north to Hebei (Beidaihe). Breeds on the Xisha Islands.
Voice: A short, noisy, hollow, and rapid *o-o-o-o* and various other notes.

Fregata ariel
白斑军舰鸟 bái bān jūn jiàn niǎo
Lesser Frigatebird

L: 66–81 cm
WS: 175–193 cm
Habitat: Occurs on tropical pelagic waters.
Behavior: Ranges over oceans looking for food all day; kleptoparasitism recorded.
Distribution: Occasional along the southeast coast; vagrant to Beijing, Shandong, Henan, Shaanxi, and Jiangxi. May breed on the South China Sea Islands.
Voice: Usually silent away from colonies, where it makes rapid, short-noted *we-we-e* whistles.

Sula dactylatra
蓝脸鲣鸟 lán liǎn jiān niǎo
Masked Booby

L: 81–92 cm
WS: 152–170 cm
Habitat: Occurs on offshore and pelagic waters.
Behavior: Pelagic bird in tropical regions; breeds on offshore islands and forages over oceans.
Distribution: Rare off Fujian and Taiwan.
Voice: Usually silent away from colonies, where it makes rapid monosyllabic calls and prolonged *hju-hju-* whistles.

Sula sula
红脚鲣鸟 hóng jiǎo jiān niǎo
Red-footed Booby

L: 66–77 cm
WS: 124–142 cm
Habitat: Occurs on offshore and pelagic waters.
Behavior: Generally flies over sea in search of fish; moves to land in breeding season. Good at flying, swimming, and diving.
Distribution: Rare off Guangdong, Hong Kong, Xisha Islands, and Taiwan; vagrant in Zhejiang.
Voice: Usually silent away from colonies, where it makes a series of short and repetitive *ga-ga-ga* calls.

Sula leucogaster
褐鲣鸟 hè jiān niǎo
Brown Booby

L: 64–74 cm
WS: 132–150 cm
Habitat: Occurs from coastal to pelagic waters in tropical regions.
Behavior: Prefers foraging for fish in flocks over oceans; great diver and swimmer.
Distribution: Occasional along China's east and south coasts, exceptionally as far north as Shandong.
Voice: Usually silent away from colonies, where it makes a rapidly repeated *e-e-e* reminiscent of Red-footed Booby but slower and lower pitched.

Christmas Island Frigatebird

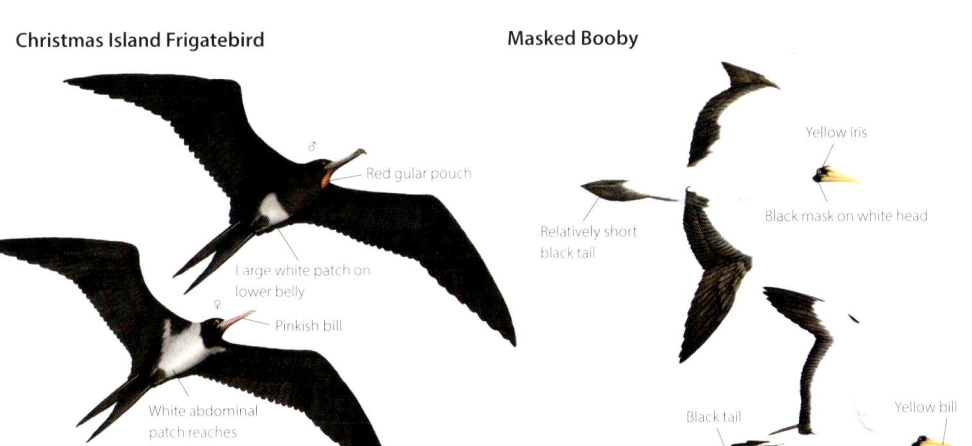

♂ Red gular pouch

Large white patch on lower belly

♀ Pinkish bill

White abdominal patch reaches lower belly

Masked Booby

Yellow iris

Black mask on white head

Relatively short black tail

Black tail

Yellow bill

Great Frigatebird

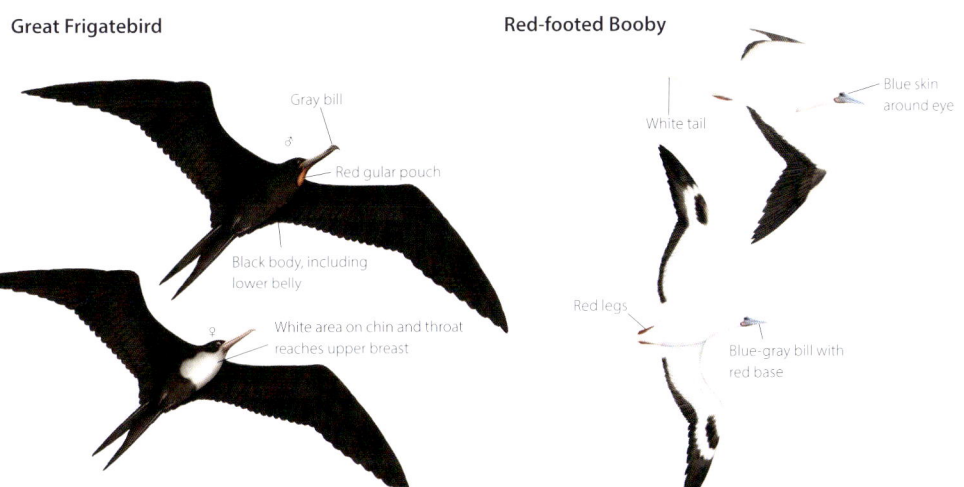

Gray bill

♂ Red gular pouch

Black body, including lower belly

♀ White area on chin and throat reaches upper breast

Red-footed Booby

White tail

Blue skin around eye

Red legs

Blue-gray bill with red base

Lesser Frigatebird

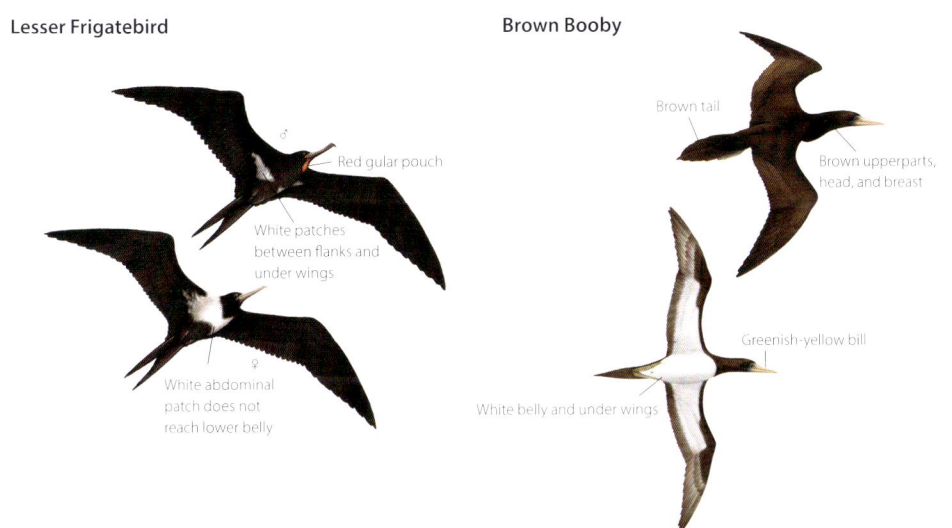

♂ Red gular pouch

White patches between flanks and under wings

♀ White abdominal patch does not reach lower belly

Brown Booby

Brown tail

Brown upperparts, head, and breast

Greenish-yellow bill

White belly and under wings

Microcarbo pygmaeus
侏鸬鹚 zhū lú cí
Pygmy Cormorant

L: 45–55 cm
WS: 75–90 cm
Habitat: Inhabits streams or small water bodies with dense reeds in flocks; rarely approaches open waters.
Behavior: Light bodied; often rests on reeds. Rarely mixes with Great Cormorant.
Distribution: Very rare vagrant or winter visitor to Xinjiang. Specimens were collected in Tian Shan Mountains in early 20th century; no record for the next hundred years. In November 2018, wintering populations were found in N Xinjiang (including Manas Wetland, Bole, Ili River, and Kalamaili Nature Reserve). The Central Asian population has clearly been expanding east in the past few years.
Voice: Gives deep groans only in nest; otherwise silent.

Microcarbo niger
黑颈鸬鹚 hēi jǐng lú cí
Little Cormorant

L: 51–56 cm
WS: 108–110 cm
Habitat: Occurs on lakes, rivers, reservoirs, ponds, and marshes.
Behavior: Generally moves about and forages in flocks. In breeding season, nests in colonies in trees. Dives to fish.
Distribution: Restricted to W Yunnan; locally common.
Voice: Largely silent away from breeding colony, where it makes a short *ke*, *kyar*, and a more prolonged *ke-e-e-e*.

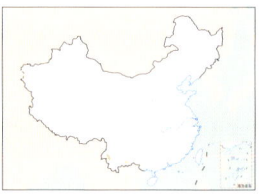

Urile pelagicus
海鸬鹚 hǎi lú cí
Pelagic Cormorant

L: 63–76 cm
WS: 91–102 cm
Habitat: Inhabits oceanic islands or coastal areas; occasionally seen on estuarine bays.
Behavior: Often rests in flocks on offshore rocky islands or cliffs. Swims and forages on the ocean's surface near oceanic islands, mainly by diving and chasing after prey.
Distribution: Seen along the east coast; locally common on the rocky islands and reefs off Liaoning, Shandong, and part of Jiangsu.
Voice: Male gives *ou-ou-ou* at breeding colony; silent in nonbreeding season.

Urile urile
红脸鸬鹚 hóng liǎn lú cí
Red-faced Cormorant

L: 71–89 cm
WS: 110–122 cm
Habitat: Inhabits oceanic islands or coastal rocky islands.
Behavior: Sometimes rests on offshore rocks or cliffs. Good at swimming and diving. Takeoff requires a run-up, where the bird flaps wings on the ocean's surface. Flies low over the water. Behaviors similar to Pelagic Cormorant.
Distribution: Vagrant to Liaodong Peninsula (historic) and Taiwan.
Voice: Unlikely to be heard in China—sometimes gives monosyllabic and deep *ai-ai*.

Phalacrocorax carbo
普通鸬鹚 pǔ tōng lú cí
Great Cormorant

L: 77–94 cm
WS: 121–149 cm
Habitat: Inhabits various kinds of water bodies, including estuaries, reservoirs, rivers, lakes, ponds, and marshes; sometimes also seen in coastal areas.
Behavior: Generally gregarious. Often rests on rocks in water or on branches. Good at swimming and diving. Dives to catch prey when fish is spotted; can stay under water for one minute. Migrates in V formation or single file, but not as organized as goose flocks.
Distribution: Widespread throughout much of China; generally common and increasing in many places. Mainly a passage migrant or summer breeder in northern China; winter visitor or resident in southern China.
Voice: Sometimes gives a deep and monotone call.

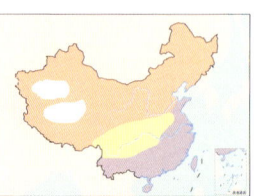

Phalacrocorax capillatus
绿背鸬鹚 (暗绿背鸬鹚) lǜ bèi lú cí (àn lǜ bèi lú cí)
Japanese Cormorant

L: 81–92 cm
WS: 150–152 cm
Habitat: Similar to Pelagic Cormorant.
Behavior: Generally rests on offshore rocks or cliffs; sometimes mixes with Pelagic Cormorant. Occasionally seen on inland estuaries during migration and winter. Difficult to differentiate from Great Cormorant. Flies low over the ocean's surface. Behaviors similar to Great Cormorant.
Distribution: Exclusively coastal and seen along China's east coast; some individuals migrate south in winter. Locally common on rocky islands and reefs off Liaoning, Shandong, and Jiangsu.
Voice: Deep guttural calls at breeding colony.

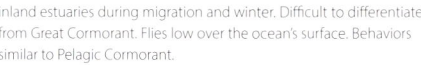

Pygmy Cormorant

Short neck; brownish head and neck

Short bill, inconspicuously hooked at tip

Small; can stand on reeds

Little Cormorant

Small

non-br.

non-br.

Only chin and throat appear whitish

Pelagic Cormorant

Relatively small crest

Small area of red bare facial skin around eye

br.

non-br.

non-br.

Relatively long slender bill

Red-faced Cormorant

Large area of red bare facial skin around eye

Prominent crest

br.

non-br.

Relatively short stout bill

Great Cormorant

non-br.

br.

Whitish head

br.

non-br.

Yellow bare skin at gape extends backward, forming an obtuse angle

Japanese Cormorant

non-br.

br.

Black body with greenish gloss; inconspicuous black feather fringe

non-br.

Yellow bare skin at gape extends backward, forming an acute angle

113

鹰形目 ACCIPITRIFORMES

Birds of prey that are mostly carnivorous and of varying size. Sexes similar in most species, but females larger in body size than males. Plumage mostly brown, white, and black. Bill strong with hooked tip; upper bill with toothlike ridges and covered with cere at base. Wings of varying shapes, equipped for different modes of flight. Legs strong, with long curved claws on feet. Tail relatively long of varying shapes. Inhabit habitats ranging from forests, deserts, open plains, and alpine cliffs to marshes, lakes, rivers, coasts, and islands. Relatively long-lived. Feed mostly on insects, fish, amphibians, reptiles, birds, mammals, or carrion; some species take plant fruits. More than half of species are migratory.

Four families, 75 genera, and 266 species recognized worldwide, with a cosmopolitan distribution. Two families, 24 genera, and 55 species recorded in China, ubiquitously distributed nationwide.

Pandion haliaetus

鹗 è

Osprey

L: 56–62 cm
WS: 147–169 cm
Habitat: Inhabits lowland rivers, lakes, reservoirs, seashores, islands, and other water bodies with abundant fish supply.
Behavior: Occurs near water year-round; feeds solely on fish; world's only raptor that can plunge all the way into water to catch fish. When foraging, searches along regular route with steady wingbeats. Plunges into water when prey is spotted. After catching a fish, carries it to a safe area to feed. Bold, unafraid of people; not highly territorial.
Distribution: Occurs throughout most of China; locally common. Summer breeder in Northeast and northwestern China; winter visitor to Taiwan; passage migrant or resident in other regions.
Voice: Seldom calls; in flight, occasionally gives soft, repeated whistle, *diu*, like a shortened begging call of a juvenile Caspian Tern.

Osprey

M-shaped wings in flight

Five "fingers"

White triangle formed by underwing coverts and belly

Black eye stripe

Brown feathers form breast band

鹗科 Pandionidae   鹰科 Accipitridae

Northern Goshawk

Osprey

Elanus caeruleus

黑翅鸢 hēi chì yuān

Black-winged Kite

L: 31–37 cm
WS: 77–92 cm
Habitat: Prefers savannas with scattered trees and open fields in arid regions; often perches on isolated trees or power lines.
Behavior: Found singly, in pairs, rarely in small flocks. In years with abundant food supply, can breed year-round. Active all day; forages more actively at dusk and dawn. Prefers rodents. Agile flier with unique wing strokes. When foraging, hovers in air and scans the ground. Flicks tail up and down when perched.
Distribution: Resident in the coastal areas in South and East China (including Hainan and Taiwan); locally common. Has been expanding north in recent years; recorded as far north as Hebei (Zhangjiakou).
Voice: Seldom calls. Persistent *bi-ou-* is contact call between pairs in breeding season.

Gypaetus barbatus

胡兀鹫 hú wù jiù

Bearded Vulture

L: 94–125 cm
WS: 235–275 cm
Habitat: Occurs on bare rocky mountains and plateaus.
Behavior: Often moves solitarily on bare rocky mountains.
Distribution: Resident in Liaoning, northwestern China, and the Qinghai-Tibet Plateau; stragglers recorded increasingly frequently in North and Central China.
Voice: Rarely heard, shrill, falconlike whistles, surprising for a bird of its size.

Neophron percnopterus

白兀鹫 bái wù jiù

Egyptian Vulture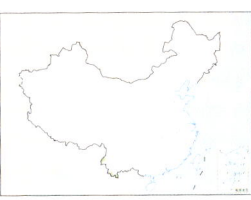

L: 55–65 cm
WS: 155–170 cm
Habitat: Inhabits mountainous and hilly areas; also in premontane deserts.
Behavior: Feeds on carcasses like other vultures; circles in sky in search of food.
Distribution: Vagrant to the Pamir Plateau and Ulugqat in SW Xinjiang, Ili in NW Xinjiang, and Gansu.
Voice: Unlikely to be heard in China and generally silent.

Sarcogyps calvus

黑兀鹫 hēi wù jiù

Red-headed Vulture

L: 76–85 cm
WS: 199–227 cm
Habitat: Inhabits open areas in tropical and subtropical zones, such as low mountains and hills, cultivated lands, and small patches of woods; occasionally seen in dense forests.
Behavior: Scavenger; often solitary or in pairs.
Distribution: Vagrant to S Yunnan.
Voice: Hoarse *wa wa*.

Black-winged Kite

Yellowish iris
Pale brown back
juv.
White spots on upperparts
Gray wing feathers and black coverts on upper wing
Red iris
Black wing feathers and white coverts on under wing
Black wing coverts
ad.
Yellow tarsi and toes

Egyptian Vulture

Black wing feathers
White wedge-shaped tail
Slender yellow bill
Yellow bare facial skin
Overall white
Slender black bill
Bluish-black bare facial skin
Overall blackish brown
ad.
juv.

Bearded Vulture

Yellow iris with red eye ring
Wedge-shaped tail
Gold head and neck
"Beard" at base of bill

Red-headed Vulture

Fleshy red head
White patch on breast
White patches on under wing
Male has pale eye
Fleshy red bare skin on head and neck
Fleshy red lappets on side of neck
Jet-black body
White patch on upper breast
White leg feathers
Pinkish-red legs

Pernis apivorus
鹃头蜂鹰 juān tóu fēng yīng
European Honey-buzzard

L: 52–59 cm
WS: 113–135 cm
Habitat: Inhabits
broadleaf, coniferous, and
mixed forests; especially
prefers woodlands and
forest edges.
Behavior: Prefers moving
solitarily at forest edges.
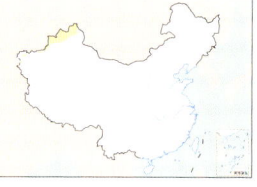
Forages mainly in trees or on the ground. Digs with claws into bee or
wasp nests for various types of food.
Distribution: Very rare in W and N Xinjiang during migration; may breed
in the western parts of Altai Mountains.
Voice: Migrants silent, so short, rising then falling, wailing whistle calls
given when breeding unlikely to be heard.

Pernis ptilorhynchus
凤头蜂鹰 fèng tóu fēng yīng
Oriental Honey-buzzard

L: 57–61 cm
WS: 121–135 cm
Habitat: Prefers forests
with abundant wasps
and bees in breeding
season; adapted to various
kinds of forests, including
primary and secondary
forests. Also forages on
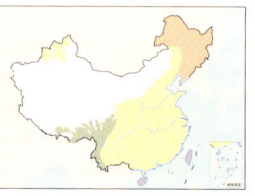
mountain bee farms or in lowland regions.
Behavior: Feeds mainly on bees and wasps; also eats other insects and
honey. In spring and fall, migrates along regular routes in large flocks.
Flies steadily with level wings during migration. In breeding season,
male exhibits display flight by raising and quivering both wings.
Distribution: *P. p. orientalis* breeds from Heilongjiang to Liaoning;
migrates through most of China, including Xinjiang, Qinghai, and Tibet;
small wintering populations in South China (including Taiwan). Common
on breeding grounds and during migration. *P. p. ruficollis* is rarer; resident
in W Yunnan and Sichuan; some individuals make local movements.
A wider variety of residency statuses have been found in recent years
(such as breeding and wintering records in eastern China and a resident
population in Taiwan), which require further research.
Voice: Male gives variety of mostly short, sometimes loud, rising and
falling, melancholy whistles in breeding season; rarely calls during
migration.

Aviceda jerdoni
褐冠鹃隼 hè guān juān sǔn
Jerdon's Baza

L: 41–48 cm
WS: 80–100 cm
Habitat: Generally seen
in montane forests and
forest edges.
Behavior: Generally
moves about solitarily in
forest habitats. Similar to
Mountain Hawk-Eagle
in flight.
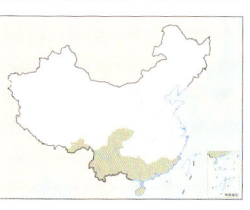
Distribution: Found in SW Yunnan, Chongqing, Guizhou, Hubei, Hunan,
SW Guangxi, and Hainan.
Voice: Lengthy, rising then falling, high-pitched whistle in display, *pi-
wheeeooo*, vaguely reminiscent of Great Eared-Nightjar.

Aviceda leuphotes
黑冠鹃隼 hēi guān juān sǔn
Black Baza

L: 28–35 cm
WS: 64–74 cm
Habitat: Prefers open
and dry broadleaf forests;
often perches on exposed
tree branches.
Behavior: Moves in pairs
in breeding season; often
migrates along regular

route in small to large flocks. Wing strokes similar to those of corvids;
holds wings level when gliding.
Distribution: Breeds in Central, East, South, and southwestern China; a
few records in Taiwan and north to Hebei during migration; wintering
records in Guangdong and Hainan.
Voice: Highly vocal when breeding, when it makes a shrill, descending,
disyllabic *dwe-tu*, with much shorter "tail" than similarly high-pitched
notes of Jerdon's Baza. Shrill gull-like calls with one to three syllables.

Aegypius monachus
秃鹫 tū jiù
Cinereous Vulture **NT**

L: 100–120 cm
WS: 250–295 cm
Habitat: Inhabits forested
hills and mountains; also
flies over grasslands and
bare mountain ridges in
Western China. Occurs up
to 4500 m.
Behavior: Requires a

run-up to take off; often glides on thermals. Usually rests on cliffs with
neck retracted.
Distribution: Occurs throughout China. Relatively common in Western
China; records in Northeast and North China mainly from winter; very
rare in East and South China.
Voice: Generally silent; occasionally gives a *ka-ka*.

European Honey-buzzard

Five "fingers"

Usually has black carpal patch

Yellow iris

Pale morph

Relatively long wings and tail

Throat lacks contrasting white patch and black streaks

Dark morph

Jerdon's Baza

Prominent crest with white tip

Six "fingers"

Prominent medial throat stripe

Broad brown bars on breast and belly

Black Baza

Black crest

White upper breast

Contrastingly black and white

Chestnut bands on belly

Relatively rounded wings

Oriental Honey-buzzard

Relatively broad wings

Six "fingers"

Unequally spaced tail bands

Dark morph ♂

Relatively pointed head

Dark streaky morph ♂

Dark morph ♀

Dark morph juv.

Pale morph ♂

Pale morph juv.

Pale streaky morph juv.

Pale morph ♀

Intermediate morph ♀

Morph mimicking Bonelli's Eagle juv.

Cinereous Vulture

Leading and trailing edges of wing almost parallel

Relatively short tail

Seven "fingers"

♂

Small head

♀

Long neck

ad.

juv.

Slender weak tarsi and talons

Gyps bengalensis

白背兀鹫 bái bèi wù jiù

White-rumped Vulture

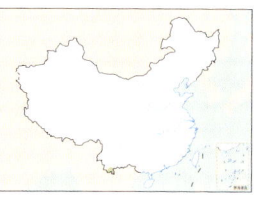

L: 75–85 cm
WS: 192–213 cm
Habitat: Generally inhabits open lowlands in tropical regions, such as wild hills and cultivated areas. Also occurs at altitudes up to 2500 m.
Behavior: Feeds on livestock and animal carcasses. Often forages solitarily; also in small flocks.
Distribution: Vagrant in W and S Yunnan.
Voice: Occasionally gives hoarse *gwa-gwa* and hissing sounds when feeding or on the nest.

Gyps himalayensis

高山兀鹫 gāo shān wù jiù

Himalayan Griffon

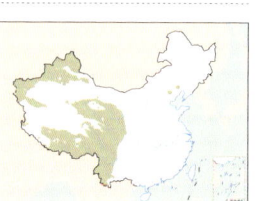

L: 103–110 cm
WS: 260–289 cm
Habitat: Inhabits grasslands, deserts, and bare rocky habitats on plateaus; also seen at forest edges at high altitudes.
Behavior: Generally searches for food while soaring; feeds gregariously on carrion. When feeding, extends head and neck, opens wings, and pursues or jumps at other scavengers.
Distribution: Widespread; often seen on the Tian Shan Mountains, Pamir Plateau, Qilian Mountains, Qinghai-Tibet Plateau, N Sichuan Mountains, and W Sichuan Plateau, all the way to the mountain ranges in NW and NE Yunnan; seen in S Yunnan during migration or in winter. Rare visitor to Liaoning, Hebei, Beijing, and C Inner Mongolia.
Voice: Occasionally gives hoarse calls.

Gyps fulvus

兀鹫 wù jiù

Eurasian Griffon

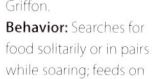

L: 93–110 cm
WS: 234–269 cm
Habitat: Inhabits open habitats on plateaus, similar to Himalayan Griffon.
Behavior: Searches for food solitarily or in pairs while soaring; feeds on carrion. Can make long-distance migration in nonbreeding season.
Distribution: Very rare in Xinjiang and SE Tibet.
Voice: Occasionally gives *gwa-gwa* or hissing sounds.

Gyps tenuirostris

细嘴兀鹫 xì zuǐ wù jiù

Slender-billed Vulture

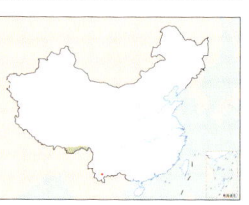

L: 93–100 cm
WS: 234–250 cm
Habitat: Inhabits open plains, cultivated areas, or villages.
Behavior: Feeds mainly on livestock carcasses; often circles in sky or follows other vultures in search of food.
Distribution: Marginally in S Tibet; vagrant in Yunnan (Pu'er). Very rare.
Voice: Gives *gaga* or grunting sounds when eating carcasses.

Spilornis cheela

蛇雕 shé diāo

Crested Serpent-Eagle

L: 65–74 cm
WS: 150–169 cm
Habitat: Inhabits open mountains and hills; occurs mainly in broadleaf forests at middle and low altitudes, as well as in orchards, along mountain roads, and in river valleys. Occurs up to 1900 m.
Behavior: Feeds mainly on snakes and other reptiles; also eats birds and small mammals. Flies steadily with slow wingbeats, wings raised when circling. Mild-tempered; not highly territorial; multiple individuals often seen in the same area. Most active during early breeding season, commonly seen flying in pairs; less active and secretive in late breeding season.
Distribution: Common. *S. c. burmanicus* occurs in southwestern China. *S. c. ricketti* is resident in southeastern and South China; stragglers recorded in Heilongjiang, Liaoning, and Beijing. *S. c. hoya* and *S. c. rutherfordi* are resident in Taiwan and Hainan, respectively. Crested Serpent-Eagle was thought to be sedentary throughout, but observations in recent years suggest that some populations of *S. c. burmanicus* and *S. c. ricketti* migrate southward in fall, probably related to food availability.
Voice: Vocal; often gives single far-carrying, plaintive whistle, *weeeearh*, sometimes combined in longer, stuttering sequence *pi-pi-weeeearch-aarrrh*.

White-rumped Vulture

White underwing coverts

White neck ruff, lower back, and rump

Black body

Slender-billed Vulture

Dark head contrasts markedly with pale body and underwing coverts

Dark gray head contrasts markedly with pale body

Long slender bill

Extremely long neck

Himalayan Griffon

Light sandy-brown underparts and underwing coverts, lighter than Eurasian Griffon

Black wing and tail feathers

Light brown body

Sparse streaks on underparts

Pale legs

Large

Crested Serpent-Eagle

Seven "fingers"

Yellow from lores to base of bill

Short stout head

Black and white crest

ad.

White bands on wings and tail

Small white spots on belly

ad.

Dark face

Pale from head to underparts

juv.

Eurasian Griffon

Pale head and neck

Paler neck feathers than Himalayan Griffon

Brown body

Brownish underparts and underwing coverts darker than Himalayan Griffon

Rounded tail

Dark legs

Circaetus gallicus
短趾雕 duǎn zhǐ diāo
Short-toed Snake-Eagle

L: 60–70 cm
WS: 166–188 cm
Habitat: Breeds in valleys or open dry woodlands on low mountains. Seen in various kinds of open habitats during migration.
Behavior: Often soars solitarily. Can hover like

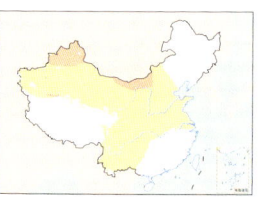

falcons; eats mainly snakes but also other reptiles.
Distribution: Breeds in northwestern China, including the Tian Shan Mountains of Xinjiang. Regular in small numbers in early fall in Beijing. Seldom seen elsewhere, but records from northern, Central, and southwestern China during migration.
Voice: Variety of whistled calls in breeding season include a short, disyllabic, plaintive *wee-arh* and short, flat, monosyllabic notes.

Nisaetus cirrhatus
凤头鹰雕 fèng tóu yīng diāo
Changeable Hawk-Eagle

L: 51–82 cm
WS: 114–150 cm
Habitat: Occurs on the edges of broadleaf and secondary forests; also seen on open farmland and woodlands.
Behavior: Often solitary. Eats various birds and small mammals.

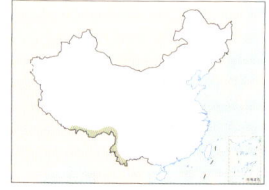

Distribution: Local and rare in S Tibet and S Yunnan.
Voice: Prolonged and shrill *kwi-kwi-kwi-kwi-kw-kweee* whistles with emphasis on final note.

Nisaetus nipalensis
鹰雕 yīng diāo
Mountain Hawk-Eagle

L: 64–84 cm
WS: 140–165 cm
Habitat: Occurs in lowland to high-altitude broadleaf, mixed, and coniferous forests.
Behavior: Generally moves around solitarily or in pairs in mature forests; forages mainly for small mammals.

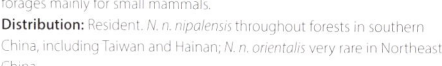

Distribution: Resident. *N. n. nipalensis* throughout forests in southern China, including Taiwan and Hainan; *N. n. orientalis* very rare in Northeast China.
Voice: Prolonged, shrill, descending whistle, *pli … pleee-ah*, with final note short and lower pitched.

Lophotriorchis kienerii
棕腹隼雕 zōng fù sǔn diāo
Rufous-bellied Eagle

L: 50–61 cm
WS: 105–140 cm
Habitat: Inhabits lowland forest edges in subtropical and tropical zones.
Behavior: Often moves around solitarily in open habitats near forest edges.
Distribution: Local resident in Hainan (rare) and W Yunnan.
Voice: Short, shrill, and rising *ki-ki-ki—*.

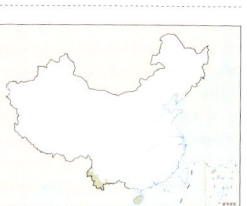

Ictinaetus malaiensis
林雕 lín diāo
Black Eagle

L: 67–81 cm
WS: 164–178 cm
Habitat: Inhabits mature evergreen broadleaf forests up to 3000 m; generally occurs at 300–2000 m.
Behavior: Feeds on small mammals and birds; often

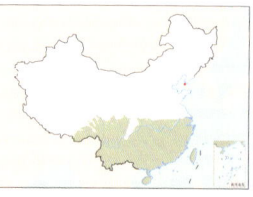

attacks nests of other bird species. Flies steadily and elegantly with occasional flaps; also flies through wooded areas or glides high in sky. In breeding season, displays with distinctive undulating flights. Often in pairs. Territorial; chases out other raptors that enter its territory.
Distribution: Common resident in East and South China (including Taiwan); several records in SW Guangxi, S Yunnan, SE Tibet, Qinghai, and Sichuan; vagrant in Liaoning (Laotie Mountain in Lvshun).
Voice: Seldom calls; gives short, rising, disyllabic whistle, *per-we*, often repeated.

Clanga clanga
乌雕 wū diāo
Greater Spotted Eagle

L: 61–74 cm
WS: 157–180 cm
Habitat: Inhabits low mountains, hills, and lowland wetlands. Breeds in forested areas. Nests mainly on pines, oaks (*Quercus dentata*), or other tall trees; nests often 8–20 m above ground.

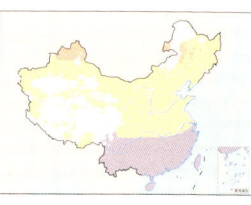
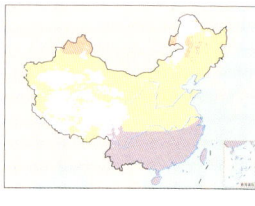

Behavior: Diurnal; forages mainly on open fields.
Distribution: An uncommon migrant in most of China, more frequently seen during migration seasons.
Voice: Repeats short shrill *kyi … kyi … kyp*.

Short-toed Snake-Eagle

Long broad wings

Short neck and rounded head, resembling large owl

Mostly pale underparts; only head is dark

Black bands on tail

Changeable Hawk-Eagle

Short crest

Grayish face, no hood

Seven "fingers"

Prominent streaks on clean white breast and belly

Mountain Hawk-Eagle

Long crest

Appears to have dark hood when viewed from distance

Broad and rounded wings

Upper breast covered with streaks, lower breast with bars

Seven "fingers"

Lacks prominent crest

N. n. nipalensis

N. n. orientalis

Rufous-bellied Eagle

Six "fingers"

juv.

ad.

Short black crest

Black head and hindneck, forming black hood

White chin, throat, upper breast

Rufous-brown belly

juv.

ad.

Black Eagle

Long, broad, flat, rectangular wings

ad.

Yellow base of bill

ad.

Relatively long tail

Seven "fingers"

juv.

Feathered tibia

Black barred tail

Wing tips reach beyond tail

Greater Spotted Eagle

ad.

subad.

juv.

juv.

ad.

Relatively long wings

Dark morph

Dark morph

Relatively short tail

Dark morph

ad.

Dark morph

ad.
Pale morph

Whitish overall

ad. Pale morph

White spots on upperparts

Dark brown overall

juv.

121

Hieraaetus pennatus

靴隼雕 xuē sǔn diāo

Booted Eagle   LC

L: 42–51 cm
WS: 110–135 cm
Habitat: Inhabits mid- to
low-altitude montane
forests and lowland
forests.
Behavior: Often solitary
or in pairs. Flies rapidly or
circles repeatedly above
forest edges.
Distribution: Uncommon. Local breeder in mid- to low-altitude forests
in Xinjiang; occasional on migration in Northeast China, Inner Mongolia,
and North, Central, and southwestern China.
Voice: Vocal, but rarely so away from nest, where it often repeats a short,
high, and thin *kle*.

Aquila nipalensis

草原雕 cǎo yuán diāo

Steppe Eagle   EN

L: 70–82 cm
WS: 175–214 cm
Habitat: Inhabits open
steppes, grassy areas,
deserts, and low hills; also
seen at high altitudes
in Western China. In
breeding season, builds
nest on cliffs or in rock
piles on mountaintops; sometimes also nests on the ground, soil
mounds, or hill slopes.
Behavior: Diurnal. Prefers perching on the ground, tree branches, or
utility poles. Forages mainly for rodents on open fields.
Distribution: Occurs mainly in Western and Northeast China; summer
breeder in northern China. Population in decline in recent years.
Voice: Occasionally a loud, rising, and increasingly gruff *kwe-ek* that
ends abruptly.

Aquila heliaca

白肩雕 bái jiān diāo

Imperial Eagle   VU

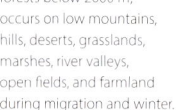

L: 68–84 cm
WS: 176–216 cm
Habitat: In summer,
often inhabits montane
forests below 2000 m;
occurs on low mountains,
hills, deserts, grasslands,
marshes, river valleys,
open fields, and farmland
during migration and winter.
Behavior: Diurnal; often solitary in nonbreeding season.
Distribution: Breeds mainly in northwestern China; few breed in
Northeast China. Uncommon in eastern China during migration; winters
in the eastern part of the Qinghai-Tibet Plateau and southwestern and
South China.
Voice: During breeding season, makes a series of hoarse, descending,
barking calls like a distant dog, *owk-owk-owk-*.

Aquila chrysaetos

金雕 jīn diāo

Golden Eagle   LC

L: 78–93 cm
WS: 190–234 cm
Habitat: Inhabits
bare rocky mountains,
montane forests, and
forest edges in summer;
also seen in open forests
and on open fields,
steppes, and wetlands
during migration and winter.
Behavior: Diurnal. Forages mainly for medium-sized to large mammals
and birds, such as blue sheep and foxes. Often stands on cliffs; dives to
catch prey when it is spotted and sometimes also forages by circling
in sky. Generally solitary or in pairs, but there are occasional records of
group hunting.
Distribution: *A. c. kamtschatica* seen in Northeast China; *A. c. daphanea*
widespread in China except for Taiwan and Hainan. Rare resident or
passage migrant in most regions.
Voice: Rarely vocalizes away from nesting area, where it makes an often
loud *yip yep yep yep*.

Aquila fasciata

白腹隼雕 bái fù sǔn diāo

Bonelli's Eagle   LC

L: 55–67 cm
WS: 143–176 cm
Habitat: Inhabits open
areas on hills and in
montane forests; nests on
cliff ledges. Juvenile often
wanders and forages in
pastures, cultivated areas,
wetlands, and other open
habitats in winter. Generally occurs in mid- and low-altitude regions,
rarely above 2000 m.
Behavior: Territorial; often soars in pairs in territory.
Distribution: Rare resident in the regions to the south of the Yangtze
River, including Taiwan and Hainan. Some juveniles wander northward
after breeding season; the northernmost record in Liaoning (Laotie
Mountain in Lvshun).
Voice: Shrill. Male gives melodious whistles in breeding season.

Booted Eagle

Small white markings on "shoulders"

Square-ended tail

Dark morph

Pale morph

White underparts and underwing coverts contrast markedly with black wing feathers

Dark morph

Pale morph

Small

Tail with sharp corners

Pale morph

Intermediate morph

Dark morph

juv.

Dark brown underparts **ad.**

White underparts

ad.

Feathered tarsi

Steppe Eagle

subad.

Prominent white band on under wing

ad.

juv.

juv.

Gape reaches behind eye

Dark brown overall

ad.

Brownish overall

ad.

juv.

Golden Eagle

ad.

juv.

juv.

ad.

White "window" on under wing

Golden nape

Dark brown overall

Gape does not reach behind eye

ad.

juv.

Relatively long wings and tail

Basal parts of the tail white

Imperial Eagle

ad.

ad.

juv.

Pale "window"

juv.

Buff nape

Dark brown overall

Buff overall

ad.

juv.

Bonelli's Eagle

Six "fingers"

Black "fingers"

Stout head

Black tail tips

ad.

juv.

Black underwing coverts contrast strongly with white belly

Finely barred tail lacks terminal black band

Third calendar year

White from throat to belly with black streaks

Brown from throat to belly

ad.

juv.

123

Accipiter trivirgatus

凤头鹰 fèng tóu yīng

Crested Goshawk   LC

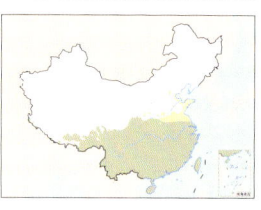

L: 40–48 cm
WS: 74–90 cm
Habitat: Adapted to various kinds of forest environments; inhabits mainly low-altitude hills. Also adapted to urban environments; can settle in small parks or green spaces.
Behavior: Most active during early breeding season, often seen in flight; circles in sky with level wings. Wing strokes fast and shallow. Adopts a unique behavior where the bird lowers and quickly shakes the wings.
Distribution: Common in the range. *A. t. indicus* is a resident in regions south of the Yangtze River (including Hainan); expanding north in recent years, with the northernmost records in Beijing. *A. t. formosae* is a resident in Taiwan.
Voice: Varied clipped, high-pitched, drongo-like calls, including a strongly disyllabic, rising *pi-wi … pi-wi … pi-wi*, and another version with stress on the rising terminal note, *pi-er-weee*, often paired.

Accipiter badius

褐耳鹰 hè ěr yīng

Shikra   LC

L: 35 cm
WS: 52–68 cm
Habitat: Prefers woodlands, open wooded areas, and farmland.
Behavior: Prefers perching on tall branches. Chases other birds with fast wingbeats; sometimes circles in sky. Forages for small mammals, lizards, rodents, and insects in forests.
Distribution: *A. b. cenchroides* is a rare breeder in desert poplar forests in W and S Junggar Basin in Xinjiang; occasionally seen in W Kashgar. *A. b. poliopsis* a rare resident in Yunnan, Tibet, Guizhou, Shaanxi, Guangxi, Guangdong, Hainan, and Macao.
Voice: Vocal. Most frequent call is a rapid-fire Ashy Drongo–like *pi-tu*. Makes various other shrill piping notes, many very similar to those of Besra, such as *kyeew*.

Accipiter soloensis

赤腹鹰 chì fù yīng

Chinese Sparrowhawk   LC

L: 25–35 cm
WS: 52–62 cm
Habitat: Breeds in forests at middle and low altitudes; rarely occurs above 1000 m. Forages in open areas.
Behavior: Darts out from perch when hunting. Forms large flocks during migration seasons; migrates along regular routes. Large flocks containing up to several thousand individuals occur in Taiwan in fall.
Distribution: Common breeder in most of central and eastern China; potentially a small wintering population in Hainan.
Voice: Varied high-pitched calls, including an accelerating, shrill *klee-klee-klee-klee-klip-kilp-ip-ip-ip* in breeding season.

Accipiter gularis

日本松雀鹰 rì běn sōng què yīng

Japanese Sparrowhawk   LC

L: 23–30 cm
WS: 46–58 cm
Habitat: Prefers woodlands on the border of low-altitude forests and open fields; forages in open areas.
Behavior: Forages mainly for small passerines. Migrates along regular routes in spring and fall. Does not gather in large flocks; generally occurs solitarily or in sporadic small flocks. Fierce and active; often provokes and attacks other raptors in sky.
Distribution: *A. g. gularis* breeds in Northeast China and migrates through eastern China; a few winter in South China, including Taiwan and Hainan. No record of *A. g. sibiricus* in China; may breed in extreme NW Inner Mongolia and migrate through northwestern and southwestern China.
Voice: Shrill *kew-kewkewkewkewkew*, descending in tone, very similar to that of Besra.

Crested Goshawk

Six "fingers," inconspicuous

Broad wings

White fluffy undertail coverts, commonly called "diapers" by Chinese birders

Short crest

Gray head

♂ ad.

Belly covered with bold bars

Strong tarsi

♀ ad.

Belly covered with relatively narrow bars

Broad and large wings

juv.

Belly densely covered with heart-shaped barring

Relatively short crest

juv.

Chinese Sparrowhawk

Four "fingers"

Black wing tips contrast strongly with white underwing coverts

Long narrow wings

Yellow iris

♀

Dark red iris, nearly black

♂

Light orange-brown belly

Japanese Sparrowhawk

Five "fingers"

Narrow wings of intermediate length

Bulging trailing edge

♂

♀

Compact silhouette

Shikra

Five "fingers"

Pale underparts with inconspicuous barring

Orangish-red iris

♂

Belly densely covered with thin fuzzy barring

Pinkish-brown flanks

Slender tarsi and toes; long middle toe

Yellow iris

♀

Brownish upperparts

Red iris

Faint medial throat stripe

Light blue-gray upperparts contrast with black primaries

A. b. poliopsis

Relatively short wings; flight feathers barely reach halfway down tail

Inconspicuous uppertail bars

Yellow iris

inconspicuous medial throat stripe

Belly densely covered with thin rufous-brown bars

inconspicuous medial throat stripe

Streaks on breast

Bars on belly

juv.

Accipiter virgatus

松雀鹰 sōng què yīng

Besra LC

L: 25–36 cm
WS: 51–70 cm
Habitat: Adapted to various kinds of wooded habitats; occurs mainly in hilly regions at low altitudes, generally not above 2200 m, but has been recorded above 3000 m in Sichuan and Tibet.

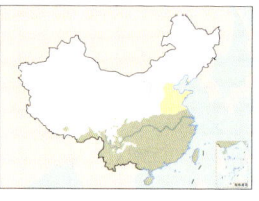

Behavior: Secretive; generally rests and forages deep in forests. Occasionally circles in sky; when circles, wings often held level, frequently flaps with interspersed short-distance gliding. Not a steady flier, often swerves or dives into forests. Afraid of people; often solitary. Highly territorial; usually initiates attacks to chase out other raptors that enter its territory.
Distribution: *A. v. affinis* a common resident in most of central, eastern, southwestern, and South China (including Hainan); a few individuals wander to North China, where the northernmost record was in Liaoning (Laotie Mountain in Lvshun). *A. v. fuscipectus* a common resident in Taiwan.
Voice: Vocal; often gives a shrill series, *kew-kewkewkewkewkew*, descending in tone.

Accipiter nisus

雀鹰 què yīng

Eurasian Sparrowhawk LC

L: 30–40 cm
WS: 60–79 cm
Habitat: Inhabits mainly montane forests, including mixed, broadleaf, and coniferous forests or forest edges; sometimes also occurs near parks and farmland.

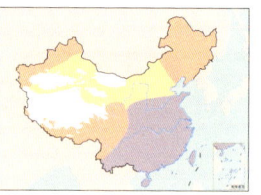

Behavior: Generally solitary in nonbreeding season; occasionally forms small loose flocks during migration. Territorial; chases out other raptors that enter its territory. Scans for prey in forests and takes prey by surprise. Agile flier; can maneuver quickly through forests.
Distribution: Widespread in China. *A. n. nisosimilis* breeds in Northeast and northwestern China; winters in North, South, southeastern, and southwestern China. *A. n. melaschistos* breeds or resides in Sichuan, Qinghai, and Tibet and winters in southwestern China.
Voice: Relatively slow but typical accipiterlike chatter, as well as more plaintive, descending whistles, *wei you-wei you*.

Accipiter gentilis

苍鹰 cāng yīng

Northern Goshawk LC

L: 47–59 cm
WS: 106–131 cm
Habitat: Generally inhabits wooded areas or forest edges; sometimes also occurs near hills, city parks, and open fields.
Behavior: Generally moves about solitarily

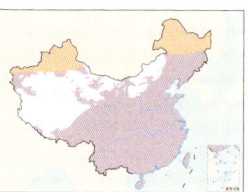

in nonbreeding season. Territorial, fierce; chases out other raptors that enter its territory, sometimes killing small raptors, such as Short-eared Owl and Eurasian Sparrowhawk. Generally hides in forests and keeps a close eye on prey. Seldom circles high in sky except during migration. Flight fast and agile; good at maneuvering through forests.
Distribution: *A. g. schvedowi* breeds in parts of Northeast, northwestern, and southwestern China; winters in most regions in northwestern, Northeast, North, and southern China. *A. g. albidus* rare in N Heilongjiang and S Liaoning. *A. g. fujiyamae* in Taiwan. *A. g. buteoides* in NW Xinjiang.
Voice: Calls like Eurasian Sparrowhawk but louder and more "fierce sounding," often with nasal quality of Black Woodpecker.

Circus melanoleucos

鹊鹞 què yào

Pied Harrier LC

L: 43–50 cm
WS: 110–125 cm
Habitat: Inhabits open habitats, including low mountains and hills, foothill plains, freshwater marshes, rivers, lakes, and grasslands; sometimes also occurs near farmland and coastal wetlands.

Behavior: Generally solitary. Often flies low above grass, wetlands, and bushes. In flight, holds wings in V shape; glides mostly, interspersed with occasional wing beats. Forages mainly by flying slowly near ground and scanning for prey; descends quickly to catch prey once it is spotted.
Distribution: Breeds in Northeast and North China; migrates through North, South, and East China. Uncommon in winter over vast areas in southern China.
Voice: Generally silent but makes a leisurely, even dreamy, nasal, Northern Lapwing–like *twee-uu* in flight display.

Besra

Orange-yellow to red iris   Gray cheek

Black medial throat stripe

Long slender yellow tarsi and toes; middle toe especially long

♀

Northern Goshawk

Gray upperparts

White underparts; belly covered with inconspicuous brown bars

ad.

Brownish upperparts

Buff underparts with brownish streaks

subad.

Eurasian Sparrowhawk

Rufous cheek

Gray upperparts

♂

Brown upperparts

♀

Pied Harrier

Trident-shaped black pattern on back and upper wing (better shown in flight)

♂

Few streaks on belly

♀

♂

♀

Circus spilonotus

白腹鹞 bái fù yào

Eastern Marsh-Harrier   LC

L: 48–58 cm
WS: 113–137 cm
Habitat: Inhabits various wetlands, including marshes, rivers, lakes, and reed ponds.
Behavior: Generally solitary or in pairs. Often flies low above water

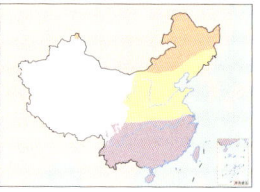

bodies or marshes while holding wings in shallow V shape. Glides often, with relatively few flaps. Generally perches on the ground or low soil mounds; tends not to perch at height, unlike other raptors.
Distribution: Breeds in Northeast and North China; rare in N Xinjiang in recent years. Migrates through North, South, and East China and winters over vast areas in southern China.
Voice: Females makes plaintive, shrill *pe-uu … pe-u … pe-uu*, and male makes more nasal note similar to, but lower pitched than, Pied Harrier. Scolding Ibisbill–like *kik-kik-kik-kik-ki* in alarm.

Circus aeruginosus

白头鹞 bái tóu yào

Eurasian Marsh-Harrier   LC

L: 43–55 cm
WS: 115–140 cm
Habitat: Occurs near open water bodies, such as lakes, rivers, marshes, and reed ponds on low mountains and plains.
Behavior: Solitary or in pairs, often hunts for prey

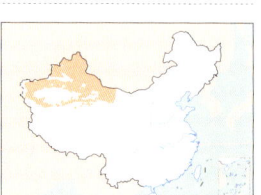

by gliding low above riparian reedbeds and marshes, wings raised in deep V shape. Perches on the ground.
Distribution: Uncommon breeder in various wetlands in Xinjiang. Migrants regular east to W Qinghai, vagrants farther east.
Voice: Vocal when displaying, when most notes similar to Eastern Marsh-Harrier; far less so in winter.

Circus pygargus

乌灰鹞 wū huī yào

Montagu's Harrier   LC

L: 39–50 cm
WS: 96–116 cm
Habitat: Inhabits low mountains and hills with bush cover, open marshes, grasslands, and farmland.
Behavior: Generally nests in reedbeds or bushes. Perches on the ground or

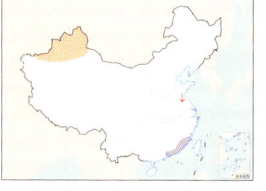

low soil mounds. Generally hunts by flying low and scanning for small animals, such as rodents, frogs, lizards, and small birds on the ground.
Distribution: Uncommon breeder in NW Xinjiang; vagrant east to Shandong, Jiangsu, Fujian, and Guangdong during migration and in winter.
Voice: Varied calls but mostly only on breeding grounds. Male gives a short nasal *kyip … kyip … kyip*; female gives a peevish Barn Owl–like hiss.

Circus cyaneus

白尾鹞 bái wěi yào

Hen Harrier   LC

L: 43–54 cm
WS: 98–124 cm
Habitat: Inhabits freshwater marshes, rivers, lakes, grasslands, and deserted fields on plains and hills; sometimes also moves to farmland, coastal wetlands, and grassy slopes.

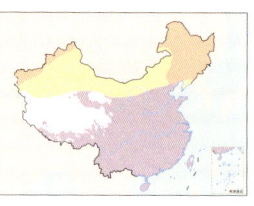

Behavior: Generally solitary or in pairs. Often flies low above wetlands or grassy areas while holding wings in V shape; when gliding, bends both wings slightly backward. Sometimes perches on the ground and watches for prey on wetlands or grass. Moves about and forages mainly during the day; most active at dusk and dawn.
Distribution: Breeds in Northeast and North China; rare records in N Xinjiang in recent years. Migrates through North, South, and East China; winters over vast areas in southern China.
Voice: Generally silent away from nest site, where it occasionally makes a repeated parakeet-like *wei jiu-wei jiu*.

Circus macrourus

草原鹞 cǎo yuán yào

Pallid Harrier   NT

L: 40–50 cm
WS: 97–118 cm
Habitat: Prefers lowland marshes and cultivated lands; also inhabits grasslands on piedmont hills and deserts near water bodies.
Behavior: Hunts by

gliding low above open fields.
Distribution: Rare breeder in NW Xinjiang; number in apparent decline in the last decade. Occasional passage migrant or winter visitor to S Xinjiang. Vagrant to Tianjin, Hebei, Inner Mongolia, Ningxia, S Tibet, Sichuan, Chongqing, Jiangxi, Jiangsu, Guangxi, and Hainan.
Voice: Generally silent; male occasionally gives a peculiar piercing *tyir´r´r´*.

Eastern Marsh-Harrier

Mainland morph ♀

Black-headed mainland morph ♂

Gray-headed mainland morph ♂

Japanese morph imm.

Japanese morph ♀

Japanese morph ♂

Mainland morph imm.

Brownish overall

Grayish-black head with streaks

♀

Brown tail without band

Black head

Black-headed mainland morph ♂

Gray-headed mainland morph

imm.

Japanese morph

imm.

Eurasian Marsh-Harrier

Pale brown head and neck with blackish-brown streaks

Brown back

Clean creamy-white crown and throat

Broad blackish-brown eye stripe

Dark brown overall

♂

♀

Hen Harrier

Five "fingers"

Gray head

White belly

Black wing tips

Heavily streaked underparts

♂

♀

Montagu's Harrier

Four "fingers"

♀

Broader white bars on secondaries

Lacks pale collar

♀

Bluish-gray upperparts and breast

Brown streaks on flanks

One black bar on wings

♂

♂

Pallid Harrier

Small wedge-shaped black patch on wing tip

No streaks on under wing

Four "fingers"

♀

Narrow white bars on secondaries

♂

Whitish head and upperparts

♂

White underparts

When perched, adult female Montagu's Harrier and Pallid Harrier are almost impossible to tell apart

129

Haliaeetus leucogaster
白腹海雕 bái fù hǎi diāo
White-bellied Sea-Eagle

L: 70–85 cm
WS: 178–218 cm
Habitat: Typical coastal bird; generally occurs near coasts. Inhabits coastal and estuarine areas, sometimes also occurs above hills and reservoirs not far from coasts.

Behavior: Generally flies solitarily or in pairs above coastal waters; wingbeats slow and powerful. Often moves about and forages during the day; especially active at dusk and dawn. Often forages by flying along coasts; sometimes also forages on land.
Distribution: Seen in coastal areas in Guangdong, Fujian, Taiwan, Hong Kong, and Hainan; stragglers exceptional north to Jiangsu.
Voice: High-pitched goose- or Oriental Pied-Hornbill–like yapping, with notes often sounding paired, *ang-ang … ang-ang … ang-ang*.

Haliaeetus leucoryphus
玉带海雕 yù dài hǎi diāo
Pallas's Fish-Eagle

L: 72–84 cm
WS: 185–215 cm
Habitat: Inhabits open water bodies, such as lakes and rivers; sometimes also moves to grassy areas and farmland.
Behavior: Generally moves around solitarily.

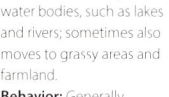

Often perches on trees or power lines for long periods to watch for prey, then quickly attacks when opportunity arises.
Distribution: Rare and local breeder in Northeast, northwestern, and southwestern China. Rare in North China during migration; vagrant in Shanghai.
Voice: Very noisy when breeding—a loud *wang ang-wang ang*.

Haliaeetus albicilla
白尾海雕 bái wěi hǎi diāo
White-tailed Eagle

L: 74–92 cm
WS: 193–244 cm
Habitat: Inhabits lakes, rivers, coasts, and estuaries; prefers water bodies near tall trees or open lakes and rivers in forested areas.
Behavior: Generally

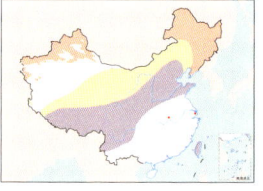

moves around solitarily or in pairs; sometimes gathers in flocks in winter. Moves about and forages mainly during the day. Prefers perching on rocks, ground, or ice surface. Often flies over lakes or seas with level wings, flapping with light wingbeats, followed by glides.
Distribution: Breeds in Northeast and northwestern China; vast winter range spans from North to southwestern China.
Voice: Usually silent away from nest, where it makes a Black Woodpecker–like slow yelp.

Haliaeetus pelagicus
虎头海雕 hǔ tóu hǎi diāo
Steller's Sea-Eagle

L: 85–105 cm
WS: 195–230 cm
Habitat: Inhabits coastal and estuarine areas; sometimes also moves inland along rivers.
Behavior: Moves about and forages mainly during the day; prefers perching

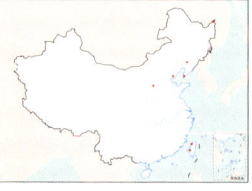

on the ground and ice surface for a long time. Often circles above lakes or seas with level wings; flies relatively slowly while gliding mostly.
Distribution: Rare winter visitor and passage migrant. Occurs in Northeast China; in recent years, has been regular at Jilin (Hunchun) in early March. Vagrant in Taiwan.
Voice: Occasional, penetrating, goose-like *klee-klee-ke*, far more raucous than White-tailed Eagle.

Haliaeetus humilis
渔雕 yú diāo
Lesser Fish-Eagle

L: 51–69 cm
WS: 130–165 cm
Habitat: Occurs in mid- and low-altitude forests near lakes, swamps, and rivers.
Behavior: Often moves about solitarily or in pairs over open wetlands in

forests; forages mainly for fish.
Distribution: Restricted to Hainan, where a rare winter visitor or vagrant.
Voice: Querulous, rising and falling, plaintive scream, *eea-a-aar*.

White-bellied Sea-Eagle

ad.

White breast, belly, and tail

ad.

ad.

Black trailing edge to wings

juv.

White-tailed Eagle

ad.

ad.

White tail

juv.

Pallas's Fish-Eagle

ad.

juv.

Pale panel on under wing

ad.

juv.

Black terminal band and white subterminal band on tail

Steller's Sea-Eagle

ad.

ad.

Stout yellow bill

juv.

Lesser Fish-Eagle

Gray head

ad.

juv.

Seven "fingers"

Plain white lower belly

ad.

Short tail

Milvus migrans
黑鸢 hēi yuān
Black Kite

L: 54–66 cm
WS: 150 cm
Habitat: Occurs near open grasslands, low mountains and hills, suburban fields, and wetlands.
Behavior: Fast and powerful flight; can circle for a long time on thermals. Hunts fast and fiercely. Nests in tall trees or cliffs.
Distribution: Common. *M. m. govinda* seen in W Yunnan; *M. m. formosanus* seen in Taiwan and Hainan; *M. m. lineatus* widespread in most of China.
Voice: Shrill, gull-like, wavering whinny or wail, *keee-aaar-a-ar*.

Haliastur indus
栗鸢 lì yuān
Brahminy Kite

L: 36–51 cm
WS: 110–125 cm
Habitat: Inhabits rivers, lakes, coastal wetlands, and fields.
Behavior: Similar to Black Kite. Feeds mainly on aquatic animals, such as fish and frogs.
Distribution: Occasionally seen in southwestern and South China; vagrant in Hubei, Qinghai, and Beijing.
Voice: Catlike rising then falling mew, *peee-ah*.

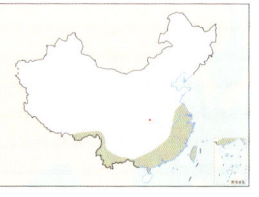

Butastur teesa
白眼鵟鹰 bái yǎn kuáng yīng
White-eyed Buzzard

L: 36–43 cm
WS: 88–100 cm
Habitat: Occurs at forest edges near open habitats, such as plains, farmland, and marshes.
Behavior: Often moves around solitarily in open wooded areas; prefers foraging for small terrestrial amphibians and reptiles.
Distribution: Rare in S Tibet.
Voice: Mournful rising then descending whistle, *kip-wheeoo*, very similar to Gray-faced Buzzard.

Butastur indicus
灰脸鵟鹰 huī liǎn kuáng yīng
Gray-faced Buzzard

L: 39–48 cm
WS: 105–115 cm
Habitat: Occurs in sparse broadleaf, mixed, and coniferous forests at middle and low altitudes.
Behavior: Moves around solitarily or in pairs at forest edges. Prefers open habitats in winter. Migrates in large flocks.
Distribution: Breeds in Northeast China; migrates through North, Central, East, and southwestern China; winters in South China, including Taiwan and Hainan.
Voice: Tremulous, whistled, rising then falling whistle *kip-weeeuh*.

Butastur liventer
棕翅鵟鹰 zōng chì kuáng yīng
Rufous-winged Buzzard

L: 35–41 cm
WS: 84–91 cm
Habitat: Occurs on the edges of low-altitude broadleaf forests near wetlands.
Behavior: Prefers moving about solitarily in open forests near wetlands.
Distribution: Restricted to SW Yunnan. Rare.
Voice: Mournful *pi-weea——* whistle, very similar to Gray-faced Buzzard and White-eyed Buzzard.

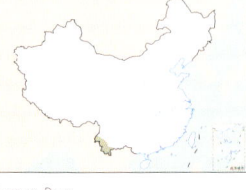

Buteo hemilasius
大鵟 dà kuáng
Upland Buzzard

L: 57–67 cm
WS: 143–161 cm
Habitat: Inhabits montane forests, foothill plains, and grasslands; also seen on edges of montane forests, open alpine steppes, and deserts. Often moves to open fields, farmland, deserted fields, or villages in winter. May occur on plateaus over 4000 m.
Behavior: Generally moves around solitarily or in pairs. Forages during the day by circling or hovering in sky. Often perches on trees, haystacks, or power lines.
Distribution: Widespread in Northeast, northwestern, North, and southwestern China as relatively common migrant or resident; uncommon winter visitor in South and East China.
Voice: Loud mewing calls similar to other buteos, perhaps especially Rough-legged Hawk. More research needed.

Black Kite

White patch on under wing

White streaks on underparts

ad.

Long tail with shallow fork

subad.

Brahminy Kite

White breast with black streaks

White head and neck

Chestnut body

Black streaks on breast

White-eyed Buzzard

White iris

White forehead, chin, and throat; white of chin extends to upper breast

Variable white pattern on wings

Gray-faced Buzzard

Prominent white supercilium

Five "fingers"

Prominent medial throat stripe

Lower underparts covered with prominent brown bars

Rufous-winged Buzzard

No supercilium

Plain gray head and neck

Lacks medial throat stripe

Rufous-chestnut wings

Rufous-brown tail

Upland Buzzard

Prominent white patch on under wing

Intermediate morph

Pale morph

Dark morph

133

Buteo refectus

喜山鵟 xǐ shān kuáng

Himalayan Buzzard

L: 45–53 cm
WS: 126–140 cm
Habitat: Inhabits
montane forests and
foothill plains; also seen
on alpine forest edges,
open mountain steppes,
and deserts. Moves to
open fields, farmland,
deserted fields, and villages in winter.
Behavior: Generally moves around solitarily or in pairs; migrates in large
flocks. Diurnal. Perches on treetops, haystacks, or utility poles. Circles or
hovers in sky to forage.
Distribution: Relatively widespread in southwestern China as common
summer breeder or resident.
Voice: Not known to differ from Eastern Buzzard or Common Buzzard.

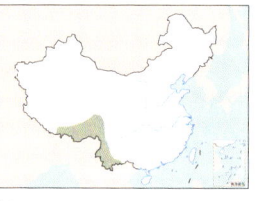

Buteo japonicus

普通鵟 pǔ tōng kuáng

Eastern Buzzard

L: 42–54 cm
WS: 122–137 cm
Habitat: Inhabits
montane forests, foothill
plains, and steppes; often
moves to open fields,
farmland, deserted fields,
and villages in winter.
Behavior: Generally
moves about solitarily or in pairs; forms large flocks during migration.
Diurnal. Often perches at height in open areas. Can hover to forage.
Distribution: Ubiquitous in eastern China. Summer breeder in
Northeast China; winter visitor to South and East China; common
throughout eastern China during migration.
Voice: Loud mewing typical of most buteos.

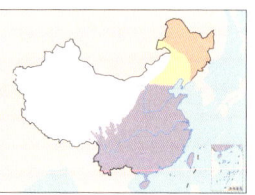

Buteo buteo

欧亚鵟 ōu yà kuáng

Common Buzzard

L: 40–48 cm
WS: 100–125 cm
Habitat: Occurs near montane forest edges in breeding season; prefers
coniferous forests. Seen on oases, steppes, farmland, deserts, and other
habitats with abundant rodents during migration and winter.
Behavior: Prefers foraging from tree branches or utility poles for rodents,

Buteo lagopus

毛脚鵟 máo jiǎo kuáng

Rough-legged Hawk

L: 45–62 cm
WS: 120–153 cm
Habitat: Breeds on tundra
in northern Eurasia.
Inhabits open plains,
low mountains and hills,
farmland, and deserted
fields in winter.
Behavior: Generally
solitary. Diurnal. Perches on trees, haystacks, or power lines. Circles or
hovers in sky to forage.
Distribution: *B. l. lagopus* seen in Xinjiang; *B. l. kamtschatkensis* is an
uncommon winter visitor to Northeast and North China. Very few
individuals occasionally winter in South China.
Voice: Loud mewing when breeding but generally silent in winter.

Buteo rufinus

棕尾鵟 zōng wěi kuáng

Long-legged Buzzard

L: 50–58 cm
WS: 130–155 cm
Habitat: Inhabits
open terrain, such as
premontane deserts, other
deserts, oases, steppes,
and mountainous areas.
Prefers dry environments;
not frequently seen in
forested areas.
Behavior: Generally moves around solitarily or in small, loose flocks on
dry plains; nests on cliff ledges or in trees.
Distribution: Relatively common breeder and passage migrant on
desert plains in Xinjiang. Uncommon in SW Xinjiang, Qinghai, C Inner
Mongolia, Ningxia, SE Gansu, S and SE Tibet, and E Yunnan during
migration and winter.
Voice: Seldom calls. Typical buteo mewing, perhaps slightly shorter and
falling than that of Eastern Buzzard.

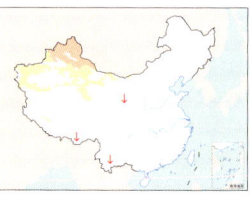

small birds, reptiles, and
large insects.
Distribution: Uncommon
breeder on the Altai
Mountains and Tian Shan
Mountains in N Xinjiang;
relatively common in
NW Xinjiang during
migration; uncommon
in SW Xinjiang. Reported
from NE Sichuan and Taiwan in winter.
Voice: Loud mewing *peeioo*.

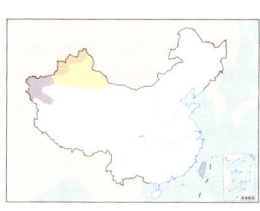

Himalayan Buzzard

Prominent dark trailing edge to wings

Rough-legged Hawk

Prominent large black carpal patches

Dark throat

Pale throat

♂

♀

Brownish carpal patches and belly

imm.

Black belly

Prominent black trailing edge

Prominent black subterminal band on tail

Eastern Buzzard

Paler underparts than Himalayan Buzzard

Brown streaks on belly

Darker tail than Himalayan Buzzard

Pale morph

Brownish overall

Brown morph

Long-legged Buzzard

Prominent large black carpal patch

Ginger-yellow tail lacks barring

Long slender neck

Tarsi not feathered; legs thick and long

Common Buzzard

Large white area at base of primaries lacks barring

Small body with large head and short neck

Black underparts and wing coverts

Relatively broad black trailing edge to flight feathers

Dark morph

Dark brown tail with prominent black subterminal band

Brown morph

Tarsi not feathered; legs thin and short

135

鸨形目　OTIDIFORMES

鸨科
Otididae

Great Bustard

Medium to large land birds. Body profile resembles ostriches, but good at flying. Plumage slightly different between sexes, mostly brown, white, and black. Head small, with short but strong bill. Neck thin and long. Wings short and rounded. Robust legs good at running. Strongly built. Flying style similar to cranes, but legs shorter than, or extending just beyond, tail. Inhabit open habitats, such as deserts and grasslands; usually in small groups. Omnivorous; feed mostly on plant buds, shoots, and seeds but also take insects and small amphibians and reptiles. Some species are migratory.

Bustards were once placed under Gruiformes but are now usually in their own order. One family, 11 genera, and 26 species recognized worldwide, distributed in the Old World and Oceania. Three genera and three species recorded in China, found in the grasslands and deserts of northern China.

Otis tarda
大鸨 dà bǎo
Great Bustard　　VU

L: 90–105 cm (♂)
75–85 cm (♀)
WS: 210–240 cm (♂)
170–190 cm (♀)
Habitat: Breeds in open dry steppes, savannas, desert steppes, and farmland in hilly regions. Occurs in winter on shallow lakes, meadows, grasslands, and wheat fields near large lakes and rivers.
Behavior: In breeding season, males gather at leks to display to females;

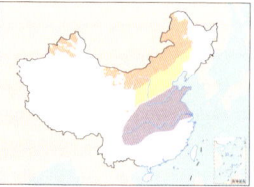

polygynous. Female solely responsible for incubation and chick rearing.
Distribution: *O. t. tarda* breeds in N Xinjiang and occasionally winters in its breeding range. *O. t. dybowskii* breeds in E Inner Mongolia, W Jilin, and SW Heilongjiang and winters in North China and the watersheds of Yellow River, Weihe River, and Yangtze River, recorded as far south as Guizhou (Caohai Lake).
Voice: Generally silent. Male gives low-pitched moans during courtship display, while both sexes give nasal bark when excited or in alarm.

Chlamydotis macqueenii
波斑鸨 bō bān bǎo
Macqueen's Bustard　　VU

L: 55–65 cm
WS: 130–150 cm
Habitat: Inhabits open desert plains and semideserts that are flat or gently undulating.
Behavior: Crepuscular. Generally moves around solitarily or in family units.

Shy and vigilant. Good at running; tends to run when in danger, rather than take off. Uses camouflage to hide among bushes; flies only when necessary.
Distribution: Uncommon breeder and passage migrant in N Xinjiang, W and C Inner Mongolia, and Gansu.
Voice: Generally silent.

Tetrax tetrax
小鸨 xiǎo bǎo
Little Bustard　　NT

L: 40–45 cm
WS: 83–91 cm
Habitat: Inhabits premontane hills, open wheat fields, alfalfa fields, harvested wet grasslands, desert steppes, and areas with low bushes.

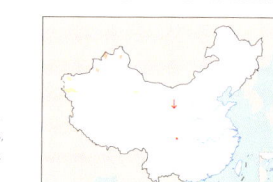

Behavior: Gregarious in nonbreeding season; forages on harvested wet grassland habitats or alfalfa fields in flocks of several dozens to hundreds of individuals. Departs breeding grounds in late October.
Distribution: Rare in N and W Xinjiang during migration; vagrant to Gansu, Qinghai, Ningxia, and NE Sichuan. In recent years, a large population has been discovered gathering in fall on grazed wet grassland habitats in the Taer Basin of Xinjiang.
Voice: Courtship call is a very dry *prrrt* reminiscent of a male Garganey but much briefer. The fourth primary produces whistling sounds in flight.

Great Bustard

No "moustache" at side of chin

Black secondaries

Dark-tipped primaries

No chestnut-brown collar on side of breast

Stout head and gray neck

♂ br.

White facial whiskers like "moustache" at side of chin

Chestnut band from base of hindneck to side of breast

Large white area on wing coverts

♂ br.

Courtship display

Macqueen's Bustard

Large black patches on secondaries contrast with large white patches at base of primaries

♂ br.

Fluffed black feathers on crown

Black plumes on side of neck

♂ br.

Breast feathers inflated

Courtship display

Little Bustard

♂ br.

Black neck in breeding season, with white V across foreneck and white collar across lower neck

Wings almost all white; large black patches only on outer four primaries

Sandy-brown upperparts with black mottling

鹤形目　GRUIFORMES

A heterogeneous order of waterbirds of varying sizes. Plumage similar between sexes, mostly white, black, brown, and red. Bill long and sharp. Neck long. Wings rounded, and most species good at flying. Tail short. Legs long. Feet variable; hind toe less developed or on a different level from the front toes. Some species with lobed feet. Inhabit primarily forests, deserts, grasslands, and marshes. Nest on ground. Feed mostly on insects, fish, and, shrimp, as well as small amphibians, reptiles, and mammals; also take plant seeds, fruits, and vegetative parts.

　　Gruiformes was once a large order consisting of diverse groups of birds but currently includes only cranes, rails, and their allies. Six families, 52 genera, and 189 species recognized worldwide, with a cosmopolitan distribution. Two families, 16 genera, and 30 species recorded in China, ubiquitously distributed nationwide.

秧鸡科
Rallidae

Brown-cheeked Rail

鹤科
Gruidae

Common Crane

Porzana porzana

斑胸田鸡 bān xiōng tián jī

Spotted Crake

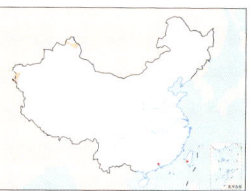

L: 22–25 cm
Habitat: Inhabits various kinds of grassy wetlands.
Behavior: Secretive; active at dusk and dawn or at night. Walks among aquatic plants or on floating vegetation; seldom flies.
Distribution: Rare breeding summer visitor to N and W Xinjiang; vagrant in winter to Guangdong and Taiwan.
Voice: Far-carrying *huwid*, repeated monotonously every 2 sec. Most vocal at night.

Rallus aquaticus

西秧鸡（西方秧鸡）xī yāng jī (xi fāng yāng jī)

Water Rail

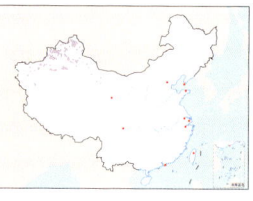

L: 23–26 cm
Habitat: Inhabits flooded fields or marshy lakes in lowlands or hilly regions; also occurs on riverside grass or bushes.
Behavior: Crepuscular. Secretive; rarely comes out of thick waterside vegetation.
Distribution: Resident and uncommon winter visitor in Xinjiang and NW Gansu; also breeds in E Qinghai. Rare visitor to SW Sichuan, Liaoning, Beijing, along the coast from Tianjin to Zhejiang, Hong Kong, and Taiwan.
Voice: Varied but most similar to call of Brown-cheeked Rail. Typical call is a series of piglike squeals, grunts, and repeated series of short *kik* notes that are slightly higher pitched and fractionally longer.

Rallus indicus

普通秧鸡 pǔ tōng yāng jī

Brown-cheeked Rail

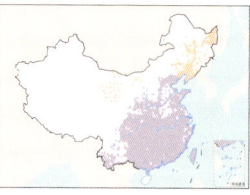

L: 23–29 cm
Habitat: Inhabits freshwater marshes, fishponds, lakeshores, and paddy fields in lowlands and hilly regions.
Behavior: Generally solitary; shy and afraid of people. Prefers thick waterside vegetation. Can walk quickly on aquatic vegetation; also good at swimming and diving. Flies rarely and fast for relatively short distances, legs dangling below body in flight.
Distribution: Locally common. Seen in Northeast, North, East, and Central China, Gansu, Qinghai, Ningxia, and most of southern China, including Hong Kong and Taiwan. The northern population is mainly summer breeders; migrate south to winter.
Voice: Hugely varied. Some sharp, rhythmic, and metallic, others squealing and unpleasant. Most similar to calls of Water Rail.

Crex crex

长脚秧鸡 cháng jiǎo yāng jī

Corn Crake LC

L: 24–27 cm
Habitat: Inhabits mountain pasture, wetlands , farmland, and forests and tall grass in river valleys at altitudes of 400–2500 m.
Behavior: Occurs on riverbanks, among lakeside tall grass, and in bushes at dusk and dawn. Calls silvery, sharp, and far-carrying. Usually heard but difficult to see.
Distribution: Locally common in NW Xinjiang in breeding season; occasionally seen in S Xinjiang and Tibet. Vagrant in Yunnan.
Voice: Distinctive loud rasping given at length at night, *krrk-krrk* or *crex crex*, like the scientific name.

Lewinia striata

灰胸秧鸡 (蓝胸秧鸡)

huī xiōng yāng jī (lán xiōng yāng jī)

Slaty-breasted Rail LC

L: 25–30 cm
Habitat: Inhabits various kinds of wetlands at low altitudes, including marshy meadows, reedbeds, paddy fields, riverbanks, mangroves, shrimp ponds, and drainage channels.
Behavior: Generally moves about solitarily or in family groups. Relatively secretive. Active in morning and at dusk; hides in grass during the day. Vigilant on the ground; quickly runs away when in danger. Good at swimming and diving, but not flying. Seldom flies; flies only for short distances.
Distribution: Uncommon resident or migrant. *L. s. albiventer* seen in SW and SE Yunnan. *L. s. jouyi* in Central, East, and South China, including Hainan; vagrant north to Beijing and Hebei. *L. s. taiwanus* in Taiwan.
Voice: Sharp *kerrek*, leisurely repeated. Also a Black-winged Stilt–like *kik* and noisy duet by pairs, *kra-ah*.

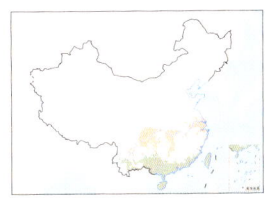

Spotted Crake

White spots from hindneck to upper back

Gray underparts with white spots

Prominent black and white bars on flanks

Water Rail

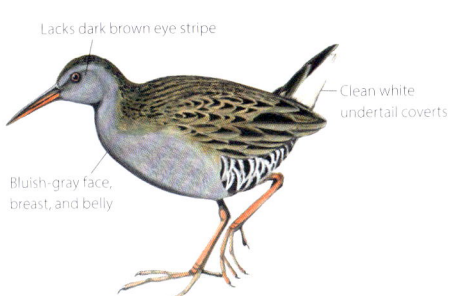

Lacks dark brown eye stripe

Clean white undertail coverts

Bluish-gray face, breast, and belly

Corn Crake

Short stout pinkish bill

Chestnut wings

Brown-cheeked Rail

Diffuse brown eye stripe, darker at lores

Lower neck and breast tinged brown

White undertail coverts with extensive dark feather centers

Lower breast and upper belly light brownish gray, not bluish gray

Slaty-breasted Rail

Prominent chestnut-brown patch from crown to hindneck

Dark brown back and wings densely covered with fine black and white barring

Pinkish bill

Dense black and white bars on flanks, belly, and undertail coverts

Bluish-gray cheek, throat, and lower breast

Grayish-black legs

Gallinula chloropus

黑水鸡 hēi shuǐ jī

Eurasian Moorhen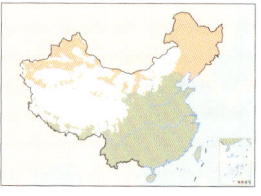

L: 30–38 cm

Habitat: Inhabits various natural or artificial freshwater wetlands; especially prefers riverbanks, ponds, marshes, reservoirs, ditches, and paddy fields with abundant reeds and other emergent plants, generally below 1700 m.

Behavior: Generally moves around solitarily or in pairs; also forms small flocks. Good at swimming and flying; requires a run-up before taking off from water. Also dives to evade predators. Often forages by walking among floating plants or riparian vegetation, and constantly flicks tail.

Distribution: Widespread and common throughout most of China. Northern populations winter further south. Populations south of the Yangtze River are mainly resident but can also disperse for short distances in response to weather and food variability.

Voice: Vocal and varied, including a loud explosive *krrrr-uk* or *kark*. Varied purrs and a rapid *kik-kik*, repeated when alarmed.

Fulica atra

白骨顶 (骨顶鸡) bái gǔ dǐng (gǔ dǐng jī)

Eurasian Coot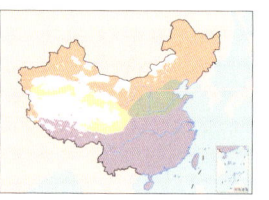

L: 36–39 cm

Habitat: Inhabits various open water bodies; especially prefers slow-flowing waters with abundant reeds, sedges, and other emergent plants, including lakes, ponds, reservoirs, estuaries, and marshes.

Behavior: Gregarious in nonbreeding season; can form flocks containing up to a few hundred individuals in winter. Also mixes with ducks. Good at swimming and diving; often dives to forage. Generally occurs on water in flocks; bobs head while swimming. Occasionally forages on land. Requires a run-up before taking off from water, where it flaps wings rapidly and loudly. Often lands after flying a short distance.

Distribution: Widespread in most of China, very common. Northerly breeding populations migrate south in winter; populations south of the Yangtze River are mostly resident but can also disperse for short distances in response to weather and food variability.

Voice: Vocal and varied—gives a brief, loud, and monosyllabic *kow* and *kick* during courtship and a disyllabic *kick-kowp*. Female's call briefer and more resonant than calls of male.

Coturnicops exquisitus

花田鸡 huā tián jī

Swinhoe's Rail

L: 12–14 cm

Habitat: Inhabits wet grasslands and marshes or tall grass near wetlands.

Behavior: Crepuscular, very secretive. When alarmed, tends to hide in grass, rather than fly.

Distribution: Breeds in Northeast China; migrates through eastern China; winters in the Yangtze Plain. Rare.

Porphyrio poliocephalus

紫水鸡 zǐ shuǐ jī

Gray-headed Swamphen

L: 41–50 cm

Habitat: Inhabits slow-flowing or calm open waters and eutrophic wetlands with dense vegetation cover below 2500 m.

Behavior: Generally moves about in pairs or family units. Mild-tempered. Forages actively at dusk and dawn. Often walks slowly and steadily on floating vegetation; can also climb reed stems. Not good at flying; flies only short distances when necessary. Flies with slow wingbeats and dangling feet; seems a bit clumsy.

Distribution: The taxonomy of Gray-headed Swamphen and related species is not well resolved. Subspecies allocation between Gray-headed Swamphen (*P. poliocephalus*) and Black-backed Swamphen (*P. indicus*) remains debated. *P. p. poliocephalus* occurs in SE Tibet and Yunnan. Classification of the subspecies in South China remains debated; in this book treated as Black-backed Swamphen; refer to Black-backed Swamphen for further details.

Voice: Male gives loud, rich, and trumpeting *chuck*; female's calls are sharper and softer. Frequently gives cackling contact calls when in flocks.

Porphyrio indicus

黑背紫水鸡 hēi bèi zǐ shuǐ jī

Black-backed Swamphen

L: 38–40 cm

Habitat: Similar to Gray-headed Swamphen.

Behavior: Similar to Gray-headed Swamphen.

Distribution: The "purple swamphens" in South China are generally referred to as the subspecies *viridis*, whose morphology is intermediate between *P. p. poliocephalus* and *P. i. indicus* and has long been listed within the *indicus* group. Recent research on the taxonomy of "purple swamphen" does not include specimens from Southeast Asia, so the specific allocation of *viridis* remains contentious. This book tentatively treats *viridis* as a subspecies of Black-backed Swamphen. Occurs in Fujian, Hainan, and Shanghai (Chongming); vagrant in Taiwan. The populations in Sichuan, W Guizhou, NE Hunan, S Hubei, Guangdong, Hong Kong, and Guangxi are generally considered to be Black-backed Swamphen, but opinions vary. Some scholars believe that both Gray-headed and Black-backed Swamphen occur in Guangdong.

Voice: Similar to Gray-headed Swamphen.

Voice: Advertising call a distinctive two-part *prit-prrrrt* given continuously mostly at dawn and dusk. Occasionally only the first syllable is given, and individuals apparently vary in pitch. Also, in breeding season, a prolonged trill, *gaga-gu*, like Ruddy-breasted Crake but often shorter, less tremulous, and less descending.

Eurasian Moorhen

Dark brown iris

Red bill and frontal shield with light yellow bill tip

Dark brown wings; remaining parts slightly bluish black

ad.

Dark red bill with light yellow tip

Red at top of tibia

Yellowish-green legs

White undertail coverts; white streaks on flanks

Light brown head, hindneck, and back

Medium to dark brown wings

White to dirty white on face, throat, foreneck, breast, and underparts

juv.

Blackish-brown tail

Gray-headed Swamphen

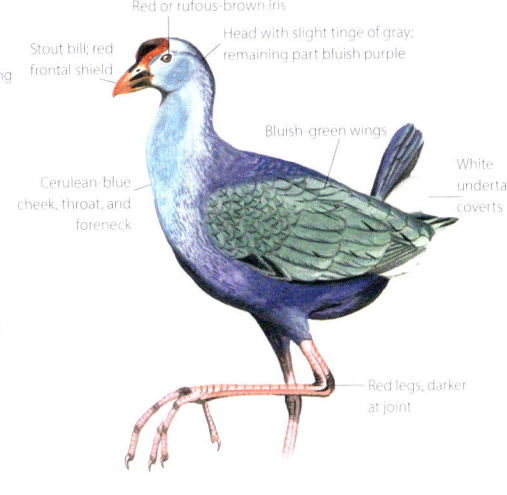

Red or rufous-brown iris

Stout bill; red frontal shield

Head with slight tinge of gray; remaining part bluish purple

Bluish-green wings

White undertail coverts

Cerulean-blue cheek, throat, and foreneck

Red legs, darker at joint

Black-backed Swamphen

Grayish-white head

Back, wings, and tail dark bluish purple to almost black

Eurasian Coot

White frontal shield and bill

Red iris

Black or dark grayish black overall

Yellowish-green or grayish-black legs

Lobed feet

Swinhoe's Rail

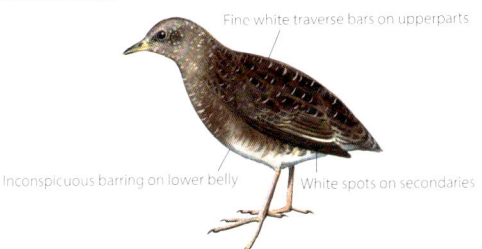

Fine white traverse bars on upperparts

Inconspicuous barring on lower belly

White spots on secondaries

Zapornia paykullii

斑胁田鸡 bān xié tián jī

Band-bellied Crake

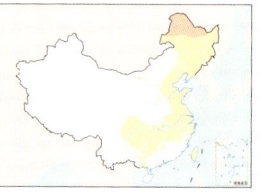

L: 22–27 cm

Habitat: Inhabits grassy lakes, water pools, and farmland; also seen on wet meadows near low-mountain forests.

Behavior: Active at dusk, dawn, and night; secretive. Occasionally forages near tussocks during the day; seeks cover in grass when alarmed.

Distribution: Breeds in Northeast and North China; migrates through Central, East, southwestern, and South China. Rare. Occasional in Taiwan.

Voice: Far-carrying territorial song a resonant amphibian-like rattle of 14–21 notes lasting about a second, mostly at night. Call a breathless *trrrup* or *krrup-up*.

Zapornia bicolor

棕背田鸡 zōng bèi tián jī

Black-tailed Crake

L: 19–25 cm

Habitat: Inhabits marshes, streams, farmland, tussocks, and scrubland near evergreen broadleaf forests at altitudes of 1000–3000 m.

Behavior: Comes out of cover to forage at dusk and dawn. Flies or runs immediately when alarmed.

Distribution: Occurs in SE Tibet, Yunnan, S Sichuan, Guizhou, Guangxi, and Guangdong. Uncommon.

Voice: Trilling song similar to Brown Crake and Ruddy-breasted Crake, as well as Little Grebe, but less faltering than Brown and more evenly paced than Ruddy-breasted. Also quiet *kik* calls.

Zapornia akool

红脚田鸡 (红脚苦恶鸟)

hóng jiǎo tián jī (hóng jiǎo kǔ è niǎo)

Brown Crake

L: 26–28 cm

Habitat: Inhabits lowlands and hilly regions; also seen on wetlands, such as ponds, lakes, riverbanks, marshes, and paddy fields with dense emergent vegetation or bushes.

Behavior: Generally moves about solitarily or in pairs. Vigilant and secretive. Occurs in dense vegetation or tussocks near water; calls and forages more actively at dusk and dawn. Good at running, wading, and swimming; bobs head and neck when walking, raising and flicking its tail. Runs to take cover among tussocks when alarmed; occasionally makes short-distance flights.

Distribution: Occurs in Central, South, and East China; locally common.

Voice: Short rising *pwi*, repeated when alarmed, or prolonged sharp trill that descends in pitch, similar to Ruddy-breasted Crake and Little Grebe.

Zapornia pusilla

小田鸡 xiǎo tián jī

Baillon's Crake

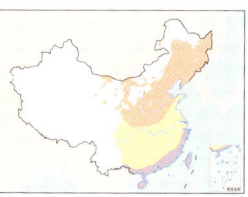

L: 15–20 cm

Habitat: Inhabits various kinds of natural or artificial grassy wetlands.

Behavior: Solitary. Hides in vegetation on the margins of water; when alarmed, takes cover among tussocks or quickly takes off and lands in cover. Occasionally swims and dives.

Distribution: Breeds from Northeast to northwestern China; uncommonly seen in most of China during migration; few winter in southwestern and southeastern China.

Voice: Song likened to prolonged frog croaking or, more vaguely, a male Garganey. Calls are a hoarse *ga*.

Zapornia parva

姬田鸡 jī tián jī

Little Crake

L: 17–19 cm

Habitat: Inhabits lakes, rivers, ponds, river channels, and deep-water marshes with abundant reeds and aquatic vegetation; especially prefers deep-water areas in large reedy wetlands with sweet flag and cattails.

Behavior: Good at swimming and diving. Crepuscular. Often swims among reeds or aquatic vegetation in water ponds; sometimes also moves outside of reeds or to areas of water that are far from banks. Forages by swimming and pecking in water; sometimes also dives to feed.

Distribution: Rare breeder in northwestern China.

Voice: Loud, far-carrying, nocturnal song an accelerating sequence of up to 20 nasal *pwa-pwa-pwa-pwa* notes that ends abruptly, often alternated with a short rapid trill.

Zapornia fusca

红胸田鸡 hóng xiōng tián jī

Ruddy-breasted Crake

L: 19–23 cm

Habitat: Inhabits reedbeds and tussocks near flooded fields, marshes, and river flats; especially prefers paddy fields and marshes.

Behavior: Shy. Generally ventures out at dusk, dawn, and night; stays among tussocks during the day. Seldom flies; seeks cover among tussocks when alarmed.

Distribution: *Z. f. erythrothorax* breeds in most of Northeast to central and eastern China; migrates through southwestern to central and eastern China; resident in Taiwan; locally common. *Z. f. bakeri* breeds in southwestern China; sometimes treated as *Z. f. fusca*.

Voice: Song is loud lengthy series of *pip* notes that gradually accelerate. Also a Brown Crake–like trill.

Band-bellied Crake

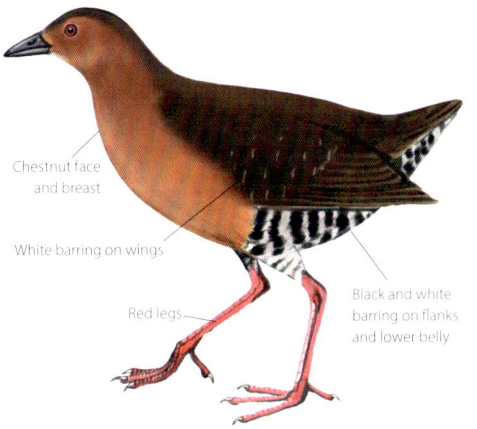

Chestnut face and breast

White barring on wings

Red legs

Black and white barring on flanks and lower belly

Baillon's Crake

Brown eye stripe

Prominent white spots

Gray from face to upper breast

White and blackish-brown bars on lower belly

Yellowish-green legs

Black-tailed Crake

Ash-gray head and breast

Bright brown hindneck, back, and wing coverts

Black tail

Red legs and toes

Little Crake

Longer primary projection and tail

Upperwing coverts lack white spots

Red base of bill

Bars only on rear flanks

Brown Crake

Dark rufous-brown iris

Slate gray on face, throat, and underparts and olive-brown upperparts; diffuse boundary between colors

Relatively thick yellowish-green bill with bright yellow base

Dark red legs

Ruddy-breasted Crake

Plain wings

Brownish-red face and breast

Narrow white bars on black lower belly

Dark red legs and toes

Rallina fasciata

红脚斑秧鸡 hóng jiǎo bān yāng jī

Red-legged Crake

L: 23–25 cm

Habitat: Generally occurs below 800 m. Inhabits open reedy marshes and wetlands, paddy fields, bushes on riverbanks, and wet areas in forests. May occur up to 1400 m during migration.

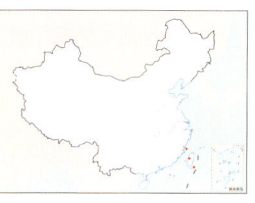

Behavior: Good at hiding. Generally hides among tussocks or bushes during the day; more active on rainy days or at dusk, dawn, and night. Calls frequently with loud and crisp voice. Runs quickly and nimbly; often evades predators by running fast.

Distribution: Vagrant; recorded at Matsu Islands, Taitung, Yunlin, and Orchid Island in summer.

Voice: Territorial vocalization is loud *gogogogok* or *girrrr*; makes courtship calls on rainy days or at night.

Rallina eurizonoides

白喉斑秧鸡 bái hóu bān yāng jī

Slaty-legged Crake

L: 21–28 cm

Habitat: Generally occurs below 700 m. Seen in dense bushes, edges, and streams within mature forests, as well as mangroves, paddy fields, and tall foothill tussocks.

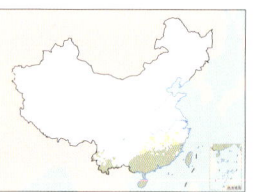

Behavior: Shy. Hides among tussocks during the day; generally comes out at dusk and dawn and in rainy weather; often calls at night. Generally solitary. Walks by raising feet high and raising and flickering tail. Immediately runs into bushes when in danger; not easily flushed. Partially nocturnal.

Distribution: *R. e. telmatophila* is a resident in southwestern, South, and southeastern China (unclear whether the Hainan population belongs to this subspecies); passage migrant to Central and East China. Breeding populations found recently in Central and East China are very secretive and rare. *R. e. formosana* is resident on Taiwan and Orchid Islands.

Voice: Series of repetitive, pulsed, paired, mechanical notes, *kek-kek*, at night; also a nasal *ow-ow*, a rhythm similar to Mountain Scops-Owl but lower pitched.

Amaurornis phoenicurus

白胸苦恶鸟 bái xiōng kǔ è niǎo

White-breasted Waterhen

L: 28–33 cm

Habitat: Generally occurs below 1500 m. Prefers marshes, wetlands, paddy fields, sugarcane fields, riverbanks, drainage ditches, mangroves, and moist forest edges with abundant reeds or tall grass. Also moves to green spaces and ponds near residential areas.

Behavior: Generally occurs solitarily or in pairs. Bolder and less afraid of people than other crakes and rails. Often moves about in open areas during the day with tail jerked up and flicking. Forages near water. Can rest on trees, swim, and occasionally dive.

Distribution: Common resident or migrant. Occurs in southwestern, Central, East, and South China, including Hainan and Taiwan; also occurs in northwestern, North, and Northeast China but in lower numbers than in southern China.

Voice: *kru-ak ... kru-ak* or *pu-ak ... ak* repeated monotonously at varying pace for up to 15 min. Also varied nasal croaks, churrs, and rattles.

Poliolimnas cinereus

白眉田鸡 bái méi tián jī

White-browed Crake

L: 20 cm

Habitat: Inhabits swampy meadows, flooded fields, and marshes; especially prefers wetlands with abundant floating vegetation.

Behavior: Bolder than its congeners. Also seen more often among floating vegetation, such as lotus leaves and water lilies, during daytime.

Distribution: Rare, but the number of records has been growing in China; recently seen in Guangxi, Hainan, Yunnan, Sichuan, Hong Kong, and Taiwan.

Voice: Small breeding-season assemblies give noisy, excited chattering and repeated yapping, *chika*. Calls with a *gleea*.

Gallicrex cinerea

董鸡 dǒng jī

Watercock

L: 40–43 cm

Habitat: Inhabits wet environments, such as marshes with abundant reeds and tussocks, flooded tall grass, and paddy fields.

Behavior: Generally occurs solitarily or in pairs. Shy. Often comes out at dusk or in cloudy weather; more secretive during the day. Occasionally moves about in open areas near field bunds or reedy ponds. Often wades in water with tail cocked and head bobbing; sometimes swims on the water's surface. When startled, runs quickly and seeks cover in nearby tussocks. Does not fly readily, but has strong flying ability; holds neck outstretched and legs stretched in flight.

Distribution: Widespread summer breeder in China; breeding records everywhere except for northwestern China, but not commonly seen.

Voice: Territorial calls a surprisingly far-carrying, low-pitched, gulped *umb* repeated every second or so and given 10–12 times.

Red-legged Crake

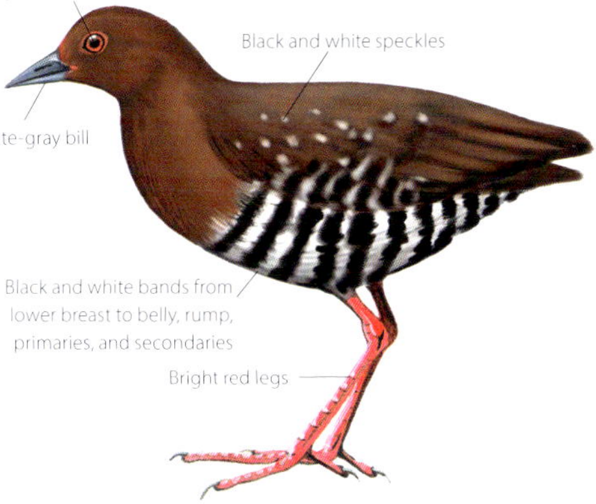

Bright red iris, eye ring, and base of bill

Black and white speckles

Slate-gray bill

Black and white bands from lower breast to belly, rump, primaries, and secondaries

Bright red legs

White-browed Crake

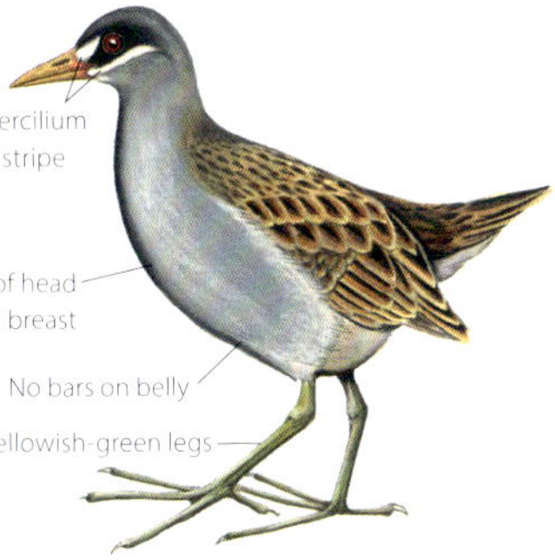

White supercilium and malar stripe

Gray side of head and breast

No bars on belly

Yellowish-green legs

Watercock

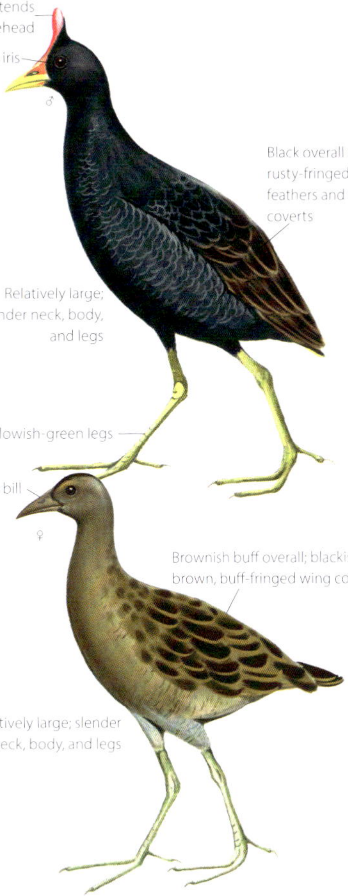

Long, sharp, red frontal shield extends to raised horn above forehead

Dark brown iris

Black overall with rusty-fringed wing feathers and wing coverts

Relatively large; slender neck, body, and legs

Yellowish-green legs

Slaty-legged Crake

Bright red iris

Bright rufous-brown head, neck, and breast

Brown to dark brown back, wings, and uppertail coverts; lacks barring

Slate-gray bill

Whitish throat

Dense black and white bands on lower breast, belly, and vent

Greenish-gray to black legs

Stout yellow bill

Brownish buff overall; blackish-brown, buff-fringed wing coverts

Relatively large; slender neck, body, and legs

White-breasted Waterhen

Yellowish-green bill; base of upper mandible reddish

Black from crown and hindneck to back, wings, and tail

White from forehead, face, and neck to lower belly

Cinnamon lower belly and undertail coverts

Yellow legs

Leucogeranus leucogeranus
白鹤 bái hè
Siberian Crane

L: 125–140 cm
WS: 210–260 cm
Habitat: The crane species with the most specialized habitat requirements: highly dependent on shallow-water wetlands. Stops over on estuarine wetlands during migration; winters on shoals and marshes of seasonally inundated lakes along the Yangtze River.
Behavior: Generally moves about solitarily, in pairs, or in family groups. Often migrates in large flocks; breaks up into smaller groups upon reaching wintering habitats; gathers in large flocks again before migration. Forages by submerging head and bill under water; walks slowly while foraging. Prefers digging roots, tubers, rhizomes, and

sprouts of aquatic vegetation. Shy and vigilant. Readily takes off when startled; flies in V formation or in single file.
Distribution: Uncommon winter visitor and passage migrant regularly found only at a few well-known wintering and stopover sites. Migrates through Northeast and North China, stopping over only briefly at a few large wetlands to rest and refuel; winters mainly in the middle reaches of the Yangtze River, where the largest wintering population is found in Jiangxi (Poyang Lake); a few winter in the lower reaches of the Yangtze River and southeastern and South China. Vagrant in Taiwan.
Voice: Flutelike *koonk, koonk*, higher pitched and more musical than call of Common Crane.

Antigone canadensis
沙丘鹤 shā qiū hè
Sandhill Crane

L: 95–120 cm
WS: 160–210 cm
Habitat: Inhabits open wetlands, shallow-water marshes, and swampy meadows. During migration, prefers stopping over on open marshes near agricultural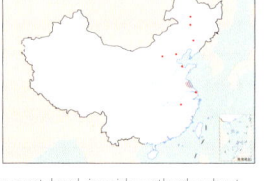
areas. Wintering habitats include coastal and riverside wetlands, wheat fields, and wet grass.
Behavior: Generally moves about in family units. In China, often occurs solitarily or in pairs and mixes often with flocks of Common Crane. Vigilant and shy; immediately takes off when danger approaches. Searches for food on land or shallow coasts.
Distribution: Rare winter visitor to eastern China, with scattered records in Shanghai, Jiangsu, Jiangxi, Shanxi, Shandong, Beijing, Hebei, and Liaoning. Regular in winter only at Yancheng NNR, Jiangsu.
Voice: Loud and deep *gar-oo-oo*, very similar to Common Crane.

Antigone vipio
白枕鹤 bái zhěn hè
White-naped Crane

L: 120–153 cm
WS: 160–210 cm
Habitat: In breeding season, inhabits open wetlands, broad river valleys, or lakeside swampy meadows; occurs on fallow agricultural areas, lakeside marshes,
and open wetlands during migration and winter.
Behavior: Moves about in pairs in breeding season, in family or small flocks after breeding season. Gregarious during migration and winter. Vigilant and afraid of people. Forages mainly by pecking while walking slowly on sand or shingly banks near lakes, or by removing topsoil with bill and pecking at seeds and tubers buried below.
Distribution: Uncommon migrant. Breeds in N Heilongjiang, Jilin, E Inner Mongolia, and (rarely) Liaoning; migrates through Northeast and North China; winters on wetlands in the Yangtze Plain. Occasionally seen along the southeast coast (including Taiwan).
Voice: Sonorous and penetrating trumpeting calls, shorter and generally higher than Common Crane and with marked changes in pitch.

Antigone antigone
赤颈鹤 chì jǐng hè
Sarus Crane

L: 152–176 cm
WS: 220–280 cm
Habitat: Inhabits lowland open grasslands, marshes, and sand or shingly banks near lakes; also forages on farmland.
Behavior: Generally moves about solitarily, in pairs, or in family groups. Walks slowly with head lowered, searching for food on wetlands or in grassy areas.
Distribution: Historical records in SW and S Yunnan; considered extinct in the wild in China.
Voice: Loud, lengthy, sonorous, penetrating trumpeting.

Anthropoides virgo
蓑羽鹤 suō yǔ hè
Demoiselle Crane

L: 90–100 cm
WS: 150–185 cm
Habitat: In breeding season, moves about on open grasslands, swampy meadows, and sandy or shingly banks near streams, lakes, and other wetlands. Can also inhabit
semidesert habitats near water sources. Can migrate through areas above 5000 m.
Behavior: Moves about in pairs in breeding season, in family or small flocks in nonbreeding season. Forms large flocks during migration and flies in V formation. Often forages by walking slowly and elegantly on relatively high-altitude meadows near shallow waters. Shy and vigilant. Keeps away from people and does not mix with other crane species.
Distribution: Uncommon migrant. Breeds in Northeast and northwestern China and on the Ordos Plateau in W Inner Mongolia; winters in S Tibet; locally common in Qinghai on migration. Rare visitor to Hebei, Beijing, Henan, and Shanxi during migration; occasionally wanders to Hubei, Jiangxi, Jiangsu, and Taiwan in nonbreeding season.
Voice: Flight calls similar to Common Crane but higher and drier.

Siberian Crane

Black primaries

Red bare facial skin from base of bill to forehead and behind eye

Light yellow iris

Dark red bill

White overall

Rusty-brown head, neck, and body

Relatively light red bill and legs

imm.

Reddish-pink legs

White-naped Crane

Sparse black downy feathers on red bare skin around head, face, and eye

Grayish-black ear spot

Light yellowish-green bill

Bright white throat, crown, and hindneck

When viewed from distance, white neck contrasts starkly with dark area on face

Slate gray overall

Dull red legs

Sarus Crane

Grayish-white crown

Red bare skin covers head, chin, throat, and upper neck

Grayish-green or yellowish-horn bill

Pale gray under wing, conspicuously blackish-brown primaries

Pale gray body tinged brown

Tail, slender

Pinkish-red legs

Sandhill Crane

Gray overall; wing feathers darker tipped

Relatively prominent red crown

Red bare facial skin on forehead, lores, and crown

Relatively short dark gray bill

Whitish face, chin, and throat

Gray body tinged brown below neck

Smaller than Common Crane

Demoiselle Crane

White ear tufts extend from behind eye to hindneck

Red iris

Black-tipped wing feathers

When viewed from distance, black neck and breast contrast starkly with white belly and under wings

Legs do not project far beyond tail

Black head and neck; black feathers on foreneck extend as chest plumes

Relatively small

Elongated secondaries and tertials cover tail

Grus japonensis

丹顶鹤 dān dǐng hè

Red-crowned Crane

L: 138–152 cm
WS: 220–250 cm
Habitat: Prefers clean and open wetlands; one of the most sensitive indicator species of wetland environmental changes. Nests on marshes or wetlands with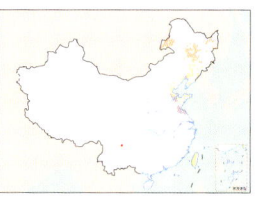
abundant reeds, sedges, and cattails in breeding season. Moves about near estuaries, freshwater wetlands, coastal saltwater marshes, tidal flats, farmland, and fallow agricultural areas during migration and winter.
Behavior: Generally moves around in pairs or family groups on breeding and wintering grounds. Forms relatively large flocks consisting of several dozen families in migration season; flocks can sometimes grow to a few dozen or over a hundred individuals. Flies in V formation. Walks slowly while gently pecking at food in water or on the ground.
Distribution: Uncommon migrant. Breeds in Northeast China and Inner Mongolia (Xilingol); winters at the Yellow River Delta and Jiangsu (Yancheng); migrates through Northeast and North China. Occasionally wanders to southeastern and South China (including Taiwan) in nonbreeding season.
Voice: Sonorous and penetrating trumpeting calls often slightly longer and more tremulous than Common Crane.

Grus grus

灰鹤 huī hè

Common Crane

L: 95–125 cm
WS: 180–200 cm
Habitat: Breeds on various kinds of shallow-water wetlands, including marshes and meadows; especially prefers open lakes and reedy marshes with abundant aquatic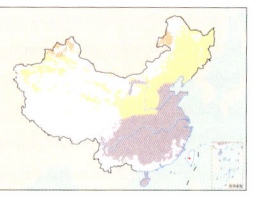
vegetation. Inhabits mainly rivers, lakes, reservoirs, or coasts during migration and winter; often forages on open farmland or fallow agricultural areas.
Behavior: Moves about in pairs or family groups of 5–10 individuals in breeding season; can form flocks containing up to 40–50 individuals in migration season and hundreds of individuals on wintering ground. Flies in V formation. Vigilant, afraid of people; one individual always stays alert when flocks move about and forage. Can forage on land and in water.
Distribution: Common migrant. Breeds in Northeast and northwestern China; winters in most of North, Central, and southwestern China. Occasionally occurs in East and South China (including Taiwan).
Voice: Sonorous, persistent, and penetrating trumpeting calls; like many cranes, juvenile makes higher-pitched, very different, plaintive piping.

Grus monacha

白头鹤 bái tóu hè

Hooded Crane

L: 91–100 cm
WS: 160–180 cm
Habitat: Prefers nesting in swamps with sparse larches and bushes, as well as forested wetlands at high altitudes. Diverse wintering habitats, including riverbanks,
shallow lakes, marshes, and paddy fields.
Behavior: Generally moves about in pairs or family groups; sometimes also seen singly or in loose flocks consisting of different family groups. Often walks and digs for food in mud. Also moves to nearby farmland to forage in winter. Roosting and foraging locations are relatively fixed if no disturbance occurs. Vigilant; often raises head to look around when moving and foraging. When danger approaches, takes off, circles in sky, and calls incessantly. In flocks, often flies in V formation or single file. Prefers digging young leaves, roots, tubers, and seeds of aquatic vegetation in shallow water.
Distribution: Uncommon migrant. A few breed in Northeast China and Inner Mongolia (Xilingol). Migrates southward to North, Central, and eastern China in winter. Occasionally occurs in Taiwan.
Voice: Sonorous and loud *kurrk*.

Grus nigricollis

黑颈鹤 hēi jǐng hè

Black-necked Crane

L: 115 cm
WS: 180–200 cm
Habitat: Inhabits boggy meadows, wetlands, and pastures at altitudes of 2950–4900 m; can descend to roughly 1375 m in winter but generally occurs on open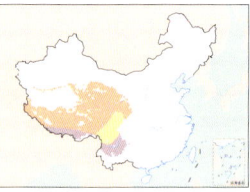
farmland, river valleys, or grassy wetlands above 3000 m.
Behavior: Highly adapted to high-altitude and cold environments, the world's only crane species that breeds on high plateaus. Generally moves around in pairs or family groups in breeding season; forms flocks of a few dozen individuals during migration and winter. In flocks, flies in V formation or single file. Spends most time foraging from dawn to dusk. Usually roosts at noon in marshes or shallow lakeshores, standing on one leg and tucking bill under back feathers. Vigilant and afraid of people; immediately takes off when approached.
Distribution: Locally common migrant. Breeds on Qinghai-Tibet Plateau and in S Gansu, SE Xinjiang, and W and NW Sichuan; winters on the Yunnan-Guizhou Plateau and in the valley along the middle reaches of the Yarlung Zangbo River.
Voice: Sonorous and penetrating trumpeting calls, sounding like *gorr-gorr-gorr* or *gage-gage*, similar to, but higher pitched than, Common Crane.

Red-crowned Crane

Black neck, secondaries, and tertials contrast strongly with otherwise-white body

Bright red crown

Dark brown iris

Secondaries and tertials black; rest of body white

Hooded Crane

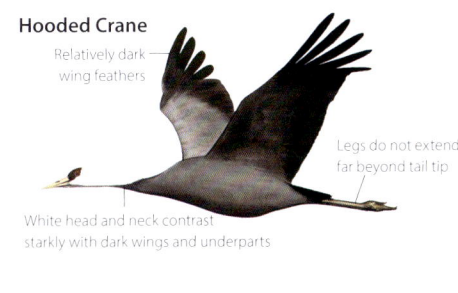

Relatively dark wing feathers

Legs do not extend far beyond tail tip

White head and neck contrast starkly with dark wings and underparts

Black forehead and lores

Small patch of red bare skin on central forehead

Deep red iris

Light yellowish-green bill

Rest of head and neck white

Rest of body slate gray

Common Crane

Dark-tipped wing feathers

Gray body

Yellow or orange iris

Red crown

Yellow bill

White band extends from behind eye to neck

Black forehead, lores, throat, foreneck, and back of head

Gray overall

Black-necked Crane

Black wing feathers and tail feathers contrast starkly with grayish-white coverts

Deep red crown and lores

Light yellow iris, with small white patch behind eye

Black head and neck

Relatively short neck

Dark wing and tail feathers clearly contrast with rest of body

Rest of body white tinged gray

鸻形目　CHARADRIIFORMES

A very heterogeneous group consisting mostly of waterbirds that associate closely with wetlands, including buttonquails, shorebirds, gulls, and alcids. Plumage similar between sexes in most species. In some species, plumage differs between sexes only during breeding season, and a handful of species exhibit marked reverse sexual dichromatism. Most species have drab plumage, dominated by black, white, gray and brown. Bill shape highly diverse, varying from short and thick to long and thin, from recurved to decurved to spatulate. Neck short to medium long. Usually very good at flying, with long and pointed wings. Legs short or long. Feet in some species partially webbed. Tail short and rounded or long and narrow. Some species good at walking and running, while some good at swimming. Inhabit almost all types of wetland habitats. Nest mostly on ground, water, or rocky areas. Omnivorous; feed mostly on mollusks, crustaceans, insects, and fish. Most species are migratory.

　　Nineteen families, 88 genera, and 386 species recognized worldwide, with a cosmopolitan distribution. Thirteen families, 49 genera, and 138 species recorded in China, ubiquitously distributed nationwide.

三趾鹑科
Turnicidae

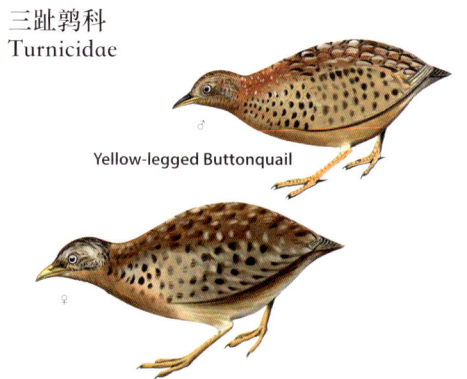

Yellow-legged Buttonquail

蛎鹬科
Haematopodidae

Eurasian Oystercatcher

反嘴鹬科
Recurvirostridae

Pied Avocet

石鸻科
Burhinidae

Eurasian Thick-knee

鹮嘴鹬科
Ibidorhynchidae

Ibisbill

鸻科
Charadriidae

Kentish Plover

彩鹬科
Rostratulidae

Greater Painted-Snipe

水雉科
Jacanidae

Pheasant-tailed Jacana

燕鸻科
Glareolidae

Oriental Pratincole

鹬科（丘鹬科）
Scolopacidae

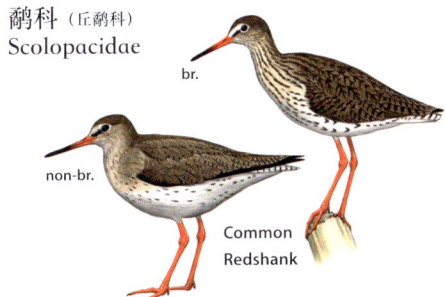

br.

non-br.

Common
Redshank

鸥科
Laridae

br.

br.

non-br.

non-br.

Black-headed Gull

贼鸥科
Stercorariidae

br.

non-br.

Pomarine Jaeger

海雀科
Alcidae

br.

Ancient Murrelet

non-br.

Turnix sylvaticus

林三趾鹑 lín sān zhǐ chún

Small Buttonquail

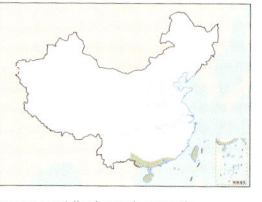

L: 13–16 cm
Habitat: Inhabits warm and dry scrubland, tussocks, meadows, and farmland on plains.
Behavior: Generally occurs solitarily or in pairs. Extremely secretive. Relatively active at dusk and dawn. Good at running; often runs rapidly through tussocks. Sometimes flushes with very fast wingbeats right at one's feet, flying only a short distance before landing among bushes or tussocks and running away.
Distribution: Rare resident in Guangdong, Guangxi, Hainan, and Taiwan.
Voice: Female's advertising call, given mostly at dusk and dawn, is unobtrusive, very low-pitched, lengthy, flat hum, hoooo, repeated every 1–3 sec; can continue for up to 30 sec. Very similar to Yellow-legged Buttonquail, but individual notes slightly longer and with slight change in pitch at end and strophes not increasing in volume.

Turnix tanki

黄脚三趾鹑 huáng jiǎo sān zhǐ chún

Yellow-legged Buttonquail

L: 15–18 cm
Habitat: Inhabits deserted fields, grassy areas, farmland, and forest-edge bushes on plains and hills below 1200 m. Nests on shallow depressions in dense grass.
Behavior: Generally occurs solitarily or in pairs. Extremely secretive. Relatively active at dusk and dawn. Good at running; often runs rapidly through tussocks. Sometimes flushes with very fast wingbeats right at one's feet, flying only a short distance before landing among bushes or tussocks and running away.
Distribution: Widespread but nowhere common. Occurs from Northeast to South and southwestern China. Summer breeder in northern China; summer breeder, passage migrant, or winter visitor to areas south of the Yangtze River.
Voice: Female gives very low-pitched, mellow series of three to eight hum calls, whooom-whooom, which last for 10–15 sec, similar to Small Buttonquail and the sound of bellows. Male gives a much quieter, easily overlooked pook–pook.

Turnix suscitator

棕三趾鹑 zōng sān zhǐ chún

Barred Buttonquail

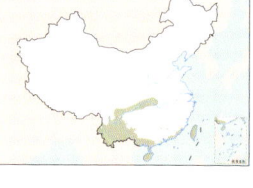

L: 14–17 cm
Habitat: Inhabits sparsely wooded areas at forest edges, scrubland, grassy slopes, paddy fields, and mountain-slope tussocks and bushes near farmland below 300 m. Also occurs on sugarcane, tea, and coffee plantations near villages.
Behavior: Generally occurs solitarily or in pairs. Extremely secretive. Relatively active at dusk and dawn. Good at running; often runs rapidly through tussocks. Sometimes flushes with very fast wingbeats right at one's feet, flying only a short distance before landing among bushes or tussocks and running away.
Distribution: T. s. blakistoni is an uncommon resident in W and SW Yunnan, S Guizhou, Hubei, Chongqing, Sichuan, Fujian, Guangdong, Hong Kong, C Guangxi, and Hainan. T. s. rostratus endemic to Taiwan; locally common.
Voice: Female often calls at night during breeding season, repeating fast, far-carrying, foghorn-like whoo-whoo-whoo, initially on even pitch but gradually rising and becoming louder, although tempo remains same, before ending abruptly. Far more "pulsed," with each shorter note repeated more rapidly than other buttonquail.

Burhinus oedicnemus

石鸻 (欧石鸻) shí héng (ōu shí héng)

Eurasian Thick-knee

L: 38–45 cm
Habitat: Seen in tamarisk bushes, desert scrubland, abandoned agricultural lands, and shingly areas along rivers near reservoirs and lakes.
Behavior: More active at dusk and night. Vigilant and shy; quickly runs and takes cover in bushes when disturbed. Does not fly readily.
Distribution: Uncommon breeder in N Xinjiang; occasional in SE Tibet in winter; vagrant in Guangdong.
Voice: A series of rapid curlew-like koo'ky' klyp koo'ky' klyp, given mostly at night.

Esacus recurvirostris

大石鸻 dà shí héng

Great Thick-knee

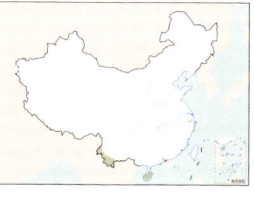

L: 53–57 cm
Habitat: Occurs on river flats or coastal beaches.
Behavior: Rests with its well-camouflaged plumage on riverbanks or beaches during the day; moves about in small flocks at dusk, dawn, and night.
Distribution: Recorded mainly in Yunnan (Tengchong, Jinghong, and Luxi) and Hainan (Haikou). Rare. Recorded once in Hong Kong.
Voice: Lengthy sequence of shrill, ascending, loud whistles given during display, often intensifying and accelerating.

Small Buttonquail

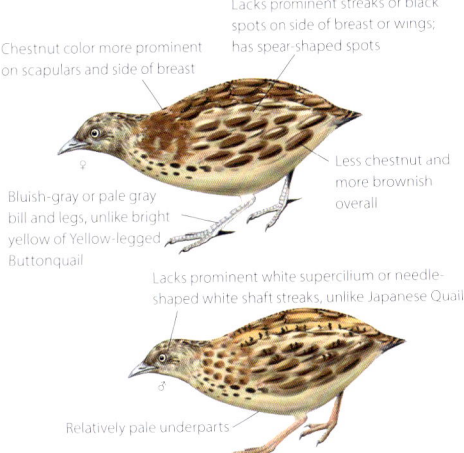

Chestnut color more prominent on scapulars and side of breast

Lacks prominent streaks or black spots on side of breast or wings; has spear-shaped spots

Less chestnut and more brownish overall

Bluish-gray or pale gray bill and legs, unlike bright yellow of Yellow-legged Buttonquail

Lacks prominent white supercilium or needle-shaped white shaft streaks, unlike Japanese Quail

♂

Relatively pale underparts

Barred Buttonquail

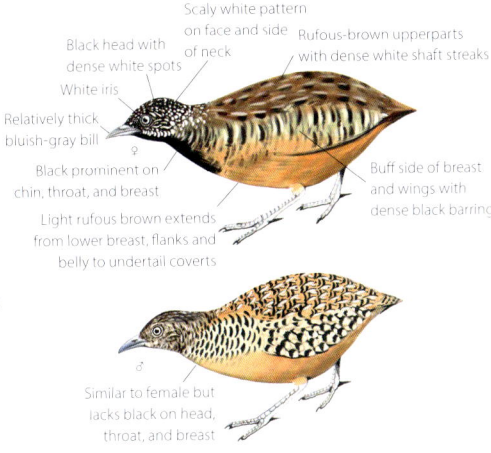

Scaly white pattern on face and side of neck

Black head with dense white spots

Rufous-brown upperparts with dense white shaft streaks

White iris

Relatively thick bluish-gray bill

Black prominent on chin, throat, and breast

Buff side of breast and wings with dense black barring

Light rufous brown extends from lower breast, flanks and belly to undertail coverts

♂

Similar to female but lacks black on head, throat, and breast

Yellow-legged Buttonquail

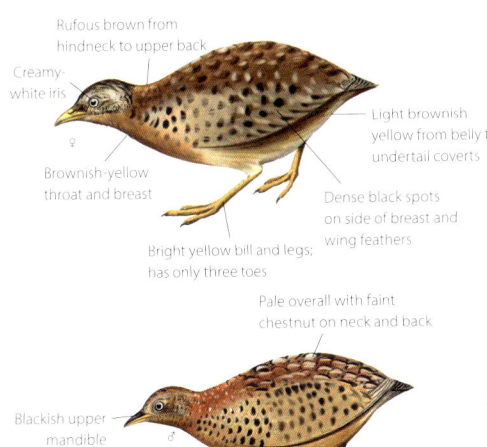

Rufous brown from hindneck to upper back

Creamy-white iris

♀

Light brownish yellow from belly to undertail coverts

Brownish-yellow throat and breast

Dense black spots on side of breast and wing feathers

Bright yellow bill and legs; has only three toes

Pale overall with faint chestnut on neck and back

Blackish upper mandible

♂

Relatively small

Eurasian Thick-knee

Yellow iris

White wing stripe bordered black

Great Thick-knee

Black wing feathers with large white patch

Yellow iris

Large head

Black lateral crown stripes, eye stripe, and short moustachial stripe contrast strongly with white supercilium

Thick bill tilted up slightly

Black and white stripes on wings

Haematopus ostralegus

蛎鹬 lì yù

Eurasian Oystercatcher　

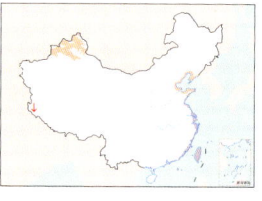

L: 40–48 cm

Habitat: In breeding season, occurs at estuaries, as well as field bunds and grassy areas near lakes or coasts. In winter, inhabits rocky, muddy, or sandy coasts and estuarine areas. Also occurs on inland lakes.

Behavior: Generally walks slowly on intertidal flats or shallow waters in flocks. Forages by probing with bill, or searches for food by sight, mainly feeding on large bivalves. Also feeds on worms and crabs.

Distribution: *H. o. osculans* breeds in Northeast and North China; a few breeding records on islands off China's east coast. Winters on the coasts of the Yellow Sea, East China Sea, and South China Sea. Locally common; occasionally recorded in wetlands on the Yangtze Plain. The population that breeds in Xinjiang and migrates through Tibet is potentially *H. o. longipes*.

Voice: Short, sharp, penetrating whistle, *weep* or *wei-wei*.

Ibidorhyncha struthersii

鹮嘴鹬 huán zuǐ yù

Ibisbill　

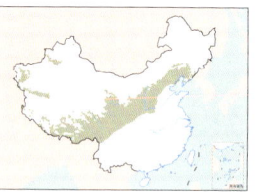

L: 39–41 cm

Habitat: Inhabits clean, pebbly streams; occasionally seen on wet grasslands.

Behavior: Moves about solitarily or in flocks of a dozen individuals. Forages mainly in the morning by walking slowly in streams; blends well with surrounding pebbles. Takes off and calls when alarmed. Makes altitudinal movements; moves deeper into mountains in summer.

Distribution: Occurs from S Liaoning, along the Yan Mountains, Taihang Mountains, and Qinling Mountains, southwest through Qinghai and Sichuan to W Yunnan, S Tibet, and W Xinjiang. Local and uncommon.

Voice: Series of short, penetrating, torrent-adapted, shrill whistles, *ji-ji-ji-ji-ji-jit*.

Himantopus himantopus

黑翅长脚鹬 hēi chì cháng jiǎo yù

Black-winged Stilt　

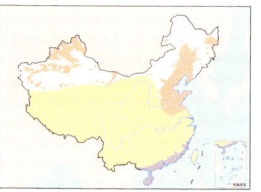

L: 35–40 cm

Habitat: Inhabits freshwater environments; prefers lakes and marshes. Also occurs on coastal saltpans and shrimp ponds. Does not show up at sea.

Behavior: Occurs in small to large flocks. Can forage in relatively deep waters. Cannot swim. Can also forage at night. Breeds in colonies. Parents fake injury to fool predators.

Distribution: Breeds in Northeast, northwestern, and North China. Migrates through most of the country; winters in South China. Common.

Voice: Loud and noisy during breeding season: *kyip* or *kik*, like Pied Avocet.

Recurvirostra avosetta

反嘴鹬 fǎn zuǐ yù

Pied Avocet　

L: 42–45 cm

Habitat: Inhabits lakes and marshes; nests on sandbanks and islets. Can also swim at sea during migration, looking gull-like from a distance.

Behavior: Gregarious. Walks swiftly, good at swimming. Forages by sweeping bill through water from side to side; can also upend to feed in water.

Distribution: Occurs everywhere except for the northern parts of Northeast China and the Qinghai-Tibet Plateau. Breeds in northwestern China and western Northeast China; winters in the Yangtze Plain and areas to its south; migrates through most of China. Very common.

Voice: Sharp, loud, piping whistles, *wii-wii*.

Eurasian Oystercatcher

White wing bar reaches primaries

Stout bill

Black-winged Stilt

Head pied. Black head pattern varies considerably; some individuals have all-white head

Long red legs

Ibisbill

Prominent white wing bar

Decurved red bill

Black band on upper breast

Pied Avocet

Contrasting black and white plumage

Distinctively upturned bill

White eye ring

Vanellus vanellus
凤头麦鸡 fēng tóu mài jī
Northern Lapwing   NT

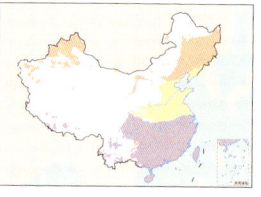

L: 28–31 cm
Habitat: Often occurs on farmland, lakes, and near marshes; does not show up near seas. Breeds in open areas without tall vegetation, such as farmland, swampy grasslands, and flat areas near wetlands.
Behavior: Generally gregarious. Runs to catch prey once detected; also forages at night. Wings appear broad in flight. When flying in flocks, flight appears undulating. Calls in flight.
Distribution: Breeds in Northeast and northwestern China; migrates through most regions except for the western part of the Qinghai-Tibet Plateau; winters in areas to the south of the Qinling-Huaihe Line.
Voice: Varied and often loud. Pleasant song, *willuch-o-weep, weep-weep*, given in exaggerated display flight and often accompanied by wing claps. Other calls include a plaintive, nasal *zi-aa*, given in alarm.

Vanellus duvaucelii
距翅麦鸡 jù chì mài jī
River Lapwing   NT

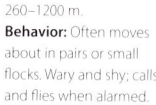

L: 29.5–31.5 cm
Habitat: Inhabits sand and shingle riverbanks and nearby farmland. Occurs at altitudes of 260–1200 m.
Behavior: Often moves about in pairs or small flocks. Wary and shy; calls and flies when alarmed.
Distribution: Previously in SE Tibet, Guangxi, Hainan, Fujian, Guangdong, Guizhou, and Yunnan; recorded only in Yunnan in recent years.
Voice: A loud, penetrating, tern- or Black-winged Stilt–like *di di di*, repeated faster when alarmed.

Vanellus cinereus
灰头麦鸡 huī tóu mài jī
Gray-headed Lapwing   LC

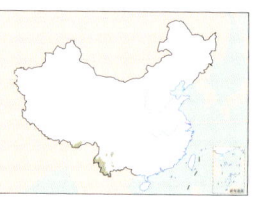

L: 34–37 cm
Habitat: Open areas near water, marshes, river flats, and wet grasslands; also occurs on farmland.
Behavior: Moves about in small flocks, forages near water, and stands on field embankments to rest.
Flight appears laborious and slow. Aggressive during breeding season; sometimes even attacks and kills chicks of other species.
Distribution: Breeds in the vast areas from eastern and southern parts of Northeast China, south to the Yangtze River Watershed and west to the eastern edge of the Qinghai-Tibet Plateau. Migrates through most of China except for Xinjiang and Tibet; winters in South China.
Voice: Noisy, rapid *kik–kik* and other notes reminiscent of Black-winged Stilt.

Vanellus indicus
肉垂麦鸡 ròu chuí mài jī
Red-wattled Lapwing   LC

L: 32–35 cm
Habitat: Inhabits open habitats, including farmland, paddy fields, marshes, river flats, cultivated lands, and grassy areas.
Behavior: Often moves about in pairs or small flocks. Wary and shy; calls and flies when alarmed. Active by day and night.
Distribution: *V. i. indicus* is rare visitor to Xinjiang; *V. i. atronuchalis* locally common in Yunnan.
Voice: Intense, loud, and strident "didhe-do-it."

Vanellus gregarius
黄颊麦鸡 huáng jiá mài jī
Sociable Lapwing   CR

L: 27–30 cm
Habitat: Breeds on dry and short grass; winters on dry farmland and deserted fields.
Behavior: Often gathers in small flocks of a few to dozens of individuals. In breeding season, prefers areas with sparse vegetation and abundant feces of herbivorous mammals.
Distribution: Historically, a summer breeder in NW Xinjiang, but no record for many years. Vagrant in Hebei (Beidaihe) in 1998.
Voice: Rarely calls outside breeding season. Gives short, quiet, and unobtrusive *kech–kech–kech* in breeding season.

Vanellus leucurus
白尾麦鸡 bái wěi mài jī
White-tailed Lapwing   LC

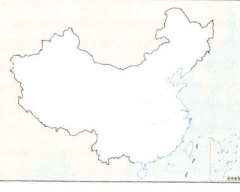

L: 28–29 cm
Habitat: Occurs on shallow lakes and slow-flowing rivers; prefers habitats with dense vegetation.
Behavior: Shy and vigilant.
Distribution: Vagrant recorded once in Xinjiang (Kashgar) in 2012.
Voice: Nasal, often quiet *pi-wheet* or *pi-ee-whee-it*, vaguely reminiscent of Northern Lapwing.

Northern Lapwing

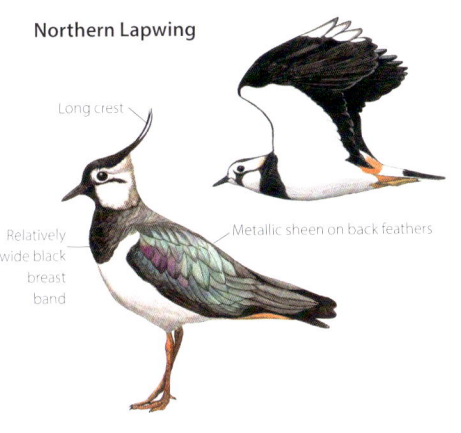

Long crest

Relatively wide black breast band

Metallic sheen on back feathers

Red-wattled Lapwing

Red wattles at lores

White patch extends from cheek to side of neck

River Lapwing

Black carpal spurs, sometimes covered by feathers

Black crown, nape, and throat

Brown back and wings

Sociable Lapwing

White-mottled black crown

♀ br.

White forehead, prominent white supercilium

Dark gray belly

♂ br.

Black belly

Streaks on throat

Scaly pattern on breast

non-br.

Gray-headed Lapwing

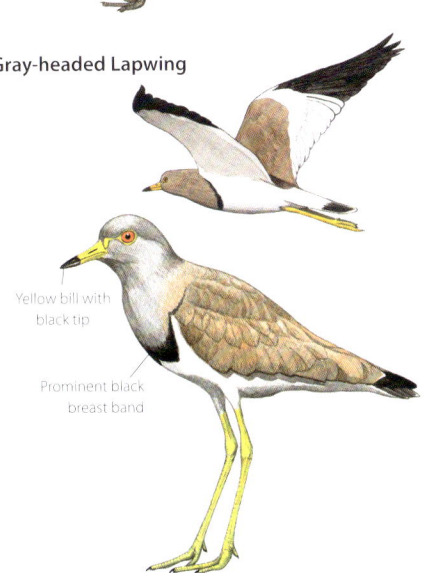

Yellow bill with black tip

Prominent black breast band

White-tailed Lapwing

All-white tail

Long legs extend well beyond tail in flight

Black bill

Pluvialis squatarola

灰鸻 (灰斑鸻) huī héng (huī bān héng)

Black-bellied Plover

L: 27–31 cm

Habitat: During migration, generally occurs gregariously on coastal tidal flats or paddy fields near coasts; a few also occur on inland wetlands. Breeds on Arctic tundra and coasts in summer. Prefers sandbars, estuaries, and tidal flats in winter.

Behavior: Generally in small flocks; can form large flocks during migration and winter. Often occurs on coastal tidal flats with other shorebirds. Makes relatively slow motions while foraging, walking a few steps to peck at food and raising head to look around.

Distribution: Widespread migrant throughout China except for the Qinghai-Tibet Plateau; a relatively common passage migrant on the east coast; uncommon on inland wetlands. Winters in coastal areas in southeastern and South China; nonbreeding summer visitors are seen in parts of the east coast.

Voice: Distinctive melancholy whistle *plee-ii-wheet*.

Pluvialis apricaria

欧金鸻 ōu jīn héng

European Golden-Plover

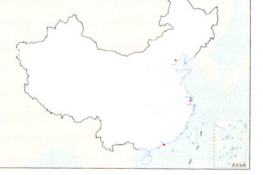

L: 25–28 cm

Habitat: Inhabits grassy areas, saltpans, and coastal flats during migration; sometimes also occurs on sandbars, estuaries, and intertidal zones.

Behavior: Usually mixes with lapwings, moving about on farmland and in grassy areas. Forages with relatively fast motions; when food is spotted, quickly walks a few steps and lowers head to peck at food.

Distribution: Vagrant in Hebei, Shanghai, and Hong Kong. One wintering record in Shanghai in February 2020.

Voice: Plaintive descending whistle *peuuu*.

Pluvialis fulva

金鸻 (金斑鸻) jīn héng (jīn bān héng)

Pacific Golden-Plover

L: 23–26 cm

Habitat: Similar to European Golden-Plover.

Behavior: Similar to European Golden-Plover. Timid; takes off immediately when in danger. Calls in flight.

Distribution: Widespread throughout China. A common passage migrant on the east coast; winters on the south coast, including Hainan and Taiwan. Nonbreeding summer visitors are seen in some parts of the east coast.

Voice: Loud ringing *chu-ii-uu* and Spotted Redshank–like *chee-vik*.

Pluvialis dominica

美洲金鸻 měi zhōu jīn héng

American Golden-Plover

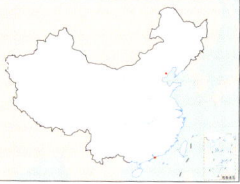

L: 24–28 cm

Habitat: Found on cultivated lands, grassy areas, estuaries, and intertidal flats during migration.

Behavior: Similar to Pacific Golden-Plover.

Distribution: Rare vagrant. Recorded in Hebei and Hong Kong during migration.

Voice: Flat disyllabic whistle, *klu-ee*.

Charadrius mongolus

蒙古沙鸻 měng gǔ shā héng

Lesser Sand-Plover

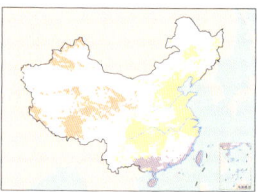

L: 18–21 cm

Habitat: Inhabits coastal intertidal flats, estuaries, and sandbars; sometimes also seen on inland rivers, marshes, and saltpans.

Behavior: Gregarious during migration and winter; often mixes with other shorebirds on coastal intertidal flats. Generally roosts in ponds inside seawalls during high tide, more active during low tide. Often walks (rather than runs) when foraging on intertidal flats, appearing less active than Kentish Plover.

Distribution: During migration, *C. m. mongolus* occurs across eastern China, including Taiwan and Hainan; *C. m. stegmanni* recorded in E Inner Mongolia and Taiwan; may also occur on the east coast during migration; both subspecies have wintering populations on the southeast coast and small nonbreeding summering population on the east coast. *C. m. pamirensis* in NW Xinjiang; *C. m. atrifrons* on the southern part of the Qinghai-Tibet Plateau; and *C. m. schaeferi* in Xinjiang and the eastern part of the Qinghai-Tibet Plateau.

Voice: Various short chattering notes: *cherk*, *trrp*, etc., all very similar to Greater Sand-Plover.

Charadrius leschenaultii

铁嘴沙鸻 tiě zuǐ shā héng

Greater Sand-Plover

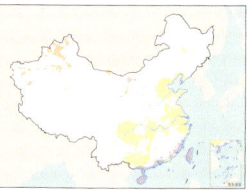

L: 22–25 cm

Habitat: Similar to Lesser Sand-Plover.

Behavior: Similar to Lesser Sand-Plover. Relatively wary and afraid of people. Sometimes runs on intertidal flats. Forages by using the foot-trembling technique to disturb prey and capture it.

Distribution: *C. l. leschenaultii* are seen everywhere except for southwestern China, mainly as a migrant; wintering population on the southeast coast and small breeding population on the east coast. *C. l. scythicus* may occur in W Xinjiang.

Voice: See Lesser Sand-Plover.

Black-bellied Plover

non-br.

Jet-black patch on breast and belly

br.

Grayish-brown streaks on breast and belly

non-br.

American Golden-Plover

Gold-spangled back

br.

br.

Jet-black patch on breast and belly

Black areas on flanks and undertail coverts more extensive than on Pacific Golden-Plover

non-br.

non-br.

European Golden-Plover

Bright white under wing in flight, compared to Pacific Golden-Plover

br.

Gold-spangled back

non-br.

Jet-black patch on breast and belly

br.

non-br.

Lesser Sand-Plover

Black forehead

br.

C. m. atrifrons

White forehead

non-br.

Relatively wide rufous breast band

br.

C. m. mongolus

Breast tinged brown

non-br.

juv.

Pacific Golden-Plover

Gray under wing, compared to European Golden-Plover

br.

Gold-spangled back

non-br.

Jet-black breast and belly

br.

non-br.

Greater Sand-Plover

non-br.

non-br.

Long heavy bill

Relatively narrow rufous breast band

br.

Usually pale legs

159

Charadrius hiaticula

剑鸻 jiàn héng

Common Ringed Plover　🄻🄲

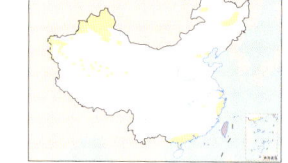

L: 18–20 cm
Habitat: Inhabits islets, coastal intertidal flats, estuaries, lake flats, reservoirs, inland rivers, farmland, swampy meadows, and grassy areas.
Behavior: Solitary or in small flocks. Vigilant; not easily approached.
Distribution: Regular but uncommon in Xinjiang; rare passage migrant or winter visitor elsewhere, with records from Heilongjiang, Beijing, Hebei, NE Inner Mongolia, Tibet, Qinghai, Jiangsu, Shanghai, Zhejiang, Jiangxi, Guangdong, Guangxi, Hong Kong, and Taiwan.
Voice: In flight, slightly melancholy, disyllabic, upward-inflected whistle, *tu–weep*.

Charadrius placidus

长嘴剑鸻 cháng zuǐ jiàn héng

Long-billed Plover　🄻🄲

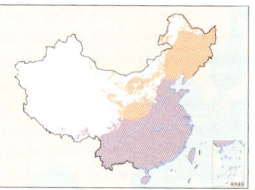

L: 19–21 cm
Habitat: Prefers shingly areas along mountain streams; also seen on rivers, lakes, flooded fields, marshes, coastal areas, or intertidal flats during migration.
Behavior: Solitary or in pairs; rarely mixes with other shorebirds. Motions slower than other small plovers; sometimes stays still and hides among rocks. Takes off suddenly when approached.
Distribution: Widespread throughout China except for Xinjiang. Mainly an uncommon migrant. Generally breeds in northern China and winters in southern China, but some individuals in northern China are resident.
Voice: Clear piping *piuw*, very similar to Little Ringed Plover.

Charadrius dealbatus

白脸鸻 bái liǎn héng

White-faced Plover　🄽🅁

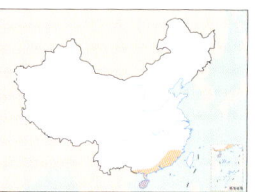

L: 15–17 cm
Habitat: Inhabits coasts and sandy intertidal flats.
Behavior: Moves swiftly; often runs for a few meters, stops briefly, and runs again. Prefers foraging and nesting on sandy coastal beaches.
Differs from the closely related Kentish Plover, which prefers foraging on coastal mudflats and nesting along seawalls, inland saltpans, and salt lakes.
Distribution: Was once considered a subspecies of Kentish Plover; recognized as a separate species only recently. Occurs on the southeast coast and Hainan.
Voice: Similar to Kentish Plover.

Charadrius alexandrinus

环颈鸻 huán jǐng héng

Kentish Plover　🄻🄲

L: 15–17 cm
Habitat: Inhabits various kinds of wetlands; often seen on coastal habitats, estuaries, sandbars, inland rivers, lakes, marshes, and saltpans.
Behavior: Gregarious in nonbreeding season, sometimes forming flocks containing up to a hundred individuals; often mixes with other small shorebirds. Moves swiftly; often runs for a few meters, stops briefly, and runs again. Forages by day and night; movements mainly shaped by tides. Roosts in ponds inside seawalls during high tide; forages on intertidal flats during low tide.
Distribution: Widespread across various wetland habitats except for the Qinghai-Tibet Plateau. Some are resident in southeastern China; individuals breeding in Northeast, northwestern, and North China generally winter in southern China.
Voice: Varied—rising nasal *pwid* and various dry rattled notes, *drrrt*.

Charadrius dubius

金眶鸻 jin kuàng héng

Little Ringed Plover　🄻🄲

L: 14–17 cm
Habitat: Prefers freshwater environments. Often seen on marshes, flooded fields, and riverbanks; sometimes also seen at estuaries and saltpans. During migration, occurs on rivers, lakes, flooded fields, marshes, coastal areas, or intertidal flats.
Behavior: Solitary or in pairs. Occasionally gathers in small flocks; seldom mixes with other shorebirds. Moves swiftly; often runs for a few meters, stops briefly, and runs again. When foraging, sometimes uses the foot-trembling technique to disturb shallow water and picks up prey that comes to the surface.

Distribution: Widespread across freshwater wetlands in China. Summer breeder in Northeast, northwestern, North, and South China, winter visitor in southwestern and southeastern China, but some individuals in northern China are resident. *C. d. curonicus* occurs everywhere except for Yunnan and Guizhou; *C. d. jerdoni* in SE Tibet, Yunnan, and SW Sichuan.
Voice: Piping *wei you*, very similar to Long-billed Plover.

Common Ringed Plover

Prominent white wing bar

br.

Lacks gold eye ring

Orange-yellow bill with black tip

juv.

Complete broad black breast band

br.

Orange legs

White-faced Plover

Plumage and especially crown relatively light colored

Lacks black eye stripe at lores, compared to breeding plumage of male Kentish Plover

Relatively long bill and tarsi

♂ br.

Rufous crown

♀

Kentish Plover

♂ br.

Brown crown

Conspicuous white hind collar, compared to sand-plovers

Rusty crown

♂ br.

♀ br.

Long-billed Plover

br.

Inconspicuous gold eye ring

Relatively long black bill

br.

Inconspicuous gold eye ring

Relatively long black bill

juv.

Little Ringed Plover

br.

Prominent gold eye ring

Black eye stripe

Black breast band

br.

Brown eye stripe and breast band

Relatively slim bill

non-br.

161

Charadrius asiaticus

红胸鸻 hóng xiōng héng

Caspian Plover

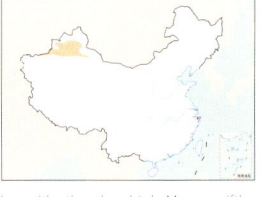

L: 18–23 cm
Habitat: Inhabits deserts, semideserts, saline-alkaline plains, and open steppes; sometimes also seen on lakes, riverbanks, and marshes.
Behavior: Moves about in pairs or small flocks in nonbreeding season; seldom mixes with other shorebirds. Moves swiftly; often runs quickly on the ground. Nests on open saline-alkaline plains or steppes; especially prefers saline-alkaline soils on mountainside deserts or semideserts.
Distribution: Occurred on the Tian Shan Mountains and Junggar Basin in Xinjiang as a rare breeder, but no record in the past 20 years. A vagrant was recorded in Zhejiang in spring 2008.
Voice: A varied nasal *chup* and soft, sharper *kik* notes.

Charadrius veredus

东方鸻 dōng fāng héng

Oriental Plover

L: 22–26 cm
Habitat: Prefers open grasslands; sometimes also seen on freshwater lakes, riverbanks, and marshes.
Behavior: Moves about in small flocks in nonbreeding season; seldom mixes with other shorebirds. Generally runs and forages swiftly in grassy areas or shallow waters. Runs and flies quickly; often makes abrupt turns in flight.
Distribution: Summer breeder in Northeast China; migrates through eastern China; a few individuals winter in Taiwan. Uncommon.
Voice: Metallic *jip*, occasionally doubled. Migrants also occasionally sing—peculiar, mechanical hissing.

Charadrius morinellus

小嘴鸻 xiǎo zuǐ héng

Eurasian Dotterel

L: 20–24 cm
Habitat: In breeding season, inhabits subalpine wetlands and grasslands at altitudes of 2000–2600 m.
Behavior: Gregarious. Produces whistling sounds in flight.
Distribution: A rare breeder on the Altai Mountains, Tarbagatai Mountains, and Saur Mountains in NW Xinjiang. Migrants rarely recorded in NE Inner Mongolia, N Heilongjiang, and the Tian Shan Mountains of Xinjiang.
Voice: Alarm calls are a clear and rhythmic *weet–weeh*; also gives deep *brroot* in flight.

Rostratula benghalensis

彩鹬 cǎi yù

Greater Painted-Snipe

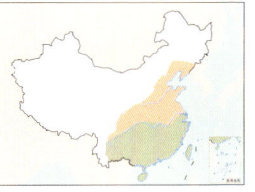

L: 23–28 cm
Habitat: Inhabits low-altitude wetlands, such as reedy ponds, marshes, wet grass, paddy fields, and mangroves.
Behavior: Moves about solitarily or in family units during breeding season; forms small flocks after breeding. Flicks tail when walking. Stays still or immediately takes cover in vegetation when alarmed; not easily flushed. When in danger, adult spreads wings in defense. Polyandrous; male responsible for incubation and chick rearing. Omnivorous. Also forages at night.
Distribution: Summer breeder in northern China, north to the Bohai Sea Rim and west to the Sichuan Basin. Resident in regions south of the Yangtze River. Locally common.
Voice: In breeding season, female gives mellow Schrenck's Bittern–like hooting, which can continue throughout the night; otherwise generally silent.

Hydrophasianus chirurgus

水雉 shuǐ zhì

Pheasant-tailed Jacana

L: 39–58 cm
Habitat: Inhabits freshwater ponds, reservoirs, lakes, marshes, and paddy fields with abundant floating vegetation; especially prefers wetlands with water chestnuts (*Eleocharis*) and prickly waterlilies (*Euryale ferox*). Occasionally occurs in coastal areas during migration.
Behavior: Moves about in small flocks. Can walk elegantly on leaves of floating vegetation. Polyandrous; male responsible for incubation and chick rearing. Nests on floating vegetation. Omnivorous.
Distribution: Locally common summer breeder or passage migrant. Breeds in North, Central, East, South, and southwestern China, including Hainan and Taiwan. Occasionally winters in southern China.
Voice: Sometimes loud, repeated, peculiar, catlike *chereeow*.

Metopidius indicus

铜翅水雉 tóng chì shuǐ zhì

Bronze-winged Jacana

L: 28–31 cm
Habitat: Inhabits small ponds or lakes with abundant floating vegetation.
Behavior: Similar to Pheasant-tailed Jacana.
Distribution: Occurs as a rare resident or migrant in S and SW Yunnan and S Guangxi. May also occur in SW Guangxi. Makes short-distance movements in response to water-level changes.
Voice: Varied but not particularly vocal—makes shrill, shrieking calls; also gives hoarse, guttural calls.

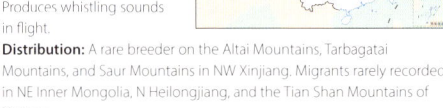

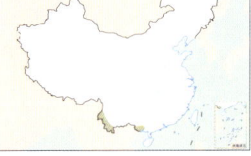

Caspian Plover

♂ br.

Distinctive rusty-red breast band with clear upper border

Oriental Plover

White crown

♂ br.

♂ br.

Distinctive rufous-red breast band

Greater Painted-Snipe

♂

Wide rounded wings

Legs dangle like a rail

♀

Chestnut or maroon head and upper breast

White patch around and behind eyes

Broad white "shoulder" band

♂

chick

Spreads wings to scare off intruders

Pheasant-tailed Jacana

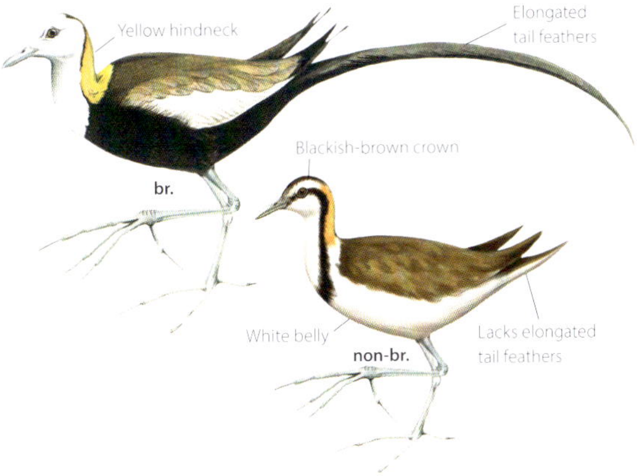

Yellow hindneck

Elongated tail feathers

Blackish-brown crown

br.

White belly

non-br.

Lacks elongated tail feathers

Eurasian Dotterel

Black crown

White supercilia meet on nape in distinctive V shape

♀ br.

White breast band

Chestnut upper belly, black lower belly

Bronze-winged Jacana

Prominent white supercilium

Dark green sheen on head and neck

Numenius phaeopus

中杓鹬 zhōng sháo yù

Whimbrel

L: 40–46 cm

Habitat: Inhabits wetlands, such as intertidal flats, coastal embankments, estuaries, and marshes. Also forages in grassy areas and cultivated fields.

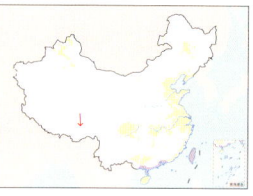

Behavior: Often solitary or in small flocks. Forages by walking and pecking; also uses bill to probe holes on mudflats.

Distribution: N. p. phaeopus migrates through Xinjiang and Tibet as an uncommon passage migrant. N. p. variegatus occurs in most of eastern China and especially coastal areas during migration as a locally common passage migrant; small nonbreeding summering populations also occur along the coast. N. p. hudsonicus, which does not have a white rump, is a vagrant but should be looked for on the east coast; sometimes treated as a separate species: N. hudsonicus.

Voice: Fast, descending, rippling, bubbly he-he-he-he-he.

Numenius madagascariensis

大杓鹬 dà sháo yù

Far Eastern Curlew

L: 53–66 cm

Habitat: Occurs on coastal and estuarine intertidal flats during migration. In winter, inhabits mainly coastal areas, such as mangroves and saltmarshes; occasionally moves to cultivated fields.

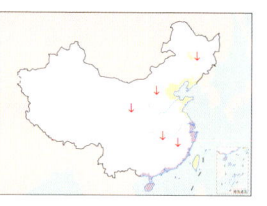

Behavior: Gathers in small flocks during migration; often associates with Eurasian Curlew. The remarkably long bill allows it to probe tidal flats for deep-burrowing prey.

Distribution: Migrates through much of China except for Tibet, Yunnan, and Guizhou; winters in Hainan. Some nonbreeding individuals spend the summer on the coastal intertidal flats of the Yellow Sea and Bohai Sea. Uncommon or occasional in most regions; population has been in drastic decline in recent years.

Voice: Loud, far-carrying cour-lee in flight, very similar to Eurasian Curlew.

Numenius minutus

小杓鹬 xiǎo sháo yù

Little Curlew

L: 28–34 cm

Habitat: Prefers dry, open, grassy areas or cultivated lands near wetlands; occasionally seen on riverbanks, marshes, or intertidal flats during migration.

Behavior: Forms flocks of a few dozen and occasionally up to a hundred individuals during migration. Forages by pecking at insects, spiders, grass seeds, and berries in grassy areas; also probes into the ground.

Distribution: Uncommon passage migrant through Northeast China and the east coast (including Taiwan). Vagrant west to Qinghai and Xinjiang.

Voice: Gives a two- to four-note te-te call, similar to Whimbrel but with fewer notes, higher pitched, and rising.

Numenius arquata

白腰杓鹬 bái yāo sháo yù

Eurasian Curlew

L: 57–63 cm

Habitat: Breeds on marshes, damp grasslands, farmland, and open fields with abundant vegetation. Moves about on coastal intertidal flats, gulfs, and estuaries in nonbreeding season; also occurs along

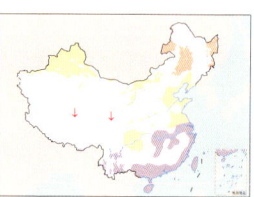

muddy banks near lakes and rivers and occasionally forages on farmland.

Behavior: Generally gregarious. Forages mainly for crabs by probing with long bill in mud; breaks off crab's legs before swallowing.

Distribution: N. a. orientalis breeds in NE Inner Mongolia, Heilongjiang, and Jilin, migrates through most of China, and winters in regions south of the Yangtze River, including Hainan and Taiwan; a few nonbreeding individuals spend summer on the intertidal flats of the Yellow Sea, Bohai Sea, and East China Sea. Uncommon; population has been in decline in recent years.

Voice: Loud cour-lee and other notes, all very similar to Far Eastern Curlew.

Arenaria interpres

翻石鹬 fān shí yù

Ruddy Turnstone

L: 21–26 cm

Habitat: Inhabits primarily coastal habitats, such as saltmarshes, reefs, seawalls, sandy beaches, mudflats, saltpans, and inland lakes. Occasionally occurs in grassy areas and cultivated fields during migration.

Behavior: Solitary or in small flocks; gathers in large groups during migration. Fast runner. When foraging, uses its strong bill to overturn stones near water to look for small invertebrates hidden underneath, hence its name.

Distribution: Migrates through most of China; some winter along the coasts of South and southeastern China, including Hainan and Taiwan. Generally uncommon.

Voice: Gives rapid metallic tuk-tuk-tuk in flight, and chattering, staccato rattle kuvi-kuvi-vitvitvitvitvit when alarmed.

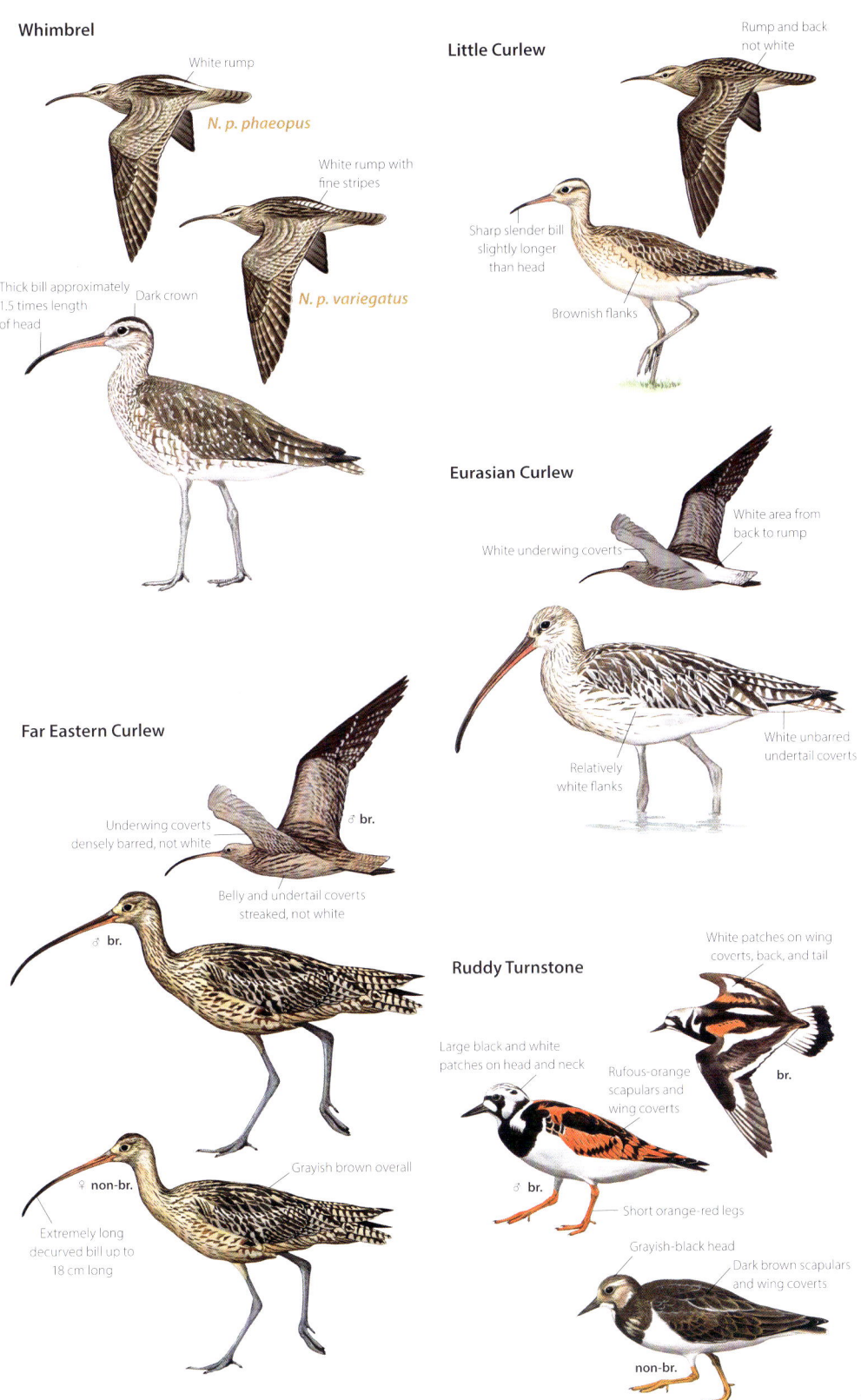

Whimbrel

White rump

N. p. phaeopus

White rump with fine stripes

N. p. variegatus

Thick bill approximately 1.5 times length of head

Dark crown

Little Curlew

Rump and back not white

Sharp slender bill slightly longer than head

Brownish flanks

Eurasian Curlew

White underwing coverts

White area from back to rump

Relatively white flanks

White unbarred undertail coverts

Far Eastern Curlew

Underwing coverts densely barred, not white

♂ br.

Belly and undertail coverts streaked, not white

♂ br.

Grayish brown overall

♀ non-br.

Extremely long decurved bill up to 18 cm long

Ruddy Turnstone

White patches on wing coverts, back, and tail

Large black and white patches on head and neck

Rufous-orange scapulars and wing coverts

br.

♂ br.

Short orange-red legs

Grayish-black head

Dark brown scapulars and wing coverts

non-br.

165

Limosa lapponica

斑尾塍鹬 bān wěi chéng yù

Bar-tailed Godwit

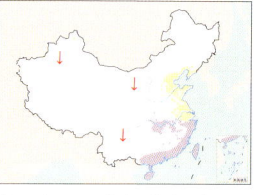

L: 37–41 cm
Habitat: Inhabits shallow intertidal zones, such as coastal tidal flats, estuaries, bays, and saltpans; seldom in freshwater habitats.
Behavior: Long-distance migrant; China's coastal wetlands are among its most important stopover sites. Often migrates in flocks of a few dozen individuals; can form large flocks containing up to several thousand individuals at staging sites. Often follows tideline in flocks on coastal tidal flats; forages by picking, pecking, and probing, feeding on clams, snails, crabs, and fish.
Distribution: During spring migration, both *L. l. baueri* and *L. l. menzbieri* migrate through Northeast China, Bohai Bay, and areas along the east coast; a few nonbreeders spend the summer in the abovementioned areas. During fall migration, adult *L. l. baueri* flies over the Pacific Ocean from the breeding region directly to wintering sites, but some juveniles stop over along the coast of China; *L. l. menzbieri* migrates through China's east coast, with a few individuals wintering on the southeast coast, including Taiwan and Hainan. Rare inland but several records in Xinjiang, Inner Mongolia, and inland southwestern China. Occurs in large numbers at major stopover sites but uncommon elsewhere.
Voice: Nasal *wheet* or more disyllabic *kit-whit*, reminiscent of Red Knot, repeated in flight.

Limosa limosa

黑尾塍鹬 hēi wěi chéng yù

Black-tailed Godwit

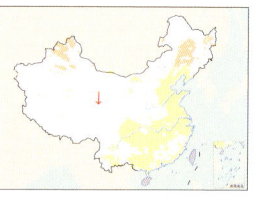

L: 37–42 cm
Habitat: Breeds on wet grasslands. In other seasons, inhabits wetlands, such as coastal tidal flats, sandy beaches, estuaries, marshes, lakes, and flooded fields. Sometimes occurs in relatively deep waters.
Behavior: Generally gregarious. Forages by probing or picking with long bill. When foraging, probes with bill deep into water; frequently raises bill to swallow food.
Distribution: Common. *L. l. melanuroides* breeds in Northeast China, Inner Mongolia, and NW Xinjiang; a few nonbreeders occur on the east coast in summer. Migrates through most of China except for Tibet; winters in the Yangtze Plain and the southeast coast, including Hainan and Taiwan. Some larger individuals that stop over in Bohai Bay belong to a newly recognized subspecies, *L. l. bohaii*.
Voice: Fast, sharp *kii* or *kio, kio* in flight; pleasant, musical, seesawing *vlee-vloe-it* repeated at speed in display.

Limnodromus semipalmatus

半蹼鹬 bàn pǔ yù

Asian Dowitcher

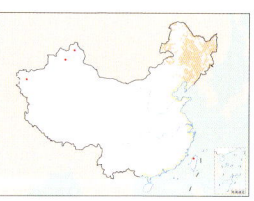

L: 33–36 cm
Habitat: Breeds on large freshwater wetlands on forest steppes. In nonbreeding season, occurs on wetlands, such as coastal tidal flats, estuaries, lagoons, marshes, saltpans and fishponds.
Behavior: Migrates in flocks; can occur in large flocks of several thousand individuals at certain stopover sites. Forages by probing, pushing and pulling bill straight in and out of mud and water; action appears mechanical, like sewing machine.
Distribution: Breeds in Northeast China and Inner Mongolia; migrates through Xinjiang, Qinghai, and the east and southeast coasts. Almost the entire population congregates in Lianyungang, Jiangsu, during spring migration.
Voice: Soft *miau* or *eouw* with froglike quality and similar purring notes in flight.

Limnodromus scolopaceus

长嘴半蹼鹬（长嘴鹬）

cháng zuǐ bàn pǔ yù (cháng zuǐ yù)

Long-billed Dowitcher

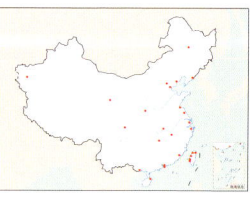

L: 24–30 cm
Habitat: Inhabits wetlands, such as inland and coastal estuaries, lakes, marshes, and fishponds.
Behavior: Generally solitary or in pairs; occurs among flocks of other large shorebirds. Forages by wading in shallow waters and using its long bill to probe for snails, small shrimp, and crabs, as well as other small invertebrates.
Distribution: Vagrant or rare migrant, recorded in Heilongjiang, Liaoning, Tianjin, Hebei, Beijing, Xinjiang, Qinghai, Sichuan, Shaanxi, Hunan, Hubei, Jiangxi, Jiangsu, Shanghai, Zhejiang, Fujian, Guangdong, Guangxi, Hong Kong, Taiwan, and Hainan.
Voice: Short sharp *kik*, often in a series; *kreek* when flushed.

Bar-tailed Godwit

L. l. baueri

Fine black bars on white rump

White from rump to back with sparse black spots

Barred tail feathers

Toes project slightly beyond tail

L. l. menzbieri

Slightly upcurved bill, pink at base

Strongly mottled scapulars

Rufous underparts

br.

Basal half of bill pink

non-br.

Asian Dowitcher

Barred rump lacks white

Rufous chestnut from head and neck to lower belly

br.

Black legs

Long, straight, relatively thick, all-black bill with slightly swollen bill tip

non-br.

Black-tailed Godwit

Contrasting black and white on wings and tail

White supercilium

Long straight bill with black tip and pinkish-yellow basal half

Fine black bars on white belly

br.

Unbarred pale gray neck

Black tail feathers

Pink basal half of bill

non-br.

Long-billed Dowitcher

White from rump to back

Blackish brown on wings

Rufous-brown underparts

br.

Greenish-yellow legs

Gray from head to chest

Long straight black bill with yellowish-green base and slightly decurved tip

non-br.

Black and white bars on undertail coverts

167

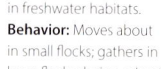

Calidris tenuirostris

大滨鹬 dà bīn yù

Great Knot

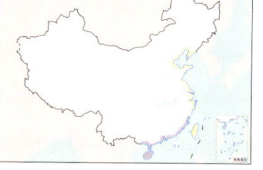

L: 26–30 cm
Habitat: Inhabits saltwater wetlands, such as coastal soft-sediment mudflats, sandy beaches, and saltpans; rarely occurs in freshwater habitats.
Behavior: Moves about in small flocks; gathers in large flocks during migration. Foraging depends mainly on the tactile receptors of bill. Forages by day and night, making relatively slow motions.
Distribution: Migrates along coastal areas of North and East China, including Taiwan, and a few individuals migrate through inland wetlands. Winters on the southeast coast and Hainan. During spring migration, thousands of individuals concentrate at Yalu Estuary, Shuangtaizi Estuary, Yellow River Delta, and Yangtze Estuary. A few also oversummer in the abovementioned areas, but almost none occur in fall. Due to development at and destruction of important stopover habitats, the population has declined sharply in recent years.
Voice: In flight, low gruff contact calls similar to Red Knot.

Calidris canutus

红腹滨鹬 hóng fù bīn yù

Red Knot

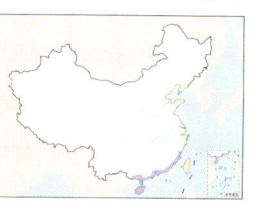

L: 23–25 cm
Habitat: Inhabits wetlands, such as coastal intertidal flats, saltpans, and estuaries.
Behavior: Occurs in small flocks; can form large groups containing up to a few thousand individuals during migration. Shy and afraid of people. Often forages by wading slowly in shallow waters or on coastal intertidal mudflats, frequently using bill to probe in mud for prey; also forages by sight. Feeds mainly on bivalves; can swallow them whole and crushes shells in its gizzard.
Distribution: Uncommon. During migration, *C. c. rogersi* and *C. c. piersmai* occur in coastal areas along the Bohai Sea, Yellow Sea, and East China Sea, including Taiwan; some oversummer in the abovementioned areas. A few individuals migrate through inland wetlands, and some winter in coastal areas in South China and Hainan. Population has been in a sharp decline in recent years due to development at and destruction of important stopover sites. Nanpu mudflat in Luannan County, Hebei Province, is the most important stopover site so far identified.
Voice: Varied contact and flight calls; most muffled *kweet-kweet*.

Calidris pugnax

流苏鹬 liú sū yù

Ruff

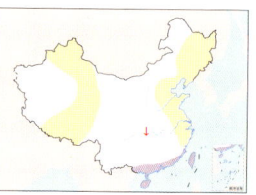

L: 26–32 cm (♂)
      20–26 cm (♀)
Habitat: Inhabits wetlands, such as coastal intertidal flats, lakes, and marshes; also forages in grassy areas, cultivated fields, and paddy fields.
Behavior: In breeding season, most males develop fluffy ruff around neck and breast; commonly seen during spring migration. Some males also mimic the body size and plumage of females.
Distribution: Migrates through N Xinjiang, the Qinghai-Tibet Plateau, and S Tibet, as well as eastern parts of Northeast China through the coastal areas along the Bohai Sea, Yellow Sea, and East China Sea, including Taiwan. A few individuals winter on the coasts of southeast and South China, including Hainan and Taiwan.
Voice: Normally silent.

Calidris falcinellus

阔嘴鹬 kuò zuǐ yù

Broad-billed Sandpiper

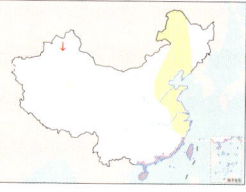

L: 15–18 cm
Habitat: Inhabits wetlands, such as coastal intertidal flats, estuaries, saltpans, lagoons, marshes, ponds, and lakes.
Behavior: Often solitary or in small flocks, mixing with other shorebirds. Forages on coastal mudflats by slowly walking and pecking. Feeds mainly on worms; also eats snails, bivalves, shrimp, crabs, and insects.
Distribution: *C. f. falcinellus* migrates through Xinjiang; *C. f. sibirica* through Northeast China and the east coast. A few individuals summer and winter along the south and southeast coast, including Hainan and Taiwan.
Voice: Flight call a distinctive rising buzzy trill, *chree-ep*, recalling Bank Swallow.

Great Knot

White U-shaped patch on rump

non-br.

Rufous-chestnut scapulars

Bill longer than head

br.

Black spots on flanks

non-br.

Ruff

White rump

Warm buff from throat to belly

Brown central tail feathers

♂

"Faeder" form

Plumage resembles female but brighter

♂ br.

"Independent" form

Black or chestnut neck ruff

♂ br.

"Satellite" form

Plain white or black and white neck ruff

"Independent" form

♂ br.

non-br.

Red Knot

Gray rump

br.

Grayish upperparts

non-br.

Gray from nape to hindneck

Extensive white area from lower belly to undertail coverts

br.

C. c. rogersi

Gray upperparts lacks black feathers or spots

Bill length equals head

Dark arrowhead-shaped markings on flanks

non-br.

Rufous brown from nape to hindneck

br.

Small diffuse white area on lower belly

C. c. piersmai

Broad-billed Sandpiper

Prominently streaked crown, nicknamed "watermelon pattern"

Decurved bill

br.

Black stripe from rump to tail

non-br.

Off-white plumage

non-br.

Dark fulvous legs

169

Calidris acuminata
尖尾滨鹬 jiān wěi bīn yù
Sharp-tailed Sandpiper LC

L: 16–23 cm
Habitat: Inhabits coastal intertidal flats, estuaries, saltpans, marshes, farmland, and grassy areas.
Behavior: Often solitary or in small flocks, mixing with other shorebirds. Forages by pecking quickly in shallow waters.
Distribution: Common passage migrant; migrates through most of central and eastern China; a few individuals winter in Taiwan.
Voice: A slightly trilling *prtt-wheet-wheet*.

Calidris melanotos
斑胸滨鹬 bān xiōng bīn yù
Pectoral Sandpiper LC

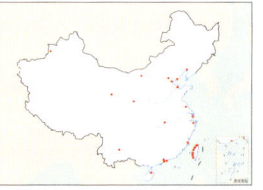

L: 19–24 cm
Habitat: Inhabits wetlands, such as coastal intertidal flats, estuaries, lakes, and marshes.
Behavior: Often solitary, mixing with flocks of other shorebirds and especially Sharp-tailed Sandpiper. Forages by wading in shallow waters and probing for prey with steady motions.
Distribution: Vagrant to Xinjiang, Qinghai, Inner Mongolia and Liaoning. Rare migrant farther east in Hebei, Beijing, Tianjin, Hubei, Anhui, Jiangsu, Shanghai, Fujian, Guangxi, Yunnan, Hainan, Hong Kong, Macao, and Taiwan.
Voice: Short Curlew Sandpiper–like trill, *kreet*, in flight and snipe-like calls when flushed.

Calidris ferruginea
弯嘴滨鹬 wān zuǐ bīn yù
Curlew Sandpiper NT

L: 18–23 cm
Habitat: Inhabits wetlands, such as coastal intertidal flats, saltpans, estuaries, and marshes. Also forages on flooded fields, lake banks, river flats, and cultivated lands.
Behavior: Generally gregarious, often mixing with flocks of other sandpipers; numbers high at some staging sites. Can also forage at night, depending on the tidal cycle.
Distribution: Common passage migrant, seen in most of China during migration; a few individuals summer and winter in coastal areas in southeastern and South China, including Hainan and Taiwan. In spring, congregates in large numbers on intertidal flats in the northern Yellow Sea and Bohai Sea.
Voice: Soft *chirrup* trills.

Calidris temminckii
青脚滨鹬 qīng jiǎo bīn yù
Temminck's Stint LC

L: 13–15 cm
Habitat: Inhabits inland freshwater wetlands, such as riverbanks, ponds, lakes, fishponds, flooded fields, and marshes; occasionally occurs on coastal intertidal flats.
Behavior: Solitary or in small flocks, rarely associates with other shorebirds. Forages by sight, quickly pecking prey from surface.
Distribution: Common passage migrant. Migrates through most of China; winters in coastal areas in southwestern, southeastern, and South China, including Taiwan.
Voice: High-pitched trill, often doubled, *tirrrr, tirrrt*.

Calidris subminuta
长趾滨鹬 cháng zhǐ bīn yù
Long-toed Stint LC

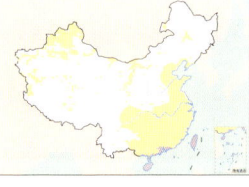

L: 13–16 cm
Habitat: Inhabits inland freshwater wetlands, such as riverbanks, ponds, lakes, fishponds, flooded fields, and marshes.
Behavior: Solitarily or in small loose flocks. Looks for prey by sight while making relatively slow motions; forages mainly for small invertebrates. Stance upright.
Distribution: Common passage migrant; migrates through most of China. A few individuals stay on the east coast in summer; winters along the coasts in South China (including Taiwan).
Voice: Silvery *chrrip*, with pronounced change in pitch in flight.

Calidris alba
三趾滨鹬 sān zhǐ bīn yù
Sanderling LC

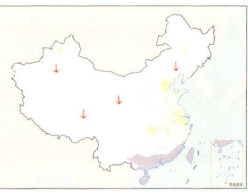

L: 19–21 cm
Habitat: Prefers open sandy beaches and sandbars; also occurs on muddy and rocky coastlines and estuaries.
Behavior: Often gathers in small flocks, mixing with flocks of other shorebirds. Usually runs and forages along tideline; walks quickly and pecks at prey on sandy or muddy coastlines.
Distribution: Uncommon passage migrant, seen in most of China on migration except for Heilongjiang, Inner Mongolia, Yunnan, and Sichuan. A few individuals winter along the southeast coast and Taiwan.
Voice: Very short sharp *chit* or *chwep*.

Sharp-tailed Sandpiper

juv.

Orange-buff wash on breast

non-br.

Black stripe from rump to tail

Inconspicuous arrowhead markings

non-br.

Prominent chestnut crown

Arrowhead-shaped streaks from side of breast to flanks

br.

Pectoral Sandpiper

Slightly decurved bill with yellowish-green base

White throat

Fine black streaks on breast form clear division from white belly

br.

non-br.

Temminck's Stint

White wing bar

Some scapulars are black with buff fringe

Black stripe from rump to tail

br.

Yellowish-green legs

Gray breast with clear lower boundary

non-br.

Long-toed Stint

Brownish-yellow crown with dark streaks

Brownish-yellow scapulars

Black stripe from rump to tail

non-br.

Long middle toes

br.

Yellowish-green legs

Brownish yellow on scapulars and crown faded, becoming lighter

non-br.

Comparison of foraging postures

Long-toed Stint

Sharp-tailed Sandpiper

Curlew Sandpiper

White wing bar

Mottled back

Plain white rump

non-br.

Rusty-red head to underparts, marbled with white streaks

br.

Prominent white supercilium

Long slender decurved bill

non-br.

Sanderling

Broad white wing bar

non-br.

Black and brown scapulars

br.

Gray scapulars

Black coverts

Black and white spots on scapulars

non-br.

juv.

Calidris pygmaea

勺嘴鹬 sháo zuǐ yù

Spoon-billed Sandpiper  **CR**

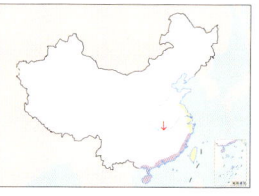

L: 14–16 cm
Habitat: Depends highly on coastal intertidal flats during migration; winters in wetland habitats, such as estuaries, lagoons, and marshes. Prefers intertidal flats consisting of relatively sandy substrates covered by a thin layer of soft mud. Often forages in shallow tide channels.
Behavior: Often gathers in small flocks; also forages with flocks of other shorebirds, especially Red-necked Stint. Forages by walking and moving the spatulate bill from side to side through the surface of shallow water or soft mud slurry.
Distribution: Rare. Migrates through the east coast, including Taiwan, and stays for a long period on the Yellow Sea intertidal flats, especially during fall migration. Small numbers winter in Fujian, Guangdong, Guangxi, Hong Kong, and Hainan. Population is now tiny. Has become a well-known shorebird in China due to education and conservation efforts.
Voice: Short, sharp Red-necked Phalarope–like *pit* in flight.

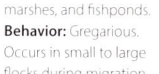

Calidris ruficollis

红颈滨鹬 hóng jǐng bīn yù

Red-necked Stint **NT**

L: 13–16 cm
Habitat: Inhabits various kinds of wetlands, including coastal intertidal flats, estuaries, lakes, marshes, and fishponds.
Behavior: Gregarious. Occurs in small to large flocks during migration, often mixing with flocks of other shorebirds. Forages on intertidal mudflats by walking and pecking at food on mud surface.
Distribution: Common passage migrant and winter visitor; migrates through most of China with a few nonbreeders summering at different places. Winters in coastal areas in southeastern and South China, including Hainan and Taiwan.
Voice: Flight call dull, *chriit*, huskier and much less sharp than Little Stint.

Calidris minuta

小滨鹬 xiǎo bīn yù

Little Stint **LC**

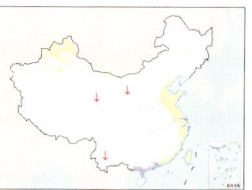

L: 14–14.5 cm
Habitat: Occurs near rivers, lakes, reservoirs, marshes, and other wetlands on open lowlands.
Behavior: Tame and gregarious; mixes with other small shorebirds. Forages by rapid pecking or picking actions; often wades in shallow waters and pecks at aquatic insects, insect larvae, small mollusks, and crustaceans.
Distribution: Commonly seen in W and N Xinjiang during migration; occasionally seen in Inner Mongolia, Qinghai, Shaanxi, Yunnan, Jilin, Tianjin, Beijing, Shandong, Jiangsu, Shanghai, Zhejiang, Guangdong, Hainan, Macao, Hong Kong, and Taiwan.
Voice: Unique short and sharp *stit* calls; also gives weak *pi, pi, pi* in flight.

Calidris alpina

黑腹滨鹬 hēi fù bīn yù

Dunlin **LC**

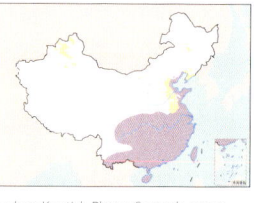

L: 16–22 cm
Habitat: Inhabits various saltwater and freshwater wetlands, such as coastal intertidal flats, estuaries, marshes, lagoons, flooded fields, and saltpans.
Behavior: Often gathers in large flocks, mixing with flocks of other shorebirds, such as Kentish Plover. Spreads out to forage but congregates when in danger and flies in close formation. Often wades in water and forages with rapid actions, pecking at food from intertidal flats using both visual and tactile cues. Foraging activities shaped by tides; sometimes forages at night.
Distribution: A common passage migrant and winter visitor. *C. a. sakhalina, C. a. kistchinski, C. a. actites,* and *C. a. arcticola* migrate through Northeast and northwestern China and the east coast; a few individuals stay in the coastal areas in summer; winters in vast coastal regions south of the Yangtze Plain, including Hainan and Taiwan.
Voice: Lengthy *krreeet* in flight.

Spoon-billed Sandpiper

Narrow white wing bar

non-br.

Black central tail

Spatulate bill

Dark brown spots
forming streaks
on breast

br.

Light brownish-gray, narrowly
fringed back and scapulars

non-br.

Clean white underparts

Mottled brown neck

Conspicuously white-fringed
back and scapulars

juv.

Red-necked Stint

Narrow white wing bar

Relatively short
stout black bill

Rusty-red fringe
on scapulars only

Red throat

Black central tail

Gray primaries

br.

Short tibia

non-br.

Black legs

Little Stint

White V on back

Rusty-red fringe on scapulars,
wing coverts, and tertials

White throat

br.

Slightly long
slender bill

Often has
inconspicuous
"braces"

non-br.

Tibia longer than on
Red-necked Stint

Dunlin

White wing bar

non-br.

Black central tail

Chestnut on scapulars

Clear-cut black
belly patch

br.

White supercilium does
not extend beyond eye

Light brownish-
gray scapulars

Long thin
decurved bill

non-br.

Calidris ptilocnemis

岩滨鹬 yán bīn yù

Rock Sandpiper

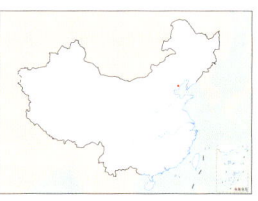

L: 20–23 cm
Habitat: Inhabits oceanic islands or coastal rocky coasts.
Behavior: Forages by walking slowly among rocks and picking mollusks, worms, crustaceans, and other small invertebrates.
Distribution: Vagrant; only recorded in Hebei (Beidaihe).
Voice: Short *chuirk*.

Calidris mauri

西滨鹬 xī bīn yù

Western Sandpiper

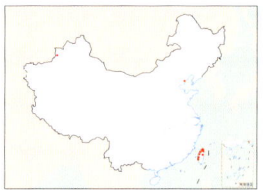

L: 14–17 cm
Habitat: Inhabits wetlands, such as coastal intertidal flats, sandy beaches, estuaries, and marshes.
Behavior: Gregarious. Records in Taiwan usually consist of one to three individuals mixed with flocks of other sandpipers (normally Dunlin and Sanderling). Runs along tideline. Uses bill to probe and peck quickly at prey on intertidal mud- and sandflats; forages using tactile and olfactory senses.
Distribution: Vagrant; recorded only in Qinghai, Hebei, Tianjin, and Taiwan. Nonbreeding plumage similar to Sanderling and may thus have been overlooked. May occur along the east coast in fall and winter.
Voice: Descending single-note *jeet* or *cheet*.

Calidris fuscicollis

白腰滨鹬 bái yāo bīn yù

White-rumped Sandpiper

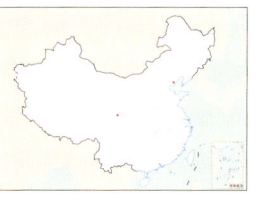

L: 15–18 cm
Habitat: Inhabits wetlands, such as coastal intertidal flats, estuaries, lakes, and marshes.
Behavior: Generally occurs solitarily, mixing with flocks of other sandpipers. Walks quickly on mudflats or in shallow waters, picking at prey on the surface.
Distribution: Vagrant; recorded only in Hebei (Beidaihe and Nanpu) and Sichuan (Zoige).
Voice: Very short, sharp, and insect- or batlike *zeet* calls in flight.

Calidris subruficollis

黄胸滨鹬 (饰胸鹬)

huáng xiōng bīn yù (shì xiōng yù)

Buff-breasted Sandpiper

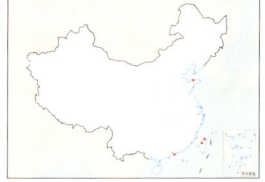

L: 16–21 cm
Habitat: Generally occurs on grass, cultivated land, and marshes near coastline.
Behavior: Generally occurs solitarily. Forages by walking rapidly on the ground, like Pacific Golden-Plover; forages for insects, spiders, and plant seeds on grass. When startled, flies a short distance and lands.
Distribution: Vagrant, recorded in Shandong (Yantai), Hong Kong, and Taiwan.
Voice: Mostly silent; gives a variety of mostly muffled calls in flight.

Calidris himantopus

高跷鹬 gāo qiāo yù

Stilt Sandpiper

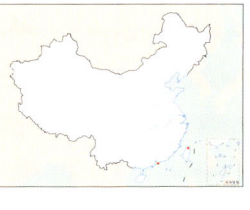

L: 18–23 cm
Habitat: Inhabits inland wetlands, such as lakes, rivers, ponds, and marshes.
Behavior: Generally occurs singly, mixing with flocks of other sandpipers. Forages by wading slowly in shallow waters and probing for prey with bill.
Distribution: Vagrant; recorded only in Hong Kong and Taiwan.
Voice: Gives various calls, *cherk* and sharp *eree*, when flushed.

Rock Sandpiper

Wide buff fringe on scapulars

Wing feathers and wing coverts in different colors

Black upper belly

br.

Yellow base of bill

Black spots on underparts

non-br.

Yellow legs

Western Sandpiper

Chestnut crown and cheeks

Black and chestnut scapulars

Arrowhead-shaped streaks on breast

br.

Resembles nonbreeding plumage of Red-necked Stint, but bill and tarsi are longer

non-br.

Webbing between middle and outer toes

White-rumped Sandpiper

Black spots on scapulars

White throat

br.

Black streaks on breast

Pale gray scapulars

Black primaries

non-br.

Buff-breasted Sandpiper

White underwing coverts

Rump not white

br.

Blackish-brown scapulars

juv.

Yellow legs

Stilt Sandpiper

Brownish-yellow cheeks and crown

White rump

No wing bar

Heavily barred underparts

br.

Light streaks on side of neck and flanks

non-br.

Yellowish-green legs

Scolopax rusticola

丘鹬 qiū yù

Eurasian Woodcock

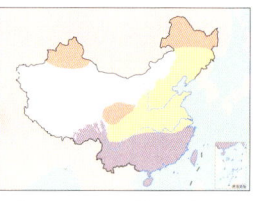

L: 33–38 cm

Habitat: Inhabits shady and moist broadleaf and mixed forests in breeding season; sometimes also occurs in marshes, wet grassy patches, and forest-edge bushes. During migration and winter, also seen among bushes on mountain slopes, streams, and farmland.

Behavior: Wings produce swishing sound during takeoff. Generally occurs solitarily; nocturnal and secretive.

Distribution: In breeding season, uncommon in N Heilongjiang, Jilin, Hebei, NW Xinjiang, Sichuan, and S Gansu. Migrates through most of China; winters mostly south of the Yangtze River.

Voice: In circling territorial flight at night, male gives two to four peculiar croaking notes, immediately followed by an explosive, sneezed, high-pitched, disyllabic note, *oort-oort-oort-oort piss-ip*.

Lymnocryptes minimus

姬鹬 jī yù

Jack Snipe

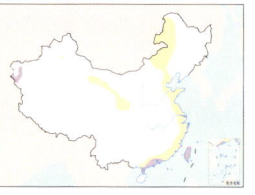

L: 18–20 cm

Habitat: Inhabits muddy river bars, marshes, lakeshores, and farmland.

Behavior: Often moves about solitarily at night or dusk; hides among tussocks by day. Generally flies a short distance and quickly lands back in tussocks. Prefers foraging on flooded muddy banks or bars; body bobs rhythmically when probing in mud with bill.

Distribution: During migration, rare in C and W Xinjiang, Gansu, NE Inner Mongolia, North, East, and South China; a few individuals winter in W Xinjiang, S Guangdong, Hong Kong, and Taiwan.

Voice: Generally silent; makes feeble *etch* when flushed.

Gallinago solitaria

孤沙锥 gū shā zhuī

Solitary Snipe

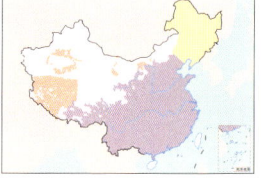

L: 29–31 cm

Habitat: Inhabits mountain rivers and swamps in breeding season. Can occur at altitudes up to 5000 m; moves to lower altitudes in winter. Occurs almost exclusively on mountain streams; prefers moving waters.

Behavior: Generally solitary; does not mix with other shorebirds. Flight slower than other snipes.

Distribution: *G. s. japonica* breeds in Northeast China; winters in mountainous regions in eastern China south of North China. *G. s. solitaria* breeds in Xinjiang, W Tibet, and W Inner Mongolia; winters in the mountains in southwestern China.

Voice: Rapidly delivered, clear *chuk-chuk-erk* and wing noise when displaying. Common Snipe–like flight call rarely heard.

Gallinago nemoricola

林沙锥 lín shā zhuī

Wood Snipe

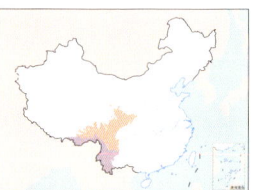

L: 28–32 cm

Habitat: Breeds on alpine meadows with dwarf, scattered bushes or in marshes with large rocks at altitudes of 3000–5000 m; prefers areas near streams. Moves to low-altitude marshes and riverbanks in nonbreeding season.

Behavior: Generally solitary. Forages more actively at dusk and dawn. Flies slowly. Forages on alpine meadows.

Distribution: Rare. Breeds in S Gansu, Sichuan, Yunnan, and S and E Tibet; winters in SE Tibet and W and NE Yunnan.

Voice: During display flight, a repeated, loud, raspy, and rhythmic *chep, chep* at rate of about three or four notes per second before running into series of faster *ip-ip-ip* notes; whole sequence lasting about 4–20 sec—*chep-chep-chep-dep-dep-de-ip-ip-ip-ip ip-ip*.

Eurasian Woodcock

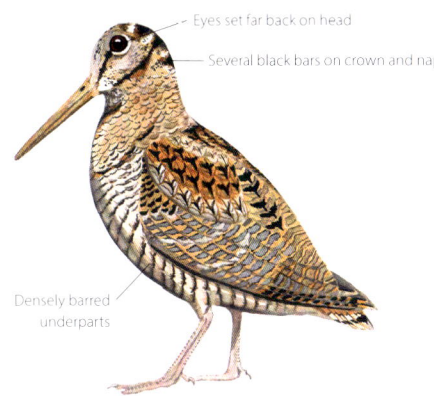

Eyes set far back on head

Several black bars on crown and nape

Densely barred underparts

Solitary Snipe

Whitish face

Heavily rufous-brown upperparts

Conspicuous narrow black and white barring on flanks

Jack Snipe

Two prominent creamy-yellow stripes on back

Small

Wood Snipe

Relatively rounded wings resemble Eurasian Woodcock

Dark underwing coverts with streaks; no white on trailing edge

Toes project well beyond tail tip

Relatively thick pale base of bill

Two brownish-yellow stripes on heavily blackish-brown back

Blackish-brown bars from breast and belly to undertail coverts

Gallinago hardwickii

拉氏沙锥 (澳南沙锥)
lā shì shā zhuī (ào nán shā zhuī)

Latham's Snipe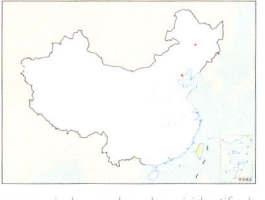

L: 28–33 cm
Habitat: Inhabits
riverbanks, marshes,
paddy fields, grassy areas,
and scrubland.
Behavior: Moves about
and forages solitarily at
dusk and dawn; probes in
soft soil for prey.
Distribution: Rare. Has short passage window and can be misidentified
easily. A small number of individuals migrate through Taiwan; possible
vagrant in Northeast and coastal eastern China.
Voice: Brief *chetch* when flushed, shorter than Common Snipe, with
quality reminiscent of Oriental Magpie.

Gallinago stenura

针尾沙锥 zhēn wěi shā zhuī

Pin-tailed Snipe

L: 25–27 cm
Habitat: Inhabits
riverbanks, paddy fields,
marshes, and grassy areas;
prefers drier environments
than Common Snipe.
Behavior: Forages
solitarily or in small
flocks at dusk and dawn,
probing bill in soft soil for prey. When flushed, flight path similar to that
of Common Snipe but shorter and lower.
Distribution: Migrates through most of China; winters in parts of South
and southwestern China and Hainan.
Voice: Strained, hoarse squawk with nasal quality when flushed, very
similar to Swinhoe's Snipe. Song, occasionally also heard from migrants
in spring, is a lengthy nasal series of asthmatic, breathless, wheezy, often
paired notes, *chiv-chiv chiv-chiv chiv-chiv*, repeated at length, gradually
accelerating to climax, followed by tail-feather winnowing, a couple of
nasal *queer* notes, and a fizzing terminus; whole sequence lasting almost
a minute.

Gallinago megala

大沙锥 dà shā zhuī

Swinhoe's Snipe

L: 27–30 cm
Habitat: Occurs
in habitats such as
riverbanks, paddy fields,
marshes, and wet grass,
similar to Common Snipe.
Behavior: Moves about
and forages solitarily or
in small flocks at dusk
and dawn, probing with bill in soft soil for prey. When flushed, makes
short-distance flight in a straight line and lands immediately; flight
relatively slow.
Distribution: May breed in some areas in Northeast China. Migrates
through most of Xinjiang and central and eastern China; winters in parts
of South China and Hainan. Passage migrant in SE Tibet.
Voice: Relatively quiet during migration. When flushed, gives abrupt
ka, similar to Pin-tailed Snipe. Flight song involves several sequence of
usually 8–30 rapid-fire (2.5 calls per second) repetitions of amphibian-
like croaking, *khre-re-re*, followed by tail-feather winnowing and varied
clicking chatter.

Gallinago gallinago

扇尾沙锥 shàn wěi shā zhuī

Common Snipe

L: 24–29 cm
Habitat: Breeds on wet
grasslands and tundra;
moves about in wet
habitats with abundant
vegetation during
migration and winter,
including lakeshores,
riversides, ponds, paddy
fields, and marshes.
Behavior: Makes display flight in breeding season, often flying high up
in sky and quickly descending. Moves about solitarily or in small flocks in
nonbreeding season. Forages actively at dusk and dawn; hides among
tussocks during the day. When alarmed, calls and breaks cover in zigzag
flight and circles high. Uses bill to probe for prey in soft mud.
Distribution: Breeds in Xinjiang and Northeast China; winters across
vast areas to the south of Yellow River. Common.
Voice: In display flight on breeding grounds, wingbeats produce a series
of pulsing, rising *huhuhuhuhu*. "Ground song" a *wicca-wicca-wicca*,
repeated at speed. Relatively quiet in nonbreeding season. Gives hoarse
ka-atch when flushed and continues to call a few times when fleeing.
Note: A few individuals of Wilson's Snipe (*G. delicata*), which occurs
mainly in North America, winter in northeastern Asia; this species may
also occur in Taiwan, but this needs confirmation.

Latham's Snipe

Outer tail feathers slightly narrower than central ones, with black and white tips and conspicuous white areas

Male has 16–18 tail feathers; female has 14–18 tail feathers, with 3–5 outer tail feathers on each side

Bill length: 64–72 mm (♂), 66–76 mm (♀)

Whitish face

Large and relatively pale overall

Relatively long wings

Relatively straight breast streaks

Tail extends well beyond primaries

Swinhoe's Snipe

Dense black streaks on underwing coverts

Toes project slightly beyond tail tip

18–26 tail feathers (generally 20–22), 7 outer tail feathers on each side (2 innermost ones showing intermediate coloration)

Outer tail feathers much narrower than central ones, with black and white tips, but white areas inconspicuous

Relatively conspicuous blackish-brown "moustache"

Bill 1.6 times the length of head

Bill length: 56–67 mm (♂), 59–74 mm (♀)

Pin-tailed Snipe

Fine dense black streaks on underwing coverts

Toes project well beyond tail tip

24–28 tail feathers (generally 26), 7–8 outer tail feathers on each side

Extremely narrow pin-shaped outer tail feathers

Bill 1.5 times the length of head

Bill length: 56–64 mm (♂), 59–68 mm (♀)

Common Snipe

Conspicuous white patch on underwing coverts, no streaks

Secondaries have white tips

12–18 tail feathers (generally 14 or 16), one outer tail feather on each side

No distinctive difference between outer and inner tail feathers

Bill 1.6–2 times the length of head

Bill length: 62–72 mm (♂), 63–75 mm (♀)

Xenus cinereus
翘嘴鹬 qiào zuǐ yù
Terek Sandpiper

L: 22–25 cm
WS: 57–59 cm
Habitat: Occurs in areas of shallow wetlands, such as coastal intertidal flats, estuaries, lakes, marshes, and saltpans.
Behavior: Often occurs

among flocks of other shorebirds. Active; locates prey by sight; when foraging, runs quickly with body held low and forward. Pecks at prey; also forages by probing with bill in soft mud or pecking at water's surface.
Distribution: Locally common; seen in many areas during migration, most commonly along the east coast, with a few summering individuals. Winters in Taiwan.
Voice: In flight, rapid, whistled, two- to five-syllable *wit–wit–wit–wit*, reminiscent of Gray-tailed Tattler in tone but faster, with individual notes shorter and lacking the tattler's initial fall in pitch.

Phalaropus lobatus
红颈瓣蹼鹬 hóng jǐng bàn pǔ yù
Red-necked Phalarope

L: 16–20 cm
WS: 30–41 cm
Habitat: During migration, inhabits various coastal water bodies, including ponds, marshes, rivers, lakes, and reservoirs. Winters at sea.
Behavior: Gregarious;

forms flocks of a few dozen to a few hundred individuals. During migration, flocks fly over seas and stop at sea to forage and rest. Forages by swimming and often spinning and pecking at prey on surface of water. Unafraid of people.
Distribution: Migrates from Xinjiang through the eastern part of the Qinghai-Tibet Plateau to Yunnan and S Tibet; or through inland wetlands from central to southern China; also migrates from northern Northeast China through the east coast and Taiwan. Winters in Hainan.
Voice: In flight, gives sharp trills, *twick*, similar to Little Stint.

Actitis hypoleucos
矶鹬 jī yù
Common Sandpiper

L: 16–22 cm
WS: 38–41 cm
Habitat: Breeds on tundra or grasslands near forests; prefers bushy areas near water bodies. Occurs in inland ponds, riverbanks, marshes, mountain streams, and various

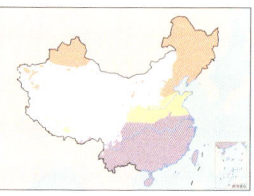

other freshwater wetlands during migration or winter; prefers rocky environments.
Behavior: Moves about solitarily; occasionally forms small flocks during migration. Often forages by walking quickly along water and constantly bobbing tail up and down; mainly a visual forager. Territorial; chases out conspecifics that enter its territory. When alarmed, calls and flies low above water with down-bowed wings and shallow wingbeats.
Distribution: Widespread and common. Breeds in vast areas across northern China and winters mostly to the south of the Yangtze River Watershed, including Hainan and Taiwan, but rarely as far north as Beijing.
Voice: Song a series of rapid, excited, twittering notes, *swee–wee–wee*. Alarm a drawn-out rising *hweeep*.

Phalaropus fulicarius
灰瓣蹼鹬 huī bàn pǔ yù
Red Phalarope

L: 20–22 cm
WS: 37–43 cm
Habitat: Occurs on various coastal water bodies, including fishponds, marshes, rivers, lakes, and reservoirs; probably occurs more often at sea.

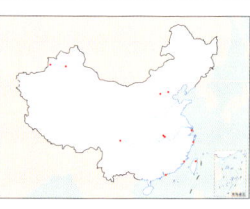

Behavior: Often occurs solitarily. Unafraid of people. Forages by swimming and often spinning and pecking at food on surface of water.
Distribution: Vagrant or rare migrant that occurs irregularly; recorded in Xinjiang, Shanxi, Beijing, Hebei, Henan, Hubei, Sichuan, Jiangxi, Jiangsu, Zhejiang, Shanghai, Fujian, Guangdong, Hong Kong, and Taiwan.
Voice: Unlikely to be heard in China, but a sharp and repetitive *pit, pit* or a brief high-pitched *zeet*.

Terek Sandpiper

Fine streaks on head, face, and side of neck

Relatively wide black feather shafts on scapulars

White trailing edge on secondaries

br.

Faded streaks on head, face, and side of neck

non-br.

Long upturned black bill with orange-yellow base

Flat standing posture

Short orange-yellow legs

Common Sandpiper

White outer tail feathers

White wing bar

Wings angled down in flight

White or buff supercilium extends behind eye

White triangle in front of closed wings

Red-necked Phalarope

1st win.

non br.

Black eye stripe

Indistinct white scapular V

Dark grayish-brown crown

Red from supercilium to neck

juv.

Dark grayish brown crown

♂ non-br.

Sharp fine bill

White throat

Rufous chestnut from supercilium to neck

White throat

br.

♂ br.

♀ br.

Dark legs and lobed toes

Red Phalarope

Relatively short stout bill with hint of yellowish-green base

Plain gray back

non-br.

br.

♂ non-br.

Buff-tinged crown

Yellow bill with black tip

Rufous brown from neck to underparts

Black crown

Gold fringe on scapulars

juv.

♂ br.

♀ br.

Pale legs and lobed toes

181

Tringa ochropus
白腰草鹬 bái yāo cǎo yù
Green Sandpiper LC

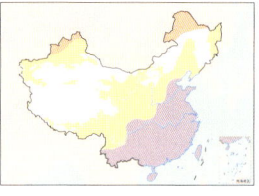

L: 21–24 cm
WS: 57–61 cm
Habitat: Breeds in damp wooded areas. Occurs in ponds, riverbanks, lakeshores, flooded fields, marshes, and other water bodies with slow-flowing water and dense grass during migration and winter; seldom occurs on coastal intertidal flats.
Behavior: Moves about solitarily. Bobs tail up and down when standing. Shy and vigilant; immediately takes off when alarmed.
Distribution: Common migrant. Breeds in NW Xinjiang, N Heilongjiang, and NE Inner Mongolia and winters mostly in the area south of the line connecting S Tibet and Bohai Bay, including Hainan and Taiwan.
Voice: Song is a lengthy rising and falling *kooi- jiji-kooi- jiji* or *kooi-kooi-kooi*; gives a clear, ringing *tlu-eet-wit-wit-it-it* in flight and a rapid and sharp *ji-ji* when alarmed.

Tringa incana
漂鹬 piāo yù
Wandering Tattler LC

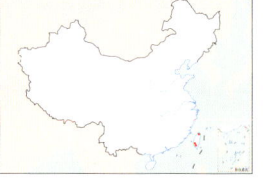

L: 26–29 cm
WS: 54–66 cm
Habitat: Prefers rocky coasts; also occurs on coastal intertidal flats and estuaries.
Behavior: Generally occurs solitarily. Rests at high points near tideline during high tide. Forages on rocks or tidal flats during low tide; occasionally probes for prey in shallow waters.
Distribution: Vagrant in Taiwan; nonbreeding plumage very similar to that of Gray-tailed Tattler and may be easily overlooked. May occur along the east coast.
Voice: Gives a 6- to 10-syllable trill, *peet*, similar to Whimbrel (but faster) and different from Gray-tailed Tattler.

Tringa glareola
林鹬 lín yù
Wood Sandpiper LC

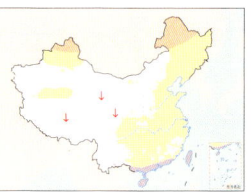

L: 19–23 cm
WS: 54–57 cm
Habitat: Breeds on marshes near coniferous forest or in open areas with abundant bushes. Prefers inland wetlands during migration; often moves about on paddy fields, ponds, marshes, and saltpans. Also occurs in grassy areas and cultivated fields; rarely occurs on intertidal flats.
Behavior: Solitary or in small flocks. Walks slowly and pecks at prey in shallow water; often bobs tail up and down when foraging. Also probes for prey with bill. Occasionally swims to forage. Shy and vigilant. Head constantly bobs forward when alarmed.
Distribution: Common visitor and passage migrant. Breeds in Northeast and northwestern China; migrates through most of China. A few individuals summer in coastal areas; some winter in coastal areas from East to South China (including Hainan and Taiwan), Yunnan, and S Tibet.
Voice: In display flight on breeding grounds, male gives loud and repeated *lior, lior, lior*; flight call a fast *chiff-if-if*.

Tringa brevipes
灰尾漂鹬 huī wěi piāo yù
Gray-tailed Tattler NT

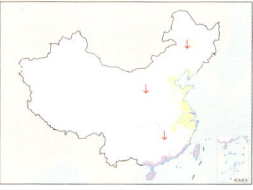

L: 23–28 cm
WS: 60–65 cm
Habitat: Inhabits coastal intertidal flats and estuaries; prefers rocky coasts.
Behavior: Generally occurs solitarily or in small flocks; forms flocks of a few dozen individuals during migration. Rests at high points near tideline during high tide. Forages by running on mudflats during low tide; also probes for prey in shallow waters.
Distribution: Uncommon. Migrates through Northeast, North, East, and South China; a few individuals stay on the east coast and islands in summer. Winters in Hainan and Taiwan.
Voice: Plaintive *to-weet, to-weet*, reminiscent of Black-bellied Plover in tone.

Green Sandpiper

Dark brown upper wings

White rump

Faint spots from head to breast

Round faintly colored spots on scapulars

non-br.

Wood Sandpiper

Black bars on white tail

White rump does not extend to back

Mottled underwing coverts

Clear white supercilium extends behind eye

Dense white spots on blackish-brown scapulars; spots more conspicuous than Green Sandpiper

Relatively short bill with olive base

br.

Wandering Tattler

Nasal grooves three-fourths the length of bill

Wandering Tattler

Nasal grooves half the length of bill

Gray-tailed Tattler

White supercilium extends behind eye

Relatively narrow white supercilium in front of eye

Dark gray flanks

Plain unbarred belly

non-br.

Dark gray back

From face, neck, and belly to undertail coverts heavily marked with wavy black pattern

br.

Gray-tailed Tattler

White from lower belly to undertail coverts, unbarred

non-br.

Relatively thick white supercilium in front of eye

Medium gray back

Inconspicuous gray on flanks

br.

183

Tringa totanus

红脚鹬 hóng jiǎo yù

Common Redshank (LC)

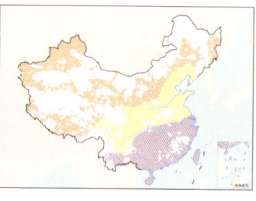

L: 26–29 cm
WS: 59–66 cm
Habitat: Breeds on wet grass or lakes. Occurs on coastal intertidal flats, saltpans, estuaries, fishponds, lakes, marshes, and paddy fields during nonbreeding season.

Behavior: Solitary, in small flocks, or mixes with flocks of other shorebirds. Forages by walking on mudflats.

Distribution: Common. *T. t. ussuriensis* breeds in N Heilongjiang. *T. t. terrignotae* breeds in S Heilongjiang and N Inner Mongolia to the east coast. *T. t. craggi* breeds in NW Xinjiang; may winter on the east coast and southwestern China. *T. t. eurhina* breeds on the Qinghai-Tibet Plateau. All subspecies may pass through most of China during southward migration; many subspecies may pass through coastal areas in southeastern and South China, including Hainan and Taiwan. Nonbreeding plumage of different subspecies is similar and not easy to differentiate.

Voice: Song a loud yodeling *tyoo … tyoo … tyoo*; repeated ringing flight calls *tiu-du*, *tiu*, or longer *diu*.

Tringa erythropus

鹤鹬 hè yù

Spotted Redshank (LC)

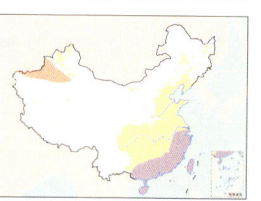

L: 26–33 cm
WS: 61–67 cm
Habitat: Breeds on marshes near forests or in open areas with abundant bushes. During migration and winter, inhabits wetlands, such as coastal intertidal flats, estuaries, ponds, lakes, flooded fields, and marshes.

Behavior: Gregarious. When foraging, probes with bill in mudflats, swipes bill from side to side through shallow waters, or submerges head in relatively deep waters; can swim.

Distribution: Common migrant. Breeding recorded in the Tian Shan Mountains in Xinjiang. Migrates through most of China; a few summering individuals recorded widely across the range. Winters in coastal areas in southeastern and South China, including Hainan and Taiwan. In winter, forms large flocks containing up to a thousand individuals in some areas.

Voice: Loud, clear, rising *chee-wik* in flight.

Tringa nebularia

青脚鹬 qīng jiǎo yù

Common Greenshank (LC)

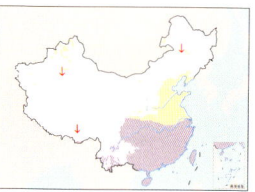

L: 30–35 cm
WS: 68–70 cm
Habitat: Inhabits various wetlands, such as coastal intertidal flats, inland marshes, rivers, lakes, ponds, saltpans, and flooded fields. Highly adaptable.

Behavior: Solitary or in small flocks; occasionally seen in large flocks containing up to several hundred individuals during migration. Often mixes with other shorebirds, egrets, or herons. Wades slowly in shallow water, or forages while walking in water.

Distribution: Common. Migrates through most of China; a few summering individuals recorded widely across the range. Winters in areas south of the Yangtze River, including Hainan and Taiwan.

Voice: Loud flight call a distinctive, ringing, fluty *tyu–tyu–tye*, descending in tone, often repeated three times but may also continue; gives monotone and sharp *chip* when alarmed.

Tringa guttifer

小青脚鹬 xiǎo qīng jiǎo yù

Nordmann's Greenshank (EN)

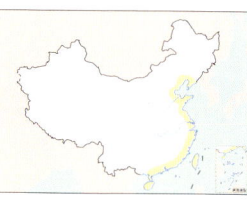

L: 29–32 cm
WS: 55 cm
Habitat: Inhabits wetlands, such as coastal intertidal flats, estuaries, and marshes; also forages on wet grass, saltpans, and paddy fields near coasts.

Behavior: In small flocks; when tide rises, often associates with medium-sized to large shorebirds, such as curlews, Black-bellied Plover, Bar-tailed Godwit, and Great Knot. Forages near the tideline when tide recedes. Prefers crabs; also feeds on other invertebrates and small fish.

Distribution: Rare passage migrant. Migrates along the coast of China, including Hainan and Taiwan; a few individuals stay on the tidal mudflats at Dongtai, Jiangsu, throughout the summer.

Voice: Short, single, flat, yipped *kyew* and a more nasal *gwak* in flight.

Common Redshank

White rump extends to back

Large white panel on trailing edge of secondaries

Relatively short bill with red base

Heavily streaked/spotted throughout body

br.

Red legs

Few streaks/spots on flanks

non-br.

Pale fringe on scapulars

Base of bill pale, barely red

juv.

Orange-yellow legs

Common Greenshank

White from rump to back

Black barring on tail

Gray-barred under wing

Relatively long upturned bill with relatively thick yellowish-green base

non-br.

Relatively long tibia

Heavily streaked from neck to side of breast

br.

Spotted Redshank

White rump extends to back

non-br.

White underparts

No white panel on trailing edge of secondaries

Black overall; body exhibits different shades of black during molt

Long slender bill with slightly drooping tip and red base of lower mandible

br.

Neck and tibia longer than Common Redshank

Nordmann's Greenshank

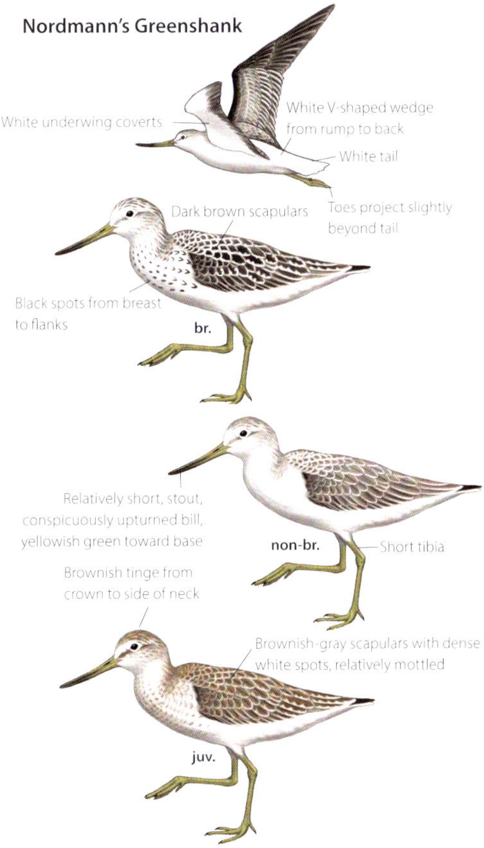

White underwing coverts

White V-shaped wedge from rump to back

White tail

Toes project slightly beyond tail

Dark brown scapulars

Black spots from breast to flanks

br.

Relatively short, stout, conspicuously upturned bill, yellowish green toward base

Brownish tinge from crown to side of neck

non-br.

Short tibia

Brownish-gray scapulars with dense white spots, relatively mottled

juv.

185

Tringa flavipes
小黄脚鹬 xiǎo huáng jiǎo yù
Lesser Yellowlegs　

L: 23–25 cm
WS: 59–64 cm
Habitat: Occurs on coastal and inland wetlands, including flooded fields, marshes, mangroves, lagoons, and intertidal flats; prefers water bodies with emergent vegetation.
Behavior: Generally occurs solitarily. Forages with elegant and slow actions.
Distribution: Vagrant in Hong Kong and Taiwan.
Voice: Loud, ringing, Common Redshank–like *diu* or *diu-diu* in flight.

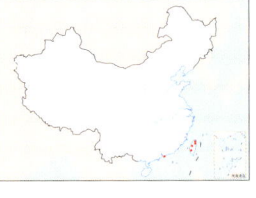

Glareola maldivarum
普通燕鸻 pǔ tōng yàn héng
Oriental Pratincole

L: 24–28 cm
Habitat: Inhabits grasslands or sandy beaches near lakes, rivers, riverbeds, and marshes; also on flooded fields, cultivated lands, and coastal beaches.
Behavior: Slowly chases prey on ground or with short run; also often forages on the wing over wetlands.
Distribution: Breeds in most regions across Northeast China to central and eastern parts of China; migrates through southwestern and South China. Common.
Voice: Little Tern–like *kirrik* and similar.

Glareola pratincola
领燕鸻 lǐng yàn héng
Collared Pratincole

L: 24–28 cm
Habitat: Generally inhabits grasslands, deserted fields, and farmland near water on open plains, also on sandflats in rivers.
Behavior: Gregarious. Forages mainly in flight, occasionally on ground.
Distribution: Uncommon breeder in N and SW Xinjiang; vagrant to W Qinghai and Hong Kong.
Voice: Little Tern–like call; gives *kik* and *kirrik* in flight, higher pitched than Black-winged Pratincole but very similar to Oriental Pratincole.

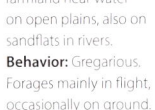

Tringa stagnatilis
泽鹬 zé yù
Marsh Sandpiper

L: 22–26 cm
WS: 55–59 cm
Habitat: Breeds on marshes among grasslands. Inhabits coastal intertidal flats, estuaries, inland lakes, and marshes during migration and winter.
Behavior: Gregarious, also mixes with other shorebirds. When foraging, pecks at food on mudflats, swipes bill from side to side through shallow water, or submerges head in relatively deep water. Capable of swimming.
Distribution: Common. Breeds in Northeast China and E Inner Mongolia. Migrates through most of China, including Hainan and Taiwan; a few individuals winter in coastal areas from East to South China and S Tibet.
Voice: Male makes display flight and gives unique and repetitive *tu-ee-u* on breeding grounds; ringing *plew* or *plew plew plew*, like Common Greenshank but higher pitched and faster.

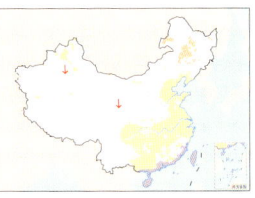

Glareola nordmanni
黑翅燕鸻 hēi chì yàn héng
Black-winged Pratincole

L: 24–28 cm
Habitat: Generally inhabits grasslands, deserted fields, and farmland on open plains at higher latitudes, usually near water. More adaptable to taller and denser vegetation than Collared Pratincole.
Behavior: Gregarious. Forages on various aerial insects in flight.
Distribution: Endemic to Central Asia; rare in N and SW Xinjiang.
Voice: Little Tern–like *ti'kik* in flight.

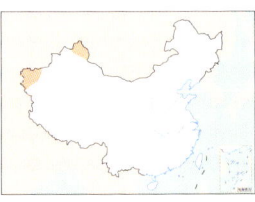

Glareola lactea
灰燕鸻 huī yàn héng
Small Pratincole

L: 15.5–19 cm
Habitat: Active near large rivers with sandbars and shingle banks, also nearby marshes and farmland.
Behavior: Gregarious. Flight agile and rapid; most active at dusk. Primarily forages in flight over water and marshes. Also chases after prey on land.
Distribution: Rare and local in S and W Yunnan and SE Tibet; occasionally seen in Hainan (Haikou).
Voice: Various; mostly short Common Tern–like flight calls *di di* or *diji diji*. Noisy when in flocks.

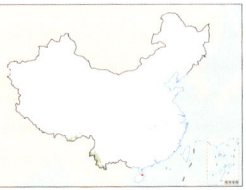

Lesser Yellowlegs

White rump does not extend to back

br.

Black bill with yellowish-brown base, relatively shorter and stouter than bill of Marsh Sandpiper

Barred white tail feathers

White spots on black scapulars

br.

Relatively long bright yellow tibia

White supercilium does not extend behind eye

Yellowish-green base of bill

juv.

Resembles Wood Sandpiper, but neck and legs longer

Marsh Sandpiper

White on rump extends to back

Toes project far beyond tail

br.

Black bars on white tail

Dense small black spots on crown and side of neck

Anchor-shaped black marks on scapulars

br.

Long, fine, straight, sharp, black bill

No black spots; white from face and neck to underparts

non-br.

Relatively long yellowish-green legs

Oriental Pratincole

Long narrow wings

Rufous underwing coverts

ad.

Warm buff throat bordered black, forming collar

ad.

Short legs; swallowlike standing position

imm.

Black-winged Pratincole

No white trailing edge on dark secondaries

Black underwing coverts

ad.

Dark brown back

Wing tips extend beyond tail tip

Collared Pratincole

White trailing edge on secondaries

Chestnut underwing coverts

Relatively pale brown back

Longer, more deeply forked tail

ad.

More red on base of lower mandible

ad.

Wing tips do not extend beyond tail tip

imm.

Small Pratincole

White secondaries with black trailing edge

ad.

Red at base of bill

No black stripes on face

ad.

chick & eggs

imm.

187

Anous stolidus

白顶玄燕鸥 bái dǐng xuán yàn ōu

Brown Noddy **LC**

L: 40–45 cm
WS: 75–86 cm
Habitat: Inhabits small oceanic islands or coastal skerries during breeding season; lives at sea in nonbreeding season.
Behavior: Oceanic species. In flocks, often

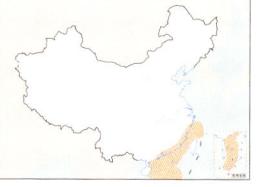

chases schools of fish. Flies with slower wingbeats than other noddies.
Distribution: Rare visitor to or vagrant in Zhejiang, Fujian, Guangdong, Hong Kong, Hainan, and Taiwan.
Voice: Unlikely to be heard in China. Loud, hoarse, purring, and guttural notes at colony, gu–a.

Anous minutus

玄燕鸥 xuán yàn ōu

Black Noddy **LC**

L: 35–39 cm
WS: 66–72 cm
Habitat: Inhabits oceanic islands; breeding colonies often far from coast. In nonbreeding season, wanders over the sea.
Behavior: Oceanic species; flies with fast wingbeats.

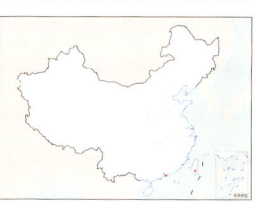

Distribution: Vagrant in Hong Kong and Taiwan (Taitung).
Voice: Unlikely to be heard in China, but at colonies, purring and guttural notes, similar to Brown Noddy but higher pitched, gar-gar.

Gygis alba

白燕鸥 bái yàn ōu

White Tern **LC**

L: 30–33 cm
WS: 70–87 cm
Habitat: Generally lives at sea solitarily or in pairs; also forms small flocks and inhabits small oceanic islands or reefs.
Behavior: Oceanic species.

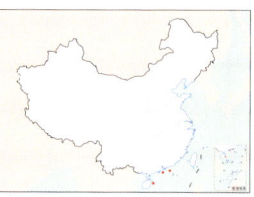

Distribution: Vagrant in Guangdong, Macao, and Hainan.
Voice: Unlikely to be heard in China, but rising and falling yips and chatters at breeding colony.

Rissa tridactyla

三趾鸥 sān zhǐ ōu

Black-legged Kittiwake **VU**

L: 37–41 cm
WS: 93–120 cm
Habitat: Inhabits inshore and coastal water areas in summer; also wanders to freshwater wetlands in winter.
Behavior: Oceanic species; breeds on coastal

and inshore islands. Forms flocks to forage while flying at sea; also follows ships.
Distribution: Uncommon migrant through Beijing and mostly along the east coast; several winter records in S Shaanxi and Sichuan. Rare in winter in Yunnan, Chongqing, and Hubei; rare migrant through Xinjiang, Gansu, Inner Mongolia, and Jilin.
Voice: Silent in the region, but noisy, often hysterical laughing, ki–wak, at breeding colonies.

Xema sabini

叉尾鸥 chā wěi ōu

Sabine's Gull **LC**

L: 27–31 cm
WS: 90–100 cm
Habitat: Commonly lives at sea during nonbreeding season; occasionally blown inshore during typhoons.
Behavior: Oceanic species. Acquires adult

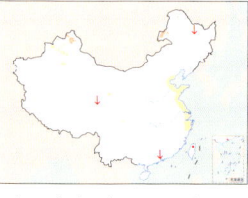

breeding plumage after six molts within 2.5 years after hatching. Flight ternlike, with frequent wingbeats and infrequent glides.
Distribution: Vagrant in Hainan and Taiwan.
Voice: Vagrants silent, but harsh Arctic Tern–like calls from breeding birds.

Hydrocoloeus minutus

小鸥 xiǎo ōu

Little Gull **LC**

L: 24–30 cm
WS: 62–69 cm
Habitat: Breeds inland on saltwater or freshwater lakes and marshes in deserts; inhabits primarily coasts, estuaries, and nearby lakes and marshes during nonbreeding season. Especially prefers water bodies with abundant aquatic plants.
Behavior: Usually gathers in flocks. Commonly flies over water; flight light and agile with shallow wingbeats.
Distribution: Rare breeder in N Xinjiang and NE Inner Mongolia; occasional during migration in Heilongjiang, Tianjin, Beijing, Hebei, Shanxi, Inner Mongolia, Qinghai, Sichuan, Jiangsu, Zhejiang, Shanghai, and Hong Kong; vagrant to Taiwan.
Voice: Soft nasal kep and similar calls, sometimes repeated.

Brown Noddy

Long slender wings

Pale under wing contrasts strongly with dark trailing edge

Long tail appears pointed when closed and slightly forked when open; same color as back

Relatively thick bill

Gray crown

Narrow incomplete white eye ring broken at lores and behind eye

Whitish forehead

Black lores

Primaries shorter than or just reaching tail

Black-legged Kittiwake

non-br.

1st win.

Thick, diffuse, black band on nape

Yellow bill

Black wing tips lack white

non-br.

Black Noddy

Tail paler than back

Relatively slender neck

Long thin bill

Primaries reach or extend beyond tail

Sabine's Gull

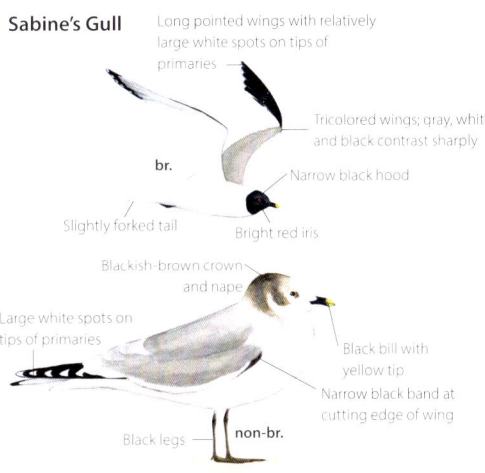

Long pointed wings with relatively large white spots on tips of primaries

Tricolored wings; gray, white, and black contrast sharply

br.

Narrow black hood

Slightly forked tail

Bright red iris

Blackish-brown crown and nape

Large white spots on tips of primaries

Black bill with yellow tip

Narrow black band at cutting edge of wing

Black legs

non-br.

White Tern

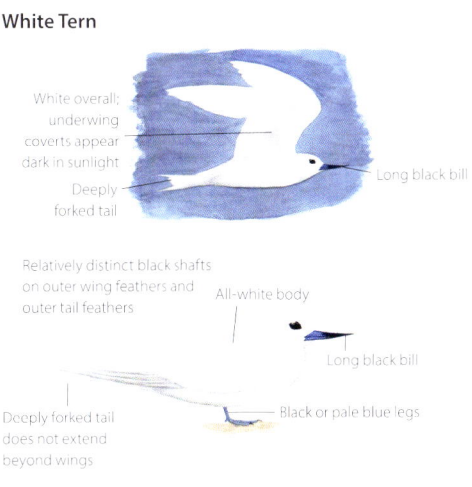

White overall; underwing coverts appear dark in sunlight

Deeply forked tail

Long black bill

Relatively distinct black shafts on outer wing feathers and outer tail feathers

All-white body

Long black bill

Deeply forked tail does not extend beyond wings

Black or pale blue legs

Little Gull

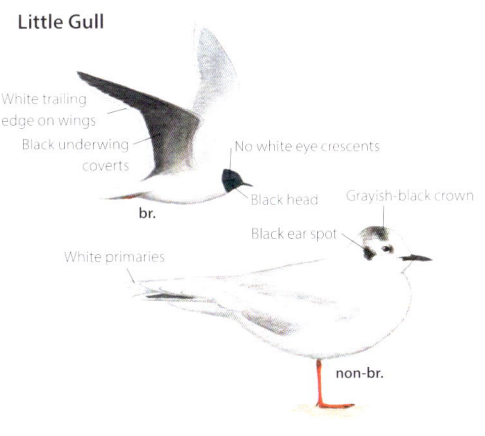

White trailing edge on wings

Black underwing coverts

No white eye crescents

Black head

Grayish-black crown

br.

Black ear spot

White primaries

non-br.

Chroicocephalus genei

细嘴鸥 xì zuǐ ōu

Slender-billed Gull   LC

L: 42–47 cm
WS: 94 cm
Habitat: Commonly
seen on coasts, marshes,
islands, and saltwater
lakes; also occasionally
seen on freshwater lakes
and rivers.
Behavior: Often in small
to large flocks; flight agile.
Distribution: Regular but uncommon in Xinjiang; possibly breeds at
Aibi (Ebi) Lake. Rare in Beijing, Tianjin, Hebei, Yunnan, Hubei, Jiangxi,
Fujian, Guangxi, Guangdong, Hong Kong, and Taiwan during migration
or winter.
Voice: Call harder and drier than that of Black-headed Gull, *krrer*.

Chroicocephalus novaehollandiae

澳洲红嘴鸥 ào zhōu hóng zuǐ ōu

Silver Gull   LC

L: 38–42 cm
WS: 91–96 cm
Habitat: Gregarious.
Generally inhabits coastal
reefs, offshore islands,
inland rivers, lakes,
farmland, and ponds.
Behavior: Gregarious;
bold scavengers.
Distribution: Vagrant in Guangdong and Taiwan.
Voice: High-pitched, descending, melancholy whistle, *pleee*, as well as
harsher Black-headed Gull–like *kreearrh* and short *kek*.

Chroicocephalus brunnicephalus

棕头鸥 zōng tóu ōu

Brown-headed Gull   LC

L: 40–46 cm
WS: 105–115 cm
Habitat: Breeds on lakes,
rivers, and marshes on
plateaus; inhabits various
wetlands in nonbreeding
season.
Behavior: Similar to
Black-headed Gull. Flocks
over wetlands.
Distribution: Common breeder on the Qinghai-Tibet Plateau and
Xinjiang; migrates through central and western regions of China; rare in
eastern China.
Voice: Very noisy at colonies. Generally similar to but lower pitched than
Black-headed Gull.

Chroicocephalus ridibundus

红嘴鸥 hóng zuǐ ōu

Black-headed Gull   LC

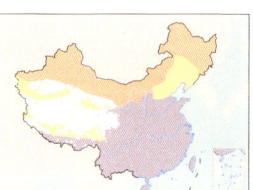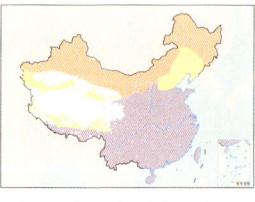

L: 36–42 cm
WS: 100–110 cm
Habitat: Inhabits various
wetlands, including lakes
in urban areas.
Behavior: Solitary or in
flocks. Wingbeats fast.
Distribution: Widely seen
across China; common.
Voice: Very noisy at colonies. Rapid grating *krreearrh* and shorter *kip*
and *kik* notes.

Saundersilarus saundersi

黑嘴鸥 hēi zuǐ ōu

Saunders's Gull   VU

L: 30–33 cm
WS: 87–91 cm
Habitat: Highly
dependent on coasts;
seldom occurs inland.
Breeds in dry, low
meadows with sparse
vegetation near coasts.
Behavior: Flight graceful;
can descend rapidly. Wings appear long and pointed. Seldom floats at
sea. Nests among coastal vegetation such as *Suaeda*.
Distribution: Breeds in coastal regions of Bohai Sea and N Yellow
Sea; winters along the coasts from Yellow Sea to South China Sea;
uncommon. Rare away from the coastline; vagrant to Beijing and
occasional in Jiangxi.
Voice: High-pitched, ternlike, piercing, chattering notes, *kip* and *khirk*,
calls often repeated. Some notes similar to Little Tern; others far less
nasal.

Rhodostethia rosea

楔尾鸥 xiē wěi ōu

Ross's Gull   LC

L: 32–36 cm
WS: 90–100 cm
Habitat: Breeds on
tundra; occurs around sea
ice and rarely along coasts
in nonbreeding season.
Behavior: Usually
depends on permanent
sea ice during
nonbreeding season.
Distribution: Vagrant in Liaoning (historic), Qinghai, and recently in
Sichuan (Deyang).
Voice: Vagrants generally silent—on breeding grounds, some notes
similar to song elements of Swinhoe's Snipe.

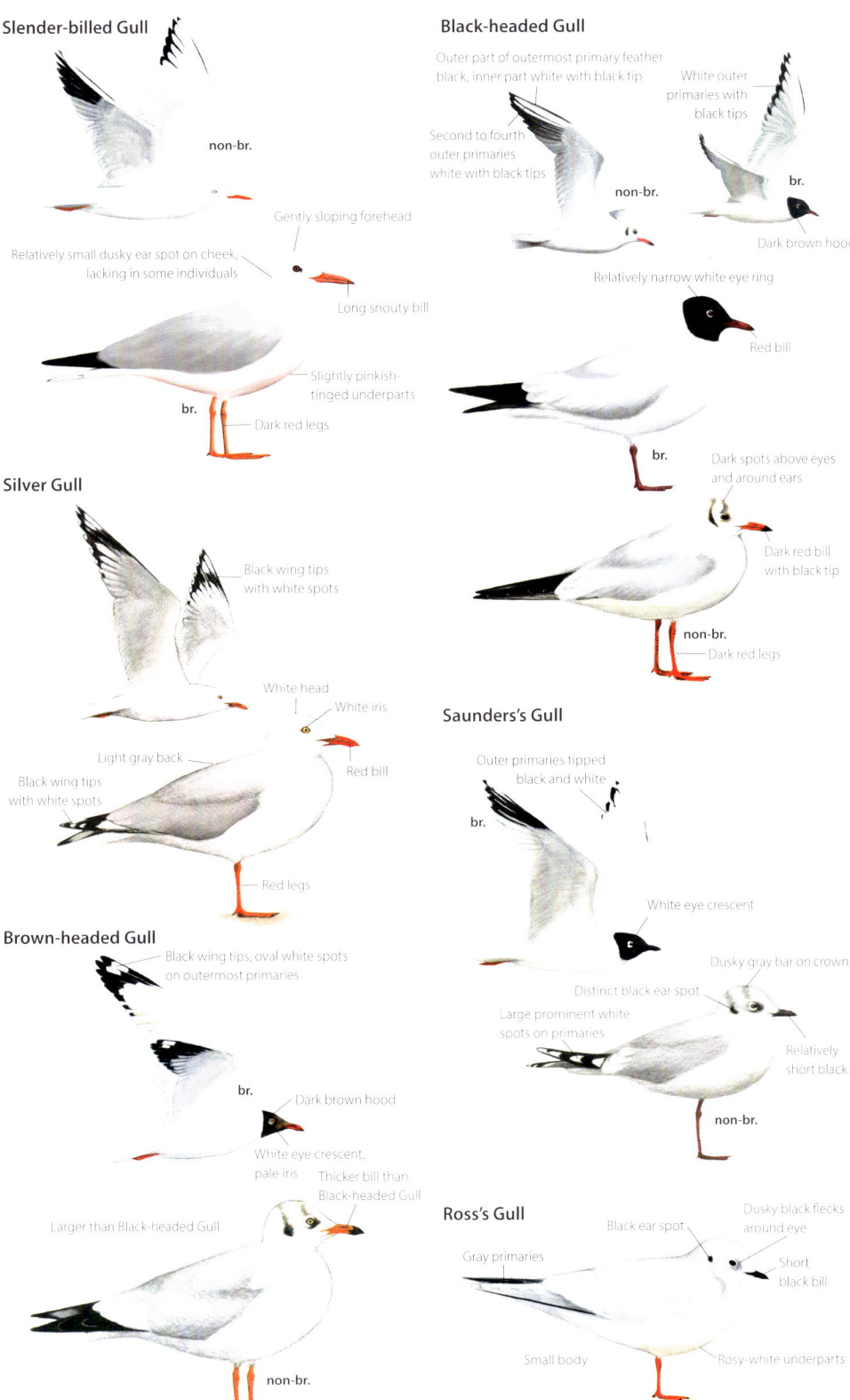

Slender-billed Gull

non-br.

Gently sloping forehead

Relatively small dusky ear spot on cheek,
lacking in some individuals

Long snouty bill

Slightly pinkish-
tinged underparts

br.

Dark red legs

Silver Gull

Black wing tips
with white spots

White head

White iris

Light gray back

Red bill

Black wing tips
with white spots

Red legs

Brown-headed Gull

Black wing tips, oval white spots
on outermost primaries

br.

Dark brown hood

White eye crescent,
pale iris

Thicker bill than
Black-headed Gull

Larger than Black-headed Gull

non-br.

Black-headed Gull

Outer part of outermost primary feather
black, inner part white with black tip

White outer
primaries with
black tips

Second to fourth
outer primaries
white with black tips

non-br.

br.

Dark brown hood

Relatively narrow white eye ring

Red bill

br.

Dark spots above eyes
and around ears

Dark red bill
with black tip

non-br.

Dark red legs

Saunders's Gull

Outer primaries tipped
black and white

br.

White eye crescent

Dusky gray bar on crown

Distinct black ear spot

Large prominent white
spots on primaries

Relatively
short black bill

non-br.

Ross's Gull

Dusky black flecks
around eye

Black ear spot

Short
black bill

Gray primaries

Small body

Rosy-white underparts

non-br.

Leucophaeus atricilla
笑鸥 xiào ōu
Laughing Gull **LC**

L: 36–40 cm
WS: 95–120 cm
Habitat: Inhabits various coastal open habitats.
Behavior: Kleptoparasitic on other waterbirds. Dark first-year juveniles often mistaken for skuas.
Distribution: Vagrant in Taiwan and Shanghai (Chongming Dongtan).
Voice: *Kiu kaw guo–guo guo guo*, like Common Gull, but flatter and lower pitched.

Leucophaeus pipixcan
弗氏鸥 fú shì ōu
Franklin's Gull **LC**

L: 32–35 cm
WS: 81–93 cm
Habitat: Breeds on prairie marshes in North America; inhabits various wetlands during migration.
Behavior: Flight graceful and agile; standing posture flat. The only gull that can have two complete molts within one year.
Distribution: Vagrants to Tianjin, Hebei, Zhejiang, Hong Kong, and Taiwan. Very rare.
Voice: Most calls shorter, rising, and higher pitched than Laughing Gull, *kowi*.

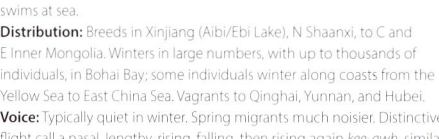

Ichthyaetus relictus
遗鸥 yí ōu
Relict Gull **VU**

L: 38–46 cm
WS: 119–122 cm
Habitat: Breeds on islands in desert lakes; wanders following precipitation. Winters at estuaries and coasts.
Behavior: Gregarious. Forages on tidal flats. Also swims at sea.
Distribution: Breeds in Xinjiang (Aibi/Ebi Lake), N Shaanxi, to C and E Inner Mongolia. Winters in large numbers, with up to thousands of individuals, in Bohai Bay; some individuals winter along coasts from the Yellow Sea to East China Sea. Vagrants to Qinghai, Yunnan, and Hubei.
Voice: Typically quiet in winter. Spring migrants much noisier. Distinctive flight call a nasal, lengthy, rising, falling, then rising again *kee-awh*, similar to the more strident, flatter calls of Pallas's Gull.

Ichthyaetus ichthyaetus
渔鸥 yú ōu
Pallas's Gull **LC**

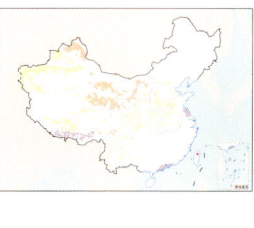

L: 58–67 cm
WS: 146–162 cm
Habitat: Breeds in lakes on plateaus. In nonbreeding season, inhabits primarily coasts, estuaries, large lakes, reservoirs, and marshes. Especially prefers water bodies with aquatic plants.
Behavior: Gregarious.
Distribution: Breeds in E Qinghai, W Inner Mongolia, and N Xinjiang. Commonly migrates through Western China, such as Xinjiang, Qinghai, and Tibet. Rare winter visitor to Northeast China and occasional in Tibet, S Xinjiang, and from Beijing and Liaoning south along the coast to Taiwan and Hong Kong.
Voice: Deep, throaty, resonant, crowlike calls, *aargh*.

Larus crassirostris
黑尾鸥 hēi wěi ōu
Black-tailed Gull **LC**

L: 46–48 cm
WS: 118–124 cm
Habitat: Inhabits estuaries and inshore habitats; breeds on oceanic islands. Very rare inland; some inland records could be misidentifications.
Behavior: Often in small flocks. Aggressive. Follows ships.

Distribution: Predominantly coastal. Very common in summer in northern China; modest numbers winter in coastal southern China. Exceptional inland.
Voice: Long call a high-pitched catlike mewing, *a–a–a*; hence the local nickname, "sea cat."

Larus canus
普通海鸥 (海鸥) pǔ tōng hǎi ōu (hǎi ōu)
Common Gull **LC**

L: 44–52 cm
WS: 105–135 cm
Habitat: Adapted to various wetlands; usually appears near coasts. Also can be seen on inland lakes.
Behavior: Often in pairs or small flocks. Flight smooth, with powerful wingbeats.

Distribution: *L. c. kamtschatschensis* is common in coastal eastern China in fall and winter; only occasionally seen inland. *L. c. heinei* is more widespread, with records from Xinjiang east to the Yellow Sea and Bohai Sea, but much rarer than *L. c. kamtschatschensis* along the coast.
Voice: High-pitched, piercing, and short *a–a–a*.

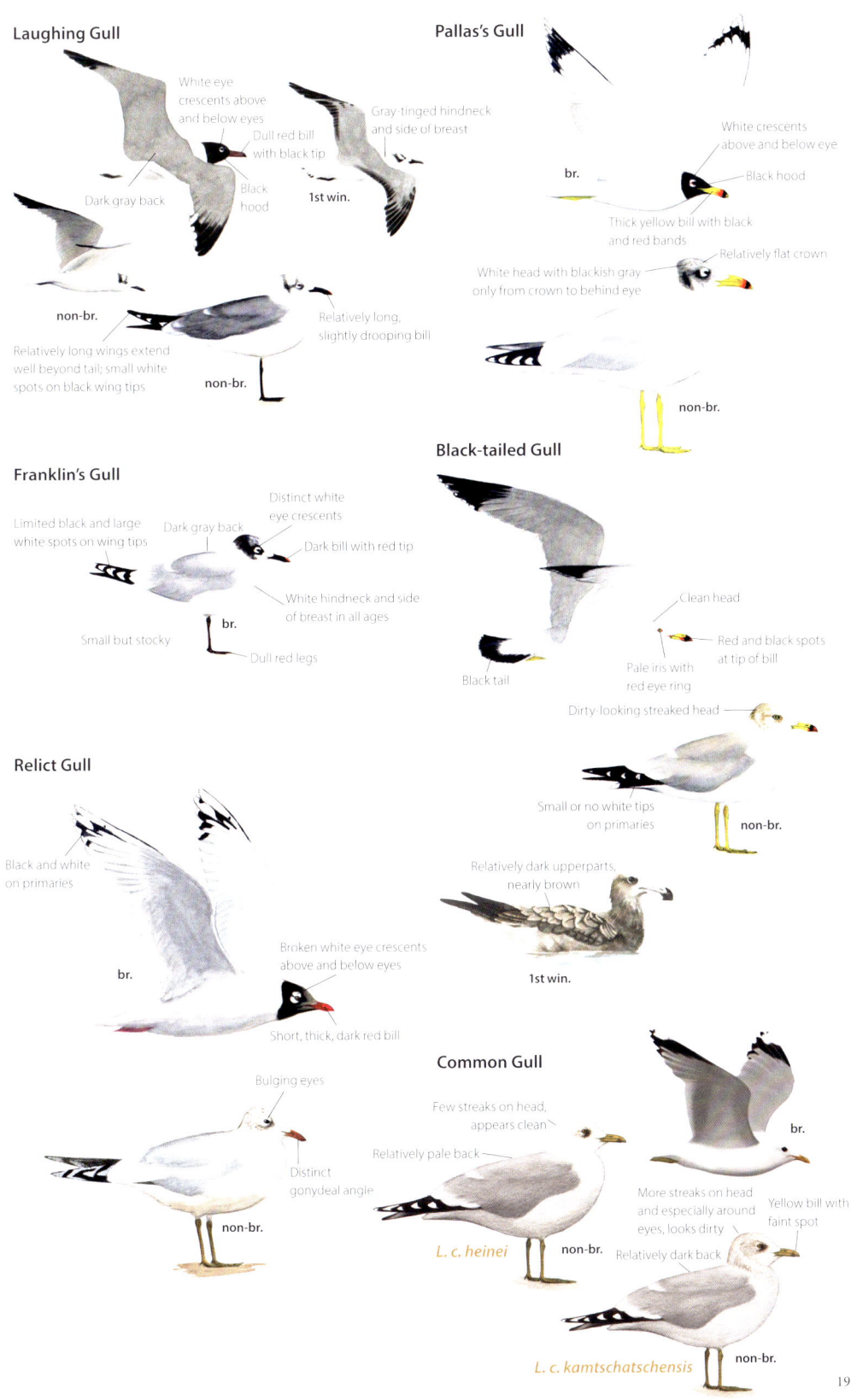

Laughing Gull

White eye crescents above and below eyes

Dull red bill with black tip

Gray-tinged hindneck and side of breast

Dark gray back

Black hood

1st win.

non-br.

Relatively long wings extend well beyond tail; small white spots on black wing tips

Relatively long, slightly drooping bill

non-br.

Pallas's Gull

White crescents above and below eye

Black hood

br.

Thick yellow bill with black and red bands

Relatively flat crown

White head with blackish gray only from crown to behind eye

non-br.

Franklin's Gull

Limited black and large white spots on wing tips

Dark gray back

Distinct white eye crescents

Dark bill with red tip

White hindneck and side of breast in all ages

Small but stocky

br.

Dull red legs

Black-tailed Gull

Clean head

Red and black spots at tip of bill

Pale iris with red eye ring

Black tail

Dirty-looking streaked head

Small or no white tips on primaries

non-br.

Relatively dark upperparts, nearly brown

1st win.

Relict Gull

Black and white on primaries

Broken white eye crescents above and below eyes

br.

Short, thick, dark red bill

Bulging eyes

Distinct gonydeal angle

non-br.

Common Gull

Few streaks on head, appears clean

Relatively pale back

br.

More streaks on head and especially around eyes, looks dirty

Yellow bill with faint spot

L. c. heinei

non-br.

Relatively dark back

L. c. kamtschatschensis

non-br.

Larus glaucescens

灰翅鸥 huī chì ōu

Glaucous-winged Gull　LC

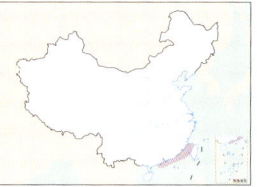

L: 60–66 cm
WS: 137–150 cm
Habitat: Inhabits wide range of coastal habitats.
Behavior: In small flocks; flies at sea, stands at shores, or slowly forages in intertidal zones.
Distribution: Very rare in China, with scattered records along the coast south of Shanghai.
Voice: Deep, slow, and lengthy *kaka*.

Larus hyperboreus

北极鸥 běi jí ōu

Glaucous Gull　LC

L: 62–70 cm
WS: 140–160 cm
Habitat: Inhabits coastal bays, estuaries, harbors, etc.
Behavior: Forages on intertidal flats alone or in small flocks; also mixes with other gulls.
Distribution: Generally rare and confined to the east coast. Regular only in Northeast China (coastal Liaoning in winter and Jilin in spring). Vagrant to many inland provinces, even west to Xinjiang and Tibet.
Voice: Harsh calls, like other large gulls.

Larus vegae

西伯利亚银鸥 xī bó lì yà yín ōu

Vega Gull　LC

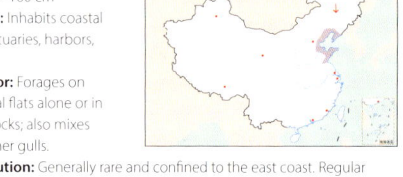

L: 55–68 cm
WS: 120–150 cm
Habitat: Inhabits large inland water bodies, such as rivers, lakes, and reservoirs, or coastal bays, intertidal zones, and reefs.
Behavior: Gregarious. Forages while flying slowly; also like to pick up food following ships.
Distribution: *L. v. mongolicus* common and widespread across much of China except for Ningxia, Tibet, and Qinghai; breeds in Northeast and northwestern China; migrates through most regions in northern China and winters in southern China. *L. v. vegae* occurs along the east coast from Shandong to Guangdong and is rare at inland wetlands in Jiangxi, Beijing, and Hubei.
Voice: Hoarse *gaga*.

Larus cachinnans

黄腿银鸥 (黄脚银鸥)

huáng tuǐ yín ōu (huáng jiǎo yín ōu)

Caspian Gull　LC

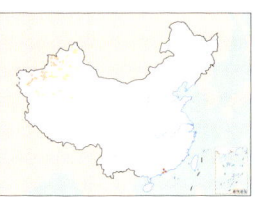

L: 55–60 cm
WS: 138–147 cm
Habitat: Occurs in grasslands, rivers, and lakes in semideserts from lowland to mountains.
Behavior: Gregarious. Forages mainly on fish and aquatic invertebrates; also catches rodents, lizards, and grasshoppers and eats animal carcasses on farmland or deserted fields near water. Forms flocks of hundreds of individuals near the edges of open water bodies during migration.
Distribution: Uncommon breeder in N and W Xinjiang; common in Xinjiang during migration; vagrant to Guangdong, Hong Kong, and Macao.
Voice: Long call higher pitched than that of Vega Gull (*L. v. vegae*) and even *L. v. mongolicus*.

Larus schistisagus

灰背鸥 huī bèi ōu

Slaty-backed Gull　LC

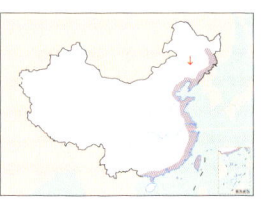

L: 60–67 cm
WS: 145–150 cm
Habitat: Often inhabits coastal tidal flats, estuaries, salt ponds, harbors, etc.
Behavior: Solitary or in small flocks. Flies over water or stands still on shores. Also forages following ships.
Distribution: Uncommon in Northeast China and vagrant to the east and south coasts. Questionable inland records.
Voice: Like other "Herring" gulls, but long call particularly lengthy.

Larus fuscus

小黑背银鸥 xiǎo hēi bèi yín ōu

Lesser Black-backed Gull　LC

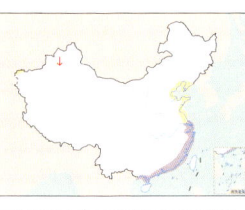

L: 51–70 cm
WS: 138–158 cm
Habitat: Inhabits coastal tidal flats, estuaries, harbors, etc.
Behavior: Similar to other "Herring" gulls.
Distribution: *L. f. heuglini* seen in Xinjiang and along the east and south coasts; locally common. *L. f. barabensis* apparently a vagrant to Hong Kong.
Voice: Like other "Herring" gulls.

Glaucous-winged Gull

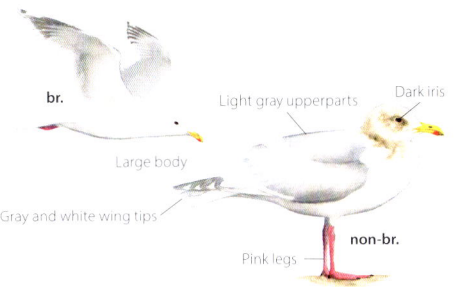

br.

Light gray upperparts

Dark iris

Large body

Gray and white wing tips

Pink legs

non-br.

Caspian Gull

br.

Dark iris

Yellow or orange legs

non-br.

Glaucous Gull

br.

Extremely pale upperparts

Pale iris

All-white wing tips

Pink legs

non-br.

Slaty-backed Gull

br.

Extremely dark slate-gray upperparts

Stocky

Dark pink legs

non-br.

Vega Gull

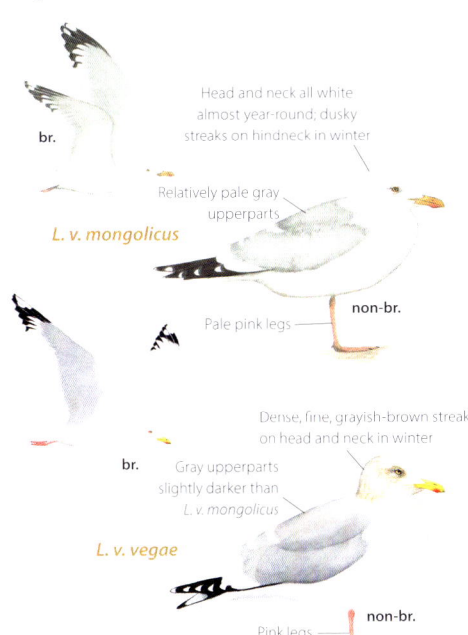

br.

Head and neck all white almost year-round; dusky streaks on hindneck in winter

Relatively pale gray upperparts

L. v. mongolicus

Pale pink legs

non-br.

br.

Dense, fine, grayish-brown streaks on head and neck in winter

Gray upperparts slightly darker than *L. v. mongolicus*

L. v. vegae

Pink legs

non-br.

Lesser Black-backed Gull

White head and neck in winter with brown shaft streaks

br.

Dark gray upperparts

L. f. heuglini

Yellowish legs

non-br.

Yellow bill with contrasting black and red on tip

br.

Black-gray upperparts

L. f. barabensis

Yellowish legs

non-br.

Hydroprogne caspia
红嘴巨燕鸥（红嘴巨鸥）
hóng zuǐ jù yàn ōu (hóng zuǐ jù ōu)
Caspian Tern   🄻🄲

L: 48–55 cm
WS: 125–140 cm
Habitat: Often inhabits coastal or estuarine wetlands; also seen on inland lakes.
Behavior: Solitary or in small flocks. Circles above water's surface with slow and light wingbeats; plunges into water for food.
Distribution: Locally common in Xinjiang and Northeast, North, East, Central, and South China.
Voice: Low-pitched, hoarse, heronlike, and intimidating *argh-ya*. Begging call of immature a pathetic, high-pitched whistle, *kyi-ve*.

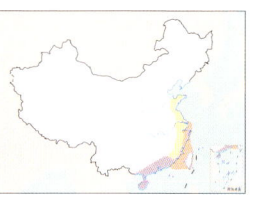

Thalasseus bergii
大凤头燕鸥 dà fèng tóu yàn ōu
Great Crested Tern   🄻🄲

L: 45–53 cm
WS: 100–130 cm
Habitat: Breeds on uninhabited oceanic islands. Wanders at sea during nonbreeding season; occasionally rests on coastal tidal flats, fishponds, and salt ponds, etc.
Behavior: Gregarious; frequently flies over sea, searching for fish in water. Bill points downward during flight; wingbeats slow. Once prey is located, flexes its wings and plunge-dives, then immediately comes out of water. Flocks fly to shallow areas near coastal tidal flats to bathe daily at specific times.
Distribution: Locally common migrant. Breeds on uninhabited oceanic islands along the coasts of southeastern China (Zhejiang, Fujian, Taiwan, and Guangdong); winters in Fujian, Guangdong, Hong Kong, Guangxi, and Hainan, occasionally to coasts in Jiangsu, Shanghai, and Shandong.
Voice: Breeding flocks make noisy *korr–korr–korr*; alarm call is a piercing *kerrak*, more powerful and deeper than Lesser Crested Tern.

Thalasseus sandvicensis
白嘴端凤头燕鸥
bái zuǐ duān fèng tóu yàn ōu
Sandwich Tern   🄻🄲

L: 36–46 cm
WS: 86–105 cm
Habitat: Forages at sea; occasionally inhabits and rests on coastal sandy beaches, rocky beaches, fishponds, intertidal zones, and shoals near harbors.

Thalasseus bengalensis
小凤头燕鸥 xiǎo fèng tóu yàn ōu
Lesser Crested Tern   🄻🄲

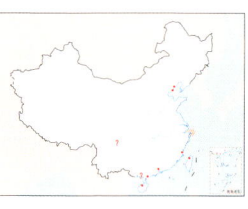

L: 35–43 cm
WS: 88–105 cm
Habitat: Inhabits tropical and subtropical oceans. Nests on uninhabited oceanic islands during breeding season; occasionally rests on shoals of coastal tidal flats, fishponds, and salt ponds.
Behavior: Similar to Great Crested Tern; often appears together with Great Crested Tern in China.
Distribution: *T. b. torresii* occasionally wanders to China's coastal regions in postbreeding season, with scattered records in Hebei, Tianjin, Zhejiang, Fujian, Guangdong, Taiwan, Guangxi, and Sichuan; very rare. Breeding recorded on islands in Zhejiang.
Voice: *kerrick* calls in flight.

Thalasseus bernsteini
中华凤头燕鸥 zhōng huá fèng tóu yàn ōu
Chinese Crested Tern   🄲🅁

L: 38–43 cm
WS: 94 cm
Habitat: Inhabits tropical and subtropical oceans. Prefers foraging in regions near coasts and with shallow sea water; rests, bathes, and mates in shallow areas on coastal tidal flats. Breeds on uninhabited oceanic islands.

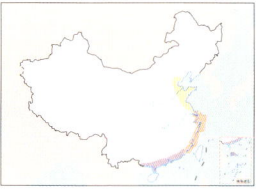

Behavior: Similar to Great Crested Tern. Usually mixes with Great Crested Tern during breeding season; migrates in single-species flocks after breeding. In winter, mixes again with flocks of Great Crested Tern.
Distribution: Extremely small population. Breeds only on several uninhabited oceanic islands in Zhejiang, Fujian, and Taiwan. Historically bred in Shandong (Qingdao), with no recent breeding records. Winters along coasts of Guangdong; may also appear in Guangxi and Hainan. Postbreeding wanderers recorded along the coasts of Shanghai, Jiangsu, Shandong, Tianjin, and Hebei.
Voice: Piercing, high-pitched *keerrick* calls, sharper than Great Crested Tern, very similar to Lesser Crested Tern and Sandwich Tern.

Behavior: Often mixes with other terns in China. Forages mainly on fishes; also feeds on crustaceans, mollusks, and other marine invertebrates.
Distribution: Vagrant in Guangdong (Zhanjiang) and Taiwan (Yilan); breeding recorded on islands in Zhejiang.

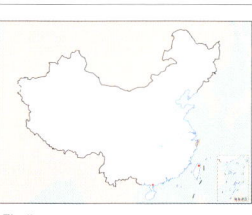

Voice: Piercing *kerrick* in flight, very similar to Lesser Crested Tern.

Caspian Tern

Black from forehead to crown

Extremely large thick red bill with black tip

br.

Relatively short, deeply forked tail

Prominent thick red bill

Slightly short crest; black streaks on forehead and crown

Black legs — non-br.

Lesser Crested Tern

br.

Eyes fully exposed, black-tinged lores

Black crest molts into white

non-br.

Black crest reaches base of bill

Relatively pale gray upperparts

Thin orange bill

Smaller than Great Crested Tern

br.

Great Crested Tern

Primaries tipped black

Pointed slender wings

br.

Prominent yellow bill

White on crown near forehead, appears mottled

Eyes fully exposed

non-br.

Black from crest to nape

White forehead and lores

Dark gray upperparts and wings

Long slender chrome-yellow bill

br.

Chinese Crested Tern

Black crest molts into white, appears mottled

Eyes fully exposed; black-tinged lores

All-white body

br.

non-br.

Black crest reaches base of bill

Distinct white on wings, back, rump, and tail

Orange bill with black tip and small white point at end

Slightly smaller than Great Crested Tern

br.

Sandwich Tern

Black on crown extends to base of bill

br.

Black crest molts into white, appears mottled

Eyes fully exposed; black-tinged lores

Gray wings

Black bill with yellow tip

non-br.

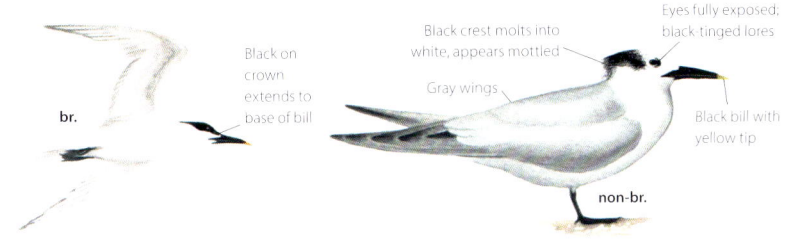

Sternula albifrons
白额燕鸥 bái é yàn ōu
Little Tern LC

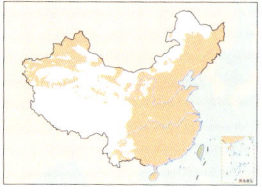

L: 20–28 cm
WS: 48–55 cm
Habitat: Inhabits a wide range of freshwater habitats or estuaries and tidal flats.
Behavior: Flight swift. Often hovers with bill pointing downward, plunge-diving into shallow water for food.
Distribution: *S. a. albifrons* seen in Xinjiang. *S. a. sinensis* ubiquitously distributed across China except for Xinjiang and Tibet. Common.
Voice: High-pitched, piercing *kyi* or *kyiet* and rapid chattering in display.

Onychoprion aleuticus
白腰燕鸥 bái yāo yàn ōu
Aleutian Tern VU

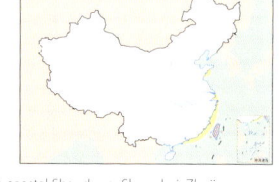

L: 32–34 cm
WS: 75–80 cm
Habitat: Lives mainly at sea; also inhabits coastal water bodies.
Behavior: Mostly oceanic. Flight graceful; rarely plunges into water for food.
Distribution: Uncommon in coastal Shandong, Shanghai, Zhejiang, Fujian, Guangdong, Hong Kong, and Taiwan.
Voice: Distinctive, multisyllabic, chittering calls reminiscent of sand-plovers, and others mellower and reminiscent of Eurasian Tree Sparrow.

Onychoprion anaethetus
褐翅燕鸥 hè chì yàn ōu
Bridled Tern LC

L: 36–41 cm
WS: 77–81 cm
Habitat: Inhabits oceanic islands and reefs; also rests on floating debris at sea.
Behavior: Oceanic species. Gathers in flocks and flies over sea; dips from the air to pluck food from water's surface; rarely plunges into water. Flight more graceful and variable than Sooty Tern.
Distribution: Breeds on uninhabited islands in the East and South China Seas.
Voice: Vagrants essentially silent. At colonies, chattering rattles, nasal *nyip*, and *kwark*, sometimes doubled.

Onychoprion fuscatus
乌燕鸥 wū yàn ōu
Sooty Tern LC

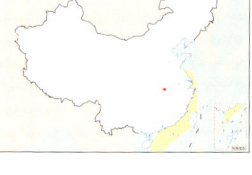

L: 38–45 cm
WS: 82–94 cm
Habitat: Seldom settles on ground except for on nest during breeding season.
Behavior: Oceanic birds. Strong wingbeats. Seldom lands on water; forages mostly over water's surface.
Distribution: Seen along coasts in East and South China, including Hainan and Taiwan; occasionally inland. Most records in China are juveniles with all-black head and neck. Rare.
Voice: Nasal *kirri-wa-ke*.

Sterna aurantia
河燕鸥 (黄嘴河燕鸥) hé yàn ōu (huáng zuǐ hé yàn ōu)
River Tern NT

L: 38–46 cm
WS: 80–85 cm
Habitat: Inhabits large rivers and surrounding marshes, ponds, reservoirs, and lakes. Nests mainly on sandy beaches.
Behavior: Often solitary or in small flocks. Can fly back and forth for a long time, patrolling over water's surface. Flight graceful and agile, with slow and easy wingbeats; when prey is spotted, rapidly plunges into water and rises back into the air.
Distribution: Local in W and S Yunnan and SE Tibet. Rare.
Voice: Loud and piercing *diu* and paired *diu diu* in flight.

Sterna dougallii
粉红燕鸥 fěn hóng yàn ōu
Roseate Tern LC

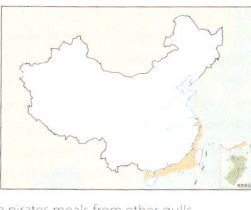

L: 31–38 cm
WS: 72–80 cm
Habitat: Breeds on oceanic islands or reefs; inhabits shores, reefs, and oceanic islands or lives at sea.
Behavior: Gregarious, similar to other terns. Flight graceful and variable. Also pirates meals from other gulls.
Distribution: Breeds on uninhabited islands in the East and South China Seas. Locally common.
Voice: Sharp disyllabic *kevick* calls, reminiscent of Spotted Redshank.

Little Tern

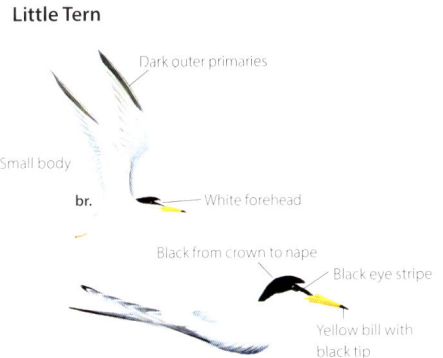

Dark outer primaries

Small body

White forehead

br.

Black from crown to nape

Black eye stripe

Yellow bill with black tip

Short yellow legs

br.

Sooty Tern

All-black primaries contrast with white underwing coverts

br.

White cutting edge of upper wing

Deeply forked tail

White forehead does not reach behind eye

Black back

Black bill

Long deeply forked tail

Black legs

br.

Aleutian Tern

Prominent black bar on trailing edge of secondaries on under wing

White forehead

White rump and tail

Deeply forked tail

br.

Black crown and nape

Black eye stripe

Gray upperparts

Black bill

Black legs

br.

Gray to white underparts

River Tern

Deeply forked tail with extremely long outer feathers

br.

White belly, unlike Black-bellied Tern

Black crown

Yellow bill

br.

Bridled Tern

White outer tail feathers

White cutting edge of wing

Deeply forked tail

br.

Black eye stripe

Black from forehead to nape

White supercilium reaches forehead and behind eye

Dusky brown upperparts and wings

Black bill

Black legs

br.

Roseate Tern

Deeply forked long tail

br.

Rosy breast

Black from forehead to nape

Extremely long outer tail feathers; tail extends well beyond wings

Red bill with black tip

Red legs

br.

Sterna sumatrana

黑枕燕鸥 hēi zhěn yàn ōu

Black-naped Tern LC

L: 30–35 cm
WS: 60–61 cm
Habitat: Inhabits coasts, reefs, or oceanic islands.
Behavior: Breeds in small flocks on uninhabited islands. Flies and forages at sea.
Distribution: Breeds on uninhabited islands in the East and South China Seas. Occasionally seen in the Yellow Sea region.
Voice: Varied; mostly short sharp chips and huskier churrs.

Sterna hirundo

普通燕鸥 pǔ tōng yàn ōu

Common Tern LC

L: 31–38 cm
WS: 77–98 cm
Habitat: Inhabits inland wetlands, such as rivers, lakes, marshes, ponds, etc. Also seen on coastal and estuarine wetlands.
Behavior: Gathers in small flocks. Flight graceful and fast. Forages by flying low over water's surface, plunging into water for prey, and rising back into the air.
Distribution: *S. h. hirundo* in Western China; *S. h. tibetana* breeds on the Qinghai-Tibet Plateau; *S. h. minussensis* breeds from Inner Mongolia south to Central China; *S. h. longipennis* widespread in eastern China. Common.
Voice: Often noisy when breeding, when it gives a variety of notes, sometimes in excited series. Some notes, such as *kiyarri*, are repeated.

Sterna acuticauda

黑腹燕鸥 hēi fù yàn ōu

Black-bellied Tern EN

L: 28–33 cm
WS: 55–60 cm
Habitat: Inhabits flat areas near wetlands, such as large rivers and marshes. Breeds on shoals and bars in large rivers.
Behavior: Often flies alone or in small flocks over rivers and ponds; seldom rests on ground.
Distribution: Historic records from Yunnan (Yingjiang and Jinghong). Probably extirpated.
Voice: High-pitched and piercing calls short, like other terns.

Chlidonias hybrida

灰翅浮鸥 (须浮鸥) huī chì fú ōu (xū fú ōu)

Whiskered Tern LC

L: 23–28 cm
WS: 74–84 cm
Habitat: Inhabits a wide range of freshwater habitats, such as lakes, marshes, ponds, and flooded fields; also seen on estuarine and coastal wetlands.
Behavior: Flies above water's surface in flocks, bill often pointing downward. Forages by surface-dipping; also hovers.
Distribution: Common and widespread.
Voice: Harsh rasping *cherk* flight calls, shorter and lower pitched than those of Black Tern.

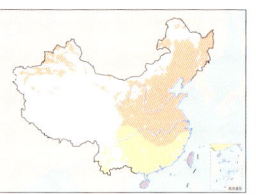

Chlidonias leucopterus

白翅浮鸥 bái chì fú ōu

White-winged Tern LC

L: 20–25 cm
WS: 63–67 cm
Habitat: Inhabits freshwater habitats, such as rivers, lakes, marshes, ponds, and flooded fields; also seen on coastal marshes.
Behavior: Often flies in flocks and constantly changes directions. Forages by surface-dipping; also hovers.
Distribution: Widespread and locally common.
Voice: *wik* flight calls lower pitched than similar short calls of Whiskered Tern.

Chlidonias niger

黑浮鸥 hēi fú ōu

Black Tern LC

L: 22–26 cm
WS: 56–62 cm
Habitat: Inhabits lakes, reservoirs, marshes, ponds, and rivers with flourishing grass.
Behavior: Often mixes with White-winged Tern. Flies low over water's surface in small flocks.
Distribution: Uncommon breeder in N Xinjiang; vagrant to Ningxia, E Inner Mongolia, Heilongjiang, Tianjin, Beijing, Hebei, Jiangsu, Hong Kong, and Taiwan.
Voice: Multisyllabic, piercing, nasal *kye-ii-ekh* flight calls longer and higher pitched than those of both Whiskered Tern and White-winged Tern.

Black-naped Tern

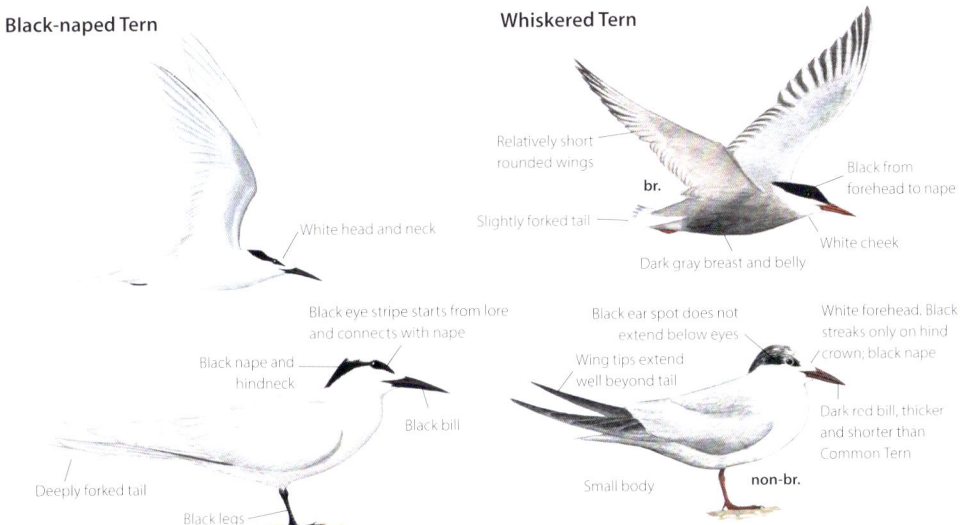

White head and neck

Black eye stripe starts from lore and connects with nape

Black nape and hindneck

Black bill

Deeply forked tail

Black legs

Whiskered Tern

Relatively short rounded wings

Slightly forked tail

br.

Black from forehead to nape

White cheek

Dark gray breast and belly

Black ear spot does not extend below eyes

Wing tips extend well beyond tail

White forehead. Black streaks only on hind crown; black nape

Dark red bill, thicker and shorter than Common Tern

Small body

non-br.

Common Tern

Black bill

S. h. longipennis

Long narrow wings

br.

Deeply forked tail

Black from forehead to nape

Slender red bill with black tip

S. h. tibetana

br.

White-winged Tern

br.

White rump and tail

Red legs

White upperwing coverts, gray wing feathers, dark outer primaries

Black head and body

Red bill

Black ear spot extends below eyes

Black streaks on crown

Whitish forehead

Black bill

non-br.

Black-bellied Tern

br.

Black crown

Orange bill

Black belly and undertail coverts

Orange legs

br.

Black Tern

Whitish-gray underwing coverts

Gray rump

br.

Black crown

Gray back

Relatively slender bill

Dark gray patches on side of breast

non-br.

Gelochelidon nilotica
鸥嘴噪鸥 ōu zuǐ zào ōu
Gull-billed Tern `LC`

L: 35–38 cm
WS: 85–100 cm
Habitat: Often inhabits coastal and estuarine wetlands; also occurs on inland lakes and marshes.
Behavior: Solitary or in small flocks. Forages in flight by dipping prey from tidal flats or water's surface; seldom plunges into water.
Distribution: *G. n. nilotica* in Xinjiang and Northeast China. *G. n. affinis* in central, eastern, and South China. Locally common.
Voice: Rising, nasal, di- or trisyllabic *kei-weik* or *ker-wee-ik* calls, with peculiar Punch and Judy–like quality.

Stercorarius pomarinus
中贼鸥 zhōng zéi ōu
Pomarine Jaeger `LC`

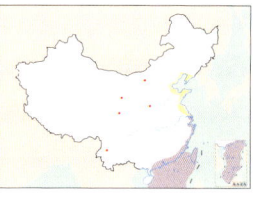

L: 47–56 cm
WS: 125–138 cm
Habitat: Lives on coasts or pelagic; occasionally appears on inland lakes and rivers.
Behavior: Often lives solitarily at sea. Pirates meals from other birds. Flight skillful.
Distribution: Wintering records from the East China Sea to the South China Sea. Migrates through the Yellow Sea regions, also inland; recorded in Beijing, Gansu, Shanxi, Henan, Sichuan, and Inner Mongolia. Uncommon migrant and vagrant.
Voice: Unlikely to be heard in China but makes noisy and piercing Common Gull–like *ya-ya* when breeding.

Stercorarius maccormicki
南极贼鸥 nán jí zéi ōu
South Polar Skua `LC`

L: 53–66 cm
WS: 130–140 cm
Habitat: Inhabits open oceans between the tropics and polar regions.
Behavior: Breeds along coasts and on islands; wanders to oceans during nonbreeding season and also comes to inshore and inland freshwater lakes and rivers. Pirates meals from or even feeds on other seabirds.
Distribution: Vagrant to South China Sea and Taiwan.
Voice: Unlikely to be heard in China but makes unpleasant chatter and *ge–ge–ge* calls, some reminiscent of braying donkey, at breeding colonies.

Rynchops albicollis
剪嘴鸥 jiǎn zuǐ ōu
Indian Skimmer `VU`

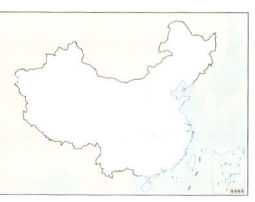

L: 40–43 cm
WS: 70–80 cm
Habitat: Occurs mainly near shoals in large rivers.
Behavior: Flies near water's surface; forages by flying with bill open like scissors and lower mandible submerged, skimming the water.
Distribution: Vagrant, with historic records along the south coast.
Voice: Mostly single-syllable, doglike, yipping flight calls, *kip* or *kyip*.

Stercorarius parasiticus
短尾贼鸥 duǎn wěi zéi ōu
Parasitic Jaeger `LC`

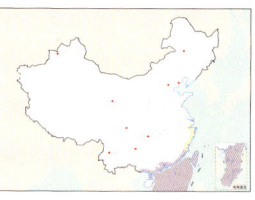

L: 41–55 cm
WS: 110–125 cm
Habitat: Inhabits open oceans between the tropics and polar regions, seldom to inshore or inlands; occasionally occurs on freshwater lakes and rivers.
Behavior: Acrobatic in flight. Wanders at sea and pirates meals from other seabirds.
Distribution: Wintering records in the East and South China Seas. Scattered records in Beijing, Hebei, Fujian, Guizhou, Zhejiang, Qinghai, and Xinjiang. Rare migrant or vagrant.
Voice: Unlikely to be heard in China but makes short *kips* and rising *e— iya-* when displaying.

Stercorarius longicaudus
长尾贼鸥 cháng wěi zéi ōu
Long-tailed Jaeger `LC`

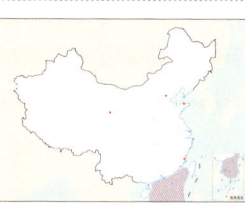

L: 48–58 cm
WS: 105–117 cm
Habitat: Inhabits open oceans between the tropics and polar regions, rivers, and lakes.
Behavior: Breeds on islands and tundra in the Arctic region; wanders at sea in winter. Also appears at inland freshwater rivers and lakes.
Distribution: Wintering records in East and South China Seas; scattered records in Beijing, Qinghai, Shandong, and Fujian. Rare migrant and vagrant.
Voice: Unlikely to be heard in China but makes short ternlike *kip* and Eurasian Coot–like *klip* and wailing *e— iya-* when displaying, higher pitched than other jaegers.

Gull-billed Tern

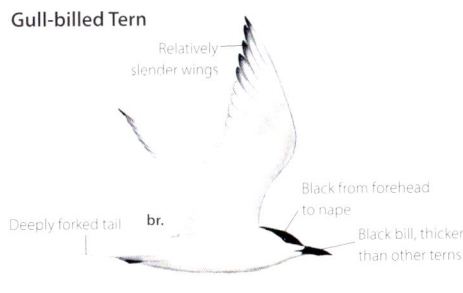

Relatively slender wings

Deeply forked tail

br.

Black from forehead to nape

Black bill, thicker than other terns

White crown, black ear spots

Relatively long wings

Stout bill

non-br.

Indian Skimmer

Black from crown to back

Long orange bill, lower mandible much longer than upper mandible

Parasitic Jaeger

White shafts of outer primaries

Prominent white patch on primaries on under wing

Slightly elongated and pointed central tail feathers

Dark brown cap

br.

Dusky breast band

non-br.

Pomarine Jaeger

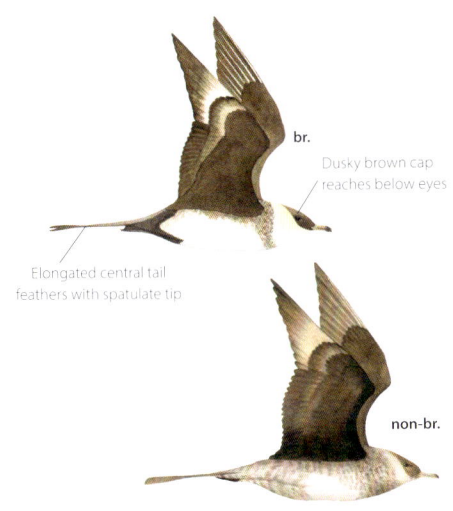

br.

Dusky brown cap reaches below eyes

Elongated central tail feathers with spatulate tip

non-br.

South Polar Skua

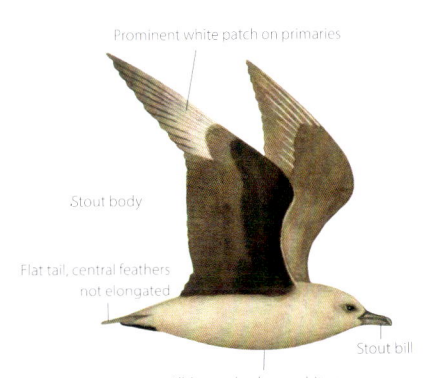

Prominent white patch on primaries

Stout body

Flat tail, central feathers not elongated

Stout bill

All-brown body, no white on throat and lower belly

Long-tailed Jaeger

Finely elongated and pointed central tail feathers

Dark brown cap

br.

non-br.

203

Uria aalge

崖海鸦 yá hǎi yā

Common Murre　LC

L: 38–46 cm
WS: 61–73 cm
Habitat: Wanders at sea in winter; seldom moves to coasts.
Behavior: Lives at sea solitarily or in small flocks in winter. Flies in straight lines with rapid wingbeats.

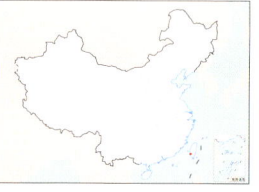

Dives to forage; feeds mainly on fish, crustaceans, and other oceanic invertebrates. Floats at sea when resting.
Distribution: Rare vagrant in Taiwan.
Voice: Silent in winter.

Brachyramphus perdix

长嘴斑海雀 cháng zuǐ bān hǎi què

Long-billed Murrelet　NT

L: 24–26 cm
WS: 43 cm
Habitat: Wanders at sea in winter; seldom approaches coasts.
Behavior: Often gathers in small flocks in winter. Strong swimmer and diver. Swims in a flat

posture with cocked tail; can dive deeper than 10 m. Feeds mainly on fish, crustaceans, and other marine invertebrates. Floats at sea when resting.
Distribution: Uncommon winter visitor in Bohai Bay and the Yellow Sea, occasionally to the East China Sea.
Voice: Silent in winter.

Synthliboramphus wumizusume

冠海雀 guān hǎi què

Japanese Murrelet　VU

L: 24–26 cm
WS: 43 cm
Habitat: Wanders at sea near subtropical regions in winter; seldom approaches coasts.
Behavior: Similar to Ancient Murrelet. Forages for various planktonic crustaceans; especially krill and larval fish.
Distribution: Vagrant or rare winter visitor in Hong Kong and Taiwan. Satellite tracking study shows birds migrate from Japan and winter off Zhejiang.
Voice: Silent in winter.

Synthliboramphus antiquus

扁嘴海雀 biǎn zuǐ hǎi què

Ancient Murrelet　LC

L: 24–27 cm
WS: 40–43 cm
Habitat: Nests on slopes of oceanic islands and rock crevices near trees during breeding season; wanders and forages offshore, occasionally moves to coasts.

Behavior: Gathers in flocks during breeding season. Does not build nest; lays eggs directly on ground. Usually gathers in small flocks during nonbreeding season. Strong swimmer and diver. Swims in a flat position without cocked tail, diving around 10–20 m deep. Forages mainly for fish, crustaceans, and other marine invertebrates; often follows shoals of fish. Floats at sea when resting. Flies straight and low.
Distribution: Breeds on islands off Shandong Peninsula; recorded in the Bohai Sea, Yellow Sea, and East and South China Seas during nonbreeding season. Rare breeder due to habitat destruction and egg harvesting in recent years.
Voice: At breeding sites, high-pitched mechanical chattering. Rarely, a very high-pitched, shorebird-like contact call, *jiu* or *kip*, given at sea.

Cerorhinca monocerata

角嘴海雀 jiǎo zuǐ hǎi què

Rhinoceros Auklet　LC

L: 35–38 cm
WS: 56–63 cm
Habitat: Wanders at sea in winter; seldom approaches coasts.
Behavior: Typical alcid behaviors. Forages mainly for fish; also catches squid, krill, and other crustaceans

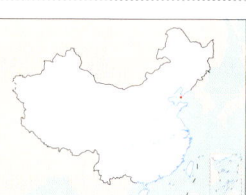

in winter. Diving depth around 30–40 m.
Distribution: Vagrant in Liaoning (Lvshun).
Voice: Silent in winter.

Common Murre

Small short wings, fast wingbeats

Blackish-brown crown, hindneck, side of neck, and upperparts

Fine line extends backward from eyes

br.

Long pointed black bill

br.

br.

Blackish brown axillaries

non-br.

non-br.

Spindle-like body

Prominent blackish-brown neck collar

Long-billed Murrelet

Short blackish-brown wings

Inconspicuous white eye ring

Blackish-brown cap and upperparts

White from chin to underparts

non-br.

Narrow white scapular bars

br.

Relatively dense black bars on breast

Ancient Murrelet

Short small wings, fast wingbeats

Spindle-like body

Black hood contrasts strongly with white underparts

Black head with narrow white lateral crown stripe

Light pink bill

Dark gray back

br.

White from side of neck to underparts

Black flanks

Narrow white lateral crown stripe disappears after molt

Back turns pale gray

No black on throat

No black on flanks

non-br.

Japanese Murrelet

Prominent white crown

br.

Black side of body contrasts strongly with white belly

Ivory bill longer and more pointed than Ancient Murrelet

White crown with black crest

br.

More black from face, chin, throat, and side of neck to flanks than Ancient Murrelet

Prominent black eye ring

No white crown

Dusky bluish-gray bill

Black molts to white on face and throat

non-br.

Rhinoceros Auklet

Appears dark sooty brown from a distance, with yellow bill

Stocky body

Triangular horn projects above bill base; size differs among individuals

Yellowish-white iris

Off-white belly

White plumes above and below eyes

Short, stout, cone-like, orange bill

br.

Dark sooty-brown body

Relatively large body, strong

No triangular projection at base of bill

Reduced plumes on side of head

Paler body with dark brown upperparts and pale brown underparts

non-br.

205

沙鸡目　PTEROCLIFORMES

Medium-sized land birds restricted to the Old World. Resemble pigeons in profile. Males noticeably more brightly patterned than females. Plumage mostly buffy, rufous, and black. Bill short and strong, resembling pheasant bills. Legs short and strong, with feathered tarsi. Wings long and pointed. Tail long, usually with elongated and pointed central tail feathers. Usually live in groups, inhabiting deserts, arid grasslands, or other arid and semiarid open habitats. Good at flying and running; usually travel long distances in search of water. Nest on ground. Feed mostly on plant seeds, fruits, buds, and insects. Some species are migratory; sometimes highly nomadic.

One family, two genera, and 16 species recognized worldwide, distributed in the dry and open habitat of the Old World. Two genera and three species recorded in China, found on the Qinghai-Tibet Plateau and the grasslands and arid zones of northern China.

Syrrhaptes tibetanus
西藏毛腿沙鸡 xī zàng máo tuǐ shā jī
Tibetan Sandgrouse

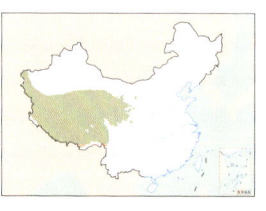

L: 39–44 cm
Habitat: Inhabits mainly grasslands, deserts, and semideserts on plateaus at altitudes of 3000–5000 m. In winter, often descends to below 4000 m. Sometimes also seen in river valleys and on lakes.
Behavior: Gregarious; can sometimes gather in flocks of hundreds. Bold. Flight agile; fast wingbeats can often be heard. Forages while walking moderately slowly; feeds mainly on grass seeds, buds, and insects.
Distribution: Uncommon resident in Qinghai, Tibet, and S Xinjiang.
Voice: Mellow, rising then falling flight call, *pluvaau*, sounds vaguely cranelike at distance.

Syrrhaptes paradoxus
毛腿沙鸡 máo tuǐ shā jī
Pallas's Sandgrouse

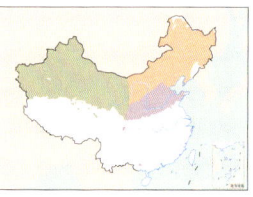

L: 39–43 cm
Habitat: Typical desert bird. Prefers deserts and semideserts; sometimes seen on farmland, grasslands, freshwater lakes, and riverbanks.
Behavior: Gregarious; may gather in large flocks of thousands of individuals. Flight agile; fast wingbeats can often be heard; lands after flying hundreds of meters. Often flies in flocks to water sources to drink at dawn and dusk; sometimes flies thousands of meters when searching for water sources. Forages while walking moderately slowly and swaying body. Feeds mainly on grass seeds, buds, and insects.
Distribution: Widespread in northern China; sometimes winters in North and southwestern China.
Voice: Various short, often Little Tern–like, *kip* notes, often doubled, *kip-kyip*, as well as mellow, soft, and nasal *kyup* or *kup-kyup* and multisyllabic, trilled *cherrk* calls in flight.

Pterocles orientalis
黑腹沙鸡 hēi fù shā jī
Black-bellied Sandgrouse

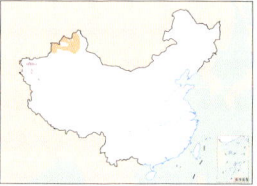

L: 30–35 cm
Habitat: Occurs near low mountains, hills, deserts, steppes, and abandoned cultivated lands.
Behavior: Often gathers in small flocks. Strong runner. Rapid flight; fast wingbeats can often be heard. Feeds on seeds, shoots, and insects on ground.
Distribution: Uncommon breeder in NW Xinjiang; occasionally wanders to SW Xinjiang in winter.
Voice: Decelerating, bubbling, lengthy purr, *durrrrr-r-r-rt*, given at intervals in flight.

Tibetan Sandgrouse

Black underwing coverts

chick & egg

Orange-yellow cheek

Off-white belly

Black-barred back, wing coverts, and tertials

Dense dark brown bars on breast

Extremely long, pointed central tail feathers

Pallas's Sandgrouse

Brownish-yellow underwing coverts

Rusty-red cheek and throat

Heavily spotted neck

Long pointed central tail feathers

Black-bellied Sandgrouse

Black primaries

chick & egg

Gray crown

Triangular black patch on lower throat

No chestnut on chin and throat

Chestnut chin and throat extends to side of neck, forming neck ring

No elongated tail feathers

Blackish-brown spots on breast

Bluish-gray breast lacks spots

Black breast band

鸽形目 COLUMBIFORMES

Small to medium-sized land birds. General profile best exemplified by feral pigeons. Plumage similar between sexes in most species, mostly bluish gray, brown, white, and green. Bill short and thick, slightly hooked and/or rounded at tip. Good at flying; wings long and pointed. Tail rounded or pointed. Legs short and strong; good at walking on ground and branches. Inhabit primarily various types of forests, but also cliffs or open habitats. Usually live in groups. Nest in trees, bushes, caves, even on buildings. Feed primarily on plant buds, shoots, seeds, and fruits, as well as insects and other small invertebrates.

One family, 49 genera, and 344 species recognized worldwide, with a cosmopolitan distribution, highly species rich in Indomalaya and Oceania. Eight genera and 33 species recorded in China, ubiquitously distributed nationwide.

鸠鸽科
Columbidae

Spotted Dove

Columba livia
原鸽 yuán gē
Rock Pigeon   LC

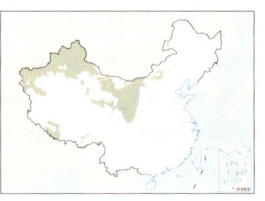

L: 30–35 cm
Habitat: Inhabits primarily vegetated plains, low mountains, hills, and cliffs.
Behavior: Cliff-dwelling species. Prefers living in flocks; populations near human residential areas hybridize with domestic or feral pigeons.
Distribution: Locally common resident in W Inner Mongolia, Ningxia, Gansu, Xinjiang, S Tibet, and Qinghai.
Voice: *oo-roo-coo*, like domestic pigeon.

Columba rupestris
岩鸽 yán gē
Hill Pigeon   LC

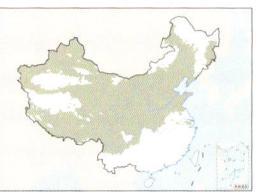

L: 30–35 cm
Habitat: Inhabits rocky high mountains or plateaus; primarily moves about in valleys and crags.
Behavior: Gathers in small flocks and forages on grasslands in valleys; also gathers in large flocks on rocky cliffs.
Distribution: *C. r. rupestris* found throughout northern China, as well as the Qinghai-Tibet Plateau and mountains in southwestern China; *C. r. turkestanica* in Xinjiang, W Qinghai, and Tibet. Locally common.
Voice: High-pitched, repeated, pulsed cooing, very similar to Rock Pigeon.

Columba leuconota
雪鸽 xuě gē
Snow Pigeon   LC

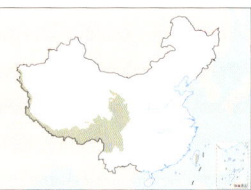

L: 28–36 cm
Habitat: Inhabits valleys, rocky cliffs, and grassy slopes at high altitudes.
Behavior: Gathers in flocks; often forages for seeds and grass roots on ground.
Distribution: *C. l. leuconota* in S Tibet; *C. l. gradaria* in W Xinjiang, Gansu, the Qinghai-Tibet Plateau, W Sichuan, and NW Yunnan. Locally common.
Voice: Clattering wing noise and a rapid *puk-uk-uk-uk-uk*, often repeated at length.

Columba oenas
欧鸽 ōu gē
Stock Dove   LC

L: 28–32 cm
Habitat: Inhabits primarily desert poplar forests, broadleaf forests in river valleys, and vegetated areas near farmland.
Behavior: Arboreal. Gregarious.
Distribution: Locally common resident in desert poplar forests in S Xinjiang (Tarim), as well as in Tian Shan Mountains and river valleys of Irtysh River in N Xinjiang; also to W Inner Mongolia and NW Gansu.
Voice: Song is a disyllabic, husky, low-pitched, flat cooing that increases in volume, *oou-o . . . oou-o . . . oo-o*.

Columba eversmanni

中亚鸽 zhōng yà gē

Yellow-eyed Pigeon VU

L: 26–30 cm
Habitat: Inhabits low mountains, hills, deserts, foothill plains, and cliffs in broadleaf forests.
Behavior: Prefers living in small flocks.

Distribution: Historically, a rare breeder in W and N Xinjiang, with passage migrants to W Gansu, but not seen in 20 years, raising concern about population status.
Voice: Low-pitched, trisyllabic song, *vr-uu-uh*, repeated fast (once per second). Alternate song a mellower, bouncy *qu-uu … qu-uu-ooh … qu-u-ooh … qu-uu-ooh … qu-uu-ooh.*

Rock Pigeon

Black bill

Slate-gray back without mottling

Two complete black bars across wings

White rump

Snow Pigeon

Dark gray head

White lower back

White underparts

Three black bars on wings

White central tail band

Hill Pigeon

Two incomplete wing bars

White central tail band

Stock Dove

Pinkish bill

Two short black bars on wings

Yellow-eyed Pigeon

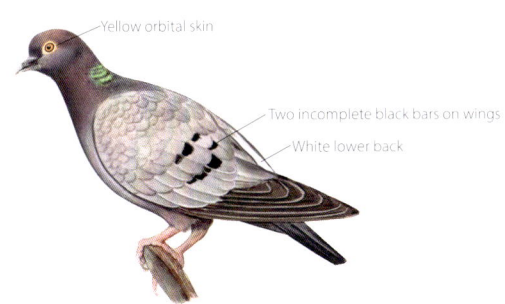

Yellow orbital skin

Two incomplete black bars on wings

White lower back

Columba palumbus

斑尾林鸽 bān wěi lín gē

Common Wood-Pigeon · LC

L: 38–43 cm
Habitat: Inhabits broadleaf forests, mixed forests, and nearby farmland shelterbelts in river valleys.
Behavior: Forest-dwelling species. Prefers gathering in flocks. Forages for seeds, grains, and berries on ground.
Distribution: *C. p. palumbus* is an uncommon resident in extreme N and NW Xinjiang; some wander to the Tian Shan Mountains and surrounding areas in winter. *C. p. casiotis* is an uncommon resident in the Tian Shan Mountains of Xinjiang.
Voice: Five-syllable, low-pitched, cooing song, with the second note emphasized and elongated and last two sounding paired, *cu–cooh–cu, coo–coo.* Loud wing clatter in alarm.

Columba hodgsonii

斑林鸽 (点斑林鸽) bān lín gē (diǎn bān lín gē)

Speckled Wood-Pigeon · LC

L: 35–40 cm
Habitat: Inhabits tall montane evergreen broadleaf forests or mixed forests.
Behavior: Primarily a forest dweller; often gathers in small flocks.
Distribution: Locally common in Gansu, Shaanxi, Sichuan, Tibet, and Yunnan.
Voice: Poorly known and clearly not very vocal—low-pitched and easily overlooked low growl (like stomach rumbling), followed by short sequence of three paired purring notes and longer terminal note, *rrrrrh wuk-oo … wuk-oo wuk-oo … arrrrh.*

Columba pulchricollis

灰林鸽 huī lín gē

Ashy Wood-Pigeon · LC

L: 32–38 cm
Habitat: Inhabits montane evergreen or deciduous broadleaf forests.
Behavior: Primarily a forest dweller; lives quietly in the canopy or in woods solitarily or in pairs. Forages in large flocks when fruits are ripe.
Distribution: Rare in Yunnan and Tibet and uncommon in Taiwan.
Voice: Low-pitched resonant *whooh-u,* with barely audible intake or breath after each note, repeated usually slowly (every second) and given three to eight times in well-separated strophes (often up to 30 sec apart).

Columba punicea

紫林鸽 zǐ lín gē

Pale-capped Pigeon · VU

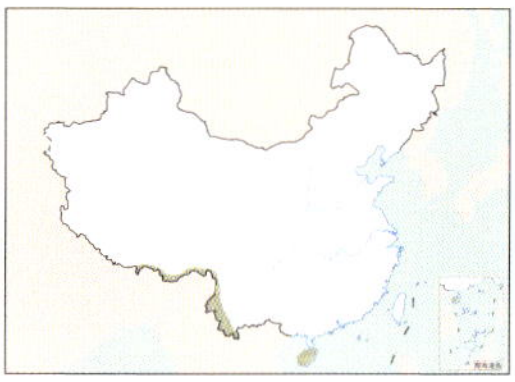

L: 35–40 cm
Habitat: Inhabits montane broadleaf forests.
Behavior: Forages in woods for fruits solitarily or in small flocks; also feeds on seeds on farmland.
Distribution: Very rare in Hainan, Yunnan, and Tibet.
Voice: Poorly known and presumably largely silent. Faint *ruhuhuhuhu* described.

Columba janthina

黑林鸽 hēi lín gē

Japanese Wood-Pigeon · NT

L: 39–43 cm
Habitat: Inhabits dense evergreen broadleaf forests on oceanic islands.
Behavior: Often lives solitarily in trees; occasionally gathers in small flocks.
Distribution: Vagrant in Shandong and Taiwan; birds caught on ships in the outer East China Sea are possible migrants.
Voice: Paired, low-pitched, mooing notes, the first slightly tremulous, the second longer, rising slightly before falling, *prrue-pruooarrhh,* repeated every 1.5 sec and typically given two or three times with lengthy pauses between strophes. Also a rattling, purring, melancholy bleat, followed by a rising seabird-like moo, *krrrrrr proooaa.*

Columba vitiensis

白喉林鸽 bái hóu lín gē

Metallic Pigeon · LC

L: 37–41 cm
Habitat: Inhabits forests or forest edges.
Behavior: Often in pairs or small flocks. Vigilant; usually stays in the tree canopy for a long time.
Distribution: Vagrant to Taiwan.
Voice: Low-pitched *wuuuu-hoo* or *vroom-hoom,* with the first note tremulous and rising.

Common Wood-Pigeon

Pale yellowish-white eye ring

Orange bill

Large white patches
on side of neck

White along
edge of wing

C. p. palumbus

Pale-capped Pigeon

Silvery gray from forehead to nape

Brownish-chestnut body;
wings and back darker and
appear purplish

Speckled Wood-Pigeon

Triangular reddish-
brown spots from
breast to belly

White spots on
median coverts

Reddish-purple scapulars
contrast with dark bluish-
gray wing coverts

Japanese Wood-Pigeon

Large body, plumage blackish with
green iridescence

Ashy Wood-Pigeon

Erect feathers with black base on
hindneck and side of neck, giving
barred impression

Cream-colored
neck collar

No spots on
underparts

Metallic Pigeon

Yellow bill
with red base

Grayish-white
throat and cheek

Greenish iridescence
on breast

Streptopelia turtur

欧斑鸠 ōu bān jiū

European Turtle-Dove

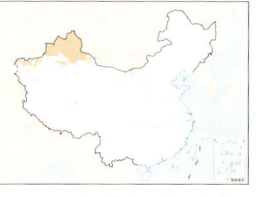

L: 25–28 cm
Habitat: Inhabits primarily open plains, deserts, and broadleaf forests on low mountains and hills. Also appears in secondary forests, residential areas, and small woods or bushes in farmland zones.
Behavior: Often nests in trees. Forages on various fruits and seeds; also feeds on a few small invertebrates.
Distribution: Uncommon breeder in Xinjiang; rarer in W Inner Mongolia, NE Gansu, W Tibet, and SW Qinghai.
Voice: Low-pitched, hesitant or nervous, paired purring, *turrrr-tororr*, repeated several times and ending abruptly.

Streptopelia orientalis

山斑鸠 shān bān jiū

Oriental Turtle-Dove

L: 28–36 cm
Habitat: Inhabits low mountains, hills, plains, montane broadleaf forests, mixed forests, agricultural farmland, and orchards; sometimes also in urban parks.
Behavior: Often in pairs or small flocks. Bold, unafraid of people. Displaying males circle in the air, fly straight up with clapping wingbeats, then glide with open wings and tail. Feeds on ground while strolling.
Distribution: Widespread throughout much of China except for the Qinghai-Tibet Plateau; mostly resident. *S. o. orientalis* in Northeast, North, East, and South China; *S. o. meena* in N and W Xinjiang; *S. o. agricola* in W and S Yunnan.
Voice: Song is a rhythmic, fast-paced phrase with two gruff notes followed by paired mellow coos, *wu-wuehr … pru-pruu*, repeated in strophes of up to 10–12 sec. Usually starts quietly, increases in volume, and ends abruptly. Song of *S. o. meena* huskier (especially the first note) and slightly slower than *S. o. orientalis* and *S. o. agricola*.

Streptopelia decaocto

灰斑鸠 huī bān jiū

Eurasian Collared-Dove

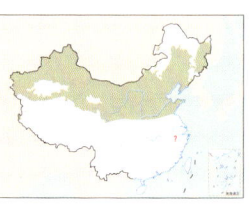

L: 25–34 cm
Habitat: Inhabits open plains, villages, agricultural farmland, and orchards; also seen in urban parks or low mountains and hills.
Behavior: Often gathers in small flocks; sometimes mixes with other doves. Prefers feeding on ground.
Distribution: Widespread; absent only from the Qinghai-Tibet Plateau. Mostly resident. Commonly seen in northern and western regions; rarely seen in East and South China. *S. d. decaocto* in Inner Mongolia, Heilongjiang, Jilin, Liaoning, Beijing, Hebei, Shandong, Shanxi, Henan, Ningxia, and Xinjiang. *S. d. xanthocycla* was reported historically from central and southeastern China, but no records in nearly a century; this subspecies has yellow eye ring, and its occurrence in China is questionable; sometimes treated as a separate species, Burmese Collared-Dove (*S. xanthocycla*).
Voice: Trisyllabic cooing song a monotonous *who-whoooo-uu*, with middle note elongated and last note short and lower. Call is a soft purring *kwerrrr*, often doubled.

Streptopelia tranquebarica

火斑鸠 huǒ bān jiū

Red Collared-Dove

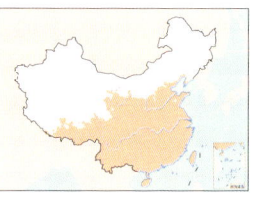

L: 20.5–23 cm
Habitat: Inhabits open plains, fields, villages, and orchards; also appears on low mountains, hills, and forest edges.
Behavior: Often in pairs or small flocks; sometimes mixes with Oriental Turtle-Dove and Spotted Dove. Usually gathers in flocks containing dozens of individuals during migration. Prefers resting on power lines or tall branches. Flight rapid.
Distribution: Widespread breeder north to Liaoning and Hebei, west to Gansu, Qinghai, W Sichuan, S Tibet, and Yunnan, east to the east coast and Taiwan, and south to Hainan. Winters in the south.
Voice: Song is a rapidly repeated, four- or five-syllable, dry, purring *cru–u–u–u–u*, often given in lengthy series.

European Turtle-Dove

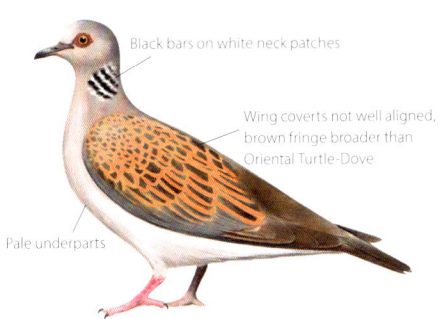

Black bars on white neck patches

Wing coverts not well aligned, brown fringe broader than Oriental Turtle-Dove

Pale underparts

Eurasian Collared-Dove

Black bar on hindneck

Oriental Turtle-Dove

Several black and white bars on neck

Grape-red lower belly

S. o. orientalis

Milky-whitish lower belly

S. o. meena

Red Collared-Dove

Gray head and rump

Thick black stripe on side of neck

Blackish-brown primaries and uppertail coverts

Purplish-pink body

Whitish vent and undertail coverts

♂

Paler on head

Grayish-brown wings

Pinkish gray from breast to underparts

Shorter tail than Eurasian Collared-Dove

♀

Spilopelia chinensis

珠颈斑鸠 zhū jǐng bān jiū

Spotted Dove

L: 27.5–30 cm
Habitat: Adapted to a wide range of habitats, including cultivated lands, gardens, parks, plantations, and secondary forests. Also in residential areas.
Behavior: Solitary or

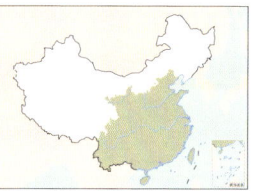

in pairs. Feeds primarily on ground. Breeds several times throughout the year. Nests simple and platelike, mainly made of loosely stacked branches. Also nests in flowerpots or on window sills in residential areas.
Distribution: Common and widespread resident. *S. c. chinensis* in most regions in southwestern, South, East, Central, North, and midwestern China, including Taiwan. *S. c. tigrina* in SW and S Yunnan. *S. c. hainana* in Hainan. *S. c. suratensis* vagrant in Xinjiang.
Voice: Widespread *S. c. chinensis* gives hoarse and repeated, two- to four-syllable, cooing *ker-wuk-keeer-kuk*, with the first note short and lower pitched, the third longest, and the last truncated. *S. c. tigrina* is like *S. c. chinensis*, while extralimital South Asian *S. c. suratensis* has much longer, more leisurely, complex, multinote strophes, where the second note is soft and the dominant one is repeated four or five times and is far less burry, and the song usually lacks an abrupt ending.

Spilopelia senegalensis

棕斑鸠 zōng bān jiū

Laughing Dove

L: 24–27 cm
Habitat: Inhabits vegetated areas, parks, artificial green space, and farmland in desert or semidesert areas.
Behavior: Feeds on ground in pairs or small flocks.
Distribution: Locally

common in Xinjiang and NW Qinghai.
Voice: Five-note laughing phrase *dodo–do–dodo*; pitch rises, then falls.

Macropygia unchall

斑尾鹃鸠 bān wěi juān jiū

Barred Cuckoo-Dove

L: 33–40 cm
Habitat: Inhabits montane evergreen broadleaf forests or bushes in secondary forests; also seen on lowland farmland.
Behavior: In pairs or small flocks. Forages primarily in trees, also in open areas in woods.

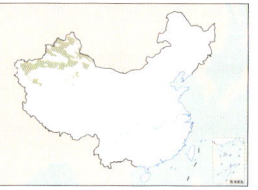

Distribution: *M. u. tusalia* occurs in the mountains in southwestern China and SE Tibet; *M. u. minor* in East and South China. Uncommon.
Voice: Low-pitched, resonant, lengthy *whoooo* that rises, then falls. At close range, a short introductory note occasionally audible, *hu-whoooo*.

Macropygia tenuirostris

菲律宾鹃鸠 fēi lǜ bīn juān jiū

Philippine Cuckoo-Dove

L: 36–40 cm
Habitat: Inhabits dense evergreen broadleaf forests.
Behavior: Timid. Gathers in small flocks in woods. Feeds on fruits or seeds.
Distribution: Locally common resident only on Orchid Island, Taiwan. Vagrant to E Taiwan and Pingtung County.
Voice: Low-pitched, resonant *cu-cuk … whoooo*.

Macropygia ruficeps

小鹃鸠 xiǎo juān jiū

Little Cuckoo-Dove

L: 30–32 cm
Habitat: Inhabits montane broadleaf forests.
Behavior: Solitary or in pairs in montane forests. Feeds on fruits and seeds. Secretive.
Distribution: Rare in S Yunnan.
Voice: Rapidly repeated, rising, nasal *pwaa.pwaa.pwaa*, vaguely reminiscent of White-breasted Waterhen.

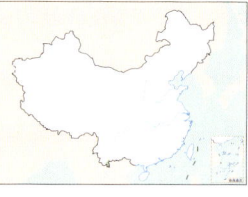

Chalcophaps indica

绿翅金鸠 lǜ chì jīn jiū

Asian Emerald Dove

L: 22–25 cm
Habitat: Inhabits montane forests at middle and low altitudes, especially in broadleaf forests.
Behavior: Often feeds on ground solitarily or in pairs.

Distribution: Resident in southwestern and South China, including Taiwan; northernmost to Zhejiang (Wenzhou). Locally common.
Voice: Low-pitched flat hum or hoot preceded by an often-inaudible, very short note, *uk-whuuu … uk-whuuu*, repeated with little variation about every 1.5 sec, often for long periods.

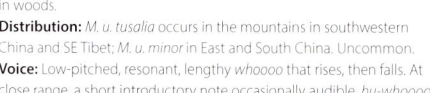

Spotted Dove

Grayish-white head

Patch of black feathers with white tips on side of neck

Grayish-brown back, wings, and tail

Pale purplish-pink neck, breast, and belly

Relatively long tail

S. c. chinensis

Philippine Cuckoo-Dove

Only female has dusky bars on hindneck and side of neck and dark spots on breast; otherwise no distinct bars or stripes

Dark underparts

Long tail

Laughing Dove

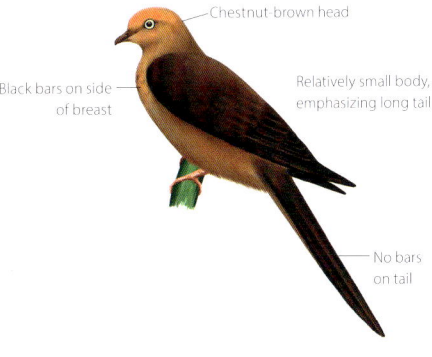

Black spots on side of neck; pink hindneck

Bluish-gray wing coverts contrast with back feathers in flight

Small body

Little Cuckoo-Dove

Chestnut-brown head

Black bars on side of breast

Relatively small body, emphasizing long tail

No bars on tail

Barred Cuckoo-Dove

Purplish-green iridescence on head and neck

Blackish-brown upperparts with fine chestnut bars

Pale underparts, scaly black bars from neck to breast

Long barred tail

Asian Emerald Dove

♂

Brownish purple from side of head to underparts

Emerald upper back and wings

Darker body; white and gray restricted to forehead and supercilium

♀

215

Treron bicinctus

橙胸绿鸠 chéng xiōng lù jiū

Orange-breasted Green-Pigeon

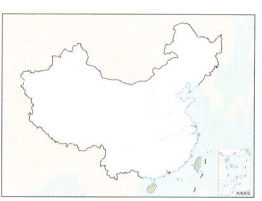

L: 24–29 cm
Habitat: Inhabits broadleaf forests at low altitudes.
Behavior: Solitary or in small flocks. Forages in fruit trees.
Distribution: Very rare resident in Hainan, Taiwan, and Hong Kong.
Voice: Comical rising nasal whistles interspersed with odd gurgling notes.

Treron phayrei

灰头绿鸠 huī tóu lù jiū

Ashy-headed Green-Pigeon

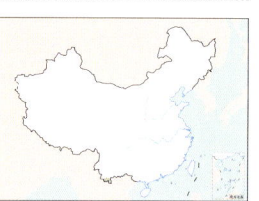

L: 24–28 cm
Habitat: Inhabits broadleaf forests at middle and low altitudes.
Behavior: Primarily arboreal; feeds in flocks in woods.
Distribution: Resident in S Yunnan. Locally common.
Voice: Typical *Treron*: attractive, gently rising and falling, short whistles. Reminiscent of Yellow-footed Green-Pigeon but lacks the clicks and harsher notes.

Treron curvirostra

厚嘴绿鸠 hòu zuǐ lù jiū

Thick-billed Green-Pigeon

L: 20–29 cm
Habitat: Inhabits broadleaf forests at middle and low altitudes.
Behavior: Often perches on branches of dead trees. Feeds in flocks in fruit-bearing trees.
Distribution: *T. c. nipalensis* in Yunnan and Guangxi; *T. c. hainanus* in Hainan and Hong Kong. Locally common.
Voice: Typical *Treron*: melodious rising and falling whistles, lower pitched and generally flatter than Ashy-headed Green-Pigeon.

Treron phoenicopterus

黄脚绿鸠 huáng jiǎo lù jiū

Yellow-footed Green-Pigeon

L: 27–34 cm
Habitat: Inhabits broadleaf forests at middle and low altitudes, especially in large trees with abundant fruits.
Behavior: Feeds in fruit trees in pairs or flocks; also visits tea plantations or farmland.
Distribution: Resident in W and S Yunnan. Rare.
Voice: Song is eclectic, ill-structured mix of rising and falling, short and long whistles interwoven with harsher notes and clicks.

Treron apicauda

针尾绿鸠 zhēn wěi lù jiū

Pin-tailed Green-Pigeon

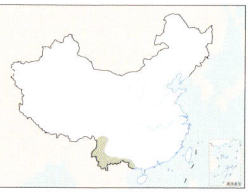

L: 31–40 cm
Habitat: Inhabits montane broadleaf forests at middle and low altitudes.
Behavior: Gathers in small flocks among clumps of tall trees. Also perches on branches of dead trees.
Distribution: *T. a. apicauda* in C and W Yunnan and W and S Sichuan; *T. a. laotianus* in E and S Yunnan and Guangxi. Locally common.
Voice: Simple poorly structured song composed of mostly flat monotone whistles of varying length, some weakly disyllabic, *whoo … whue … who-whoo*, repeated.

Treron seimundi

白腹针尾绿鸠 bái fù zhēn wěi lù jiū

Yellow-vented Green-Pigeon

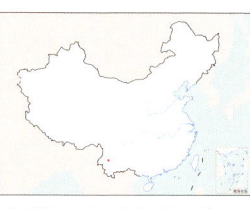

L: 31–33 cm
Habitat: Inhabits evergreen broadleaf forests and forest edges.
Behavior: Often solitary or in pairs; records also exist of mixing with other green-pigeons. Wanders far when foraging.
Distribution: Vagrant. First record in China is a banded individual from Yunnan (Nanjian County of Dali) on October 10, 2020.
Voice: Particularly melancholy, slow, and hesitant song usually starts with an unpleasant Common Hill Myna–like note and is slightly higher pitched than some green-pigeons, *cherk po-popo-yooyoo-po*.

Orange-breasted Green-Pigeon

Bluish-gray nape and upper back

Mauve-pink band and broad orange-yellow band on upper breast

Bright yellow fringe on wing coverts, forming streaks

Pale bluish-gray nape and upper back

Yellow fringe on wing coverts

Pale reddish-brown uppertail coverts

Ashy-headed Green-Pigeon

Bluish gray from crown to bill

Purplish-chestnut scapulars, back, and wing coverts

Yellow fringe on wing coverts, forming stripes

Green upperparts, yellow fringe on wing coverts

Thick-billed Green-Pigeon

Gray from forehead to crown

Yellow bill with red base

Purplish-chestnut back and wing coverts

Reddish-brown undertail coverts

Yellow fringe and conspicuous yellow stripes on wing coverts

Yellow fringe on wing coverts, forming stripes

Yellow-footed Green-Pigeon

Bluish-gray head

Yellow band on hindneck, upper back, and upper breast

Yellow legs

Pin-tailed Green-Pigeon

Prominent elongated central tail feathers

Rufous-brown undertail coverts

Yellow-vented Green-Pigeon

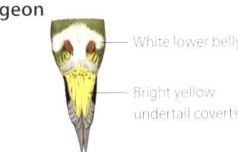

White lower belly

Bright yellow undertail coverts

No reddish-purple patch on wings

Azure bare orbital skin

Relatively long central tail feathers

Reddish-purple patch on wings

217

Treron sphenurus

楔尾绿鸠 xiē wěi lǜ jiū

Wedge-tailed Green-Pigeon

L: 28–33 cm
Habitat: Inhabits montane broadleaf or mixed forests.
Behavior: Forages in trees in small flocks; especially prefers large trees with abundant fruits.
Distribution: Resident in southwestern China. Locally common.

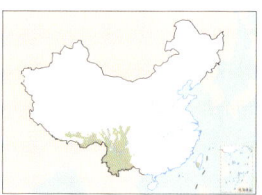

Voice: Flutelike song typically preceded by two or three short whoops followed by series of tremulous or guttural whistles, including one series that accelerates and ends in three drawn-out whistles, wo-whu … wo-whu … whoooeeeao … whu-whu … wuwuwuwuwuwuwuwu …whoo-u … uwhoo …whohoohoo.

Treron sieboldii

红翅绿鸠 hóng chì lǜ jiū

White-bellied Green-Pigeon

L: 21–33 cm
Habitat: Inhabits broadleaf or mixed forests; also seen in bushes in parks or on cultivated lands.
Behavior: Gathers in small flocks. Some subspecies migrate.

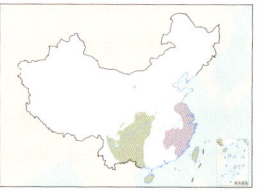

Distribution: T. s. sieboldii in East China; T. s. fopingensis in Hubei, Chongqing, Sichuan, S Shaanxi, and Inner Mongolia (Helan Mountains); T. s. murielae in Guangxi, Guizhou, and Hainan; T. s. sororius in Taiwan. Locally common.
Voice: Hesitant, slowly delivered series of often short whistles, each with sudden changes in volume and pitch, wu–oh–wu, repeated at length.

Treron formosae

红顶绿鸠 hóng dǐng lǜ jiū

Whistling Green-Pigeon

L: 33–35 cm
Habitat: Inhabits lowland broadleaf forests on tropical and subtropical islands.
Behavior: Solitary or in small flocks in the canopy.
Distribution: Uncommon in E and S Taiwan; vagrant in Hong Kong.

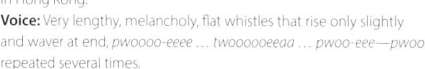

Voice: Very lengthy, melancholy, flat whistles that rise only slightly and waver at end, pwoooo-eeee … twoooooeeaa … pwoo-eee—pwoo, repeated several times.

Ptilinopus leclancheri

黑颏果鸠 hēi kē guǒ jiū

Black-chinned Fruit-Dove

L: 26–28 cm
Habitat: Inhabits broadleaf forests and coastal shelterbelts at low altitudes.
Behavior: Moves about solitarily in the canopy; also gathers in small flocks when fruits are ripe.

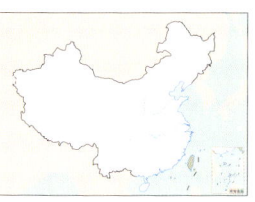

Distribution: P. l. taiwanus in Taiwan; P. l. longialis on Orchid Island, Taiwan. Rare.
Voice: Series of simple, very low-pitched, flat notes, wuu, at 1 sec intervals repeated three or four times every 8–15 sec.

Ducula aenea

绿皇鸠 lǜ huáng jiū

Green Imperial-Pigeon

L: 42–45 cm
Habitat: Inhabits lowland broadleaf forests and mangroves.
Behavior: Solitary or in pairs in the canopy; seldom on the ground.
Distribution: Resident in S Yunnan, SW Guangdong, and Hainan. Very rare.

Voice: Very low-pitched disyllabic purring, uk-rrhhuuu, with emphasis on second note, repeated every 4–8 sec.

Ducula badia

山皇鸠 shān huáng jiū

Mountain Imperial-Pigeon

L: 43–51 cm
Habitat: Inhabits montane evergreen broadleaf forests.
Behavior: Moves about in pairs or small flocks in the canopy; also perches on dead branches of large trees.

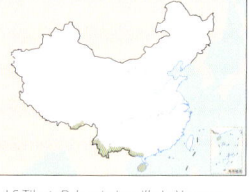

Distribution: D. b. insignis in E and S Tibet; D. b. griseicapilla in Yunnan, Guangxi, and Hainan. Locally common.
Voice: Very low-pitched, resonant, paired booming, usually with short, frequently inaudible, introductory note, ut-whoom-whoom.

Wedge-tailed Green-Pigeon

♂
Orange-tinged breast
Rufous-chestnut wing coverts
Yellowish-green underparts

♀
Green breast
Grayish-green wing coverts

White-bellied Green-Pigeon

♂
Rufous-chestnut wing coverts
White belly

♀
Dusky green wing coverts

Whistling Green-Pigeon

♂
Orange crown
Purplish-red scapulars and wing coverts
Narrow yellowish-white fringe on wing coverts

♀
Olive belly

Black-chinned Fruit-Dove

White head, neck, and upper breast
Black chin
Emerald-green upperparts
Broad dark band between breast and belly
Reddish-brown undertail coverts

Green Imperial-Pigeon

Dark green upperparts
Pale pink underparts
Large body

Mountain Imperial-Pigeon

Gray crown and cheek
Dark purple upperparts
Gray underparts
Large body
D. b. griseicapilla

Pink crown and cheek
D. b. insignis

鹃形目 CUCULIFORMES

Medium-sized perching birds. Body long and slender. Plumage similar between sexes in most species; coloration highly variable. Bill relatively long and slightly decurved. Wings usually long and pointed; flight style resembles small hawks. Legs short, with zygodactyl feet. Tail long. Inhabit a wide range of habitats, including forests, scrubland, deserts, and reedbeds. Calls loud and distinctive. Most species are brood parasites, laying eggs in nests of other species. Insectivorous; feed primarily on caterpillars and other insects. Most species are migratory.

One family, 33 genera, and 149 species recognized worldwide, with a cosmopolitan distribution. Nine genera and 20 species recorded in China, ubiquitously distributed nationwide.

杜鹃科
Cuculidae

Common Cuckoo

Centropus sinensis
褐翅鸦鹃 hè chì yā juān
Greater Coucal

L: 47–56 cm
Habitat: Usually occurs at altitudes below 1200 m. Inhabits secondary forests, tall grasslands, scrubland, bamboo thickets, scrub near cultivated lands, mangroves, paddy fields, and gardens.

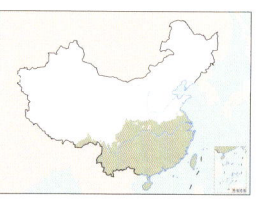

Behavior: Often walks, hops, or chases prey on ground or in bushes. Not brood parasitic; nests in grass, shrubs, or bamboo thickets.
Distribution: Resident; locally common. *C. s. sinensis* occurs in S Tibet, E Yunnan, Guizhou, Guangxi, Guangdong, Hong Kong, Fujian, Hubei, and Hunan, with few records in Jiangxi, Zhejiang, and Henan. *C. s. intermedius* in W and S Yunnan and Hainan.
Voice: Primate-like song is a series of 3–30 low-pitched *hoop* notes that gradually slow. Call an unpleasant, grating *skaaitch*. Also, in alarm, a breathless hissing and a series of rapid, nasal *kokokokokok* notes reminiscent of White-breasted Waterhen.

Centropus bengalensis
小鸦鹃 xiǎo yā juān
Lesser Coucal

L: 34–38 cm
Habitat: Occurs at altitudes below 1500 m. Inhabits tall grass, reedbeds, marshes, bamboo thickets, secondary forests, open country, or cultivated lands; generally in habitats

closer to water and more open than those preferred by Greater Coucal.
Behavior: Often solitary or in pairs. Generally feeds on ground. Sometimes flies for short distances. Often rests on bushtops. Vigilant and secretive; when disturbed, immediately runs and hides in dense bushes or grass. Not brood parasitic; nests in bushes or on branches of small trees.
Distribution: Seen in S Henan, Yunnan, S Guizhou, Hubei, Hunan, S Anhui, Jiangxi, Jiangsu, Shanghai, Zhejiang, Fujian, Guangdong, Hong Kong, Guangxi, Hainan, and Taiwan; occurs primarily as summer breeder in Jiangsu and as locally common resident in other regions. Recorded as summer breeder in Shandong (Jiaodong Peninsula) and increasingly regular north to Beijing and Hebei.
Voice: Three to five resonant *wup* notes in an accelerating sequence (shorter and "fuller" than the *hoop* of Greater Coucal), followed by a similar number of clear, multisyllabic, ringing notes, *wup-wup-wup-wup … kitituk-kitituk kitituk-kitituk.*

Clamator coromandus
红翅凤头鹃 hóng chì fèng tóu juān
Chestnut-winged Cuckoo

L: 38–46 cm (tail 16–24 cm)
Habitat: Inhabits open woodlands and scrubland on low mountains, hills, and piedmont plains below 1500 m; also in mixed broadleaf-coniferous forests.

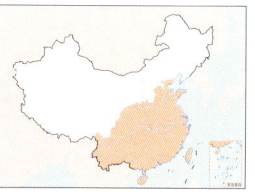

Behavior: Usually solitary or in pairs. Secretive, but does not hide in dense leaves like *Cuculus* cuckoos. Often forages in low vegetation. Flight rapid, with crest flattened. Brood parasite; the major hosts are laughingthrushes.
Distribution: Common summer breeder in East, Central, southwestern, and South China, including Hainan; also recorded in North China in recent years.
Voice: In breeding season, gives far-carrying, ringing, metallic territorial song of paired flat toots, *thu-thuu.* Also hoarse, scolding, woodpecker-like chatter, *crititick.*

Clamator jacobinus
斑翅凤头鹃 bān chì fèng tóu juān
Pied Cuckoo

L: 31–34 cm (tail 15–17 cm)
Habitat: Inhabits a wide range of forest habitats below 2000 m, such as open woodlands, bamboo thickets, and scrubland.
Behavior: Often solitary or in pairs. Usually on

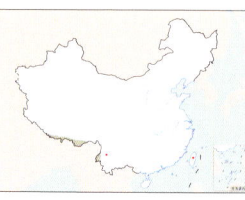

tall and exposed branches. Flight slow. Generally forages in trees and bushes, sometimes on the ground. Brood parasite; the major host species are babblers.
Distribution: Fringes of S Tibet and SW Yunnan; very rare. Vagrant to Yunnan (Dali) and Taiwan.
Voice: Loud, descending, ringing *kleeuw-kleeuw-kleeuw,* repeated at length. Call a loud chatter, like Chestnut-winged Cuckoo.

Greater Coucal

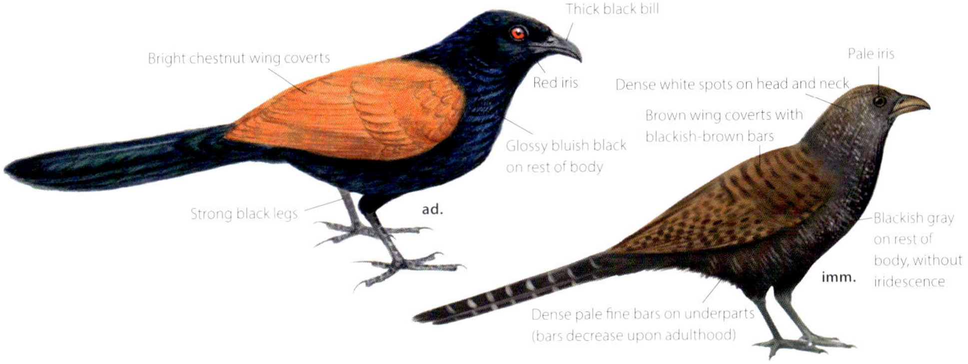

Thick black bill

Bright chestnut wing coverts

Red iris

Glossy bluish black on rest of body

Strong black legs

ad.

Pale iris

Dense white spots on head and neck

Brown wing coverts with blackish-brown bars

Blackish gray on rest of body, without iridescence

imm.

Dense pale fine bars on underparts (bars decrease upon adulthood)

Lesser Coucal

Blackish-brown iris

Fine white streaks on upperparts

Black bill

Brown wings

Blackish brown from head to belly

Blackish-brown tail with slight greenish iridescence

Black legs

ad.

Short, stout, decurved, pink bill

Dense pale streaks on body

imm.

Chestnut and brown wings with blackish-brown bars

Chestnut-winged Cuckoo

Black head with prominent crest

Pale ochre to reddish-brown throat (differs among individuals)

Chestnut wing coverts contrast sharply with black back

Metallic glossy black tail

Pied Cuckoo

Prominent black crest

Black and white body with black head, tail, and wings; remaining parts white

Small white wing patch

Relatively broad white band on tail tip

Phaenicophaeus tristis

绿嘴地鹃 lǜ zuǐ dì juān

Green-billed Malkoha　LC

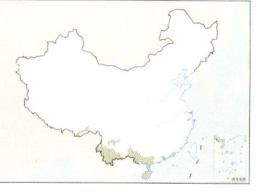

L: 43–60 cm (tail 27–36 cm; males have larger body size)
Habitat: Occurs at altitudes below 1600 m. Inhabits primary, secondary, and artificial forests; also adapted to a wide range of habitats, such as dense scrub, cultivated lands, rubber plantations, montane forests, and bamboo thickets.
Behavior: Often solitary or in pairs; hides in bushes, bamboo thickets, rattan, or branches near ground. Usually feeds on ground. Quietly rests on low branches near ground; makes rapid and short-distance flight when disturbed.
Distribution: Resident in SE Tibet, Yunnan, SW Guangxi, Guangdong, and Hainan; locally common.
Voice: Song is a measured, descending, trogonlike *tiaup*. Call a bad-tempered-sounding clucking or froglike croaking, *ko, ko, ko, ko*, reminiscent of Blue-bearded Bee-eater.

Eudynamys scolopaceus

噪鹃 zào juān

Asian Koel　LC

L: 39–46 cm (tail 17–20 cm)
Habitat: Occurs at altitudes below 1000 m. Inhabits mangroves, secondary forests, gardens, and artificial forests.
Behavior: Often solitary. Hides in flourishing leaves of dense canopy. Far more often heard than seen; difficult to spot if the bird does not call. Feeds on a wide range of food. Brood parasite; the major hosts are species in the Corvidae (e.g., Oriental Magpie and Red-billed Blue-Magpie) and Sturnidae (e.g., Black-collared Starling).
Distribution: Common summer breeder. *E. s. chinensis* widespread in Central, southwestern, East, South, and southeastern China, resident in S Yunnan; *E. s. harterti* in Hainan, resident. Spreading north.
Voice: Loud, noisy, piercing, and onomatopoeic *kiau … kiau … kiau … kiau* in summer, often repeated 5–10 times, rising then falling in tone. Female gives an equally loud, rising, bubbling, White-throated Kingfisher–like *kyik-kyik-kyik-kyik-kyik*.

Chrysococcyx maculatus

翠金鹃 cuì jīn juān

Asian Emerald Cuckoo　LC

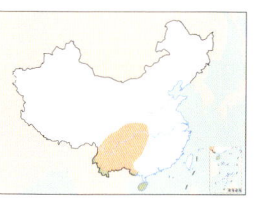

L: 17–18 cm
Habitat: Inhabits evergreen broadleaf forests, bushes within forests, and forest edges.
Behavior: Often solitary or in pairs. Gathers when food is abundant. Brood parasite; the major host species are passerines.
Distribution: Found in Yunnan, Sichuan, Chongqing, Guizhou, Hubei, Hunan, Guangdong, Hong Kong, and Hainan.
Voice: Song a sharply enunciated *tu-wi-uk*, repeated four to six times in flight. Call a rapid, slowing chatter, *dididi*, similar to Violet Cuckoo but shorter and not descending.

Chrysococcyx xanthorhynchus

紫金鹃 zǐ jīn juān

Violet Cuckoo　LC

L: 16 cm
Habitat: Inhabits evergreen broadleaf forests on low mountains, forests on piedmont plains, woodlands at forest edges, and scrubland.
Behavior: Often solitary or in pairs. Brood parasite; the major host species are passerines, such as sunbirds.
Distribution: SW and S Yunnan.
Voice: Song is an intense, high-pitched, sharp, penetrating, and disyllabic *khis-ik … khis-ik … khis-ik* in flight, similar to Asian Emerald Cuckoo but missing the introductory note and slightly more lisping. Also a lengthy, descending, swiftlet-like chatter similar to Asian Emerald Cuckoo.

Cacomantis sonneratii

栗斑杜鹃 lì bān dù juān

Banded Bay Cuckoo　LC

L: 22–24 cm
Habitat: Inhabits evergreen broadleaf forests, secondary forests, scrubland, bamboo thickets, and clumps of trees near villages; occurs from lowlands to 1600 m.
Behavior: Often solitary or in pairs. Sometimes calls and perches on open branches. Brood parasite.
Distribution: SW Sichuan, NE Guangxi, and S Yunnan.
Voice: Four flat whistled notes in rapid succession, *di-di-di-diu*, "chalky-pepper." Also an intensifying alternate song with three or four slow whistled notes followed by three to six shorter notes delivered faster before suddenly stopping. Also a hurried, purring *ki-ri-ki-krrr*.

Green-billed Malkoha

Red bare skin around eye

Very thick pale green bill

Extremely long, glossy, dusky bluish-green or dusky green tail

White tips on tail feathers, forming several well-aligned large white spots

Asian Emerald Cuckoo

Glossy emerald head

Red bare skin around eye

Glossy emerald back and wings

♂

White belly with green bars

Pale yellow head

♀

Bronze and green back

Finely barred white throat

Black wing feathers

Asian Koel

Ivory or pale green bill

Bright red iris

All black with dark bluish iridescence

♂

Violet Cuckoo

Red bare skin around eye

Dark purple body

♂

White belly with purple bars

♀

White face, throat, and belly with fine brown bars

Brown upperparts with dense yellowish-buff and brown spots

♀

Yellowish-buff underparts with dense dark brown bars

Banded Bay Cuckoo

Chestnut eye stripe

Chestnut upperparts with black bars

White from throat to underparts with black bars

Cacomantis merulinus

八声杜鹃 bā shēng dù juān

Plaintive Cuckoo LC

L: 21–25 cm
Habitat: Inhabits open forests, bushes, orchards, and parks.
Behavior: Perches solitarily in trees and calls. Brood parasite.
Distribution: Locally common in southwestern, East, and South China, including Hainan and Taiwan.
Voice: Series of high-pitched, short, and descending whistles ending in a trill, *phi phi phi phi-pipipi*.

Surniculus lugubris

乌鹃 wū juān

Square-tailed Drongo-Cuckoo LC

L: 24–28 cm
Habitat: Inhabits broadleaf forests or scrubland at low altitudes.
Behavior: Often perches solitarily and calls in the upper and middle stories of forest. Brood parasite.
Distribution: Uncommon summer visitor to southwestern, Central, East, and South China, including Hainan.
Voice: Rapid-fire ascending series of five to seven short *pi-pi-pi-pi* notes.

Hierococcyx sparverioides

大鹰鹃 （鹰鹃） dà yīng juān (yīng juān)

Large Hawk-Cuckoo LC

L: 38–42 cm
Habitat: Inhabits broadleaf forests from plains to mountains.
Behavior: Secretive; hides and calls in forests, hard to find. Brood parasite.
Distribution: Widespread across China except for Northeast China, Xinjiang, W Qinghai, and N Tibet. Common.
Voice: Song an intense whistle, "brain fe-ver," that increases in intensity (speed and pitch) to a climax. Similar to song of Common Hawk-Cuckoo. Also an intense *preu-pr … preu-pr … preu-pr*, repeated 7–10 times on a rising scale.

Hierococcyx varius

普通鹰鹃 pǔ tōng yīng juān

Common Hawk-Cuckoo LC

L: 33–37 cm
Habitat: Inhabits lowland deciduous forests or secondary forests and bushes.
Behavior: Secretive. Brood parasite.
Distribution: Uncommon in SE Tibet.
Voice: Song like Large Hawk-Cuckoo but more shrill and with last note lower in pitch.

Hierococcyx hyperythrus

北棕腹鹰鹃 （北鹰鹃）
běi zōng fù yīng juān (běi yīng juān)

Northern Hawk-Cuckoo LC

L: 28–30 cm
Habitat: Inhabits broadleaf forests at middle and low altitudes.
Behavior: Secretive; often solitary and perches in the canopy. Brood parasite.
Distribution: Breeds in Northeast and North China; migrates through East and South China, including Taiwan. Uncommon.
Voice: Peculiar nasal whistle given up to 16 times in intensifying sequences, *ju-yichi-jlui*, occasionally ending with a rapid splutter of chittering swiftlet-like notes.

Hierococcyx nisicolor

棕腹鹰鹃 （霍氏鹰鹃）
zōng fù yīng juān (huò shì yīng juān)

Hodgson's Hawk-Cuckoo LC

L: 28–30 cm
Habitat: Inhabits broadleaf forests at middle and low altitudes.
Behavior: Secretive; often perches solitarily in the canopy. Brood parasite.
Distribution: Breeds in southwestern, Central, East, and South China. Uncommon.
Voice: Shrill *gee-whiz* repeated 4–10 times and intensifying, and a chittering swiftlet-like chatter, *trrrrr-titititititirrrtrrr*.

Plaintive Cuckoo

No bars on body

Chestnut-brown underparts

No supercilium

Small body

♂

♀ **Rufous morph**

Densely barred underparts

Common Hawk-Cuckoo

Less black on chin

No black streaks on throat and breast

Brown underparts with pale bars

Square-tailed Drongo-Cuckoo

All-black drongo-like bird but with relatively slender bill and different toe arrangement

Fine bars on outer tail feathers and undertail coverts

Northern Hawk-Cuckoo

Medium to large white patch on hindneck

No prominent difference in color between head and back

Distinct white spots on tertials

Almost no streaks on underparts

Narrow dark bars on upper tail

Reddish-brown tail tip

Large Hawk-Cuckoo

No patch on hindneck; dark gray head contrasts sharply with dusky grayish-brown back

Relatively small black patch on chin

Reddish-brown breast and side of neck with broad blackish-brown streaks

Bars on lower breast and belly

White tail tip

Hodgson's Hawk-Cuckoo

Lacks or has extremely small white patch on hindneck

Both head and back gray, no contrast

Gray patch on chin

Brownish-pink streaks on underparts

Reddish-brown tail tip

Cuculus poliocephalus
小杜鹃 xiǎo dù juān
Lesser Cuckoo **LC**

L: 24–26 cm
Habitat: Inhabits primarily montane forests; also occurs in various forests and bushes during migration.
Behavior: Secretive; often calls solitarily in the canopy without being spotted. Brood parasite.
Distribution: Widespread and locally common in China except for Xinjiang and Ningxia.
Voice: Rapid five-note, six-syllable song with notes rising and falling in pitch, like "that's your cho-ky pepper," repeated at length.

Cuculus micropterus
四声杜鹃 sì shēng dù juān
Indian Cuckoo **LC**

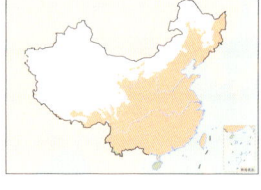

L: 31–34 cm
Habitat: Inhabits forests or trees on farmland at low altitudes.
Behavior: Solitary or in pairs; more often heard than seen. Brood parasite.
Distribution: Seen across China except for Xinjiang. Locally common.
Voice: Repeated, leisurely, four-syllable *kwer-kwah-kwah-kurh*, likened to "crossword puzzle," with each note lower than the previous. Females make bubbling trill.

Cuculus canorus
大杜鹃 dà dù juān
Common Cuckoo **LC**

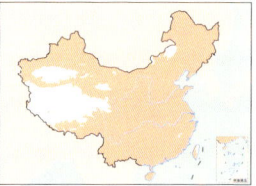

L: 32–35 cm
Habitat: Inhabits wooded areas, open farmland, or trees on wetlands at middle and low altitudes.
Behavior: Occurs and calls in trees; also seen on power lines. Bold. Brood parasite.
Distribution: *C. c. canorus* in Northeast, North, and northwestern China, including Taiwan; *C. c. subtelephonus* in C Inner Mongolia and C and W Xinjiang; *C. c. bakeri* in North, East, Central, South, and southwestern China. Common.
Voice: Familiar disyllabic *cuck-oo*, with the first note higher than the second. Female makes a loud bubbling chuckle that rises, then falls, *wik-wik-wik-wik-wik-wik*, and is vaguely like Whimbrel.

Cuculus optatus
东方中杜鹃 (北方中杜鹃)
dōng fāng zhōng dù juān (běi fāng zhōng dù juān)
Oriental Cuckoo **NR**

L: 25–34 cm
Habitat: Inhabits montane forests.
Behavior: Often calls solitarily in the canopy; hard to see. Brood parasite.
Distribution: Breeds in Northeast China, Inner Mongolia, Xinjiang, and Taiwan. Locally common.
Voice: Disyllabic *huuhuu*, reminiscent of Eurasian Hoopoe. (See Himalayan Cuckoo.) Females make bubbling trill.

Cuculus saturatus
中杜鹃 zhōng dù juān
Himalayan Cuckoo **LC**

L: 25–34 cm
Habitat: Inhabits montane forests; seen in wooded areas on plains during migration.
Behavior: Often perches and calls solitarily in the upper and middle stories of tall trees. Brood parasite.
Distribution: Locally common from southwestern and South China to central and eastern China, and north to Beijing and NE Hebei.
Voice: Song similar to Oriental Cuckoo but invariably more than two notes (usually three or four), with the first note typically higher pitched than the others and the entire song subtly higher pitched than that of Oriental Cuckoo, *hoop, hoop-hoop-hoop*, repeated often for long periods. Females make bubbling trill.

Lesser Cuckoo

Small body

Fewer and wider-spaced bars on underparts

♂

♀

Rufous morph

Oriental Cuckoo

Closely resembles Himalayan Cuckoo, best identified by voice

♂

Rufous morph

♀

Indian Cuckoo

Dusky brown iris, yellow eye ring

♂

Gray head contrasts with dusky brown back

♀

Relatively broad black subterminal band on tail

Himalayan Cuckoo

Brown iris, yellow eye ring

♂

Bars on underparts broader and more widely spaced than Common Cuckoo

Rufous morph

♀

These *Cuculus* cuckoos are best distinguished by voice.

Common Cuckoo

Yellow iris and eye ring

Relatively dense and fine bars on belly

227

鸮形目　STRIGIFORMES

Birds of prey known for their unique nocturnal habits. Cryptic plumage similar between sexes, mostly brown and gray. Body cylindrical and ovoid and strongly built; usually "stands" straight when perched. Most species with distinctive facial disk, some with "ear tufts." Eyes large and forward facing. Inhabit primarily tropical and temperate forests but also open habitats, including deserts and grasslands, solitarily or in pairs. Mostly nocturnal; hunt by hearing and vision. Fly noiselessly. Feed mostly on rodents, birds, and other small animals. A few species are migratory.

Two families, 28 genera, and 248 species recognized worldwide, with a cosmopolitan distribution. Two families, 12 genera, and 32 species recorded in China, ubiquitously distributed nationwide.

草鸮科
Tytonidae

Barn Owl

鸱鸮科
Strigidae

Long-eared Owl

Tyto alba
仓鸮 cāng xiāo
Barn Owl

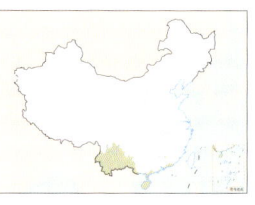

L: 33–39 cm
Habitat: Prefers human habitations; in the daytime or breeding season, usually nests or roosts in ruins, lofts, and bridge piers, as well as in crevices and cavities of walls, trees, and rocks in countryside and towns.
Behavior: Nocturnal. Often solitary. Usually perches for a long time on buildings or open branches, waiting for prey.
Distribution: *T. a. javanica* in S Yunnan; *T. a. stertens* in C Yunnan, Guangxi, Guangdong, and Hainan. Uncommon resident.
Voice: Variety of harsh screeches and hisses; clicks bill as alarm call.

Tyto longimembris
草鸮 cǎo xiāo
Australasian Grass-Owl

L: 32–38 cm
Habitat: Occurs on desolate tall grasslands, such as swamps or reedbeds.
Behavior: Nocturnal. Often solitary. Usually circles above deserted grasslands in search of prey. Nests on ground among dense grass.
Distribution: *T. l. chinensis* widespread from Yunnan to Hubei and Shandong, with vagrants north to Hebei; *T. l. pithecops* in Taiwan. Uncommon resident.
Voice: Clicking, almost churring notes, more reminiscent of a sand-plover than an owl.

Phodilus badius
栗鸮 lì xiāo
Oriental Bay-Owl

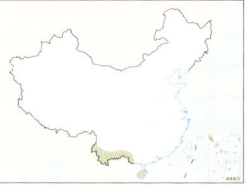

L: 23–29 cm
Habitat: Occurs in evergreen broadleaf forests and dense bamboo thickets.
Behavior: Nocturnal. Often solitary or in pairs. Not too afraid of people, but usually hides deep among leaves.
Distribution: Rare resident in W and S Yunnan, Guangxi, and Hainan.
Voice: Series of four to seven eerie, melancholy whistles that rise and fall slightly in pitch and often start loud and gradually fade away, *du-duo … duo … duo … duo.*

Otus lettia

领角鸮 lǐng jiǎo xiāo

Collared Scops-Owl

L: 23–25 cm

Habitat: Occurs in deciduous forests, evergreen forests, secondary forests, open scrubland near cultivated lands, villages, and urban parks at altitudes below 2200 m.

Behavior: Nocturnal. Hides in dense trees and bamboo thickets during daytime; flies low to hunt for food at night.

Distribution: Common resident. *O. l. erythrocampe* in most regions in southwestern, South, and East China; *O. l. lettia* in SW and S Yunnan and SE Tibet; *O. l. umbratilis* in Hainan; *O. l. glabripes* in Taiwan.

Voice: Monosyllabic, mellow, and flat *buuo*, repeated at intervals of 12–20 sec, often for minutes on end; female higher pitched than male.

Otus semitorques

北领角鸮 běi lǐng jiǎo xiāo

Japanese Scops-Owl

L: 21–26 cm

Habitat: Inhabits montane broadleaf and mixed forests; sometimes also seen on margins of montane forests, forests near villages, or urban parks.

Behavior: Nocturnal. Often hides in dense trees during daytime, active at night. Nests in natural tree cavities, sometimes uses abandoned cavities of woodpeckers, occasionally uses nests of Oriental Magpie.

Distribution: Found in Northeast, North, Central, and East China; primarily a summer breeder in Northeast China. Migrates to North and East China in winter.

Voice: Low-pitched, unobtrusive, buttonquail-like, pulsed booming, *woo … woo … woo*, with a short (0.19 sec), low-pitched, flat note, starting quietly and repeated metronomically every 0.6 sec for over a minute. Additional vocalizations have been described, but many apparently in error. More research needed.

Barn Owl

Crown paler than Australasian Grass-Owl

White facial disk

Black and white spots on back

Relatively short tarsi

Australasian Grass-Owl

Crown darker than Barn Owl

Generally yellowish, sometimes whitish, facial disk

White spots on back

Relatively long tarsi

Oriental Bay-Owl

Dark chestnut with white spots from head to back

Pinkish-chestnut facial disk

Pinkish-chestnut breast and belly with black spots

Collared Scops-Owl

Grayish-brown or brown body with distinctive erect ear tufts

Dark brown iris

Line of triangular pale yellowish-buff spots on scapulars

Distinctly dark brownish-rimmed facial disk

Thin dark streaks on belly

Unfeathered toes

Japanese Scops-Owl

Red iris

Distinct blackish-brown shaft streaks and wavy pale brown bars on underparts

Feathered tarsi and toes

Otus spilocephalus

黄嘴角鸮 huáng zuǐ jiǎo xiāo

Mountain Scops-Owl

L: 18–20 cm

Habitat: Occurs in dense evergreen broadleaf forests, montane forests, valleys, and forest edges at altitudes of 600–2600 m.

Behavior: Nocturnal. Often hides in the dark among trees or in caves during daytime, moves about in dense canopy at night; hard to find.

Distribution: Uncommon resident. *O. s. hambroecki* locally common in Taiwan; *O. s. latouchi* in S and SW Yunnan, Guangxi, Hainan, Guangdong, Fujian, Jiangxi, and Zhejiang.

Voice: Clear, penetrating, high-pitched, disyllabic whistle, *plu-plu*, repeated metronomically every 3–10 sec.

Otus brucei

纵纹角鸮 zòng wén jiǎo xiāo

Pallid Scops-Owl

L: 20–22 cm

Habitat: Inhabits primarily desert poplar forests in deserts and other arid areas; seldom occurs in other broadleaf forests.

Behavior: Nocturnal.

Distribution: Uncommon breeder in desert poplar forests of SW and NW Xinjiang.

Voice: Very low, metronomic *hoop* or *who* notes that are made at rate of just over one note per second and start quietly and increase in volume, very different from Eurasian Scops-Owl.

Otus elegans

优雅角鸮 (琉球角鸮)

yōu yǎ jiǎo xiāo (liú qiú jiǎo xiāo)

Ryukyu Scops-Owl

L: 18–22 cm

Habitat: Inhabits evergreen broadleaf and mixed forests on low mountains; sometimes seen in forests or orchards near residential areas.

Behavior: Often solitary or in pairs. Nocturnal; usually hides in dense forests during daytime, active at night. Nests in coconut trees and other natural tree cavities.

Distribution: Restricted to Orchid and Green Islands off S Taiwan, where locally common residents. Vagrant to Hong Kong.

Voice: Low-pitched, disyllabic *wu wu* repeated every 2 sec. Pairs often duet, female replying with lower, nasal *nyet*.

Otus sunia

红角鸮 hóng jiǎo xiāo

Oriental Scops-Owl

L: 16–22 cm

Habitat: Inhabits broadleaf forests and mixed forests on mountains or plains; sometimes seen at forest edges, forests near residential areas, or urban parks.

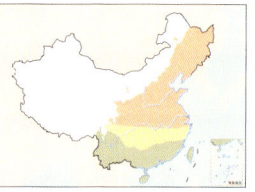

Behavior: Nocturnal. Hides in forests during daytime, active at night. Nests in natural tree cavities or old woodpecker cavities; occasionally nests in rock crevices or artificial nest boxes. A few individuals reuse old nests of Corvidae.

Distribution: Widespread in eastern China, primarily as summer breeder and passage migrant. *O. s. stictonotus* rare in E Inner Mongolia, Heilongjiang, Jilin, Liaoning, Hebei, Shandong, Gansu, and Henan; *O. s. malayanus* in Guangdong, Guangxi, Yunnan, Sichuan, Guizhou, Hubei, Anhui, Jiangxi, Jiangsu, Zhejiang, and Fujian. *O. s. japonicus* occurs in Taiwan during migration.

Voice: Trisyllabic, ringing *toi … toi-toik*, with longer pause between the first and second notes. Rhythm varies geographically.

Otus scops

西红角鸮 xī hóng jiǎo xiāo

Eurasian Scops-Owl

L: 19–21 cm

Habitat: Inhabits broadleaf forests, farmland shelterbelts, and urban parks on plains and mountains. Avoids desert poplar forests.

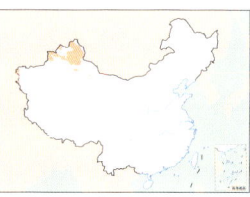

Behavior: Nocturnal.

Distribution: Locally common summer breeders in N and W Xinjiang.

Voice: Simple deep *toik* calls, repeated metronomically at intervals of about 3 sec.

Mountain Scops-Owl

Brown upperparts, yellowish-brown facial disk, more brownish plumage than Oriental Scops-Owl

Line of triangular white spots on scapulars

Ivory to lemon-yellow bill, paler than Oriental Scops-Owl

Pale brown-tinged underparts with faint streaks

Oriental Scops-Owl

Gray morph

Yellow iris

Rufous morph

Distinct blackish-brown shaft streaks on underparts

Pallid Scops-Owl

Line of buff spots on scapulars

Streaks on underparts with little branching

Wing tips do not extend beyond tail tip

Eurasian Scops-Owl

Gray morph

Very similar to Oriental Scops-Owl and best distinguished by voice

Line of white spots on wing coverts

Rufous morph

Wing tips extend beyond tail tip

Streaks on underparts, mostly branched

Ryukyu Scops-Owl

Yellow iris

Ketupa zeylonensis

褐渔鸮 hè yú xiāo

Brown Fish-Owl

L: 51–55 cm
Habitat: Inhabits streams, rivers, and ponds in forests and open woodlands.
Behavior: Partially nocturnal. Often flies and forages solitarily above water's surface.
Distribution: *K. z.*

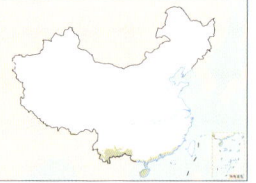

leschenaulti occurs in W Yunnan; *K. z. orientalis* in Yunnan and South China, including Hainan. Uncommon.
Voice: Very low-pitched, typically paired, resonant hooting, *oom-uum*, not far-carrying and easily overlooked.

Ketupa flavipes

黄腿渔鸮 huáng tuǐ yú xiāo

Tawny Fish-Owl

L: 48–55 cm
Habitat: Occurs near rivers, streams, fishponds, and lakes in forests at middle and low altitudes.
Behavior: Partially nocturnal. Perches in large trees or rocks near water and waits for prey.

Distribution: Uncommon in southwestern, East, Central, and South China, including Taiwan.
Voice: Long, descending, strained, fine whistle, *sreeeooouu*.

Ketupa nipalensis

林雕鸮 lín diāo xiāo

Spot-bellied Eagle-Owl

L: 51–63 cm
Habitat: Occurs in evergreen broadleaf forests and secondary forests.
Behavior: Nocturnal. Often solitary. Hides in dense trees during daytime. Not too afraid of people.

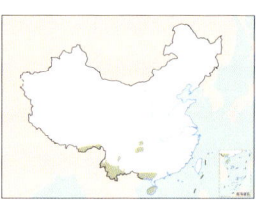

Distribution: Rare in SE Tibet, W Yunnan, Sichuan, Chongqing, Guangxi, and Hainan.
Voice: Melancholy, eerie whistle that rises and falls in pitch, *klueeuw*.

Bubo bubo

雕鸮 diāo xiāo

Eurasian Eagle-Owl

L: 59–73 cm
Habitat: Inhabits a wide range of habitats, such as montane forests, plains, deserted fields, forest-edge bushes, woodlands, bare high mountains, and cliffs; capable of living at altitudes above 3000 m.

Usually occurs in remote areas.
Behavior: Usually solitary except during breeding season. Nocturnal; often rests in trees, cliffs, and withered grass during daytime. Excellent hearing; immediately opens eyes when approached; flies away immediately if approached too close. Flight slow and silent, often low.
Distribution: Widespread across China but absent in Hainan and Taiwan; often as uncommon resident. *B. b. hemachalanus* in Inner Mongolia, Gansu, Xinjiang, Tibet, Qinghai, W Yunnan, and W Sichuan; *B. b. tibetanus* in SW Gansu, Tibet, Qinghai, W Sichuan, and NW Yunnan; *B. b. kiautschensis* in Shandong, Henan, Shaanxi, Chongqing, Guizhou, Hubei, Jiangxi, Zhejiang, Fujian, and Hong Kong; *B. b. turcomanus* in Xinjiang; *B. b. tarimensis* in NE Xinjiang; *B. b. ussuriensis* in Heilongjiang, Jilin, Beijing, Hebei, and Shandong; *B. b. yenisseensis* in N Xinjiang.
Voice: Deep, booming *ooo-uu*, the first syllable stressed and audible at great range. Alarm a yapped *yak-ak*, occasionally in longer sequence.

Ketupa blakistoni

毛腿雕鸮 （毛腿渔鸮）

máo tuǐ diāo xiāo (máo tuǐ yú xiāo)

Blakiston's Fish-Owl

L: 67–77 cm
Habitat: Inhabits rivers, streams, and fishponds in forests on low mountains.
Behavior: Nocturnal. Often solitary. Feeds primarily on fish.
Distribution: Very rare in forests in northwestern

Northeast China. Population has declined rapidly due to massive loss of suitable habitat.
Voice: Low-pitched, short, resonant booms, *bu-yu … boo* or *bu-yu … boo-bu*.

Brown Fish-Owl

Long ear tufts

Thick black streaks
on upperparts

Fine brown bars
on underparts

Eurasian Eagle-Owl

Prominent ear tufts

Orange-red iris

Tawny Fish-Owl

Long ear tufts

Slightly larger than
Brown Fish-Owl

Orange-brown
body

Blakiston's Fish-Owl

Relatively long ear tufts

Inconspicuous facial disk

Enormous body

Spot-bellied Eagle-Owl

Prominent long ear tufts

Dark brown back

Heart-shaped spots on belly

Bubo scandiacus

雪鸮 xuě xiāo

Snowy Owl

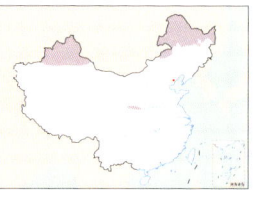

L: 55–64 cm
Habitat: Prefers snow-covered open fields or plains in winter; sometimes also seen on open woodlands or forest edges. Inhabits primarily Arctic tundra in summer.
Behavior: Diurnal. Often stands in snow or on stakes for a long time during the day; once prey is located, flies low to approach and pounce on the prey.
Distribution: Restricted to northwestern and Northeast China. Uncommon winter visitor; rare in North and Central China in winter.
Voice: Usually silent in winter. Occasionally *kre-kee-ke* given in flight, like a large gull.

Strix leptogrammica

褐林鸮 hè lín xiāo

Brown Wood-Owl

L: 39–55 cm
Habitat: Inhabits evergreen broadleaf forests in subtropical mountainous areas, sometimes near villages.
Behavior: Nocturnal. Often solitary. Usually hides in dense trees during daytime.
Distribution: *S. l. caligata* in Taiwan and Hainan; *S. l. ticehursti* in Sichuan, Yunnan, and most of southern China to the east; *S. l. newarensis* in S and SE Tibet.
Voice: Eerie, wailing scream that rises, then falls, longer and higher pitched than similar wail of Spot-bellied Eagle-Owl, *perwieuu*. Also low-pitched hoot *huturu*.

Strix nivicolum

灰林鸮 huī lín xiāo

Himalayan Owl

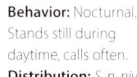

L: 37–40 cm
Habitat: Inhabits forests at middle and high altitudes.
Behavior: Nocturnal. Stands still during daytime, calls often.
Distribution: *S. n. nivicolum* in southwestern, Central, East, and South China; *S. n. yamadae* in Taiwan; *S. n. ma* in Xinjiang and North and Northeast China.
Voice: Simple song is a clear, resonant, short pair of low-pitched hoots, *boo-boo*, delivered in rapid succession with the second note sometimes slightly longer and lower pitched. Repeated every few seconds, it varies little but is occasionally trisyllabic. *S. n. ma* and birds in N Xinjiang have very different vocalization; its more complex "compound hooting" is reminiscent of Tawny Owl (*S. aluco*), *haoou … hu, hww'O'o'u'u*. The first note descends, there is a lengthy pause before the brief second note, and then the final note is tremulous or wavering.

Strix uralensis

长尾林鸮 cháng wěi lín xiāo

Ural Owl

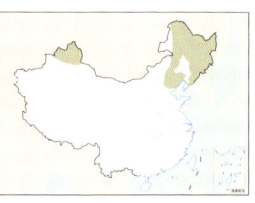

L: 45–54 cm
Habitat: Inhabits montane coniferous, mixed, and broadleaf forests; especially common in the latter two types of forests. Sometimes seen in secondary forests at forest edges or woodlands.
Behavior: Often solitary except during breeding season. Nests in natural tree cavities; occasionally nests on ground under tree roots or on rocky cliffs. Mainly nocturnal; usually in deep forests during daytime, resting close to tree trunks, sometimes also feeds during the day. Flight buoyant, silent, and undulating, with deep wingbeats.
Distribution: *S. u. yenisseensis* in N Xinjiang; *S. u. nikolskii* in Northeast China and northern North China. Resident; local and uncommon.
Voice: Not especially vocal. Song a deep, intimidating *wo-hu … wuhu uhwo-ho*, with distinct pause after the first pair of notes. Makes harsh *krief* when threatened.

Strix davidi

四川林鸮 sì chuān lín xiāo

Sichuan Wood-Owl

L: 54 cm
Habitat: Inhabits mixed broadleaf-coniferous forests and coniferous forests from subalpine to alpine zones.
Behavior: Often solitary. Inhabits forests at high altitudes. Nocturnal and secretive.
Distribution: Chinese endemic restricted to S Gansu, W Sichuan, and SE Qinghai.
Voice: Similar to Ural Owl.

Strix nebulosa

乌林鸮 wū lín xiāo

Great Gray Owl

L: 56–65 cm
Habitat: Prefers primary coniferous forests and mixed broadleaf-coniferous forests composed mainly of larch, birch, and aspen.
Behavior: Similar to Ural Owl. Nests in trees; nest similar to that of magpie but simpler. Sometimes occupies the nests of other raptors or large birds. Excellent hearing; can hear rodents under snow. Flight slow and silent, with gentle wingbeats.
Distribution: Resident in N Xinjiang and northern Northeast China; local and uncommon.
Voice: Occasionally gives prolonged series of 9–14 low-pitched booming hoots, *wu … wu … wu … wu*, at less than 1 sec intervals. Variety of other calls include descending nasal *chiep* and similar but more strained, almost hissing, begging notes of fledglings and bill clicking.

Snowy Owl

All-white plumage with few bars

More bars on upper- and underparts

Ural Owl

Prominent streaks on belly

Brown Wood-Owl

V-shaped white supercilium

Dark brown back

Pale brown breast and belly with dense thin black bars

Sichuan Wood-Owl

Blackish-brown back with white spots

Conspicuous streaks on breast and belly; relatively pale bars hardly visible

Himalayan Owl

Rufous morph

Smaller than Brown Wood-Owl

Dark streaks and bars on underparts

Gray morph

Paler than Brown Wood-Owl

Great Gray Owl

Barring in concentric circles on facial disk

Relatively large body

235

Surnia ulula

猛鸮 měng xiāo

Northern Hawk Owl

L: 34–40 cm

Habitat: Inhabits primary coniferous or mixed broadleaf-coniferous forests; also seen on forest tundra and forests on plains. Especially prefers open areas in forests.

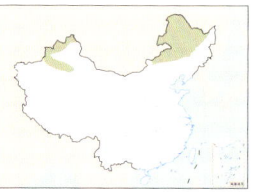

Behavior: Usually solitary except during breeding seasons. Nests on treetops or in cavities of dead trees; sometimes reuses old nests of Corvidae. Forages primarily during daytime; crepuscular. Flight rapid, alternating between wing flapping and gliding. Often perches on treetops during daytime; can hover in search of prey. Rapidly pounces on prey if located.

Distribution: *S. u. tianschanica* in N Xinjiang; *S. u. ulula* in northern parts of Northeast China. Local and uncommon resident.

Voice: Very fast, lengthy, bubbling trill, *lulululululululu* (higher pitched and longer than Boreal Owl) and quarrelsome cackling. Makes fast falconlike *kik-kik-kik* in alarm.

Athene noctua

纵纹腹小鸮 zòng wén fù xiǎo xiāo

Little Owl

L: 20–26 cm

Habitat: Inhabits farmland, deserts, or around villages; also occurs on low mountains and hills, forests on plains, and forest-edge bushes.

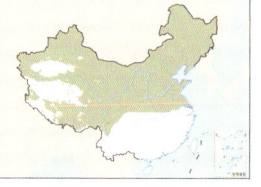

Behavior: Usually solitary except during breeding seasons. Nests in cavities in abandoned buildings, crevices and caves in rocks. Crepuscular and nocturnal. Usually stands on roofs or power lines in the morning and evening. Hunts for prey with rapid flight.

Distribution: Uncommon resident. *A. n. impasta* in Gansu, Qinghai, and N Sichuan; *A. n. ludlowi* on the eastern edge of the Qinghai-Tibet Plateau; *A. n. orientalis* in C and N Xinjiang; *A. n. plumipes* in Northeast and North China.

Voice: Varied; includes a mellow *kee-ew* or *klie-uw* that rises sharply at the end and is repeated; makes a series of short, high-pitched *kiwi-kiwi-kiwi* notes in alarm.

Athene brama

横斑腹小鸮 héng bān fù xiǎo xiāo

Spotted Owlet

L: 19–22 cm

Habitat: Often inhabits hills, low mountains, plains, farmland, and woodlands near villages; also visits orchards.

Behavior: Solitary or in pairs. Often nests in tree cavities. Diurnal and crepuscular.

Distribution: Rare resident in SE Tibet and SW Yunnan.

Voice: Unpleasant cat-fighting cacophony of screeches, chatters, and chuckles. Also hissing.

Glaucidium passerinum

花头鸺鹠 huā tóu xiū liú

Eurasian Pygmy-Owl

L: 15–19 cm

Habitat: Inhabits coniferous and mixed broadleaf-coniferous forests; also seen in open wooded areas and birch and poplar forests.

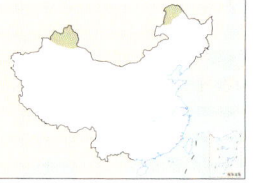

Behavior: Nocturnal and crepuscular. Often nests in tree cavities. Reuses old nests of woodpeckers; sometimes builds nests with tree branches. Capable of hunting for prey larger than itself; forages primarily for mice. Caches food in winter.

Distribution: *G. p. passerinum* in N Xinjiang; *G. p. orientale* in Northeast China. Local and rare resident. Some move south in winter.

Voice: Song a far-carrying, piping, Eurasian Scops-Owl–like *deu* given about once every 1.5 sec and repeated at length. Also a scale-ascending series of 5–10 strained whistles. Female gives short, high-pitched, rising whistle, like a passerine alarm call.

Taenioptynx brodiei

领鸺鹠 lǐng xiū liú

Collared Owlet

L: 15–17 cm

Habitat: Often inhabits montane forests, open forest edges, and scrubland at altitudes of 1350–2500 m.

Behavior: Primarily diurnal; also active at night. Prefers perching

and calling in dense canopy, swaying tail from side to side. The calls often elicit mobbing by small passerines.

Distribution: Locally common resident. *T. b. brodiei* widespread south of the Qinling-Huaihe Line. *T. b. paradalotum* in Taiwan.

Voice: Male's territorial song is a continuously repeated, mellow, four-syllable toot, *hū hū-hū hū*, with the middle notes grouped closer. Often heard during day.

Glaucidium cuculoides

斑头鸺鹠 bān tóu xiū liú

Asian Barred Owlet

L: 22–26 cm

Habitat: Inhabits open habitats at altitudes below 2700 m; adapted to a wide range of habitats, from forests to villages.

Behavior: Similar to Collared Owlet. Often makes undulating flight from perch.

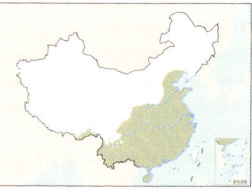

Distribution: Locally common resident. *G. c. austerum* in SE Tibet; *G. c. brugeli* in S Yunnan; *G. c. persimile* in Hainan; *G. c. rufescens* in SW Yunnan; *G. c. whitelyi* widespread in southern China.

Voice: Lengthy winnowing trill, *wowowowowowowowo*, that can last 6–7 sec. Also combative, accelerating *puk-loo … puk-lu puk-lu holo, holo* that continues for 7–14 sec.

Northern Hawk Owl

Spots on forehead

Conspicuous bars on belly

Eurasian Pygmy-Owl

Spots on crown

Clear streaks on belly

Collared Owlet

Brown body with pale spots on crown

Inconspicuous facial disk, no ear tufts

Brown bars on breast extend to flanks

Two prominent black nape spots, creating "false face"

White belly streaked with teardrop-shaped brown feathers

Little Owl

Thick white supercilium

Distinct streaks on belly

Asian Barred Owlet

Dense bars on crown, no white spots

No "false eyes" on nape

Clear bars from breast to upper belly

White lower belly with diffuse dark brown streaks

Spotted Owlet

Conspicuous broad white band on throat

Distinct brown spots on belly

Relatively larger body and longer tail compared to Collared Owlet

Ninox scutulata

鹰鸮 yīng xiāo

Brown Boobook LC

L: 26–31 cm

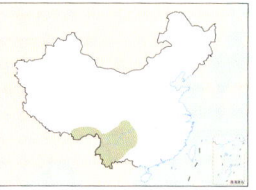

Habitat: Often inhabits broadleaf forests; sometimes occurs in mixed forests, foothill plains, orchards, and villages on low mountains and hills.

Behavior: Often solitary; in pairs during breeding season. Usually nests in natural tree cavities. Nocturnal; mainly hides and rests in forests during daytime. Flight rapid and agile.

Distribution: Uncommon resident in southwestern China. *N. s. burmanica* in S and W Yunnan, Sichuan, and Chongqing; *N. s. lugubris* in SE Tibet.

Voice: Mellow, melancholy, upwardly inflected, and slightly disyllabic *whooup … whooup … whooup*, hoots repeated at intervals greater than a second in series of three to seven notes.

Ninox japonica

日本鹰鸮 (北鹰鸮) rì běn yīng xiāo (běi yīng xiāo)

Northern Boobook LC

L: 27–33 cm

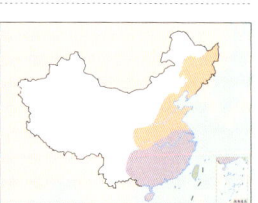

Habitat: Often inhabits mixed broadleaf-coniferous and broadleaf forests; especially prefers river valleys in forests. Sometimes occurs on low mountains and hills, foothill plains, orchards, and urban parks.

Behavior: Often solitary; in pairs during breeding season. Usually nests in natural tree cavities; also reuses old nests of Mandarin Duck. Nocturnal and crepuscular; often hides in trees during daytime. Flight rapid and agile.

Distribution: Widespread in Northeast, North, East, and South China, mostly as summer breeder in northern China. *N. j. japonica* in Heilongjiang, Jilin, Liaoning, Hebei, Beijing, Shanghai, Shandong, and Hubei. *N. j. totogo* occurs in Taiwan; resident.

Voice: Repetitive, resonant, low-pitched, paired *wup-wup* notes, separated by about 0.4 sec and repeated about every 0.7 sec. Notes lower pitched, shorter, and flatter than those of Brown Boobook.

Asio otus

长耳鸮 cháng ěr xiāo

Long-eared Owl LC

L: 33–40 cm

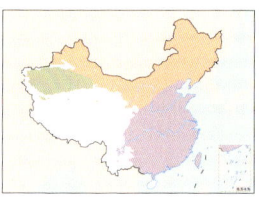

Habitat: Often inhabits coniferous forests and mixed broadleaf-coniferous forests; sometimes also seen in broadleaf forests, woodlands at forest edges, urban parks, orchards, and in proximity of villages.

Behavior: Sometimes gathers in small flocks during migration and winter. Often nests in old nests of magpies or raptors; sometimes also in natural tree cavities. Nocturnal and crepuscular; often rests in forests during daytime.

Distribution: Widespread throughout China but absent from the Qinghai-Tibet Plateau and Hainan. Breeds in northern China and winters in the south. Resident in C Xinjiang.

Voice: Adults not particularly vocal—male gives deep, rising, then falling hoot, *whoo*, every 2.5 sec; female gives a harsher, more disyllabic version. Fledglings often noisy—a sad, plaintive, descending whistle, *plee-ah*.

Asio flammeus

短耳鸮 duǎn ěr xiāo

Short-eared Owl LC

L: 35–40 cm

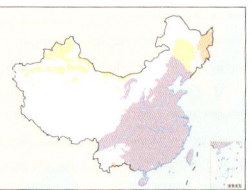

Habitat: Often inhabits varied habitats, such as plains, grasslands, deserts, marshes, low mountains, and hills; especially prefers open grasslands on plains or grass near lakes.

Behavior: Gathers in small flocks in winter. Usually nests in grass on ground; sometimes also breeds in tree cavities. Crepuscular; hides in grass during daytime and hunts for food at dusk and night. Flies slowly, alternating between flapping and gliding.

Distribution: Widespread but absent on the Qinghai-Tibet Plateau. Breeds in northern Northeast China; passage migrant and winter visitor in most regions.

Voice: Usually silent. Occasionally a low-pitched, strained *chreeee-wif*.

Aegolius funereus

鬼鸮 guǐ xiāo

Boreal Owl LC

L: 23–26 cm

Habitat: Often inhabits coniferous and mixed broadleaf-coniferous forests; especially prefers mixed forests of pine, birch, and poplar. Sometimes also on farmland and villages.

Behavior: Often solitary except during breeding season. Nests in natural tree cavities; also reuses old nests of woodpeckers. Nocturnal; hides in dense forests during daytime. Bold; occasionally visits residential areas

in fall and winter. Flight rapid, undulating.

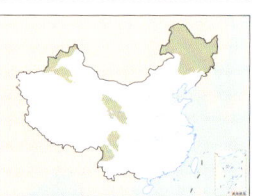

Distribution: Uncommon resident in northwestern, Northeast, and southwestern China. *A. f. beickianus* in S Gansu, E Qinghai, NW Yunnan, and Sichuan; *A. f. pallens* in the Tian Shan Mountains of Xinjiang; *A. f. sibiricus* in northernmost Xinjiang, Heilongjiang, Jilin, and NE Inner Mongolia.

Voice: Tremulous Common Snipe–like winnowing, *wu wu wu-wu-wu*, shorter and lower pitched than Northern Hawk Owl. Also a loud *kyiak*.

Brown Boobook

Barred belly

N. s. lugubris

Northern Boobook

Streaked belly

N. j. japonica

Boreal Owl

Distinct white X
between eyes

Long-eared Owl

Conspicuous
ear tufts

Orange iris

Distinct white
X between
eyes

Short-eared Owl

Relatively
short ear tufts

Yellow iris

夜鹰目　CAPRIMULGIFORMES

Small to medium-sized nocturnal perching birds, resembling hawks in flight. Plumage similar between sexes, mostly brown, white, and black; patterned similar to owls. Head flat. Bill appears small but with very large gape, usually with whiskerlike feathers on head or well-developed rictal bristles. Wings long and pointed or short and rounded; fly without sound. Tail relatively long, squared or rounded; some with elongated tail feathers. Legs short with syndactyl feet. Inhabit primarily forests, but also open habitats. Nocturnal; feed mostly on insects, but some feed primarily on plant fruits. Nest on ground or tree branches. Most species are nonmigratory.

Four families, 25 genera, and 122 species recognized worldwide, distributed mainly in the tropics and subtropics, some extending to temperate zones. Two families, three genera, and seven species recorded in China, widely distributed except on the Qinghai-Tibet Plateau.

蛙口夜鹰科 (蟆口鸱科)
Podargidae

Hodgson's Frogmouth

夜鹰科
Caprimulgidae

Gray Nightjar

Batrachostomus hodgsoni

黑顶蛙口夜鹰 (黑顶蟆口鸱)
hēi dǐng wā kǒu yè yīng (hēi dǐng má kǒu chī)

Hodgson's Frogmouth LC

L: 22–27 cm
Habitat: Inhabits evergreen deciduous forests and plantations at forest edges.
Behavior: Often solitary or in pairs. Nocturnal. During nighttime, uses its large mouth to hawk insects in forests or at forest edges. Stays still, like a stake, during daytime; very well camouflaged.
Distribution: Resident in W and S Yunnan.
Voice: Male gives a nasal rising then falling *whaaeee*, often repeated multiple times; female gives a rising *perweeuu* or *diu diudiu*.

Lyncornis macrotis

毛腿夜鹰 (毛腿耳夜鹰)
máo tuǐ yè yīng (máo tuǐ ěr yè yīng)

Great Eared-Nightjar LC

L: 31–40 cm
Habitat: Inhabits evergreen broadleaf forests and bushes and bamboo thickets at forest edges.

Caprimulgus jotaka

普通夜鹰 pǔ tōng yè yīng
Gray Nightjar LC

L: 24–29 cm
Habitat: Breeds in open scrubland, broadleaf forests, mixed broadleaf-coniferous forests, woodland bushes, bamboo thickets, and farmland on plains or hills. Also becomes increasingly adaptable to urban environments and often rests on buildings. *C. j. hazarae* occurs up to 3400 m.
Behavior: Nocturnal. Often perches motionlessly on branches, tree trunks, or on the ground in forests; hard to find. Flies out to forage at dusk and night, especially active at dusk. Flight rapid and silent; usually glides after flapping wings. Circles and calls in the air, catching various insects, such as moths and true bugs. Also calls from perch.
Distribution: Common. *C. j. hazarae* is resident in SE Tibet and S Yunnan; *C. j. jotaka* is widely distributed from Northeast to southwestern China and regions to its east, including Hainan and Taiwan, as summer breeder and passage migrant.
Voice: Short percussive notes repeated 5–15 times in fast-paced, loud sequence, *chuk-chuk-chuk-chuk*. Silent in winter.

Behavior: Often solitary or in pairs. Nocturnal. Often rests on grassy patches in forests or perches on shaded branches during daytime; flies and forages over open areas at night. Large body size. Flies like a harrier.
Distribution: Rare in S Yunnan.
Voice: Cheerful, clear, lengthy, descending whistle, *per-weeeoooo*.

Hodgson's Frogmouth

Well-developed whiskerlike feathers on forehead and crown

Well-developed whiskerlike feathers on forehead and crown

Grayish-white body

Reddish-brown body

Great Eared-Nightjar

Distinct ear tufts

White patch on throat

Gray Nightjar

Fine streaks on crown

White submoustachial streak

Densely barred throat with prominent white patches on side

Grayish-brown body with splotches

Brown spots on wings

Barred tail

White tail tip

White patch on primaries (yellow patch on female)

Caprimulgus europaeus

欧夜鹰 ōu yè yīng

Eurasian Nightjar　LC

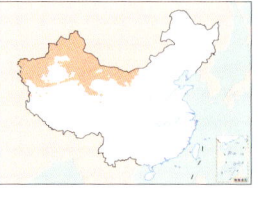

L: 24–28 cm
Habitat: Inhabits desert edges, desert grasslands, farmland, desert fields, and wooded areas on plains.
Behavior: Often solitary or in pairs; gathers in small flocks during migration.
Nocturnal; active mainly at dusk and night, often rests on branches or ground in forests during daytime. Usually calls at night. Displaying male glides with wings open in V shape and tail fanned. Flight rapid, agile, and silent. Catches prey in flight using large mouth; feeds on insects such as mosquitoes, blackflies, beetles, and moths.
Distribution: *C. e. europaeus* breeds in Altai Mountains in N Xinjiang; *C. e. unwini* breeds in SW Xinjiang and NW Gansu; *C. e. plumipes* breeds in E Xinjiang to NW Gansu, W Inner Mongolia, and Ningxia.
Voice: Song an intense purring or reeling with changes in pitch but few "intakes of breath," errrrrrrrrrrr … urrrrrrrrrrrrrr … errrrrr, often continuing for minutes and ending with wing claps. Nasal *quoik* contact calls.
Note: Vaurie (1960) described a new species, *Caprimulgus centralasicus* (Vaurie's Nightjar 中亚夜鹰), based on a specimen collected in 1929 in Pishan County, SW Xinjiang. There were no subsequent encounters. Schweizer et al. (2020) conducted molecular and morphological analyses on the type specimen, and their results indicated that the type specimen was a subadult Eurasian Nightjar. Therefore, *Caprimulgus centralasicus* is no longer considered a valid species but a synonym of *C. e. plumipes*.

Caprimulgus aegyptius

埃及夜鹰 āi jí yè yīng

Egyptian Nightjar　LC

L: 24–27 cm
Habitat: Inhabits primarily dry plains, grasslands, semideserts, and deserts with bushes. Especially prefers deserts and semideserts near rivers or lakes.
Behavior: Active mainly at dusk and night. Rests on ground or in shadows under bushes during daytime; frequently flies and hunts over water or ground at dusk. Flight rapid; usually flies low, close to the ground or water's surface.
Distribution: One specimen record from W Xinjiang in 1876; could be a misidentification. Cheng (1994) questioned the reliability of this record.
Voice: Pulsating series of purring *krorr* notes with pauses, as if the bird is easily distracted, and where tempo drops toward end. Short, dry *etch* notes after being flushed.

Caprimulgus macrurus

长尾夜鹰 cháng wěi yè yīng

Large-tailed Nightjar　LC

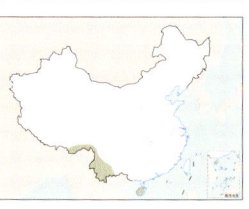

L: 25–29 cm
Habitat: Adapted to a wide range of habitats, including bamboo thickets, scrubland, monsoon forests, secondary forests, open villages, cultivated lands, etc. Usually active at altitudes below 2700 m.
Behavior: Nocturnal. Rests in dense forests or shadows on ground near forest edges during daytime; flies out and forages at dusk and night, especially active at dusk. Calls from perch or ground for as long as half an hour. Catches prey in flight; forages for flying insects, such as moths, crickets, wasps, earwigs, and beetles. Sometimes feeds on insects attracted by streetlights.
Distribution: Resident in SE Tibet, SW and S Yunnan, and Hainan; locally common.
Voice: Resonant *tonk … tonk … tonk … tonk*, repeated hesitantly in series of two to eight leisurely notes.

Caprimulgus affinis

林夜鹰 lín yè yīng

Savanna Nightjar　LC

L: 20–26 cm
Habitat: Inhabits grasslands, open forests, cultivated lands, and stony hillsides below 1500 m. The subspecies in Taiwan, *C. a. stictomus*, prefers broad riverbeds with mixed sand and stone and bare stony beaches near riverbeds at the middle and lower reaches of rivers. Becoming more adaptable to rural and urban environments; often rests on open ground of roads, airports, industrial zones, construction sites, or campuses. Also rests and breeds on terraces of buildings.
Behavior: Nocturnal. Often rests in shadows on ground during daytime; flies out and feeds at dusk and night, especially active at dusk. Flies high when chasing prey; flight slow, buoyant, and silent. Also usually waits for insects attracted by artificial light at fixed locations and flies out for food when prey appears. Relies on sight when hunting; thus makes use of light in cities at night to forage.
Distribution: Uncommon resident or migrant. *C. a. amoyensis* in southwestern and South China and coastal southeastern China and Hainan; breeding records in Taiwan. *C. a. stictomus* is endemic to Taiwan.
Voice: Male makes loud, piercing, slightly nasal, and rising *chweep* call, repeated several times at intervals of 0.8–3 sec; one individual can call repeatedly and continuously for 45 min. Calls all night; peaks one hour before sunrise and after sunset. Female seldom calls; makes only low-pitched calls in flight with males when other nightjars intrude into territory in breeding season.

Eurasian Nightjar

Prominent black streaks on upperparts

Broad buff fringe on scapulars

Grayish-brown body

C. e. europaeus

Large-tailed Nightjar

♂

Slender body; tail accounts for large proportion of body length

White-tipped outermost tail feathers (yellow tipped on female)

White patch on primaries (yellow patch on female)

White submoustachial stripe

Brownish-yellow neck collar

Small white throat patch

Blackish-barred tail

Grayish-brown body with dense brownish-white and blackish-brown spots

Egyptian Nightjar

Sandy-gray upperparts, spots inconspicuous

Scapular pattern different from female Eurasian Nightjar *C. e. plumipes*; Egyptian Nightjar also has fewer streaks

Savanna Nightjar

♂

Brown and black bars on wing feathers

Whitish outer tail feathers

White patch on black primaries (yellow patch on female)

Pale brown eye ring

Brownish-yellow feathers on scapulars and wing coverts

Two relatively small white patches on side of throat

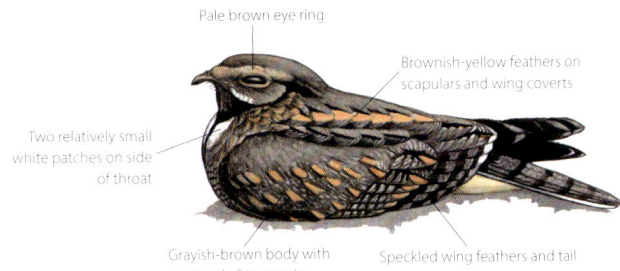

Grayish-brown body with extremely fine streaks

Speckled wing feathers and tail

雨燕目　APODIFORMES

Small forest-dwelling perching birds, including swifts and hummingbirds. Plumage of swifts is drab, and is similar or differs slightly between sexes in most species. Hummingbirds are usually brightly colored; plumage differs between sexes in most species. Bill of swift appears small but with large gape and well-developed rictal bristles; in hummingbirds, bill extremely long and fine. Good fliers with long and pointed wings. Legs short and weak. Tail long, squared or forked, but highly variable in hummingbirds. Swifts feed mostly on insects, while hummingbirds feed almost exclusively on nectar. Some species are nonmigratory, while others undertake extraordinary migrations.

Four families, 127 genera, and 486 species recognized worldwide, with a cosmopolitan distribution (hummingbirds exclusively in the Americas). Two families, six genera, and 15 species recorded in China, ubiquitously distributed nationwide, with high species diversity in southern China.

凤头雨燕科 (树燕科)
Hemiprocnidae

Crested Treeswift

雨燕科
Apodidae

Common Swift

Hemiprocne coronata

凤头雨燕 (凤头树燕)
fèng tóu yǔ yàn (fèng tóu shù yàn)

Crested Treeswift   **LC**

L: 23–25 cm
Habitat: Inhabits evergreen broadleaf forests, pine forests, mixed broadleaf-coniferous forests, shrubby meadows, bamboo thickets, etc. Occurs at altitudes of 1065–2565 m.
Behavior: Often feeds solitarily; also in pairs or small flocks. Usually active on ground or in understory bushes; occasionally ascends to branches covered with mosses and lichens.
Distribution: S and W Yunnan and SE Tibet.
Voice: Percussive *yip-yap* and Small Pratincole–like chatter in flight.

Crested Treeswift

Gray cheek

Erect crest on forehead

Red cheek

Slender wings cross and protrude when perched

All-gray plumage

Aerodramus brevirostris

短嘴金丝燕 duǎn zuǐ jīn sī yàn

Himalayan Swiftlet

LC

L: 13–14 cm
Habitat: Inhabits cliffs of limestone caves from low to high altitudes.
Behavior: Nests colonially on vertical rock faces and in caves on rocky mountains; catches prey in flight.

Distribution: *A. b. brevirostris* in SE Tibet and NW Yunnan; *A. b. innominatus* in Guangdong, Hainan, Shanghai, Hubei, Hunan, Chongqing, Sichuan, Guizhou, and Yunnan; *A. b. rogersi* in SW Yunnan. Increasingly regular along the east coast north to Hebei. Vagrant in Beijing.
Voice: Calls include a dry repeated *chit … chit* and a drawn-out reeling rattle. In display, an extremely rapid, buzzy, musical twittering mixed with clicks and squeaky twittering. *A. b. innominatus* echolocates.

Aerodramus germani

戈氏金丝燕 gē shì jīn sī yàn

Germain's Swiftlet

NR

L: 12–13 cm
Habitat: Inhabits solutional caves on coasts and oceanic islands.
Behavior: Nests colonially in coastal caves; can get in and out of dark caves freely.

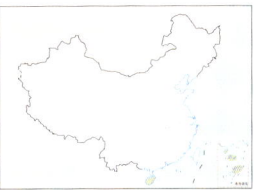

Distribution: Coastal South China; breeding populations recorded in Hainan (Dazhou Island in Wanning).
Voice: High-pitched, buzzy chittering, like Himalayan Swiftlet.

Himalayan Swiftlet

Tail more deeply forked than Germain's Swiftlet

Relatively pale brown rump

Blackish-brown body, paler on belly

Germain's Swiftlet

Paler brown or grayish-white rump

Wings shorter than Himalayan Swiftlet

Tail more shallowly forked than Himalayan Swiftlet

Hirundapus caudacutus

白喉针尾雨燕 bái hóu zhēn wěi yǔ yàn

White-throated Needletail　LC

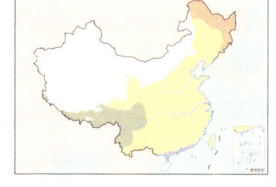

L: 19–21 cm
Habitat: Inhabits mixed broadleaf-coniferous forests and coniferous forests at middle to high altitudes; also occurs in river valleys in forests.
Behavior: Solitary or in small flocks. High-speed flight. Larger and more agile than other swifts. Calls in flight.
Distribution: *H. c. caudacutus* breeds in Northeast China and widespread during migration in Northeast, North, and Central China; *H. c. nudipes* breeds in southwestern China.
Voice: High-pitched, insectlike trilling, *tr-tr-tr.*

Hirundapus cochinchinensis

灰喉针尾雨燕 huī hóu zhēn wěi yǔ yàn

Silver-backed Needletail　LC

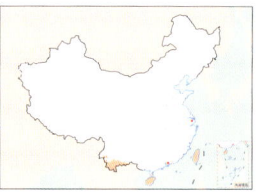

L: 20–22 cm
Habitat: Occurs in montane forests on oceanic islands, coasts, and mainland.
Behavior: Often solitary or in small flocks; forages for insects on the wing over forests and open areas.
Distribution: Breeds in S Yunnan, occasionally also South and East China, including Hainan and Taiwan.
Voice: High-pitched chittering and trills similar to other needletails.

Hirundapus giganteus

褐背针尾雨燕 hè bèi zhēn wěi yǔ yàn

Brown-backed Needletail　LC

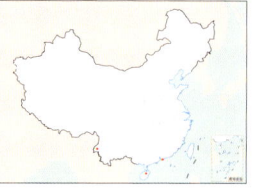

L: 21–26 cm
Habitat: Occurs in lowland forests and secondary forests near rivers and lakes.
Behavior: Often gathers in small flocks; forages for insects on the wing over forests.
Distribution: Rare in W Yunnan, Hong Kong, and Hainan.
Voice: High-pitched chittering and trills similar to other needletails, although a markedly disyllabic *zreee-it* might be diagnostic.

Hirundapus celebensis

紫针尾雨燕 zǐ zhēn wěi yǔ yàn

Purple Needletail　LC

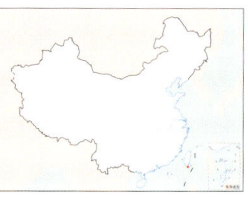

L: 24–25 cm
Habitat: Occurs in mature and secondary forests on low mountains and open farmland.
Behavior: Solitary or in small flocks; forages on the wing over open forests.
Distribution: Vagrant in Hong Kong and Taiwan.
Voice: No information.

Cypsiurus balasiensis

棕雨燕 zōng yǔ yàn

Asian Palm Swift　LC

L: 11–12 cm
Habitat: Occurs at forest edges, farmland, towns, and villages on low mountains and plains.
Behavior: Often gathers in large flocks. Nests on palm fronds. Forages in flight.
Distribution: Locally common in W and SW Yunnan and Hainan.
Voice: Excited high-pitched reedy trills, *chit-trt-trr.*

Apus melba

高山雨燕 gāo shān yǔ yàn

Alpine Swift　LC

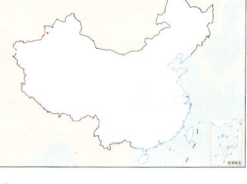

L: 20–23 cm
Habitat: Occurs in large river valleys or open areas with cliffs.
Behavior: Gathers in flocks; forages on the wing.
Distribution: One record in the Ili River valley of NW Xinjiang. Likely also occurs in Tibet.
Voice: High-pitched trills that usually slow and drop in pitch.

White-throated Needletail

Grayish-white forehead

Small white tips on tertials

Horseshoe-shaped white mark from flanks to undertail coverts

H. c. caudacutus

Pure white throat

H. c. nudipes

Purple Needletail

Bluish-purple upper back

White on tertials

White lores

Bluish-black throat

Silver-backed Needletail

No white on tertials

Forehead lacks white spots

Gray throat

Asian Palm Swift

Lacks pale rump

Deeply forked tail

Wings more slender than swiftlets

Alpine Swift

Wings longer and broader than Common Swift

White breast and belly

White throat

Broad brown breast band

Pale brown upperparts

White throat

Brown-backed Needletail

More distinct white lores

Inconspicuous small white throat patch

Pale brown back

Apus apus

普通雨燕 (普通楼燕)

pǔ tōng yǔ yàn (pǔ tōng lóu yàn)

Common Swift

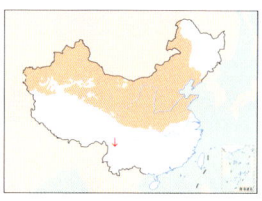

L: 16–19 cm
Habitat: Seen in deserts, grasslands, and even cities; often nests on cliffs in the wild or in ancient buildings in cities. Also found in recent years breeding in crevices of overpass piers.
Behavior: Often gathers in flocks. Hunts for insects on the wing in high-speed flight. Collects food in the back of the throat, making it bulge like a ball.
Distribution: Breeds in Northeast, North, and northwestern China, south to N Sichuan, W Hubei, and N Jiangsu; rare migrant through several regions in southwestern China.
Voice: Loud, piercing, screaming flight calls, srreeeeerrr.

Apus pacificus

白腰雨燕 bái yāo yǔ yàn

Pacific Swift

L: 17–20 cm
Habitat: Often inhabits cliffs, forests, and tundra near streams and reservoirs.
Behavior: Gathers in small flocks and forages on the wing.
Distribution:
A. p. pacificus in northern, East, and South China; A. p. kanoi in eastern northwestern, southwestern, Central, East, and South China.
Voice: Loud and piercing flight calls similar to Common Swift but very slightly lower pitched and more often slowing and descending.

Apus salimalii

青藏白腰雨燕 qīng zàng bái yāo yǔ yàn

Salim Ali's Swift

L: 17–20 cm
Habitat: Inhabits cliffs and forests near rivers.
Behavior: Prefers flying and hunting in flocks above open habitats.
Distribution: Endemic to Western China; found in E Tibet and W Sichuan at the eastern edge of the Qinghai-Tibet Plateau.
Voice: Screaming flight calls very similar to Pacific Swift, perhaps very slightly lower pitched, flatter, and less descending.

Apus acuticauda

暗背雨燕 àn bèi yǔ yàn

Dark-rumped Swift

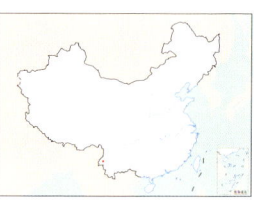

L: 16–17 cm
Habitat: Occurs in tropical forests and woodlands in lowlands with plenty of river valleys.
Behavior: Solitary or gathers in small flocks; forages in flight over forest edges and open areas.
Distribution: Reported in W Yunnan.
Voice: Noisy and very high-frequency chittering at colony, zi–zi–zi–zi–zi.

Apus cooki

印支白腰雨燕 yìn zhī bái yāo yǔ yàn

Cook's Swift

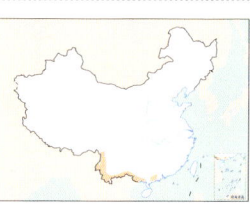

L: 17–20 cm
Habitat: Occurs in forests, woodlands, or cliffs near water.
Behavior: Gathers in flocks and forages on the wing.
Distribution: Found in W and SE Yunnan, S Guangxi, and NW Guangdong.
Voice: Poorly known but probably similar to Pacific Swift.

Apus nipalensis

小白腰雨燕 xiǎo bái yāo yǔ yàn

House Swift

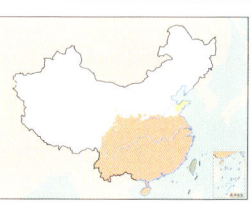

L: 11–15 cm
Habitat: Seen in a wide range of habitats, including forest edges, open areas, towns, cliffs, and caves.
Behavior: Often gathers in flocks and forages on the wing, usually mixed with other swifts; calls in flight.
Distribution: A. n. kuntzi in Taiwan; A. n. nipalensis in the Yangtze River Watershed and regions to its south, and vagrant north to Shandong. The populations in southern China are sometimes treated as A. n. subfurcatus.
Voice: High-pitched, fast, descending twitter.

Common Swift

Scaly belly

White throat

Dark-rumped Swift

Blackish-brown upperparts without white rump

Prominent scaly pattern on underparts

Dense scaly pattern on throat, does not appear white

Pacific Swift

White rump narrower than House Swift but broader than other white-rumped swifts

Relatively dull upper back

Deeply forked tail

Scaly lower belly with distinct fringe

White throat

Cook's Swift

Relatively narrow white rump

Grayish-white throat with denser fine bars

Black underwing coverts, darker than other white-rumped swifts

Salim Ali's Swift

Extremely narrow white rump, distinctly narrower than other white-rumped swifts

Tail longer than other white-rumped swifts

Relatively bright upper back

Scaly lower belly with indistinct fringe

Grayish-white throat

House Swift

Relatively broad white rump

Square and slightly notched tail

Blackish-brown underparts

White throat

咬鹃目　TROGONIFORMES

Brightly colored, medium-sized, forest-dwelling birds. Plumage differs between sexes. Bill short and thick, with slightly hooked tip and well-developed rictal bristles. Wings short and rounded. Legs short and weak, with heterodactyl feet. Tail long and squared. Inhabit dense broadleaf forests, usually roosting or perching on branches, often in pairs. Feed mostly on insects, amphibians, reptiles, and mollusks; also take plant fruits. All species are residents.

One family, seven genera, and 43 species recognized worldwide, distributed in tropical and subtropical forests. One genus and three species recorded in China.

Harpactes erythrocephalus

红头咬鹃 hóng tóu yǎo juān

Red-headed Trogon

L: 31–35 cm
Habitat: Inhabits relatively intact evergreen broadleaf forests and mixed bamboo thickets. Prefers shady and moist river valleys at altitudes up to 2400 m.
Behavior: Often solitary or in pairs. Usually perches quietly on branches. Forages in forests in short-distance flights.
Distribution: *H. e. helenae* in SE Tibet and W Yunnan; *H. e. yamakanensis* in SE Sichuan and Guizhou, east to Fujian; *H. e. erythrocephalus* in SW Yunnan; *H. e. intermedius* in SE Yunnan and SW Guangxi; *H. e. hainanus* in Hainan.
Voice: Pleasant, mellow, flat, and relaxed *tyaup … tyaup … tyaup*, repeated four to eight times at rate of almost two notes per second. Also a mellow purring, given as a contact call, and a loud chatter, given in alarm.

Harpactes oreskios

橙胸咬鹃 chéng xiōng yǎo juān

Orange-breasted Trogon

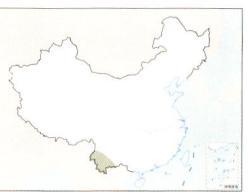

L: 25–31 cm
Habitat: Active in lowland evergreen broadleaf forests, tropical monsoon forests, and bamboo thickets.
Behavior: Often solitary or in pairs. Usually in trees and occasionally forages on ground. Often perches on branches for a long time. Sometimes bold.
Distribution: Rare and local in S Yunnan.
Voice: Fast song is 3–10 repetitions of *tau-tau-tau* notes, sometimes slightly accelerating. Also makes scolding chatter in alarm, like Chestnut-winged Cuckoo.

Harpactes wardi

红腹咬鹃 hóng fù yǎo juān

Ward's Trogon

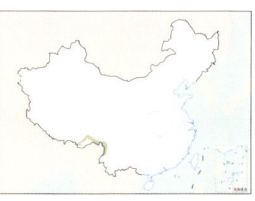

L: 35–38 cm
Habitat: Inhabits pristine subtropical alpine forests and temperate evergreen broadleaf forests, usually with bamboo thickets and vines. Mostly at altitudes of 1500–3200 m; occasionally descends to 300 m.
Behavior: Solitary or in pairs. Often perches still on branches and waits for insects; also forages for fruits on treetops. Flying ability poor; mostly makes short-distance flights.
Distribution: Rare resident in SE Tibet and in the Gaoligong Mountains in W and NW Yunnan.
Voice: Lengthy series of over 20 flat *klew* notes repeated rapidly in 4 sec sequence, often starting quietly, slightly accelerating and lengthening before slowing and shortening toward end. Call a harsh spluttering purr, *whirrurur*.

Red-headed Trogon

Red head, breast, and belly

Brown back

White breast band

White breast band

Brown head and breast without white line on breast

Brown head and breast

Brown-tinged breast

H. e. intermedius

Narrow white breast band

H. e. yamakanensis

Orange-breasted Trogon

Olive-yellowish head and breast

Ochre-red back, rump, and tail

Orangish-yellow breast and belly

Dense narrow white barring on black wings

Grayish head

Grayish back

Brown barring on upperwing coverts

Ward's Trogon

Red forehead and bill

Dark red head, breast, and back

Bright pinkish-red belly

Yellow forehead and bill

Brown head, breast, and back

Yellow belly

佛法僧目   CORACIIFORMES

Brightly colored small to medium-sized perching birds. Vivid plumage often resulting from structural colors is similar or differs slightly between sexes. Head big on thick neck. Bill long and strong. Wings long and broad. Legs short with syndactyl feet. Tail usually squared or rounded, some with unique elongated central tail feathers. Inhabit primarily rivers, lakes, forests, and open plains. Nest in cavities. Feed mostly on fish, shrimp, amphibians, reptiles, insects, and plant seeds and fruits. Most species are nonmigratory.

Six families, 35 genera, and 178 species recognized worldwide, mostly distributed in tropical and subtropical regions, some in temperate zones. Three families, 11 genera, and 23 species recorded in China, found mostly in southern China.

佛法僧科
Coraciidae

Indochinese Roller

蜂虎科
Meropidae

Blue-tailed Bee-eater

翠鸟科
Alcedinidae

Common Kingfisher

Coracias affinis

棕胸佛法僧 zōng xiōng fó fǎ sēng

Indochinese Roller   LC

L: 30–35 cm
Habitat: Inhabits open areas at forest edges; also seen on farmland and deserted fields on plains.
Behavior: Often occurs solitarily or in pairs in open habitats; pounces on prey on ground or in the air.

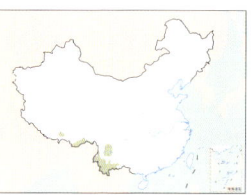

Distribution: Uncommon resident in Yunnan, S Tibet, and S Sichuan.
Voice: Variety of unpleasant, grating, scolding, chatter notes, *char* or *chak*, given at rest, but mostly in flight.

Coracias garrulus

蓝胸佛法僧 lán xiōng fó fǎ sēng

European Roller   LC

L: 29–32 cm
Habitat: Inhabits wooded areas on plains, premontane hills, deserts, wetlands, vegetated areas, and farmland.
Behavior: Solitary or in pairs. Nests in tree cavities, cliffs, or riverbanks. Often

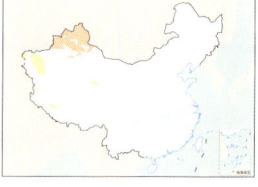

perches high on trees or power lines; searches for potential prey on ground.
Distribution: Breeds in N Xinjiang, more common at border regions in western parts. Rare migrant in S Xinjiang and W Tibet. Vagrant east to Beijing.
Voice: Various harsh crow- or pond-heron-like notes and clicking *krik* and hoarse marsh tern–like *chack-ack*, with the first note stressed.

Eurystomus orientalis

三宝鸟 sān bǎo niǎo

Dollarbird   LC

L: 26–32 cm
Habitat: Inhabits open areas at the edges of broadleaf and mixed broadleaf-coniferous forests.
Behavior: Often solitary or in pairs; prefers perching and waiting for

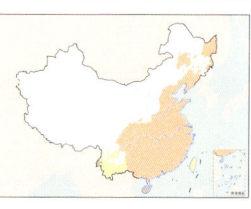

prey on top of tall trees at open forest edges at middle and low altitudes.
Distribution: Common breeder and passage migrant. Widespread in the east but absent in Xinjiang, Qinghai, and Tibet.
Voice: Series of short rasping *rak* or *kak-ak-ak*.

Nyctyornis amictus

赤须蜂虎 (赤须夜蜂虎)

chì xū fēng hǔ (chi xū yè fēng hǔ)

Red-bearded Bee-eater ⓛⒸ

L: 32–34.5 cm
Habitat: Inhabits evergreen broadleaf forests at middle and low altitudes.
Behavior: Often perches quietly on horizontal branches in the middle and lower layers of trees or hunts insects on the wing.
Distribution: One in Yunnan (Ruili), but far from its known range (Malay Peninsula to islands in Southeast Asia), possibly an escaped bird.
Voice: Noisy, scolding Indochinese Roller–like chatter, *jie-guo-guo-guo*.

Nyctyornis athertoni

蓝须蜂虎 (蓝须夜蜂虎)

lán xū fēng hǔ (lán xū yè fēng hǔ)

Blue-bearded Bee-eater ⓛⒸ

L: 29–35 cm
Habitat: Seen in montane forests at middle and low altitudes.
Behavior: Usually solitary in the canopy; prefers forests more than other bee-eaters.
Distribution:
N. a. athertoni in Yunnan and Guangxi; *N. a. brevicaudatus* in Hainan.
Voice: A series of bad-tempered, grumbling, grouchy croaks, cackles, and rattles interspersed with harsher churrs, *ga-ga-ga-ga-kchar-ke-char*.

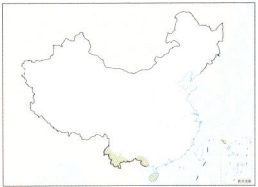

Red-bearded Bee-eater

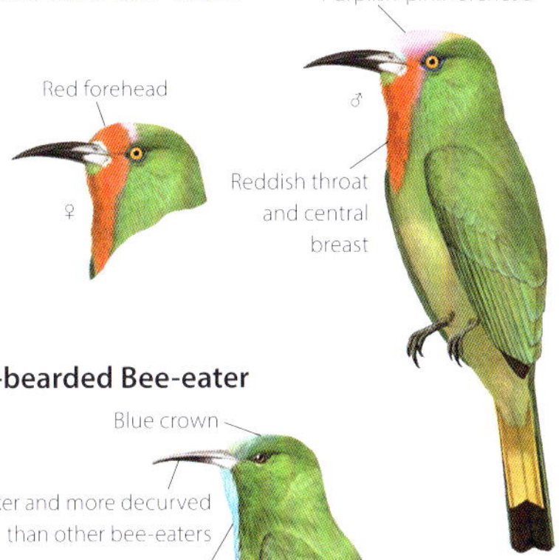

Purplish-pink forehead

Red forehead

♀

♂

Reddish throat and central breast

Indochinese Roller

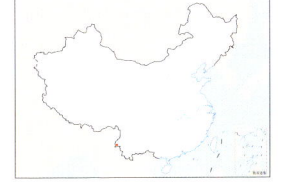

Blue crown

Grayish brown from face to breast

Blue-bearded Bee-eater

Blue crown

Bill thicker and more decurved than other bee-eaters

Blue "beard" feathers on breast

Broad green streaks on belly

Square green tail without elongated central feathers

European Roller

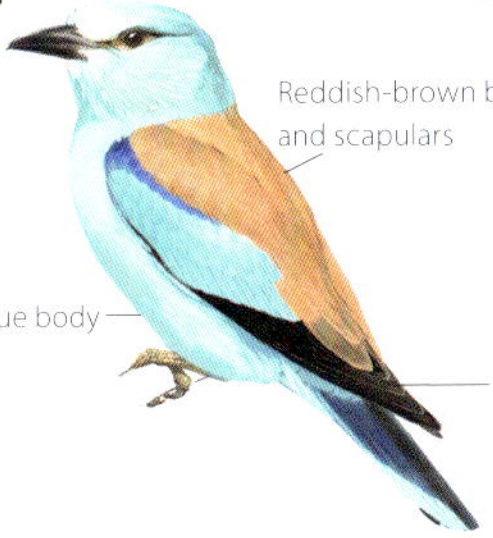

Reddish-brown back and scapulars

Pale blue body

Black wing feathers

Asian Green Bee-eater

Rufous-chestnut head

Bluish-green throat

All-green body

Narrow black band

Dollarbird

Relatively short, thick, broad, red bill

Pale blue patches at base of primaries in flight

Elongated black central tail feathers

Merops orientalis

绿喉蜂虎 lù hóu fēng hǔ

Asian Green Bee-eater ⓛⒸ

L: 18–20 cm
Habitat: Inhabits open gardens, farmland, riverbanks, woodlands, and bamboo thickets.

Behavior: Often gathers in small flocks and moves about in dry open habitats; forages for insects in flight.
Distribution: S and W Yunnan, north to S Sichuan.
Voice: Rolling, burry *tr–tr–tr–ri* and more staccato *ti-it*.

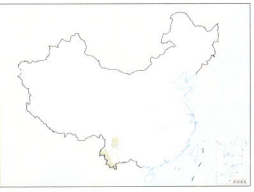

Merops persicus

蓝颊蜂虎 lán jiá fēng hǔ

Blue-cheeked Bee-eater

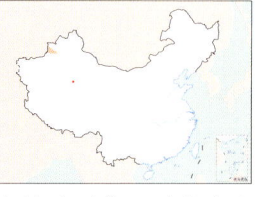

L: 28–32 cm
Habitat: Inhabits edges of lowland deserts with sparse trees and nearby river valleys.
Behavior: Gathers in small flocks. Digs nests on sand dunes or earthen cliffs. Often perches on branches or power lines; flies and catches insects from perch. Also hunts for insects on the wing; especially prefers butterflies.
Distribution: Rare in Xinjiang. A small but stable breeding population found in Xinjiang (Ili) in recent years; vagrant in SE Xinjiang (Altun Mountains).
Voice: Trilling *priip priip*, higher pitched and drier than European Bee-eater.

Merops philippinus

栗喉蜂虎 lì hóu fēng hǔ

Blue-tailed Bee-eater

L: 25–36 cm
Habitat: Occurs on edges of woodlands, open farmland, and riverbanks.
Behavior: Moves about in flocks in open dry areas; hunts for insects on the wing.
Distribution: Found in Yunnan, SW Sichuan, Guangxi, Guangdong, Hong Kong, Fujian, Taiwan, and Hainan.
Voice: A series of dry *kwr–kwr–kwr*.

Merops ornatus

彩虹蜂虎 cǎi hóng fēng hǔ

Rainbow Bee-eater

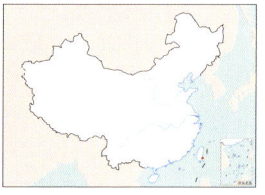

L: 19–21 cm
Habitat: Dry open farmland, woodlands, scrubland, and riverbanks.
Behavior: Often gathers in flocks. Fond of bees. Inhabits dense bushes or trees. Utters gentle calls in flight.
Distribution: Vagrant to Taiwan.
Voice: Penetrating dry trills and purrs, *gr–gr–gr–gr*.

Merops viridis

蓝喉蜂虎 lán hóu fēng hǔ

Blue-throated Bee-eater

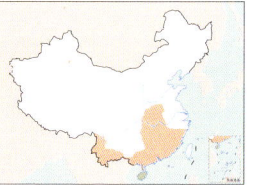

L: 21–32 cm
Habitat: Inhabits woodlands, riverbanks, farmland, soil slopes, and gardens.
Behavior: Often gathers in small flocks in open habitats; forages primarily for bees.
Distribution: Locally common breeder in S Henan, SE Yunnan, Hubei, Hunan, Jiangxi, Zhejiang, Fujian, Guangdong, Hong Kong, Guangxi, and Hainan.
Voice: Twangy *trrurrip … trrurip … trrurip* and harsher *kip* notes.

Merops leschenaulti

栗头蜂虎 lì tóu fēng hǔ

Chestnut-headed Bee-eater

L: 20–23 cm
Habitat: Occurs in open wooded areas at middle and low altitudes.
Behavior: Prefers gathering in small flocks at open forest edges; forages on the wing.
Distribution: Locally common breeder in S and W Yunnan.
Voice: Similar to Asian Green Bee-eater but slightly coarser and lower pitched, with many notes slightly longer.

Merops apiaster

黄喉蜂虎 huáng hóu fēng hǔ

European Bee-eater

L: 25–29 cm
Habitat: Inhabits primarily earthen cliffs, steep slopes, and river valleys at foothills and open plains with trees; especially prefers habitats around lakes, reservoirs, and other wetlands.
Behavior: Gregarious. Excavates burrows in earthen cliffs. Often perches on branches or power lines; hunts from perch when insect approaches. Also hawks various insects; especially fond of bees.
Distribution: Locally common breeder in NW Xinjiang; rare in SW Xinjiang.
Voice: Soft, melodious, liquid, nasal purring, *prrit*, often repeated.

Blue-cheeked Bee-eater

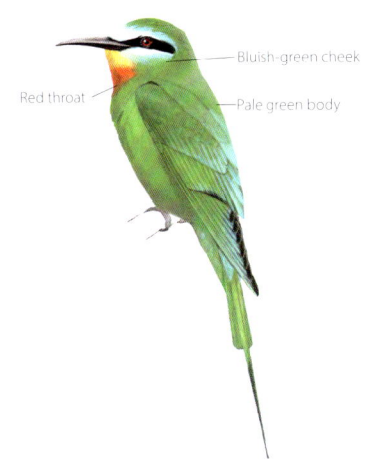

Red throat

Bluish-green cheek

Pale green body

Blue-throated Bee-eater

Chocolate-brown head and upper back

Blue throat

Blue rump and tail

Extremely elongated central tail feathers

Blue-tailed Bee-eater

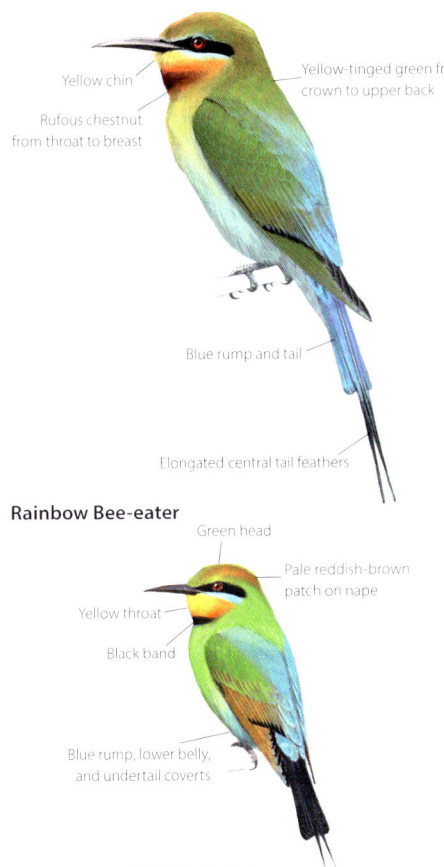

Yellow chin

Yellow-tinged green from crown to upper back

Rufous chestnut from throat to breast

Blue rump and tail

Elongated central tail feathers

Chestnut-headed Bee-eater

Yellow throat and chin

Chestnut-brown hood extends to upper back

Narrow black band

Blue rump

Green tail without elongated central feathers, slightly notched tail tip

Rainbow Bee-eater

Green head

Pale reddish-brown patch on nape

Yellow throat

Black band

Blue rump, lower belly, and undertail coverts

Black tail with elongated central feathers

European Bee-eater

Yellow cheek, chin, and throat

Chestnut head and back

Black band

Pale blue underparts

255

Halcyon coromanda
赤翡翠 chì fěi cuì
Ruddy Kingfisher

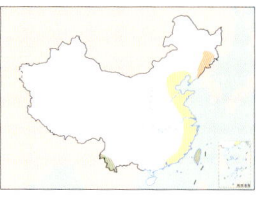

L: 25–27 cm
Habitat: Breeds in dense evergreen broadleaf forests or groves and secondary forests near mountain streams below 1800 m. *H. c. major* often occurs on oceanic islands and coastal forests during migration.
Behavior: Often solitary or in pairs during breeding season. Usually secretive during daytime, or flies quickly while calling and perches on bare branches. Often hunts from perch, descending to the ground or diving into water.
Distribution: *H. c. coromanda* is resident in S and SW Yunnan, more common in summer; also possible in SE Tibet. *H. c. major* breeds in Northeast China and migrates through coastal regions from East China to South China (including Taiwan); rare. *H. c. bangsi* is an uncommon resident in Taiwan (including Orchid Island).
Voice: Song is a loud, descending, slow, trilling *quirrr-r-r-r-r*, repeated in sequences of four to five notes. Lower pitched, longer, and less staccato than similar whinnying trill of White-throated Kingfisher.

Halcyon smyrnensis
白胸翡翠 bái xiōng fěi cuì
White-throated Kingfisher

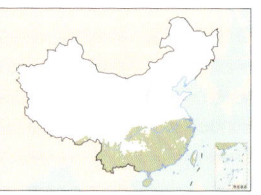

L: 26.5–29.5 cm
Habitat: Occurs on ponds, reservoirs, swamps, paddy fields, fishponds, rivers on plains, lakeshores, coasts, mangroves, or water bodies near villages at altitudes up to 1200 m; sometimes also seen in areas far from water.
Behavior: Usually solitary. Prefers perching on power lines, branches, or rocks near water. Watches the water for a long time and immediately plunges to the ground or dives into the water if prey is located. Flight rapid, in straight line.
Distribution: Common resident. *H. s. smyrnensis* in SE Tibet. *H. s. perpulchra* in SE Tibet, NW to S Yunnan, and islands in the South China Sea, possibly in SW Guangxi. *H. s. fokiensis* widespread in southern China, including Hainan; occasional in Taiwan.
Voice: Noisy. Calls include a loud piercing series of descending *pe-pe-pe-pe-pe-pe-pe* notes, a repeated simple *chake ake ake-ake ake-ake* (very similar to Black-capped Kingfisher), and a descending trill, like Ruddy Kingfisher.

Pelargopsis capensis
鹳嘴翡翠 guàn zuǐ fěi cuì
Stork-billed Kingfisher

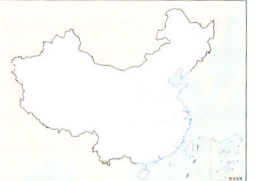

L: 35–41 cm
Habitat: Occurs near lowland waters, such as streams, canals, rivers, reservoirs, lakes, and forest edges on coasts, at altitudes up to 870 m.
Behavior: Usually solitary or in pairs. Often stands on branches or rocks near water for a long time and waits for prey.
Distribution: Rare in S Yunnan (Xishuangbanna) and W Yunnan (Yingjiang).
Voice: Song a loud, descending, cackling laugh, *kya-kya-kya-ka*; also a more musical *di-di-du … dididu*, repeated.

Halcyon pileata
蓝翡翠 lán fěi cuì
Black-capped Kingfisher

L: 26–31 cm
Habitat: Inhabits wetlands, such as rivers and reservoirs, on plains and low mountains.
Behavior: Often stands on power lines or branches near water and stares at the water's surface, waiting for small aquatic prey, such as fish or shrimp, to approach. Nests in cavities on earthen shores.
Distribution: Common in a wide range of areas, from Northeast China to southwestern and South China, including Taiwan and Hainan.
Voice: Repeated monosyllabic *jiu-jiu-jiu* or piercing *ga-jiu, ga-jiu*, both similar to White-throated Kingfisher.

Todiramphus chloris
白领翡翠 bái lǐng fěi cuì
Collared Kingfisher

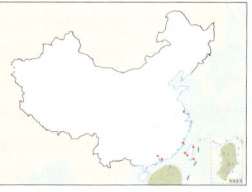

L: 22.5–25 cm
Habitat: Often occurs in coastal mangroves and vegetation, also open areas along rivers or lakes.
Behavior: Prefers perching on trees or rocks for a long time, plunging to the ground or into water once prey is located.
Distribution: *T. c. armstrongi* in Jiangsu, Fujian, Guangdong, and Hong Kong; *T. c. collaris* in Taiwan. Vagrant or rare passage migrant; likely resident on islands in the South China Sea.
Voice: Song is a short series of three to six loud, penetrating, nasal or husky yips, *kyip-kyip-kyip-kyip*. Calls similar to song but often in lengthy wavering series and then more of a hesitant chatter.

Ruddy Kingfisher

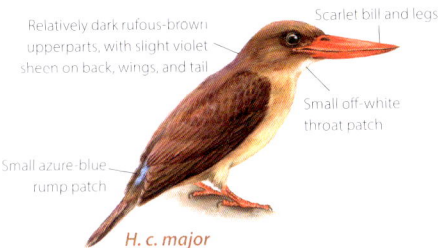

Scarlet bill and legs

Relatively dark rufous-brown upperparts, with slight violet sheen on back, wings, and tail

Small off-white throat patch

Small azure-blue rump patch

H. c. major

Purple tinge on head, upperparts, breast, and tail more prominent on wings and tail

Small azure-blue rump patch

Plumage darker than *H. c. major*

H. c. bangsi

Darkest overall plumage, relatively large blue rump patch

H. c. coromanda

Stork-billed Kingfisher

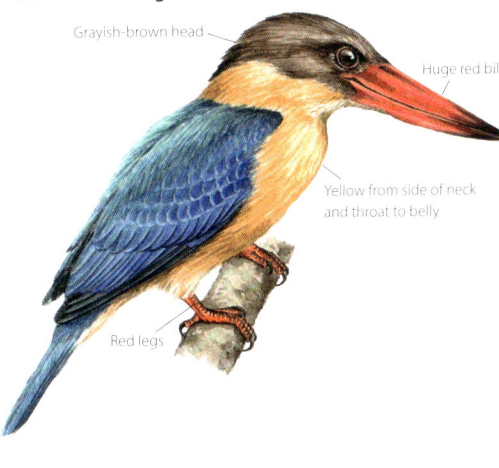

Grayish-brown head

Huge red bill

Yellow from side of neck and throat to belly

Red legs

Black-capped Kingfisher

Black head

White neck and breast

Chestnut-yellow flanks

White-throated Kingfisher

Long, thick, bright red bill sometimes appears dark red

White from throat to breast

Black median coverts

Glossy bright blue wings, back, and tail

Bright red legs

Collared Kingfisher

Narrow white supercilium does not extend behind eye

Glossy blue head, back, wings, and tail sometimes appear bluish green

Pale base of lower mandible

White throat, neck, breast, lower belly, and undertail coverts

Alcedo meninting
蓝耳翠鸟 lán ěr cuì niǎo
Blue-eared Kingfisher

L: 15.5–17 cm
Habitat: Occurs near streams, swamps, and river mouths in evergreen broadleaf forests. More often in forests than Common Kingfisher.
Behavior: Usually solitary or in pairs. Often perches on branches or rocks near water for a long time, waiting for fish, insects, small amphibians, and reptiles to approach.
Distribution: Restricted to S Yunnan (Xishuangbanna), possibly extirpated.
Voice: Penetrating, high-pitched *tjiet*, similar to, but typically longer than, calls of Oriental Dwarf-Kingfisher and with tremulous quality.

Alcedo atthis
普通翠鸟 pǔ tōng cuì niǎo
Common Kingfisher

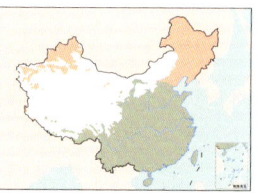

L: 15–17 cm
Habitat: Inhabits wetlands, such as open lakes, rivers, and ponds.
Behavior: Feeds primarily on fish and shrimp; often plunges rapidly into water for food from branches or rocks near water. Excavates nests on streamside or lakeside earthen banks.
Distribution: *A. a. atthis* in Xinjiang; *A. a. bengalensis* widespread across China except for Xinjiang. Very common.
Voice: High-pitched metallic *zir*, slightly lower pitched than both Blue-eared Kingfisher and Oriental Dwarf-Kingfisher.

Alcedo hercules
斑头大翠鸟 bān tóu dà cuì niǎo
Blyth's Kingfisher

L: 22–23 cm
Habitat: Highly selective. Occurs only near streams in evergreen broadleaf forests on low mountains and hills from 200 to 1200 m, or foothill rivers surrounded by forests.
Behavior: Usually solitary or in pairs; shy. Often perches on rocks in streams, streamside bamboo, or low branches of bushes, quietly watching the water's surface; plunges into water for fish once spotted. Also forages for crustaceans and various aquatic insects and insect larvae.
Distribution: Rare resident; fragmented distribution in SE Tibet, S and SW Yunnan, Guangxi, Guangdong, Hainan, Fujian, and Jiangxi.
Voice: Piercing, loud, flat *pseet* in flight, louder and fractionally higher pitched than Common Kingfisher, very similar to Blue-eared Kingfisher and Oriental Dwarf-Kingfisher.

Ceyx erithaca
三趾翠鸟 sān zhǐ cuì niǎo
Oriental Dwarf-Kingfisher

L: 12.5–14 cm
Habitat: Inhabits gully streams in tropical monsoon forests. Sometimes also occurs in ponds and estuaries at forest edges. From lowlands to 1300 m.
Behavior: Often solitary or in pairs. Prefers understory; perches on branches or rocks in streams for a long time waiting for prey. Often bobs head or flicks tail.
Distribution: *C. e. erithaca* is rare in S and W Yunnan and Hainan; vagrant to Matsu Islands. *C. e. rufidorsa* is vagrant in Taiwan; sometimes treated as a color morph of the current species; others consider it a separate species, Rufous-backed Drawf-Kingfisher (*C. rufidorsa*).
Voice: Gives piercing flight calls, *ji*, like Blue-eared Kingfisher but more variable (some rising); other calls often shorter, less tremulous, higher pitched than Common Kingfisher.

Megaceryle lugubris
冠鱼狗 guān yú gǒu
Crested Kingfisher

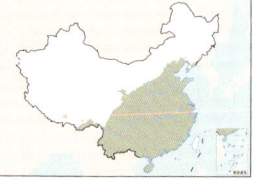

L: 37–42 cm
Habitat: Inhabits rivers and streams on mountains and plains. Especially prefers clear, fast-flowing, and boulder-strewn rivers.
Behavior: Forages primarily for fish and shrimp; often perches on small trees or large rocks near water and plunges into water rapidly to hunt for prey. Excavates nests on banks along rivers, lakes, or fields.
Distribution: *M. l. guttulata* is widespread in eastern China, common; *M. l. lugubris* is historically recorded in Liaoning (Liaoyang).
Voice: Piercing *ket-ket-ket* calls in flight with a nervous quality and reminiscent of River Lapwing, sometimes slow, sometimes fast, singly and in series. Also a more nasal *yit* or *kyit*.

Ceryle rudis
斑鱼狗 bān yú gǒu
Pied Kingfisher

L: 27–31 cm
Habitat: Inhabits streams and lakes on plains and low mountains.
Behavior: Often flies slowly back and forth close to the water's surface looking for food; rapidly dives into water with wings folded once prey is detected. Nests in earthen cavities on banks.
Distribution: *C. r. insignis* is mainly south of the Yangtze River; occasionally farther north; vagrant to North China. *C. r. leucomelanurus* in Yunnan and Guangxi.
Voice: Piercing, rapid, repeated *ji-ji-ji-ji*, sometimes with trills.

Blue-eared Kingfisher

Blue ear coverts

Black bill

White patches on side of neck

Dark blue head, back, and wings

♂

♀

Red lower mandible

Common Kingfisher

Large white patches on side of neck

Bluish-green head and back

White throat

Small turquoise-blue spots on wing coverts

Oriental Dwarf-Kingfisher

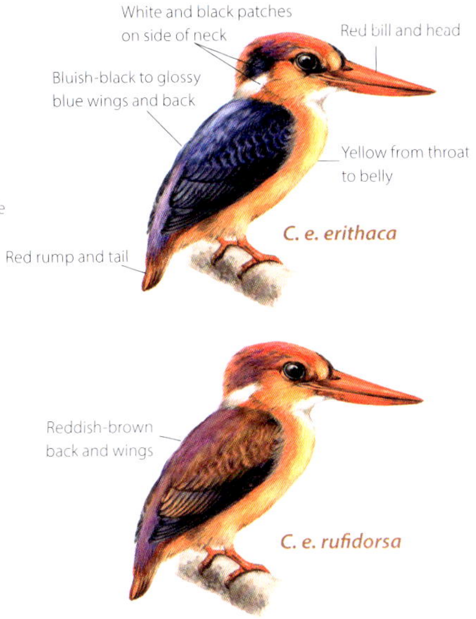

White and black patches on side of neck

Red bill and head

Bluish-black to glossy blue wings and back

Yellow from throat to belly

C. e. erithaca

Red rump and tail

Reddish-brown back and wings

C. e. rufidorsa

Blyth's Kingfisher

Ear coverts lack orange brown

Dark blue head and wings

♂

Black bill

Red lower mandible

♀

Relatively large body

Red legs

Pied Kingfisher

♀

No breast band

Crested Kingfisher

Well-developed crest

Large white patch on cheek

Black-and-white upperparts

Light-colored bill tip

Broad black eye stripe

♂

Two black breast bands

犀鸟目　BUCEROTIFORMES

Medium-sized to large perching birds. Plumage similar between sexes in most species, mostly black, white, and brown. Bill long and curved, usually with a casque on upper bill. Some species developed long erectile crest on head. Legs strong. Tail long. Inhabit dense forests or open plains. Nest in cavities. Hornbills are known to exhibit unique breeding behavior, where females are sealed in cavities to incubate eggs and rear chicks, while males collect and deliver food for females and chicks. Feed mostly on plant fruits, especially figs, but also plant shoots, amphibians, reptiles, and small birds and mammals; some also take insects. Most species are residents, while some are migratory.

　　Four families, 19 genera, and 74 species recognized worldwide, distributed in the Old World and N Oceania. Two families, six genera, and six species recorded in China: Eurasian Hoopoe is widely distributed across the country, while hornbills are found in tropical and subtropical forests of southwestern China.

犀鸟科
Bucerotidae

Buceros bicornis
双角犀鸟 shuāng jiǎo xī niǎo
Great Hornbill　**VU**

L: 95–105 cm
Habitat: Inhabits large, intact tropical monsoon forests, tropical rain forests, and forests with tall trees. Occurs up to 2000 m.
Behavior: In breeding season, only male is out and about, while female stays in nest cavity; female and chicks fed in nest by male. In nonbreeding season, often in pairs or with juveniles. Sometimes also gathers in places with abundant food.
Distribution: Resident in S and W Yunnan and SE Tibet.
Voice: Very loud, resonant, gruff, barked *kyok*, singly or in sequence. Also an intimidating growling *krrrohh*.

Aceros nipalensis
棕颈犀鸟 zōng jǐng xī niǎo
Rufous-necked Hornbill　**VU**

L: 90–100 cm
Habitat: Inhabits intact dense evergreen broadleaf forests, at altitudes up to 2900 m.
Behavior: Similar to Great Hornbill.
Distribution: Local resident in SE Tibet and vagrant to westernmost Yunnan.
Voice: Barking *kup* or *kuk*, singly, paired, or in series, higher pitched and less fierce than Great Hornbill.

Rhyticeros undulatus
花冠皱盔犀鸟 huā guān zhòu kuī xī niǎo
Wreathed Hornbill　**LC**

L: 75–85 cm
Habitat: Inhabits large areas of intact evergreen broadleaf forests. Occurs from lowlands to 2600 m.

Anthracoceros albirostris
冠斑犀鸟 guān bān xī niǎo
Oriental Pied-Hornbill　**LC**

L: 55–60 cm
Habitat: Inhabits well-vegetated forests, such as tropical monsoon forests and tropical rain forests; also occurs in artificial and secondary forests. Adaptable.
Behavior: Similar to Great Hornbill.
Distribution: Found in S and W Yunnan, S Guangxi, and SE Tibet.
Voice: Thin-sounding, cackling *kek-ek-ek-kek-ek-ek* and occasional single *kek* notes.

Anorrhinus austeni
白喉犀鸟 bái hóu xī niǎo
Brown Hornbill　**NT**

L: 60–65 cm
Habitat: Inhabits well-vegetated forests, such as tropical monsoon forests and tropical rain forests; also occurs in rubber plantations at forest edges.
Behavior: Similar to Great Hornbill.
Distribution: Extremely rare in S Yunnan.
Voice: High-pitched, penetrating yapping and screaming sounds.

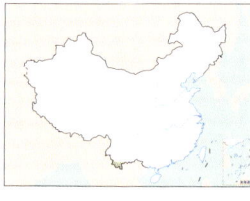

Behavior: Similar to Great Hornbill.
Distribution: W Yunnan and possibly SE Tibet.
Voice: Dog-yelping, slightly rising *u-gyar*, longer than Rufous-necked Hornbill and with an "intake of breath" at the start. Also in series, *ga ga o*, or comical and harsh *ge ge*.

戴胜科
Upupidae

Upupa epops
戴胜 dài shèng
Eurasian Hoopoe (LC)

L: 25–31 cm
Habitat: Forages on open low grasslands, farmland, and deserted fields; nests in tree cavities or cliff crevices.
Behavior: Solitary or gathers in small flocks. Forages on ground. Hunts for worms by probing with bill in soft soil; prey is thrown into the air and swallowed. When disturbed, flies for a short distance and stops or flies into nearby trees. Crest opens when excited or startled. Flight undulating at low speed.
Distribution: Widespread but absent in S Xinjiang; very common in northern and western China; conspicuous. Occasionally seen in South and East China.
Voice: Trisyllabic oop-oop-oop, reminiscent of Oriental Cuckoo.

Great Hornbill

Yellow casque bulges on both sides

Red front and back of casque

White iris, red bare skin around eyes

Dark red iris

Black face, throat, and base of bill contrasts with yellow bill and neck

Black body with white wing patch and white fringe on wing feathers

Black subterminal band on white tail

Eurasian Hoopoe

Rufous-necked Hornbill

Many black stripes on upper mandible

Blue bare skin around eyes

Brown from head to neck and belly

Inflatable red bare skin on throat

Black back and wings

Mostly black body

Black tail with white distal half

Slightly smaller than male

Oriental Pied-Hornbill

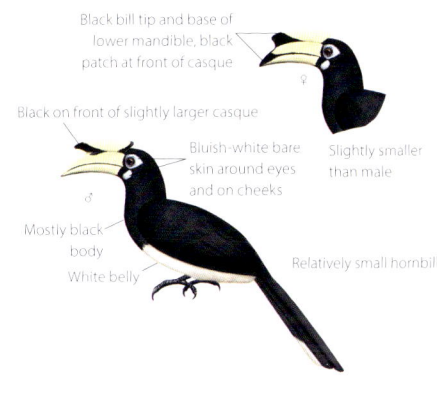

Black bill tip and base of lower mandible, black patch at front of casque

Black on front of slightly larger casque

Bluish-white bare skin around eyes and on cheeks

Slightly smaller than male

Mostly black body

White belly

Relatively small hornbill

Wreathed Hornbill

Several ridges on both mandibles and casque

Bright red iris, red bare skin around eyes

White face and neck

Inflatable yellow bare throat skin with black central bar

Slightly fluffy dark brown feathers on hindneck

White tail

Inflatable blue bare throat skin with black central bar

Mostly black body

Slightly smaller body

Brown Hornbill

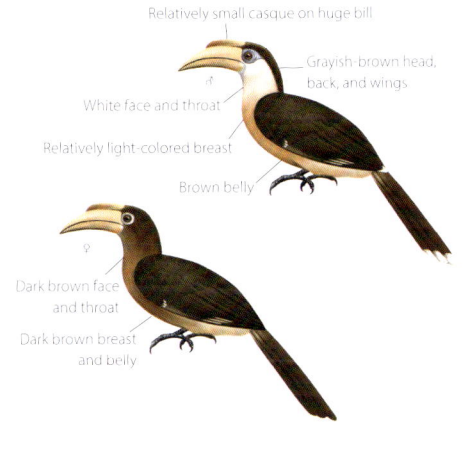

Relatively small casque on huge bill

Grayish-brown head, back, and wings

White face and throat

Relatively light-colored breast

Brown belly

Dark brown face and throat

Dark brown breast and belly

261

啄木鸟目 (鴷形目)   PICIFORMES

The most species-rich group of perching birds, consisting of small to medium-sized species. Plumage similar or slightly different between sexes in most species. Bill strong and pointed. Wings mostly short and rounded. Legs short but strong, with zygodactyl feet, well adapted to climbing or clinging onto vertical surfaces. Tail relatively long, usually pointed or squared; some groups have strong and stiff tail feathers to keep body in vertical position. Inhabit forests from tropics to temperate zones. Most cavity-nesting species make their own tree cavities; honeyguides are brood parasites. Feed mostly on insects and plant seeds and fruits, while honeyguides feed mostly on honey and beeswax. A few species are migratory.

Nine families, 71 genera, and 445 species recognized worldwide, with a near-cosmopolitan distribution except for Oceania. Three families, 19 genera, and 43 species recorded in China: Woodpeckers are widely distributed across the country, while barbets and Yellow-rumped Honeyguide are found in forests of southern China.

拟啄木鸟科
Megalaimidae

响蜜鴷科
Indicatoridae

啄木鸟科
Picidae

Great Barbet

Yellow-rumped Honeyguide

Great Spotted Woodpecker

Psilopogon virens
大拟啄木鸟 dà nǐ zhuó mù niǎo
Great Barbet   LC

L: 30–35 cm
Habitat: Inhabits broadleaf and mixed broadleaf-coniferous forests at middle and low altitudes.
Behavior: Moves about solitarily or in small flocks in tall trees. Resonant but simple calls. Easily found in montane forests.
Distribution: Seen in various regions in southern China, except for Hainan and Taiwan.
Voice: Loud descending *keearr … keearrr*, with character of a distant large gull, repeated once every second. Querulous, grating, and unpleasant *kyarr-h* call.

Psilopogon lineatus
绿拟啄木鸟 lǜ nǐ zhuó mù niǎo
Lineated Barbet   LC

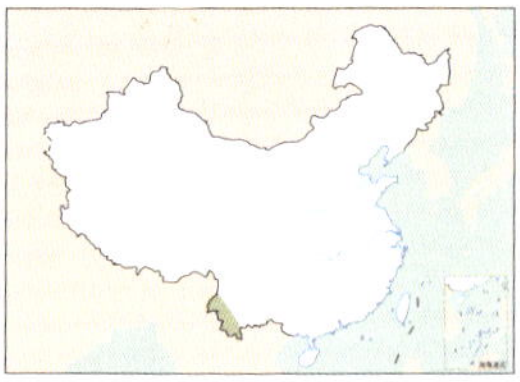

L: 26–30 cm
Habitat: Seen on small mountains at low altitudes; also in forests, gardens, and wooded areas on farmland on plains.
Behavior: Often solitary or in pairs. Forages in upper and middle stories; seldom comes to ground.
Distribution: S and W Yunnan.
Voice: Loud, percussive, disyllabic *pu-tuk* at rate of more than one call per second, repeated metronomically. Also short, resonant, bubbling purr that starts slowly, *pr-pr-prrrrrrrt … curutuk … curutuk … curutuk*, with gradual, stuttering start, then a long strophe followed by about 15 fading *curutuk* notes.

Psilopogon faiostrictus

黄纹拟啄木鸟 huáng wén nǐ zhuó mù niǎo

Green-eared Barbet

L: 24–27 cm
Habitat: Inhabits evergreen broadleaf forests, woodlands, and secondary forests on plains and low mountains.
Behavior: Often solitary or in pairs in open forests.
Distribution: Extreme southern South China, including the Leizhou Peninsula.
Voice: Series of rising *poid* notes, with quality of Asian Barred Owlet, delivered slowly; song a lively, rapid, ringing, four-syllable *tu-ku-cu-tuk*, repeated at rate of more than one call per second.

Psilopogon franklinii

金喉拟啄木鸟 jīn hóu nǐ zhuó mù niǎo

Golden-throated Barbet

L: 20.5–23.5 cm
Habitat: Inhabits evergreen broadleaf forests at altitudes of 900–2600 m, slightly higher than other barbets.
Behavior: Usually several individuals gather at places with abundant food. Often calls continuously from perch.
Distribution: *P. f. franklinii* is resident in SE Tibet, S, W, and SE Yunnan, and S Guangxi; *P. f. ramsayi* possibly occurs in SW Yunnan.
Voice: Resonant disyllabic *qu-kuk … qu-kuk*, repeated about once per second.

Great Barbet

Bright yellow bill, brown tip on upper mandible

Dark blue head

Bright red undertail coverts

Lineated Barbet

Yellow bare skin around eyes

No green, but pale brown streaks from head and neck to lower breast

Yellow legs

Green-eared Barbet

Green cheek and ear coverts

Red patches on side of neck

Green breast

Golden-throated Barbet

More black and white streaks from ear coverts to eyebrow

P. f. ramsayi

Gold crown, red forehead and nape

Broad black eyebrow

Grayish white from ear coverts to lower throat

Gold chin

Green body with glossy blue on edge of wings

P. f. franklinii

263

Psilopogon faber
黑眉拟啄木鸟 hēi méi nǐ zhuó mù niǎo
Chinese Barbet

L: 20–22 cm

Habitat: Inhabits montane evergreen broadleaf forests and mixed broadleaf-coniferous forests at middle and low altitudes; also seen in woodlands and at forest edges.

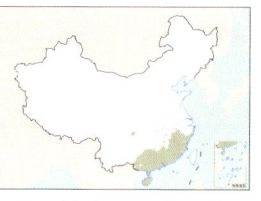

Behavior: Often solitary or gathers in small flocks; moves about in the canopy. Often makes simple calls, easily detected.

Distribution: Locally common in mountains of southern China. *P. f. sini* in Yunnan, Chongqing, Guizhou, Hunan, Zhejiang, Fujian, Guangxi, and Guangdong; *P. f. faber* in Hainan.

Voice: Stuttering, rapid series, *tu–tu–tu–tututututututu*. *P. f. sini* has more notes that are lower pitched and delivered more slowly than *P. f. faber*, which sounds hurried and frenetic.

Psilopogon nuchalis
台湾拟啄木鸟 tái wān nǐ zhuó mù niǎo
Taiwan Barbet

L: 20–22 cm

Habitat: Inhabits subtropical broadleaf forests at middle and low altitudes.

Behavior: Often solitary in upper and middle stories in forests. Not agile. Very vocal, thus easily spotted.

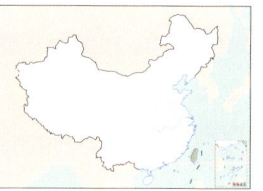

Distribution: Chinese endemic, restricted to Taiwan.

Voice: Similar to Chinese Barbet but usually with fewer (six) notes in each phrase, and with neighboring phrases delivered alternately fast and slow, with each phrase gradually becoming louder, then quieter. Rapid *tu–tu–tu–tu–tu–tu–tu*, intermediate in pace between fast *P. f. faber* and slower *P. f. sini*.

Psilopogon asiaticus
蓝喉拟啄木鸟 lán hóu nǐ zhuó mù niǎo
Blue-throated Barbet

L: 22–23 cm

Habitat: Inhabits various habitats, such as evergreen broadleaf forests, plantations, parks, and orchards. Usually at altitudes up to 2000 m.

Behavior: Usually gathers at places with abundant food. Often calls continuously from perch.

Distribution: *P. a. asiaticus* in W Yunnan and SE Tibet; *P. a. davisoni* in S and SE Yunnan and Guangxi.

Voice: Rapid, resonant, three- or four-note *tu-ku-tuk* or *u-tu-ku-tuk*, reminiscent of Green-eared Barbet but shorter, faster, and with individual notes difficult to discern.

Psilopogon duvaucelii
蓝耳拟啄木鸟 lán ěr nǐ zhuó mù niǎo
Blue-eared Barbet

L: 16–17 cm

Habitat: Inhabits evergreen broadleaf forests, secondary forests, and tea plantations. Usually below 1000 m.

Behavior: Usually gathers at places with abundant food. Often calls continuously from perch.

Distribution: S and W Yunnan.

Voice: Double *tu-tuk*, repeated rapidly (1.5–2 times per second), sometimes for minutes on end; also series of up to 25 whistled *plow* notes.

Psilopogon haemacephalus
赤胸拟啄木鸟 chì xiōng nǐ zhuó mù niǎo
Coppersmith Barbet

L: 15–17 cm

Habitat: Inhabits various habitats, such as evergreen broadleaf forests, plantations, gardens, orchards, and cultivated areas. Usually at altitudes below 1500 m.

Behavior: Usually gathers

at places with abundant food. Often calls continuously from perch.

Distribution: S and W Yunnan and SE Tibet.

Voice: Ringing *tonk … tonk … tonk*, repeated metronomically at high speed, 1.5–2 times per second.

Indicator xanthonotus
黄腰响蜜䴕 huáng yāo xiǎng mì liè
Yellow-rumped Honeyguide

L: 15–16 cm

Habitat: Inhabits wooded rocky gorges and streams, with large colonies of honeybees on cliffs. Primarily at altitudes of 1500–2600 m.

Behavior: Usually solitary or in pairs. Feeds primarily

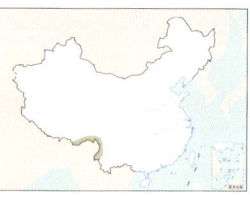

on beeswax and honey from rock bee nests on rock faces.

Distribution: Rare in SE Tibet and Gaoligong Mountains in W Yunnan.

Voice: Very high-pitched, intense, penetrating, descending *sip*, with quality of alarm call of Common House-Martin, and milder, repeated *seet*.

Chinese Barbet

Black eyebrow and crown
Red on upper nape

Red forehead
Black eyebrow, yellow crown
Yellow throat

P. f. faber

P. f. sini

Taiwan Barbet

Yellow forehead
Blue crown
Black eyebrow
Red lores
Red patch on hindneck
Yellow throat

Blue-throated Barbet

Blue on central crown

Black on central crown
Red forehead and crown
Black lateral crown stripe
P. a. davisoni

Blue from face to throat

Green body

P. a. asiaticus

Blue-eared Barbet

Blue crown
Black forehead
Blue ear coverts with two red patches above and below

Blue area on throat smaller than on Blue-throated Barbet

Coppersmith Barbet

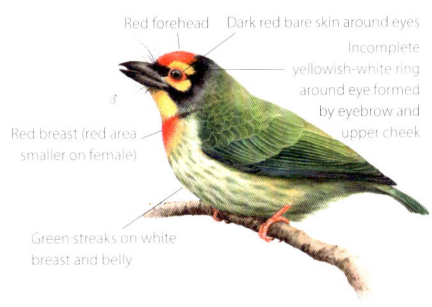

Red forehead
Dark red bare skin around eyes
Incomplete yellowish-white ring around eye formed by eyebrow and upper cheek

Red breast (red area smaller on female)

Green streaks on white breast and belly

Yellow-rumped Honeyguide

Yellow forehead and lower cheek

Dark streaks on grayish-white belly
White inner webs of tertials

Yellow on forehead and lower cheek paler than male
Yellow rump

Jynx torquilla
蚁䴕 yǐ liè
Eurasian Wryneck

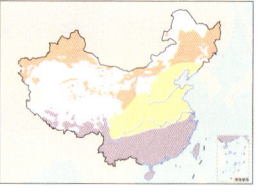

L: 16–19 cm
Habitat: Often inhabits open woodlands on low mountains, hills, and plains; especially prefers broadleaf and mixed broadleaf-coniferous forests. Sometimes also active in forest-edge bushes, river valleys, orchards, and urban parks.
Behavior: Often solitary; in pairs only in breeding season. Usually nests in tree cavities; also reuses old nests of other woodpeckers. Sometimes breeds in natural cavities on rotten stumps. Flight undulating and rapid. Mainly hops and forages on the ground for ants and their eggs.
Distribution: Widespread. Summer breeder in northern China, winter visitor in southern China, and migrates through much of China.
J. t. torquilla is widespread; the populations in E Asia (including those in China) are sometimes treated as subspecies *J. t. chinensis*. *J. t. himalayana* in S Tibet.
Voice: Loud, piercing, falconlike song, *ji-ji-ji-ji-ji*, similar to that of Lesser Spotted Woodpecker but lesser stereotyped and with whining tone. Otherwise silent.

Picumnus innominatus
斑姬啄木鸟 bān jī zhuó mù niǎo
Speckled Piculet

L: 9–10 cm
Habitat: Often inhabits evergreen broadleaf forests on foothill plains, low mountains, and hills; sometimes also in subalpine mixed forests. Especially prefers woodlands, bamboo thickets, and forest-edge bushes in open areas.
Behavior: Usually solitary. Nests in tree cavities. Often joins other small passerines in mixed-species flocks in short trees or bushes.
Distribution: Widespread in southern China; uncommon resident.
P. i. innominatus in E Tibet; *P. i. malayorum* in S and W Yunnan; *P. i. chinensis* in Shandong, Jiangxi, Jiangsu, Shanghai, Zhejiang, Fujian, Guangdong, Hong Kong, S Henan, S Shanxi, S Shaanxi, S Gansu, S Sichuan, Chongqing, and Guizhou.
Voice: Pulsed, intense, metallic *zi zi zi zi*. Calls similar but weaker and erratic. Often drums very loudly.

Sasia ochracea
白眉棕啄木鸟 bái méi zōng zhuó mù niǎo
White-browed Piculet

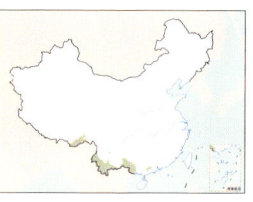

L: 8–9 cm
Habitat: Often inhabits evergreen broadleaf forests, bamboo thickets, woodlands, and scrubland on low mountains, hills, and foothill plains; sometimes occurs in reedbeds on riverbanks and sugarcane fields.
Behavior: Occurs solitarily. Usually nests in tree cavities. Prefers short trees or tall bushes. Sometimes joins mixed-species flocks. Often climbs along branches to forage. Seldom comes to the ground.
Distribution: Local in southwestern China; uncommon resident.
S. o. ochracea in SE Tibet; *S. o. reichenowi* in Yunnan; *S. o. kinneari* in SE Yunnan, Guizhou, Guangxi, and Guangdong.
Voice: Loud, slightly nasal, intense *tsiii*. Tremulous, chittering, descending trill, *chi-rrrrrrrr*.

Picoides tridactylus
三趾啄木鸟 sān zhǐ zhuó mù niǎo
Eurasian Three-toed Woodpecker

L: 21–24 cm
Habitat: Inhabits primarily coniferous or mixed broadleaf-coniferous forests on mountains or plains; especially prefers untouched primary coniferous forests. Typical forest-dwelling birds.
Behavior: Usually in upper and middle stories of forests; prefers foraging on dead branches.
Distribution: Uncommon. *P. t. tridactylus* in NW Heilongjiang, Jilin, NE Inner Mongolia, and Altai Mountains in N Xinjiang. *P. t. tianschanicus* in Tian Shan Mountains in C and SW Xinjiang. *P. t. funebris* in Gansu, E Tibet, S and E Qinghai, NW Yunnan, and W Sichuan; sometimes treated as a separate species, *P. funebris* (Dark-bodied Three-toed Woodpecker, 暗腹三趾啄木鸟).
Voice: *kik* is subtly lower pitched and softer than call of Great Spotted Woodpecker. Drumming often slightly slower and longer (1.2 sec.) than Great Spotted (often 0.6 sec.), with slight acceleration at very end.

Eurasian Wryneck

Clear brown eye stripe

Distinctly vermiculated throat

Sparse black spots on relatively pale belly

J. t. torquilla

White-browed Piculet

Relatively short white supercilium

Rufous forehead

Brown underparts

Speckled Piculet

White supercilium

Olive forehead

Brown spots on breast

Eurasian Three-toed Woodpecker

Yellow forehead

Black crown

Grayish-black underparts, white bars on flanks

P. t. funebris

Yellow forehead

Black crown

White underparts, black bars on flanks and under tail

P. t. tridactylus

Yungipicus canicapillus

星头啄木鸟 xīng tóu zhuó mù niǎo

Gray-capped Pygmy Woodpecker

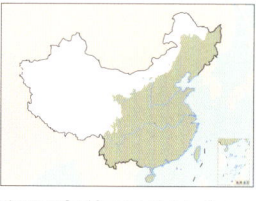

L: 14–17 cm

Habitat: Often inhabits various habitats on plains or mountains, such as broadleaf forests, coniferous forests, mixed broadleaf-coniferous forests, bamboo thickets, secondary forests, urban parks, and orchards; sometimes also in artificial forests and near villages.

Behavior: Often solitary or in pairs; occurs in family unit when juveniles fledge. Usually nests in tree cavities 3–15 m from ground. Flight rapid and undulating. Prefers upper and middle stories of trees; forages by climbing on branches. Occasionally comes to the ground.

Distribution: Widespread in eastern and central parts of China; common resident in some regions. *Y. c. doerriesi* in Heilongjiang, E Jilin, E Liaoning, NE Inner Mongolia; *Y. c. scintilliceps* in SW Liaoning, Hebei, Beijing, Shandong, Shanghai, Zhejiang, Henan, Ningxia, and Anhui; *Y. c. szetschuanensis* in S Shaanxi, Ningxia, S Gansu, and N and C Sichuan; *Y. c. omissus* in NW and W Yunnan and C and SW Sichuan; *Y. c. nagamichii* in E and SE Yunnan, Guizhou, Jiangxi, Fujian, Guangdong, and Guangxi; *Y. c. swinhoei* in Hainan; *Y. c. obscurus* in S Yunnan.

Voice: Familiar *kik*, similar to call of Great Spotted Woodpecker but shorter, higher pitched, and more intense (and extremely similar to that of Fulvous-breasted Woodpecker). Sometimes gives a series of monosyllabic calls, *zhi zhi zhi*. Rapid (15 notes in 0.8 sec), tinny drumming very different from Fulvous-breasted.

Yungipicus kizuki

小星头啄木鸟 xiǎo xīng tóu zhuó mù niǎo

Japanese Pygmy Woodpecker

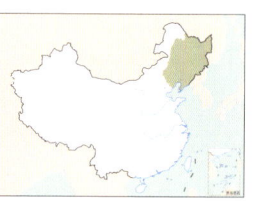

L: 14–18 cm

Habitat: Often inhabits montane coniferous, mixed broadleaf-coniferous, and broadleaf forests; sometimes also comes to birch or secondary forests.

Behavior: Often solitary except during breeding season; occurs in family unit when juveniles fledge. Usually nests in tree cavities. Flight rapid and undulating. Often flies from the top of one tree to the lower part of another tree and then climbs up to forage.

Distribution: Uncommon resident in E Heilongjiang, Jilin, E Liaoning, and N Hebei.

Voice: A series of peculiar, buzzy *zzzzrer* notes. Makes fast (20 calls in 2.9 sec) sequence of intense, penetrating *tititititititi* notes as song. Drumming speed (18 notes in 1.3 sec) similar to, but much longer than, Gray-capped Pygmy Woodpecker and doesn't fade.

Dendrocoptes auriceps

褐额啄木鸟 hè é zhuó mù niǎo

Brown-fronted Woodpecker

L: 19–20 cm

Habitat: Inhabits montane coniferous and mixed coniferous-oak forests at altitudes of 1200–3000 m.

Behavior: Forages on tree trunks or in bushes for caterpillars, berries, and pine nuts. Often solitary or in pairs; also joins tits and minivets in mixed-species flocks.

Distribution: Rare in Tibet (Gyirong).

Voice: *kik* call very similar to that of Great Spotted Woodpecker, occasionally in intense, lengthy, excited, chattering series. Also a mellower version and a plaintive, nasal, whining *pleeoo-u*.

Dryobates cathpharius

赤胸啄木鸟 chì xiōng zhuó mù niǎo

Crimson-breasted Woodpecker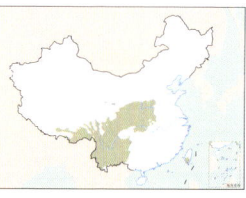

L: 16–18 cm

Habitat: Inhabits montane broadleaf or mixed forests at altitudes of 700–2800 m.

Behavior: Moves about on small branches or rotten woods. Forages for insects and insect larvae; also feeds on nectar. Occasionally in mixed-species flocks.

Distribution: Subspecies group *cathpharius*, including *D. c. cathpharius* and *D. c. ludlowi*, in S and SE Tibet; *pernyii* group, including *D. c. pernyii*, *D. c. tenebrosus*, and *D. c. innixus*, in C and N Yunnan, N and W Sichuan, E Chongqing, SE Gansu, W Hubei, and S Shaanxi.

Voice: Sharp *kik* shorter and clearer than call of Great Spotted Woodpecker (but virtually identical to Darjeeling Woodpecker). Also gives a series of rapid and accelerating *chikchikchikchik* when agitated (clearer and more ringing than Great Spotted, less tinny than Rufous-bellied Woodpecker) and often followed by a nervous, nasal, flickerlike *wicca-wicca-wicca*.

Dryobates minor

小斑啄木鸟 xiǎo bān zhuó mù niǎo

Lesser Spotted Woodpecker

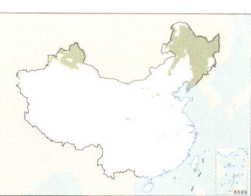

L: 14–18 cm

Habitat: Occurs in broadleaf or mixed forests on hills and foothill plains at altitudes of 500–1300 m. Also appears in woodlands, farmland, and orchards in fall and winter.

Behavior: Often in the canopy. Forages and calls on branches; rarely on tree trunks.

Distribution: *D. m. kamtschakensis* in N Xinjiang; *D. m. amurensis* in Heilongjiang, Jilin, Liaoning, NE Inner Mongolia, and S Gansu. Vagrant south to Beijing.

Voice: Series of 10–14 high-pitched and essentially identical *pee-pee-pee-pee* calls; drumming often more rapid and tinny than Great Spotted Woodpecker.

Gray-capped Pygmy Woodpecker

Small red patch on head

Black streaks on pale brown belly

White patches on back

Brown-fronted Woodpecker

Brown forehead

Gold crown

Red patch on nape

Yellowish-brown crown and nape

♂

♀

Japanese Pygmy Woodpecker

Regular white bars on back

Crimson-breasted Woodpecker

♂

Red breast patch with diffuse boundary

Red patch on nape extends to side of neck and connects to black malar stripe

D. c. cathpharius

Yellowish-brown to grayish-white breast and belly

♀

No red patch on nape

Buff breast and belly

cathpharius Group

Lesser Spotted Woodpecker

Red crown

Buff-tinged white forehead and cheek

Black crown

♂

Red patch on nape does not extend to side of neck

Buff breast and belly

Grayish-white underparts, with black streaks on flanks

D. m. amurensis

D. c. ludlowi

♂ White belly, with almost no streaks

♀

White undertail coverts

D. m. kamtschakensis

Red breast patch with clear boundary

Red patch on nape does not extend to side of neck

Buff-tinged white breast and belly

White breast and belly

D. c. pernyii

pernyii Group

Dendrocopos hyperythrus
棕腹啄木鸟 zōng fù zhuó mù niǎo
Rufous-bellied Woodpecker　 LC

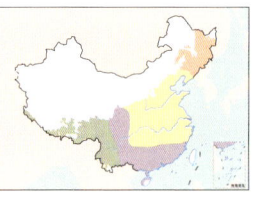

L: 19–23 cm
Habitat: Inhabits montane coniferous or mixed forests. Different subspecies are distributed at different altitudes. *D. h. hyperythrus* and *D. h. marshalli* in the Himalayas and the Hengduan Mountains can be found at altitudes over 2500 m.
Behavior: Occurs and forages mostly in the canopy.
Distribution: *D. h. hyperythrus* and *D. h. marshalli* are residents in southwestern China; seen in Tibet, Sichuan, and W Yunnan. *D. h. subrufinus* is a migrant; breeds in Heilongjiang, winters in Guangdong, Guangxi, Guizhou, Sichuan, and migrates through eastern parts of China.
Voice: Varied; lengthy rattling *chit-chit-chit-r-r-r-r-h* and faster, very intense chatter *ptikitititititititit*, sometimes slightly wavering (speeding and slowing with changes in pitch). When agitated, makes rapid, tinny, scolding *twicca-twicca-twicca*. Drumming is short (five to eight notes lasting just 0.5–0.6 sec), accelerating, fading, and resonant—similar to sounds produced by hardwood Chinese temple blocks.

Dendrocopos macei
纹腹啄木鸟 (茶胸斑啄木鸟)
wén fù zhuó mù niǎo (chá xiōng bān zhuó mù niǎo)
Fulvous-breasted Woodpecker　 LC

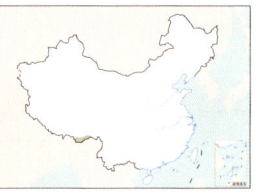

L: 18–20 cm
Habitat: Inhabits mixed forests, secondary forests, and open areas at forest edges up to 3000 m.
Behavior: Usually occurs in the canopy of tall trees; seldom comes to the understory of forests.
Distribution: Rare in extreme SE Tibet.
Voice: Call either a short, sharp, high-pitched *kik* (shorter and sharper than Great Spotted Woodpecker but extremely similar to Gray-capped Pygmy Woodpecker) or a more complex *chi-ter-cher*, sometimes with a fourth note, *chi-ter-cher-chik*. Drumming is short (seven to nine notes over 0.6 sec), fading and accelerating toward the end, like a dropped hard ball.

Dendrocopos atratus
纹胸啄木鸟 wén xiōng zhuó mù niǎo
Stripe-breasted Woodpecker　 LC

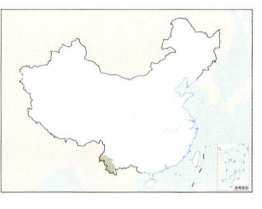

L: 18–22 cm
Habitat: Inhabits evergreen or deciduous broadleaf forests at altitudes below 1500 m.
Behavior: Forages primarily for beetles, butterflies, moths, and caterpillars. Occasionally feeds on fruits or seeds.
Distribution: Resident in W and S Yunnan.
Voice: Short *kik* calls, subtly more muffled than most *Dendrocopos* woodpeckers; also short sequences of excited multisyllabic chattering, *zha-zha-zha*. Drumming like Darjeeling Woodpecker.

Dendrocopos darjellensis
黄颈啄木鸟 huáng jǐng zhuó mù niǎo
Darjeeling Woodpecker　 LC

L: 21–24 cm
Habitat: Inhabits montane coniferous or mixed forests at altitudes of 1000–3000 m.
Behavior: Often occurs in middle and lower layers of trees. Climbs and forages along tree trunks; seldom seen in the canopy.
Distribution: Resident in SE Tibet, SW Sichuan, and W Yunnan.
Voice: Short sharp *kik*, indistinguishable from Crimson-breasted Woodpecker; also an intense chatter, *twick-wik-wik-wik-wik*, that soon develops into a rolling trill—less tinny than Rufous-bellied Woodpecker, faster than Great Spotted Woodpecker, and, unlike both, accelerating. Lengthy but evenly paced drumming (more than 20 notes in a second), like Stripe-breasted Woodpecker.

Dendrocopos leucopterus
白翅啄木鸟 bái chì zhuó mù niǎo
White-winged Woodpecker　 LC

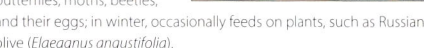

L: 22–24 cm
Habitat: Prefers large desert poplar forests in deserts; also inhabits forests on plains, river valleys, and artificial forests.
Behavior: Forages for butterflies, moths, beetles, and their eggs; in winter, occasionally feeds on plants, such as Russian olive (*Elaeagnus angustifolia*).
Distribution: Locally common in desert poplar forests in S Xinjiang; uncommon in broadleaf forests in S and W Junggar Basin in N Xinjiang.
Voice: Varied vocalizations include relatively soft *kik*, like Eurasian Three-toed Woodpecker; multisyllabic *chiterrk*; flickerlike notes reminiscent of Himalayan Flameback; and excited, nervous chatter. Drumming extremely rapid (18 notes in 0.6 sec) and fading.

Dendrocopos major
大斑啄木鸟 dà bān zhuó mù niǎo
Great Spotted Woodpecker　 LC

L: 20–25 cm
Habitat: Inhabits broadleaf forests on plains, hills, and mountains; also occurs in urban parks.
Behavior: Flight slow, undulating. Good at foraging for insects under bark. Nests by excavating cavities in tree trunks.
Distribution: Widespread and common. *D. m. brevirostris* in northern Northeast China and N Xinjiang; *D. m. japonicus* in central and southern Northeast China; *D. m. cabanisi* in eastern North China to East China; *D. m. wulashanicus* in W Inner Mongolia; *D. m. beicki* in Ningxia, Gansu, and E Qinghai; *D. m. stresemanni* in southwestern China; *D. m. mandarinus* in the Yangtze River Watershed and regions to its south; *D. m. hainanus* in Hainan.
Voice: Gives monosyllabic or short *zha-*, *zha-*. In breeding season, male drums on trees rapidly and makes a series of *dududu* calls.

Rufous-bellied Woodpecker

Black and white spots on crown

Red from crown to hindneck

White cheek

♂

Brown breast and belly

♀

Darjeeling Woodpecker

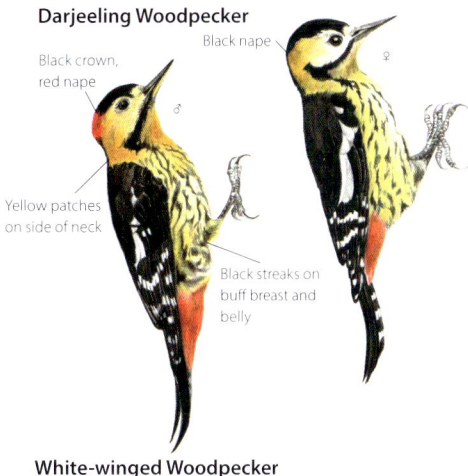

Black crown, red nape

Black nape

♀

Yellow patches on side of neck

♂

Black streaks on buff breast and belly

Fulvous-breasted Woodpecker

Black crown

♀

Red crown

♂

Black nape

Thin pale black streaks on belly and flanks

White-winged Woodpecker

Black nape

Red patch on nape

♂

♀

Large white wing patch

Black streaks on mostly white primaries

Stripe-breasted Woodpecker

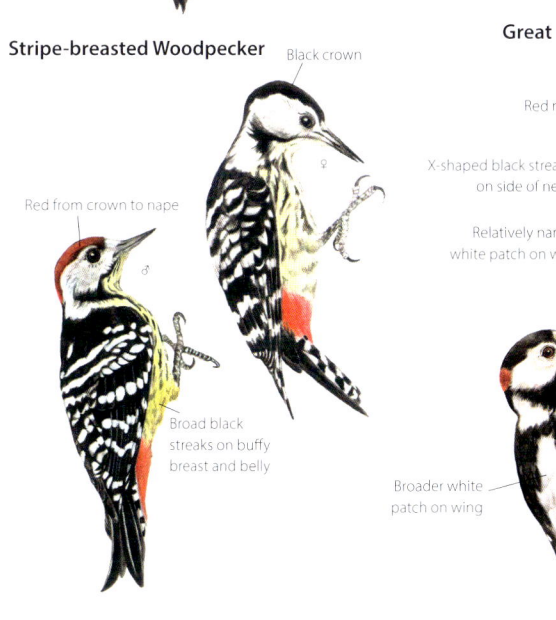

Black crown

♀

Red from crown to nape

♂

Broad black streaks on buffy breast and belly

Great Spotted Woodpecker

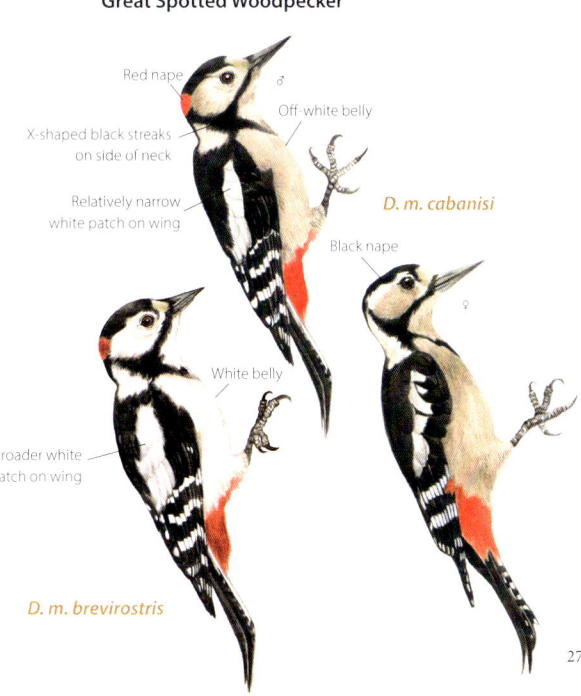

Red nape

♂

Off-white belly

X-shaped black streaks on side of neck

Relatively narrow white patch on wing

D. m. cabanisi

Black nape

♀

White belly

Broader white patch on wing

D. m. brevirostris

Dendrocopos leucotos
白背啄木鸟 bái bèi zhuó mù niǎo
White-backed Woodpecker

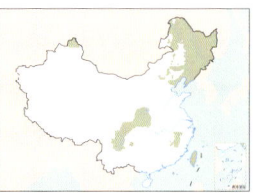

L: 25–30 cm
Habitat: Inhabits primarily various forests at altitudes of 500–2000 m; especially common in primary mixed broadleaf-coniferous forests and broadleaf forests. Also occurs and forages on edges of secondary forests and woodlands.
Behavior: Often solitary or in pairs. Forages by creeping up tree trunks; pecks continuously when grubs are detected in decaying wood; does not move to another tree until all pests are eaten in one tree. Flight undulating. Sometimes also feeds on ants or other insects on fallen wood or in soil mounds on ground.
Distribution: Locally common but disjunctly distributed. *D. l. leucotos* in Northeast China, south to Beijing and Hebei and the Altai Mountains in N Xinjiang; *D. l. fohkiensis* in NW Fujian and NE Jiangxi; *D. l. tangi* in S Shaanxi to C Sichuan and Chongqing; *D. l. insularis* in Taiwan.
Voice: *kik* call very similar to Three-toed Woodpecker (slightly lower pitched than Fulvous-breasted Woodpecker); *wicca* when excited. Drumming starts strong and slow but accelerates and weakens (like bouncing ball).

Chrysophlegma flavinucha
大黄冠啄木鸟 dà huáng guān zhuó mù niǎo
Greater Yellownape

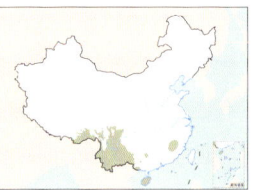

L: 31–36 cm
Habitat: Inhabits broadleaf forests at altitudes of 500–2000 m.
Behavior: Climbs and forages primarily along tree trunks; comes occasionally to the ground.
Distribution: Resident in E Tibet, S and W Yunnan, SW Sichuan, Jiangxi, Fujian (Wuyi Mountains), S Guangdong, and Hainan.
Voice: Variable but loud and barely disyllabic *kiyaep*, *ki-yep*, or series of notes, *klaeyp-yep-kep*. In flight, an excited sequence of shorter notes, *ki-ki-ki-kae-yep kayep*. Drumming relaxed, evenly paced, slow (27 notes in 1.5 sec), and mellow, like drumming on a bucket.

Dryocopus javensis
白腹黑啄木鸟 bái fù hēi zhuó mù niǎo
White-bellied Woodpecker

L: 42–48 cm
Habitat: Inhabits montane broadleaf, mixed broadleaf-coniferous, and coniferous forests.
Behavior: Often solitary or in pairs in forests at middle and low altitudes; also seen at high altitudes. Prefers tall dead trees. Produces loud drumming sounds.
Distribution: Rare and local resident in Yunnan, SW Sichuan, W Inner Mongolia, and C Fujian.
Voice: Loud *kiyow* calls; loud, smooth, and moderately long or long *you*. Resonant slow (more than 20 notes in over 2.5 sec) drumming that starts slowly and accelerates slightly.

Picus vittatus
花腹绿啄木鸟 huā fù lǜ zhuó mù niǎo
Laced Woodpecker

L: 30–33 cm
Habitat: Inhabits open forests, bamboo thickets, and orchards on low mountains and plains.
Behavior: Often forages at base of tree trunks or on ground.
Distribution: Very rare in Xishuangbanna, S Yunnan.
Voice: *Dendrocopos*-like *kip* call or disyllabic *kee-ip*, reminiscent of Greater Yellownape. Low-pitched, resonant drumming is fast and lengthy (28 notes in 1.7 sec) and evenly paced and sounds like a distant motorcycle engine.

Picus chlorolophus
黄冠啄木鸟 huáng guān zhuó mù niǎo
Lesser Yellownape

L: 23–27 cm
Habitat: Inhabits montane broadleaf forests and mixed forests at altitudes of 500–2000 m. Range often overlaps with Greater Yellownape.
Behavior: Similar to Greater Yellownape. Sometimes mixes with other birds.
Distribution: Resident in E Tibet and W Yunnan, Jiangxi, Fujian (Wuyi Mountains), and Hainan.
Voice: Lengthy, mournful, raptorlike mewing call, *peeee-u*, sometimes more nasal or more strongly disyllabic. Reminiscent of the call of Rufous Woodpecker but almost invariably given singly. Rarely drums, but slow and rhythmic.

Dryocopus martius
黑啄木鸟 hēi zhuó mù niǎo
Black Woodpecker

L: 45–55 cm
Habitat: Inhabits mainly primary coniferous forests and mixed broadleaf-coniferous forests at altitudes below 1800 m; sometimes also appears in broadleaf forests and at forest edges of secondary forests.
Behavior: Flight undulating. Often solitary. Forages primarily on tree trunks, thick branches, and dead wood; chisels large holes. Also forages for ants or other insects on ground or in decaying fallen wood. Usually pecks and drums on tree trunks when foraging and makes far-reaching *kuang kuang* drumming sounds.
Distribution: *D. m. martius* in Northeast China south to Shanxi and N Xinjiang, historically also Beijing and Hebei. *D. m. khamensis* in S and W Gansu, E Tibet, S and E Qinghai, W Yunnan, and W Sichuan.
Voice: Gives loud and melodious *kwee-kwee-kwee-kwee-kwee-kwee-kwee* in spring. Loud and resonant drumming at even tempo.

White-backed Woodpecker

Red crown

Black crown

♀

Fine black streaks on breast and belly

White back

Red vent

D. l. leucotos

Laced Woodpecker

Red crown and nape

Yellowish-green throat and breast

Black crown and nape

♀

Black and white streaks on belly

Greater Yellownape

♀

Brown chin

♂

Yellow chin

Short black and white streaks on throat

Prominent yellow nape

Olive-gray breast and belly

Lesser Yellownape

Discontinuous red patch on side of head broadens and extends to nape

Red patch extends from behind eye to nape

White "moustache" stripe

♀

Bright yellow crest

♂

Olive-brown breast

Grayish-white bars on belly

White-bellied Woodpecker

Prominent red crest

Blackish-gray bill

No red malar stripe

♂

♀

Red malar stripe

White belly

Black Woodpecker

Red only on nape

♀

Red on forehead, crown, and nape

♂

All-black body

Picus xanthopygaeus
纹喉绿啄木鸟 wén hóu lǜ zhuó mù niǎo
Streak-throated Woodpecker

L: 28–30 cm
Habitat: Inhabits montane broadleaf forests or bamboo thickets at altitudes below 1500 m.
Behavior: Often forages for ants or other insects on the ground or tree trunks.
Distribution: Very rare in SW Yunnan.
Voice: High-pitched *kik* subtly longer, tinnier, and higher pitched than Laced Woodpecker.

Picus squamatus
鳞腹绿啄木鸟 lín fù lǜ zhuó mù niǎo
Scaly-bellied Woodpecker

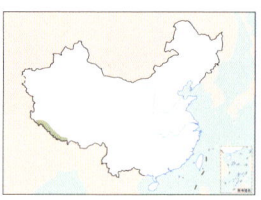

L: 30–36 cm
Habitat: Inhabits montane forests at altitudes below 3000 m.
Behavior: Forages primarily for ants, mainly on dead branches, decaying wood, and the ground.
Distribution: Restricted to S Tibet (Gyirong and Yadong).
Voice: Loud, ringing, one-, two-, or three-syllable *kle-glu* or *kee-glu-gu*, with the stress on the first syllable, similar in tone to Gray-headed Woodpecker. Short, reasonably rapid drumming (nine notes in 1.1 sec) that accelerates and descends.

Picus rabieri
红颈绿啄木鸟 hóng jǐng lǜ zhuó mù niǎo
Red-collared Woodpecker

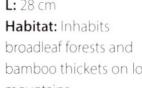

L: 28 cm
Habitat: Inhabits broadleaf forests and bamboo thickets on low mountains.
Behavior: Similar to other *Picus* green woodpeckers.
Distribution: Very rare in Yunnan (SE Honghe).
Voice: *kik* calls similar to Streak-throated Woodpecker but less tinny.

Picus canus
灰头绿啄木鸟 huī tóu lǜ zhuó mù niǎo
Gray-headed Woodpecker

L: 26–31 cm
Habitat: Inhabits forests and forest edges on middle and low altitudes, groves near farmland and villages, and urban green spaces and gardens.
Behavior: Often occurs at the middle and lower parts of tree trunks; climbs up spirally to the forks and then flies to the base of another tree. Forages primarily for insects, also berries and seeds.
Distribution: Subspecies group *canus* (the subspecies of this group in China is *P. c. jessoensis*) in N Xinjiang, Northeast China, and south to Beijing, Hebei, and Shandong; *guerini* group (including *P. c. kogo, P. c. guerini, P. c. sobrinus, P. c. tancolo, P. c. sordidior*, and *P. c. hessei*) is common in North, East, Central, South, and southwestern China. The two subspecies groups are sometimes treated as two separate species.
Voice: Nervous *chk* when agitated; series of six to nine loud descending whistles (yaffle), *kiu-kiu-kiu-kliu-kliu—kliu*, reminiscent of hearty laughter. Lengthy, fast, and evenly paced drumming.

Dinopium shorii
喜山金背啄木鸟 xǐ shān jīn bèi zhuó mù niǎo
Himalayan Flameback

L: 30–32 cm
Habitat: Inhabits primarily lowland primary forests at altitudes below 1200 m.
Behavior: Occasionally mixes with other birds.
Distribution: Rare in SE Tibet.
Voice: Lengthy nervous, rapid, loud series, *tibittbitibitibitiibit*, similar to very similar Greater Flameback but less tinny, longer, and with shorter notes that decelerate.

Dinopium benghalense
小金背啄木鸟 xiǎo jīn bèi zhuó mù niǎo
Black-rumped Flameback

L: 26–29 cm
Habitat: Inhabits montane moist broadleaf forests at altitudes below 1800 m.
Behavior: Feeds primarily on ants, beetles, and spiders. Occasionally searches for ant nests on the ground.
Distribution: Rare in SE Tibet.
Voice: Typical wavering trill that accelerates, rises, and then falls in pitch.

Streak-throated Woodpecker

Black upper mandible

White supercilium

Red crown

Black streaks on throat ♂

Black crown ♀

Scaly black and white pattern on breast and belly

Gray-headed Woodpecker

guerini Group

Black spots on gray forehead ♀

Black from hindcrown to nape ♂

Olive breast and belly

Gray from hindcrown to nape ♂

Gray head ♀

canus Group

Scaly-bellied Woodpecker

Black crown ♀

Yellow bill

Red crown

Grayish-green throat ♂

Thick scaly black pattern on breast and belly

Himalayan Flameback

White spots on black crown and crest ♀

Red crown

Thick black eye stripe extends to neck

Red on yellowish back; red rump

Scaly black pattern on belly

Red-collared Woodpecker

White eye ring ♂

Bright red crown, nape, and neck collar

Grayish-green cheek

Grayish-green crown ♀

Black-rumped Flameback

White spots on black crown ♀

Red crown ♂

White spots on black chin and throat

White spots on black wing coverts

Black streaks on belly

275

Dinopium javanense

金背三趾啄木鸟
jīn bèi sān zhǐ zhuó mù niǎo

Common Flameback

LC

L: 28–31 cm
Habitat: Inhabits open secondary forests, sparse woodlands, plantations, and parks on low mountains.
Behavior: Moves about and forages on tree trunks, branches, and fallen woods on ground. Calls in undulating flight.
Distribution: Uncommon resident in SE Tibet and S Yunnan.
Voice: Typical trill is slightly coarser, lower pitched, and of more even tempo than other flamebacks.

Chrysocolaptes guttacristatus

大金背啄木鸟 dà jīn bèi zhuó mù niǎo

Greater Flameback

LC

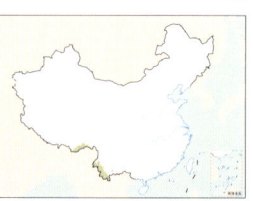

L: 30–34 cm
Habitat: Inhabits open evergreen broadleaf forests or deciduous tropical monsoon forests, forest edges, and nearby villages; usually prefers habitats with large trees (including dead trees).
Behavior: Often in pairs or family groups. Climbs on tree trunks and forages by pecking and drumming. Calls from time to time.
Distribution: Resident in S and W Yunnan and SE Tibet.
Voice: Typical trill is tinnier and more evenly paced than Himalayan Flameback but not as coarse or uniform as Common Flameback.

Gecinulus grantia

竹啄木鸟 zhú zhuó mù niǎo

Pale-headed Woodpecker

LC

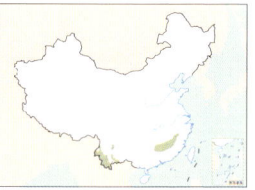

L: 23–25 cm
Habitat: Inhabits bamboo thickets or mixed forests with bamboo at altitudes below 1200 m.
Behavior: Often occurs in pairs or small flocks in the understory of forests; forages for ants, beetles, and insect larvae on bamboo or tree trunks.
Distribution: *G. g. grantia* in W Yunnan; *G. g. indochinensis* in S and SE Yunnan; *G. g. viridanus* in Jiangxi, Fujian (Wuyi Mountains), and N Guangdong. Local and occasional.
Voice: *kittituk-ti-tuk* chatter very similar to Bay Woodpecker but often slower. Also a whining, plaintive, nasal *kleek kleek-kleek* given four to five times and similar to that of Bay Woodpecker, but all notes equally stressed (that of Bay fades), more evenly paced (Bay's accelerates), and less flat. Lengthy, rapid, very resonant drumming (27 notes in 2 sec), often on bamboo, slows markedly.

Blythipicus pyrrhotis

黄嘴栗啄木鸟 huáng zuǐ lì zhuó mù niǎo

Bay Woodpecker

LC

L: 25–32 cm
Habitat: Inhabits montane broadleaf forests at altitudes below 2000 m; migrates to low-altitude regions in winter.
Behavior: Forages for various kinds of insects, occasionally berries. Noisy. Often mixes with other birds.
Distribution: Occasionally seen in Yunnan, Sichuan, Chongqing, Guizhou, Guangxi, Hunan, Fujian, Zhejiang, and Hainan.
Voice: Hoarse series, *kwa-kwa-kwa-kwa-*, decreases in pitch and is slow at first, then accelerates. Chatter very similar to that of Pale-headed Woodpecker. Not known to drum.

Micropternus brachyurus

栗啄木鸟 lì zhuó mù niǎo

Rufous Woodpecker

LC

L: 21–24 cm
Habitat: Inhabits montane broadleaf forests and bamboo thickets at altitudes below 1500 m.
Behavior: Forages primarily for ants; often occurs near ant nests. In pairs or small flocks; sometimes mixes with other birds.
Distribution: Occasional in S Yunnan, Guangxi, Guangdong, Jiangxi, S Zhejiang, S Hunan, C Fujian, and Hainan.
Voice: Rapid nasal whinny of 4–16 *kwee-kwee-kwee-kwee-* notes that intensify before accelerating and fading slightly. Calls with a flat, nasal, whining *pleet-it*, sometimes repeated four times and similar in tone to the longer, invariably single note of Lesser Yellownape. Drumming soft and decelerating.

Mulleripicus pulverulentus

大灰啄木鸟 dà huī zhuó mù niǎo

Great Slaty Woodpecker

VU

L: 45–50 cm
Habitat: Inhabits open evergreen broadleaf forests or deciduous tropical monsoon forests, forest edges, and nearby villages; usually prefers habitats with large trees (including dead trees).
Behavior: Often in small family groups. Drums, forages, and climbs on tree trunks, calls from time to time. Exhibits a unique behavior during which birds in a small family group call and flap wings around tree trunks.
Distribution: Local resident in W and S Yunnan and SE Tibet.
Voice: High-pitched, excited, even hysterical, whinnying cackle and laughter, *twoik-twoik-twoik-woik*, and more subdued, squeaky *twik-wik*.

Common Flameback

White streaks on black crown and crest

Red crown and crest

Only one black moustachial line

Black hindneck

Only one toe facing backward

♀

Bay Woodpecker

Yellow bill

Lacks red patch on nape

Red nape

♂

♀

Dense black bars on brown wings

Greater Flameback

White spots on black crest

Four toes

Red crest

♂

Two moustachial lines on cheek, connected at side of neck

♀

Gold wings and back

Rufous Woodpecker

Cheek lacks red patch

Red patch on cheek

Black bill

Black spots on throat

Dense black bars on wings, back, and tail feathers

♂

♀

Great Slaty Woodpecker

Cheek lacks red patch

♀

Red patch on cheek

Yellow throat

♂

Short dense feathers on head

Gray plumage; relatively large body

Pale-headed Woodpecker

Pale red crown

Grayish-green head

♂

Pale brown bars on brown wings

Crown lacks red patch

隼形目　FALCONIFORMES

Small to medium-sized diurnal birds of prey. Plumage similar or slightly different between sexes in most species, mostly black, white, gray, and rufous. Bill short, strong, with hooked tip; upper bill with cere and toothlike ridges. Body slender and cone shaped. Wings long and pointed, enabling fast and powerful flight. Tail long, rounded or wedge shaped. Inhabiting forest edges and open habitats. Feed by hunting in flight or diving on ground targets, mostly insects, birds, and rodents. Nest in crevices, tree cavities, or nests built by other birds (especially those by corvids). Some species are migratory.

    Falconiformes once included not only falcons but also eagles, hawks, vultures, and their allies, but recent studies suggest that the similarities between the two types of diurnal birds of prey result from convergent evolution while falcons are more closely related to parrots. One family, 11 genera, and 66 species recognized worldwide, with a cosmopolitan distribution. Two genera and 12 species recorded in China, ubiquitously distributed nationwide.

隼科
Falconidae

Microhierax caerulescens
红腿小隼 hóng tuǐ xiǎo sǔn
Collared Falconet   LC

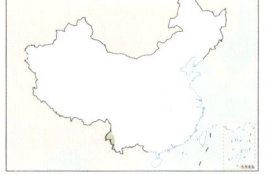

L: 14–18 cm
WS: 28–34 cm
Habitat: Inhabits deciduous forests, artificial forests, and evergreen forest edges, usually near rivers or streams. At altitudes up to 2000 m, but usually lower than 900 m.
Behavior: Usually gathers in small family groups near nest at dawn and dusk and returns to nest to rest every night. Hunts separately during daytime, but family members stay close together. Usually stands on exposed branches waiting for prey and bobs head before hunting.
Distribution: W Yunnan, possibly also in SE Tibet.
Voice: High-pitched, multisyllabic chittering, *kilkilkilkillii*, with nasal tone reminiscent of Crested Treeswift.

Microhierax melanoleucos
白腿小隼 bái tuǐ xiǎo sǔn
Pied Falconet   LC

L: 15–19 cm
WS: 35–37 cm
Habitat: Often inhabits low mountains and hills at altitudes up to 2000 m; especially prefers broadleaf forests and forest edges of woodlands.
Behavior: Often in pairs or gathers in small flocks. Usually nests in cavities abandoned by woodpeckers. Flight rapid and agile; often hunts for prey on the wing. Prefers resting or searching for food on tall trees during daytime and returns to the cavity to rest at night.
Distribution: Patchily distributed in southern China; uncommon. Relatively stable breeding populations around Wuyuan, Jiangxi.
Voice: Shrill, piercing, short whistles, *kleek* and *kik*, lacking the nasal quality of Collared Falconet.

Falco naumanni
黄爪隼 huáng zhǎo sǔn
Lesser Kestrel

L: 29–34 cm
WS: 61–66 cm
Habitat: Often inhabits open fields, deserts, river valleys, and forest edges; especially prefers rocky areas.
Behavior: Often in pairs or gathers in small flocks.
Usually nests in crevices on mountain cliffs; some individuals also breed in cavities of large trees. Flight rapid; glides frequently.
Distribution: Local summer breeder in northern China; locally common in N Xinjiang. Occasionally seen in eastern and central China during migration.
Voice: At colony, a husky, often unobtrusive *wei-wei-wei*, more like Gray Partridge than a falcon. Other notes very similar to Eurasian Kestrel.

Falco tinnunculus
红隼 hóng sǔn
Eurasian Kestrel

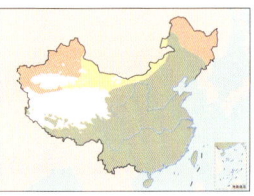

L: 31–38 cm
WS: 69–74 cm
Habitat: Often inhabits farmland, villages, montane forests, forest edges, grasslands, and fields. Highly adaptable; some individuals also breed in cities.
Behavior: Often solitary. Usually nests in trees in old nests of magpies, crows, or other birds; sometimes also nests on cliff faces, rock crevices, or earthen burrows. Hovers in the air to watch the ground; dives rapidly to catch mice and other prey.
Distribution: Widespread and common. *F. t. intertinctus* nationwide; *F. t. tinnunculus* in Heilongjiang, N Inner Mongolia, N Xinjiang, and Beijing.
Voice: Piercing *qi qi qi qi qi*.

Collared Falconet

Black eye stripe on white face

Black head, back, and wings

Rufous chin and throat

Rufous lower belly, vent, and leg feathers

Male slightly smaller than female

Pied Falconet

White breast and belly

Lesser Kestrel

Nearly unspotted pale under wings

Tail generally wedge shaped

Few streaks on rufous-brown belly

Yellow claws

Eurasian Kestrel

Densely spotted under wings

Dark brown streaks on belly

Black claws

Tail generally flat

Falco vespertinus

西红脚隼 xī hóng jiǎo sǔn

Red-footed Falcon NT

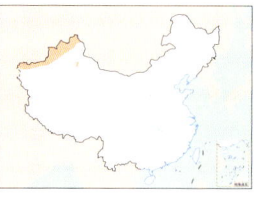

L: 27–33 cm
WS: 65–76 cm
Habitat: Often inhabits open plains, woodland edges, fields, and river valley woodlands.
Behavior: Often in pairs or small flocks. Usually nests in trees in old nests of magpies, crows, or other birds; sometimes also builds its own nest in trees or tree cavities. Rests mostly on trees or utility poles. Flight rapid; also glides or hovers.
Distribution: Primarily in N and W Xinjiang; uncommon summer breeder.
Voice: Chattering *yi yi yi yi*, like Eurasian Kestrel or Eurasian Hobby.

Falco amurensis

红脚隼 hóng jiǎo sǔn

Amur Falcon LC

L: 25–30 cm
WS: 63–71 cm
Habitat: Often inhabits plains, low mountain woodlands, hills, forest edges, open fields, farmland, and cultivated lands. Also seen in cities during migration.
Behavior: Often gathers in flocks during nonbreeding season. Usually nests on top of tall trees in sparse woodlands; sometimes also occupies old nests of magpies, crows, or other birds. Perches on trees or utility poles when resting. Flight rapid; forages for insects on the wing. Also searches for food by hovering in the air.
Distribution: Widespread but absent in Xinjiang and the Qinghai-Tibet Plateau. Summer breeder in northern China and passage migrant in southern China.
Voice: Chattering *yi yi yi*, like Eurasian Kestrel or Eurasian Hobby.

Falco columbarius

灰背隼 huī bèi sǔn

Merlin 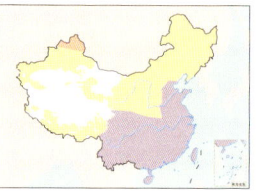 LC

L: 27–32 cm
WS: 64–73 cm
Habitat: Often inhabits open plains, deserted fields, and farmland; sometimes also comes to low mountains, hills, and tundra.
Behavior: Often solitary. Usually uses old nests of magpies, crows, and other birds in trees, on cliff faces, or on ground. Rests on trees, soil mounds, or cow dung. Flight rapid; prefers flying low, close to the ground, and chases prey at high speed.
Distribution: A few breeding records in Xinjiang. Passage migrant and winter visitor in northwestern, Northeast, North, East, southwestern, and southeastern China. *F. c. insignis* in Heilongjiang, Jilin, Hebei, Beijing, Shandong, Shanghai, Sichuan, Anhui, Qinghai, Gansu, Fujian, and Taiwan; *F. c. lymani* in Qinghai and Xinjiang; *F. c. pacificus* in Hebei and Inner Mongolia; *F. c. pallidus* in S Tibet.
Voice: Generally silent away from nest site, where it makes an intense, scolding chatter, *yi yi yi*.

Falco subbuteo

燕隼 yàn sǔn

Eurasian Hobby LC

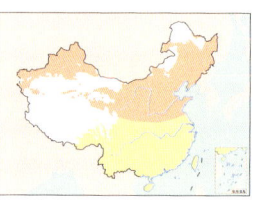

L: 29–35 cm
WS: 69–78 cm
Habitat: Often inhabits open plains, woodlands, deserted fields, and farmland; sometimes also comes to low mountains, hills, and forest edges.
Behavior: Often solitary or in pairs. Usually occupies nests of magpies and crows; nests in trees in woodlands or at forest edges. Rests on trees, utility poles, or soil mounds. Flight rapid and agile; chases after swallows and dragonflies in the air.
Distribution: Widespread. Summer breeder in northern China; seen throughout China on migration. *F. s. subbuteo* in Heilongjiang, Beijing, Hebei, Henan, Shandong, Shanxi, Xinjiang, Tibet, and Qinghai; *F. s. streichi* in Yunnan, Sichuan, Chongqing, Guizhou, Hubei, Jiangxi, Zhejiang, Shanghai, Guangxi, and Taiwan.
Voice: Intense scolding very similar to Eurasian Wryneck, *yi yi yi*.

Red-footed Falcon

Dark gray under wings

Orange-red head, breast, and belly

Merlin

Bright gray upperparts

Brown streaks on rufous-brown underparts

Thick brown streaks on underparts

Amur Falcon

Pale gray underwing coverts contrast strongly with black wing feathers

Many black spots on nearly white underwing coverts

Eurasian Hobby

Prominent black moustachial stripe

Dark brown streaks on belly

281

Falco severus

猛隼 měng sǔn

Oriental Hobby

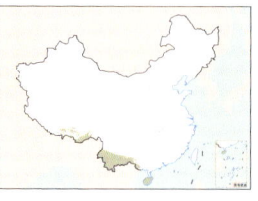

L: 24–30 cm
WS: 61–71 cm
Habitat: Usually prefers forest openings in coastal mangroves and deciduous or evergreen broadleaf forests. Hunts in bushes, grasslands, tea plantations, or paddy fields.
Behavior: Usually solitary or in pairs; active in small family groups when subadults fledge. Often stands on exposed branches waiting for prey. Flight rapid; forages primarily for small birds.
Distribution: Local resident in S and W Yunnan, SE Tibet, and Hainan.
Voice: Intense scolding very similar to Eurasian Hobby, but shorter notes in faster, more piercing sequence.

Falco cherrug

猎隼 liè sǔn

Saker Falcon

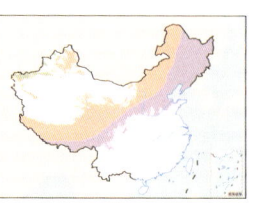

L: 42–60 cm
WS: 106–129 cm
Habitat: Often inhabits open foothill plains, deserted fields, farmland, and cultivated fields; especially prefers rocky hills with few trees and open fields.
Behavior: Often solitary or in pairs. Usually nests on cliff faces or in trees; sometimes also breeds in old nests of other birds. Often rests on soil mounds or trees. Flight rapid and powerful; chases prey at high speed. Fierce; sometimes also captures other medium-sized raptors.
Distribution: Breeds in northwestern and Northeast China; passage migrant and winter visitor in North, East, and southwestern China. *F. c. cherrug* in Xinjiang; *F. c. milvipes* in Jilin, Hebei, Beijing, Shandong, Gansu, Qinghai, Sichuan, and also Xinjiang. Escaped "Altai Falcons," Gyrfalcon x Saker Falcon hybrids kept by falconers, create identification problems in Xinjiang.
Voice: Usually silent away from nest, where it calls *qi qi qi qi*, hoarser than Peregrine Falcon.

Falco rusticolus

矛隼 máo sǔn

Gyrfalcon

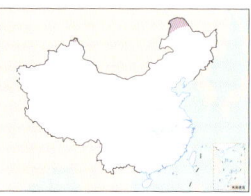

L: 53–63 cm
WS: 110–130 cm
Habitat: Often inhabits open foothill plains, deserted fields, rocky mountains, rocky coasts, etc.
Behavior: Often solitary. Usually nests in trees on cliffs or tundra. Often perches on trees and utility poles, resting and watching for prey; dives rapidly to hunt for prey, primarily in low flight. Fierce; often drives away other invading medium-sized to large raptors.
Distribution: Rare winter visitor or vagrant to N Xinjiang and northern Northeast China. Mostly dark-morph individuals.
Voice: Silent away from nest, where it makes a deep, guttural *ang ang ang*.

Falco peregrinus

游隼 yóu sǔn

Peregrine Falcon

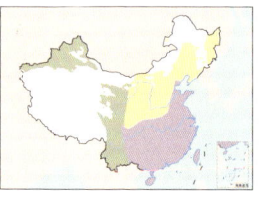

L: 41–50 cm
WS: 84–120 cm
Habitat: Inhabits varied habitats, such as mountains, hills, deserts, grasslands, swamps, lakes, coasts, farmland, and cultivated fields; some individuals also stay and breed in cities.
Behavior: Often solitary. Usually nests on cliff faces; sometimes breeds in trees and buildings. Also breeds in old nests of other birds. Often perches on clifftops and trees, resting and watching the surroundings; sometimes patrols in the air. Dives rapidly and chases after prey once detected; sometimes grabs prey with feet in the air.
Distribution: Widespread. *F. p. calidus* in Heilongjiang, Jilin, Hebei, Beijing, Shandong, Shanghai, Hubei, Anhui, and Taiwan; *F. p. japonensis* in Shandong, Jiangsu, Zhejiang, and Fujian; *F. p. peregrinator* in Yunnan, Sichuan, Chongqing, Guizhou, Hubei, Hunan, and Jiangxi and increasingly north to Beijing; *F. p. peregrinus* in Heilongjiang, Jilin, and Xinjiang; and *F. p. ernesti* in S Yunnan. *F. p. babylonicus*, in Ningxia, Qinghai, and Xinjiang, was once treated as a subspecies of "Barbary Falcon" (*F. pelegrinoides*) that is now merged with Peregrine Falcon.
Voice: Husky, repetitive chatter, *ga ga ga ga*.

Oriental Hobby

Black hood with flat lower boundary

Black from back and wings to tail

Pale yellowish-white throat

Rufous-brown belly

ad.

Heavy black streaks on belly

juv.

Saker Falcon

juv.

Clear spots on breast

ad.

Gyrfalcon

White morph

Mostly white plumage

ad.

Peregrine Falcon

Prominent moustachial stripe on face

Clear bars on belly

ad.

F. p. calidus

ad.

Streaks on belly

juv.

Reddish belly

F. p. peregrinator

From cheek to hindneck slightly tinged brown

Pale gray back

"Barbary Falcon"
F. p. babylonicus

鹦鹉目（鹦形目） PSITTACIFORMES

Brightly colored perching birds of the tropics. Plumage similar or slightly different between sexes in most species. Bill short, thick, and strong; upper bill curved and hooked, with cere at base. Zygodactyl feet well equipped for grabbing or clinging to vertical surfaces. Most species have long and thin tails. Inhabit a variety of habitats, ranging from desert scrubland to tropical rain forests, but most dwell in forests. Active and loud. Commonly gather in large groups. Herbivorous; feed mostly on plant fruits, nectar, and seeds. Nest in tree cavities, holes, or crevices. Most species are residents, and some form large nomadic groups in nonbreeding seasons.

Three families, 93 genera, and 389 species recognized worldwide, distributed in tropical and subtropical areas around the world. One family, three genera, and nine species recorded in China, found in Hainan and southwestern China.

Psittinus cyanurus
蓝腰鹦鹉 lán yāo yīng wǔ
Blue-rumped Parrot

L: 18–19.5 cm
Habitat: inhabits primary tropical rain forests, mangroves, logged rain forests, forest edges, open areas, and oil palm, rubber, and coconut plantations. Usually at altitudes up to 700 m. One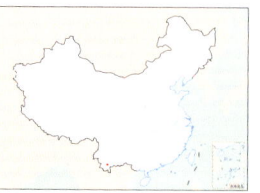
record in evergreen broadleaf forests in Yunnan (Pu'er).
Behavior: Usually gathers in small flocks. Forages for palm fruits in the canopy. Nests in tree cavities.
Distribution: S Yunnan (Pu'er).
Voice: Two or three piercing, harsh Black Drongo–like notes, *chi, chi, chi, whee-chi-chi,* or *chew-ee.*

Psittacula finschii
灰头鹦鹉 huī tóu yīng wǔ
Gray-headed Parakeet

L: 36–40 cm
Habitat: Inhabits evergreen broadleaf forests, open mixed deciduous forests, teak forests, secondary forests, and plantations. Especially prefers edges of cultivated lands, such as cornfields.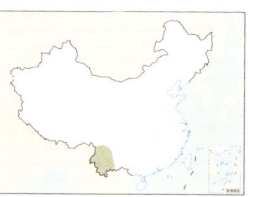
Mostly in foothills at altitudes of 600–1200 m, sometimes up to 3800 m.
Behavior: Usually gregarious, sometimes in flocks containing hundreds of individuals. Gathers noisily and forages in the canopy. Nests in tree cavities.
Distribution: Resident in Yunnan and SW Sichuan.
Voice: Loud, piercing, rising, markedly disyllabic *ju ju* calls. When resting, calls mixed with some short *qiu qiu* notes.

Psittacula himalayana
青头鹦鹉 qīng tóu yīng wǔ
Slaty-headed Parakeet

L: 39–41 cm
Habitat: Inhabits subtropical evergreen broadleaf forests, deciduous forests, and coniferous forests; prefers large cedar forests. At altitudes of 600–2500 m, mostly above 1300 m.
Behavior: Usually gregarious, sometimes in flocks containing hundreds of individuals. Gathers noisily and forages in the canopy. Nests in tree cavities.
Distribution: S Tibet, close to the border with Nepal.
Voice: Flatter, longer, and more monosyllabic flight calls than Gray-headed Parakeet. When resting, calls mixed with some short *qiu qiu* notes.

Psittacula roseata
花头鹦鹉 huā tóu yīng wǔ
Blossom-headed Parakeet

L: 30–36 cm
Habitat: Inhabits open forests and open areas at forest edges at altitudes up to 1000 m.
Behavior: Gregarious. Gathers noisily and forages in the canopy. Nests in tree cavities.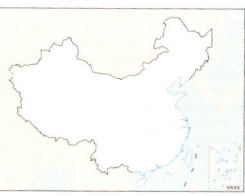
Distribution: W Yunnan (Yingjiang).
Voice: Slightly shorter, flat, and less piercing flight calls than other parakeets, with more melodious short notes.

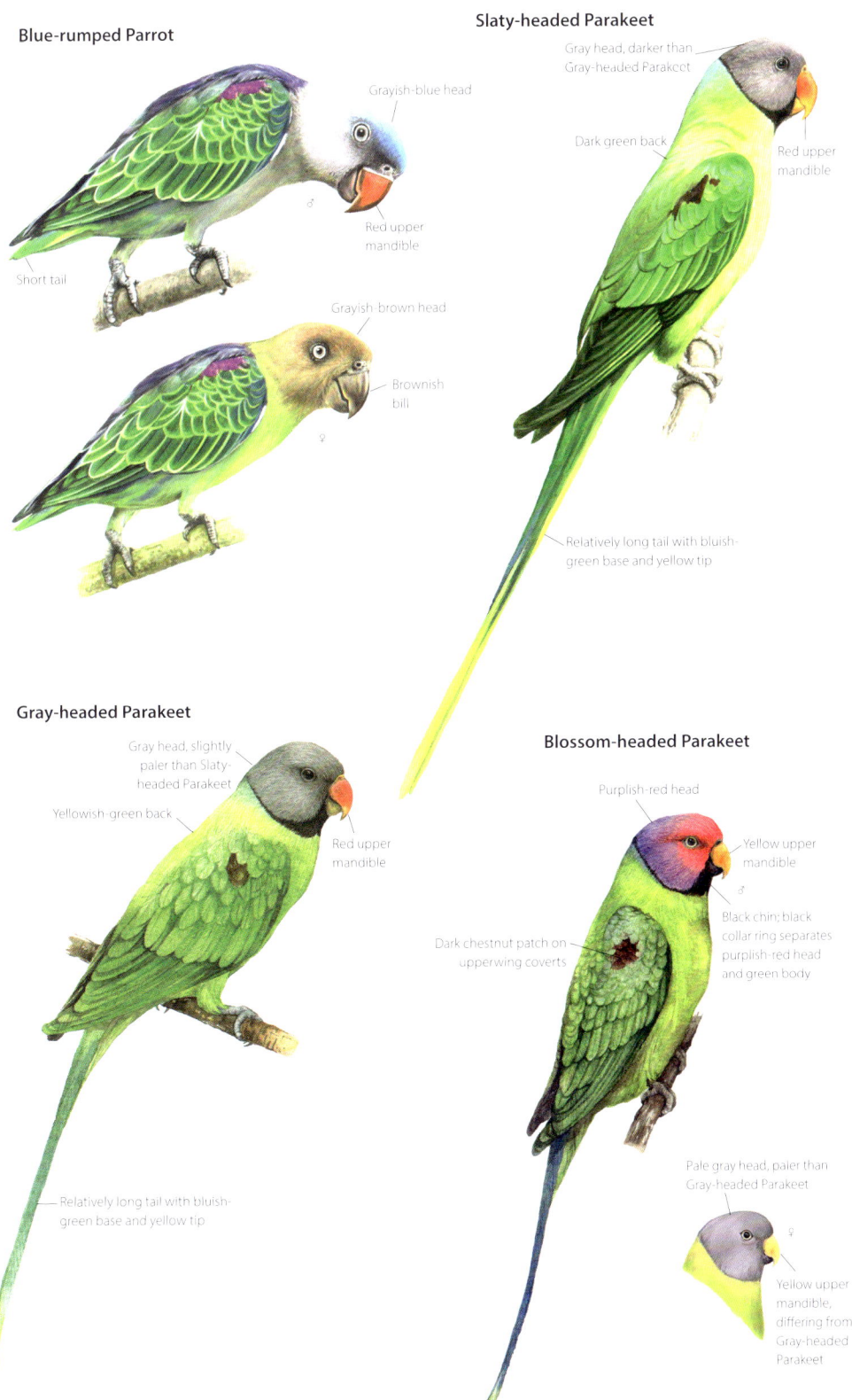

Blue-rumped Parrot

Grayish-blue head

Red upper mandible

Short tail

Grayish-brown head

Brownish bill

Slaty-headed Parakeet

Gray head, darker than Gray-headed Parakeet

Dark green back

Red upper mandible

Relatively long tail with bluish-green base and yellow tip

Gray-headed Parakeet

Gray head, slightly paler than Slaty-headed Parakeet

Yellowish-green back

Red upper mandible

Relatively long tail with bluish-green base and yellow tip

Blossom-headed Parakeet

Purplish-red head

Yellow upper mandible

Black chin; black collar ring separates purplish-red head and green body

Dark chestnut patch on upperwing coverts

Pale gray head, paler than Gray-headed Parakeet

Yellow upper mandible, differing from Gray-headed Parakeet

Psittacula alexandri

绯胸鹦鹉 fēi xiōng yīng wǔ

Red-breasted Parakeet

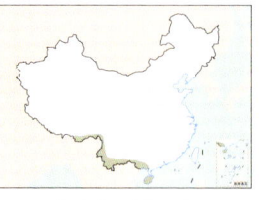

L: 33–38 cm
Habitat: Inhabits
evergreen broadleaf
forests, deciduous
forests, secondary forests,
mangroves, teak forests,
and coconut plantations
in foothills and lowland
regions; also occurs
in cultivated areas and villages near woodlands. Especially prefers
cultivated fields with corn or other crops near forest edges.
Behavior: Usually gregarious, sometimes in flocks of four or five
thousand birds. Gathers noisily and forages. Nests in tree cavities.
Distribution: Resident in W Yunnan, SE Tibet, S Guangxi, and Hainan.
Voice: Flat, descending, raucous flight call, lower pitched and huskier
than other parakeets. Very noisy when in flocks.

Psittacula derbiana

大紫胸鹦鹉 dà zǐ xiōng yīng wǔ

Derbyan Parakeet

L: 37–50 cm
Habitat: Inhabits
montane forests at middle
and high altitudes.
Behavior: Gregarious;
flies often.
Distribution: Locally
common in SE Tibet,
Yunnan, W Sichuan, and
SW Guangxi. Rapid population decline due to illegal hunting and habitat
fragmentation.
Voice: Short, husky, descending, and tremulous flight calls.

Psittacula eupatria

亚历山大鹦鹉 yà lì shān dà yīng wǔ

Alexandrine Parakeet

L: 50–58 cm
Habitat: Inhabits open
woodlands at low
altitudes, especially mixed
forests.
Behavior: In pairs or small
flocks, forages in tall fruit-
bearing trees.
Distribution: Local in
W Yunnan. Rare.
Voice: Harsh, low-pitched, rising, then descending, tremulous,
screaming *kyah* or *keeh*, more raucous and less obviously disyllabic than
calls of Rose-ringed Parakeet.

Psittacula krameri

红领绿鹦鹉 hóng lǐng lù yīng wǔ

Rose-ringed Parakeet

L: 38–42 cm
Habitat: Inhabits mixed
forests, open woodlands,
and farmland edges at
low altitudes.
Behavior: In pairs or small
flocks.
Distribution: Rare in
SW Yunnan; established
populations in Hong Kong and Macao from released and escaped cage
birds.
Voice: Short, noisy, penetrating calls with a noticeable drop in pitch,
ga–ga.

Loriculus vernalis

短尾鹦鹉 duǎn wěi yīng wǔ

Vernal Hanging-Parrot

L: 13–15 cm
Habitat: Inhabits
evergreen broadleaf
forests, deciduous
seasonal rain forests,
secondary forests,
abandoned farmland,
bamboo thickets,
orchards, tall scrubland,
and coastal wooded areas. At altitudes up to 1800 m.
Behavior: Usually gregarious. Often climbs; forages by hanging upside
down from trees. Nests in tree cavities.
Distribution: Resident in SE Tibet, S and W Yunnan.
Voice: Gives piercing and continuous disyllabic or trisyllabic *gigi* calls;
sometimes makes more hurried and continuous *zhizhi*.

Red-breasted Parakeet

Grayish-white head

Bright red upper mandible

Black throat

Pink breast, differing from purple breast of Derbyan Parakeet

Blue tail

Rose-ringed Parakeet

Green head, pale blue nape

Pink neck collar

Little or no blue on nape

Bluish-green central tail feathers with extremely small yellow tip

Green tail

Derbyan Parakeet

Black stripe on forehead

Red bill

Relatively broad black moustachial stripe extends to chin and hindneck

Purple underparts

Black bill

Vernal Hanging-Parrot

Red upper and lower mandibles

Pale blue throat

Red rump and uppertail coverts

Short tail

Alexandrine Parakeet

Large-looking head

Pink neck collar

Lacks neck collar

Red patch on upperwing coverts

Chin lacks black

Relatively long central tail feathers with yellowish tip

阔嘴鸟科 Eurylaimidae

Small to medium-sized songbirds of tropical and subtropical forests. Plumage similar or slightly different between sexes; usually brightly colored. Large eyes on a big head. Neck short and thick. Bill flat and thick, with hooked tip; rictal bristles less developed. Wings short and rounded. Usually inhabit dense forests near streams, in pairs or groups. Shy and not active. Weave pear-shaped hanging nests. Primarily insectivorous; some also take fruits. All species are residents and nonmigratory.

Seven genera and nine species recognized worldwide. Only one species is in Africa, and four are monotypic genera in E Himalayas and Southeast Asia. Two genera and two species recorded in China, found in W and S Yunnan, W Guangxi, SW Guizhou, and Hainan.

Long-tailed Broadbill

Unique yellow and black pattern on head

Yellow spot on side of head

Long blue tail

Silver-breasted Broadbill

Black lore connects with supercilium

White breast band on female

S. l. polionotus

Yellow base of bill reaches lores

Broad black supercilium

Brown upper back

Grayish-black supercilium

Dark gray back

S. l. rubropygius

S. l. elisabethae

Black tail

Psarisomus dalhousiae
长尾阔嘴鸟 cháng wěi kuò zuǐ niǎo
Long-tailed Broadbill

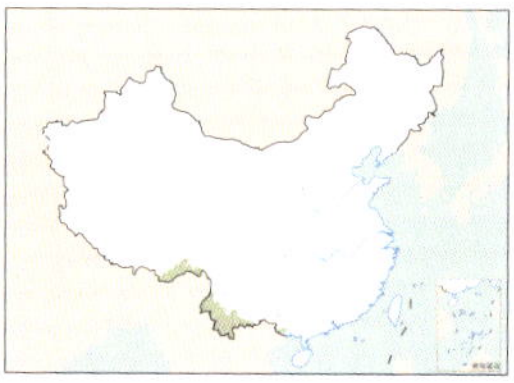

L: 20–28 cm
Habitat: Occurs in tropical and subtropical evergreen broadleaf forests at middle and low altitudes; also seen at forest edges.
Behavior: Often in pairs or small flocks. Occurs in dense forests near streams or river valleys; forages in forests.
Distribution: Resident in SE Tibet, S and W Yunnan, SW Guizhou, and SW Guangxi.
Voice: Series of 5–10 piercing, high-pitched, descending whistles, *triue-triieu-triiu-triu-triu*, that become shorter and more rapid, frequently repeated. Rarely heard call is a rising, raspy *tsweeep*.

Serilophus lunatus
银胸丝冠鸟 yín xiōng sī guān niǎo
Silver-breasted Broadbill

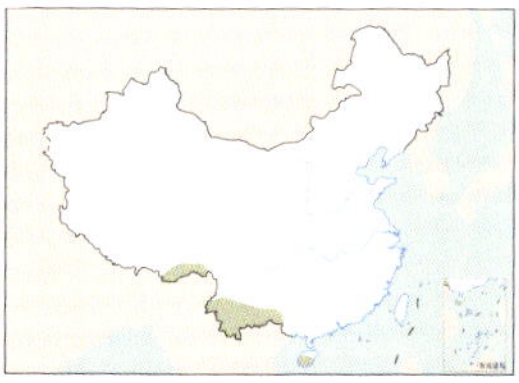

L: 15–18 cm
Habitat: Occurs near streams and forest edges in subtropical and tropical forests at middle and low altitudes.
Behavior: Often in small flocks. Occurs in forests or below the canopy of forest edges. Quiet but not shy.
Distribution: *S. l. elisabethae* in Yunnan and SW Guangxi; *S. l. polionotus* in Hainan. *S. l. rubropygius*, in SE Tibet, sometimes treated as a separate species, *S. rubropygius* (Gray-browed Broadbill, 灰眉丝冠鸟).
Voice: Soft, unmusical, melancholy, descending *tiiuu*. Sometimes markedly disyllabic *tiu-uu* and shorter, more piping versions, *tiu* or *tyu*, as well as short clicks repeated at intervals.

八色鸫科　Pittidae

Brightly colored, medium-sized, ground-dwelling songbirds. Body profile short and rounded. Plumage similar between sexes except for *Hydrornis* species. Bill strong, resembling those of thrushes. Neck short and thick. Legs long and strong. Tail extremely short. Inhabit primarily forests, scrubland, and mangroves, solitarily or in pairs. Nest on ground or in bushes. Most species are residents; very few are migratory.

Three genera and 42 species recognized worldwide, distributed in Indomalaya, Oceania, and the Afrotropics. Two genera and eight species recorded in China, widely distributed in southern China.

蓝八色鸫属
Hydrornis

八色鸫属
Pitta

Blue Pitta　　**Hooded Pitta**

Eared Pitta

White supercilium extends backward, forming "braid"

Few black spots on brown belly

Shorter "braid"

Dense black spots on pale brown belly

Blue Pitta

Yellowish-brown forehead　Red nape

Blue from back to tail

♂ ad.

Less red on nape

Dense black spots and bars on pale blue belly

Brown back

Less pale blue on belly　juv.

Hydrornis phayrei

双辫八色鸫 shuāng biàn bā sè dōng

Eared Pitta　LC

L: 20–24 cm
Habitat: Inhabits lowland tropical rain forests, secondary forests, and mixed bamboo thickets; especially prefers dry areas. Mostly at altitudes below 900 m but recorded up to 1830 m in Myanmar, Thailand, and China.
Behavior: Usually in the understory; forages on ground with fallen leaves and decaying tree trunks. Very quiet and vigilant; often stands still for a long time.
Distribution: S and W Yunnan.
Voice: Makes 1 sec long, almost monotone whistle that wavers in pitch slightly toward end, *weeeei-oo-whit*.

Hydrornis cyaneus

蓝八色鸫 lán bā sè dōng

Blue Pitta　LC

L: 23–24 cm
Habitat: Active in evergreen broadleaf forests and bamboo thickets with steep ravines. Also occurs in moist forests adjacent to streams and dense ground vegetation, or in dry and less dense forests. At altitudes of 60–2000 m, usually the lower range.
Behavior: Usually solitary or in pairs on the forest floor. Forages by turning fallen leaves and digging with bill in decaying tree trunks. Very quiet and vigilant; often stands still for a long time.
Distribution: Rare and local in S and W Yunnan.
Voice: Disyllabic *pleeou-wit*, duller than Rusty-naped Pitta and with more complex introductory note than Rusty-naped's clearly enunciated *tu-wit*.

Hydrornis nipalensis
蓝枕八色鸫 lán zhěn bā sè dōng
Blue-naped Pitta

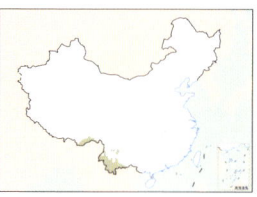

L: 22–25 cm
Habitat: Inhabits tropical and subtropical rain forests, secondary forests, bamboo thickets, and woodlands with dense understory, usually in river valleys or slopes near water at altitudes up to 2150 m.
Behavior: Usually solitary or in pairs on the forest floor. Forages by turning fallen leaves and digging with bill in decaying tree trunks. Very quiet and vigilant; often stands still for a long time.
Distribution: *H. n. hendeei* in W and S Yunnan; *H. n. nipalensis* in SE Tibet.
Voice: Lengthy monosyllabic whistle, *puuuwhet*, rising at the end. See Rusty-naped Pitta.

Hydrornis soror
蓝背八色鸫 lán bèi bā sè dōng
Blue-rumped Pitta

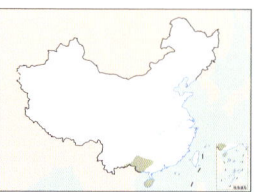

L: 22–24 cm
Habitat: Inhabits a wide range of habitats: primary evergreen forests, secondary or logged forests, and mixed forests with bamboo, as well as moist river areas, dry steep mountain slopes, and craggy limestone ground. From lowlands to 1700 m.
Behavior: Similar to Blue-naped Pitta.
Distribution: *H. s. tonkinensis* in SE Yunnan, S Guangxi; *H. s. douglasi* in Hainan.
Voice: Short resonant single *diu* repeated about every 8 sec.

Hydrornis oatesi
栗头八色鸫 lì tóu bā sè dōng
Rusty-naped Pitta

L: 23–25 cm
Habitat: Prefers dense primary or secondary montane forests, often with dark and moist ravines. At altitudes of 380–2600 m.
Behavior: Similar to Blue-naped Pitta.
Distribution: *H. o. oatesi* in S and W Yunnan; *H. o. castaneiceps* in SE Yunnan.
Voice: Percussive, strongly disyllabic, almost whiplash-like *du-wit*, repeated every 4–5 sec. Call of Blue-naped Pitta is the most similar but is monosyllabic, very slightly lower pitched, and inflected upward at end.

Pitta sordida
绿胸八色鸫 lǜ xiōng bā sè dōng
Hooded Pitta 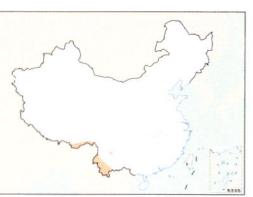

L: 16–19 cm
Habitat: Inhabits a wide range of habitats, including primary or secondary evergreen broadleaf forests, tropical monsoon forests, plantations, parks, and orchards, as well as bushes and grassy areas in forests. At altitudes up to 1000 m.
Behavior: Usually solitary or in pairs on the forest floor. Forages by turning fallen leaves and digging with bill in decaying tree trunks. Very quiet; often stands still for a long time. Also perches on branches. Sometimes bold.
Distribution: *P. s. cucullata* in S and W Yunnan, SE Tibet, and Sichuan; *P. s. sordida* is vagrant to Taiwan.
Voice: Short, paired, rising, whistled notes, *whew-whew*.

Pitta nympha
仙八色鸫 xiān bā sè dōng
Fairy Pitta

L: 16–20 cm
Habitat: Inhabits moist forests with dense bushes on lowlands and hills, especially near streams. Also occurs in plantations or parks during migration season. From lowlands to altitudes up to 1200 m.
Behavior: Usually solitary or in pairs on the forest floor. Forages by turning fallen leaves and digging with bill in decaying tree trunks. Very quiet; often stands still for a long time. Also perches on branches.
Distribution: Breeds in East, Central, and South China, including Hainan and Taiwan. Rare overshoots on migration along coastal regions in North and Northeast China.
Voice: Lengthy, complex, multisyllabic whistle, *puworh-puwee*.

Pitta moluccensis
蓝翅八色鸫 lán chì bā sè dōng
Blue-winged Pitta

L: 16–20 cm
Habitat: Occurs in a wide range of habitats, including moist or dry, dense or sparse primary forests, mixed deciduous forests, secondary forests, scrubland, and bamboo thickets. Also occurs in plantations and parks. From lowlands to altitudes up to 1800 m.
Behavior: Similar to Fairy Pitta.
Distribution: S Yunnan; vagrant to Shanghai, Guangdong, Guangxi, Hainan, and Taiwan.
Voice: Husky, whistling, disyllabic *tai-wuw*, with shorter, less complex notes than Fairy Pitta.

Blue-naped Pitta

- More blue on nape than female
- Greenish back
- ♂
- Less blue on nape
- Brownish-green back
- ♀

Hooded Pitta

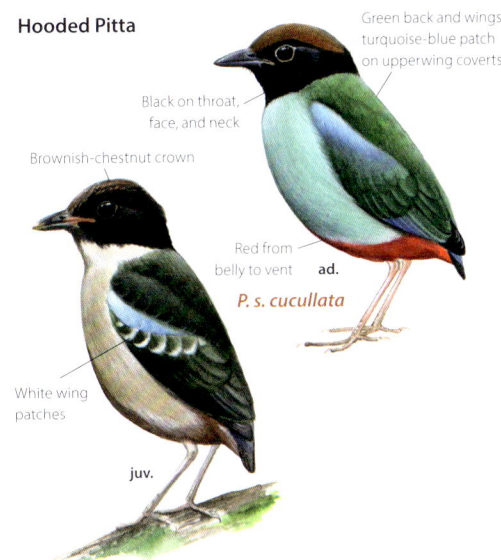

- Green back and wings, turquoise-blue patch on upperwing coverts
- Black on throat, face, and neck
- Brownish-chestnut crown
- Red from belly to vent
- *P. s. cucullata*
- ad.
- White wing patches
- juv.

Blue-rumped Pitta

- Blue crown
- ♂
- Relatively bright blue rump
- Brown belly
- *H. s. tonkinensis*

Fairy Pitta

- Chestnut crown
- White supercilium and thick black eye stripe
- Greenish-blue back and wings with glossy blue patch on upperwing coverts
- Off-white underparts
- Scarlet from belly to vent

Rusty-naped Pitta

- Brownish-chestnut head
- ♂
- Green back and tail with blue wash
- Brownish-chestnut belly

Blue-winged Pitta

- Buff-brown crown
- Thick black eye stripe
- Green back with bright blue wings
- Brownish-yellow underparts
- Scarlet from belly to vent

291

钩嘴䴕科 *Vangidae*

Small to medium-sized forest-dwelling songbirds. Plumage similar or slightly different between sexes in most species, mostly black, white, and brown. Bill in some species resembles that of shrikes: strong with hooked tip. Inhabit primarily lowland and montane forests. Insectivorous. All species are residents.

Twenty-one genera and 39 species recognized worldwide, distributed primarily in Africa, with a few in Indomalaya. Two genera and two species recorded in China, found in South and southwestern China.

Bar-winged Flycatcher-shrike

- Black head
- Brown back
- Long white wing bar
- Yellowish-brown underparts
- Blackish-brown head, paler than male
- Pale grayish-white underparts

Large Woodshrike

- Gray crown
- Broad black eye stripe extends from lores to ear coverts
- White rump
- Short tail
- Brownish gray from crown to neck

Hemipus picatus
褐背鹟䴕 hè bèi wēng jú
Bar-winged Flycatcher-shrike   LC

L: 14–15 cm
Habitat: Occurs at lowland forest edges, including disturbed forests, secondary forests, bamboo thickets, scrubland, artificial forests, and areas far from large forests. At altitudes of 600–1800 m.

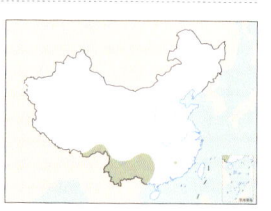

Behavior: Usually gathers in small flocks; also mixes with other birds. Forages mostly on top of trees or bushes; occasionally comes to the ground.
Distribution: Resident in S Tibet, Yunnan, C and S Guizhou, Jiangxi, and SW Guangxi.
Voice: Noisy, high-pitched, excited, repetitive, multisyllabic chattering, *tidititididit.*

Tephrodornis virgatus
钩嘴林䴕 gōu zuǐ lín jú
Large Woodshrike   LC

L: 18.5–23 cm
Habitat: Inhabits forest edges and clearings in evergreen broadleaf forests.
Behavior: In pairs during breeding season. Gathers in small flocks in nonbreeding season,

often in mixed-species flocks with other passerines. Noisy. Flies through tree crowns, making short sallies after disturbed insects. Often hunts from perch; also catches insects off water's surface.
Distribution: Uncommon resident. *T. v. latouchei* in E Yunnan, Guizhou, Guangxi, N Guangdong, SW Jiangxi, C and E Fujian, S Zhejiang; *T. v. jugans* in SW Yunnan; *T. v. hainanus* in Hainan.
Voice: Song is 1–3 sec rapid repetition of 5–16 short *wit-wit-wit-wit* notes, all on same pitch but increasing in volume. Calls include short *chip* and nasal *chrk,* as well as explosive grating.

燕鵙科 Artamidae

Small to medium-sized forest-dwelling songbirds. Body profile streamlined. Plumage similar between sexes, mostly gray, white, black, and brown. Bill thick, wide, and strong. Wings long and pointed, triangular. Tail short and rounded. Inhabit primarily the open areas adjacent to forests on plains or low mountains, usually in groups. Excellent fliers whose profile and behavior resemble swallows. Feed on insects in flight; some also take nectar. All species are residents.

One genus and 11 species recognized worldwide, primarily in Oceania, with only two species in Indomalaya. One genus and one species recorded in China, found in South and southwestern China.

Ashy Woodswallow

Large thick
bluish-gray bill

ad.

juv.

Pinkish-gray
underparts

White rump

Triangular wings

Flat tail

Artamus fuscus

灰燕鵙 huī yàn jú

Ashy Woodswallow

L: 16–19 cm
Habitat: Inhabits open country at altitudes up to 1500 m. Prefers perching on bare branches or power lines. Usually flies low.
Behavior: Often circles in the air hawking insects; sometimes forages on water's surface. Glides like swallows. Sometimes besieges and chases away passing raptors or crows. Individuals usually perch close together

and preen one another with bill.
Distribution: Locally common resident in S, W, and SE Yunnan, Guizhou, S Guangxi, S Guangdong, and Hainan.
Voice: Nasal *cherk*, often repeated in short intense series of up to six chattering notes; also makes even more nasal *nyrrt*, often doubled. Rambling twittering song includes many of these same notes.

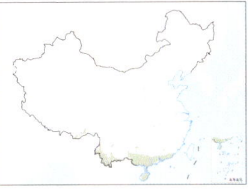

雀鹛科 *Aegithinidae*

Small forest-dwelling songbirds. Plumage similar between sexes, mostly olive, black, and dark green, while males usually darker than females. Bill strong, resembling those of thrushes. Legs relatively short but strong. Inhabit broadleaf forests and forest edges on plains and at lower altitudes, solitarily or in pairs. Usually silent but active. Mostly insectivorous but also take plant fruits and seeds. All species are residents.

Ioras constitute this recently recognized family, formerly grouped with leafbirds and fairy-bluebirds in Irenidae. One genus and four species recognized worldwide, endemic to Indomalaya. One genus and two species recorded in China, found in SE Tibet, W and S Yunnan, and SW Guangxi.

Common Iora

Great Iora

Two conspicuous white bars on relatively black wings

Black-tinged back and wings

Lacks wing bar

Black tail

Yellowish-white wings

Yellowish-green tail

Green back and wings

Aegithina tiphia

黑翅雀鹛 hēi chì què bēi

Common Iora 🄻🄲

L: 12.5–13.5 cm
Habitat: Occurs at lowland forest edges, including disturbed forests, secondary forests, bamboo thickets, scrubland, artificial forests, and messy habitats near villages. Primarily at altitudes up to 1000 m.

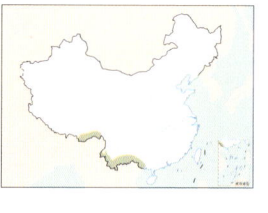

Behavior: Usually solitary or in pairs. Forages and hops back and forth in trees.
Distribution: Resident in SE Tibet, S and W Yunnan, and S Guangxi.
Voice: Highly vocal, with large vocabulary. Makes variety of intense descending whistles, trills, and continuous chattering, *dididiidididiu*, or pleasant disyllabic *didiu*.

Aegithina lafresnayei

大绿雀鹛 dà lǜ què bēi

Great Iora 🄻🄲

L: 13.6–15.4 cm
Habitat: Occurs at lowland forest edges; occasionally seen in secondary forests, bamboo thickets, scrubland, artificial forests, and habitats near villages. Primarily at lowlands.

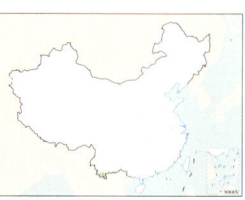

Behavior: Usually solitary or in pairs. Forages while hopping back and forth in trees. Sometimes joins mixed-species flocks.
Distribution: S Yunnan.
Voice: Spluttering chatter and an intensifying series of 4–10 Common Tailorbird–like whistles, *tiu-tiu-tiu-tu-tiu*.

山椒鸟科　Campephagidae

Small to medium-sized forest-dwelling songbirds, including brightly colored minivets and gray-colored cuckooshrikes. Plumage of most minivets differs between sexes and is mostly red, yellow, black, and gray. The plumage of cuckooshrikes is mostly similar between sexes, some resembling cuckoos. Bill short and strong, slightly hooked at tip. Wings long and pointed, enabling powerful flights. Tail medium to long; some constitute half of total body length. Inhabit primarily lowland and montane forests, in pairs or groups. Insectivorous, but also take plant fruits and seeds. Most species are residents; some are migratory.

Vangas, helmetshrikes, and their allies were once grouped with cuckooshrikes and minivets, but only the latter two groups are currently included in the family. Eleven genera and 93 species recognized worldwide, endemic to Indomalaya. Three genera and 11 species recorded in China, found mostly in southern China, while some ranges extend to the northern regions.

山椒鸟属
Pericrocotus

鸣鹃䴗属
Lalage

鸦鹃䴗属
Coracina

Gray-chinned Minivet

Black-winged Cuckooshrike

Large Cuckooshrike

Pericrocotus solaris

灰喉山椒鸟 huī hóu shān jiāo niǎo

Gray-chinned Minivet

L: 17–19 cm

Habitat: Inhabits low mountains, hills, and montane forests; often forages at forest edges. Adapted to a wide range of altitudes, usually at 500–3300 m; wanders at low altitudes in winter.

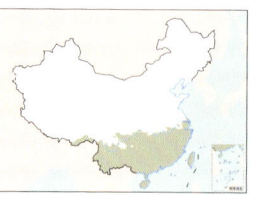

Behavior: In pairs during breeding season; in flocks during nonbreeding season. Forages in the canopy; active. Good at calling.

Distribution: Common resident. *P. s. solaris* in SE Tibet and S Yunnan; *P. s. montpellieri* in N and NW Yunnan; *P. s. griseogularis* in southwestern, South, Central, and southeastern China, including Hainan and Taiwan.

Voice: Distinctive thin two-note whistle, *twis-wees.*

Pericrocotus divaricatus

灰山椒鸟 huī shān jiāo niǎo

Ashy Minivet

L: 18–21 cm

Habitat: Inhabits wooded areas at low altitudes. Adapted to varied forests, including evergreen, broadleaf, and secondary forests, etc. Seen in various kinds of habitats in migration season.

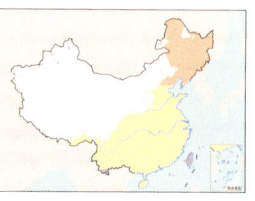

Behavior: Often in pairs in breeding season; gathers in small to medium-sized flocks in nonbreeding season. Active in the canopy; often perches on power lines, dead branches, or branches with few leaves. Calls in undulating flight.

Distribution: Breeds in Northeast China and migrates south, mostly along east coast, rarely as far west as Shaanxi and C Sichuan; wintering individuals found in Taiwan and S Yunnan. Huge numbers seen in recent years at Laotie Mountain in Lvshun, Liaoning.

Voice: Calls a series of metallic *gi-li-li* notes. Songs are similar to calls but with more trills.

Pericrocotus tegimae

琉球山椒鸟 liú qiú shān jiāo niǎo

Ryukyu Minivet

L: 18–21 cm

Habitat: Inhabits broadleaf forests at middle and low altitudes.

Behavior: Similar to other gray-colored minivets. Makes short-distance migrations in winter.

Distribution: An island species split from Ashy Minivet in recent years; breeding in Ryukyu Islands. Vagrant to Taiwan, Shanghai, Zhejiang (Hangzhou), and Jiangsu (Wuxi and Suzhou).

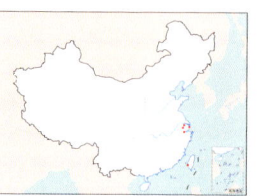

Voice: Similar to Ashy and Brown-rumped Minivets, but some sequences begin with a distinct, indrawn, breathless *e-uff*, while others include shorter notes that are delivered faster and at slightly higher pitch. Some vocalizations reminiscent of trilling waxwing.

Pericrocotus brevirostris

短嘴山椒鸟 duǎn zuǐ shān jiāo niǎo

Short-billed Minivet

L: 19–20 cm

Habitat: Inhabits deciduous forests and montane secondary forests from 1000 to 2800 m; wanders at low altitudes in winter.

Behavior: Often in pairs. Behaviors similar to other minivets.

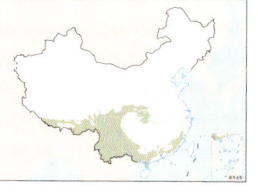

Distribution: Uncommon resident. *P. b. brevirostris* in SE Tibet and NW Yunnan; *P. b. affinis* in S Sichuan, W and SW Yunnan; *P. b. anthoides* in SE Yunnan, Guizhou, Guangxi, and N Guangdong.

Voice: Short, plaintive, monosyllabic, descending whistle, *tiuu*, reminiscent of whistle call of rubythroats.

Pericrocotus cantonensis

小灰山椒鸟 xiǎo huī shān jiāo niǎo

Brown-rumped Minivet

L: 18–19 cm

Habitat: Inhabits forests on hills and plains; prefers broadleaf forests with tall trees.

Behavior: Similar to Ashy Minivet.

Distribution: Predominantly south of S Henan to SE Gansu, and east of C Sichuan; has recently bred in Beijing. Rare migrant in Shandong and vagrant to Liaoning.

Voice: Trill a series of *zii-zii-zii* notes, very similar to Ashy Minivet but very slightly higher pitched, more "open," with last note often rising. Sequences subtly more muffled, with a less bouncing or ringing quality.

Pericrocotus roseus

粉红山椒鸟 fěn hóng shān jiāo niǎo

Rosy Minivet

L: 18–20 cm

Habitat: inhabits montane secondary broadleaf forests, mixed forests, and coniferous forests from 300 to 1800 m. Also forages in parks and orchards on low mountains in winter.

Behavior: In pairs in breeding season; gathers in small flocks in nonbreeding season. Less active than other minivets.

Distribution: Uncommon. Resident or summer breeder in Yunnan, SW Sichuan, S and SE Guizhou, S Guangxi, and SW Guangdong; vagrant to Hong Kong and Matsu Islands.

Voice: Flight calls similar to Ashy, Ryukyu, and Brown-rumped Minivets but shorter and with fewer notes (typically just four or five).

Gray-chinned Minivet

Gray throat

Gray throat clearly set off from breast ♂

Comma-shaped orange-red wing patch

Yellow belly

Comma-shaped yellow wing patch

Orange-red belly

P. s. griseogularis

Orange-tinged throat, diffusely set off from breast ♂

Yellow-tinged throat ♀

P. s. solaris

Short-billed Minivet

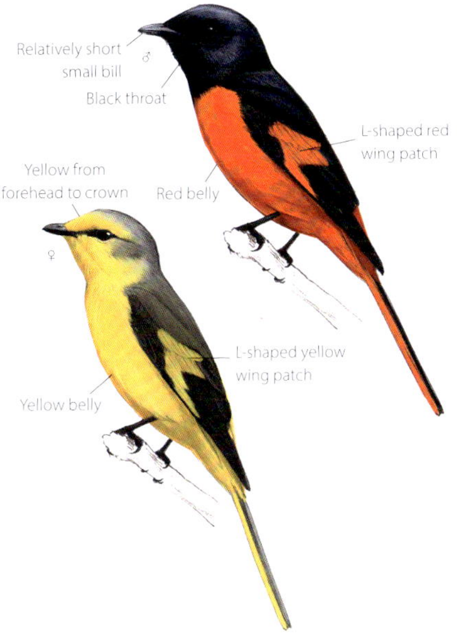

Relatively short small bill ♂

Black throat

Yellow from forehead to crown

Red belly

L-shaped red wing patch

L-shaped yellow wing patch

Yellow belly

Ashy Minivet

White on forehead ends above eyes ♂

Gray-tinged forehead

♀

Clean-looking pale gray underparts

Gray secondaries

Gray rump

Brown-rumped Minivet

White on forehead extends into relatively broad supercilium behind eyes ♂

Brown-tinged underparts, looks dirty

Gray-tinged forehead

Brown-tinged secondaries

Pale brown rump

Ryukyu Minivet

White from forehead to supercilium ♂

Consistent ashy gray from crown to upperparts

Crescent-shaped patch on lower cheek

Pale gray head ♀

Rosy Minivet

White throat ♂

Gray head

Pink-tinged belly and rump

♀

Pale yellow on belly

Pericrocotus ethologus

长尾山椒鸟 cháng wěi shān jiāo niǎo

Long-tailed Minivet

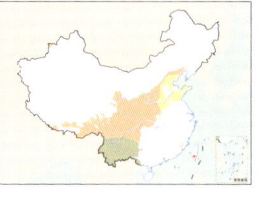

L: 17–20 cm
Habitat: Inhabits mountainous areas at middle and low altitudes. Adapted to habitats with various types of vegetation, including broadleaf forests, coniferous forests, and mixed forests.
Behavior: Often gathers in small to large flocks; forages in the canopy. If one individual flies away, the rest of the flock will follow. Often calls in flight.
Distribution: *P. e. ethologus* in mountainous areas of southwestern China, through the Qingling Mountains and Taihang Mountains to northwestern and northern North China; vagrant to Taiwan; *P. e. laetus* in S Tibet; *P. e. yvettae* in W Yunnan.
Voice: Short series of rising notes (some almost flat) of variable length, similar to Scarlet Minivet.

Pericrocotus speciosus

赤红山椒鸟 chì hóng shān jiāo niǎo

Scarlet Minivet

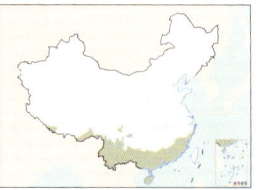

L: 17–22 cm
Habitat: Inhabits dense broadleaf forests at altitudes of 500–1500 m; wanders at low altitudes in winter.
Behavior: Forages in flocks in the canopy; active. Good at calling.
Distribution: Locally common resident. *P. s. speciosus* in SE Tibet; *P. s. fraterculus* in Yunnan and Hainan; *P. s. fohkiensis* in South and southeastern China, north to S Zhejiang.
Voice: Three to 11 piercing, thin, rising whistles, *weep-weep-weep*, similar to Long-tailed Minivet but in longer series and higher pitched and with change of pitch.

Coracina macei

大鹃鵙 dà juān jú

Large Cuckooshrike

L: 29–33 cm
Habitat: Inhabits open woodlands at altitudes up to 1500 m.
Behavior: Often solitary or in pairs; often perches on top of protruding tall branches at forest edges.
Distribution: Uncommon resident. *C. m. siamensis* in Yunnan and S Guizhou; *C. m. rexpineti* in southeastern China north to S Zhejiang, including Taiwan; *C. m. larvivora* in Hainan.
Voice: Querulous *queee-ee*.

Lalage nigra

斑鹃鵙 (黑鸣鹃鵙) bān juān jú (hēi míng juān jú)

Pied Triller

L: 16.5–18 cm
Habitat: Inhabits coastal broadleaf forests.
Behavior: Often in pairs or small flocks in its normal range; usually single birds recorded in China. Active in the canopy.
Distribution: Vagrant to Taiwan.
Voice: Varied. Song a mellow, descending, seesawing Common Iora–like warble. Calls include an Ashy Woodswallow–like nasal chatter and indrawn whistles.

Lalage melaschistos

暗灰鹃鵙 àn huī juān jú

Black-winged Cuckooshrike

L: 19.5–24 cm
Habitat: Inhabits open woodlands and bamboo thickets up to 2450 m. In winter, occurs in woodlands, orchards, and river valleys at low altitudes.
Behavior: Often solitary or in pairs; also joins mixed-species flocks when foraging. Usually forages in the canopy and active.
Distribution: Uncommon or locally common. *L. m. melaschistos* breeds in SE Tibet and NW Yunnan. *L. m. avensis* breeds in Sichuan, Chongqing, Yunnan, S Guizhou, and SW Guangxi. *L. m. intermedia* breeds in northwestern China (E Gansu) to most regions in North, East, Central, and South China; winters in South China (including Taiwan). *L. m. saturata* is resident in Hainan. Range expanding north.
Voice: Slow, haunting, variable, fluty, descending, two- to four-syllable whistles, *ti-tiuu* or *ti-ti-teeeuw*.

Long-tailed Minivet

Small area of pale yellow on forehead

Black throat

Wing patch shaped like upside-down U

♂

♀

Pied Triller

White supercilium

Black back

Gray from crown to upper back

Gray rump

♂

♀

Scarlet Minivet

Black throat

Yellow from forehead to crown

Red belly

Scattered red wing patches

♂

Yellow belly

Scattered yellow wing patches

♀

P. s. fohkiensis

Orangish underparts

P. s. speciosus

Black-winged Cuckooshrike

Black wings

♂

Dense fine bars on belly

♀

Large Cuckooshrike

Large thick bill

Dark face

♂

Pale gray underparts

Clean undertail coverts lack spots

Fine bars on belly

♀

C. m. siamensis

Dark gray underparts

♂

C. m. rexpineti

299

伯劳科　Laniidae

Ferocious, small to medium-sized, forest-dwelling songbirds. Plumage is similar or differs slightly between sexes, mostly gray, black, white, and rufous. Bill thick, strong, and hooked at tip, resembling bills of raptors, with well-developed rictal bristles. Tail relatively long. Legs strong with sharp claws. Inhabit primarily scrubland, open forests, and forest edges on plains, in deserts, or in low mountains. Ferocious hunters. Uniquely among other songbirds, diet dominated by meat; feed mostly on insects, as well as small amphibians and reptiles. Most species are migratory.

Four genera and 33 species recognized worldwide, mostly in the Old World except two species in North America. One genus and 14 species recorded in China, widely distributed except on C Qinghai-Tibet Plateau.

伯劳属
Lanius

Long-tailed Shrike

Lanius tigrinus
虎纹伯劳 hǔ wén bó láo
Tiger Shrike　**LC**

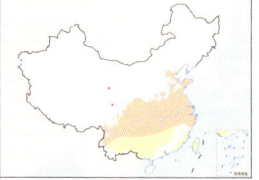

L: 17–18.5 cm
Habitat: Inhabits forested areas at low altitudes, usually below 1000 m.
Behavior: Usually solitary or in pairs. Often forages at forest edges; hides in trees or bushes, keeping a close watch on prey.
Distribution: Breeds from Northeast China to Central, East, and southwestern China; occasionally winters in South China. Few migration records in Taiwan. Vagrant west to Qinghai.
Voice: Harsh *chick* or repeated scolding chatter, *tcha-tcha-tcha* in alarm. Erratic, usually quiet, scratchy song mixes varied squeaky and harsh notes with skilled mimicry of other species.

Lanius bucephalus
牛头伯劳 niú tóu bó láo
Bull-headed Shrike　**LC**

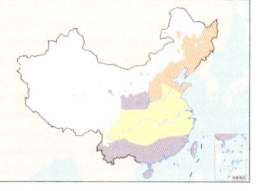

L: 19–20 cm
Habitat: Breeds primarily in montane broadleaf forests up to 1800 m. Winters at low altitudes in montane forest edges and countryside. Adapted to secondary vegetation and farmland.
Behavior: Usually solitary or in pairs. When foraging, perches on branches, power lines, or walls and observes; rapidly hunts prey if located. Carnivorous; forages mainly for invertebrates. Also eats plant seeds.
Distribution: Uncommon. *L. b. bucephalus* breeds in Northeast and North China. Winters in eastern and southeastern China, including Taiwan; migrates through various regions in East and Central China. *L. b. sicarius* is rare and local resident in C and S Gansu, C and N Sichuan, and SW Shaanxi.
Voice: Rapid and variable songs, often with imitations of other birds' vocalizations, resembling reed warblers. Calls variable; some spluttering and harsh, others mellower, softer chattering. Many given in lengthy sequence.

Lanius cristatus
红尾伯劳 hóng wěi bó láo
Brown Shrike　**LC**

L: 17–20 cm
Habitat: Inhabits open areas at forest edges; prefers bushy habitats with many small trees.
Behavior: Often solitary; gathers in small flocks during migration. Territorial; chases away

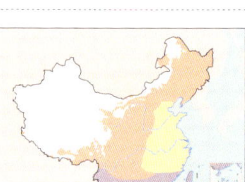

other intruding shrikes. Usually observes the surroundings and hunts from perch; often wags tail in circles when perched.
Distribution: Locally common. *L. c. cristatus* is seen only in migration in Central China and eastern China, including Taiwan. *L. c. confusus* breeds in Northeast China; migrates through eastern China (including Taiwan). *L. c. lucionensis* breeds from North to East China; migrates through eastern China and winters along the southeast coast and Taiwan. *L. c. superciliosus* winters in South China and Hainan; migrates through eastern China (including Taiwan).
Voice: Coarse chattering *ga, ga, ga, ga*; also gives mellow and gentle songs in spring. Like many shrikes, males are skilled mimics.

Lanius collurio
红背伯劳 hóng bèi bó láo
Red-backed Shrike　**LC**

L: 16–18 cm
Habitat: Inhabits moist and open woodlands, forest edges, forest clearings, scrubland, and riverside groves on low mountains, hills, and foothill plains; also comes to orchards and farmland.
Behavior: Often solitary or in pairs.
Distribution: Summer breeder in N Xinjiang (Altai Mountains); locally common. Occasionally in the Tian Shan Mountains; vagrant to Hong Kong and Taiwan.
Voice: Varied scolding notes.

Tiger Shrike

Bars on brown wings, back, and tail

Strong thick bill

White underparts

Pale lores

Barred underparts

Brown Shrike

White supercilium narrows toward forehead

Pale rufous brown or brownish gray from crown to back

Brown tail

L. c. cristatus

White supercilium broadens toward forehead

Brownish gray from crown to back

White band on forehead and relatively broad white supercilium

Relatively dark brown from crown to back

L. c. confusus

L. c. superciliosus

Gray crown

Grayish-brown back

Pinkish-brown flanks

L. c. lucionensis

Bull-headed Shrike

Brown head

Gray back

White wing patch

Brown head

Densely and finely barred underparts

Red-backed Shrike

Dark gray head

Black eye stripe

Rufous-chestnut back

Black tail

Grayish-brown head

Brown patch behind eye

Brown back

Vermiculated bars on underparts

Blackish-brown tail

301

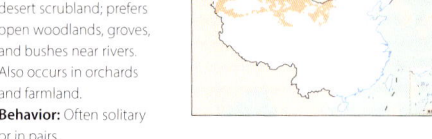

Lanius isabellinus
荒漠伯劳 huāng mò bó láo
Isabelline Shrike　LC

L: 16–18 cm
Habitat: Inhabits primarily low mountains, hills, and desert scrubland; prefers open woodlands, groves, and bushes near rivers. Also occurs in orchards and farmland.
Behavior: Often solitary or in pairs.
Distribution: Common in northwestern China. *L. i. isabellinus* breeds in Inner Mongolia, N Ningxia, C Gansu, and NE Xinjiang; *L. i. arenarius* breeds in S Xinjiang to NW Gansu, and Ningxia; *L. i. tsaidamensis* breeds in Qinghai and SE Xinjiang.
Voice: Calls similar to Brown and Red-backed Shrikes. Song includes skilled mimicry.

Lanius collurioides
栗背伯劳 lì bèi bó láo
Burmese Shrike　LC

L: 19–21 cm
Habitat: Inhabits open woodlands at middle and low altitudes; also seen in green spaces, and bushes and trees near farmland.
Behavior: Solitary or in pairs. Perches on branches or power lines waiting to hunt for insects.

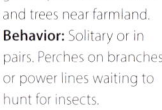
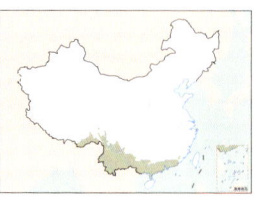

Distribution: Locally common from southwestern China to Guangdong.
Voice: Common call a loud scolding chatter, *etch-etch-etch*. Songs are scratchy and varied; start with titlike *jiji-* followed by some musical notes and often end with trills.

Lanius phoenicuroides
棕尾伯劳 zōng wěi bó láo
Red-tailed Shrike　LC

L: 16–19 cm
Habitat: Inhabits primarily moist low mountains, hills, and foothill plains; prefers open woodlands, forest edges, forest clearings, and groves and bushes near rivers. Also inhabits orchards and farmland.
Behavior: Often solitary or in pairs.
Distribution: Common breeder in N Xinjiang; vagrant to Qinghai.
Voice: Variable, with *zhizhi* calls or hoarse gasps and imitations of other birds' vocalizations.

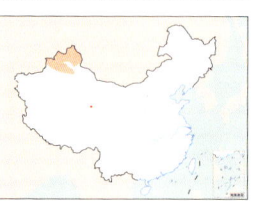

Lanius vittatus
褐背伯劳 hè bèi bó láo
Bay-backed Shrike　LC

L: 16–17 cm
Habitat: Inhabits open woodlands and bushes near farmland.
Behavior: Solitary or in pairs. Hunts for insects.
Distribution: Vagrant to Sichuan (Ma'erkang).
Voice: Calls include a nasal scolding, not dissimilar to other shrikes.

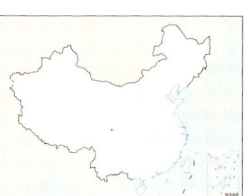

Lanius schach
棕背伯劳 zōng bèi bó láo
Long-tailed Shrike　LC

L: 20–25 cm
Habitat: Inhabits plains and hills. Adapted to varied habitats, including farmland, deserted fields, woodlands, and nurseries. Often near villages. At altitudes up to 1600 m.
Behavior: Often solitary.

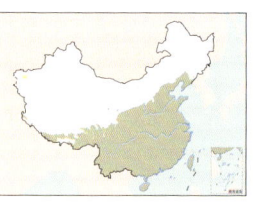

Territorial; chases away other intruding shrikes. Often perches high on tall branches and power lines in open areas, observing the surroundings and waiting to hunt; wags tail in a circular motion when perched. Carnivorous and fierce; with a wide variety of prey items, including fish.
Distribution: *L. s. schach* is common resident in Central, East, South, and southeastern China, spreading north in recent years; stable populations in northwestern and North China; dark morph is locally common from South China to southeast coast. *L. s. tricolor* is common resident in Yunnan and SE Tibet. *L. s. erythronotus* is restricted to SW Xinjiang (Kashi); uncommon resident or passage migrant; *L. s. nasutus* is vagrant to Taiwan (Orchid Island).
Voice: Variable, mostly hoarse, and piercing calls; also makes melodious and variable songs. Able to imitate songs of several passerines. Abundance of song types possibly used to attract females.

Lanius minor
黑额伯劳 hēi é bó láo
Lesser Gray Shrike　LC

L: 20–23 cm
Habitat: Inhabits river-valley forests, deserts, and farmland shelterbelts on open plains, and premontane hills with sparse trees.
Behavior: Solitary or in pairs. Nests in broadleaf forests or bushes. Prefers perching on trees, bushtops, or power lines.
Distribution: Uncommon breeder in N Xinjiang.
Voice: Simple and piercing calls, some reminiscent of Eurasian Magpie.

Isabelline Shrike

Sandy-brown head

Eye stripe paler at lores

Pale brown underparts

Relatively small white wing patch

Pale brown tail

♂

L. i. arenarius

Bay-backed Shrike

Pale gray head and nape

Dark rufous-chestnut back

White wing patch

Gray rump and uppertail coverts

White outer tail feathers

Burmese Shrike

Dark gray head and nape

Rufous-chestnut back and rump

Small white wing patch

♂

White outer tail feathers

Upperparts paler than on male

White-tinged forehead and lores

♀

Long-tailed Shrike

Black eye stripe reaches forehead

All-black face and throat

Consistent pale yellowish brown on back and flanks

Black wings

Dark morph

L. s. schach

Narrow black eye stripe does not reach forehead

Gray head often tinged pale brown

Pale brownish back contrasts with pale yellowish-brown flanks

Black from head to upper back

Gray back

L. s. erythronotus

L. s. nasutus

Degree of black varies; face usually black, different degrees of black on other parts of head

L. s. tricolor

Red-tailed Shrike

Rufous-brown head

Relatively broad white supercilium

Black eye stripe

♂

Prominent white wing patch

Brown head and upper back

Brown tail

♀

Inconspicuous wing patch

Small, fine, scaly, black pattern on underparts

Lesser Gray Shrike

Black forehead

Pale pinkish-brown underparts

Relatively long white wing patch

Relatively long primaries

303

Lanius tephronotus

灰背伯劳 huī bèi bó láo

Gray-backed Shrike

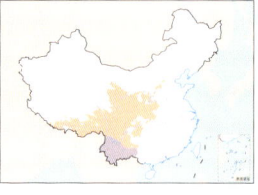

L: 21–23 cm

Habitat: Inhabits open farmland, fields, and woodlands. Prefers scrubby forest edges; often comes to habitats near villages. At altitudes up to 4500 m in the Himalayas; in winter at lower altitudes.

Behavior: Similar to Long-tailed Shrike.

Distribution: Locally common resident or migrant. *L. t. tephronotus* breeds in northwestern, Central, and southwestern China; migrates to southern regions or low altitudes in winter; vagrant east to Beijing and Hebei. *L. t. lahulensis* breeds in W Tibet; winters at low altitudes.

Voice: Similar to Long-tailed Shrike.

Lanius borealis

灰伯劳 huī bó láo

Northern Shrike

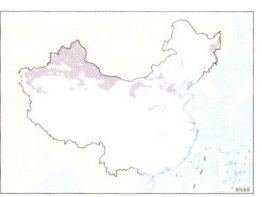

L: 22–26 cm

Habitat: *L. b. funereus* inhabits montane coniferous forests and bushes. *L. b. mollis* and *L. b. sibiricus* inhabit primarily open regions with sparse trees and bushes, and forest-edge bushes in winter.

Behavior: Often perches on protruding branches or power lines, flies down from perch to the ground to pounce on prey, then flies back to branches. Prey is carried to special impaling or wedging structures on trees and is pulled apart using bill.

Distribution: *L. b. mollis* winters in E Liaoning, N Hebei, C Inner Mongolia, NW Gansu, and NE Xinjiang; *L. b. sibiricus* winters in Heilongjiang, Jilin, Liaoning, Beijing, Tianjin, Hebei, Shanxi, and NE Inner Mongolia; *L. b. funereus* breeds in N Xinjiang.

Voice: Gives piercing and clear *schrreea* calls and elongated nasal *eeh*.

Lanius sphenocercus

楔尾伯劳 xiē wěi bó láo

Chinese Gray Shrike

L: 25–31 cm

Habitat: Often inhabits low mountains, hills, deserts, grasslands, and cultivated fields in open areas with sparse trees or scrub.

Behavior: Often solitary or in pairs. Usually perches on branches or bushtops, watching the surroundings; also hovers in the air and observes the situation on ground. Fierce; rapidly pounces on prey if located.

Distribution: Widespread but absent in Xinjiang. Uncommon breeder in northern regions; passage migrant and winter visitor to the regions south of North China. *L. s. sphenocercus* in Heilongjiang, Jilin, Inner Mongolia, Hebei, Beijing, Shandong, Zhejiang, Hubei, Anhui, Taiwan, Qinghai, and Gansu. *L. s. giganteus* in E Qinghai, E Tibet, and N and W Sichuan; sometimes treated as a separate species, Giant Shrike (*L. giganteus*).

Voice: Highly vocal. Calls include a disyllabic *tch-ick* or *kerr-ick*.

Lanius excubitor

西方灰伯劳 xī fāng huī bó láo

Great Gray Shrike

L: 21–25 cm

Habitat: *L. e. pallidirostris* inhabits saxaul forests, tamarisk bushes, desert poplar forests, and deserted fields in deserts. *L. e. homeyeri* inhabits open areas with sparse trees and bushes in winter.

Behavior: Often solitary or in pairs.

Distribution: *L. e. pallidirostris*, formerly treated as a separate species, "Steppe Gray Shrike" (草原灰伯劳), breeds in Xinjiang, NW Gansu, and N Ningxia. *L. e. homeyeri* winters in NW and S Xinjiang.

Voice: Variable, including hoarse gasping calls, *zhi zhi* notes, and imitations of other birds' vocalizations.

Gray-backed Shrike

atively pale gray
upperparts with
ght brown tinge

Relatively short bill

L. t. lahulensis

Relatively dark gray
upperparts

Brown hue to
wings and tail

Relatively thick bill

ad.

L. t. tephronotus

juv.

Fine scaly bars
on body

Chinese Gray Shrike

Gray rump

Relatively broad
white supercilium

Relatively
pale back

Larger white
wing patch than
Northern Shrike

Black eye stripe
broader behind
eye, relatively
narrow white
supercilium

L. s. sphenocercus

Relatively dark back

Relatively dark back

Longer tail than
Northern Shrike

L. s. giganteus

Northern Shrike

Small white patches at
base of primaries

Dark gray head and upper back

Dusky brown bars on
white underparts

L. b. sibiricus

Prominent white
wing patches

Dense scaly dusky-brown
bars on underparts

Inconspicuous
wing patches

L. b. mollis

Great Gray Shrike

Large broad white patch on
primaries and secondaries

Relatively gray
head and upper
back

Prominent white
supercilium

White fringe
on tertials

White underparts
lack bars

Two white
wing patches

L. e. homeyeri

Long white patch at
base of primaries

Pale gray head
and upper back

White
underparts
tinged
brownish red

Relatively long white
wing patch

Relatively long primaries

L. e. pallidirostris

莺雀科　Vireonidae

Small forest-dwelling songbirds. Plumage of most but not all species similar between sexes. Bill fine and pointed, or short and thick. Both wings and tail medium in length. Legs moderately strong. Usually make short flights among trees. Habitats range from forest canopy to understory bushes; some also found in mangroves. Nest on tree branches. Feed mostly on insects and other arthropods but also plant fruits. Some temperate-breeding species are migratory.

Six genera and 64 species recognized worldwide, mostly in the New World and Southeast Asia. Two genera and six species recorded in China, found in the montane forests in Central, South, and southwestern China.

绿凤鹛属
Erpornis

White-bellied Erpornis

鸥鹛属
Pteruthius

White-browed Shrike-Babbler

Erpornis zantholeuca

白腹凤鹛 bái fù fèng méi

White-bellied Erpornis　LC

L: 11–13 cm
Habitat: Inhabits broadleaf forests and secondary forests on mountains and hills. Usually seen at altitudes of 250–1600 m; recorded up to 3300 m in Taiwan, descending to lower altitudes in winter.

Behavior: Active and agile. Often gathers in small flocks and forages in the canopy, also in bushes. Often joins other passerines, such as fulvettas and babblers, in mixed-species flocks.
Distribution: Locally common or uncommon resident. *E. z. zantholeuca* in S and W Yunnan; *E. z. tyrannulus* in SE Yunnan and Hainan; *E. z. griseiloris* from South China to southeastern China, including Taiwan.
Voice: High-pitched, titlike, nasal chattering, trills, and scolding notes, *see see-see-se-drt.*

Pteruthius rufiventer

棕腹鸥鹛 zōng fù jú méi

Black-headed Shrike-Babbler　LC

L: 18–21 cm
Habitat: Occurs in primary evergreen broadleaf forests and secondary forests at middle altitudes.
Behavior: Gathers in small flocks in upper and middle stories of trees; mixes with other birds. Sometimes also seen in understory bushes.
Distribution: Uncommon resident in SE Tibet, W and NW Yunnan.
Voice: Song a far-carrying, ringing, trisyllabic *kip-chu-weeik* or *wi-tu-tuik*, with stress on final syllable, repeated every 3–4 sec. Calls include a scolding gruff chatter, *rrrrt-rrrrt*, and loud grating.

Pteruthius aeralatus

红翅鸥鹛 hóng chì jú méi

White-browed Shrike-Babbler　LC

L: 14–15 cm
Habitat: Inhabits broadleaf forests, coniferous forests, and rhododendron forests from 350 to 3050 m; migrates to lower altitudes in winter.
Behavior: Solitary, in pairs, or in small flocks, forages for insects in the canopy; often mixes with other birds.
Distribution: Locally common resident. *P. a. validirostris* in S and SE Tibet and NW Yunnan; *P. a. ricketti* in NE Yunnan, Sichuan, Chongqing, S Guizhou, Hunan, NE Jiangxi, Zhejiang, Fujian, Guangdong, and Hainan.
Voice: Song is a loud, hurried, strident series of rhythmic, percussive notes, *ip-chu chu chu-chu-chu*, or shorter *tiu-tiu … tiu-tiu*, repeated every 2 sec, often incessantly. Grating churrs when agitated.

Pteruthius xanthochlorus

淡绿鸥鹛 dàn lǜ jú méi

Green Shrike-Babbler　LC

L: 12–13 cm
Habitat: Inhabits evergreen broadleaf forests, subalpine mixed forests, and coniferous forests from 210 to 3600 m; migrates to lower altitudes in winter.
Behavior: Usually solitary or in pairs; often joins tits, leaf warblers, and other babblers in mixed-species flocks. Forages while moving slowly.
Distribution: Uncommon resident. *P. x. xanthochlorus* in SE Tibet; *P. x. pallidus* in Central, southwestern, South, and southeastern China.
Voice: Rapid, monotonous, even metronomic, repetition of clear, fairly high-pitched, one-, two-, and occasionally three-syllable notes, *tiu … tiu … tiu, whi-tu … whi-tu … whi-tu-tu*, every 2–8 sec, and a faster *tweedleeeddelee*. Calls include a nasal, wheezy *nyeep* and chattering.

Pteruthius melanotis

栗喉鵙鹛 lì hóu jú méi

Black-eared Shrike-Babbler

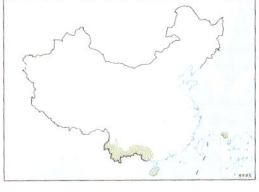

L: 10–12 cm

Habitat: Occurs in upper and middle stories of trees in montane evergreen broadleaf forests at middle and low altitudes.

Behavior: Solitary or gathers in small flocks in upper and middle stories of trees; mixes with other small birds. Relatively less active and easy to observe.

Distribution: Resident in SE Tibet, W Guangxi, and NW, W, S, and SE Yunnan.

Voice: Song a rapidly repeated series of 4–14 short *twi–twi–twi–twi–twi–twi–twi–twi* notes. Scolding, peevish *chwik-wik-wik.*

Pteruthius intermedius

栗额鹀鹛 lì é jú méi

Clicking Shrike-Babbler

L: 10–12 cm

Habitat: Occurs in montane evergreen broadleaf forests at middle and low altitudes.

Behavior: Often solitary or in pairs. Joins mixed-species flocks in lowland forests. Moves about slowly. Not shy.

Distribution: Resident in W, S, and SE Yunnan, Guangxi, and Hainan.

Voice: Song a rapid (two notes per second) repetition of short rising notes, *chwe.*

Black-eared Shrike-Babbler

Bright yellow forehead

Crescent-shaped black rear ear coverts

Chestnut brown on chin and throat extends to upper breast

Chestnut brown on side of throat

White-bellied Erpornis

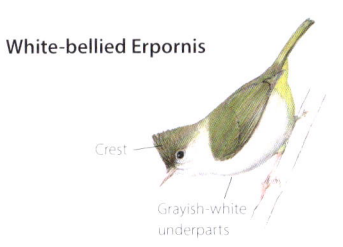

Crest

Grayish-white underparts

Black-headed Shrike-Babbler

Black crown and ear coverts

Gray from head to throat

Olive back

Short, thick, hooked bill, resembling shrike

Black hood

White to grayish-white throat

Rufous-chestnut back

Pale reddish brown from belly to undertail coverts

Clicking Shrike-Babbler

Chestnut-brown forehead

Lacks crescent-shaped black rear ear coverts

Chestnut lighter on forehead, lacks yellow bar of male

Grayish-green underparts

White-browed Shrike-Babbler

Rufous-chestnut tertials

White from throat to breast

Gray hood

Green tertials

P. a. validirostris

Green tail

P. a. ricketti

Gray from throat to breast

Gold or orange tertials

Green Shrike-Babbler

Lacks distinct white eye ring

Green wings

White eye ring

P. x. xanthochlorus

Gray outer primaries

P. x. pallidus

307

黄鹂科　Oriolidae

Medium-sized forest-dwelling songbirds. Plumage similar or slightly different between sexes, mostly yellow, black, olive, and gray. Bill thick and strong, resembling bills of thrushes. Good at flying, with long and pointed wings. Tail medium in length, usually rounded. Legs moderately strong. Inhabit primarily broadleaf forests, especially active in canopy. Nest among tree branches. Feed mostly on insects and other small invertebrates but also plant seeds and fruits. Some species are migratory.

Four genera and 38 species recognized worldwide, mostly in the Old World and Oceania. One genus and seven species recorded in China, ubiquitously distributed nationwide.

黄鹂属
Oriolus

Black-naped Oriole

Oriolus oriolus
金黄鹂 jīn huáng lí
Eurasian Golden Oriole

L: 22–25 cm

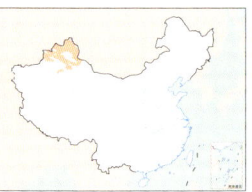

Habitat: Inhabits primarily broadleaf forests in river valleys; especially prefers natural poplar and birch forests. Also appears in trees in countryside and farmland.
Behavior: Arboreal. Loud and melodious calls. Primarily in the canopy of tall trees; seldom comes to the ground.
Distribution: Uncommon breeder in N Xinjiang.
Voice: Loud, fluty, whistled song, *oh wheela whee*, with strophes well separated. Nasal, almost Eurasian Jay–like, scolding *skaaaa*.

Oriolus kundoo
印度金黄鹂 yìn dù jīn huáng lí
Indian Golden Oriole

L: 22–25 cm

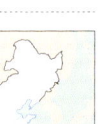

Habitat: Inhabits broadleaf forests, shelterbelts of farmland, orchards, and large trees or groves in urban parks on mountains and foothill plains.
Behavior: Arboreal. Melodious calls. Primarily in the canopy of tall trees; seldom comes to the ground.
Distribution: Uncommon to locally common breeder in S Xinjiang and SW Tibet; overlaps with Eurasian Golden Oriole on the northern slope of Tian Shan Mountains in N Xinjiang. Rare east to Qinghai.
Voice: Loud, fluty, whistled song similar to, but more complex than, that of Eurasian Golden Oriole.

Oriolus xanthornus
黑头黄鹂 hēi tóu huáng lí
Black-hooded Oriole

L: 23–25 cm

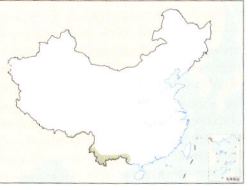

Habitat: Occurs in lowland evergreen broadleaf forests, semideciduous tropical monsoon forests, and forest edges; occasionally occurs in secondary forests, bamboo thickets, scrubland, artificial forests, and habitats around villages. Primarily at altitudes up to 1200 m.
Behavior: Usually solitary or in pairs in breeding season; occasionally in small flocks of several individuals in nonbreeding season. Sometimes also joins mixed-species flocks. Forages by hopping among leaves. Feeds on insects; also favors nectar.
Distribution: Local and uncommon in W Yunnan and SW Guangxi.
Voice: Loud, liquid, fluty, whistled song, -u-u-lu-kwe-ut or similar, with stress on last note and with introductory notes only audible at close range. Makes *yiyou* calls and harsh whining mewing, *cheeah* or nasal *kwaak*.

Eurasian Golden Oriole

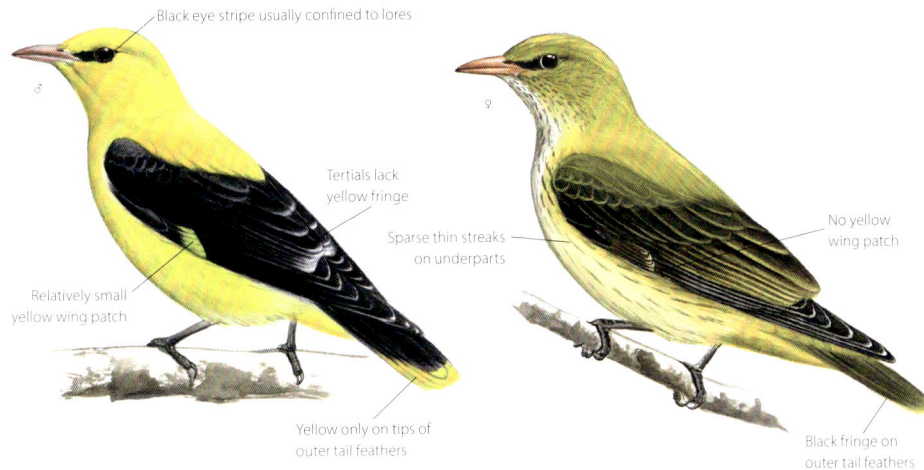

Black eye stripe usually confined to lores

Tertials lack yellow fringe

Sparse thin streaks on underparts

No yellow wing patch

Relatively small yellow wing patch

Yellow only on tips of outer tail feathers

Black fringe on outer tail feathers

♂

♀

Indian Golden Oriole

Black eye stripe extends behind eye

Yellow-tinged fringe on tertials

Yellow-tinged fringe on tertials

Small yellow wing patch

Relatively large yellow wing patch

Relatively thick dense streaks on underparts

Yellow fringe on outer tail feathers

Yellow fringe on outer tail feathers

♂

♀

Black-hooded Oriole

Yellow forehead

Black head

Black streaks on pale white throat

ad.

juv.

Oriolus chinensis

黑枕黄鹂 hēi zhěn huáng lí

Black-naped Oriole

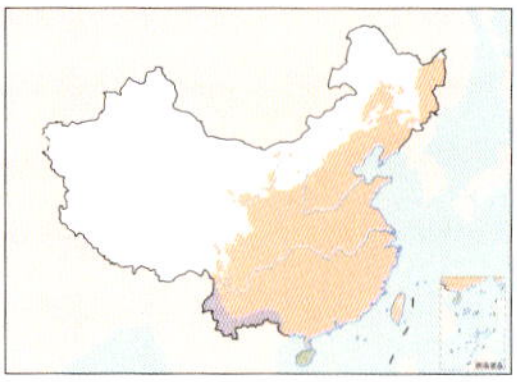

L: 23–28 cm
Habitat: Occurs in various open forests on low mountains, hills, and plains; also seen in tall trees near farmland and parks.
Behavior: Often solitary or in pairs. Flight undulating. Usually perches in canopy and does not come to the ground. Easy to find by calls.
Distribution: Widespread but absent in Tibet, Xinjiang, Qinghai, and W Gansu. Fewer records in Northeast China. More commonly seen in North China and regions to its south.
Voice: Various melodious whistled songs, usually with short strophes, *ou-li-liooou*. Makes catlike mewing calls.

Oriolus traillii

朱鹂 zhū lí

Maroon Oriole

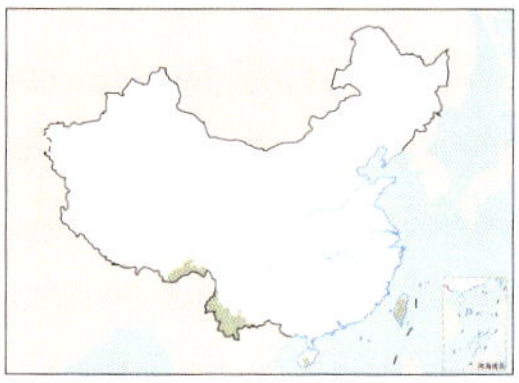

L: 23–28 cm
Habitat: Inhabits dense woodlands at low altitudes; especially prefers broadleaf forests and bamboo thickets.
Behavior: Usually solitary or in pairs in upper and middle stories of trees; often appears at forest edges.
Distribution: *O. t. traillii* in S Tibet, Yunnan, Guizhou; relatively common in suitable habitats. *O. t. nigellicauda* in Hainan, S Guangxi, SE Yunnan; very rare. *O. t. ardens* is endemic to Taiwan.
Voice: Song is mellow, melodic, slow, fluty whistling, incorporating longer whistles than Black-hooded Oriole. Gives *pii-ga-ga-gagagaga* and a mewing *kee-ah* call.

Oriolus tenuirostris

细嘴黄鹂 xì zuǐ huáng lí

Slender-billed Oriole

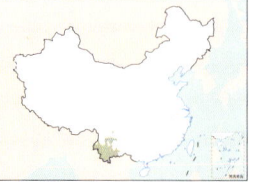

L: 22–26 cm
Habitat: Inhabits natural broadleaf forests and artificial coniferous forests on hills and plains at low altitudes; also seen in urban parks.
Behavior: Often solitary or in pairs in wooded areas with tall broadleaf trees. Melodious and distinct songs.
Distribution: Local resident in S and W Yunnan and S Sichuan.
Voice: Melodious and long songs, often sounding more hurried than those of Black-naped Oriole, with multiple syllables. Makes hoarse *zia*.

Oriolus mellianus

鹊鹂（鹊色鹂） què lí (què sè lí)

Silver Oriole

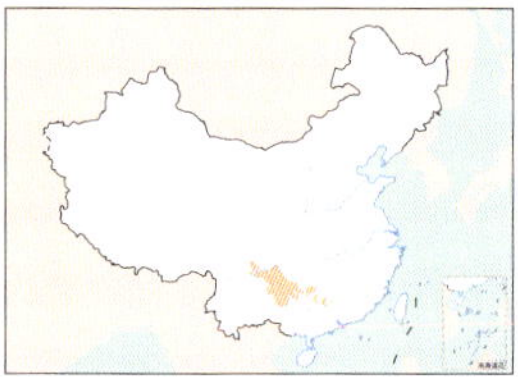

L: 24–28 cm
Habitat: Inhabits montane broadleaf forests at middle and low altitudes; usually seen in secondary forests and sparse woodlands.
Behavior: Usually in pairs in the canopy of trees; seldom joins mixed-species flocks.
Distribution: Rare breeder in southwestern China to western South China.
Voice: Poorly known. Song melodious and fluty. Call is catlike mewing.

Black-naped Oriole

Relatively thin bill

Relatively thick black eye stripe

♂

♀ Back tinged olive brown

juv.

Streaks on belly

Maroon Oriole

♂ Bright red back

♀

Streaked breast

O. t. ardens

♂ Dark brownish-red back

O. t. traillii

♂ Dark red back

O. t. nigellicauda

Slender-billed Oriole

Relatively thin bill

Relatively narrow black band on nape

♂

Yellow back tinged olive

♀ Darker olive back

juv.

Slender black bill

Black streaks on belly and relatively whitish breast

Silver Oriole

♂ Milky-white iris

Grayish-white body

Black wings

♀ Grayish-brown body

311

卷尾科   Dicruridae

Medium-sized forest-dwelling songbirds. Plumage similar between sexes, mostly gray and black, usually with iridescence. Bill thick and strong, resembling bills of magpies, with well-developed rictal bristles. Usually perch in upright positions with short legs. Tail long and characteristically forked, some with long ornamental feathers. Inhabit primarily montane forests or scrubland. Often territorial. Insectivorous, catching insects in flight or on ground. Most species are residents, while some more northerly breeding species are migratory.

One genus and 29 species recognized worldwide, mostly in Indomalaya and Afrotropics. One genus and seven species recorded in China, found mostly in southern regions and some extending to North and Northeast China.

卷尾属
Dicrurus

Black Drongo

Dicrurus leucophaeus
灰卷尾 huī juǎn wěi
Ashy Drongo **LC**

L: 26–28 cm

Habitat: Inhabits various broadleaf forests at middle and low altitudes; recorded up to 4000 m in Yunnan. Moves about at forest edges or in clearings in woodlands.

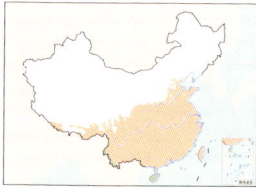

Behavior: Often perches on conspicuous or exposed branches; hawks insects from perch or dives rapidly to capture prey. Flight acrobatic, sometimes undulating when flying a long distance.

Distribution: *D. l. leucogenis* restricted to south of the Qinling-Huaihe Line, west to S Gansu, Sichuan, and E Yunnan; *D. l. salangensis* in South and southwestern China; *D. l. hopwoodi* in southwestern China and western South China, common in suitable habitats. Vagrant north to Shandong, Hebei, and Beijing.

Voice: Songs flutier and less jarring than Black Drongo, more leafbird-like, but still include rapid, jumbled, chattering phrases and more often mimicry. Calls include raptorlike whistles and *wee-peepee*.

Dicrurus macrocercus
黑卷尾 hēi juǎn wěi
Black Drongo **LC**

L: 24–30 cm

Habitat: Inhabits open farmland and forest edges on plains and low altitudes.

Behavior: Often perches on branches or power lines. Hawks insects from perch; also glides in the air. Fierce; attacks raptors.

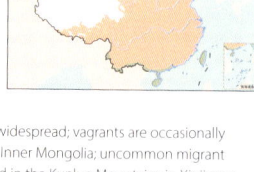

Distribution: *D. m. cathoecus* widespread; vagrants are occasionally seen in Heilongjiang, Jilin, and Inner Mongolia; uncommon migrant to Qinghai, and only one record in the Kunlun Mountains in Xinjiang; common in other regions. *D. m. albirictus* in SE Tibet. *D. m. harterti* in Taiwan.

Voice: Often sings early in morning. Variable but often repetitive, raspy squawks, chatters, and whistles, occasionally including mimicry. Harsh *zhaazhaa* calls.

Dicrurus annectens
鸦嘴卷尾 yā zuǐ juǎn wěi
Crow-billed Drongo **LC**

L: 27–32 cm

Habitat: In breeding season, occurs in mature evergreen broadleaf forests up to 1200 m; winters in broadleaf forests at middle and low altitudes, usually below 800 m.

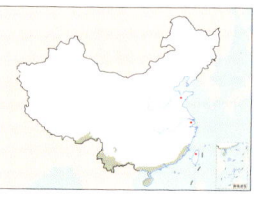

Behavior: Usually solitary or in pairs; also gathers in small flocks in winter. Hawks insects from perch.

Distribution: Status uncertain; uncommon resident or migrant. Distributed in E Tibet, S and W Yunnan, Guangxi, Guangdong, Fujian, Hong Kong, Macao, and Hainan; vagrant to Shandong (Weifang), Shanghai (Nanhui Dongtan), and Taiwan, possibly wandering juveniles.

Voice: Typical *Dicrurus*, with loud musical or discordant whistles combined with metallic trills; noisy in flocks.

Black Drongo

Deep crimson iris

All-black body

D. m. cathoecus

Prominent white spot at gape

D. m. albirictus

Ashy Drongo

Distinct white patch around eye with sharply defined boundary

Relatively pale body

D. l. leucogenis

Indistinct yellow-tinged white patch around eye with diffuse boundary

Relatively dark body

D. l. salangensis

Slightly dark patch around eye

Deep crimson iris

Dark gray, nearly black, body

Underparts paler than back

D. l. hopwoodi

Crow-billed Drongo

Thick base of bill

Glossy blackish-blue wings

juv.

ad.

White spots from belly to undertail coverts

Relatively shallowly forked tail compared with Black Drongo

Dicrurus aeneus

古铜色卷尾 gǔ tóng sè juǎn wěi

Bronzed Drongo　**LC**

L: 22–25 cm
Habitat: Inhabits broadleaf forests and secondary forests on mountains and hills at middle and low altitudes, up to 2000 m. In Taiwan, *D. a. braunianus* prefers mature forests.

Behavior: Often solitary or in pairs. Perches on open branches at forest edges and hawks insects from perch. Also joins other passerines (in Taiwan, often Gray-chinned Minivet) in mixed-species flocks and forages in the canopy. May initiate attacks on raptors and crows.
Distribution: Locally common resident. *D. a. aeneus* in SE Tibet, Yunnan, SW to C Guangxi, S Guizhou, N Guangdong, Macao, and Hainan; *D. a. braunianus* in Taiwan.
Voice: Noisy. Song loud accelerating combination of musical whistles, harsh notes, and rapid chatters. Calls include penetrating *nyip* and scolding shrike-like chatter.

Dicrurus remifer

小盘尾 xiǎo pán wěi

Lesser Racket-tailed Drongo　**LC**

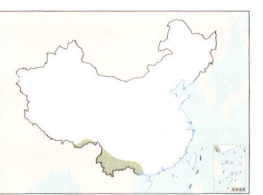

L: 25–27 cm, excluding outermost tail feathers (45–58 cm, including outermost tail feathers)
Habitat: Inhabits primarily broadleaf forest edges, sparse secondary forests, and bamboo thickets at middle and low altitudes.
Behavior: Occurs in open areas at forest edges. Hawks passing insects.
Distribution: Locally common in SE Tibet, W to S Yunnan, and SW Guangxi.
Voice: Highly varied and complex repertoire that includes metallic sounds, whistles, and complex trills, some high pitched, others low and resonant. Strophes generally short, with variable rhythm that at times recalls a thrush.

Dicrurus hottentottus

发冠卷尾 fà guān juǎn wěi

Hair-crested Drongo　**LC**

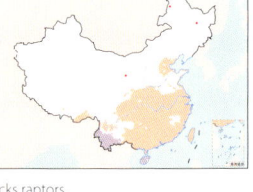

L: 29–34 cm
Habitat: Inhabits broadleaf forests on mountains and hills at middle and low altitudes.
Behavior: Inhabits sparse woodlands. Hawks passing insects. Sometimes gathers in small flocks and calls. Fierce; attacks raptors.
Distribution: *D. h. brevirostris* breeds in North, Central, South, and southwestern China; wintering in SW and S Yunnan and Hainan; vagrants to Heilongjiang, Ningxia, and Qinghai. *D. h. hottentottus* in W Yunnan. Not rare in suitable habitats.
Voice: Variable songs usually start with loud, incisive, short *tsip* notes and become melodious bell-like or jangling notes at different pitches and tempos, often repetitive. Calls include various whistles and harsh discordant notes.

Dicrurus paradiseus

大盘尾 dà pán wěi

Greater Racket-tailed Drongo　**LC**

L: About 35 cm, excluding outermost tail feathers (length of outermost tail feathers is about 50 cm on *D. p. grandis* and about 41.5 cm on *D. p. johni*)
Habitat: Occurs in moist evergreen broadleaf forests up to 1400 m; migrates to lower altitudes in winter.
Behavior: Often in pairs; sometimes displays in flocks. Hawks insects from perch. Also joins other passerines in mixed-species flocks and forages in the middle story and understory of trees. Forages for insects and small invertebrates.
Distribution: Locally common or uncommon resident. *D. p. grandis* in SE Tibet, S and W Yunnan, and SW Guangxi; *D. p. johni* in Hainan.
Voice: Noisy, talented vocalist. Songs include various whistles, bell-like ringing, and hoarse trills, usually repeated several times. Skilled mimic.

Bronzed Drongo

Short blunt bill

Plumage strongly glossed metallic blackish blue

Forked tail does not curl upward

Hair-crested Drongo

Filoplumes rise from forehead

Iridescent bluish-green sheen on wing coverts

Tail curls upward and inward

Lesser Racket-tailed Drongo

Lacks crest

Bill mostly covered by feathers, appears short

Square tail

Outermost pair of tail feathers extremely elongated, width of outer and inner vanes almost equal

Greater Racket-tailed Drongo

Crest on crown

Forked tail

Greatly reduced inner vane of outermost tail feather tip, outer vane broad and twisted

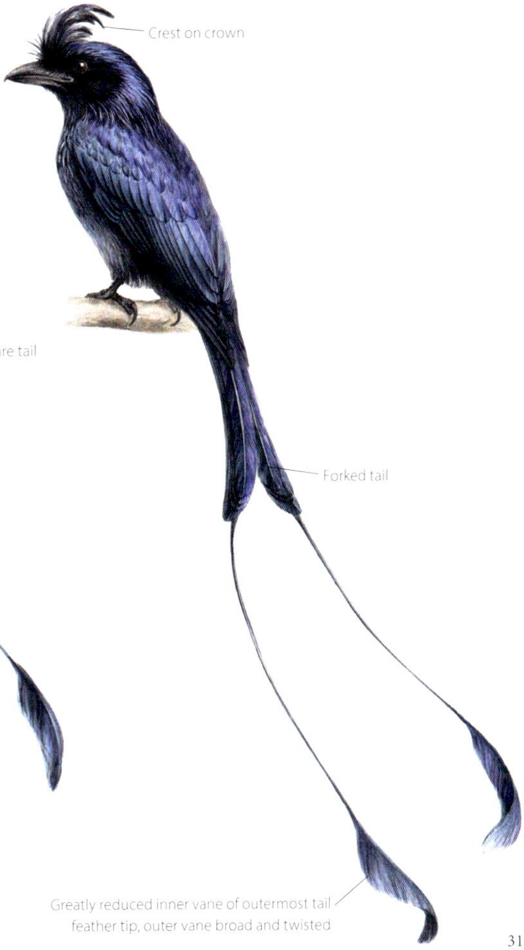

扇尾鹟科 Rhipiduridae

Small forest-dwelling songbirds. Body shape resembles Old World flycatchers, but with longer tail and often perching in horizontal position. Plumage similar between sexes, mostly white, blue, rufous, and dark gray. Inhabit primarily forests and scrubland on plains and at lower altitudes. Insectivorous; active. Common name in Chinese refers to their regularly cocked and fanned tail. All species are resident.

Three genera and 52 species recognized worldwide, distributed in Oceania and South and Southeast Asia. One genus and three species recorded in China; found in Taiwan and southwestern regions.

White-throated Fantail

Thin white supercilium

White on throat extends to side of neck

Slaty black overall

White-browed Fantail

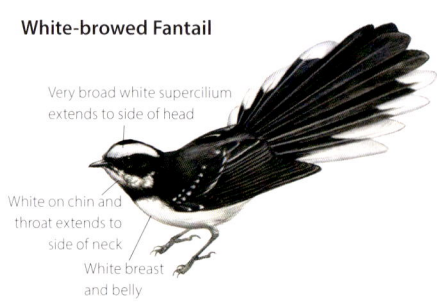

Very broad white supercilium extends to side of head

White on chin and throat extends to side of neck

White breast and belly

Philippine Pied-Fantail

Long broad protruding tail, four pairs of outer feathers broadly tipped white

Relatively broad white supercilium

Black from crown to forehead

Broad black eye stripe

White underparts with black band between throat and breast

Rhipidura albicollis

白喉扇尾鹟 bái hóu shàn wěi wēng

White-throated Fantail

L: 17–20 cm

Habitat: Occurs in broadleaf forests, mixed forests, secondary forests, and bamboo thickets at middle and high altitudes; also seen in artificial woodlands, scrubland, and farmland.

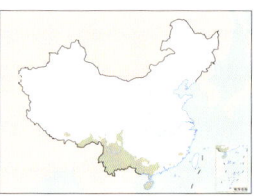

Behavior: Often solitary or in pairs at forest edges and in bushes. Active; often joins mixed-species flocks.

Distribution: Resident in southwestern China, Guangdong, Guangxi, and Hainan.

Voice: Halting, hesitant song of four to nine short notes, first notes rising, later ones falling, *zu–zu–zu–ze–ze–da–di–da*. Call short penetrating *itch* or *kitch*, like paradise-flycatchers.

Rhipidura aureola

白眉扇尾鹟 bái méi shàn wěi wēng

White-browed Fantail   LC

L: 17 cm

Habitat: Occurs in woodlands, scrubland, farmland, and orchards on low mountains and plains at middle and low altitudes.

Behavior: Usually solitary or in pairs at forest edges and in open areas. Active. Not shy.

Distribution: Specimens collected from W Yunnan (Yingjiang and Lushui); no other records.

Voice: Song similar to that of White-throated Fantail—halting and hesitant, with four to nine short notes, first notes rising, later ones falling, but lower pitched. Various short metallic calls, *chip* or *trrk*, very different from White-throated Fantail.

Rhipidura nigritorquis

菲律宾斑扇尾鹟 fēi lù bīn bān shàn wěi wēng

Philippine Pied-Fantail

LC

L: 18–19 cm

Habitat: Inhabits green spaces, parks, bamboo thickets, mangroves, etc.

Behavior: Active. Solitary or in pairs. Prefers to perch with tail cocked and fanned.

Distribution: Vagrant to Taiwan.

Voice: Song reminiscent of White-throated Fantail but far less stereotyped; incorporates harsher clicking notes and is, in some way, more bulbul-like. Calls more similar to White-throated Fantail but coarser, longer, and without the rising tone.

王鶲科　Monarchidae

Small to medium-sized forest-dwelling songbirds with a well-proportioned profile. Plumage similar or slightly different between sexes, but note that males of some species have two or more color morphs. Bill in most species flat and wide, resembling bills of Old World flycatchers. Good fliers with wide wings. Tail medium in length; most with a flat end, but central tail feathers extremely elongated in some species. Inhabit primarily the middle story and understory in broadleaf forests. Some also occur in grasslands and mangroves, and a handful of species are ground dwellers. Nest on tree branches. Feed mostly on insects and insect larvae. Some species are migratory.

Sixteen genera and 102 species recognized worldwide, including monarch flycatchers, paradise-flycatchers, and magpie-larks, distributed in Africa, Indomalaya, Oceania, and Pacific islands. Two genera and five species recorded in China; mostly found in eastern and southern China.

Hypothymis azurea
黑枕王鶲 hēi zhěn wáng wēng
Black-naped Monarch　(LC)

L: 14–16 cm
Habitat: Inhabits various types of forests at low altitudes; also on farmland or in forests, scrubland, and bamboo thickets near villages.
Behavior: Solitary or in pairs, flies back and forth in trees, hunting for insects; also joins mixed-species flocks.
Distribution: *H. a. oberholseri* is endemic to Taiwan. *H. a. styani* in southwestern and South China (including Hainan); vagrant to Shanghai (Nanhui Dongtan and Minhang) and Zhejiang (Wenzhou). Locally common.
Voice: Song a fast three-note whistle, *wee–wee–wee*. Calls include strained nasal *chweek* or *chweek-werk*.

寿带属
Terpsiphone

Amur Paradise-Flycatcher

王鶲属
Hypothymis

Black-naped Monarch

Black-naped Monarch

Black patch on nape

Azure-blue upperparts, throat, and breast

Black breast band

Lacks black patch on nape

Dull blue upperparts

Lacks black breast band

Terpsiphone paradisi
印度寿带 <small>(印缅寿带)</small>
yìn dù shòu dài <small>(yìn miǎn shòu dài)</small>

Indian Paradise-Flycatcher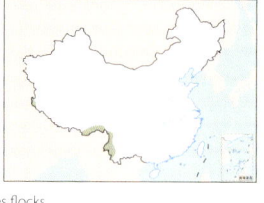

L: 20 cm (excluding male's
central tail feathers of up
to 30 cm)
Habitat: Inhabits
broadleaf forests at middle
altitudes.
Behavior: Occurs in
upper and middle stories
of trees. Hawks passing
insects. Also joins mixed-species flocks.
Distribution: *T. p. leucogaster* in SW Tibet, rare. *T. p. saturatior* in
W Yunnan, locally common; also recorded in S Tibet; this subspecies is
sometimes treated as a subspecies of Blyth's Paradise-Flycatcher.
Voice: Similar to Amur Paradise-Flycatcher.

Terpsiphone incei
寿带 shòu dài

Amur Paradise-Flycatcher

L: 20–22 cm (♂; excluding
central tail feathers of up
to 30 cm); 17.5–21 cm (♀)
Habitat: Inhabits lowland
evergreen broadleaf
forests up to 1200 m.
Nests in wooded areas
near water.
Behavior: Solitary or in
pairs. Usually forages in upper and middle stories of trees; hawks insects
from perch. Sometimes also forages in the understory.
Distribution: Widespread in China but absent in Inner Mongolia,
Qinghai, Xinjiang, and Tibet; common breeder and passage migrant.
Voice: Song a fast (four or five notes per second) monotonous repetition
of usually three (rarely just one) rising notes, *weep ... weep* or *tu-we
weep*. Calls are rasping and husky *jouey*.

Terpsiphone affinis
东方寿带 <small>(中南寿带)</small>
dōng fāng shòu dài <small>(zhōng nán shòu dài)</small>

Blyth's Paradise-Flycatcher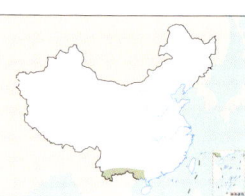

L: 19–23 cm (excluding
male's central tail feathers
of up to 27 cm)
Habitat: Inhabits
broadleaf forests and
bamboo thickets at
middle and low altitudes.
Behavior: Similar to
Amur Paradise-Flycatcher.
Perches in tall trees or flies through trees hunting for insects. Also joins
mixed-species flocks.
Distribution: Locally common in S Yunnan to S Guangxi.
Voice: Similar to Amur Paradise-Flycatcher.

Terpsiphone atrocaudata
紫寿带 zǐ shòu dài

Japanese Paradise-Flycatcher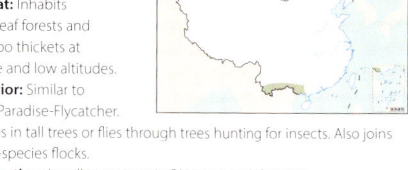

L: 18.5–20 cm
(♂; excluding central tail
feathers of up to 25 cm);
17–18 cm (♀)
Habitat: *T. a. atrocaudata*
usually occurs in coastal
shelterbelts or nurseries
during migration.
T. a. periophthalmica
prefers mature broadleaf forests on Orchid Island in Taiwan.
Behavior: Similar to Amur Paradise-Flycatcher.
Distribution: *T. a. atrocaudata* migrates through coastal regions in South
and East China, including Taiwan. *T. a. periophthalmica* breeds on Orchid
Island in Taiwan; some individuals are resident.
Voice: Mellow, melodious, multisyllabic, whistling song alternates high-
and low-pitched, flat notes, *wu-ti-ti-wu-ti-twit*, each strophe lasting one
second, very different from songs of other regional paradise-flycatchers.
Harsh *jouey* call very similar to Amur and other paradise-flycatchers but
slightly shorter, huskier, and flatter.

Indian Paradise-Flycatcher

Relatively short flat crest

Relatively dense black shaft streaks on upperparts

Bluish-gray throat

White morph

T. p. saturatior

Long crest

Relatively light-colored

Grayish-white throat, white breast and belly

T. p. leucogaster

Amur Paradise-Flycatcher

All white below head

White morph

Blue eye ring

Rufous morph

Clear boundary between colors of throat and breast

Diffuse boundary between colors of breast and belly

Blyth's Paradise-Flycatcher

White morph male very rare

Dark chestnut upperparts

Bluish-gray throat

Whitish belly

Japanese Paradise-Flycatcher

Black from head to breast

Purple back

Dark bluish-black tail

T. a. atrocaudata

Blackish back

Black lower breast and flanks

T. a. periophthalmica

Diffuse boundary between colors of throat and breast

Relatively distinct boundary between colors of breast and belly

鸦科　Corvidae

Medium-sized to large forest-dwelling or ground-dwelling songbirds. Plumage similar or slightly different between sexes, mostly black, white, green, blue, and rufous, usually with structural colors. Bill in most species strong and cone shaped, slightly hooked at tip. Wings short and rounded. Legs strong. Tail moderately long and squared, rounded, or wedge shaped; tail feathers in some species especially elongated. Inhabit a variety of habitats, including open plains, dry scrubland, alpine meadows, various types of forests, and urban areas. Usually gather in flocks. Nest in trees, crevices, tree holes, or artificial cavities. Omnivorous; feed mostly on insects and other arthropods, as well as small amphibians, reptiles, birds, and mammals. Also take plant seeds and fruits. Most species are resident, while some are migratory.

Twenty-five genera and 134 species recognized worldwide, with a cosmopolitan distribution. Thirteen genera and 31 species recorded in China, ubiquitously distributed nationwide.

噪鸦属
Perisoreus

Siberian Jay

松鸦属
Garrulus

Eurasian Jay

灰喜鹊属
Cyanopica

Azure-winged Magpie

蓝鹊属
Urocissa

Red-billed Blue-Magpie

绿鹊属
Cissa

Common Green-Magpie

树鹊属
Dendrocitta

Gray Treepie

塔尾树鹊属
Temnurus

Ratchet-tailed Treepie

喜鹊属
Pica

Oriental Magpie

地鸦属
Podoces

Xinjiang
Ground-Jay

星鸦属
Nucifraga

Eurasian Nutcracker

山鸦属
Pyrrhocorax

Red-billed Chough

鸦属
Corvus

Carrion Crow

寒鸦属
Coloeus
(merged with *Corvus* in
eBird/Clements Checklist)

Daurian Jackdaw

Perisoreus infaustus

北噪鸦 běi zào yā

Siberian Jay

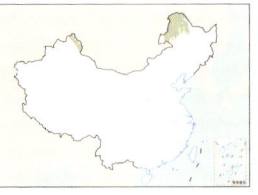

L: 26–31 cm
Habitat: Typical bird of taiga forests. Often inhabits coniferous and mixed broadleaf-coniferous forests; especially prefers forests of spruce and fir.
Behavior: Often in pairs or small flocks. Active. Noisy when calling. Shy; hides quietly in trees observing the surroundings when people are present. Flight slow and silent; tail often fanned in flight.
Distribution: Restricted to N Xinjiang and northern Northeast China; uncommon resident. *P. i. maritimus* in NE Heilongjiang and NE Inner Mongolia; *P. i. opicus* in N Xinjiang.
Voice: Usually quiet. Occasional subdued twittering and chattering; rising buzzardlike mew, *mio*, and short scolding *kiau*.

Perisoreus internigrans

黑头噪鸦 hēi tóu zào yā

Sichuan Jay

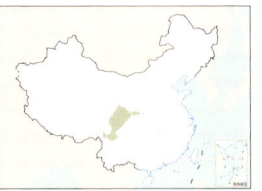

L: 29–32 cm
Habitat: Occurs in coniferous forests at high altitudes.
Behavior: Mostly in family units in alpine and subalpine coniferous forests; social animal that forages and lives together.
Distribution: Chinese endemic, restricted to the eastern edge of the Qinghai-Tibet Plateau.
Voice: Buzzardlike plaintive mewing; short *kip-kip-kip*. Much of its vocabulary mimicked by Giant Laughingthrush.

Garrulus glandarius

松鸦 sōng yā

Eurasian Jay

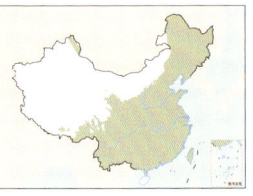

L: 30–36 cm
Habitat: Prefers montane forests and plains near mountains; also occurs in urban parks.
Behavior: Usually inhabits tree canopies. Solitary or in pairs in breeding season, gathers in small flocks in nonbreeding season. Flight variable; undulating when flying a short distance, in queue when flying over open areas. Forages mostly on ground; caches food.
Distribution: Multiple subspecies. Widespread across China but absent from the Qinghai-Tibet Plateau, S Xinjiang, W Inner Mongolia, and Hainan; not rare in suitable habitats. Sometimes treated as three species: *G. leucotis*, including *G. g. leucotis* in S and W Yunnan; *G. bispecularis*, including *G. g. interstinctus* in S Tibet (Zayu, Bome, Nyingchi, Chamdo), *G. g. sinensis* in eastern and central China, and *G. g. taivanus* in Taiwan; and *G. glandarius*, including *G. g. brandtii* in Northeast China and N Xinjiang, *G. g. kansuensis* in Gansu and Qinghai, and *G. g. pekingensis* in North China to E Gansu.
Voice: Varied; includes a buzzardlike plaintive mewing, Northern Goshawk–like cackling, and grating *skkaip*. Vocal differences between groups insufficiently studied.

Cyanopica cyanus

灰喜鹊 huī xǐ què

Azure-winged Magpie

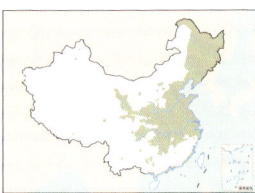

L: 31–40 cm
Habitat: Occurs in varied habitats on plains; also common in residential areas. Seen in the canopy of tall trees, scrubland, and on ground. Uncommon in mountainous regions.
Behavior: Gathers in small to large flocks. Noisy; usually maintains contact with calls. Not shy, but when people get too close, makes alarm calls, and the entire flock flies away. Caches food. Chases raptors less frequently than Oriental Magpie.
Distribution: Widespread but absent from Tibet. *C. c. cyanus* in northern Northeast China; *C. c. stegmanni* in Northeast China; *C. c. pallescens* in N Heilongjiang; *C. c. kansuensis* in Gansu and N Qinghai; *C. c. interposita* in North China and eastern northwestern China; *C. c. swinhoei* in Central, East, and South China. All the populations mentioned above are sometimes treated as *C. c. cyanus*. Very common in northeastern, northwestern, and North China; common or occasional in the remaining regions. Range is possibly expanding, occasionally aided by local introductions as a pest-control measure.
Voice: Varied; includes an up-slurred, nasal scolding, *vrrruuee*, repeated often; a lengthier, faint, almost hissing version of the same; and, when agitated, a more rattled *k-r-r-r-r-r*.

Siberian Jay

Gray back

Distinct orange patch
on upperwing coverts

P. i. maritimus

Brownish back

P. i. opicus

Sichuan Jay

Black head contrasts
strongly with smoky-
gray body

Yellow to horn bill

Smoky-black to
smoky-gray plumage

Azure-winged Magpie

White tip on
central tail feathers

Rufous-brown
forehead and crown

Black around nostrils
inconspicuous

Lacks white wing patch

G. g. sinensis

Eurasian Jay

Relatively thick black
streaks on crown

Relatively large white
wing patch

White rump

G. g. brandtii

Thin black streaks on crown

Small white
wing patch

G. g. pekingensis

Distinct black
around nostrils

G. g. taivanus

Black nape

White forehead

White cheek and throat

G. g. leucotis

Urocissa caerulea
台湾蓝鹊 tái wān lán què
Taiwan Blue-Magpie

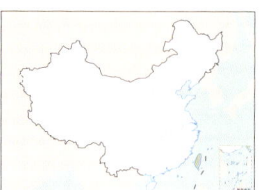

L: 63–68 cm
Habitat: Inhabits deciduous forests from 300 to 1200 m.
Behavior: Gathers in small flocks of several to a dozen individuals. Fierce. Usually flies in a single file.
Distribution: Chinese endemic; locally common resident in Taiwan .
Voice: Typical corvid *gaga*; sometimes also gives *didi*.

Urocissa flavirostris
黄嘴蓝鹊 huáng zuǐ lán què
Yellow-billed Blue-Magpie

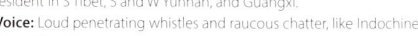

L: 45–69 cm
Habitat: Inhabits broadleaf and mixed forests at middle and high altitudes; seldom seen at forest edges and open areas in coniferous forests.
Behavior: Usually occurs in pairs and small flocks in the middle story and understory of trees. Noisy and bold.
Distribution: S Tibet and W Yunnan.
Voice: Varied vocabulary includes several notes reminiscent of Red-billed Blue-Magpie, such as a staccato, scolding chatter, *rak-rak-rak-rak*, that is slightly higher pitched and slower, and distinctive loud screams, such as *kik plear*.

Urocissa erythroryncha
红嘴蓝鹊 hóng zuǐ lán què
Red-billed Blue-Magpie

L: 42–60 cm
Habitat: Inhabits broadleaf and mixed broadleaf-coniferous forests on low mountains and plains.
Behavior: Moves about in small flocks of several to a dozen individuals. Forages primarily in trees, sometimes on ground. Often glides quietly. Fierce; seldom mixes with other birds.
Distribution: *U. e. brevivexilla* in northern China north to Liaoning and west to Ningxia. *U. e. erythroryncha* in eastern and southern China west to W Sichuan and south to Hainan; populations in Taiwan are introduced. *U. e. alticola* in N Yunnan. Common in suitable habitats.
Voice: Noisy, varied, and complex. Includes simple, grating, scolding notes, as well as more melodious and variable songs (with some mimicry).

Urocissa whiteheadi
白翅蓝鹊 bái chì lán què
White-winged Magpie

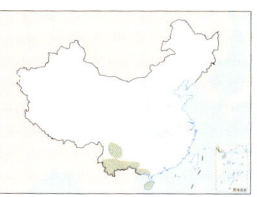

L: 45–47 cm
Habitat: Inhabits tropical and subtropical evergreen forests in lowlands; prefers areas near streams.
Behavior: Moves about in flocks, with flock size up to more than 20 individuals. Active in forests.
Distribution: *U. w. whiteheadi* rare endemic in montane rain forests in Hainan. *U. w. xanthomelana* in S Yunnan, SW Guangxi, and SW Sichuan; rare. The two subspecies are sometimes treated as two separate species.
Voice: Various calls. Most common is perhaps a raucous *zhu-reek* or *reek-reek-reek*. Other calls include more nasal notes, slightly reminiscent of Azure-winged Magpie, and some similar to Red-billed Blue-Magpie.

Cissa chinensis
蓝绿鹊 lán lǜ què
Common Green-Magpie

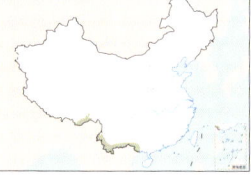

L: 36–38 cm
Habitat: Inhabits broadleaf forests, bamboo thickets, and forest-edge bushes at low altitudes.
Behavior: Solitary or in pairs. Hides in trees, but calls noisily.
Distribution: Uncommon resident in S Tibet, S and W Yunnan, and Guangxi.
Voice: Loud penetrating whistles and raucous chatter, like Indochinese Green-Magpie.

Cissa hypoleuca
黄胸绿鹊（印支绿鹊）
huáng xiōng lǜ què (yin zhi lǜ què)
Indochinese Green-Magpie

L: 31–34 cm
Habitat: Inhabits broadleaf forests and bamboo thickets at middle and low altitudes.
Behavior: Solitary, in pairs, or in small flocks in the middle story and understory of trees, with well-camouflaged feathers and loud calls.
Distribution: *C. h. jini* in SE Sichuan and Guangxi, uncommon; *C. h. katsumatae* in Hainan, rare.
Voice: Loud penetrating whistles and raucous chatter, very similar to calls of Common Green-Magpie and equally variable.

Taiwan Blue-Magpie

Black nape
Yellow iris
Black breast

White-winged Magpie

U. w. xanthomelana

White tip on
black tail

White tip on gray tail with
relatively narrow black
subterminal spots

U. w. whiteheadi

**Yellow-billed
Blue-Magpie**

White on nape does
not extend to crown
Yellow bill
Yellow legs

Common Green-Magpie

White tips and black subterminal
spots on secondaries

Relatively long tail

**Red-billed
Blue-Magpie**

White on nape
extends to crown
Red bill
ad.
Red legs

Orange-red bill
juv.

Indochinese Green-Magpie

Pale green secondaries lack white
tip and black subterminal spots

Relatively short tail

Dendrocitta vagabunda

棕腹树鹊 zōng fù shù què

Rufous Treepie

L: 35–45 cm
Habitat: Inhabits broadleaf forests, secondary forests, and artificial forests at middle and low altitudes.
Behavior: Solitary, in pairs, or in small flocks. Flight undulating.
Distribution: Rare in W Yunnan.
Voice: Noisy raucous chatter higher pitched and faster than that of Gray Treepie (but much slower than Oriental Magpie), *clee-clee-clee-cle*, more intense when agitated; nasal *klu-ee* and multisyllabic *kle-ar-rit*.

Dendrocitta formosae

灰树鹊 huī shù què

Gray Treepie

L: 36–40 cm
Habitat: Inhabits montane broadleaf forests, secondary forests, and artificial forests at middle and low altitudes.
Behavior: Gathers in small flocks, hops in trees, or glides downward from one tree and flies up to another.
Distribution: *D. f. himalayana* in W Yunnan; *D. f. sinica* in southwestern China to East and South China; *D. f. sapiens* in Sichuan; *D. f. formosae* in Taiwan; *D. f. insulae* in Hainan. Common.
Voice: Harsh chatter, *clee-clee-clee-clee*, slower, lower pitched, and harsher than Rufous Treepie. Ringing "song" a multisyllabic *cog-u-ba-lik-zree-tut*, more staccato than similar-pitched song of Rufous.

Dendrocitta frontalis

黑额树鹊 hēi é shù què

Collared Treepie

L: 35–39 cm
Habitat: Inhabits various types of wooded areas at middle and low altitudes; especially prefers broadleaf forests and bamboo thickets.
Behavior: In pairs or in small flocks. Glides between trees.
Distribution: Uncommon resident in SE Tibet and W Yunnan.
Voice: Varied and complex. Chatter shorter, much higher pitched, faster, and far less grating than other treepies. Resonant *u-tik* or *tuk-up*; ringing bell-like *kraink* reminiscent of Greater Racket-tailed Drongo; and remarkable, musical, mellow ratchet calls, *pli-ur-r-r-r-r-r-r-r*.

Pica bottanensis

青藏喜鹊 qīng zàng xǐ què

Black-rumped Magpie

L: 45–54 cm
Habitat: Occurs near forests and scrubland up to 5500 m.
Behavior: Often solitary or in family units; sometimes gathers in small flocks containing at most seven or eight individuals in winter.
Distribution: Monotypic. Common in E Tibet, NW Yunnan, Qinghai, W Sichuan, and SW Gansu.
Voice: Very similar to Oriental Magpie. Chatter call is faster than Oriental Magpie, with simpler notes and fewer harmonics, but slower than Eurasian Magpie.

Pica pica

欧亚喜鹊 ōu yà xǐ què

Eurasian Magpie

L: 44–50 cm
Habitat: Adapted to various habitats; seen from villages to city centers but seldom occurs in dense forests.
Behavior: Similar to Oriental Magpie. Often gathers in flocks of 5–20 individuals; flock size larger in winter.
Distribution: *P. p. bactriana* in Xinjiang and NW Tibet; *P. p. leucoptera* in NE Inner Mongolia. Locally common.
Voice: Very similar to Oriental Magpie. Common chatter call is faster, with more notes per second.

Pica serica

喜鹊 xǐ què

Oriental Magpie

L: 40–50 cm
Habitat: Adapted to various habitats, especially human habitations, but seldom occurs in dense forests or open areas without trees nearby. Usually builds nest with branches or wires on tall trees or transmission towers.
Behavior: Solitary or gathers in flocks; forms large flocks of several dozen to a hundred individuals in winter. Hops and walks on the ground. Omnivorous and bold; often initiates attacks on raptors. Tail often leaves a trail when walking on snow.
Distribution: Widespread in eastern China but absent in NE Inner Mongolia; common in northern China, fewer in the south. West to a geographical line connecting Gansu (Tiantangsi), Qinghai (Xining, Banma), Sichuan (Kangding), and Yunnan (Deqin).
Voice: Wide variety of vocalizations; mostly unmusical and harsh. Harsh, spluttering, scolding chatter, *tsche-tsche-tsche*, made when agitated, is perhaps the most familiar. This is slower than the similar chatter of Eurasian Magpie, and individual notes are more complex and with harmonics.

Rufous Treepie

Blackish gray from head to upper breast

Dark rufous-brown upperparts

Grayish-white upperwing coverts

Brownish-yellow underparts

Black tip on long gray tail

Black-rumped Magpie

Black rump

Tail shorter than Oriental Magpie

Gray Treepie

Gray from crown to nape

Rest of head and throat black

Brown upperparts

Small white patch on black wings

Central tail feathers mostly gray

D. f. himalayana

Central tail feathers all black

D. f. sinica

Eurasian Magpie

More white on primaries

Glossy green wings

White or grayish-white rump

Collared Treepie

Grayish-white nape, neck, and breast

Rufous-chestnut back, scapulars, and belly

Black on throat and front of head

Oriental Magpie

Grayish-white rump

Glossy blue wings

Temnurus temnurus

塔尾树鹊 tǎ wěi shù què

Ratchet-tailed Treepie 🟢LC

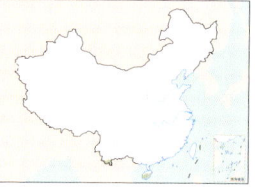

L: 32–35 cm
Habitat: Occurs in tropical rain forests.
Behavior: Often solitary or in pairs in the center of canopy. Secretive. Flight relatively heavy. Sometimes joins leafbirds, barbets, and Greater Racket-tailed Drongos in mixed-species flocks; forages in flowering trees.
Distribution: Rare in S Yunnan and locally common in rain forests of C Hainan.
Voice: Varied; includes loud, tremulous, rising purr, *preeeee*, and a harsher *graak*. Both are more reminiscent of drongos or green-magpies than other treepies.

Podoces hendersoni

黑尾地鸦 hēi wěi dì yā

Mongolian Ground-Jay 🟢LC

L: 28–31 cm
Habitat: Inhabits premontane desert, low mountains and hills, and other deserts and semideserts with sparse bushes.
Behavior: Often in family units, with stable territories. Picks up garbage along roadsides.
Distribution: Local and uncommon resident in Ningxia, NW Gansu, W Inner Mongolia, Xinjiang, and N Qinghai.
Voice: Most common vocalization is rapid spluttering trill of 4–22 intense, far-carrying notes repeated at intervals of 15 sec or more and slowing very slightly toward the end. Xinjiang Ground-Jay has a similar trill, but Mongolian Ground-Jay's individual notes are shorter, rising, and with more harmonics; those of Xinjiang more of a diphthong. Presumed contact call a penetrating, intense, high-pitched *twi-sk* or *tiss-ik*, with the stress on the first note and more reminiscent of a ground squirrel (*Spermophilus*) than a bird.

Podoces biddulphi

白尾地鸦 bái wěi dì yā

Xinjiang Ground-Jay 🟢NT

L: 29–30 cm
Habitat: Inhabits desert hinterlands with sparse bushes, desert poplar forests along the Tarim River, and tamarisk bushes at the edges of desert oases.
Behavior: Typical desert-dwelling species. Occurs only on soft sands. Legs long and powerful; strong runner on sandy ground. Often in family units, with stable territories. Sometimes picks up food along roadsides.
Distribution: Uncommon endemic to Taklamakan Desert in S Xinjiang. Also recorded in W Gansu.
Voice: Intense, far-carrying trill similar to Mongolian Ground-Jay but not usually slowing and with more of a whinnying horselike quality. Contact calls unknown.

Nucifraga caryocatactes

星鸦 xīng yā

Eurasian Nutcracker 🟢LC

L: 29–34 cm
Habitat: Commonly seen in open coniferous forests; occasionally seen in mixed broadleaf-coniferous forests.
Behavior: Solitary or in pairs, occasionally gathers in small flocks. Flight straight and flat. Often perches on top of coniferous trees or in the canopy; seldom comes to the ground. Caches food.
Distribution: Found in Northeast, North, northwestern, and southwestern China and Taiwan. The current species is sometimes treated as two species: *N. caryocatactes*, including *N. c. macrorhynchos* (Northeast China, northern parts of North China, and N Xinjiang) and *N. c. rothschildi* (Xinjiang), with characteristic large, dense spots on the belly, extending to lower belly; and *N. hemispila*, including *N. c. hemispila* (S Tibet), *N. c. macella* (from S Shanxi to southwestern China), *N. c. interdicta* (southern parts of Northeast China to North China), and *N. c. owstoni* (Taiwan), with characteristic small belly spots that do not extend to the lower belly.
Voice: Dry, grating, mechanical scolding, *kraaaaark*, often repeated in rapid succession.

Ratchet-tailed Treepie

Eurasian Nutcracker

White spots on upperparts extend to rump

Black and white tail

White undertail coverts

Relatively dense, large, triangular-shaped, white spots on underparts extend to lower belly

N. c. macrorhynchos

Relatively small, round, white spots on underparts extend to upper belly

N. c. hemispila

Few white spots on belly

N. c. interdicta

Mongolian Ground-Jay

No black cheek patch

Black tail

Xinjiang Ground-Jay

Black cheek patch

White tail

329

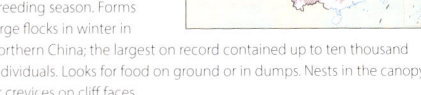

Pyrrhocorax pyrrhocorax

红嘴山鸦 hóng zuǐ shān yā

Red-billed Chough　LC

L: 36–40 cm
Habitat: Inhabits rocky mountains, alpine meadows, open river shoals, and cultivated lands.
Behavior: Gregarious. Uses bill to dig and turn vegetation or rocks on ground. Also mixes with other corvids. Sometimes turns or dives abruptly in flight.
Distribution: *P. p. brachypus* in southern Northeast China and North China; *P. p. himalayanus* at middle and high altitudes in northwestern and southwestern China; *P. p. centralis* in W Xinjiang. Not rare in suitable habitats.
Voice: Short, piercing, and high-pitched *chiaa*, vaguely reminiscent of Daurian Jackdaw.

Pyrrhocorax graculus

黄嘴山鸦 huáng zuǐ shān yā

Yellow-billed Chough　LC

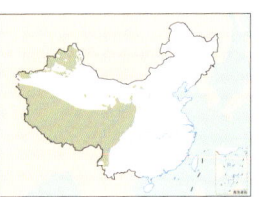

L: 37–39 cm
Habitat: In breeding season, inhabits alpine pastures above tree line, rocky valleys, and cliffs, often at higher altitudes than Red-billed Chough; descends to slightly lower altitudes in winter.
Behavior: Lives in flocks; large flocks split into several small flocks when foraging. Walks energetically on ground; occasionally hops. Forages for invertebrates on ground or in rock crevices; also feeds on berries and discarded human food.
Distribution: High altitudes in Western China, NW Yunnan, and W Sichuan; not rare in suitable habitats.
Voice: Piercing *zirrr*, higher pitched and very different from Red-billed Chough.

Corvus splendens

家鸦 jiā yā

House Crow　LC

L: 40–42 cm
Habitat: Occurs in various habitats, such as cities and towns, villages and farmland.
Behavior: Not shy. Boldly walks on ground looking for food; also actively chases and attacks raptors.
Distribution: S Tibet and S and W Yunnan; quite a few records in the past, but now rare. Vagrant to Macao, Taiwan, and Yunnan (Kunming). A small established population in Hong Kong.
Voice: Hoarse, dry, muffled *kaaa-kaaa*.

Corvus dauuricus

达乌里寒鸦 dá wū lǐ hán yā

Daurian Jackdaw　LC

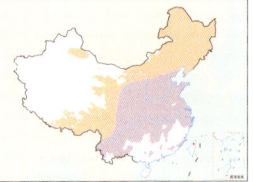

L: 29–37 cm
Habitat: Inhabits sparse woodlands and open villages; often forages on farmland.
Behavior: Gregarious; may also congregate in breeding season. Forms large flocks in winter in northern China; the largest on record contained up to ten thousand individuals. Looks for food on ground or in dumps. Nests in the canopy or crevices on cliff faces.
Distribution: Broadly distributed across China except for Hainan, W Qinghai, W Tibet, and N and W Xinjiang. Common resident or breeder in northern China; occasional winter visitor in southern China; vagrant to Taiwan.
Voice: Short, piercing *kyarr* and intense *kyak*, extremely similar to call of Eurasian Jackdaw.

Corvus monedula

寒鸦 hán yā

Eurasian Jackdaw　LC

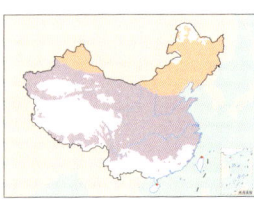

L: 30–34 cm
Habitat: Inhabits primarily low mountains, hills, and plains in breeding season; prefers farmland near cities and countryside in winter.
Behavior: Gregarious; often gathers in small flocks. Noisy. Usually mixes with Rooks.
Distribution: Locally common resident or wintering birds in Xinjiang, W Tibet, and southwestern China. Vagrant to Beijing.
Voice: Wide variety of calls used in different situations, including a simple repeated *kyar*; sometimes noisy.

Corvus frugilegus

秃鼻乌鸦 tū bí wū yā

Rook　LC

L: 46–47 cm
Habitat: Prefers open areas near farmland, grassy areas, and wetlands with tall trees nearby; also occurs in villages.
Behavior: Highly gregarious even in breeding season; sometimes joins other crows or starlings in mixed-species flocks. Forages on the ground, probing with bill in ground for seeds and other food items. Chases and attacks raptors, but not as persistent as other crows.
Distribution: *C. f. frugilegus* in N and W Xinjiang. *C. f. pastinator* in Northeast, North, northwestern, and East China and eastern South China; previously very common but now much rarer; vagrant to Taiwan and Hainan. The individuals in C Xinjiang are possibly also *C. f. pastinator*.
Voice: Deep caws and croaks, generally low pitched and powerful, *craa-craa-craa*.

Red-billed Chough

Orange bill more slender than Yellow-billed Chough

juv.

Flat tail relatively short

Relatively long "fingers"

Slightly decurved red bill

Daurian Jackdaw

Dark iris

White collar

Relatively short bill

White belly

Dark iris

Fine white streaks

Relatively short bill

1st win.

Eurasian Jackdaw

Grayish-white hindneck

White iris

Yellow-billed Chough

Relatively short thick yellow bill

Rook

Distinct "fingers"

Bare skin on lores and chin

C. f. frugilegus

House Crow

Relatively flat and rounded forehead

Dark gray nape and neck

Long narrow wings

Slender pointed cone-shaped bill with grayish-white base

Slightly bulging crown

Feathered lores and chin

C. f. pastinator

331

Corvus corone

小嘴乌鸦 xiǎo zuǐ wū yā

Carrion Crow

L: 48–56 cm
Habitat: Adapted to a
wide range of habitats,
such as wooded areas,
farmland, river flats, and
cities, at altitudes up to
3600 m.
Behavior: Lives in flocks;
also mixes with other

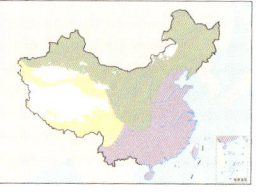

crows. Wingbeats slow; calls in flight. Around cities in northern China,
often forages on the ground in the countryside during daytime and flies
to city centers at night; perches on tall trees or buildings. Chases raptors.
Distribution: Widespread but absent in Tibet, Guizhou, and Guangxi.
Very common in northern China; occasional or rare in southern China.
Voice: Typical crow calls, *kraaa-kraaa*, varied and less "musical" than
Rook.

Corvus cornix

冠小嘴乌鸦 guān xiǎo zuǐ wū yā

Hooded Crow

L: 44–51 cm
Habitat: Inhabits city
gardens, farmland in the
countryside, and reservoir
wetlands in winter.
Behavior: Often mixes
with other crows in the
wild.
Distribution: Common in
winter in W Xinjiang.
Voice: Calls indistinguishable from Carrion Crow.

Corvus pectoralis

白颈鸦 bái jǐng yā

Collared Crow

L: 47–55 cm
Habitat: Inhabits open
lowlands, such as river
shoals, farmland, and
edges of villages; prefers
areas near water.
Behavior: Similar to other
crows, but gathers in
smaller flocks. Vigilant, less
aggressive than Large-billed Crow.
Distribution: In central and eastern China, south of N Hebei and east of
the Sichuan Basin. Population much declined. Rare in the north but still
locally common in South China.
Voice: Varied. Typical *kraa-kraa*, very similar to call of Carrion Crow but
high pitched. Also more resonant, lower-pitched notes.

Corvus macrorhynchos

大嘴乌鸦 dà zuǐ wū yā

Large-billed Crow

L: 45–57 cm
Habitat: Adapted to a
wide range of habitats,
including cities, villages,
deserted fields, and forest
edges, at altitudes up to
5000 m or higher.
Behavior: Moves about in
flocks during nonbreeding

season; usually gathers in large flocks in northern China, solitary or in
smaller flocks in southern China. Mixes with other crows. Omnivorous;
usually forages on the ground. Chases and attacks raptors.
Distribution: Widespread but absent in C and N Tibet and C Xinjiang.
Very common in Northeast and North China; common or occasional
in other regions. *C. m. intermedius* in SW Xinjiang, S and W Tibet;
C. m. tibetosinensis along the eastern edge of the Qinghai-Tibet Plateau;
C. m. mandshuricus in Northeast China; *C. m. colonorum* in most regions
in central and eastern parts of China. *C. m. levaillantii*, in S Tibet and
W Yunnan, smaller than other subspecies and making higher-pitched
calls, is sometimes treated as "Jungle Crow," *C. levaillantii* (丛林鸦).
Voice: Varied, extensive repertoire includes a typically hoarse *kraa-kraa*.

Corvus corax

渡鸦 dù yā

Common Raven

L: 63–70 cm
Habitat: Adapted to a
wide range of habitats,
including deserted fields,
grasslands, and river
valleys, at altitudes up to
5000 m or higher. Seldom
occurs in cities.
Behavior: Often in small

flocks; sometimes gathers in large flocks after breeding season. Sways
when walking on the ground. Wingbeats slow; sometimes rolls in the air.
Actively attacks or even kills raptors.
Distribution: *C. c. kamtschaticus* in western Northeast China and
northwestern China; *C. c. tibetanus* in eastern northwestern China, the
Qinghai-Tibet Plateau, and its adjacent regions.
Voice: Varied and complex; includes harsh, low-pitched, hollow-
sounding *pruk*, *kwop*, and *korrp*.

Large-billed Crow

Carrion Crow

Flat forehead

Bill thinner and smaller than Large-billed Crow

Bill smaller than other subspecies

C. m. levaillantii

Distinct bulging on forehead

Thick strong bill

C. m. colonorum

Hooded Crow

Grayish-white back

Black breast

Grayish-white underparts

Common Raven

Long wings, like raptors, with relatively narrow wing tips

Almost wedge-shaped tail

Nasal bristles reaching half of exposed culmen

Flat head

Thick strong bill

Collared Crow

White breast band

太平鸟科 Bombycillidae

Small forest-dwelling songbirds with a compact profile. Plumage similar between sexes, mostly pinkish brown, red, yellow, and black. Bill short and moderately wide. Head has prominent crest and conspicuous black eye stripe. Good fliers, with long and pointed wings. Inhabit primarily coniferous, mixed, and deciduous forests. Nest among tree branches. Feed mostly on plant fruits, seeds, and shoots, as well as insects. Migratory. Sometimes gather in large flocks during winter.

One genus and three species recognized worldwide, in the Arctic and temperate zones of Northern Hemisphere. One genus and two species recorded in China, found mostly in Northeast, northwestern, North, and East China.

Bohemian Waxwing

Crest lacks black

Black bib has sharper boundary

More red waxy tips on secondaries

Yellow-tipped tail

Yellow outer vane of primaries forms distinct yellow stripe

Diffuse black boundary

Indistinct red waxy tips on secondaries

Relatively narrow yellow fringe on primaries

Japanese Waxwing

Thick black eye stripe, lower part of crest also black

Red-tipped tail

White tips on primaries form several white bars

1st win.

White only on outer vane of primaries, forms one white stripe

---

Bombycilla garrulus
太平鸟 tài píng niǎo
Bohemian Waxwing **LC**

L: 19–23 cm
Habitat: Usually inhabits tall trees; prefers cypresses and pagoda trees. Occurs in broadleaf forests, secondary forests, and city gardens.
Behavior: Gathers in flocks, often in trees; mixes with Japanese Waxwing and thrushes. Flight rapid. Nomadic in winter; generally does not stay at one place for a long time.
Distribution: Winter visitor to Northeast and North China and W Xinjiang. Numbers fluctuate markedly; major irruption occurs every two to three years, during which also found in Central and East China.
Voice: High-pitched tremulous trills, *chi-lilili* or *lilili*.

Bombycilla japonica
小太平鸟 xiǎo tài píng niǎo
Japanese Waxwing **NT**

L: 18–20 cm
Habitat: Breeds in forests; prefers coniferous forests. Wintering habitats similar to those of Bohemian Waxwing.
Behavior: Gathers in flocks, often in trees. Fond of cypress cones and various berries. Mixes with Bohemian Waxwing and Naumann's Thrush. Individuals stand close together when resting.
Distribution: Winter visitor in China, seen in Northeast and North China. Number of individuals fluctuates largely from year to year; more commonly seen in East China and also possibly in South and southwestern China during major irruptions.
Voice: Many vocalizations similar to those of Bohemian Waxwing. Also a fine, descending "boiling kettle" whistle, *treee*, with a quality recalling penduline-tit.

玉鹟科 (仙莺科)　Stenostiridae

Small forest-dwelling songbirds. Body profile resembles Old World flycatchers, with short, broad, and pointed bill. Plumage similar between sexes. Inhabit primarily broadleaf, mixed, or bamboo forests and understory bushes near streams and valleys. Nest among branches. Feed primarily on insects and insect larvae; often forage in flight ("flycatching") or on ground. A few species are migratory.

Fairy flycatchers were previously placed in Muscicapidae but now constitute this newly recognized family. Four genera and nine species recognized worldwide, with six species in Africa and three species in Indomalaya. Two genera and two species recorded in China, found in montane forests from Central to southwestern China.

Yellow-bellied Fairy-Fantail

Yellow supercilium

Yellow underparts

Gray-headed Canary-Flycatcher

Gray head contrasts strongly with yellowish-green body

Chelidorhynx hypoxanthus

黄腹扇尾鹟 huáng fù shàn wěi wēng

Yellow-bellied Fairy-Fantail　LC

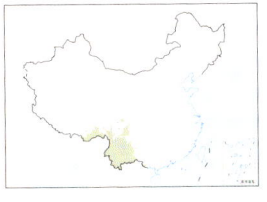

L: 11.5–12.5 cm

Habitat: Inhabits mountains and hills from 800 to 3700 m; migrates to lower altitudes in winter. Adapted to a wide range of wooded habitats, but mostly in middle story of moist evergreen broadleaf forests and low bushes.

Behavior: Very active; continuously changes postures and opens or raises tail to flush insects and catch them. Often joins mixed-species flocks in winter.

Distribution: Locally common resident in S and SE Tibet, Yunnan, W and SW Sichuan.

Voice: Repeats intense *tsip* call. Song a rambling, discordant jumble of high-pitched trills, short warbles, and occasional call notes.

Culicicapa ceylonensis

方尾鹟 fāng wěi wēng

Gray-headed Canary-Flycatcher　LC

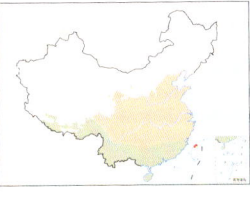

L: 12–13 cm

Habitat: At altitudes of 100–2000 m, inhabits evergreen and deciduous broadleaf forests, bamboo thickets, mixed forests, and forest-edge bushes. Recorded up to 3100 m in Tibet; migrates to lower altitudes in winter.

Behavior: Usually solitary; in family units during breeding season. Forages at all levels in forests. Active; often flicks tail when perched. Territorial.

Distribution: Resident and migrant in Central, southwestern, southeastern, and South China (including Hainan) and SE Tibet; vagrants to Taiwan. Common in southwestern China, uncommon in other regions. Recorded along the east coast during migration in recent years; presumably from unknown breeding populations on mountains in East to North China.

Voice: Song is loud, sweet, and distinctive, a multisyllabic series of rising and falling fast whistles, *chi-wi ee-whi-cheee*, repeated with little variation every few seconds. Call is high-pitched, accelerating series of short notes, *chit-tit-tit-tit-tirrr-h*, like a ball bearing being dropped.

山雀科　Paridae

Small forest-dwelling songbirds with a short and rounded profile. Plumage similar or slightly different between sexes, mostly olive, gray, black, yellow, and white. Cone-shaped bill short, thick, and strong. Wings short and rounded, enabling agile flight. Tail squared, moderately long. Inhabit primarily forests, scrubland, parks, and plantations. Very active. Nest in tree cavities or crevices. Omnivorous; feed in all forest layers and on the ground, mostly on plant seeds and fruits, as well as insects and insect larvae, but occasionally also take small amphibians, reptiles, even other birds. Very few species are migratory.

Fourteen genera and 64 species recognized worldwide, widely distributed in the Old World and North and Central America. Twelve genera and 24 species recorded in China, ubiquitously distributed nationwide.

火冠雀属
Cephalopyrus

Fire-capped Tit

林雀属
Sylviparus

Yellow-browed Tit

冕雀属
Melanochlora

Sultan Tit

黑冠山雀属
Periparus

Coal Tit

黄腹山雀属
Pardaliparus
(merged with *Periparus* in eBird/Clements Checklist)

Yellow-bellied Tit

冠山雀属
Lophophanes

Gray-crested Tit

杂色山雀属
Sittiparus

Varied Tit

高山山雀属
Poecile

Marsh Tit

蓝山雀属
Cyanistes

Azure Tit

地山雀属
Pseudopodoces

Ground Tit

山雀属
Parus

Japanese Tit

黄山雀属
Machlolophus

Yellow-cheeked Tit

Cephalopyrus flammiceps

火冠雀 huǒ guān què

Fire-capped Tit

L: 8–11 cm
Habitat: Inhabits montane coniferous or mixed broadleaf-coniferous forests; also seen in scrubland at high altitudes.
Behavior: Often in flocks in nonbreeding season; also joins mixed-species flocks.
Distribution: Locally common. *C. f. flammiceps* in SW Tibet; *C. f. olivaceus* in SE Tibet, Yunnan, Sichuan, Chongqing, Guizhou, Guangxi, W Hubei, S Shaanxi, Ningxia, and SE Gansu.
Voice: Song a short high-pitched warble, *zi-zi-zi*, interspersed with intense trills and occasional Yellow-browed Warbler–like *tsooest* notes.

Sylviparus modestus

黄眉林雀 huáng méi lín què

Yellow-browed Tit

L: 9–10 cm
Habitat: Inhabits various forests, bamboo thickets, secondary forests, and scrubland in mountains at middle and low altitudes.
Behavior: Solitary or gathers in small flocks; joins mixed-species flocks. Forages and hops about in trees.
Distribution: Locally common from southwestern to South China.
Voice: Calls a high-pitched Goldcrest-like *zi* or *zee*.

Fire-capped Tit

Red forehead
Olive upperparts
Grayish-green belly
Yellowish-green throat and breast

Yellow-browed Tit

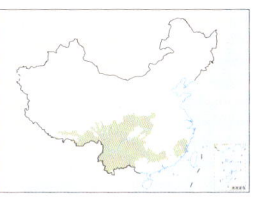

Short thick yellow supercilium, sometimes covered and inconspicuous
Short crest
One wing bar
All body yellowish green

Sultan Tit

Long bright citrine-yellow crest
Short crest
Blackish gloss on rest of body
Dusky brown or blackish brown, sometimes greenish
Bright yellow breast and underparts

Melanochlora sultanea

冕雀 miǎn què

Sultan Tit

L: 20–21 cm
Habitat: Inhabits deciduous forests, mixed forests, evergreen forests, and secondary forests on lowland and mountains below 1220 m.
Behavior: Often gathers in small flocks or forages in the canopy with other passerines in mixed-species flocks. Active and noisy.
Distribution: Locally common. *M. s. sultanea* in S and SE Tibet, W to S Yunnan; *M. s. seorsa* in SW Guangxi, Guangdong, Jiangxi, Fujian, north to S Zhejiang; *M. s. flavocristata* in Hainan.
Voice: Varied. Songs include a resonant explosive *chí-dip ... tri-trip*; a series of three to five melodious, whistled *chew* notes, often repeated for long periods; and a shrill, rapid *chit-ter-der chit-ter-der*. Calls include a scolding, metallic *tji-jup* and a harsh *krssh-krssh*.

Periparus rufonuchalis

棕枕山雀 zōng zhěn shān què

Rufous-naped Tit　LC

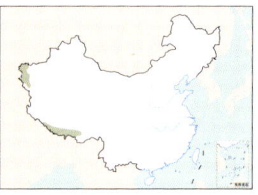

L: 12–13 cm
Habitat: Inhabits alpine forests from 2800 to 3500 m.
Behavior: Solitary or in pairs in breeding season; otherwise often gathers in flocks. Often in the canopy; also moves and forages in understory bushes or on the ground.
Distribution: Locally common in SW Xinjiang; also in S Tibet.
Voice: Jaunty, variable songs are a Coal Tit–like series of repeated, slightly hoarse, simple, disyllabic whistles, *wi- tee … wi- tee* or *pitchuee*, and a multisyllabic *pit-chu-pi-chu-eer*. Calls include a plaintive *tseep*, short *sip*, and rapid *trrrr*.

Periparus rubidiventris

黑冠山雀 hēi guān shān què

Rufous-vented Tit　LC

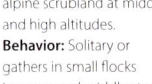

L: 12–13 cm
Habitat: Inhabits mixed broadleaf-coniferous forests, coniferous forests, bamboo thickets, and alpine scrubland at middle and high altitudes.
Behavior: Solitary or gathers in small flocks in upper and middle stories of trees; often joins other small birds in mixed-species flocks.
Distribution: *P. r. rubidiventris* in SW Tibet. *P. r. beavani* (sometimes treated as *P. r. whistleri*) in southwestern to central China.
Voice: Song is a repeated *wich-uu … wich-uu-wich-uu*. Calls include a thin rising *pweeit*, a high-pitched, rapid, scolding *si-si-si-si-si*, a hoarse rattling *zizizizizi*, and a churring *dzeee*.

Periparus ater

煤山雀 méi shān què

Coal Tit　LC

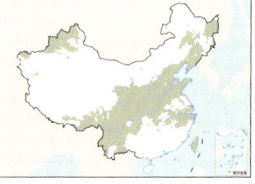

L: 9–12 cm
Habitat: Inhabits various wooded areas on mountains at middle altitudes; also in scrubland.
Behavior: Often gathers in small flocks or joins other tits in mixed-species flocks. Hops about and forages among leaves or branches.
Distribution: Locally common. *P. a. ater* in Northeast China and N Xinjiang; *P. a. aemodius* (sometimes treated as *P. a. eckodedicatus*) from southwestern to central China; *P. a. rufipectus* in C and W Xinjiang; *P. a. pekinensis* in North China; *P. a. insularis* in W Liaoning and NE Hebei; *P. a. kuatunensis* from East China to NW Fujian; *P. a. ptilosus* in Taiwan.
Voice: Song is higher pitched and finer than song of Japanese Tit, a repeated *sichuu-sichuu-sichuu* or *pi-chu-eit*. Calls include various short buzzy and nasal notes and a plaintive *tseuu*, similar to Yellow-browed Warbler.

Periparus venustulus

黄腹山雀 huáng fù shān què

Yellow-bellied Tit　LC

L: 9–11 cm
Habitat: Inhabits various wooded areas at middle and low altitudes.
Behavior: Gathers in large flocks in winter. Hops about and forages in trees. Also mixes with other tits in mixed-species flocks.
Distribution: Widespread but absent from northwestern China. Locally common. Some populations increasing and spreading north.
Voice: Varied. Song a repeated, silvery, ringing, Coal Tit–like *wi-chu-uu*. Calls include various short buzzy notes, a nasal *tchay*, and a di- or trisyllabic, high-pitched, fine *ti-di* or *ti-di-di*.

Rufous-naped Tit

Brown patch behind nape

Chestnut-brown vent

Black bib more extensive

Coal Tit

Short crest on black head

White nuchal patch

White cheek

Two wing bars

P. a. kuatunensis

No crest

P. a. insularis

Short crest

Pale brown belly

P. a. rufipectus

Rufous-vented Tit

Prominent black crest

No wing bar

Yellowish-brown undertail coverts

Dusky gray breast and belly

Yellow-bellied Tit

Black head

White nuchal and cheek patches

♂ br.

Two wing bars

Bright yellow underparts

♂ non-br.

Yellow throat

Grayish-green upperparts

Two wing bars

Grayish-white cheek patch

♀

Yellowish-green underparts

juv.

Poecile superciliosus
白眉山雀 bái méi shān què
White-browed Tit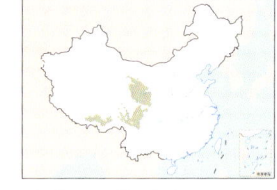

L: 13–14 cm
Habitat: Inhabits mixed broadleaf-coniferous forests, coniferous forests, and alpine scrubland and meadows at high altitudes.
Behavior: Often in pairs or small flocks at forest edges at high altitudes; also seen in villages and forest edges near farmland. Not shy.
Distribution: Endemic to midwestern China; found in S and E Tibet, N and NW Sichuan, W to NW Gansu, and E Qinghai.
Voice: Varied, often loud, far-carrying vocalizations include a ringing, multisyllabic *ci–ci–piu–piu*, *plu-plu-plu-plu*, and *tii-lu-lu*.

Poecile palustris
沼泽山雀 zhǎo zé shān què
Marsh Tit

L: 12–13 cm
Habitat: Inhabits montane woodlands; also seen in parks and orchards in winter.
Behavior: Often gathers in small flocks outside breeding season; also joins other tits in mixed-species flocks. Forages among branches and leaves in the canopy.
Distribution: Locally common. *P. p. brevirostris* in Northeast China and N Xinjiang; *P. p. jeholicus* from North China to W Henan; *P. p. hellmayri* in East China. The latter two subspecies are sometimes treated as synonyms.
Voice: Song a repeated monosyllabic *chup-chup-chup-chup-chup* (more disyllabic versions inviting confusion with Japanese Tit) or *weeta-weeta-weeta*. Calls include a cheerful, often loud, or almost peevish *pitchay*.

Poecile montanus
褐头山雀 hè tóu shān què
Willow Tit

L: 11–13 cm
Habitat: Inhabits montane forests at middle and high altitudes, especially coniferous and mixed broadleaf-coniferous forests.
Behavior: Gathers in small flocks and forages in the lower part of the canopy; also mixes with other tits.
Distribution: Locally common. *P. m. baicalensis* in Northeast China and N Xinjiang; *P. m. songarus* in the Tian Shan Mountains in Xinjiang; *P. m. affinis* in Ningxia, S Gansu, E Qinghai, N Sichuan, and SW Shaanxi; *P. m. stoetzneri* in North China.
Voice: Song a series of penetrating, flat whistles, *tiu-tiu-tiu-tiu* or a more jolly *ti-du-tidu-ti-du*, repeated at length or recombined with shorter calls. Most distinctive call a buzzy *chairr*, often given in series that are longer and more nasal than Marsh Tit.

Poecile davidi
红腹山雀 hóng fù shān què
Pere David's Tit

L: 11–13 cm
Habitat: Inhabits coniferous forests, mixed broadleaf-coniferous forests, and bamboo thickets at middle to high altitudes.
Behavior: Prefers gathering in small flocks in upper and middle stories of trees; also comes down to bushes and ground. Mixes with other small birds in nonbreeding season.
Distribution: Endemic to mountains in central China. Seen in W Hubei, N to SW Sichuan, S Shaanxi, and SW Gansu.
Voice: Silver-throated Tit–like purr and unobtrusive, short *sit* contact calls. Others very similar to Willow Tit: *zi-zi* … *tchaay*, *chip-ip* … *tschay*, *tr-trip-it-tchay*, and powerful *tchay-chay-chat* when agitated.

Poecile hypermelaenus
黑喉山雀 hēi hóu shān què
Black-bibbed Tit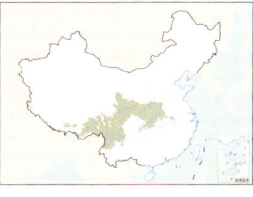

L: 11–12 cm
Habitat: Inhabits montane coniferous or broadleaf forests at middle and high altitudes.
Behavior: Gathers in small flocks; forages among branches and leaves or on the ground.
Distribution: Uncommon. *P. h. hypermelaenus* in S Gansu, S Shaanxi, W Hubei, S Shanxi, and W Henan; *P. h. dejeani* in Sichuan, W Guizhou, and NW Yunnan. The two subspecies are sometimes treated as synonyms.
Voice: Song a fast, ringing, repeated, almost jarring *wichitu-wichitu-wichitu* and a bouncy Japanese Tit–like *ter-dididit-er-wu-ter-didi-er-wu*. Calls include a strong *tchiu*, often combined with scolding chatter; other calls very similar to Marsh Tit: *pitchu* or *tit-pitchu*.

Poecile weigoldicus
四川褐头山雀 (川褐头山雀)
sì chuān hè tóu shān què (chuān hè tóu shān què)
Sichuan Tit

L: 11–13 cm
Habitat: Inhabits coniferous forests and forest-edge bushes at high altitudes.
Behavior: Gathers in small flocks. Insectivorous.
Distribution: Locally common in SE Tibet, NW Yunnan, and Sichuan.
Voice: Poorly known but variable. One song type a ringing *ti-di-du-di*, very much like Willow Tit, another a markedly buzzier *we-cheer* … *we-cheer*. Nasal, buzzy calls very similar to Willow Tit.

White-browed Tit

Distinct white supercilium

Pinkish-brown cheek

Black crown, eye stripe, and throat

Pere David's Tit

Rufous-chestnut neck ring

Black from throat to upper breast forms triangular bib

Rufous chestnut from breast to undertail coverts

Marsh Tit

White patch at base of upper mandible

Inconspicuous pale fringe on wing feathers

Black-bibbed Tit

Peak or short crest on black cap

Brown tinge on olive back

Black bib extends to upper breast, with clear boundary

Warm brown underparts

Willow Tit

Black cap

Gray upperparts

Small black bib

P. m. baicalensis

Conspicuous pale fringe on wing feathers

Brownish cap

Relatively large black bib

P. m. affinis

inconspicuous pale fringe on wing feathers

Relatively large black bib

P. m. stoetzneri

Sichuan Tit

Relatively dark blackish-brown cap; upperparts more dusky grayish brown, differing from dusky brown on Willow Tit (*P. m. affinis*)

No crest

Black bib smaller than Black-bibbed Tit, with diffuse lower border

More dark gray than brown on flanks, differing from Willow Tit (*P. m. affinis*) and Black-bibbed Tit

Lophophanes dichrous
褐冠山雀 hè guān shān què
Gray-crested Tit

L: 10–12 cm
Habitat: Inhabits montane mixed broadleaf-coniferous forests, coniferous forests, bamboo thickets, and alpine scrubland at high altitudes.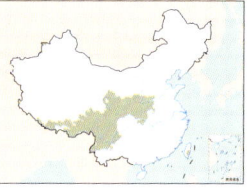
Behavior: Often in pairs or small flocks in middle story of trees. Active. Often joins other small tits and warblers in mixed-species flocks.
Distribution: *L. d. dichrous* in SE Tibet; *L. d. dichroides* in W Hubei, S Shaanxi, S Gansu, N Sichuan, S and E Qinghai, and N Tibet; *L. d. wellsi* in NW Yunnan and Sichuan.
Voice: Song starts with high-pitched fine notes followed by more faltering ones, *wee-wee-trzz-i-tr-trtt*, repeated. Calls include high-pitched Goldcrest-like notes, similarly high-pitched fine trills reminiscent of Bohemian Waxwing, and other more titlike sounds, such as *tsiti-sit-sit*.

Sittiparus varius
杂色山雀 zá sè shān què
Varied Tit

L: 12–14 cm
Habitat: Inhabits various woodlands at middle and low altitudes.
Behavior: Gathers in small flocks in trees; also mixes with other tits.
Distribution: Irruptive migrant in Northeast China and along the coast; a separate population in Guangdong (Nanling Mountains). Locally common.
Voice: One song type is high pitched, fine, and like Japanese Tit; another is a higher, tinkling version, vaguely reminiscent of Mugimaki Flycatcher's song. Calls include two to four unremarkable, high-pitched, rising contact calls, *tsee-tsee-tsee* or *si-si-wi*, and other harsher, grating calls, *tchay-chay-chay*, like an excited Willow Tit.

Sittiparus castaneoventris
台湾杂色山雀 tái wān zá sè shān què
Chestnut-bellied Tit

L: 12–14 cm
Habitat: Inhabits montane broadleaf forests at middle and low altitudes.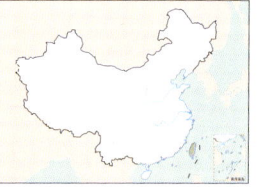
Behavior: Often in upper and middle stories of trees; also joins mixed-species flocks.
Distribution: Chinese endemic; uncommon resident in Taiwan .
Voice: Some vocalizations similar to Varied Tit (more research needed), but accelerating sequence of three to eight very short notes, *ti-di-dit-it*, some ending with very short, harsher *tchay*, are more similar to Yellow-bellied Tit than to Varied.

Cyanistes cyanus
灰蓝山雀 huī lán shān què
Azure Tit

L: 12–13 cm
Habitat: Inhabits broadleaf forests and mixed forests on mountains and plains. Especially common in trees and bushes on the banks of streams, rivers, and lakes. Also occurs in 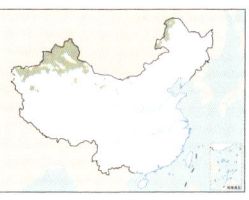 reed marshes, and bushes and willows at desert edges.
Behavior: Solitary or in pairs during breeding season; otherwise often gathers in flocks. Active. Often hops about or flies in trees, sometimes hangs on branches and forages for insects, such as beetles, butterflies, moths, caterpillars, and their eggs. Sometimes also eats plants.
Distribution: *C. c. tianschanicus* in Heilongjiang, NE Inner Mongolia, and Xinjiang; *C. c. berezowskii* (once considered a separate species, Yellow-breasted Tit "黄胸山雀") extremely rare and possibly extirpated in Gansu and NE Qinghai. *C. c. flavipectus* and *C. c. carruthersi*, also part of the "Yellow-breasted Tit" group, possibly occur in W Xinjiang.
Voice: Contact calls are a slurred tutting and nasal *tsee-tsee-dze-dze*. Alarm calls are a scolding *chr-r-r-r-rit*. Songs are pleasant short verses, *tis-tis-tsheuw … tis-tis-tsheuw*, often repeated with little variation. Wide range of appealing calls, including repeated *te-te-cher*, bubbling trills, and short chatters.

Pseudopodoces humilis
地山雀 dì shān què
Ground Tit

L: 14–17 cm
Habitat: Inhabits grasslands and grassy slopes at high altitudes; often in habitats with pikas.
Behavior: Gathers in flocks and forages on ground; great hopper. Seldom flies a long distance.
Distribution: Locally common on the Qinghai-Tibet Plateau and in Ningxia, Gansu, Xinjiang, and Sichuan.
Voice: Plaintive, occasionally intense, short, descending whistles, *jiu*, often paired and slightly reminiscent of Robin Accentor, and nasal chattering, sometimes combined in rapid sequence, *jee … jee … tu-didi-di diu*.

Gray-crested Tit

Erect gray crest

Rufous-brown iris

Gray back lacks black

L. d. wellsi

Brownish-gray throat differs from brown underparts

Contrasting colors on throat and breast; other subspecies have same color

L. d. dichrous

Brown-tinged back

Same color on throat and underparts

L. d. dichroides

Azure Tit

Black band on hindneck connects with blue-black eye stripe

Bluish-gray upperparts

Distinct white wing bar

C. c. tianschanicus

Relatively blue crown

Relatively dull back

Yellow-tinged breast

C. c. berezowskii

Ground Tit

Slender decurved bill

Earth-brown upperparts

Whitish underparts

Varied Tit

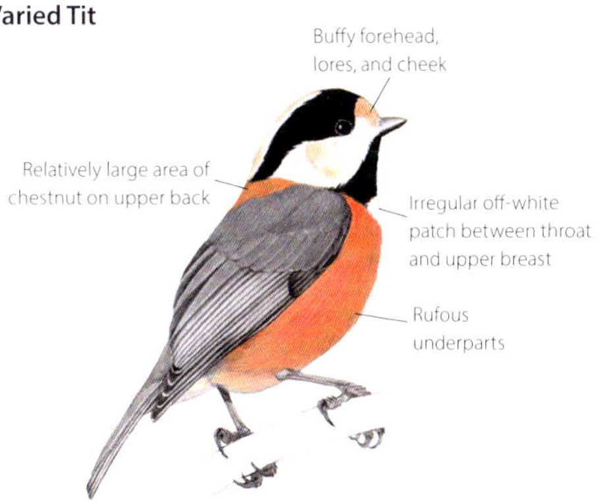

Buffy forehead, lores, and cheek

Relatively large area of chestnut on upper back

Irregular off-white patch between throat and upper breast

Rufous underparts

Chestnut-bellied Tit

White forehead, lores, and cheek

Small chestnut patch on upper back

Rich chestnut underparts

Parus major

欧亚大山雀（大山雀）
ōu yà dà shān què (dà shān què)
Great Tit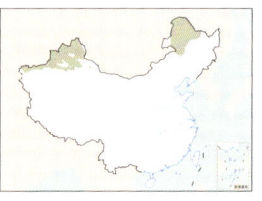

L: 13–15 cm
Habitat: Inhabits various types of woodlands and scrubland at middle and low altitudes.
Behavior: Solitary or gathers in small flocks. Forages in trees and bushes. Not shy.
Distribution: Locally common. *P. m. kapustini* in Xinjiang, N Heilongjiang, and NE Inner Mongolia; *P. m. turkestanicus* in N Xinjiang.
Voice: Large vocabulary, with many notes not readily separable from those of Japanese Tit.

Parus minor

大山雀（远东山雀）
dà shān què (yuǎn dōng shān què)
Japanese Tit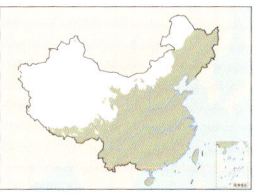

L: 12–14 cm
Habitat: Inhabits various types of woodlands.
Behavior: Solitary or gathers in small flocks. Hops about in trees and bushes; occasionally catches insects in flight or on the ground.
Distribution: Common across much of China. *P. m. minor* in Northeast, North, northwestern, Central, and East China; *P. m. tibetanus* (including *P. m. subtibetanus*) in Tibet, Qinghai, Yunnan, W Guizhou, N and W Sichuan; *P. m. commixtus* in southern China except for Hainan. *P. m. hainanus* in Hainan; this subspecies is sometimes placed in *P. cinereus* (Cinereous Tit, 苍背山雀).
Voice: Large vocabulary. Some songs cheerful, including the bouncing *tea-cha … tea-cha … tea-cha*; others timid, and still others intense, scolding, or inquiring.

Parus monticolus

绿背山雀 lǜ bèi shān què
Green-backed Tit

L: 12–15 cm
Habitat: Inhabits montane woodlands at middle and low altitudes; migrates to lower altitudes in winter.
Behavior: In pairs or small flocks; also joins mixed-species flocks. Insectivorous.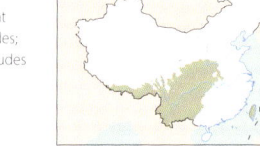
Distribution: Locally common. *P. m. monticolus* in S and SE Tibet; *P. m. yunnanensis* from southwestern to Central China; *P. m. insperatus* in Taiwan.
Voice: Large vocabulary. Many notes including ringing, strident song *tutta-seeta-seeta*. Chatter and contact notes similar to those of Japanese Tit but generally simpler, with slower, shorter repetitions and more modest frequency range.

Machlolophus holsti

台湾黄山雀 tái wān huáng shān què
Taiwan Yellow Tit

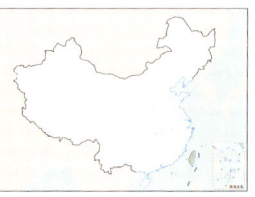

L: 12–13 cm
Habitat: Inhabits broadleaf forests, mixed forests, or secondary forests at middle and high altitudes.
Behavior: Solitary or in pairs in the canopy of tall trees; also joins mixed-species flocks.
Distribution: Chinese endemic; uncommon resident in Taiwan.
Voice: Variable ringing songs, most similar to Japanese Tit but generally simpler, with shorter repetitions, *tziu-du … tziu-du* and *ti-ti-ti-du … tiu-du* and *t-wichi-tu … t-wichi-tu*. Scolding, raucous, almost woodpecker-like chattering.

Machlolophus xanthogenys

眼纹黄山雀 yǎn wén huáng shān què
Himalayan Black-lored Tit

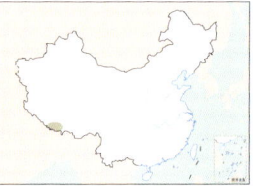

L: 12–15 cm
Habitat: Inhabits open woodlands, bamboo thickets, and forest-edge bushes at middle altitudes.
Behavior: Arboreal. In pairs or small flocks. Insectivorous.
Distribution: Uncommon resident in S Tibet (Zhangmu and Gyirong).
Voice: Large vocabulary, with many notes similar to Yellow-cheeked Tit, but songs higher pitched, with greater frequency range, longer, with fewer staccato notes, and frequently include a lengthy *tsieuw* and other falling notes. Ringing *tsi-pit-tsu-tsip-it*.

Machlolophus spilonotus

黄颊山雀 huáng jiá shān què
Yellow-cheeked Tit

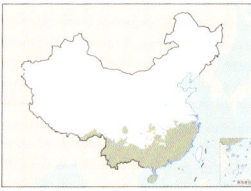

L: 13.5–15.5 cm
Habitat: In breeding season, at altitudes of 1000–3000 m; prefers open temperate and subtropical deciduous forests, mixed forests, montane evergreen forests, secondary forests, bamboo thickets, and forest edges. Migrates to altitudes of 350–700 m in nonbreeding season.
Behavior: Active. Often in pairs or flocks; usually joins other passerines in mixed-species flocks. Forages primarily in the middle story of trees or in bushes.
Distribution: Locally common resident. *M. s. spilonotus* in S and SE Tibet; *M. s. subviridis* in W to S Yunnan; *M. s. rex* in E, C, and N Yunnan, S Guizhou, S Sichuan, S Hunan, C Guangxi, Guangdong, C and NW Fujian, and Jiangxi, north to S Zhejiang.
Voice: Song is a ringing, clear, fast repetition of mostly flat notes, with no great changes in pitch, *chiu-chiu-piu … chiu-chiu-piu, ch* or *dzi-dzi-pu … dzi-dzi-p*. Calls include *tsee-tsee-du-de* or *witch-a witch-a witch-a* and nasal or whining chatter when agitated. (See Himalayan Black-lored Tit.)

Great Tit

Disconnected black on throat and neck

Yellowish-green back

One wing bar

P. m. turkestanicus

P. m. kapustini

Broad black ventral line on yellow underparts

Japanese Tit

Greenish upperparts

Broad black ventral line on white underparts

Gray upperparts

P. m. minor

Broad black ventral line on white underparts

P. m. hainanus

Green-backed Tit

Yellowish-green back

Two wing bars

Broad black ventral line on yellow underparts

Taiwan Yellow Tit

Black crown

Long crest, white nuchal patch extends to tip of crest feathers

Dark bluish-gray upperparts

Yellow base of forehead, cheek, and underparts

White undertail coverts

Himalayan Black-lored Tit

Long black crest

Black lores

Olive upper back lacks streaks or spots

Two wing bars

Broad black ventral line on yellow underparts

Yellow-cheeked Tit

Slightly grayish-blue upperparts and tail

Broadest black ventral line among subspecies; black on breast connects to hindneck

M. s. rex

No yellow below head; side of underparts gray

No black on forehead

Olive upper back

Broad black ventral line

M. s. subviridis

Bright yellow becomes paler from cheek to lower belly

Dense yellow spots on black upper back

Black on chin and throat extends to lower belly, broadest on breast

M. s. spilonotus

White vent

White outer tail feathers

No black ventral line; throat and breast slightly black-tinged

攀雀科　Remizidae

Small forest-dwelling songbirds. Profile resembles tits, but with finer and more pointed bill. Plumage similar or slightly different between sexes, mostly gray, yellow, and rufous. Inhabit primarily forest patches, reedbeds, or bushes near open habitats, including deserts, plains, lakes, and agricultural fields. Apart from Verdin in North America, all species make delicate bag-shaped nests that hang from tree branches. Primarily insectivorous, but also take plant shoots, seeds, and nectar. Eurasian species are mostly migratory, while those in Africa and North America usually do not undertake long-distance migrations.

Three genera and 11 species recognized worldwide: four in Eurasia, six in Africa, and one in North America. One genus and three species recorded in China, found mostly in northern regions, while Chinese Penduline-Tit winters down to the Yangtze River Watershed.

White-crowned Penduline-Tit

White crown
White neck collar
Brown upperparts
♂

Chinese Penduline-Tit

White supercilium
Gray from head to hindneck
Broad black eye stripe
♂

Pale brown supercilium
Brown eye stripe
Grayish brown from head to hindneck
♀

Black-headed Penduline-Tit

All-black head
Chestnut-brown upper back
Grayish-black throat

Remiz coronatus

白冠攀雀 bái guān pān què

White-crowned Penduline-Tit LC

L: 10–12 cm
Habitat: Inhabits broadleaf forests and bushes near lakes and rivers; also prefers reedbeds during migration.
Behavior: In pairs in breeding season; often gathers in flocks during migration or winter. Prefers nesting on branches of poplars and willows; delicate teapot-shaped nest woven with plant fibers and wool. Active and agile.
Distribution: *R. c. coronatus* in N Xinjiang (Ili, Bozhou, and Tacheng); *R. c. stoliczkae* in N Ningxia and N Xinjiang. Uncommon in S Xinjiang during migration and in winter.
Voice: Call very similar to that of Chinese Penduline-Tit and Black-headed Penduline-Tit but "flatter" than both and not so "dreamy," and typically longer than Black-headed. Song slightly more staccato; gives faint tutting.

Remiz consobrinus

中华攀雀 zhōng huá pān què

Chinese Penduline-Tit LC

L: 10–11 cm
Habitat: Inhabits woodlands or broadleaf forests, reedbeds, cattails, and tall grasslands on plains.
Behavior: Builds pouch-like nest on broadleaf trees near water. Gathers in small flocks in winter.
Distribution: Widespread in eastern, central, and southwestern China. Seasonally common.
Voice: Call a fine, descending, breezy whistle, *tsseiu*, reminiscent of Reed Bunting but lacking the vigor of that species. Compare Yellow-browed Warbler. Shorter versions, combined with fine twitters and trills, given in flight and in songs. Calls average slightly longer than both other penduline-tits.

Remiz macronyx

黑头攀雀 hēi tóu pān què

Black-headed Penduline-Tit LC

L: 10–12 cm
Habitat: Inhabits willows and reedbeds near rivers and lakes.
Behavior: Same as White-crowned Penduline-Tit.
Distribution: Probably a rare breeder in Ili River valleys of N Xinjiang. The individuals observed so far have not looked typical of the species and possibly represent intermediate types between White-crowned Penduline-Tit and Chinese Penduline-Tit.
Voice: Vocalizations similar to White-crowned Penduline-Tit and Chinese Penduline-Tit, but calls slightly shorter than both and not as "flat" as White-crowned. Song a repeated *tsee-diu* or *tseiu … dz-du*.

文须雀科　Panuridae

Small songbirds living in wetlands of the Palearctic. Plumage differs between sexes. Common name in Chinese refers to the characteristic "black moustache" extending from lores to lower cheek in males. Prefer reed marshes along rivers or in lakes. Usually in pairs or small groups. Omnivorous; feed on the seeds of reed and other grasses but also take insects. Nomadic during nonbreeding seasons.

Bearded Reedling represents a unique clade. It was once grouped with tits, chickadees, and titmice in Paridae, but recent studies suggested it constitutes an independent family, more closely related to Alaudidae (larks). One genus and one species recognized worldwide, distributed in Europe and northern Asia. One genus and one species recorded in China, widely distributed in wetland habitats of northern China.

百灵科　Alaudidae

Small to medium-sized ground-dwelling songbirds, mostly sandy to brown colored. Plumage similar between sexes. Cone-shaped bill fine and long or strong and thick. Head usually crested. Good fliers, with long and pointed wings. Tail medium in length. Most species with well-developed hind claws. Inhabit primarily grasslands and various types of deserts; also in riverbeds, marshes, agricultural fields, and alpine meadows. Highly variable and melodic vocalizations; sing often in flight. Nest on ground or in bushes. Feed mostly on insects, grass seeds, and other plant seeds or shoots. Some species are migratory.

Twenty-one genera and 98 species recognized worldwide, distributed in the Old World, Oceania, and North and Central America; highly species rich in Africa. Seven genera and 15 species recorded in China, ubiquitously distributed nationwide.

Bearded Reedling

Black "moustache" ♂

Yellowish-brown head

No "moustache" ♀

歌百灵属
Mirafra

Horsfield's Bushlark

云雀属
Alauda

Eurasian Skylark

凤头百灵属
Galerida

Crested Lark

角百灵属
Eremophila

♂

Horned Lark ♀

短趾百灵属
Calandrella

Mongolian Short-toed Lark

百灵属
Melanocorypha

Mongolian Lark

沙百灵属
Alaudala

Asian Short-toed Lark

Panurus biarmicus

文须雀 wén xū què

Bearded Reedling　🟢 LC

L: 15–18 cm

Habitat: Often inhabits reedbeds in lakes or along rivers; sometimes also comes to reedbeds in urban parks.

Behavior: Often gathers in small flocks. Active. Fond of calling. Often hops about in reedbeds or climbs reed stems. Flight and wingbeats slow, flicking the long tail. Forages primarily for insects and grass seeds.

Distribution: Widespread in northern China. Uncommon breeder or winter visitor.

Voice: Far-carrying, very distinctive *ping* and, less commonly, a short *tutt*.

Mirafra javanica

歌百灵 gē bǎi líng

Horsfield's Bushlark　 LC

L: 14 cm
Habitat: Often inhabits dry grassy areas and scrubby plains and low mountains; prefers paddy fields and open areas with short grass.
Behavior: Often walks on ground or makes undulating flight. Great singer; sings in flight. Songs melodious and ringing. Perches in thickets.
Distribution: Rare resident in C Yunnan, Guangdong, and Guangxi. Vagrant to Hong Kong.
Voice: Chatlike, hesitant. Short song starts slowly, often with one to five sparrowlike introductory chirps, before accelerating into a variable warble and trill that frequently includes mimicry of others species, such as Richard's Pipit, Blyth's Pipit, Red-throated Pipit, and white-eyes.

Alauda gulgula

小云雀 xiǎo yún què

Oriental Skylark　LC

L: 14–16 cm
Habitat: Inhabits open plains, grassy areas, small hills, riverbanks, sandy beaches, grasslands, graveyards, deserted hillsides, farmland, and coastal plains.
Behavior: In pairs during breeding season; often in flocks outside breeding season. Good at running, primarily on ground. Often takes off suddenly from the ground and performs aerial displays. Difficult to distinguish from Eurasian Skylark in the wild; best identified by song.
Distribution: *A. g. lhamarum* in SW Tibet; *A. g. weigoldi* in Central and East China; *A. g. inopinata* in northwestern China, and S and E Qinghai-Tibet Plateau; *A. g. vernayi* in southwestern China; *A. g. coelivox* in southeastern China and Hainan; *A. g. wattersi* in Taiwan.
Voice: Song similar to Eurasian Skylark but includes dry, mechanical *drzz* or *bazz bazz* call notes.

> ### An alternative species delineation of Oriental Skylark–Eurasian Skylark complex
>
> The subspecies of the Oriental Skylark–Eurasian Skylark complex occurring in China are sometimes grouped as follows: *A. a. dulcivox* still in Eurasian Skylark (*A. arvensis*); *A. a. kiborti*, *A. a. intermedia*, *A. a. pekinensis*, and *A. a. lonnbergi* treated as "Pekin Skylark" (*A. pekinensis*); *A. a. japonica* and the subspecies of Oriental Skylark *A. g. wattersi* treated as "Japanese Skylark" (*A. japonica*); the subspecies of Oriental Skylark *A. g. lhamarum*, *A. g. weigoldi*, and *A. g. coelivox* (including *A. g. sala* as a synonym of *A. g. coelivox*) still in Oriental Skylark (*A. gulgula*); and *A. g. inopinata* and *A. g. vernayi* grouped as *A. vernayi*.

Alauda leucoptera

白翅百灵 bái chì bǎi líng

White-winged Lark　LC

L: 17–19 cm
Habitat: Inhabits dry grassy areas and low grasslands, especially on saline-alkaline soils with sparse and short vegetation.
Behavior: Often walks on ground or makes undulating flight. Great singer; sings in flight. Songs melodious and ringing.
Distribution: Rare in N Xinjiang.
Voice: Song similar to Eurasian Skylark but drier, more hesitant, stuttering, or halting. Includes mimicry, like Eurasian Skylark. Calls a dry *drrrt* or *drrrt-it-it*.

Galerida cristata

风头百灵 fēng tóu bǎi líng

Crested Lark　LC

L: 17–19 cm
Habitat: Inhabits dry plains, open fields, semideserts, and desert edges; also occurs in open areas, such as cultivated fields, abandoned fields, sandy or rocky mountain slopes, and lowlands.
Behavior: Often in flocks outside breeding season. Bold; unafraid of people. Good at running on ground; sometimes perches on bushes and soil mounds to rest.
Distribution: *G. c. magna* in W Inner Mongolia, Ningxia, NW Gansu, Xinjiang, and NE Qinghai; *G. c. leautungensis* in southwestern Northeast China, E Inner Mongolia, North and Central China, Qinghai, Gansu, S Tibet, and N Sichuan.
Voice: Flight song slower, shorter, and clearer than that of Eurasian Skylark; best recognized by inclusion of call notes. Simple song is a desolate, fluty *twee-tee-tuu*. Call a rising, nasal *djuuee*.

Alauda arvensis

云雀 yún què

Eurasian Skylark　LC

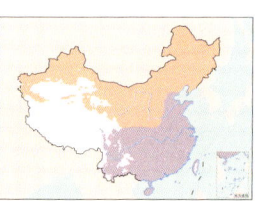

L: 16–18 cm
Habitat: Inhabits open plains, grasslands, marshes, cultivated fields, and coasts; also occurs on mountains with sparse trees and at forest edges. Especially prefers grassy areas near water.
Behavior: Similar to Oriental Skylark.
Distribution: *A. a. dulcivox* breeds in NW Xinjiang; *A. a. kiborti* in North and eastern Northeast China; *A. a. intermedia* in Northeast, North, Central, South, and eastern northwestern China; *A. a. pekinensis* in North and western Northeast China; *A. a. lonnbergi* vagrant in Jiangsu; *A. a. japonica* ("Japanese Skylark," 日本云雀) recorded during migration along Jiangsu coast.
Voice: Exuberant song delivered skillfully from ground or in towering flight: a very lengthy series of trills, whistles, and chirrups, including considerable repetition and some mimicry. Calls are variable, a dry *chirrup* or *prreet*.

Horsfield's Bushlark

Brown on wings

White-winged Lark

Rufous-brown crown lacks streaks

White wing patch

Rufous-brown upperwing coverts

Oriental Skylark

Buff-white outer tail feathers

Buff-white fringe on secondaries

Relatively short primary projection

Relatively short tail

Crested Lark

Long spiky crest

Ear coverts lack brown

Long decurved bill

Eurasian Skylark

Relatively long primary projection

Relatively long tail

Pure white outer tail feathers

Relatively broad white fringe on secondaries

Eremophila alpestris
角百灵 jiǎo bǎi líng
Horned Lark **LC**

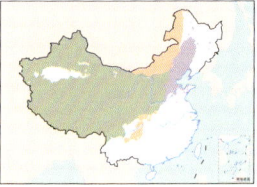

L: 16–19 cm
Habitat: Inhabits dry grassland habitats in alpine cold deserts, plateau grasslands, alpine meadows, deserts, and semideserts; also occurs in coastal regions, roadsides, and farmland in winter.
Behavior: Gathers in large flocks during nonbreeding season. Strong short-distance runner; flies a short distance only when danger approaches.
Distribution: *E. a. flava* winters in Heilongjiang, Liaoning, Beijing (very rare), Hebei, Inner Mongolia, and N Xinjiang; *E. a. brandti* in N and E Xinjiang, Inner Mongolia, Beijing, Hebei, Ningxia, E Qinghai, N Gansu, N Shaanxi, and N Shanxi; *E. a. albigula* in W Xinjiang; *E. a. argalea* in SW Xinjiang and W Tibet; *E. a. elwesi* in E Tibet, S Qinghai, and NW Sichuan; *E. a. przewalskii* in NW Qinghai; *E. a. teleschowi* in S Xinjiang; *E. a. khamensis* in Gansu, Tibet, Qinghai, and S and W Sichuan; *E. a. nigrifrons* in NE Qinghai.
Voice: Flight calls are variable but invariably short, fine, and high pitched, *tsiu*, *tis-i*, or *tsi-i-tiu*, and White Wagtail–like *chis-uu*. Song is a short jumbled repetition of a small number of high-pitched notes with a stumbling, tentative beginning and faster, "open-ended," jolting chatter, tone similar to calls.

Calandrella acutirostris
细嘴短趾百灵 xì zuǐ duǎn zhǐ bǎi líng
Hume's Lark **LC**

L: 13–14 cm
Habitat: Inhabits dry environment with stunted vegetation, such as dry steppes, plateaus, gravelly mountains, saline-alkaline deserts, and semideserts.
Behavior: Ground dweller; good at running and hopping. When disturbed, takes off immediately and flies a short distance (about 50 m). Often forages for insects and grass seeds on saline-alkaline grasslands and sandy fields with needle grasses near plateau lakes.
Distribution: *C. a. acutirostris* breeds in C and W Xinjiang, migrates through Tibet; *C. a. tibetana* breeds in W Inner Mongolia, Ningxia, N Sichuan, S Gansu, E Qinghai, S and SW Xinjiang, and Tibet.
Voice: Short and explosive song similar to Greater Short-toed Lark but usually slower, with shorter strophes, that begins quietly with two to five similar notes, then increases in volume before more rapid chatter and often a more gradual end. Includes whistles, harsh notes, and mimicry. Adjacent strophes sometimes the same but otherwise far more variation than in Greater Short-toed. Most common call a dry, rasping, or even buzzy *chrrrrrr* or *chrrrrrp*, very different from Mongolian Short-toed and Greater Short-toed Lark.

Alaudala cheleensis
短趾百灵 (亚洲短趾百灵)
duǎn zhǐ bǎi líng (yà zhōu duǎn zhǐ bǎi líng)
Asian Short-toed Lark **LC**

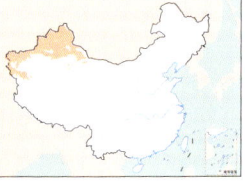

L: 13–14 cm
Habitat: Inhabits steppes, grasslands, and semideserts; especially prefers wet gravelly grasslands and grassy areas near water sources, such as lakes and rivers. Also occurs on dry steppes, foothill plains, and gravelly deserts with sparse vegetation and bushes.
Behavior: In pairs during breeding season; otherwise often gathers in flocks. Primarily on ground. Fast runner; often stops intermittently. Not shy. Usually makes vertical flights and sings during breeding season.
Distribution: *A. c. seebohmi* in Xinjiang; *A. c. kukunoorensis* in N Qinghai to SE Xinjiang; *A. c. tangutica* in NE Tibet to S Qinghai; *A. c. stegnmnni* in NW Gansu; *A. c. beicki* in W Inner Mongolia, Ningxia, N Gansu, and E Qinghai; *A. c. cheleensis* in Northeast and North China, N Ningxia, Shaanxi, Sichuan, Jiangsu, Zhejiang, and Taiwan.
Voice: Typical flight calls are characteristic soft, dry rattle, *prrrt* or *prrr-rrr-rrr*. Variable and melodic fast-paced song delivered during descending or circling flight; often includes rattles, trills, whistles, churrs, call notes, and considerable mimicry.

Calandrella dukhunensis
蒙古短趾百灵 měng gǔ duǎn zhǐ bǎi líng
Mongolian Short-toed Lark **LC**

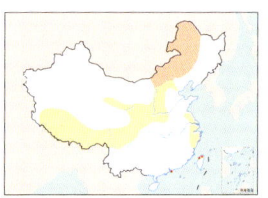

L: 14–15 cm
Habitat: Inhabits open dry steppes, deserts, and semideserts, especially dry sandy steppes and deserts with sparse vegetation and low bushes; also occurs in grassy areas near water and farmland.
Behavior: Often walks on ground, hides, and stays still when disturbed; difficult to find due to cryptic plumage. Flies and sings high in the air when displaying.
Distribution: A recently recognized species previously treated as a subspecies (*C. b. dukhunensis*) of Greater Short-toed Lark. Occurs in steppe and desert regions from western Northeast China to North China and the Qinghai-Tibet Plateau, as well as in Jiangsu, Shanghai, and Zhejiang; occasionally seen in Hong Kong and Taiwan.
Voice: Flight calls lower pitched, drier, more complex, and more muffled than Greater Short-toed Lark.

Calandrella brachydactyla
大短趾百灵 dà duǎn zhǐ bǎi líng
Greater Short-toed Lark **LC**

L: 14–15 cm
Habitat: Similar to Mongolian Short-toed Lark.
Behavior: Similar to Mongolian Short-toed Lark.
Distribution: *C. b. longipennis* in Xinjiang; subtle race *C. b. orientalis* presumed vagrant to W Heilongjiang and NW Jilin.
Voice: Song short (less than 2 sec and fewer than 15 notes) and explosive, with faltering start, rapid acceleration, and abrupt end, repeated with little variation. Most common call a dry *chirrup* or *dreet*, often repeated; also *teeoo* or *trilp*. Flight calls similar to *tjirp* of sparrows and *drelt* of skylarks.

Horned Lark

Short or inconspicuous crest

♀

E. a. brandti

Small horn-shaped black crest

♂

Broad distinct black breast band

♂

Black on side of head and breast connects

E. a. albigula

Yellow-tinged neck, cheek, and throat

♂

E. a. flava

Hume's Lark

Yellow bill with black tip

Grayer above than Mongolian Short-toed and Greater Short-toed Lark

Longest tertial cloaks primaries like Mongolian Short-toed and Greater Short-toed Lark

Black lores

Asian Short-toed Lark

Side of breast lacks black patch

Prominent primary projection

Sparsely streaked breast

Mongolian Short-toed Lark

Black patch on side of breast

Dark brown upperparts

Primaries cloaked by tertials

Relatively dark brown breast

Greater Short-toed Lark

Black patch on side of breast

Relatively pale brown upperparts

Primaries cloaked by tertials

Buff-tinged breast

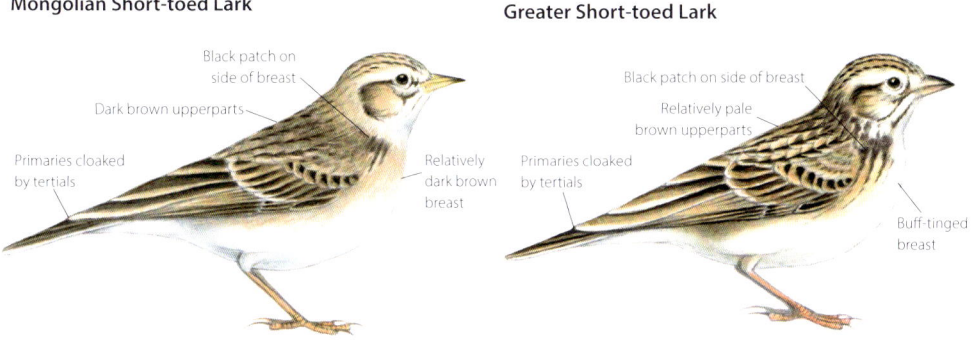

Melanocorypha bimaculata

双斑百灵（二斑百灵）
shuāng bān bǎi líng (èr bān bǎi líng)
Bimaculated Lark

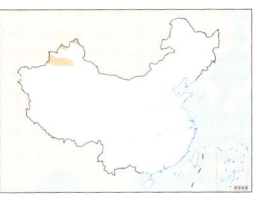

L: 16–18 cm

Habitat: Inhabits open steppes with sparse vegetation and river valleys; also on foothill plains of gently sloping mountains. Especially prefers semideserts with sparse vegetation and some gravel.

Behavior: Shy and alert. Strong runner and flier.

Distribution: *M. b. bimaculata* is an uncommon breeder in N Xinjiang.

Voice: Song very similar to Calandra Lark. Flight calls are a hoarse and loud *trrelit*, also *dre-lit*.

Melanocorypha calandra

草原百灵 cǎo yuán bǎi líng
Calandra Lark

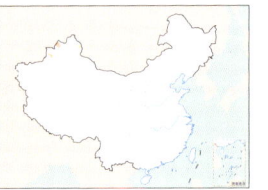

L: 18–20 cm

Habitat: Inhabits open areas, such as mountain grasslands, farmland in hilly regions, needle grass fields, and deserts with mugwort.

Behavior: Often in flocks during nonbreeding season. Usually stays on ground. Melodious and ringing calls in flight; also calls from ground.

Distribution: Breeder in NW Xinjiang, locally common.

Voice: Song similar to Eurasian Skylark (with even more mimicry) and includes distinctive buzzy notes.

Melanocorypha yeltoniensis

黑百灵 hēi bǎi líng
Black Lark

L: 18–21 cm

Habitat: Inhabits farmland and needle grass fields on foothill plains. Seen in desert steppes in winter.

Behavior: Gathers in flocks. Usually forages on ground. Males display with strong wingbeats, resembling large black bats. Males and females gather in separate flocks in winter.

Distribution: A stable small breeding population in NW Xinjiang (Tacheng). Wintering populations in N and W Junggar Basin fluctuate considerably in numbers from year to year.

Voice: Song similar to Eurasian Skylark but more "rambling," with frequent and erratic changes of tempo and pitch. Flight calls unremarkable, like Eurasian Skylark.

Melanocorypha mongolica

蒙古百灵 měng gǔ bǎi líng
Mongolian Lark

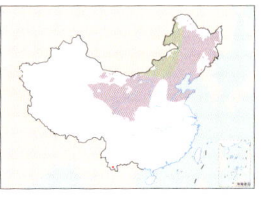

L: 17–22 cm

Habitat: Inhabits open areas, such as grasslands and semideserts; often occurs on saline-alkaline grasslands near water. Especially prefers moist areas with dense herbaceous plants. Sometimes also along roads or near human residential areas in winter. Avoids dry desert steppes.

Behavior: Sings often. Gathers in flocks in winter.

Distribution: Breeds in Inner Mongolia, SW Heilongjiang, W Jilin, Ningxia, W Gansu, and E Qinghai; winters south to Beijing, Tianjin, N Hebei, Shandong, and N Shaanxi. Vagrant to Yunnan.

Voice: Complex, highly accomplished, and variable songs, for which the species has often been illegally hunted and traded.

Melanocorypha maxima

长嘴百灵 cháng zuǐ bǎi líng
Tibetan Lark

L: 20–23 cm

Habitat: Inhabits marshes and meadow grasslands around lakes, at altitudes of 3200–4800 m. Easiest to find around lakes, bays, and river flats.

Behavior: Often walks on ground or makes gently undulating flight. Rises high into the air; strong runner on ground. Often hides and stands still when disturbed; difficult to find due to cryptic plumage. Forages for insects and seeds on ground.

Distribution: Resident in Tibet, S Gansu, S Xinjiang, NW Sichuan, and Qinghai.

Voice: Simple, slow, hesitant song, with short strophes and very considerable mimicry. (Some sequences consist almost exclusively of mimicry of Eurasian Skylark, Common Snipe, Horned Lark, Rosy Pipit, Chinese Gray Shrike, Common Redshank, or Common Tern, with very little "original" material.) Mellow flight calls, *djue-du-du*, reminiscent of Crested Lark.

Mongolian Lark

Bimaculated Lark

Black lores

Thick blunt bill

Pale yellowish-brown crown with chestnut border

Black neck patch

Black neck patch

Primary projection almost reaches tail tip

Short tail

Distinct white wing patch

Calandra Lark

Black underwing coverts

Broad white trailing edge on wing in flight

White lores

Thick blunt bill

Black neck patch

Tibetan Lark

Long slightly decurved bill

Distinct white tips on tertials and secondaries

Diffuse black patch on neck

Long tail

Primary projection does not reach tail tip

Black Lark

Nearly yellow bill

All-black body ♂

♀

Black wing feathers

Black legs

鹎科　Pycnonotidae

Small to medium-sized songbirds, found mostly in forests. Plumage similar between sexes, mostly olive, brown, black, white, and chestnut. Bill moderately long and thick. Wings mostly short and rounded. Tail moderately long, rounded or squared. Legs appear thin and weak. Inhabit primarily rain forests, broadleaf forests, scrubland, plantations and parks. Nest on trees or in bushes. Feed on insects and plant fruits. A few northerly breeding species are migratory.

　　Twenty-seven genera and 154 species recognized worldwide, distributed in the Old World and islands of N Oceania. Ten genera and 23 species recorded in China, found mostly in tropical and subtropical regions, a few species extending to North and Northeast China.

冠鹎属
Alophoixus

White-throated Bulbul

绿鹎属
Alcurus
(merged with *Pycnonotus* in eBird/Clements Checklist)

Striated Bulbul

伊俄勒短脚鹎属
Iole

Gray-eyed Bulbul

灰短脚鹎属
Hemixos

Chestnut Bulbul

纹胸鹎属
Ixos

Mountain Bulbul

短脚鹎属
Hypsipetes

Black Bulbul

黑头鹎属
Brachypodius

Black-headed Bulbul

黄鹎属
Rubigula

Black-crested Bulbul

雀嘴鹎属
Spizixos

Collared Finchbill

鹎属
Pycnonotus

Light-vented Bulbul

White-throated Bulbul

Grayish-white throat

Grayish-white cheek patch

Yellowish-green breast and belly

Alophoixus flaveolus

黄腹冠鹎 huáng fù guān bēi

White-throated Bulbul   LC

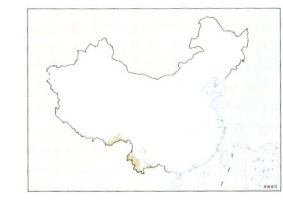

L: 19–24 cm
Habitat: Inhabits montane broadleaf forests or secondary forests at altitudes of 600–1800 m.
Behavior: Gathers in small flocks to feed on berries and insects in middle story of trees or in bushes. Often joins babblers and minivets in mixed-species flocks.
Distribution: Locally common in S and W Yunnan and SE Tibet.
Voice: Vocalizations include a nasal *bia* or *twerk*, given either singly or in a rapidly repeated sequence, *bia-bia-bia*.

Alophoixus pallidus

白喉冠鹎 bái hóu guān bēi

Puff-throated Bulbul   LC

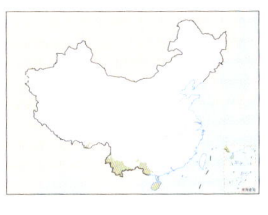

L: 20–25 cm
Habitat: Inhabits montane broadleaf forests and open woodland scrub at altitudes below 1500 m.
Behavior: Feeds primarily on berries in the middle story and understory of trees and in bushes; also feeds on insects and nectar.
Distribution: *A. p. henrici* in S and W Yunnan and S Guangxi; *A. p. pallidus* in Hainan. Locally common.
Voice: Raucous, staccato chattering, *chutt-chutt-chutt-*; sometimes a repeated disyllabic *ga-ga-*.

Puff-throated Bulbul

White throat feathers puffed out like "beard"

Dark gray face

Olive-brown breast and belly

Pycnonotus striatus

纵纹绿鹎 zòng wén lǜ bēi

Striated Bulbul   LC

L: 20–24 cm
Habitat: Inhabits montane forests at altitudes of 1000–2500 m; occasionally seen in forest-edge bushes.
Behavior: Prefers gathering in small flocks in the canopy of tall trees; seldom occurs in open scrub. Forages for berries and insects.
Distribution: *P. s. striatus* in S and W Yunnan and S Guangxi; *P. s. arctus* in SE Tibet. Sometimes placed in genus *Alcurus*.
Voice: Song a ringing, clear, disyllabic *tic-up* or *twi-wi*, repeated leisurely, a ringing, strongly inflected *chu-wick*, and a harsh, slurred *djrrri*.

Olive from crown to nape with short white streaks

Striated Bulbul

Fine white streaks on back

Broad white streaks on breast and belly

Iole propinqua
灰眼短脚鹎 huī yǎn duǎn jiǎo bēi
Gray-eyed Bulbul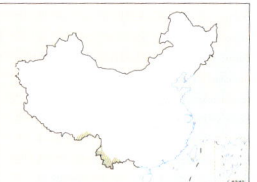

L: 18–21 cm
Habitat: Inhabits broadleaf forests on low mountains and plains at altitudes below 1500 m; also seen in open woodlands, scrubland, and orchards.
Behavior: Gathers in small flocks and forages on fruiting trees and bushes at forest edges and gaps; seldom on ground.
Distribution: *I. p. propinqua* is occasional in S and W Yunnan; *I. p. aquilonis* is occasional in SE Yunnan and SW Guangxi.
Voice: Distinctive, nasal, catlike mewing, *uuu-wit* or *brerrit*.

Hemixos flavala
灰短脚鹎 huī duǎn jiǎo bēi
Ashy Bulbul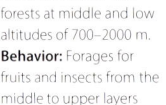

L: 19–22 cm
Habitat: Inhabits evergreen broadleaf forests and ravine rain forests at middle and low altitudes of 700–2000 m.
Behavior: Forages for fruits and insects from the middle to upper layers of trees; often gathers in small flocks. Forages in cultivated fields and orchards in nonbreeding season.
Distribution: *H. f. flavala* is occasional in SE Tibet and W and SW Yunnan; *H. f. bourdellei* is occasional in S Yunnan and SW Guangxi.
Voice: Stereotyped, nasal, rapid, and perky three- to five-note song, *phe-ttle-dot* or *skrink-er-rink*, interspersed with short whistles, churrs, and unobtrusive buzzy notes.

Hemixos castanonotus
栗背短脚鹎 lì bèi duǎn jiǎo bēi
Chestnut Bulbul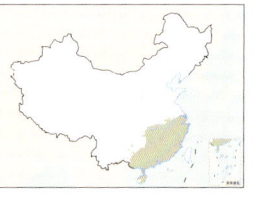

L: 19.5–21.5 cm
Habitat: Inhabits evergreen broadleaf forests, forest edges, and secondary forests on low mountains and hills at altitudes below 1000 m; migrates to lower altitudes during nonbreeding season.
Behavior: Often solitary or in pairs during breeding season; gathers in large flocks during nonbreeding season. Active and noisy. Often joins other bulbuls in mixed-species flocks in the canopy; also moves about and forages in understory bushes and small trees.
Distribution: Locally common resident. *H. c. canipennis* in southwestern, Central, South, southeastern, and East China, expanding north in recent years; *H. c. castanonotus* in Hainan.
Voice: Cheerful, simple song very similar to that of Ashy Bulbul.

Hypsipetes amaurotis
栗耳短脚鹎 lì ěr duǎn jiǎo bēi
Brown-eared Bulbul

L: 27–29 cm
Habitat: Occurs in deciduous forests and mixed forests; also occurs in urban parks.
Behavior: Not shy. Solitary or in small flocks. Moves about in canopy or at forest edges; flight undulating.
Distribution: Two subspecies occur in China: *H. a. amaurotis* is locally common in eastern Northeast China in winter, occasional in eastern North China, and rare in coastal regions of Shandong, Jiangsu, Shanghai, and Zhejiang; common passage migrant to Taiwan. *H. a. nagamichii* is endemic to China and found on the islands in E Taiwan, such as Orchid Island.
Voice: Noisy. High-pitched, piercing, scolding calls, *whee-wi-wi* or *weesp*. Song a rapidly repeated *wiccu-wiccu-wiccu*, heard far less frequently.

Hypsipetes leucocephalus
黑短脚鹎 hēi duǎn jiǎo bēi
Black Bulbul

L: 23.5–26.5 cm
Habitat: Occurs in evergreen broadleaf forests, mixed forests, and forest edges from 500 to 3000 m during breeding season; migrates to lower altitudes during nonbreeding season, wandering on low mountains and plains. Often occurs in broadleaf, mixed, and secondary forests and green spaces on low mountains. Occasionally on oceanic islands.
Behavior: Usually only in the canopy. Gathers in large flocks with hundreds of birds in winter; noisy. Black-headed morph and white-headed morph sometimes form mixed flocks, but more often in separate flocks.
Distribution: Locally common resident. *H. l. psaroides* in S and SE Tibet; *H. l. ambiens* in W Yunnan (Gongshan); *H. l. concolor* in W, SW, and S Yunnan; *H. l. leucothorax* breeds in S Shaanxi, SW and C Sichuan, Guizhou, and Hubei and migrates south in winter; *H. l. stresemanni* in N Yunnan—mountains in Lijiang and those between Lancang (Mekong) River and Nujiang (Salween) River; *H. l. sinensis* in NW Yunnan (Zhongdian, Weixi, Bijiang, and Hekou); *H. l. leucocephalus* in S Anhui, S Hunan, and Zhejiang, south to Guangdong, Guangxi, and Hong Kong; *H. l. nigerrimus* in Taiwan; and *H. l. perniger* in Hainan and SW Guangxi.
Voice: Slightly nasal, very simple song, with two alternating notes, *snyur-khwi*, repeated rapidly. Varied calls include a piercing *nyeee* and a catlike mewing, as well as a cacophonous chatter from flocks.

Gray-eyed Bulbul

Short brown crest

Clean olive-brown body almost without streaks

Brown-eared Bulbul

Chestnut ear coverts

Gray breast

Brown on breast connects with ear coverts

Relatively long tail

H. a. amaurotis

H. a. nagamichii

Ashy Bulbul

Blackish-gray crest

Grayish-brown or brown ear coverts

White throat contrasts strongly with gray breast and belly

Large yellowish-green wing patch

Black Bulbul

Black from forehead to crown, black patch below eye

Dark gray body

Bright red bill and legs

Black body with gray wing feathers

H. l. psaroides

Lacks white-headed morph

Black tail

White-headed morph

H. l. nigerrimus

White head and breast

Black-headed morph

H. l. leucothorax

H. l. ambiens

Darkest subspecies, body all black

All-black body

Chestnut Bulbul

Puffed, raised, blackish-brown crest

Chestnut upperparts contrast strongly with white underparts

Whitish throat and vent

Broad diffuse pale gray breast band

Blackish-brown or dark brown wings and tail

H. c. canipennis

Relatively pale chestnut upperparts

Olive wings

H. c. castanonotus

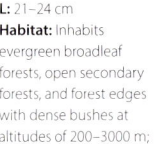

Ixos mcclellandii
绿翅短脚鹎 lǜ chì duǎn jiǎo bēi
Mountain Bulbul　LC

L: 21–24 cm
Habitat: Inhabits evergreen broadleaf forests, open secondary forests, and forest edges with dense bushes at altitudes of 200–3000 m; migrates to lower altitudes in winter.
Behavior: Often gathers in small flocks of three to five or a dozen individuals. Usually hops about or flies in the canopy or bushes and makes noisy calls. Often captures aerial insects from perch.
Distribution: Locally common resident. *I. m. mcclellandii* in S and SE Tibet; *I. m. similis* in S and W Yunnan and Hainan; *I. m. holtii* in most regions of Central, East, South, and southeastern China, east to S Jiangsu and north to the southern edge of the Yellow River Watershed.
Voice: Calls are a penetrating, high-pitched, descending, metallic *tsiu*; often as paired *chirrut-chewt* when repeated in lengthy, excited sequence, or perhaps used as a song.

Spizixos semitorques
领雀嘴鹎 lǐng què zuǐ bēi
Collared Finchbill　LC

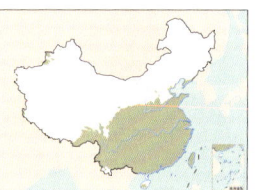

L: 21–23 cm
Habitat: Inhabits open wooded areas, secondary forests, scrubland, hills, and edges of towns from 400 to 1500 m; adapted to various types of secondary habitats.
Behavior: Often solitary or in pairs during breeding season. Usually perches on bare branches or power lines; flies into dense bushes when disturbed. Prefers gathering in flocks during nonbreeding season.
Distribution: Common resident. *S. s. semitorques* in central, southern, and eastern China, with apparent northward expansion in recent years. Records in North China increasing, but many are probably escapes. *S. s. cinereicapillus* in Taiwan.
Voice: Spluttering, staccato, slow chattering song, *chuwichu-chuwichu-chuwichu-chuw*. Calls are a rippling *wrrr*, often repeated in lengthy series.

Brachypodius melanocephalos
黑头鹎 hēi tóu bēi
Black-headed Bulbul　

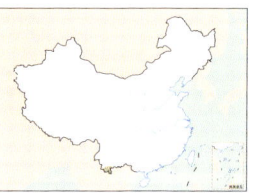

L: 17–19 cm
Habitat: Inhabits evergreen broadleaf forests on low mountains and open scrub at forest edges.
Behavior: Often gathers in small flocks. Prefers figs and other fruits; also hawks insects.
Distribution: Restricted to Yunnan (Xishuangbanna).
Voice: Simple song is a 3–8 sec series of well-spaced, short whistles or chirps that start slowly and meander erratically in pitch initially but soon become repetitive, *diu diu diu tit-titichitchittititititititit*. Calls are a scolding, explosive *jiu*.

Rubigula flaviventris
黑冠黄鹎 hēi guān huáng bēi
Black-crested Bulbul　

L: 18–21 cm
Habitat: Inhabits low mountains and hills, secondary forests, forest-edge bushes, farmland, and orchards.
Behavior: Often gathers in small flocks; also joins mixed-species flocks. Feeds on berries in the understory bushes of open woodlands. Seldom occurs in tall and dense forests.
Distribution: *R. f. flaviventris* in W Yunnan and SE Tibet; *R. f. vantynei* in S Yunnan and S Guangxi, common; *R. f. johnsoni* in W Guangxi, rare.
Voice: Considerable geographic variation in songs. *R. f. flaviventris* sings a pleasant, short, simple, musical ditty that starts with an abbreviated note followed by multisyllabic whistling, *we-twee-wee-u-wi*. Song of *R. f. vantynei* is less accomplished, short, and simple. Song of *R. f. johnsoni* is far less musical, flatter, and also short. Call is soft churr, *prrt*.

Spizixos canifrons
凤头雀嘴鹎 fèng tóu què zuǐ bēi
Crested Finchbill　

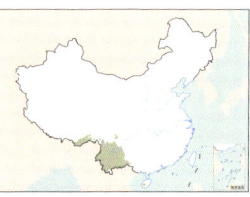

L: 18–22 cm
Habitat: Inhabits montane broadleaf forests, secondary forests, and forest-edge bushes from 1200 to 3000 m.
Behavior: Often in small flocks; gathers in large flocks in winter. Forages for fruits, seeds, and insects in the middle story of trees or bushes.
Distribution: Resident in C, W, and S Yunnan and SW Sichuan.
Voice: Songs are simple, dry, bubbling trills, at lower frequency for the first two or three notes, higher for the next three or four notes, *prrt, prr-t-ttit*. Short versions, *prrt* or *trrt*, given as contact calls, occasionally as lengthy chatter.

Mountain Bulbul

Slightly erect brown crest

Relatively slender black bill

Brownish-gray scapulars and back

Dense white streaks on gray throat

Pale brown face, side of neck, breast, and flanks

Bright olive wings and tail

I. m. holtii

Brownish-yellow vent and undertail coverts

Olive scapulars and back

Pale brownish-yellow lower mandible

Gray scapulars and back

I. m. mcclellandii

I. m. similis

Black-headed Bulbul

Black head lacks crest

Blue iris

Black subterminal band on yellow-tipped tail

Black-crested Bulbul

Black crest

White iris

R. f. flaviventris

Prominent crest

Red throat

Grayish-black tail tip

R. f. johnsoni

Collared Finchbill

Dark gray head

S. s. cinereicapillus

Black head

Cone-shaped ivory bill with white spot at base

Fine white streaks on cheek and ear coverts

Characteristic white neck collar

Brown-tipped tail

S. s. semitorques

Black-tipped tail

Crested Finchbill

Erect black crest

Yellow bill

Gray cheek

Pycnonotus finlaysoni
纹喉鹎 wén hóu bēi
Stripe-throated Bulbul　LC

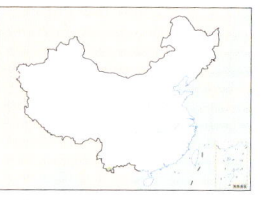

L: 19 cm
Habitat: Inhabits secondary forests, orchards, farmland, and villages on low mountains and plains.
Behavior: Omnivorous. Gathers in small flocks in the middle story and understory of trees or in orchards. Forages for berries and insects. Capable of flying a long distance, chasing after insects such as butterflies and moths. Also joins other bulbuls in mixed-species flocks.
Distribution: Rare in SW Yunnan.
Voice: Explosive scolding. Song a jumble of slightly hoarse chattering notes, *wit-chu wich-ik*, rising and falling in pitch. Contact calls are short, repeated, mellow churrs, *pur-whip-purr*.

Pycnonotus flavescens
黄绿鹎 huáng lù bēi
Flavescent Bulbul　LC

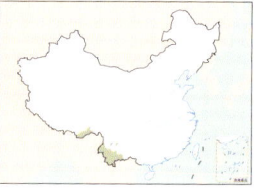

L: 19–22 cm
Habitat: Inhabits open secondary forests, forest-edge bushes, farmland, and orchards on low mountains and plains at altitudes of 400–2000 m.
Behavior: Gathers in small flocks. Prefers foraging for berries and insects in dense bushes and the canopy. Quieter than other bulbuls.
Distribution: Locally common in S and W Yunnan and S Guangxi.
Voice: Song is an abrupt series of four to six rising hoarse whistles, *brr, brr … zink-zink-zink*. Calls include a lengthy spluttering or scolding chatter.

Pycnonotus xanthorrhous
黄臀鹎 huáng tún bēi
Brown-breasted Bulbul　LC

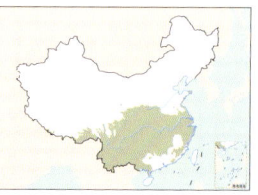

L: 19–21 cm
Habitat: Often occurs in secondary forests, scrubland, tall grasslands, and gardens at altitudes of 600–2300 m; recorded at altitudes up to 4275 m in N Yunnan. Migrates to lower altitudes in winter.
Behavior: Prefers gathering in flocks.
Distribution: Locally common resident. *P. x. xanthorrhous* in southwestern China, east to C Guangxi; *P. x. andersoni* in most regions in South, East, and Central China.
Voice: Songs are short, simple, rapid, and repeated, *churr-wee* or *churr-wi-uu*. Calls are a mellow purring or rapid *brzzp-brzzp-brzz*, similar in tone to the song.

Pycnonotus sinensis
白头鹎 bái tóu bēi
Light-vented Bulbul　LC

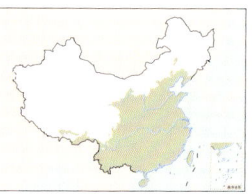

L: 18–20 cm
Habitat: Occurs in various habitats, including open secondary forests, cultivated fields, forest edges, orchards, gardens, scrubland, cities, countryside, and oceanic islands. Highly adaptable.
Behavior: The most common passerine in many cities. Active and noisy. Often gathers in small flocks of three to five individuals during breeding season; hops about and flies in trees. Gathers in flocks of dozens of individuals during nonbreeding season. Usually wanders and forages in trees; also comes down to bushes. Bold and unafraid of people.
Distribution: Common and widespread resident. *P. s. sinensis* is common in most regions of southwestern, South, southeastern, Central, East, and North China, expanding to northern regions in recent years; already to Northeast China. Populations in northern China migrate south after breeding season. *P. s. formosae* in Taiwan; *P. s. hainanus* in S Guangdong, S Guangxi, and Hainan.
Voice: Typical bulbul song—a simple, mellow, repetitive *wut-wut-wut … tik-er-ti-do*, repeated at leisure. Some geographic variation in its delivery. Calls varied, some chattering, scolding, and often loud, *tocc-tocc-toc*.

Pycnonotus taivanus
台湾鹎 tái wān bēi
Styan's Bulbul　VU

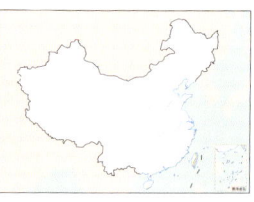

L: 18–19 cm
Habitat: Inhabits lowland secondary forests, farms, and villages in coastal regions.
Behavior: Often solitary or in pairs in trees during breeding season; gathers in large flocks during nonbreeding season.
Distribution: Chinese endemic; locally common resident in Taiwan. Often hybridizes with the sympatric subspecies of Light-vented Bulbul (*P. s. formosae*) in Taiwan.
Voice: A variety of vocalizations, with short, choppy, cheerful song similar to Light-vented Bulbul.

Stripe-throated Bulbul

Yellowish-green forehead

Short black streaks on yellow throat

Yellowish-green vent

Light-vented Bulbul

Small crest on blackish-brown crown

Large white patch behind eye extends to nape, patch smaller in northern populations

Brown ear coverts

Black moustachial stripe

White from throat to upper breast

Blackish-brown crest lacks white

P. s. sinensis

P. s. hainanus

Flavescent Bulbul

Thick short white supercilium

Black lores

Pale yellow vent

Styan's Bulbul

Blackish-brown crest and forehead

White ear coverts

Characteristic small red spot at base of bill

Pale grayish brown from hindneck to back

Blackish-brown moustachial stripe

From side of neck to breast slightly tinged pale pinkish gray

Brown-breasted Bulbul

Black hood

Pale brown breast band

Bright yellow vent

Dark brown tail

Pycnonotus jocosus

红耳鹎 hóng ěr bēi

Red-whiskered Bulbul

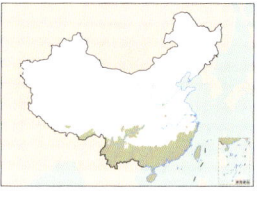

L: 18–20.5 cm
Habitat: Usually occurs at middle to low altitudes; adapted to various habitats, including montane secondary forests, parks, orchards, roads, reedbeds, villages, and human residential areas.
Behavior: Prefers gathering in flocks. Active. Often occurs in trees and bushes.
Distribution: Locally common resident. *P. j. monticola* in SE Tibet and W Yunnan. *P. j. jocosus* in S Yunnan, S Guizhou, N and W Guangdong, Hong Kong, SW Guangxi, Fujian, and S Zhejiang. *P. j. hainanensis* in S Guangdong; the populations on Hainan Island are believed to be introduced from South China. In recent years, many places in Central and North China have records of escaped or released individuals.
Voice: Songs are usually short, excited, and musical; the second note is often stressed (loudest and highest pitched), then the pitch descends. Also makes long songs *wit-ti-waet*, *queep kwil-ya*, or *queek-ka*. Calls are often fragments of songs, such as *queee-kwu*.

Pycnonotus aurigaster

白喉红臀鹎 bái hóu hóng tún bēi

Sooty-headed Bulbul

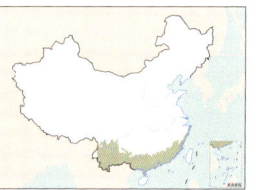

L: 19–21 cm
Habitat: Usually occurs in scrubland, secondary forests, meadows, towns, gardens, and cultivated fields at low altitudes.
Behavior: Often in pairs during breeding season; gathers in small flocks during nonbreeding season. Also joins Red-whiskered Bulbul and Brown-breasted Bulbul in mixed-species flocks. Active. Good at singing; often sings high on trees or bushes. Individuals maintain contact and communicate through calls. Forages in trees; sometimes comes down to bushes or ground.
Distribution: Locally common resident. *P. a. latouchei* in W to S Yunnan, Guizhou, SW Sichuan, S to E Hunan, and NE Guangxi; *P. a. resurrectus* in W Guangdong and from SE to SW Guangxi; *P. a. chrysorrhoides* in N and E Guangdong, Hong Kong, Fujian, Jiangxi, and S Zhejiang.
Voice: Songs are a loud and percussive *whiit-wit-i-wit*. Calls are usually mellow and simple but variable and typical of genus, *pu-tuu*, some with a tone reminiscent of Asian Green Bee-eater.

Pycnonotus cafer

黑喉红臀鹎 hēi hóu hóng tún bēi

Red-vented Bulbul

L: 19–23 cm
Habitat: Inhabits wooded areas on low mountains and hills, areas surrounding farmland and villages, and urban parks at altitudes below 1000 m.
Behavior: Prefers gathering in flocks. Noisy. Roosts in flocks at night. Omnivorous.
Distribution: Common in SE Tibet, S and W Yunnan.
Voice: Variety of notes, with the stress on the first and last notes, more subdued and less musical than Red-whiskered Bulbul. Calls are a simple and repetitive *dwee-dwee*.

Pycnonotus goiavier

白眉黄臀鹎 bái méi huáng tún bēi

Yellow-vented Bulbul

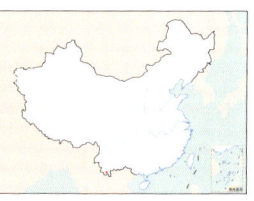

L: 19–21 cm
Habitat: Inhabits woodland scrub, reedy wetlands, urban parks, and farmland at altitudes below 1500 m.
Behavior: Omnivorous. Diet includes fruits, seeds, nectar, and a large number of invertebrates.
Distribution: First recorded in Yunnan (Xishuangbanna) in November 2019. Possibly resulting from the northward expansion from Southeast Asia.
Voice: Melodious fluty song an erratic, lengthy (up to 7 sec) jumble of short notes at various pitches, reminiscent of Stripe-throated Bulbul but considerably longer, faster paced, and "flatter." Calls are a short noisy *churt* or similar.

Pycnonotus leucogenys

白颊鹎 bái jiá bēi

Himalayan Bulbul

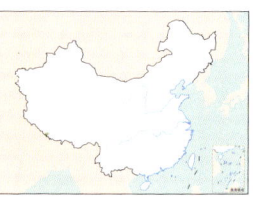

L: 19–21 cm
Habitat: Inhabits montane ravine forests and forest-edge bushes at altitudes of 300–2500 m.
Behavior: Often gathers in small flocks in bushes and open woodlands; forages for berries or small invertebrates.
Distribution: Occasional in S Tibet (Gyirong, Zhangmu).
Voice: Short lively song similar to Red-vented Bulbul.

Red-whiskered Bulbul

Erect crest on black head

Bright red spot below eye

White cheek and throat

Dark brown lightens to pale brown from side of neck to side of breast

P. j. jocosus

Red vent

Heavy uniform blackish brown from side of neck to side of breast

White tip on tail conspicuous in flight

P. j. monticola

Red-vented Bulbul

Short crest on black head

Grayish-brown belly

Red vent

Yellow-vented Bulbul

Brown crown

Broad white supercilium

Black eye stripe

Yellowish-green vent

Sooty-headed Bulbul

Black cap

Buff white from ear coverts to throat

Grayish-brown wings

Bright red vent

White tip on tail conspicuous in flight

Himalayan Bulbul

Gray crest curves forward

White cheek

Yellowish-green vent

燕科　Hirundinidae

Small agile songbirds with an elegant body profile. Plumage similar between sexes. Bill short, flat, and moderately wide. Wings long, pointed, and triangular; excellent fliers. Tail medium in length, emarginated or forked in most species. Legs short and thin, usually covered with feathers. Inhabit primarily cliffs, riverbanks, valleys, and agricultural fields; usually in flocks. Some species well adapted to human-dominated landscapes. Nest on cliffs, walls and roofs of buildings, or holes on sandy banks. Feed mostly on insects while flying. Most species are migratory.

　Eighteen genera and 88 species recognized worldwide, with a cosmopolitan distribution. Six genera and 14 species recorded in China, ubiquitously distributed nationwide.

沙燕属
Riparia

Bank Swallow

燕属
Hirundo

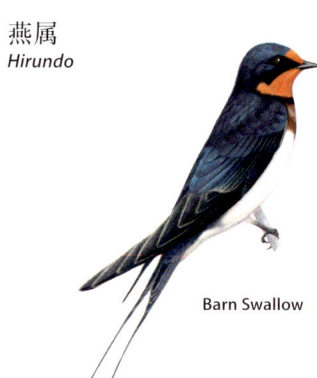

Barn Swallow

岩燕属
Ptyonoprogne

Eurasian Crag-Martin

毛脚燕属
Delichon

Asian House-Martin

金腰燕属
Cecropis

Red-rumped Swallow

石燕属
Petrochelidon

Streak-throated Swallow

Riparia chinensis

褐喉沙燕 hè hóu shā yàn

Gray-throated Martin

L: 10–13 cm

Habitat: Inhabits sandy beaches near wetlands at middle and low altitudes.

Behavior: Flies in flocks and forages on the wing.

Distribution: Locally common in S Yunnan, Guangxi, Hong Kong, and Taiwan.

Voice: Soft, easily overlooked, weak, twittering song. Calls include a purring *trrrr* and various shorter *chit* or *chut* notes, more intense and sharper when agitated. Some notes vaguely like house-martin; others similar to Bank Swallow.

Riparia riparia

崖沙燕 yá shā yàn

Bank Swallow

L: 12–13 cm

Habitat: Inhabits sandy beaches or banks near wetlands; also perches on branches or power lines.

Behavior: Flies in flocks. Hawks insects over water. Joins other martins and swallows in mixed-species flocks.

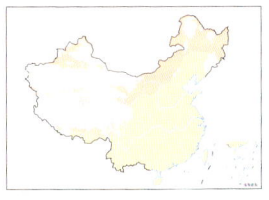

Distribution: *R. r. taczanowskii* breeds in Northeast China and migrates through North, Central, and South China; *R. r. riparia* breeds in Xinjiang and migrates through Western China. Seasonally common.

Voice: Dry, hoarse rasping, *ji–j.*

Riparia diluta

淡色崖沙燕 dàn sè yá shā yàn

Pale Sand Martin

L: 12–13 cm

Habitat: Active near wetlands.

Behavior: Similar to Bank Swallow. Nests on earthen cliffs; forages on the wing in large flocks over wetlands.

Distribution: *R. d. diluta* breeds in Xinjiang and N Qinghai; *R. d. tibetana* breeds in SE Tibet, Qinghai, and N Sichuan; *R. d. fohkienensis* breeds in Central, South, and southwestern China. Seasonally common.

Voice: Similar to Bank Swallow. Noisy at nest site.

Gray-throated Martin

Brown upperparts

Relatively uniform pale grayish brown on throat and upper breast

Whitish on remaining underparts

Bank Swallow

Clear boundary of ear coverts

Grayish-brown upperparts

Clear breast band

Pale Sand Martin

Grayish-brown upperparts

Breast band with diffuse boundary

Hirundo rustica

家燕 jiā yàn

Barn Swallow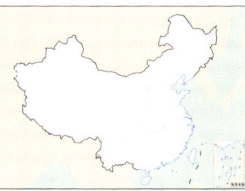

L: 17–19 cm

Habitat: Inhabits farmland and fields; especially prefers habitats near water. Also adapted to cities, but modern construction often cannot provide high-quality nest sites for it, resulting in population declines in cities.

Behavior: One of the most familiar birds. Collects mud and builds nests near water, usually under the eaves of buildings. Fledglings often perch on power lines waiting to be fed by parents. Adults hunt for insects on the wing in agile flight.

Distribution: *H. r. gutturalis* is a common and widespread breeder in most regions; winter visitor or resident in South China. *H. r. rustica* breeds in Xinjiang and Tibet. *H. r. tytleri* and *H. r. mandschurica* breed in Northeast China and migrate through eastern China; possibly winter in Yunnan.

Voice: Highly vocal, with wide variety of notes, including high-pitched twittering song, a high-pitched *ziz*, and sharper notes, such as *chi-chitt* or *flittt*, given in alarm.

Hirundo tahitica

洋燕 (洋斑燕) yáng yàn (yáng bān yàn)

Pacific Swallow

L: 16–18 cm

Habitat: Forages on the wing over villages, farmland, or forests; nests on mountain cliffs or walls and pillars of buildings.

Behavior: Often gathers in small flocks; forages on the wing, flying low.

Prefers perching on power lines or dead branches. Hovers near nest, rather than landing, when feeding nestlings.

Distribution: *H. t. namiyei* is common in Taiwan; *H. t. javanica* is occasionally seen on Green Island and Orchid Island, Taiwan.

Voice: Varied. Gives nasal *nyit* and churrs in contact, and shrill, piercing *ti-ti* (surprisingly reminiscent of Gray-headed Lapwing) in alarm.

Hirundo smithii

线尾燕 xiàn wěi yàn

Wire-tailed Swallow

L: 14–21 cm

Habitat: Inhabits open areas, grasslands, and villages near rivers and lakes; also occurs over flooded fields.

Behavior: Flight rapid, usually above water; sometimes flies low over water.

Distribution: Restricted to Yunnan (Dehong), where uncommon and local resident.

Voice: High-pitched, squeezed, twittering songs include lengthy trills and rattle notes, *chirrickwee*. Calls include *dili* and *chit-chit*.

Ptyonoprogne rupestris

岩燕 yán yàn

Eurasian Crag-Martin

L: 14–15 cm

Habitat: Inhabits primarily mountain valleys; especially prefers valleys with rivers or water sources. Nests and rests on cliff faces or in rock crevices.

Behavior: Solitary or gathers in loose small flocks. Prefers foraging on the wing above water, with variable flight speed. Individuals in some northern regions of China do not migrate in winter; instead, they hide in rock crevices during bad weather and come out to forage only when the weather improves.

Distribution: Widespread west of a line connecting S Liaoning, Taihang Mountains, Qinling Mountains, C Sichuan, and the Hengduan Mountains, but absent from N Xinjiang and central Qinghai-Tibet Plateau; locally common.

Voice: Gives short unobtrusive *prrt* flight calls.

Ptyonoprogne concolor

纯色岩燕 chún sè yán yàn

Dusky Crag-Martin

L: 13–14 cm

Habitat: Nests on cliff faces. Often occurs in mountain valleys and open areas near mountains and above rivers; occasionally above villages.

Behavior: Solitary or gathers in small loose flocks. Flight speed variable.

Distribution: Very rare in S Yunnan and SW Guangxi.

Voice: Makes gentle *chi* and *trrrt* calls in flight, the latter more forceful than Eurasian Crag-Martin.

Barn Swallow

Small area of chestnut on forehead

Chestnut- or brown-tinged central breast band with diffuse boundary

White or buff-yellow belly

H. r. gutturalis

Long outermost tail feathers

Bluish-black breast band has clear upper border

H. r. rustica

Pale salmon-red belly

H. r. mandschurica

Relatively narrow bluish-black breast band

Rufous-chestnut belly paler than throat

H. r. tytleri

Wire-tailed Swallow

juv.

Rufous-chestnut crown

White throat

White throat

Tail lacks long filamentous feathers; flat tail differs from Barn Swallow

ad.

Elongated outer tail feathers form filaments

Eurasian Crag-Martin

White spots on tail, except for outermost tail feathers

Relatively flat tail

Grayish-white throat and breast; lacks breast band

Pacific Swallow

Large area of chestnut on forehead reaches above eye

No breast band

Generally relatively short outermost tail feathers

Dusky Crag-Martin

Dusky brown throat and breast, color similar to belly

Delichon urbicum

毛脚燕（白腹毛脚燕）

máo jiǎo yàn (bái fù máo jiǎo yàn)

Common House-Martin

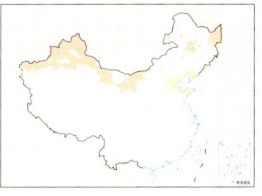

L: 13–14 cm
Habitat: Occurs in forests, wetlands, open farmland, and villages and towns. Nests on cliff faces or in buildings. Seldom rests in reedbeds during migration, unlike Barn Swallow.
Behavior: Gathers in flocks. Behaviors resemble other martins and swallows; usually forages high in the air.
Distribution: *D. u. lagopodum* breeds in Northeast China and migrates through most regions in eastern and central China; common or occasional. *D. u. urbicum* in Xinjiang and W Tibet; common. The two subspecies are sometimes treated as two separate species, *D. urbicum* and *D. lagopodum*.
Voice: Soft, dry twitters, *prrt* or *chi-ch*.

Delichon dasypus

烟腹毛脚燕 yān fù máo jiǎo yàn

Asian House-Martin

L: 11–13 cm
Habitat: Usually inhabits mountain valleys above 1500 m; also occurs in villages at middle and high altitudes. Occurs in various types of open areas during migration.
Behavior: Gathers in flocks; flocks especially large before precipitation. Flight rapid, wheeling in irregular routes.
Distribution: *D. d. dasypus* breeds in northern Northeast China and migrates through eastern China. *D. d. cashmeriense* breeds in northern North China, W Gansu, E Qinghai, W Hubei, and southwestern China; migrates through most regions in central China. *D. d. nigrimentale* is resident in Zhejiang, Fujian, Guangdong, and Taiwan.
Voice: Flight call, *prree*, similar to, but less harsh than, Common House-Martin.

Delichon nipalense

黑喉毛脚燕 hēi hóu máo jiǎo yàn

Nepal House-Martin

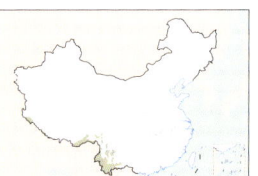

L: 11–13 cm
Habitat: Inhabits mountains with streams at altitudes above 1000 m; especially frequent near waterfalls.
Behavior: Colonial; nests on cliff faces or in rocky caves. Flies and forages along mountain ridges or treetops.
Distribution: *D. n. nipalense* in SW Tibet; *D. n. cuttingi* in C and W Yunnan.
Voice: Flight calls far "buzzier" than calls of other house-martins.

Cecropis daurica

金腰燕 jīn yāo yàn

Red-rumped Swallow

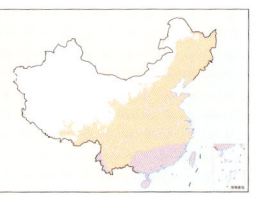

L: 16–20 cm
Habitat: Occurs in cities, farmland, and open areas near rivers.
Behavior: Typical swallow behaviors. Closely associated with human habitations. Sometimes mixes with Barn Swallow, but flight speed relatively slower. Bottle-shaped nests, more delicate than those of Barn Swallow.
Distribution: Widespread but absent from W Inner Mongolia, W Gansu, and C and W Qinghai-Tibet Plateau. *C. d. japonica* is a common breeder in eastern China, a wintering visitor or resident in S Guangdong, S Guangxi and Hainan, and a rare passage migrant to Taiwan. *C. d. daurica* is occasional visitor or very rare breeder in Xinjiang and northern Northeast China. *C. d. gephyra* in W Gansu, extreme E Qinghai, E Tibet to W Yunnan and Sichuan; this subspecies sometimes treated as a synonym of *C. d. daurica*. *C. d. nipalensis* in SE Tibet, Guangxi, and W Yunnan.
Voice: Flight calls more nasal than Barn Swallow—a rising *twerk*. Song much slower and hoarse.

Cecropis striolata

斑腰燕 bān yāo yàn

Striated Swallow

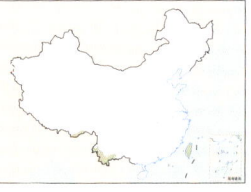

L: 17–19 cm
Habitat: Inhabits open areas in cultivated fields.
Behavior: Similar to Red-rumped Swallow.
Distribution: *C. s. striolata* in Taiwan; *C. s. stanfordi* in S and W Yunnan. Locally common. This species is sometimes treated as a subspecies of Red-rumped Swallow.
Voice: Nasal flight calls similar to Red-rumped Swallow. Song occasionally includes an intense, twangy, descending, chittering rattle.

Petrochelidon fluvicola

黄额燕 huáng é yàn

Streak-throated Swallow

L: 11–12 cm
Habitat: Inhabits open areas, piedmont hills, farmland, and villages; especially prefers habitats near water.
Behavior: Similar to other swallows.
Distribution: Vagrant. Only one record from Beijing (Changping) in 2014.
Voice: Variety of short, dry, rasping *trrrt* calls.

Common House-Martin

D. u. lagopodum

Pure white underparts

Deeply forked tail

Pure white uppertail coverts

Uppertail coverts with black spots

D. u. urbicum

Red-rumped Swallow

Chestnut on side of neck conspicuous

Relatively fine streaks on breast and belly

C. d. daurica

Orange rump lacks fine streaks

Yellow-tinged undertail coverts

Heavily streaked breast and belly

Brownish-yellow underparts

C. d. japonica

Asian House-Martin

Slightly forked tail

Smoky-gray or grayish-white belly

Striated Swallow

Chestnut side of neck inconspicuous

Fine streaks on yellow rump

Relatively thick streaks on breast and belly

Contrasting black thigh patch (not illustrated)

Whitish undertail coverts

Nepal House-Martin

Black throat

Flat tail, appears protruding in flight

White belly

Black undertail coverts

Streak-throated Swallow

Chestnut from forehead to crown extends to nape

Streaked throat and breast

White belly

369

鳞胸鹪鹛科   Pnoepygidae

Small ground-dwelling songbirds. Profile resembles wren-babblers, with extremely short tail. Plumage similar between sexes, mostly yellowish brown, black, and white, with scaly patterns on breast and belly. Neck short. Bill straight and thin, similar to babblers but weaker. Wings short and rounded. Strong legs and feet good for running and hopping. Prefers wet broadleaf, mixed, or bamboo forests and rhododendron thickets at middle and high altitudes. Behavior similar to wrens. Secretive; presence often revealed by high-pitched vocalizations. Nest in bushes, among rocks, in low tree branches, or on ground. Feed mostly on insects and insect larvae; also take plants seeds and fruits.

Until recently, cupwings were placed in Timaliidae, but recent phylogenetic studies suggest that they are only distantly related to the current members of Timaliidae and Sylviidae and therefore constitute this newly recognized family. One genus and five species recognized worldwide, endemic to Indomalaya. China is the core area of cupwing distribution, with all five species recorded; found in the Yangtze River Watershed and regions to its south, including Taiwan and Hainan.

Pnoepyga albiventer
鳞胸鹪鹛 lín xiōng jiāo méi
Scaly-breasted Cupwing   LC

L: 9–10 cm
Habitat: Inhabits damp shady alpine forests at altitudes of 2400–3800 m.
Behavior: Often hops and forages on ground covered with moss and fallen trees. Secretive and quiet.
Distribution: Occasional in W Yunnan and SE Tibet.
Voice: Song is an intense, powerful, hurried, explosive jumble of slightly shrill whistles that gradually decrease in pitch. Calls an intense, jolting *dzik*, like Pygmy Cupwing.

Pnoepyga mutica
中华鹪鹛 zhōng huá jiāo méi
Chinese Cupwing   NR

L: 9–10 cm
Habitat: Inhabits dense forests near streams and forest edges. Occurs at 1500–3000 m.
Behavior: Similar to Scaly-breasted Cupwing.
Distribution: Chinese endemic. Seen in W Hubei, S Shaanxi, N and W Sichuan.
Voice: Song is very similar to that of Scaly-breasted Cupwing; the highest frequency attained is well short of that of Scaly-breasted. Calls are indistinguishable from Scaly-breasted and Pygmy Cupwings.

Pnoepyga formosana
台湾鹪鹛 tái wān jiāo méi
Taiwan Cupwing   LC

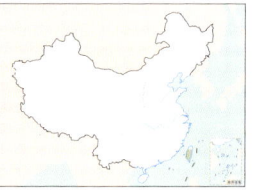

L: 8.5–9 cm
Habitat: Inhabits mid- to high-altitude broadleaf and mixed forests from 1200 to 2700 m.
Behavior: Prefers dense undergrowth with abundant leaf litter. Hops about in search for food. Occasionally flies short distances and moves rapidly among large trees.
Distribution: Chinese endemic; restricted to Taiwan.
Voice: Simple, distinctive song is a series of six or seven high-pitched, mostly flat, short whistles, *tz ... ti ... tchiti-tzu-wi*, with a long pause after the first note, a slightly shorter pause between the second and third, and then a fast conclusion. Call equally distinctive—a wheezy, loud, hissing *shhhhhh*.

Pnoepyga immaculata
尼泊尔鹪鹛 ní bó ěr jiāo méi
Immaculate Cupwing   LC

L: 8.5–9 cm
Habitat: Inhabits montane forests from 1600 to 3100 m. Prefers forest-edge bushes near streams where mosses and herbs are abundant.
Behavior: Often hops around to forage in bushes, on twigs, and on the ground.
Distribution: Rare in S Tibet (Zhangmu).
Voice: Very different from songs of Scaly-breasted Cupwing or Chinese Cupwing, song is a slightly longer series of six to eight flat whistles given in a slower, gradually descending sequence and lacking the intensity and changes in pitch and speed.

Pnoepyga pusilla
小鳞胸鹪鹛 xiǎo lín xiōng jiāo méi
Pygmy Cupwing   LC

L: 8.5–9 cm
Habitat: Inhabits mid- to high-altitude broadleaf forests and mixed forests from 1200 to 3000 m. Commonly seen in shady moist forests.
Behavior: Often moves about and forages on ground covered by moss and fallen trees. Calls as it moves. Easier heard than seen.
Distribution: Locally common in Sichuan, Guizhou, Yunnan, S Shaanxi, Anhui, Jiangxi, Fujian, and Hainan.
Voice: Distinctive song is a slow simple repetition of paired monosyllabic notes, the first higher pitched than the second, *ti ... tu*, occasionally trisyllabic. *Dzik* call very similar to Scaly-breasted Cupwing.

Scaly-breasted Cupwing

Pale scaly spots on head

Faint buff spots on wings

Dark morph

Scaly buff breast and belly

Tail about 2 cm, slightly longer than Pygmy Cupwing

Pale morph

Scaly white breast and belly

Taiwan Cupwing

Ochreous-buff mottling on head

Relatively thick black scaly pattern on underparts

Dark morph

Pale morph

Chinese Cupwing

Intense olive-brown back

Pygmy Cupwing

Inconspicuous scaly pattern on head

Two lines of clear brownish-yellow spots on wings

Extremely short tail, about 1 cm

Dark morph

Immaculate Cupwing

Almost no spots on back and wings

Slightly long bill

Pale morph

371

树莺科　Cettiidae

Small songbirds consisting of groups of species that differ markedly in morphology. Plumage similar between sexes, mostly olive, brown, yellow, black, and chestnut. Usually good fliers, with relatively long and pointed wings. Tail short or long. Strong legs and feet good for running and hopping. Inhabit primarily mid-level and lower vegetation in montane forests, bamboo forests, bushes, and grasses. Nest among branches. Feed primarily on insects and insect larvae. Some more northerly breeding species are migratory, and some perform altitudinal migration.

　　Cettiidae is a newly recognized family whose members were previously placed in Sylviidae. Seven genera and 32 species recognized worldwide, distributed in the Old World and N Oceania. Seven genera and 19 species recorded in China, found in eastern and southern regions.

拟鹟莺属
Abroscopus

Rufous-faced Warbler

拟缝叶莺属
Phyllergates

Mountain Tailorbird

暗色树莺属
Horornis

Brownish-flanked Bush Warbler

宽嘴鹟莺属
Tickellia

Broad-billed Warbler

树莺属
Cettia

Cetti's Warbler

地莺属
Tesia

Slaty-bellied Tesia

淡脚树莺属
Urosphena

Pale-footed Bush Warbler

Abroscopus superciliaris

黄腹鹟莺 huáng fù wēng yīng

Yellow-bellied Warbler (LC)

L: 9–10 cm
Habitat: Inhabits evergreen broadleaf forests, secondary forests, or surrounding bamboo thickets and tall bushes.
Behavior: Generally found in bamboo thickets. Often catches insects on the wing.
Distribution: *A. s. drasticus* in SE Tibet; *A. s. superciliaris* in SE Tibet, S Yunnan, S Guangxi, and Guangdong. Locally common.
Voice: Jaunty, halting song of three or four short flat whistles in an ascending sequence, *ter-ti-du-di*. Soft, slightly nasal churring when agitated.

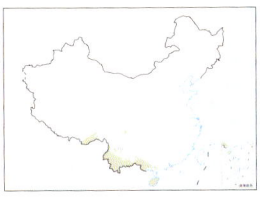

Yellow-bellied Warbler

Gray head
White supercilium
Olive upperparts
Grayish-white throat and breast
Bright yellow belly

Abroscopus albogularis

棕脸鹟莺 zōng liǎn wēng yīng

Rufous-faced Warbler (LC)

L: 8–9 cm
Habitat: Inhabits evergreen broadleaf forests, coniferous forests, bamboo thickets, and bushes on tropical and subtropical low hills. Found mostly at forest edges.
Behavior: Often solitary or in small flocks; feeds on insects.
Distribution: Common resident. *A. a. albogularis* in S and SW Yunnan. *A. a. fulvifacies* in Gansu, Shaanxi, Hubei, Jiangsu, and regions to the south, including Hainan and Taiwan; vagrant to Beijing and Hebei (Nanpu).
Voice: Song is a high-pitched, tremulous whistle of just less than a second, *riiiiiiiiiiii*, repeated every 3–5 sec. Short hard *chrt* calls repeated rapidly (at rate of over two calls per second) and in lengthy series (over 30 sec) when agitated.

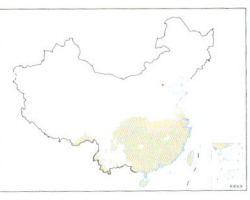

Rufous-faced Warbler

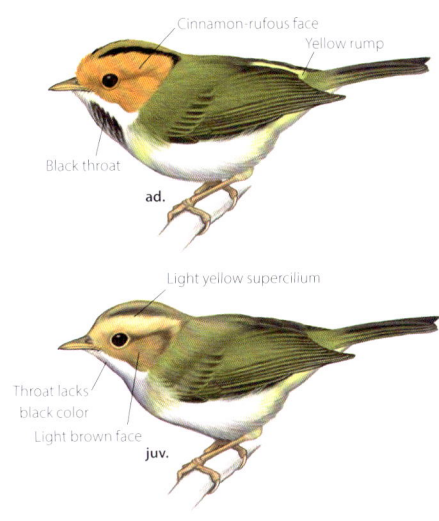

Cinnamon-rufous face
Yellow rump
Black throat
ad.

Light yellow supercilium
Throat lacks black color
Light brown face
juv.

Abroscopus schisticeps

黑脸鹟莺 hēi liǎn wēng yīng

Black-faced Warbler (LC)

L: 9–10 cm
Habitat: Inhabits mid-altitude evergreen broadleaf forests.
Behavior: Often gathers in small flocks, moving quickly through bushes. Also joins mixed-species flocks.
Distribution: *A. s. flavimentalis* (entirely yellow throat) in S Tibet and NW Yunnan. *A. s. ripponi* (yellow throat with a central dusky patch) in W and S Yunnan, Sichuan, and Guangxi. Locally common.
Voice: High-pitched, nervous, ethereal, somewhat tremulous, tinkling song, *tsirrririr tiririr-tsii*. Call similar but short and more obviously descending.

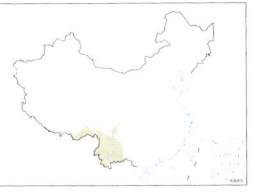

Black-faced Warbler

Wide black eye stripe
Gray crown, nape, and upper breast
Yellow supercilium and throat

373

Phyllergates cucullatus
栗头织叶莺 (金头缝叶莺)
lì tóu zhī yè yīng (jīn tóu féng yè yīng)
Mountain Tailorbird

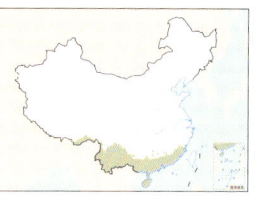

L: 12–13 cm
Habitat: Inhabits the understory or forest-edge bushes of evergreen broadleaf forests and bamboo thickets at middle and low altitudes.
Behavior: Often hides in dense undergrowth; secretive.
Distribution: Uncommon in Yunnan, Sichuan, Hunan, Guangdong, and Hainan.
Voice: Song a very thin, high-pitched series of four to seven notes, *pe-pe-prree-pe-de*, with slightly higher-pitched ending. Calls include a dry, nasal chatter, an often-descending trill, and, in alarm, a low scolding *chrt*.

Tickellia hodgsoni
宽嘴鹟莺 kuān zuǐ wēng yīng
Broad-billed Warbler

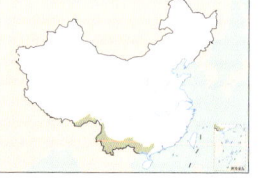

L: 10–11.5 cm
Habitat: Inhabits bamboo thickets and evergreen broadleaf forests at middle altitudes.
Behavior: Moves about in the lower and middle parts of undergrowth. Often joins mixed-species flocks.
Distribution: *T. h. hodgsoni* in SE Tibet; *T. h. tonkinensis* in SE Yunnan and Guangxi. Rare.
Voice: Song is very high-pitched series of about eight fine whistles given with very short pauses, so that they almost merge into one, *si-seeee-ee-eee*. Only slightly higher pitched but far less tremulous and juddering than song of Rufous-faced Warbler. Intense *tsik* calls often combined with strained "song," *twisip-wispi-tsik-tlip*, when agitated.

Horornis diphone
短翅树莺 (日本树莺)
duǎn chì shù yīng (rì běn shù yīng)
Japanese Bush Warbler

L: 14–18 cm
Habitat: Inhabits bushes and tall grass near forest edges. Often occurs in woodlands or scrubland on oceanic islands or coastal areas during migration and winter.
Behavior: Moves about solitarily or in pairs, hopping around in the understory of trees, in bushes, or on the ground. Feeds on insects.
Distribution: *H. d. cantans* is an uncommon winter visitor in Taiwan. *H. d. riukiuensis* occurs in small numbers along the east coast (including Taiwan) during migration and winter; residence status unclear.
Voice: Song like Manchurian Bush Warbler. Occasionally gives short repeated *ze-, ze-* during migration and winter. Calls are clear and crisp, unlike the mumbling calls of Manchurian.

Horornis canturians
远东树莺 yuǎn dōng shù yīng
Manchurian Bush Warbler

L: 15–18 cm
Habitat: Inhabits sparse broadleaf forests and scrubland below 1500 m in breeding season; prefers bushes or tall grassy areas near forest edges; also breeds in bushes on oceanic islands off the east coast. Often occurs in open forests or bushes on plains or hills during migration and winter.
Behavior: Similar to Japanese Bush Warbler. Often sings on prominent branches of bushes in breeding season.
Distribution: Relatively common. *H. c. canturians* breeds in Shaanxi, Gansu, Chongqing, Sichuan, Hunan, Hubei, and East China and winters from East to South China, with a few records from southwestern China. *H. c. borealis* breeds from Northeast to North China and migrates south to overwinter, with migration and wintering areas possibly overlapping with *H. c. canturians*.
Voice: Loud haunting song begins with a series of hollow, guttural sounds, followed by a series of very fast whiplash whistles, *guuuuuuu-guuuuuuu-lu-fenqiu*. Contact call a mellow *trrrt*, most similar to Hume's Bush Warbler but lower pitched, "fuller," and coarser. Alarm calls are a muffled, slowing, very lengthy series sounding like "teacup … teacup … teacup."

Horornis fortipes
强脚树莺 qiáng jiǎo shù yīng
Brownish-flanked Bush Warbler

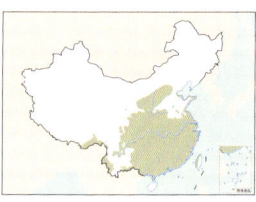

L: 11–12.5 cm
Habitat: Breeds in broadleaf forests and scrubland below 2400 m; descends to low altitudes in nonbreeding season, in orchards, tea plantations, and bamboo thickets at foothills, and green spaces, agricultural lands, and villages on plains.
Behavior: Often solitary. Moves through dense bushes and hops around constantly; far easier to hear than see in breeding season. Forages for various insects and some plant items.
Distribution: Common resident. *H. f. fortipes* in SE Tibet; *H. f. davidianus* widespread in Central, South, southeastern, and southwestern China south of Beijing, Hebei, Shaanxi, and Gansu; *H. f. robustipes* in Taiwan. *H. f. davidianus* and *H. f. robustipes* are sometimes treated as one separate species.
Voice: Clear, loud, three- or four-syllable song, starting with a long whistle, *weeeeeeee*, followed by two or three sharply divergent notes, *weeeeeee-chi-yuu*. Call a harsh *chuk*, often doubled and very similar to that of Aberrant Bush Warbler but slightly "softer." Call is a soft monosyllabic *tsk, tsk* or *trrt, trrt*.

Mountain Tailorbird

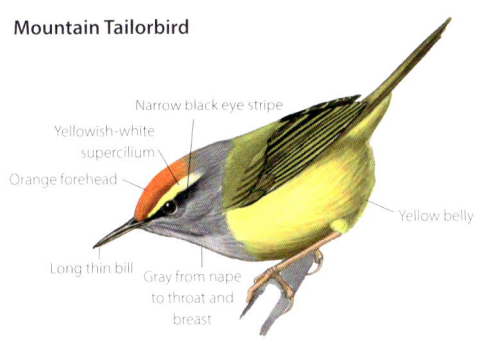

Narrow black eye stripe

Yellowish-white
supercilium

Orange forehead

Long thin bill

Gray from nape
to throat and
breast

Yellow belly

Manchurian Bush Warbler

Cinnamon-brown crown
contrasts with nape and back

Body significantly larger than
Brownish-flanked Bush Warbler,
with relatively long tail

Broad-billed Warbler

Relatively
short bill

Chestnut crown

Grayish supercilium
and eye stripe

White outer
tail feathers

Brownish-flanked Bush Warbler

Brownish overall

H. f. davidianus

Relatively strong
legs and toes

Japanese Bush Warbler

Olive-brown crown, slightly
darker than nape and back

Slightly brown-tinged upperparts

H. d. cantans

Crown and upperparts duller than *H. d. cantans*,
sometimes grayish green

H. d. riukiuensis

H. f. robustipes

Distinctive brown tone
from flanks to vent

H. f. fortipes

Horornis brunnescens

喜山黄腹树莺 (休氏树莺)
xǐ shān huáng fù shù yīng (xiū shì shù yīng)

Hume's Bush Warbler

L: 11–12 cm
Habitat: Inhabits bamboo thickets and dense understory bushes at middle and high altitudes.
Behavior: Very secretive. Often solitary under bushes.
Distribution: Uncommon in S Tibet.

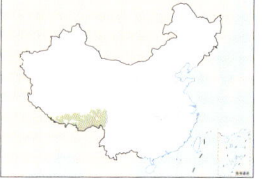

Voice: Remarkable manic song varies more than the similar song of Yellowish-bellied Bush Warbler. It usually consists of two parts, starting with three to five even, protracted, slightly strangulated, thin whistles on ascending scale, wiiiiiiii wiiiiiiii wiiiiiiii wiiiiiiii, changing abruptly to second part: a long series of very loud, rattling, hard, paired trills, such as witit-rr-r, repeated mechanically for up to 15 sec. Call a thin dry rattle.

Horornis acanthizoides

黄腹树莺 huáng fù shù yīng

Yellowish-bellied Bush Warbler

L: 10–11 cm
Habitat: Inhabits scrubland and bamboo thickets at middle to high altitudes.
Behavior: Secretive. Often solitary in bushes or grassy areas.
Distribution: *H. a. acanthizoides* in the mid- to high-altitude mountains in Shaanxi, S Gansu to Central, East, southwestern, and South China. *H. a. concolor* in Taiwan. Locally common.

Voice: See Hume's Bush Warbler for comparison of songs. Calls higher pitched and softer—more like Aberrant and Brownish-flanked Bush Warbler or even Eurasian Wren.

Horornis flavolivaceus

异色树莺 yì sè shù yīng

Aberrant Bush Warbler

L: 11–12 cm
Habitat: Inhabits understory bushes, bamboo thickets, and tall grass at middle and high altitudes.
Behavior: Solitary. Active, but often hides in dense thickets or tall grass.
Distribution: *H. f. flavolivaceus* in SE Tibet. *H. f. intricatus* in Shandong, S Shanxi, Shaanxi, Yunnan, and Sichuan. Uncommon.

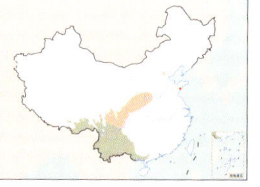

Voice: Pleasant whistled song comprises several extremely short, rising syllables (often inaudible), followed by a long whistle, t-t-t- we-weeeeeuu. Contact call very similar to Brownish-flanked Bush Warbler.

Tesia cyaniventer

灰腹地莺 huī fù dì yīng

Gray-bellied Tesia

L: 9–9.5 cm
Habitat: Inhabits understory bushes, bamboo thickets, and tall grass among mid-altitude evergreen broadleaf forests. Often occurs near streams.
Behavior: Often solitary or in pairs. Active in understory bushes or on the ground.
Distribution: Uncommon in S Tibet, S Sichuan, W and S Yunnan, and Guangxi.

Voice: Attractive, halting song is surprisingly loud and mellow. Starts with up to four short, high-pitched, deliberately spaced notes, followed by a sequence of lower-pitched whistles. Call a loud rattling trrrrrrk, lower pitched than Slaty-bellied Tesia.

Tesia olivea

金冠地莺 jīn guān dì yīng

Slaty-bellied Tesia

L: 9–9.5 cm
Habitat: Inhabits understory bushes in ravine forests and mid-altitude evergreen broadleaf forests.
Behavior: Often solitary or in pairs. Moves about near the ground. Secretive.
Distribution: Uncommon in SE Tibet, S and W Yunnan, Guizhou, and Guangxi.

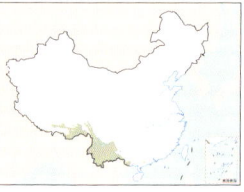

Voice: Loud song starts with a few alternating high- and low-pitched notes, ji-ju-ji-ju, that gradually accelerate to an intense, high-speed, explosive, jumbled crescendo. Principal calls a sharp tchirik and a spluttering rattle, trrrrt trrrrt trrrrt.

Cettia cetti

宽尾树莺 kuān wěi shù yīng

Cetti's Warbler

L: 13–14 cm
Habitat: Inhabits scrubland, reedbeds, and tall grass along riverbanks and marshy wetlands.
Behavior: Often forages on the lower parts of reeds and bushes, with tail occasionally cocked.

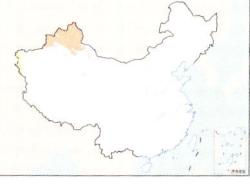

Feeds on insects and insect larvae, spiders, crustaceans, and other small aquatic invertebrates.
Distribution: Uncommon or locally common breeder in NW Xinjiang. Also uncommonly seen in S Xinjiang during migration or winter.
Voice: Distinctive, simple, stereotyped song is a loud staccato explosion of rich notes that begin with a short chip, followed by a heavy trill phrase, chuti-chuti chuti and its variations, and ending with a trill. Calls include an equally penetrating spik and intense rattle.

Hume's Bush Warbler

Long slender bill, almost entirely black

Narrow white supercilium

Light brown upperparts with olive-tinged back

Whitish throat and breast

Pale yellowish from belly and flanks to undertail coverts

Gray-bellied Tesia

Yellowish-green supercilium

Yellow lower mandible with black tip

Light gray underparts

Yellowish-bellied Bush Warbler

White supercilium

Rufous-brown crown and wings

Tail shorter than Brownish-flanked Bush Warbler

Grayish throat and breast

Pale yellowish belly

Slaty-bellied Tesia

Crown more yellowish than back

Orange lower mandible

Dark gray underparts

Aberrant Bush Warbler

Larger than Yellowish-bellied Bush Warbler

Earth-brown supercilium

Olive-brown upperparts

Pale throat and breast

Buff belly

H. f. intricatus

Cetti's Warbler

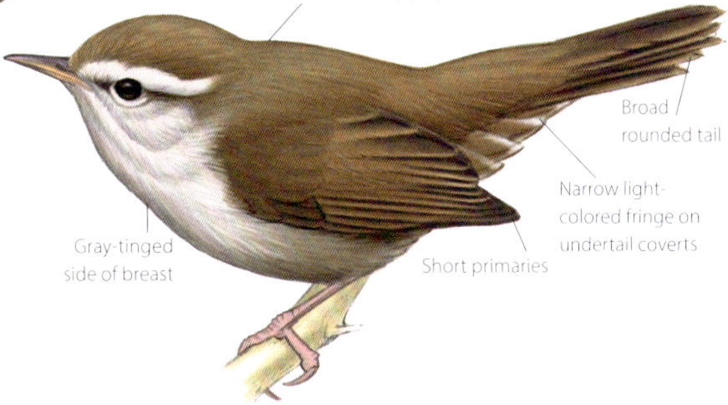

Rufous-brown upperparts

Broad rounded tail

Narrow light-colored fringe on undertail coverts

Gray-tinged side of breast

Short primaries

Cettia major
大树莺 dà shù yīng
Chestnut-crowned Bush Warbler   LC

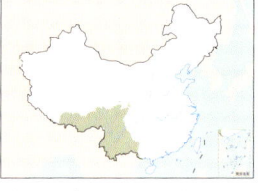

L: 12.5–13 cm
Habitat: Inhabits understory bushes, bamboo thickets, and rhododendron thickets at middle and high altitudes.
Behavior: Solitary or in pairs. Active, often in bushes.
Distribution: Uncommon in SE Tibet, W and NW Yunnan, and W Sichuan.
Voice: Song is an intense rapid warble that starts with a couple of short introductory notes and is followed by three to six powerful warbled notes, *i-i-iwi-wi-wirri-i*. Call is extremely similar to that of Gray-sided Bush Warbler—a sharp buntinglike *tzik*.

Cettia brunnifrons
棕顶树莺 zōng dǐng shù yīng
Gray-sided Bush Warbler   LC

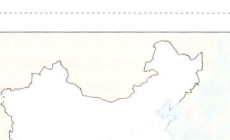

L: 10.5–11.5 cm
Habitat: Inhabits bushes in the understory or at forest edges at middle and high altitudes.
Behavior: Active. Often solitary or in pairs.
Distribution: Uncommon in S Tibet, W and NW Yunnan, and Sichuan.
Voice: Loud, wheezy, repetitive song starts with a strong, short, high-pitched note, followed by three prolonged notes, and ends with a few softer nasal notes, *twiss-wiss-is-is-is … neyu-neyu*. Call like Chestnut-crowned Bush Warbler.

Cettia castaneocoronata
栗头树莺 (栗头地莺) lì tóu shù yīng (lì tóu dì yīng)
Chestnut-headed Tesia   LC

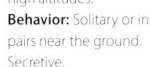

L: 8–8.5 cm
Habitat: Inhabits understory bushes, bamboo thickets, and tall grassy areas at middle and high altitudes.
Behavior: Solitary or in pairs near the ground. Secretive.
Distribution: *C. c. castaneocoronata* in S Tibet, W Yunnan, Sichuan, and NW Guizhou. *C. c. ripleyi* in SE Tibet, NW Yunnan, and Guangxi. Uncommon.
Voice: Loud high-pitched *tis tisu-eeet*. Calls a single sharp *tis*. Some geographic variation in both song and calls.

Urosphena squameiceps
鳞头树莺 lín tóu shù yīng
Asian Stubtail   LC

L: 9.5–10.5 cm
Habitat: Inhabits broadleaf and mixed forests below 1000 m in breeding season. Prefers primary mixed forests along streams; often forages in forest undergrowth, tall grass, fallen trees, and near streams.
Behavior: Particularly active when foraging, hopping around constantly and moving briskly.
Distribution: Breeds in mountainous areas in Northeast and North China. Often seen in most coastal areas during migration, and a few winter in S to SE Yunnan and coastal areas in South and southeastern China, including Hainan and Taiwan.
Voice: High-pitched insectlike song is a rising and intensifying series of short buzzing notes, *see-see-see-see*. Calls include a single *zik* and multisyllabic *chip-chip-chip*.

Urosphena pallidipes
淡脚树莺 dàn jiǎo shù yīng
Pale-footed Bush Warbler   LC

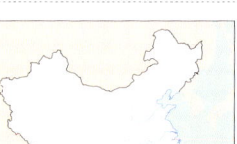

L: 9.2–9.6 cm
Habitat: In breeding season, inhabits evergreen broadleaf forests, secondary forests, bamboo thickets, and forest-edge bushes on hills and low mountains from 1000 to 1500 m. Moves to low-altitude scrubland in winter.
Behavior: Solitary or in pairs. Secretive; far easier to hear than see. Very active when foraging, hopping about constantly.
Distribution: Uncommon resident. *U. p. pallidipes* in W, SW, and S Yunnan; *U. p. laurentei* in SE Yunnan, C, W, and SW Guangxi, and N Guangdong. Occasional in Hong Kong and Macao in winter.
Voice: Song a loud, rich, and explosive *zip … zip-tschuk-o-tschuk*, similar to Slaty-bellied Tesia but lacks that species' introductory whistles. Call a sharp *chip-chip* or *chick-chick*. Spluttering ratchet call similar to that of Cetti's Warbler but lower pitched. Also a hard churr.

Chestnut-crowned Bush Warbler

Supercilium rufous buff at front

Distinct brown crown

Whitish underparts

Asian Stubtail

Fine scaly pattern on crown

Creamy-white supercilium from base of bill to neck

Extremely short tail

Gray-sided Bush Warbler

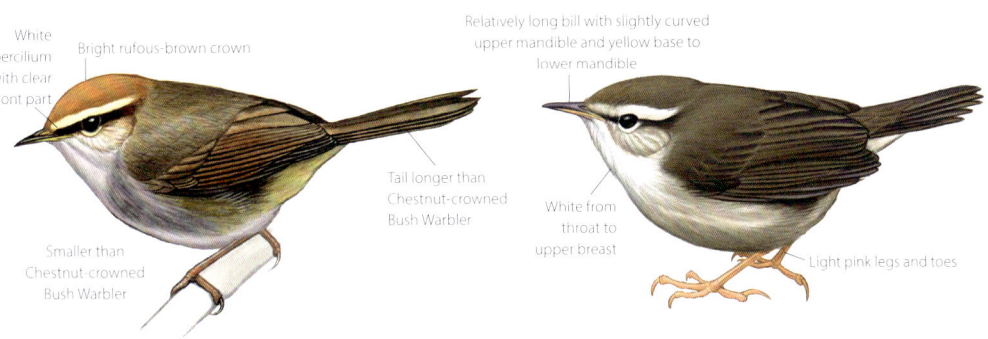

White supercilium with clear front part

Bright rufous-brown crown

Smaller than Chestnut-crowned Bush Warbler

Tail longer than Chestnut-crowned Bush Warbler

Pale-footed Bush Warbler

Relatively long bill with slightly curved upper mandible and yellow base to lower mandible

White from throat to upper breast

Light pink legs and toes

Chestnut-headed Tesia

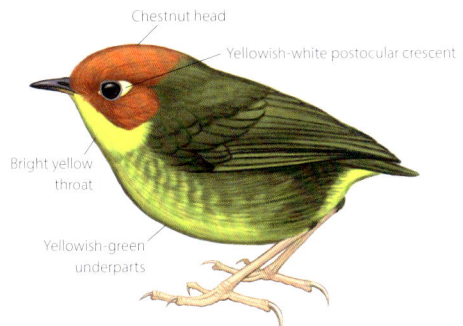

Chestnut head

Yellowish-white postocular crescent

Bright yellow throat

Yellowish-green underparts

长尾山雀科　Aegithalidae

Small songbirds whose morphology and behavior resemble tits, chickadees, and titmice. Plumage similar between sexes, mostly white, black, brown, and red. Body small and compact. Bill short and thick. Wings short and rounded. Legs and feet relatively strong. Tail long and narrow, with squared or rounded end. Inhabit primarily montane broadleaf, mixed, or bamboo forests and forest-edge bushes. Active. Nest among branches. Feed mostly on insects (including larvae and eggs); also take plant seeds and fruits in winter. Most species are resident, forming nomadic flocks during nonbreeding seasons.

Long-tailed tits and bushtits were previously placed in Paridae. Three genera and 13 species recognized worldwide, mostly in Eurasia except one species in North and Central America. Two genera and eight species recorded in China, ubiquitously distributed nationwide, with high richness in southwestern China.

长尾山雀属
Aegithalos

Black-throated Tit

雀莺属
Leptopoecile

White-browed Tit-Warbler

Aegithalos caudatus

北长尾山雀 běi cháng wěi shān què
Long-tailed Tit　🔵LC

L: 13–16 cm
Habitat: Inhabits broadleaf and mixed forests with rich bushes; sometimes seen in urban parks.
Behavior: Gathers in small or large groups; also joins mixed-species flocks.

Active from bushes to the canopy; rarely on the ground. Often hangs upside down from branches to feed. Forages only briefly on one tree before flying to the next; usually does not fly far.
Distribution: Common resident in Northeast China and N Xinjiang. Rare in winter south to N Hebei and Beijing.
Voice: Highly vocal. Calls include low purring churrs and higher-pitched trills and chatty clicking.

Aegithalos glaucogularis

银喉长尾山雀 yín hóu cháng wěi shān què
Silver-throated Tit　🔵LC

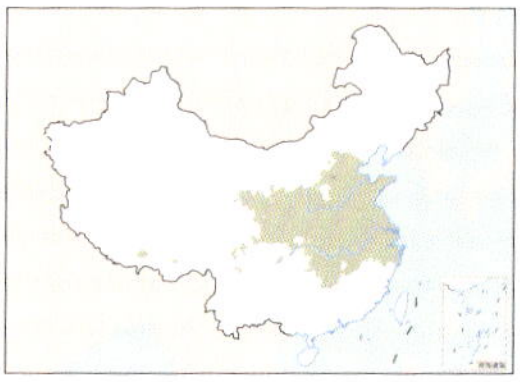

L: 13–16 cm
Habitat: Inhabits broadleaf forest edges and scrubland; also in urban parks.
Behavior: Similar to Long-tailed Tit.
Distribution:

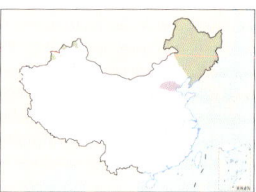

A. g. vinaceus is commonly seen in North China and ranges westward to C Gansu, E Qinghai, C Sichuan, and NW Yunnan. A. g. glaucogularis in Central and East China.
Voice: Similar to Long-tailed Tit, but churrs perhaps fractionally lower pitched, sii-sii.

Aegithalos concinnus

红头长尾山雀

hóng tóu cháng wěi shān què

Black-throated Tit

LC

L: 9–12 cm
Habitat: Inhabits
deciduous forest edges;
also occurs in urban green
spaces and residential
areas. Very common.
Behavior: Highly
gregarious; sometimes
join mixed-species flocks.
Very active; constantly hops among branches and rarely stays in one tree
for long period.
Distribution: *A. c. concinnus* widely distributed south of the Qinling-
Huaihe Line except for Hainan and southwestern China; also in S Gansu,
C to S Shaanxi, S Shanxi, and Henan; vagrant or escapes north to Beijing
and Shandong. *A. c. talifuensis* in southwestern China. *A. c. iredalei* in
S and SE Tibet; this subspecies is sometimes treated as a separate
species, *A. iredalei*.
Voice: Similar to other *Aegithalos*, but particularly clear, dry, purring trills
of continuous, fine *zii-zii-zii* sounds.

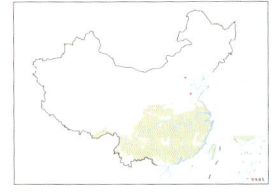

Aegithalos iouschistos

棕额长尾山雀 zōng é cháng wěi shān què

Rufous-fronted Tit

LC

L: 10–11 cm
Habitat: Inhabits
coniferous forests, mixed
forests, and alpine
scrubland from 2000 to
3000 m.
Behavior: Often joins
mixed-species flocks.
Feeds on insects, plant
seeds, and fruits in the canopy or bushes.
Distribution: Locally common in S and SE Tibet.
Voice: Short three- or four-syllable, fricative, metallic sound, like *zir-zir-
heer-*, sometimes with trills.

Long-tailed Tit

White head

Rufous-fronted Tit

Cinnamon-brown
forehead

Grayish-white
throat

Rufous-brown
belly

Silver-throated Tit

Black lateral
crown stripe

ad.

Rufous-brown
breast and cheek

juv.

Black-throated Tit

Incomplete
rufous-brown
breast band

White belly

A. c. concinnus

Broad white
supercilium

Lacks breast
band

Light brown
belly

A. c. iredalei

381

Aegithalos bonvaloti

黑眉长尾山雀 hēi méi cháng wěi shān què

Black-browed Tit

L: 10–11 cm
Habitat: Inhabits alpine coniferous and mixed forests from 1500 to 3500 m.
Behavior: Often gathers in small flocks; forages and moves through forests quickly, making frequent contact calls.

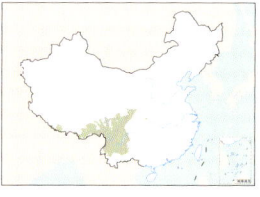

Distribution: *A. b. obscuratus* in C Sichuan; *A. b. bonvaloti* in W Sichuan, NW Guizhou, Yunnan, and SE Tibet. Locally common.
Voice: Similar to Rufous-fronted Tit but with slightly more muffled trills.

Aegithalos fuliginosus

银脸长尾山雀 yín liǎn cháng wěi shān què

Sooty Tit

L: 10–11 cm
Habitat: Inhabits subalpine broadleaf forests and rhododendron thickets at 1000–2600 m. Commonly found in oak forests and mixed forests with oak trees.
Behavior: Often in flocks;

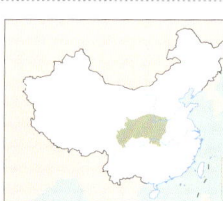

makes short hops within the canopy to forage.
Distribution: Chinese endemic. Distributed in S Shaanxi, S Gansu, C and N Sichuan, and W Hubei.
Voice: A pleasant tri- or multisyllabic song similar to high-pitched bell sounds. Alarm call is a *terr-terr-* trill, similar to Japanese Tit.

Leptopoecile sophiae

花彩雀莺 huā cǎi què yīng

White-browed Tit-Warbler

L: 9–12 cm
Habitat: Inhabits alpine dwarf forests, rhododendron thickets, and cold desert zones above tree line from 2500 to 5000 m. Darker-bodied *L. s. sophiae* and *L. s. obscurus* prefer high-

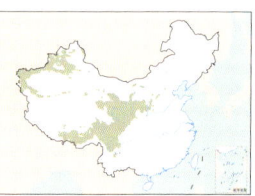

altitude moist forests, while lighter-colored *L. s. major* and *L. s. stoliczkae* favor low-altitude arid forests. The two groups are sometimes treated as two separate species.
Behavior: Forages mostly in small flocks. Moves very fast, catching insects in various ways. Sometimes joins mixed-species flocks. Often found in open scrub at forest edges, rarely in dense forests.
Distribution: *L. s. sophiae* in N Gansu, S Qinghai, and Xinjiang; *L. s. obscurus* in S Gansu, SE Tibet, E Qinghai, and Sichuan. *L. s. major* in W Xinjiang and W Qinghai; *L. s. stoliczkae* in Xinjiang, W Tibet, and W Qinghai.
Voice: Most common call, a buzzy *psrt*, frequently doubled, is identical to that of Crested Tit-Warbler. Also gives song that Crested doesn't match: a high-pitched descending *seeee* 0.2–0.3 sec long and similar to Bohemian Waxwing.

Leptopoecile elegans

凤头雀莺 fèng tóu què yīng

Crested Tit-Warbler

L: 9–10 cm
Habitat: Inhabits alpine coniferous forests, krummholz, and sparse scrubland from 3000 to 4000 m.
Behavior: Often moves from twig to twig. Foraging behavior similar

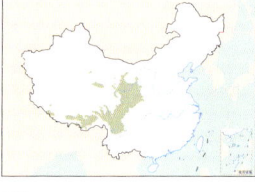

to that of nuthatches and Goldcrest.
Distribution: Chinese endemic. *L. e. meissneri* in SE Tibet; *L. e. elegans* in Qinghai, W and N Sichuan, and S Gansu. The species is sometimes treated as monotypic.
Voice: Most common call a buzzy *psrt*, identical to White-browed Tit-Warbler. Song is a distinctive series of three to five intense, very short, slightly rising, high-pitched, pulsed notes, *t-t-t-sip-sip*.

Black-browed Tit

White median crown stripe

Conspicuous white neck ring

Rufous-brown breast and flanks

Grayish-white belly

Sooty Tit

Silver-gray face

White throat and neck ring

Brown breast band

White belly

White-browed Tit-Warbler

Rufous-brown crown

White supercilium

Iridescent purplish-blue throat, breast, and rump

Violet-blue belly

L. s. obscurus

Pale chestnut crown

Buff belly

L. s. sophiae

Grayish-brown throat, breast, and upper belly

Wine-red crown

Buff breast, extending to throat

Whitish belly

L. s. stoliczkae

Crested Tit-Warbler

Pale purplish-gray crown with short crest

Turquoise back with metallic sheen

Rufous-chestnut face, neck, throat, and breast

Conspicuous black eye stripe

Grayish-brown crown

Rufous-brown flanks

Grayish-brown breast and belly

柳莺科　Phylloscopidae

A highly species-rich family of small insectivorous songbirds. Plumage similar between sexes, mostly olive, brown, and yellow. Bill thin and short. Wings short and rounded to long and pointed. Legs and feet relatively weak. Species similar in size and shape; best identified by plumage and voice. Prefers broadleaf, mixed, or coniferous forests and forest-edge bushes. Active across the canopy and in the lower and middle layers of vegetation. Nest among branches, rocks, or bushes. Feed primarily on insects, insect larvae, and other arthropods. Northerly breeding species are migratory.

Leaf warblers were previously placed in Sylviidae; now they constitute this newly recognized family. One genus and 80 species recognized worldwide, distributed in the Old World and islands of N Oceania. One genus and 51 species recorded in China, ubiquitously distributed nationwide.

柳莺属
Phylloscopus

Yellow-browed Warbler

Wood Warbler

Relatively broad yellow supercilium

Well-defined border between yellow breast and white belly

No wing bar

Long primary projection, similar in length to tertials

Buff-barred Warbler

Diffuse median crown stripe, indistinct on forehead

Dark olive upperparts

Pale yellow rump

White outer tail feathers

Orange wing bars

Ashy-throated Warbler

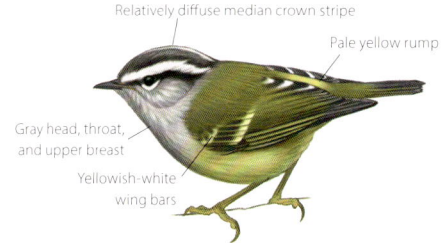

Relatively diffuse median crown stripe

Pale yellow rump

Gray head, throat, and upper breast

Yellowish-white wing bars

Hume's Warbler

Grayish overall, with hint of green

Pale fringe on tertials

Wing bar formed by lesser coverts less distinct

P. h. humei

Black bill and legs, differing from Yellow-browed Warbler

Indistinct median crown stripe

Grayish brown, instead of pale olive, from crown to upperparts, duskier than *P. h. humei*

P. h. mandellii

Yellow-browed Warbler

Diffuse median crown stripe

Yellow base of lower mandible

Pale fringe on tertials

Two wing bars

fresh

Yellowish-brown legs

worn

Phylloscopus sibilatrix
林柳莺 lín liǔ yīng
Wood Warbler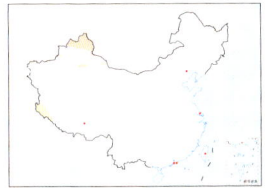

L: 11–12.5 cm
Habitat: In breeding season, inhabits coniferous, mixed broadleaf-coniferous, and broadleaf forests below 1500 m. Also occurs in willows in river valleys and vegetated plains during migration.
Behavior: Sings actively in breeding season. When singing, hops among branches or flies up from perch to midair and then drops back down; sometimes also sings on branches. Forages among branches in the undergrowth and canopy.
Distribution: Rare breeder in the W Altai Mountains and E Tian Shan Mountains in Xinjiang; occasional migration in N Xinjiang, SW Tibet, and Yunnan. Fall vagrant in Beijing, Jiangsu, Guangdong, Hong Kong, Taiwan, and Tibet.
Voice: Calls include a melancholy *siuh*, given singly or in series. Song an accelerating, descending, shimmering trill, *sip sip sip sip-sip-srrrrrrrrrrrrrr*, often likened to a coin spinning to a stop on a marble floor.

Phylloscopus pulcher
橙斑翅柳莺 chéng bān chì liǔ yīng
Buff-barred Warbler

L: 11–11.5 cm
Habitat: Inhabits coniferous forests, evergreen broadleaf forests, and rhododendron thickets at middle to high altitudes.
Behavior: Often solitary or in pairs. Active; moves about in the canopy or understory bushes.
Distribution: Locally common in S Shaanxi, W Inner Mongolia, Gansu, S Tibet, S Qinghai, Yunnan, and Sichuan.
Voice: Song an intensifying, gradually slowing, descending, insectlike trill lasting over 5 sec and often preceded by one or two call notes, reminiscent of the songs of Lemon-rumped Warbler and Sichuan Leaf Warbler but continuous and not in two parts. Calls an intense, short, sharp *tsip*, often repeated, similar to Ashy-throated Warbler, but very slightly lower pitched.

Phylloscopus maculipennis
灰喉柳莺 huī hóu liǔ yīng
Ashy-throated Warbler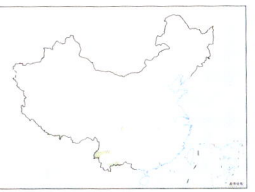

L: 9–9.5 cm
Habitat: Inhabits evergreen broadleaf forests and bamboo thickets at middle and high altitudes.
Behavior: Often solitary or in pairs. Very active.
Distribution: Locally common in S Tibet, Yunnan, and Sichuan. Vagrant north to Beijing.
Voice: Simple song a pleasant high-pitched *wi-tsi-wi-tsi-wi*, repeated. Call very similar to Buff-barred Warbler.

Phylloscopus humei
淡眉柳莺 dàn méi liǔ yīng
Hume's Warbler

L: 10–11 cm
Habitat: Inhabits coniferous forests, subalpine pine forests, birch krummholz, and scrubby alpine grasslands from 1000 to 3500 m. Particularly common in rhododendron thickets, pine forests, and birch krummholz; sometimes even found near snowline. Inhabits gullies, riverine broadleaf forests, orchards, sparsely wooded grassy slopes, and scrubby grasslands during migration and winter.
Behavior: Often solitary or in pairs; also gathers in small, loose flocks during nonbreeding season. Hops incessantly among branches and bushes; also moves about and forages near the ground.
Distribution: Locally common. *P. h. humei* breeds in northwestern China (Junggar Basin, Turpan, Kashgar, and Tian Shan Mountains) and winters in S Tibet. *P. h. mandellii* breeds from NW Yunnan to Sichuan, Qinghai, Gansu, Ningxia, S Shaanxi, SE Shanxi, and Hebei and winters in S Yunnan and SE Tibet; vagrant in Shanghai, Hong Kong, and Taiwan.
Voice: Distinctive song often starts with a call note, followed by a drawn-out, lengthy (1.5 sec), descending, nasal European Greenfinch–like buzzy wheeze, *zweeeeee*. Songs of both subspecies similar, but song of *P. h. mandellii* falls more markedly in pitch. Typical calls differ— *P. h. humei* is a clearly disyllabic *dsu-weet*, with second note rising, while *P. h. mandellii* is usually a stronger, more lisping, vaguely trisyllabic *dsu-uu-wit*, with an abrupt, falling ending.

Phylloscopus inornatus
黄眉柳莺 huáng méi liǔ yīng
Yellow-browed Warbler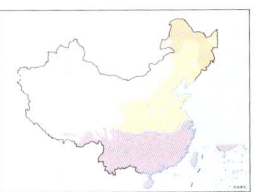

L: 10–11 cm
Habitat: In breeding season, inhabits coniferous forests, mixed broadleaf-coniferous forests, willow thickets, and forest-edge bushes from 1000–2400 m. Moves about in various wooded areas and bushes on plains and hilly areas in nonbreeding season.
Behavior: Often solitary or in small flocks; sometimes forms large groups during migration. Active; constantly hops among branches and bushes to forage.
Distribution: Common migrant. Breeds in Northeast China and winters from southwestern China and the Yangtze Plain to South China (including Taiwan); found in most parts of China during migration.
Voice: Calls a high-pitched, penetrating *tsooest*, similar to Coal Tit. Song is essentially an extended version of the calls.

Phylloscopus yunnanensis
云南柳莺 yún nán liǔ yīng
Chinese Leaf Warbler   LC

L: 9–10 cm
Habitat: Inhabits montane forests from 1000 to 2800 m in breeding season. Especially prefers mixed broadleaf-coniferous forests dominated by coniferous trees; also breeds in secondary evergreen broadleaf forests at middle and low altitudes. Inhabits broadleaf forests from 400 to 1800 m in nonbreeding season.
Behavior: Often occurs solitarily or in pairs in the upper parts of trees; very active when foraging.
Distribution: Locally common. Breeds in E Qinghai, S Gansu, Sichuan, Shaanxi, Shanxi, to N Hebei. Winters in S and SW Yunnan; migrates through Central and southwestern China. Vagrant in Shanghai, Hong Kong, and Taiwan.
Voice: In breeding season, males often sing from the top branches of pine trees. Song loud, crisp, mechanical chanting, often repeated for minutes on end, *tsiri-tsiri-tsiri*. Intense rising call note *twis* can be repeated rapidly as a second song type.

Phylloscopus chloronotus
淡黄腰柳莺 dàn huáng yāo liǔ yīng
Lemon-rumped Warbler   LC

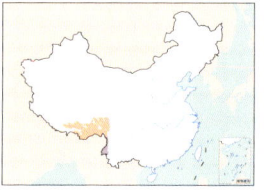

L: 10–10.5 cm
Habitat: Inhabits coniferous and mixed forests at middle and high altitudes.
Behavior: Active. Often gathers in small groups; hovers among branches and leaves.
Distribution: Locally common in E and S Tibet.
Voice: Two very different song types. A thin, drawn-out, straight rattle, immediately followed by a lower-pitched, even longer, straight rattle, such as *tsirrrrrrrrrrsisisisisisisisisisisisisisisisisisisi* (generally shorter than Sichuan Leaf Warbler), and an extraordinarily protracted, stuttering, "false starting" series of notes that include short trills and multiple different call notes, *tsetse tsirrp tsitsi tsetsetsetsetse*. Call, a disyllabic *chwis*, is shorter than that of Sichuan Leaf Warbler.

Phylloscopus forresti
四川柳莺 sì chuān liǔ yīng
Sichuan Leaf Warbler   LC

L: 9.5–10 cm
Habitat: Inhabits coniferous or mixed forests at middle and high altitudes.
Behavior: Often solitary or in pairs; sometimes joins other warblers and tits in mixed-species flocks.
Distribution: Locally common in S Shaanxi, S Gansu, S Qinghai, N Yunnan, Sichuan, and Chongqing.
Voice: Sings a descending two-part trill, similar to Lemon-rumped Warbler, but the opening sequence is higher pitched and descending, while the second part, the lengthy straight rattle, is slower paced. Call similar to that of Lemon-rumped but longer.

Phylloscopus kansuensis
甘肃柳莺 gān sù liǔ yīng
Gansu Leaf Warbler   LC

L: 9–9.5 cm
Habitat: Coniferous or mixed forests at middle and high altitudes.
Behavior: Solitary or in pairs in upper and middle stories.
Distribution: Locally common breeder in W and S Gansu and NE Qinghai; rare in southwestern China during migration.
Voice: Shimmering song is like Wood Warbler—a series of intense *tsip-tsip-tsip* call notes, followed by a rapid trill, like a coin spinning to a stop on a marble floor. Calls a multisyllabic *chi-si-wi*.

Phylloscopus proregulus
黄腰柳莺 huáng yāo liǔ yīng
Pallas's Leaf Warbler   LC

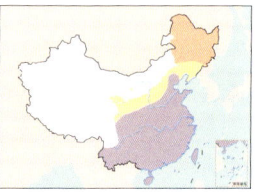

L: 9–10 cm
Habitat: In breeding season, inhabits coniferous and mixed forests, from foothills to tree line, up to 1700 m; sometimes also in broadleaf forests. During migration and winter, mostly in nurseries, orchards, gardens, or woodland bushes on plains and hills.
Behavior: Solitary or in pairs in the canopy; often joins small passerines, such as Yellow-browed Warbler and Goldcrest, in mixed-species flocks. Active and agile; often hops among branches in search of food. The lemon-yellow rump is clearly visible when hovering.
Distribution: Common migrant. Breeds in Heilongjiang, E and N Jilin, and N Inner Mongolia. Widespread in North, Central, East, South, southeastern, and southwestern China during migration and winter.
Voice: Males often stand between branches at the top of coniferous trees and sing loudly. Attractive, lively, highly varied song with accomplished trills and complex warbled phares, *tivi-tivi-tivi*, almost wrenlike in quality. Calls are a disyllabic, nasal, upward-inflected *dju-ee*, very different from call of Yellow-browed Warbler.

Phylloscopus cantator
黄胸柳莺 huáng xiōng liǔ yīng
Yellow-vented Warbler   LC

L: 10.5–11 cm
Habitat: Inhabits broadleaf forests at low and middle altitudes.
Behavior: Active in middle story; often joins mixed-species flocks.
Distribution: Uncommon in SE Tibet, SW and S Yunnan, and S Guangxi.
Voice: Simple, high-pitched, fine song is repeated every 8–10 sec and usually starts with a couple of short notes, followed by a pleasant sequence of four or, rarely, five notes, the first three of which are rising and the final one is falling, *tsi ... tsi-weet-weet-weet-seeu*. Rarely heard call, given singly or in extended series, is an intense call note, *tzis*, that rises at the very start but is otherwise flat.

Chinese Leaf Warbler

Upper mandible blackish brown, lower mandible yellow with blackish-brown tip

Pale crown with relatively diffuse median crown stripe

Gray-tinged green body lacks yellow tone

Pale fringe on tertials

Pale yellow rump, difficult to see in field

Two wing bars; front one relatively pale

Wing bars lack black

Yellowish-brown legs

Gansu Leaf Warbler

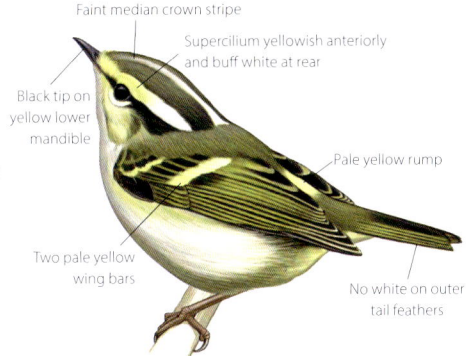

Faint median crown stripe

Supercilium yellowish anteriorly and buff white at rear

Black tip on yellow lower mandible

Pale yellow rump

Two pale yellow wing bars

No white on outer tail feathers

Lemon-rumped Warbler

Relatively light-colored median crown stripe

Pale yellow median crown stripe, supercilium, wing bars, and rump

Almost entirely black bill

No white on outer tail feathers

Pallas's Leaf Warbler

Well-defined median crown stripe

Supercilium relatively thick, bright yellow at front

Olive upperparts with yellowish-green wings

Pale fringe on tertials

Lemon-yellow rump

Two well-defined, yellow-tinged wing bars

Sichuan Leaf Warbler

Extremely similar to Lemon-rumped Warbler, best distinguished by vocalizations

Yellow-vented Warbler

Bright yellow median crown stripe and supercilium

Dark green lateral crown stripe and eye stripe

Two wing bars

Bright yellow throat, upper breast, and undertail coverts

White lower breast and belly

Phylloscopus armandii
棕眉柳莺 zōng méi liǔ yīng
Yellow-streaked Warbler

L: 12–14 cm

Habitat: In breeding season, inhabits mountains from 1200 to 3500 m and forests on foothill plains. Particularly prefers coniferous forests, poplar and birch forests, forest edges, and riverine bushes. Often forages on low scrubland and ground. Active in forests below 1500 m in nonbreeding season.

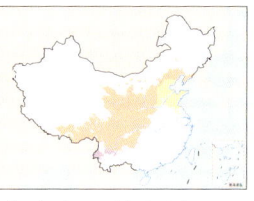

Behavior: Often solitary or in pairs; sometimes also forms small loose flocks. Hopping and foraging among branches and bushes.

Distribution: Locally common. *P. a. armandii* breeds in western Northeast China, North China, and E Qinghai-Tibet Plateau; *P. a. perplexus* breeds in Ningxia, Hubei, and southwestern China and winters in W and S Yunnan and Guizhou.

Voice: Complex, highly variable song lasts 1.1–1.4 sec; usually starts with one or two calls and skillfully mixes trills with more complex short warbles, such as *zik ... zik-trr-dwitt-dwitt-dwitt-dwitt-dwi*. Song repeated every 3–5 sec and is more reminiscent of Gray-crowned Warbler than Radde's Warbler. Distinctive, sharp, buntinglike call, *zik*.

Phylloscopus schwarzi
巨嘴柳莺 jù zuǐ liǔ yīng
Radde's Warbler

L: 12.5–13.5 cm

Habitat: Inhabits low mountains and hills below 1400 m and foothill plains; occurs mostly at the edges of broadleaf and mixed forests in breeding season. Forages in the understory bushes and low branches of broadleaf forests or in grassy areas at forest edges.

Behavior: Solitary or in pairs. Vigilant. In breeding season, male often perches on top of bushes or low trees and sings all day; especially active in the early morning. Hops around frequently near the ground.

Distribution: Locally common migrant. Breeds in the Greater and Lesser Xing'an (Khingan) Mountains in Northeast China; some winter in S Yunnan, Guangdong, and Hong Kong. Seen across eastern China during migration.

Voice: Often-loud song typically starts with one or two tentative, nervous notes but is mostly powerfully delivered, repeated notes delivered faster and more "explosively" than Dusky Warbler, *jiao-jiao-jiao*. Call a nervous, nasal *prrp*, far less sharp and clear than the "smack, smack" of Dusky Warbler.

Phylloscopus fuligiventer
烟柳莺 yān liǔ yīng
Smoky Warbler

L: 10–11 cm

Habitat: Inhabits alpine scrubland and gully grasslands from 3000 to 4500 m. Migrates to lower altitudes in winter.

Behavior: Often moves about solitarily or in pairs in low bushes or on the ground; flicks wings and tail when foraging.

Phylloscopus griseolus
灰柳莺 huī liǔ yīng
Sulphur-bellied Warbler

L: 11–12 cm

Habitat: Inhabits open forests close to tree line and adjacent alpine and upland scrubland; particularly prefers rocky slopes and gully bushes. Also inhabits forest edges and grasslands with sparse trees or bushes.

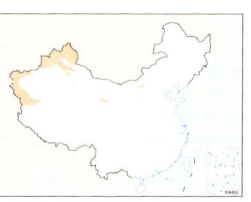

Behavior: Prefers moving around in rock crevices or under bushes; sometimes forages in shrubs or on tree trunks. Mostly rests on ground or rocks.

Distribution: Uncommon breeder in C Inner Mongolia, Xinjiang, and N Qinghai.

Voice: The song is about a second long, is introduced by a call note, and consists of four to five notes at the same pitch, *tsi-tsi-tsi-tsi-tsi*, similar to song of Tickell's Leaf Warbler but slower (so that all notes can be counted), mellower, and slightly more "muffled." Call is a characteristic, soft *stup* note, with a nasal tone reminiscent of Radde's Warbler yet harder. Much more muffled and mellower than *quip* of Tickell's Leaf Warbler.

Phylloscopus fuscatus
褐柳莺 hè liǔ yīng
Dusky Warbler

L: 11–12 cm

Habitat: Occurs from foothill plains to montane forests and alpine scrubland above tree line, up to 4500 m. Especially prefers sparse and open forest edges of broadleaf, mixed, and coniferous forests, and woodlands and scrubland along streams; during migration and winter, inhabits farmland, orchards, woodlands, and deserted fields on plains and hills. Feeds mostly in forest understory and forest edges, also in scrub and grass along streams.

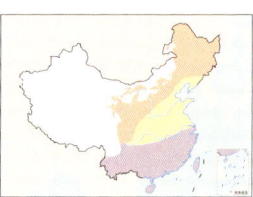

Behavior: Often solitary or in pairs. Active; hopping among branches in search of food.

Distribution: Fairly common. *P. f. fuscatus* breeds in Northeast China and overwinters in most of southeastern, South, and southwestern China, including Hainan and Taiwan. *P. f. robustus* breeds in Gansu, Inner Mongolia, and Sichuan; found in most of China during migration.

Voice: Simple song (with usually just one and, very occasionally, two notes, repeated) sounds drier, slower, and more repetitive than Radde's Warbler, *chett, chett, chett, chett*—and the tentative introductory notes are usually absent; variable. Call is a sharp, stone-clicking, or Lesser Whitethroat–like *check* or *chack*.

Distribution: *P. f. weigoldi* is locally common in S and E Tibet, E Qinghai, and W Sichuan; *P. f. tibetanus* in E Tibet and *P. f. fuligiventer* in W Tibet.

Voice: Song and calls of *P. f. weigoldi* very like Dusky Warbler. Call of *P. f. fuligiventer* a mellow, unobtrusive *trrt-trrt*, reminiscent of Aberrant Bush Warbler. Song of the latter is a short *P. f. weigoldi*-like trill, with multiple *tsli-tsli* or *trrssuu* notes admixed.

Yellow-streaked Warbler

Pale buff supercilium; yellow-tinged at front

Green-tinged olive-brown upperparts (including crown, neck, back, rump, and uppertail coverts), no wing bars

Slightly curved dark brown upper mandible, yellow lower mandible

Pale buff undertail coverts

Sulphur-bellied Warbler

Bright yellow supercilium, paler at rear

Cold brown upperparts

No wing bar

Sulphur-colored underparts

Radde's Warbler

Resembles Yellow-streaked Warbler, but supercilium more yellowish and not well defined at front

Bill and legs stronger

Heavy brown-buff undertail coverts

Dusky Warbler

Supercilium white with well-defined front part and brown-tinged at rear

Bill and legs thinner than Radde's Warbler

Brown or olive-brown upperparts

Pale brown undertail coverts, not as intense as Radde's Warbler

Smoky Warbler

Dark brown from crown to back

Olive-brown breast and belly tinged smoky gray

Blackish-brown tail

Phylloscopus affinis

黄腹柳莺 huáng fù liǔ yīng

Tickell's Leaf Warbler　LC

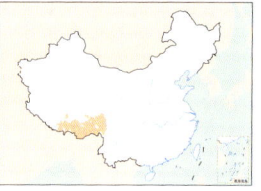

L: 11–11.5 cm
Habitat: Inhabits open scrubland and tall grass at middle and high altitudes.
Behavior: Solitary or in small flocks in bushes near the ground.
Distribution: Locally common in S Tibet.
Voice: Call a sparrowlike *trrp, chep,* or *chit.* Song a short rapid trill, typically preceded by one or two call notes, *trp, whi-whi-whi-whi-whi,* similar to song of Sulphur-bellied Warbler.

Phylloscopus occisinensis

华西柳莺 huá xī liǔ yīng

Alpine Leaf Warbler　NR

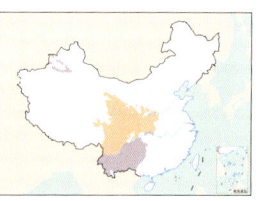

L: 10–11 cm
Habitat: Inhabits forests and agricultural lands above 2000 m and alpine scrubland above 4500 m.
Behavior: Solitary or in pairs; also found in small flocks of three to five individuals during nonbreeding season. Active, agile, hopping constantly to feed in bushes near the ground. Sometimes sallies from branches to catch insects.
Distribution: Breeds in Gansu, Qinghai, Sichuan, Shaanxi, Ningxia, W Hubei, W Hunan, and Chongqing, makes short altitudinal migrations after breeding season, and winters in Yunnan.
Voice: Vocalizations very similar to those of Tickell's Leaf Warbler and doubtfully separable in the field.

Phylloscopus subaffinis

棕腹柳莺 zōng fù liǔ yīng

Buff-throated Warbler　LC

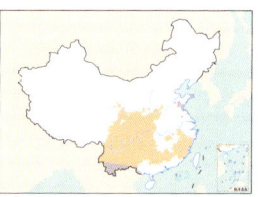

L: 10.5–11 cm
Habitat: Inhabits montane coniferous forests and forest-edge bushes from 900 to 2800 m in breeding season. Also inhabits coniferous forests or broadleaf woodlands, scrubland, and scrubby meadows on low mountains, hills, and foothill plains. Migrates south to lower altitudes in nonbreeding season.
Behavior: Often solitary or in pairs; sometimes in small, loose flocks in nonbreeding season. Moves among branches. Active when foraging, hopping around constantly.
Distribution: Breeds in Central, East, southeastern, and southwestern China. Migrates south in winter.
Voice: Song is even slower and slightly longer than Sulphur-bellied Warbler and much slower than Tickell's Leaf Warbler or Alpine Leaf Warbler, with a much smaller range of frequencies and sounding more "muffled." Calls a soft *chrrup,* longer, lower pitched, and more complex than Tickell's and Sulphur-bellied.

Phylloscopus trochilus

欧柳莺 ōu liǔ yīng

Willow Warbler　LC

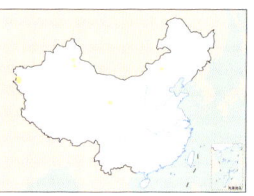

L: 11–12.5 cm
Habitat: In breeding season, inhabits dense deciduous forests with sparse undergrowth below 1500 m. Also found in premontane bushes and vegetated plains during migration.
Behavior: Forages mostly among branches in the undergrowth and canopy; also catches insects on the wing.
Distribution: Occasionally seen in Inner Mongolia, N Xinjiang, and Qinghai during migration. Vagrant farther east, with records in Taiwan, Hong Kong, and coastal Hebei.
Voice: Pleasant song (unlikely to be heard in China) is a series of accelerating, descending warbles, *sisi-sisii-sisisi-suy-suy-suy-sui-sui-sui tuuy tuuy tuuy si-si-sviiy-sui.* Call an up-slurred, disyllabic *hoo-eet!*

Phylloscopus sindianus

中亚叽喳柳莺 (东方叽喳柳莺)

zhōng yà jī zhā liǔ yīng (dōng fāng jī zhā liǔ yíng)

Mountain Chiffchaff　LC

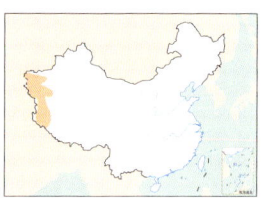

L: 10–12 cm
Habitat: Inhabits alpine coniferous forests, broadleaf forests, river valley bushes, scrubby semiarid barren slopes, grassy areas, and other open habitats above 2500 m. Also inhabits alpine tundra grasslands and gullies.
Behavior: Forages mostly among bushes and on the ground.
Distribution: Breeds in SW Xinjiang and W Tibet. Locally common.
Voice: Song is a *tissiyun, tissiyun,* higher and faster than Common Chiffchaff. Call a plaintive, flat, monosyllabic whistle very similar to Common Chiffchaff.

Phylloscopus collybita

叽喳柳莺 jī zhā liǔ yīng

Common Chiffchaff　LC

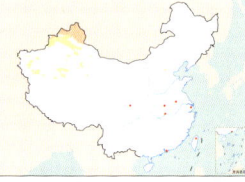

L: 10–11.5 cm
Habitat: Inhabits montane forests in breeding season; especially prefers coniferous forests with dense understory, and groves and willows in river valleys and along streams. Found also in desert scrubland, willow thickets in river valleys, reedbeds, grasses, and vegetated plains during migration.
Behavior: Often solitary or in pairs. Active, agile; constantly hops about or flies among trees and bushes.
Distribution: Breeds in the Altai Mountains of Xinjiang. Common throughout Xinjiang during migration; regular in W Qinghai, rare farther east. Vagrant to Beijing, Henan, Anhui, Shanghai, Jiangsu, Hubei, Sichuan, Hong Kong, and Taiwan.
Voice: Song is low and similar to *chi-vi, chi, vi, chi, vi.*

Tickell's Leaf Warbler

Yellow supercilium, front part of eye stripe indistinct

Grayish-brown upperparts tinged olive

Relatively long bill, lower mandible almost entirely orange

Short-looking tail

Yellowish-green underparts lack distinct breast band

Willow Warbler

Relatively narrow pale yellow supercilium

Yellowish-green upperparts

Primary projection similar in length to tertials

No wing bar

Buff underparts paler on lower belly

Yellowish-brown legs

Alpine Leaf Warbler

Grayish-brown crown and upperparts

No wing bar

Slender bill, lower mandible entirely yellow

Pale yellow from supercilium, throat, and breast to underparts, pale on flanks

Mountain Chiffchaff

Supercilium whiter and broader than Common Chiffchaff

Feathers on upperparts lack olive fringe

Base of lower mandible usually tinged yellow

White throat

Buff-throated Warbler

Upperparts olive brown or olive tinged brown

No wing bar

Yellow lower mandible with black on tip or anterior half

Pale brown or buff from throat to upper breast, belly pale buff

Common Chiffchaff

All-black bill

Buff supercilium

Off-white throat

Wing and tail feathers generally fringed olive

Phylloscopus coronatus

冕柳莺 miǎn liǔ yīng

Eastern Crowned Warbler LC

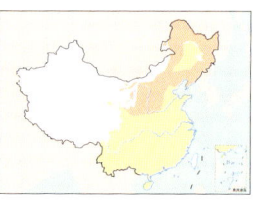

L: 11–12 cm
Habitat: Inhabits montane coniferous forests, mixed broadleaf-coniferous forests, broadleaf forests, and their marginal areas below 2000 m in breeding season; prefers foraging in sparsely wooded scrubland and river valleys. Occurs in broadleaf forests on plains and hilly areas during migration; found mostly in foothills, gardens, or coastal woods.
Behavior: Often solitary or in pairs; gathers in small loose flocks and sometimes joins other warblers in mixed-species flocks during migration. Found mostly in the canopy; often forages in the canopy of broadleaf trees, such as maple, linden, and Manchurian ash. Active; hops about constantly among branches to forage.
Distribution: Common migrant. Breeds in Northeast, North, Central, and southwestern China, south to Shaanxi, Chongqing, and Sichuan; migrates through most of eastern China.
Voice: Simple song is repeated *chip-chip … wheez, jia-jia-whizuji—*, or *ts-chits-chip ts-chits-chip-sii*, all having a longer, rising, terminal note. Flat, nasal, bullfinch-like call note, *djue*, rarely heard.

Phylloscopus ijimae

日本冕柳莺 (饭岛柳莺)
rì běn miǎn liǔ yīng (fàn dǎo liǔ yīng)

Ijima's Leaf Warbler VU

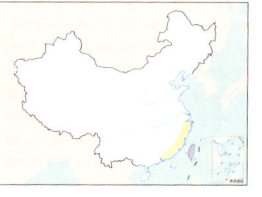

L: 11–12 cm
Habitat: Occurs in broadleaf woods near coasts during migration.
Behavior: Similar to Eastern Crowned Warbler. Often hops about and forages actively from understories to upper stories of dense groves.
Distribution: Rare vagrant or passage migrant in China. Found in Taiwan during migration; scattered records in Hong Kong. May also occur in coastal areas in southeastern China.
Voice: Song strophes are short and repetitive *swee-swee-swee-swee*, with a rhythm that resembles the song of the Arctic Warbler and a quality similar to that of Eastern Crowned Warbler. Call is a single or repeated short, flat, melancholy *twee*, reminiscent of Common Chiffchaff.

Phylloscopus emeiensis

峨眉柳莺 é méi liǔ yīng

Emei Leaf Warbler LC

L: 11–12 cm
Habitat: Inhabits evergreen broadleaf forests at middle altitudes.
Behavior: Solitary or in pairs. Does not flick wings alternately.
Distribution: Uncommon in S Shaanxi, Yunnan, Sichuan, and N Guangdong. Vagrant in Hong Kong.
Voice: Song is a series of lengthy, rapid, Arctic Warbler–like trills, *zhe-zhe-zhe*. Call is a short, sparrowlike, mono- or disyllabic *chit-chit*.

Phylloscopus plumbeitarsus

双斑绿柳莺 shuāng bān lǜ liǔ yīng

Two-barred Warbler LC

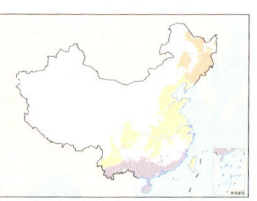

L: 11.5–12 cm
Habitat: In breeding season, inhabits coniferous forests, mixed broadleaf-coniferous forests, and groves of birch and poplar from 400 to 4000 m; occurs on secondary forest edges, green spaces, and bushes on plains and hilly areas during migration.
Behavior: Often solitary or in pairs; sometimes gathers in small scattered groups during migration. Active; constantly hops about and forages in the canopy.
Distribution: Uncommon migrant. Breeds in Northeast China and winters in South China, Yunnan, and Hainan. Passage migrant in central and eastern China.
Voice: Song is very similar to Greenish Warbler but with longer strophes. Sharp call, *chi-ree-wee*, is subtly more trisyllabic than the equally Eurasian Tree Sparrow–like call of Greenish Warbler.

Phylloscopus trochiloides

暗绿柳莺 àn lǜ liǔ yīng

Greenish Warbler LC

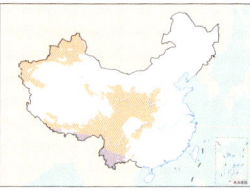

L: 11–12 cm
Habitat: Inhabits various wooded areas and forest-edge bushes from middle to high altitudes.
Behavior: Solitary or in pairs in the upper and middle stories of trees. Active.
Distribution: *P. t. viridanus* in Xinjiang; *P. t. trochiloides* in S Shaanxi, S Gansu, S and SE Tibet, S Qinghai, Yunnan, Sichuan, Hubei, and Jiangxi; *P. t. obscuratus* in W Inner Mongolia, Ningxia, S and E Tibet, Qinghai, Yunnan, and Hainan. Locally common.
Voice: Song is an intense, erratic, mostly high-pitched, jerky warble, reminiscent of a wren. Call a disyllabic White Wagtail–like *chiwi*.

Phylloscopus magnirostris

乌嘴柳莺 wū zuǐ liǔ yīng

Large-billed Leaf Warbler LC

L: 12–12.5 cm
Habitat: Inhabits evergreen broadleaf forests or mixed forests near mountain streams at middle to high altitudes.
Behavior: Solitary or in pairs in the upper and midstories of trees. Active.
Distribution: Found in North, Central, southwestern, and midwestern China. Locally common.
Voice: Loud slow song is a five-syllable, descending, Willow Tit–like series of clear, far-carrying, whistled, musical, flutelike whistles, with an introductory note followed by two paired notes, *tee … ti-ti … tu-tu*, running down the scale; may also begin song with *zi zi, zu zu* notes. Call equally distinctive—a rising *der-t-t-ti*.

Eastern Crowned Warbler

- Dark crown
- White median crown stripe, clear at nape
- Relatively long thick bill, orange lower mandible
- Relatively bright olive upperparts
- Pure white from throat to underparts
- One or two wing bars
- Pale yellow undertail coverts

Two-barred Warbler

- Supercilium extends to base of bill
- Dark olive from crown to upperparts
- Relatively stout bill; yellow lower mandible lacks dusky spot, differing from Arctic Warbler
- Two broad well-defined wing bars
- Primary projection shorter than Arctic Warbler
- Off-white undertail coverts
- Compact body, dark legs

Ijima's Leaf Warbler

- Gray crown differs from Arctic Warbler
- Resembles Eastern Crowned Warbler but with flatter head and no median crown stripe
- Relatively short faint eye stripe contrasts nicely with distinct pale eye ring
- Lower mandible lacks dusky spot
- Grayish overall, especially head and back

Greenish Warbler

- No median crown stripe
- Yellowish-green supercilium
- Olive upperparts
- Pale lower mandible
- One or two wing bars

Emei Leaf Warbler

- Two wing bars
- Relatively pale median crown stripe
- Orange lower mandible
- Whitish underparts

Large-billed Leaf Warbler

- Long thick bill with dark lower mandible
- Dark olive upperparts
- One or two wing bars

Phylloscopus intermedius

白眶鹟莺 bái kuàng wēng yīng

White-spectacled Warbler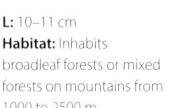

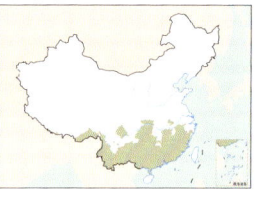

L: 10–11 cm
Habitat: Inhabits broadleaf forests or mixed forests on mountains from 1000 to 2500 m.
Behavior: Prefers dense primary forests or dense multilayered secondary forests. Often catches insects on the wing; sometimes feeds in the canopy. Very active; often mixes with other birds.
Distribution: *P. i. zosterops* is found in SE Tibet; *P. i. intermedius* has a "sky island" distribution among the mid- to high-altitude mountains in Sichuan, Guizhou, Hunan, Hubei, Jiangxi, Guangdong, Fujian, and S Zhejiang.
Voice: Song often a loud whistle and trills, vaguely reminiscent of White-tailed Robin and very similar to Gray-cheeked Warbler, but with fewer trills and generally at lower pitch. Calls are highly variable, with two hiccupping notes in the C Himalayas; a ringing, almost Large-billed Leaf Warbler–like (but rising) three notes, *der-teee-t*, with stress on the middle syllable or *wi-tu-chik* in the E Himalayas; a slightly stuttering *p-r-r-chut* in Myanmar and E Yunnan; a flat multisyllabic *chisiwit* in Sichuan; and a different *tu-tu-prrrt* in southeastern China. More research needed.

Phylloscopus poliogenys

灰脸鹟莺 huī liǎn wēng yīng

Gray-cheeked Warbler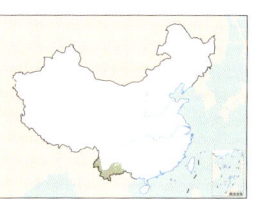

L: 9–10 cm
Habitat: Inhabits evergreen broadleaf forests and understory bushes below 2500 m.
Behavior: Active, with a relatively stable range throughout the year. Feeds mainly on insects and insect larvae in tall trees and understory bushes.
Distribution: Locally common in SW and SE Yunnan. .
Voice: Song similar to White-spectacled Warbler. Call a rising *twees* or a disyllabic *chi-wees*, vaguely reminiscent of Yellow-browed Warbler.

Phylloscopus burkii

金眶鹟莺 jīn kuàng wēng yīng

Green-crowned Warbler

L: 11–12 cm
Habitat: Inhabits montane evergreen broadleaf forests at 2000–2600 m, lower altitudes than Whistler's Warbler.
Behavior: Often hawks insects in forest understory.
Distribution: Rare in S and SE Tibet.
Voice: Song includes varied whistled phrases, some ending in rapid trills, which separate them from Whistler's Warbler. Call a soft, whipping *huit*, somewhat reminiscent of a high-pitched Common Chaffinch.

Phylloscopus tephrocephalus

灰冠鹟莺 huī guān wēng yīng

Gray-crowned Warbler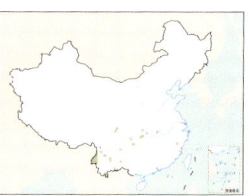

L: 10–11 cm
Habitat: Inhabits montane evergreen broadleaf forests and secondary forests from 1200 to 1900 m.
Behavior: Similar to Green-crowned Warbler.
Distribution: Found in Yunnan, Sichuan, Guizhou, Shaanxi, Hubei, Zhejiang, Jiangxi, and Fujian.
Voice: Short, subdued, soft *trrup* or *turup* call note. Song a series of short phrases and whistles; presence of trills and tremolos distinguish this species from Bianchi's Warbler and Alstrom's Warbler, while the higher overall frequency distinguishes it from Martens's Warbler.

Phylloscopus whistleri

韦氏鹟莺 wéi shì wēng yīng

Whistler's Warbler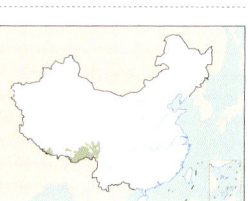

L: 11–12 cm
Habitat: Inhabits mixed deciduous forests and evergreen broadleaf forests in mountainous areas from 2800 to 3300 m.
Behavior: Similar to Green-crowned Warbler.
Distribution: Occasional in S and SE Tibet.
Voice: Song closely similar to that of Bianchi's Warbler but higher pitched; more easily separated from Green-crowned and Gray-crowned Warblers by lack of trills. Call a soft *chip* or *tiu*.

Phylloscopus valentini

比氏鹟莺 bǐ shì wēng yīng

Bianchi's Warbler

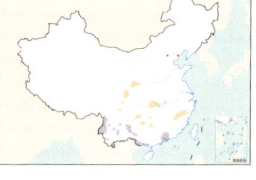

L: 11–12 cm
Habitat: Inhabits mixed forests and evergreen broadleaf forests from 1700 to 3100 m. Distribution overlaps with Martens's Warbler and Gray-crowned Warbler, but at higher altitudes than Alstrom's Warbler.
Behavior: Similar to Green-crowned Warbler.
Distribution: *P. v. valentini* is found in S Shaanxi, S Gansu, Yunnan, and Sichuan; *P. v. latouchei* is found in Central, East, and South China. Locally common. Vagrant north to Hebei and Beijing.
Voice: Call a short *tiu*, occasionally doubled. Sings a pleasant whistle with one or two initial syllables, and its overall frequency is lower than the song of the Gray-crowned Warbler and Martens's Warbler.

White-spectacled Warbler

White eye ring broken above eye

Bluish-gray crown with black lateral crown stripe

Yellow wing bar

P. i. zosterops

Yellow eye ring broken above eye

Bluish-gray crown

Olive cheek

P. i. intermedius

Gray-cheeked Warbler

Broad white eye ring broken above eye

Relatively squared bluish-gray head

One well-defined wing bar

Green-crowned Warbler

Well-defined black lateral crown stripe

Yellow eye ring broken at rear

Olive cheek

Almost no wing bar

Gray-crowned Warbler

Gray supercilium

Distinct bluish-gray median crown stripe

Yellow eye ring broken at rear

Whistler's Warbler

Lateral crown stripe clear blackish at rear, fades away on forehead

Short pale yellow wing bar

Complete yellow eye ring

Bianchi's Warbler

Gray below lateral crown stripe relatively pale, lacks contrast with olive face

Yellow wing bar

Complete yellow eye ring

Phylloscopus soror

淡尾鹟莺 dàn wěi wēng yīng

Alström's Warbler

L: 11–12 cm
Habitat: Inhabits evergreen broadleaf and secondary forests from 600 to 1500 m, at lower altitudes than Bianchi's Warbler.
Behavior: Similar to Green-crowned Warbler.
Distribution: Common in S Shaanxi, Sichuan, Guizhou, Guangxi, Jiangxi, and Fujian. Breeds north to Beijing and Hebei.
Voice: High-pitched fine song consists of different groups of syllables, each consisting of the same syllable repeated two or three times. Call a short high-pitched *tsi-dit* or single *tsrit*.

Phylloscopus omeiensis

峨眉鹟莺 é méi wēng yīng

Martens's Warbler

L: 11–12 cm
Habitat: Inhabits montane evergreen broadleaf forests from 1200 to 2300 m. Distribution overlaps with Gray-crowned Warbler and Bianchi's Warbler.
Behavior: Similar to Green-crowned Warbler.
Distribution: Found in N Yunnan, Sichuan, S Shaanxi, Guizhou, W Hubei, and S Gansu.
Voice: Sings a series of continuous rapid thrills, *zhe-zhe-zhe*. Song contains short, varied, whistled notes, often ending in trills. Calls a short faint *chup* and *chu-du*.

Phylloscopus castaniceps

栗头鹟莺 lì tóu wēng yīng

Chestnut-crowned Warbler

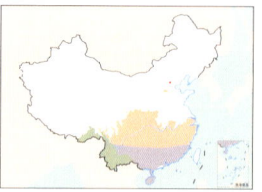

L: 9–10.5 cm
Habitat: Inhabits broadleaf forests and forest-edge woodland and bushes in foothills and low mountains below 2000 m.
Behavior: Often solitary or in pairs in breeding season; gathers in small flocks in nonbreeding season.
Distribution: Locally common. *P. c. castaniceps* in S and SE Tibet, NW, W, and SW Yunnan. *P. c. laurentei* in SE Yunnan and S Guangxi. *P. c. sinensis* in Central, southwestern, East, and South China; occasionally wanders to North China.
Voice: Piercingly high-pitched, very fine song of four to seven notes that descend slightly, *we te see- see- see-see*. Call is a multisyllabic dry trill, *chrrk*, or longer, more purring rattle, *trrrrrrt*.

Phylloscopus calciatilis

灰岩柳莺 huī yán liǔ yīng

Limestone Leaf Warbler

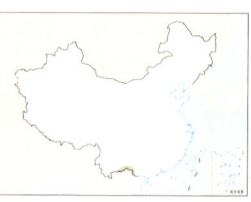

L: 10.5–11 cm
Habitat: Inhabits evergreen broadleaf forests in limestone landscapes.
Behavior: Solitary or in pairs in the middle story of trees. Active. Does not migrate.
Distribution: Uncommon in S Guangxi and SE Yunnan.
Voice: Appealing high-pitched song lasts about 1.5 sec and starts with three or four lengthening rising notes and ends with a short three-note warble, *wi-wii-wee-weee-tui-wis-uu*, reminiscent of, but clearer and slower than, that of Sulphur-breasted Warbler.

Phylloscopus ricketti

黑眉柳莺 hēi méi liǔ yīng

Sulphur-breasted Warbler

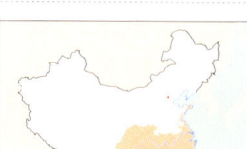

L: 10–11 cm
Habitat: In breeding season, inhabits low-mountain broadleaf forests and secondary forests below 2000 m; also in mixed forests, coniferous forests, forest-edge bushes, and orchards. Occurs in open woods on plains and hills during migration.
Behavior: Moves about singly or in pairs in breeding season; solitary during migration. In breeding ground, mixes with other passerines while foraging. Active; hops about and flies among leaves and branches. Catches flying insects from perch; also forages in understory bushes. Often flicks wings alternately, presumably to startle hiding insects.
Distribution: Uncommon. Breeds in S Gansu, Shaanxi, Sichuan, Chongqing, Guizhou, Hubei, Hunan, Jiangxi, Fujian, Zhejiang, Guangdong, and N Guangxi; migrates through most of southwestern and southeastern China; a few records in the coastal areas of North and East China.
Voice: Jolly, descending, seven- or eight-syllable song, *si ... sre-sree-sree-pitch-ee-you*, reminiscent of Limestone Leaf Warbler but faster; introductory notes barely rise. Whole impression is of a subtly flatter, more muted song.

Alström's Warbler

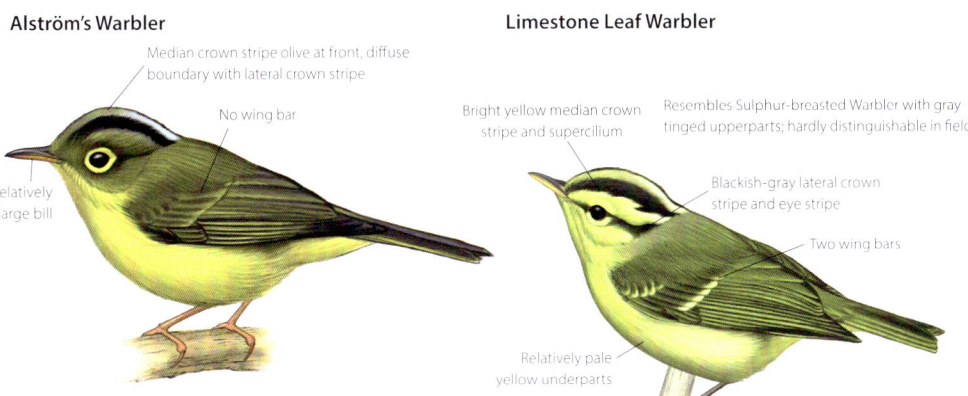

Median crown stripe olive at front, diffuse boundary with lateral crown stripe

No wing bar

Relatively large bill

Limestone Leaf Warbler

Bright yellow median crown stripe and supercilium

Resembles Sulphur-breasted Warbler with gray tinged upperparts; hardly distinguishable in field

Blackish-gray lateral crown stripe and eye stripe

Two wing bars

Relatively pale yellow underparts

Martens's Warbler

Gray-tinged olive below lateral crown stripe

Well-defined gray median crown stripe and black lateral crown stripe

No wing bar

Sulphur-breasted Warbler

Black lateral crown stripe

Pale yellowish-green median crown stripe

Olive upperparts

Yellowish-green wings

One or two well-defined yellow wing bars

Bright yellow from supercilium, throat, breast, and belly to undertail coverts

Chestnut-crowned Warbler

White eye ring

Rufous-chestnut crown

Black lateral crown stripe

Gray neck, side of neck, and back

Grayish white from throat to upper breast

Bright yellow belly

P. c. sinensis

Grayish-white central belly

P. c. castaniceps

Phylloscopus borealoides
萨岛柳莺 (库页岛柳莺)
sà dǎo liǔ yīng (kù yè dǎo liǔ yīng)
Sakhalin Leaf Warbler

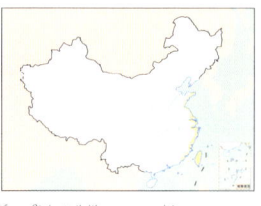

L: 11–12 cm
Habitat: Inhabits plains, hills, coastal secondary woodlands, green spaces, and scrubland during migration.
Behavior: Active in the lower and middle stories of woods or bushes; hops about actively when foraging. Often flicks tail, like some robins.
Distribution: Uncommon passage migrant. Migrates through the coastal areas of East and South China. Hard to distinguish morphologically from Pale-legged Leaf Warbler; more easily identified by sound. Likely misidentified or overlooked.
Voice: Call a single, sharp, but slightly flat note between 4.5–5.2 kHz, slightly lower pitched than Pale-legged Leaf Warbler's *tink*, which is between 5.6–5.9 kHz. Very distinctive song is vaguely reminiscent of Large-billed Leaf Warbler—a clear, drawn-out series of three or four whistled phrases beginning with *hee-tsoo-kee.*

Phylloscopus tenellipes
淡脚柳莺 dàn jiǎo liǔ yīng
Pale-legged Leaf Warbler

L: 10–11 cm
Habitat: Breeds in broadleaf forests and woodlands below 1800 m; especially prefers dense forests near rivers. Inhabits secondary woodlands, green spaces, and bushes on plains, hills, and coastal areas during migration. Winters in lowland forests.
Behavior: In lower and middle stories of woods or bushes, hops about and flicks tail when foraging, like some robins.
Distribution: Common migrant. Breeds in Northeast China; migrates through coastal areas of East and South China, with a few individuals wintering in Hainan.
Voice: Sings an insectlike *tiriririrririri* that ends abruptly. Call similar to Sakhalin Leaf Warbler.

Phylloscopus xanthoschistos
灰头柳莺 huī tóu liǔ yīng
Gray-hooded Warbler

L: 10–11 cm
Habitat: Inhabits various forests at middle altitudes.
Behavior: Solitary or in small flocks in the upper and middle stories of trees; often joins mixed-species flocks.
Distribution: Uncommon in SW and S Tibet.
Voice: Short (0.8 sec), cheerful, fast, slightly lisping song rises, then descends, with strophes repeated, with minimal variation, every 5 sec or so, *tis-lissi-liu-whee-oo.* Multiple different calls include sharp *sik, chis,* and *twiss.*

Phylloscopus examinandus
堪察加柳莺 kān chá jiā liǔ yīng
Kamchatka Leaf Warbler

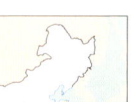

L: 13 cm
Habitat: Inhabits secondary woodlands, parks, or coastal green spaces on open plains and hills during migration.
Behavior: Similar to Arctic Warbler.
Distribution: Uncommon passage migrant. Migrates along the island chains in W Pacific; mostly recorded in Taiwan, with a few records from the east coast and Hong Kong, north to Liaoning. Once treated as a subspecies of Arctic Warbler.
Voice: Song is like that of Arctic Warbler but more hesitant, pulsed or pumping, *t'treet't't reet't'treet't'treet.* Crackling multisyllabic call is slightly longer and has greater frequency range than Arctic Warbler—*trrrt,* a faster *trrt,* or doubled *trr-trrt.*

Phylloscopus borealis
极北柳莺 jí běi liǔ yīng
Arctic Warbler

L: 12–13 cm
Habitat: In breeding season, inhabits coniferous forests, sparse broadleaf forests, mixed broadleaf-coniferous forests, and forest-edge bushes from 400 to 1200 m. Seen in secondary forests, forest edges, artificial forests, orchards, gardens, and other green spaces during migration.
Behavior: Often solitary or in pairs in breeding season; gathers in small flocks during migration. Sometimes joins other warblers in mixed-species flocks to forage in tree canopy. Active, fast, and agile, often hops about and forages among branches and leaves; also forages in bushes and low parts of trees. Feeds mainly on insects.
Distribution: Common migrant. Breeds in Northeast and northwestern China and winters in South China (including Taiwan); found in most parts of eastern China during migration.
Voice: Powerful, insectlike, dry, rattling song, *tzik-tzik-tzik-tzik-tzik-tzik-tzik,* has only occasional changes in pitch and tempo. Call is a short dipperlike *dzit.*

Phylloscopus xanthodryas
日本柳莺 rì běn liǔ yīng
Japanese Leaf Warbler

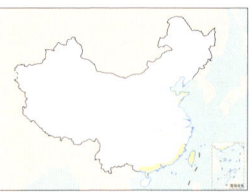

L: 12–13 cm
Habitat: Similar to Kamchatka Leaf Warbler.
Behavior: Similar to Arctic Warbler.
Distribution: Uncommon passage migrant. Migrates through island chains; found mostly in Taiwan (possibly its wintering grounds), with a few records along the east coast (including Hong Kong). Once treated as a subspecies of Arctic Warbler.
Voice: Song is slower and lower pitched than that of Arctic Warbler or Kamchatka Leaf Warbler, *tree'diret-tree'diret-tree'diret.* Each repeated song lasts 1.5–2 sec. Calls markedly lower pitched, *dzyrrt* or similar.

Sakhalin Leaf Warbler

Resembles Pale-legged Leaf Warbler but with slightly larger body and longer bill, upperparts more greenish

Two relatively faint, indistinct wing bars

Pinkish to dark pink legs

Kamchatka Leaf Warbler

Not safely separable from Arctic or Japanese Leaf Warbler except by vocalizations but fractionally larger

P10

Kamchatka Leaf Warbler

Pale-legged Leaf Warbler

Gray-tinged head and neck contrast strongly with olive back and wings

Two wing bars

ostly blackish .wer mandible with light-colored tip

.ight colored, nearly whitish, from throat to underparts

Extremely pale pink legs

Arctic Warbler

Supercilium does not reach base of bill

Flat head

Olive or grayish-green upperparts

Two thin wing bars, worn in some individuals

Relatively long primaries

Relatively slender pointed bill, yellowish-brown lower mandible with dusky spot

Off-white underparts

Slender body

Arctic Warbler

P10

Gray-hooded Warbler

Pale gray head, nape, and upper back

White supercilium

Yellow underparts

Japanese Leaf Warbler

Not safely separable from Kamchatka or Arctic Leaf Warbler except by vocalizations or, occasionally, yellow hues on face and throat

P10

Japanese Leaf Warbler

Phylloscopus ogilviegranti

白斑尾柳莺 bái bān wěi liǔ yīng

Kloss's Leaf Warbler

L: 10.5–11 cm
Habitat: Inhabits broadleaf forests at middle and low altitudes.
Behavior: Flicks wings simultaneously, unlike Claudia's Leaf Warbler.
Distribution: *P. o. disturbans* in Gansu, Shaanxi, Sichuan, Chongqing, Guizhou, Hunan, and Guangdong; *P. o. ogilviegranti* in Jiangxi, Fujian, and Guangdong. Uncommon.
Voice: Very similar to Claudia's Leaf Warbler.

Phylloscopus intensior

云南白斑尾柳莺 yún nán bái bān wěi liǔ yīng

Davison's Leaf Warbler

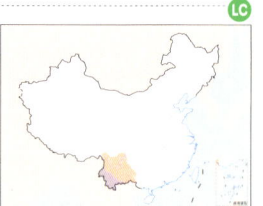

L: 10.5–11 cm
Habitat: Inhabits broadleaf forests at middle and low altitudes.
Behavior: Highly mobile; often moves quickly among branches. Flicks wings alternately.
Distribution: Uncommon in Yunnan.
Voice: Similar to Kloss's Leaf Warbler.

Phylloscopus hainanus

海南柳莺 hǎi nán liǔ yīng

Hainan Leaf Warbler **VU**

L: 10–11 cm
Habitat: Inhabits secondary forest edges from 640 to 1500 m; seldom occurs in primary forests.
Behavior: Solitary or in pairs in breeding season; mixes with other passerines in nonbreeding season. Often moves about in canopy.
Distribution: Locally common resident. Found only in some mountainous areas in C, W, and S Hainan. Sometimes treated as a subspecies of Davison's Leaf Warbler.
Voice: Song similar to Davison's Leaf Warbler, *tsitsitsui-tsitsitsui … titsu-titsui-titsui*. Call is a high-pitched *pitsiu, pitsiu*.

Phylloscopus reguloides

西南冠纹柳莺 xī nán guān wén liǔ yīng

Blyth's Leaf Warbler

L: 11–11.5 cm
Habitat: Inhabits broadleaf forests at middle and low altitudes.
Behavior: Prefers climbing on tree trunks or thick branches; also hangs upside down like nuthatch. Flicks wings alternately.
Distribution: *P. r. reguloides* in S Tibet; *P. r. assamensis* in SE Tibet, SW and S Yunnan, and SW Sichuan. Locally common.
Voice: Song begins with essentially an extended version of the species' call, *chi-wi-chi-chi*, before an intense wrenlike trill.

Phylloscopus claudiae

冠纹柳莺 guān wén liǔ yīng

Claudia's Leaf Warbler

L: 10 cm
Habitat: In breeding season, inhabits various types of forests and forest-edge bushes below 3500 m; migrates to low mountains or foothill plains in nonbreeding season.
Behavior: Solitary or in pairs in breeding season; gathers and forages in small groups or joins other species in mixed-species flocks in nonbreeding season. Often moves about in the canopy; also forages in bushes. Flicks wings alternately.
Distribution: Locally common migrant. Breeds in the mountainous areas of Central and North China, overwinters in Yunnan, and migrates through most of Central and southwestern China, with a few individuals recorded in coastal areas in East China.
Voice: Song a rapid and wrenlike *chi chi pit-chew pit-chew*. Calls a repeated disyllabic *pit-cha* or a trisyllabic *pit-chew-a*, like Blyth's Leaf Warbler.

Phylloscopus goodsoni

华南冠纹柳莺 huá nán guān wén liǔ yīng

Hartert's Leaf Warbler

L: 10.5–12 cm
Habitat: Similar to Claudia's Leaf Warbler.
Behavior: Similar to Claudia's Leaf Warbler. Often hangs upside down from the underside of branches or forages on tree trunks.
Distribution: Locally common resident. *P. g. fokiensis* breeds in W Hubei, Guizhou, Guangxi, Anhui, Jiangxi, and Zhejiang; migrates to the southern parts of the breeding range in winter. *P. g. goodsoni* breeds in Guangdong and Guangxi, possibly wintering in Yunnan, with a few wintering records in Hong Kong and Hainan.
Voice: Sings a titlike *whee-cheet-a whee cheet-a*, which is delivered quickly and repeatedly. Calls a repeated disyllabic *pit-cha*, similar to Claudia's Leaf Warbler.

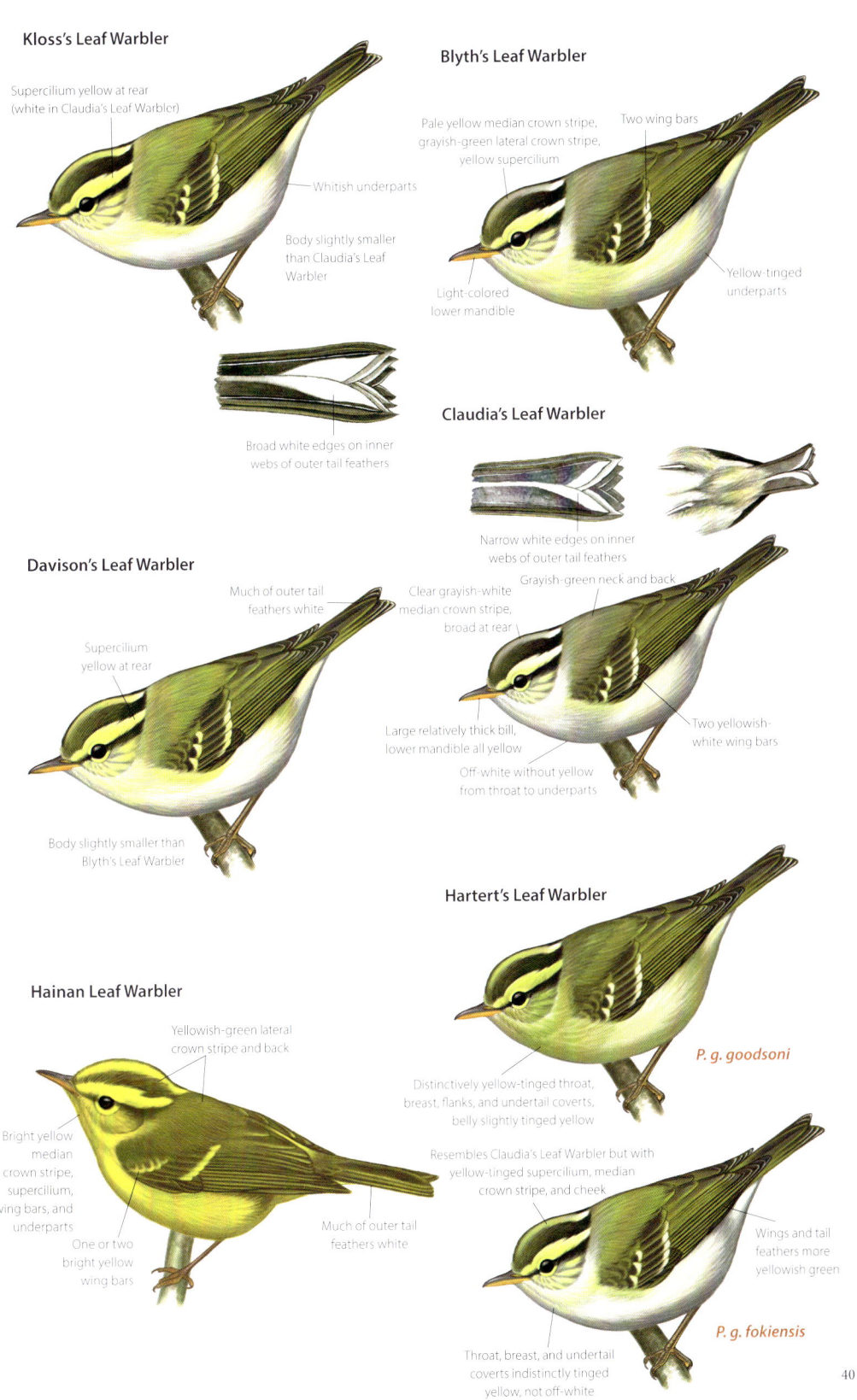

Kloss's Leaf Warbler

Supercilium yellow at rear (white in Claudia's Leaf Warbler)

Whitish underparts

Body slightly smaller than Claudia's Leaf Warbler

Broad white edges on inner webs of outer tail feathers

Blyth's Leaf Warbler

Pale yellow median crown stripe, grayish-green lateral crown stripe, yellow supercilium

Two wing bars

Light-colored lower mandible

Yellow-tinged underparts

Claudia's Leaf Warbler

Narrow white edges on inner webs of outer tail feathers

Grayish-green neck and back

Clear grayish-white median crown stripe, broad at rear

Large relatively thick bill, lower mandible all yellow

Two yellowish-white wing bars

Off-white without yellow from throat to underparts

Davison's Leaf Warbler

Much of outer tail feathers white

Supercilium yellow at rear

Body slightly smaller than Blyth's Leaf Warbler

Hartert's Leaf Warbler

Distinctively yellow-tinged throat, breast, flanks, and undertail coverts, belly slightly tinged yellow

P. g. goodsoni

Resembles Claudia's Leaf Warbler but with yellow-tinged supercilium, median crown stripe, and cheek

Hainan Leaf Warbler

Yellowish-green lateral crown stripe and back

Bright yellow median crown stripe, supercilium, wing bars, and underparts

One or two bright yellow wing bars

Much of outer tail feathers white

Wings and tail feathers more yellowish green

P. g. fokiensis

Throat, breast, and undertail coverts indistinctly tinged yellow, not off-white

401

苇莺科　Acrocephalidae

Small to medium-sized insectivorous songbirds; body size slightly larger than other Old World warblers. Plumage similar between sexes, mostly olive brown, white, and yellow. Bill thin and pointed, but powerful looking. Good fliers, with relatively long and pointed wings. Tail medium to long, rounded or slightly graduated. Inhabit primarily reedbeds, tall grasses, bushes, or forest edges along rivers, lakes, or coasts. Nest in middle and lower layers of vegetation. Feed primarily on insects and other invertebrates, occasionally plant seeds and fruits. Most species are migratory.

Seven genera and 62 species recognized worldwide, distributed in the Old World, Australia, and Pacific islands. Three genera and 16 species recorded in China, widely distributed except the Qinghai-Tibet Plateau.

苇莺属
Acrocephalus

Oriental Reed Warbler

厚嘴苇莺属
Arundinax

Thick-billed Warbler

篱莺属
Iduna

Booted Warbler

Acrocephalus arundinaceus
大苇莺 dà wěi yīng
Great Reed Warbler　

L: 19–20 cm
Habitat: Inhabits dense reedbeds or bushes in lakes, reservoirs, rivers, ponds, swamps, and other water bodies.
Behavior: Often moves clumsily among reeds; rarely comes down to the ground. In breeding season, usually sings loudly from top of reeds near its nest.
Distribution: Common breeder in water bodies throughout Xinjiang; occasionally seen in C Inner Mongolia, Gansu, and S Yunnan during migration.
Voice: Rhythmic, loud, and guttural song with repetition of notes, *karra-karra-karra … krie-krie-krie … kra-kra … kieh*, slightly lower pitched and faster than that of Oriental Reed Warbler. Calls include a hard *tack* and a *churr*, all very similar to Oriental Reed Warbler.

Acrocephalus orientalis
东方大苇莺 dōng fāng dà wěi yīng
Oriental Reed Warbler

L: 17–19 cm
Habitat: Inhabits low mountains, hills, and foothill plains in breeding season; prefers foraging and nesting in grasslands and reedbeds near lakes, ponds, streams, riverbanks, and reed swamps, and in swamps and wet grasslands with dense vegetation.
Behavior: Often solitary or in pairs in breeding season. Active, frequently hopping about or climbing along grass stems or bush branches. In the early morning, often sings from top of reeds near its nest or on nearby twigs.
Distribution: Common migrant. Breeds in Northeast, North, Central, and East China, west to Gansu and E Qinghai. Seen in most parts of China during migration. Winters in Taiwan.
Voice: Loud, intense, lengthy, raspy song, such as *ga-ga-ji … kiruk- kiruk-kiruk, jee- jee-jee*. Calls include a short *kirr*.

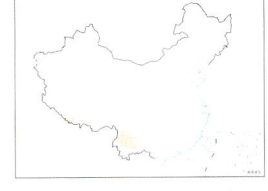

Acrocephalus stentoreus

噪苇莺 (噪大苇莺) zào wěi yīng (zào dà wěi yīng)

Clamorous Reed Warbler 🔵LC

L: 18–20 cm
Habitat: Inhabits reedbeds, thickets, and moist grass near water bodies, such as lakes, rivers, ponds, and swamps, on low mountains and plains from 400 to 900 m.
Behavior: Solitary or in pairs in breeding season. Moves quickly through reedbeds and thickets. In breeding season, often perches on top of reeds or other emergent plants or bushes and calls for long periods, especially in the early morning and at dusk.
Distribution: Breeds in SE Tibet, southwestern China (S Sichuan, C and SW Guizhou, Yunnan), and C Gansu. Not uncommon in suitable habitats.
Voice: Very noisy during breeding season. Loud song is very similar to that of Oriental Reed Warbler but often less intense or hurried, with more pauses and hesitation and less rigidly repetitive. Calls include a short hoarse *chack*.

Acrocephalus melanopogon

须苇莺 xū wěi yīng

Moustached Warbler 🔵LC

L: 12–13.5 cm
Habitat: Inhabits dense reedbeds or sweet flag in rivers, ponds, marshes, and other water bodies.
Behavior: Skulks in reedbeds or sweet flag near water; often comes down to the ground to feed near water.
Distribution: A small breeding population was found in Urumqi and the Ili River valley in N Xinjiang.
Voice: Song uniformly paced; resembles that of Common Reed Warbler but with lower-pitched notes and usually including a gradually increasing sequence, *du- du- du*. Call a stonechat-like *chack* and a rattling *trrt-trrt*.

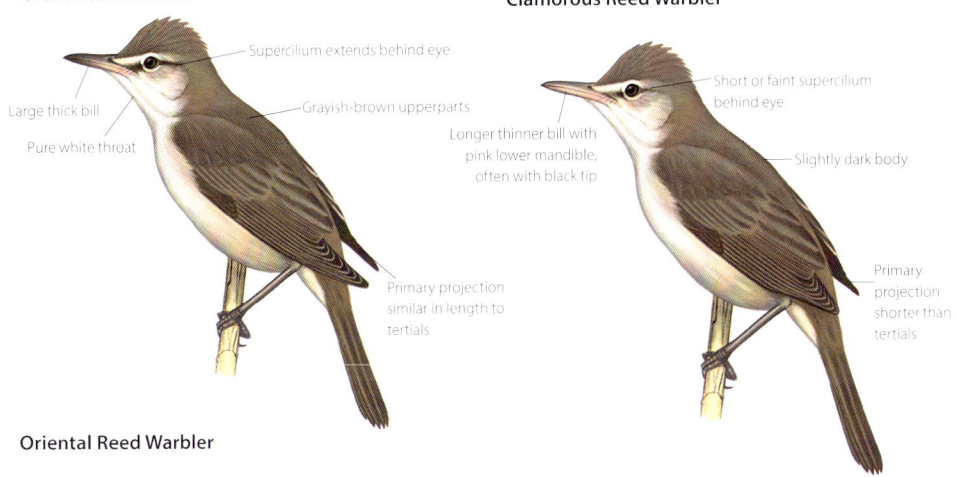

Great Reed Warbler

Supercilium extends behind eye
Large thick bill
Pure white throat
Grayish-brown upperparts
Primary projection similar in length to tertials

Clamorous Reed Warbler

Short or faint supercilium behind eye
Longer thinner bill with pink lower mandible, often with black tip
Slightly dark body
Primary projection shorter than tertials

Oriental Reed Warbler

Large thick bill with pinkish-brown lower mandible
Distinctive orange-red color inside bill
Crown slightly erect when perched
Blackish color shows when feathers on side of neck are fluffed
Creamy white from throat to breast
Finely streaked breast
Primary projection shorter than tertials
Dark bluish-gray or pale gray legs

Moustached Warbler

Dense black streaks on crown
White supercilium broadens behind eye
Brownish upperparts
Dusky ear coverts
Relatively short primary projection

Acrocephalus bistrigiceps

黑眉莺莺 hēi méi wěi yīng

Black-browed Reed Warbler　LC

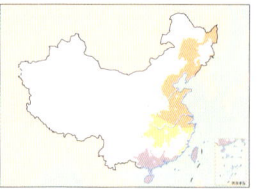

L: 13.5–14 cm
Habitat: Inhabits bushes and reedbeds near water in breeding season; also in woods or bushes far from water during migration.
Behavior: Often solitary or in pairs in breeding season. Sings from top of small bushes or tips of mugwort; flies a short distance into dense thickets when disturbed. Agile and active when foraging, hopping and weaving among reed stems and leaves or climbing up and down.
Distribution: Breeds in Northeast, North, Central, and East China. A few overwinter in southeastern and South China, including Hainan and Taiwan. Migrates through most central and eastern parts of China.
Voice: Rapidly delivered song similar to that of Manchurian Reed Warbler and Blunt-winged Warbler, but delivery is faster, with more short notes and a greater frequency range. Songs contain remarkable amounts of mimicry (in some strophes, almost 100 percent) and considerable repetition of adjacent notes. Call a short *chur, chur*.

Acrocephalus concinens

钝翅莺莺 dùn chì wěi yīng

Blunt-winged Warbler　LC

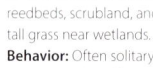

L: 13–14 cm
Habitat: Breeds in bushes and grassy areas at forest edges from 500 to 1450 m; also occurs in reedbeds, scrubland, and tall grass near wetlands.
Behavior: Often solitary or in pairs in breeding season, perching and singing on the top of reeds or grass.
Distribution: Uncommon migrant. Breeds in East, North, and Central China, Guangxi, and Guizhou, migrates through North, East, South, and southwestern China, and possibly overwinters in S and SW Yunnan.
Voice: Complex and varied song is similar to both Black-browed Reed Warbler and Manchurian Reed Warbler, but with slower delivery, slightly shorter strophes, less repetition, and a more modest frequency range than either reed warbler and with far less mimicry than Manchurian. Call is a soft *tschask*.

Acrocephalus agricola

稻田莺莺 dào tián wěi yīng

Paddyfield Warbler　LC

L: 12–13.5 cm
Habitat: Inhabits reedbeds and tamarisk bushes near water banks.
Behavior: Often in the lower part of reedbeds, hopping quickly back and forth among grass stems. Distinctive behaviors include flicking tail rapidly, cocking tail, and raising crown feathers.
Distribution: Common breeder in Xinjiang. A few records in southwestern and southeastern China (including Yunnan, Shanghai, Hong Kong, and Taiwan) during migration.
Voice: Sings a long smooth song, unlike the hoarse song of other reed warblers. Call is a shrill *chik-chik*, with a muffled *zack-zack* or a hoarse trill.

Acrocephalus tangorum

远东莺莺 yuǎn dōng wěi yīng

Manchurian Reed Warbler　VU

L: 13–14.5 cm
Habitat: Inhabits large areas of reeds in lakes or marshes in breeding season; winters on tall grasslands or reedbeds near wetlands.
Behavior: Similar to Blunt-winged Warbler. Often cocks tail.
Distribution: Uncommon migrant. Breeds in Northeast China (Hulun Lake in Inner Mongolia and the Songhua and Nenjiang River basins in Heilongjiang) and exceptionally south to Hebei. Also found in coastal areas in Northeast, North, East, and South China during migration, with a possible wintering population in Taiwan.
Voice: Song similar to that of both Black-browed Reed Warbler and Blunt-winged Warbler but with longer strophes than the latter and far less mimicry than the former. Some repetition, with occasional notes repeated up to 10 times, but adjacent notes repeated far less often than they are by Black-browed. Most songs also have a higher proportion of longer notes and fewer churrs than Black-browed. Call a mellower *tac* than Black-browed.

Acrocephalus dumetorum

布氏莺莺 bù shì wěi yīng

Blyth's Reed Warbler　LC

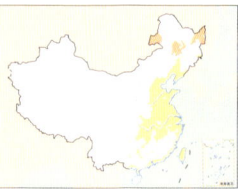

L: 12.5–14 cm
Habitat: Prefers scrubland to reedbeds. Inhabits mainly bushes and wooded areas near wetlands; sometimes occurs in mountain bushes and urban parks far from water. Also found in reeds and other vegetation in wetlands during migration.
Behavior: Often solitary. Active and vigilant, moves about and forages throughout the day; frequently fans and flicks tail.
Distribution: Uncommon to common breeder in NW Xinjiang; uncommon in SW Xinjiang during migration. Vagrant in Sichuan, Hong Kong, Beijing, and Taiwan.
Voice: Slow song includes considerable mimicry, extensive repetition of phrases, occasional call notes, and strophes that climb up the scale. Calls a harsh *thik*, hard *chak*, and rubbing *cherr*.

Acrocephalus scirpaceus

芦莺 (芦莺莺) lú yīng (lu wěi yīng)

Common Reed Warbler　LC

L: 12.5–14 cm
Habitat: Inhabits reedbeds and bushes near lakeshores, reservoirs, and marshes.
Behavior: Often solitary or in pairs. Prefers roosting on reed stems and small willow branches. Active, agile, and vigilant; often hides in bushes and reeds.
Distribution: Uncommon breeder in reed wetlands throughout Xinjiang and vagrant in Jiangsu and Yunnan.
Voice: Song is generally slow and sedate, with notes repeated two to four times. Call is a calm *che* or weak *churr*.

Black-browed Reed Warbler

Heavy blackish-brown lateral crown stripe

Long thick creamy-white supercilium

Relatively long primary projection

White from throat to underparts

Flanks tinged pale brown

Manchurian Reed Warbler

Relatively long tail

Relatively thin dark brown lateral crown stripe

Relatively thick bill

Pale brown rump and uppertail coverts

Brown iris

Relatively short primary projection

Whitish throat contrasts with buff-yellow breast and belly

Body heavily tinged brownish yellow

Blunt-winged Warbler

Creamy-white supercilium short, thin, and faint, inconspicuous behind eye

From upper edge of supercilium to forehead slightly tinged dark brown, inconspicuous in some individuals

Brown iris

White from throat to upper breast

Buff or brownish from lower breast to belly and flanks

Relatively short primary projection

Blyth's Reed Warbler

Thin white supercilium, inconspicuous behind eye

Lower mandible with black tip

White throat

Grayish-brown rump and upperparts lack warm tones

Short primary projection

Paddyfield Warbler

White supercilium bordered above by diffuse short dark line

Lower mandible with black tip

Brown rump

Short primary projection

Common Reed Warbler

Yellowish-white supercilium ends at lores

Square headed

Pinkish lower mandible

Rump tinged brown

Relatively long primary projection

Acrocephalus schoenobaenus

蒲苇莺（水蒲苇莺）pú wěi yīng (shuǐ pù wěi yīng)

Sedge Warbler

L: 11.5–13 cm
Habitat: Inhabits dense sweet flag or reedbeds near wetlands.
Behavior: Vigilant. Often skulks and forages in grass. Male perches and sings on tall grass in breeding season and sometimes makes courtship flights.
Distribution: Uncommon breeder in NW Xinjiang.
Voice: Song is a loud medley of hard and sweet notes, faster than Common Reed Warbler, with marked changes in speed and more mimicry. Song can last for several minutes without interruption. Call includes a hoarse chattering *tue* and a scolding beep.

Acrocephalus sorghophilus

细纹苇莺 xì wén wěi yīng

Streaked Reed Warbler

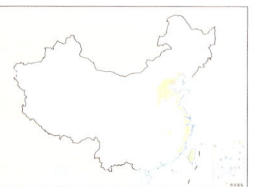

L: 12–13 cm
Habitat: Inhabits wetlands and nearby bushes, reedbeds, and farmland during migration.
Behavior: Often solitary during migration.
Distribution: Rare and mysterious; very little distribution information is available. Presumably breeds in wetlands in Liaoning and Hebei. Migrates through East and North China, with a few records in Liaoning, Hebei, Beijing, Hubei, Jiangsu, Fujian, and Taiwan.
Voice: Song is essentially undocumented but apparently often includes trills.

Arundinax aedon

厚嘴苇莺 hòu zuǐ wěi yīng

Thick-billed Warbler

L: 18–21 cm
Habitat: Often forages in woodland bushes and grassy areas near river valleys in breeding season; inhabits bushes and grass near forest edges or water during migration and winter.
Behavior: Solitary or in pairs in breeding season. Male often sings on bushtops near nest. Skulking and agile when foraging.
Distribution: Uncommon. *A. a. aedon* breeds in Inner Mongolia and may winter in SE Tibet and W Yunnan; *A. a. rufescens* breeds in Northeast China and may winter in S Yunnan and SW Guangxi. Both subspecies can be seen in most areas from Northeast and North China to southwestern and southeastern China during migration. The two subspecies are assumed to use different migration routes: *A. a. aedon* in the west and *A. a. rufescens* in the east.
Voice: Pleasant song is loud and rich in mimicry and often starts with *tschok-tschok* calls, followed by a pleasant whistling phrase. Call is a short, raspy *chack, chack*.

Iduna caligata

靴篱莺 xuē lí yīng

Booted Warbler

L: 11–12.5 cm
Habitat: Inhabits bushes, grassy areas, and forest edges of woodlands; especially prefers wet tall grasslands.
Behavior: Often solitary or in pairs. Hides in dense bushes or grass. Moves quickly, hard to find. Usually perches on grass stems and hops up and down agilely along the stems.
Distribution: Breeds in W Inner Mongolia and N Xinjiang; uncommon or locally common. Occasionally found in S Xinjiang during migration, and vagrant in Liaoning (Dalian), Fujian (Xiamen), Taiwan, and Yunnan (Dali).
Voice: Exuberant, enchanting song starts slowly but builds rapidly into fast warbling, with short notes that lack a clear pattern or repetition. Call a soft *clik*.

Iduna rama

赛氏篱莺 sài shì lí yīng

Sykes's Warbler

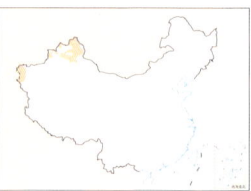

L: 11.5–13 cm
Habitat: The population in N Xinjiang inhabits desert poplar forests, saxaul forests, and tamarisk bushes at desert edges or within deserts. The population in SW Xinjiang prefers sea buckthorn bushes and desert scrubland in river valleys. Prefers more arid habitats than Booted Warbler.
Behavior: Often solitary or in pairs. Mostly hidden in dense bushes. Male perches and sings from branches in breeding season.
Distribution: Locally common breeder in W and N Xinjiang.
Voice: Compared to Booted Warbler, song contains some mimicry and with many notes repeated three to five times. (Booted Warbler's slightly more relaxed song lacks this repetition of adjacent notes.) Calls include a dry tongue-clicking *chek* and a fuller *tslek*, similar to those of Booted but shorter, sharper, and less "full."

Iduna pallida

草绿篱莺 cǎo lù lí yīng

Eastern Olivaceous Warbler

L: 12–13.5 cm
Habitat: Inhabits thickets of vegetation in desert and semidesert regions; also occurs at forest edges, sparsely wooded scrubland, or tall grass far from water.
Behavior: Solitary or in pairs. Vigilant but bold; hops frequently among bush branches. In breeding season, male often perches on treetops and sings and makes courtship flights. Diagnostic downward tail flicking.
Distribution: Very rare breeder in NW Xinjiang.
Voice: Song is a monotonous "babbling" that is repeated cyclically, with notes going up and down the scale in a repetitive pattern, with hoarse, scratchy, squeaky, nasal notes and few whistles, with poor articulation and notes often blurring together. Calls are a slightly nasal *tick-tick-tick*.

Sedge Warbler

Relatively thin streaks on crown

Relatively broad buff-white supercilium

Pale brown ear coverts

Pale brown upperparts

Relatively long primary projection

Streaked Reed Warbler

Buff supercilium slightly tinged brown

Fine streaks on crown

Relatively narrow blackish-brown line above supercilium, with diffuse boundary

Fine blackish-brown streaks on back

Thick-billed Warbler

Relatively thick bill

Pale brown body

A. a. rufescens

Relatively large body with long tail

Booted Warbler

Long broad supercilium extends behind eye

Relatively small bill with black tip

Upperparts more brownish

Relatively short primary projection

Buff-white underparts

Dark toes contrast with pale legs, like wearing boots

Sykes's Warbler

Faint supercilium extends behind eye

Relatively long bill, lower mandible either pure yellow or with diffuse black spot in middle

Less brown on upperparts

Relatively short primary projection

Pure white underparts

Eastern Olivaceous Warbler

Short supercilium usually ends at lore

Relatively long bill with yellow lower mandible

Grayish-brown upperparts tinged olive

White fringe on wing feathers

White underparts

Relatively long primary projection

A. a. aedon

Relatively pale body, with more olive color

蝗莺科　Locustellidae

Small insectivorous songbirds. Plumage similar between sexes, mostly brown, white, and buffy; some species heavily streaked. Bill thin and pointed. Good fliers, with long and pointed wings. Most species good at walking, with long and strong legs. Tail medium to long. Inhabit primarily marshes, scrubland, reedbeds, tall grasses, rhododendron thickets, and forest understory. Nest on ground or in lower vegetation. Feed mostly on insects and other invertebrates. Most species are migratory.

Grassbirds, grasshopper warblers, and bush warblers were previously placed in Sylviidae but now constitute this newly recognized family. Eleven genera and 63 species recognized worldwide, distributed in the Old World, Australia, and Pacific islands. Three genera and 18 species recorded in China, widely distributed except on the Qinghai-Tibet Plateau.

小蝗莺属
Helopsaltes

Pallas's Grasshopper Warbler

蝗莺属
Locustella

Lanceolated Warbler

大尾莺属
Cincloramphus

Striated Grassbird

Helopsaltes amnicola
库页岛蝗莺 kù yè dǎo huáng yīng
Sakhalin Grasshopper Warbler　LC

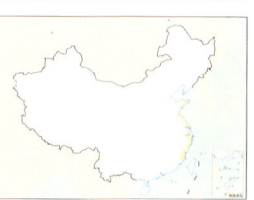

L: 16.5–18 cm
Habitat: During migration, inhabits woodland bushes and grassy areas in foothills and near coasts.
Behavior: Similar to Gray's Grasshopper Warbler.
Distribution: Very uncommon passage migrant. Migrates along island chains, passing through Taiwan and, to a lesser extent, the eastern coastal areas.
Voice: Song sounds slower, more mellow, and more relaxed than that of Gray's Grasshopper Warbler. Call also like Gray's.

Helopsaltes fasciolatus
苍眉蝗莺 cāng méi huáng yīng
Gray's Grasshopper Warbler　LC

L: 16.5–18 cm
Habitat: Inhabits forest edges and river valleys on low mountains in breeding season; prefers reedbeds and bushes in swampy areas. Occurs in bushes in open woodlands during migration.
Behavior: Often solitary or in pairs in breeding season; also gathers in small flocks during migration. Usually hides in dense grass. Extremely vigilant; seldom moves to the upper layer of bushes and rarely flies. Creeps, runs, and hops in the undergrowth.
Distribution: Uncommon migrant. Breeds in NE Inner Mongolia, Heilongjiang, and Jilin; migrates through Liaoning, Hebei, and the east and southeast coasts of China, including Taiwan.
Voice: Song is loud, far-carrying, rich, fluty, mellow, bulbul-like *tryt-to tryt-to trytorytoryt*. Call is a dry *terreck-terreck*.

Helopsaltes pryeri

斑背大尾莺 bān bèi dà wěi yīng

Marsh Grassbird

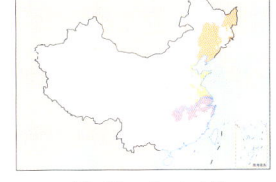

L: 12–14 cm
Habitat: Inhabits tall grasslands and reedbeds in lakes, rivers, streams, and coasts on plains; highly dependent on reedbeds.
Behavior: Often solitary or in pairs. Prefers perching on bushtops. Good hopper but not good flier. When disturbed, flies only a few meters before dropping back into reeds. Nests in reedbeds; nests are usually made of withered reed leaves supported by reed stems. Male also uses reed stems as the main perch for singing, courtship, and warning. In breeding season, often perches on top of reeds or grass, flies into the air while singing, and then drops back into the reeds.
Distribution: Breeds in lakes and other wetlands in Northeast, North, and Central China, as well as along the east coast, south to Poyang Lake in Jiangxi; wanders south in winter, often to the coastal regions.
Voice: Song is an often-intense, unmusical, sputtering, swiftlet-like chatter of up to 50 rapidly repeated, short clicking notes between 2–8.6 kHz that vary in intensity, often starting quietly; chatter lasts 2–3 sec and is repeated every 6–13 sec. Also a shorter series of mellower notes, with Radde's Warbler–like nasal character, *werrrrk … twerrrk … twrrrrrr*, given when agitated, and a scolding *tschick-tschick-tschick*, repeated in intense rapid sequence when alarmed.

Helopsaltes certhiola

小蝗莺 xiǎo huáng yīng

Pallas's Grasshopper Warbler

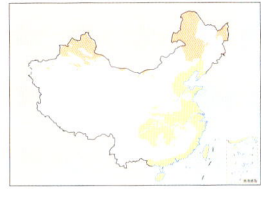

L: 12–14 cm
Habitat: Inhabits forest edges, reedbeds, or scrubland near water.
Behavior: Often solitary, secretive. Rarely calls outside breeding season. Usually flies through dense bushes or forages on the ground.
Distribution: Uncommon migrant. *H. c. certhiola* breeds in Northeast China; *H. c. centralasiae* breeds in northwestern China (Tian Shan Mountains, Qaidam Basin, and Ordos Plateau); *H. c. minor* breeds in Northeast China; *H. c. rubescens* has also been recorded in Northeast China. A few records from Taiwan and Hong Kong in winter. During migration, found in most areas of central, eastern, and southern China: *H. c. minor* and *H. c. rubescens*, which breed in Northeast China, migrate through the east coast, while the migration routes of the other two subspecies align toward the central regions. *H. c. sparsimstriatus*, which breeds in S Siberia, may also pass through China during migration.
Voice: Loud, strident, explosive song lasts 3–4 sec and often starts with one or two *dzik* call notes before *tri-tri- rich-rich-rich-che-che-che-che-che*. Calls include a mellow *trrrt*, often extended into lengthy series, and a penetrating *dzik* that is also often extended into a ratcheting splutter.

Sakhalin Grasshopper Warbler

Resembles Gray's Grasshopper Warbler closely but has more brownish and reduced olive tones on head, back, wings, and rump

Gray's Grasshopper Warbler

Olive-brown upperparts

Large thick bill

Tinged gray from cheek and throat to breast

Relatively large body, rounded tail

ad.

Buff supercilium and cheek

Dark brown from crown to upperparts

Dark brown spots from chin to breast and belly

Buff underparts, differing from juvenile Sakhalin Grasshopper Warbler

juv.

Yellowish-brown flanks

Marsh Grassbird

Black streaks from head and neck to back

Brown upperparts

White underparts

Black shaft streaks on uppertail coverts and tail

Pallas's Grasshopper Warbler

Relatively pale body, more olive brown on upperparts

Relatively fine and dense streaks on back

H. c. certhiola

Black streaks on crown and back

Body slightly tinged rufous brown, more conspicuous than other subspecies

Rufous-brown rump and uppertail coverts, distinctive in flight

H. c. rubescens

Helopsaltes pleskei

东亚蝗莺 dōng yà huáng yīng

Pleske's Grasshopper Warbler

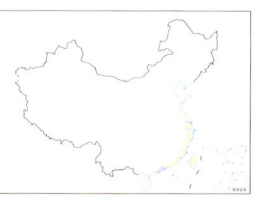

L: 16–17 cm

Habitat: Breeds in bushes, grassy areas, or willow thickets near woods on islands; inhabits mangroves, bushes, and grass near coasts during migration.

Behavior: Often solitary or in pairs. Secretive. Moves about in dense vegetation near the ground.

Distribution: Uncommon migrant. Breeds only on the offshore islands of Qingdao in Shandong; migrates through the east and southeast coasts and Taiwan, with a few wintering records in Hong Kong.

Voice: Sings frequently on branches in the breeding season. Song is loud and explosive, like Pallas's Grasshopper Warbler but slightly simpler, with fewer repetitions and fewer notes, *chip-chit-chir-chit-chi-schwee-scheee-scheee*, lasting between 1.2–3.2 sec. Contact call a short penetrating *chit*.

Helopsaltes ochotensis

北蝗莺 běi huáng yīng

Middendorff's Grasshopper Warbler

L: 13.5–14.5 cm

Habitat: Prefers wetter environments than Pleske's Grasshopper Warbler. Often inhabits river valleys on low mountains and in foothills. Also in dense bushes and tall grass near wetlands.

Behavior: Similar to Pleske's Grasshopper Warbler but climbs more skillfully in tall grass.

Distribution: Uncommon passage migrant. Migrates through Liaodong Peninsula and the east and southeast coasts; winters in Taiwan.

Voice: Pallas's Leaf Warbler–like song consists of two parts, *drrrt-chit chit-chit-cherwee-cherwee*, and contains less frequent and less rigid repetition. Calls include a short *chit*.

Locustella lanceolata

矛斑蝗莺 máo bān huáng yīng

Lanceolated Warbler

L: 12–13.5 cm

Habitat: Inhabits reed ponds and bushes near water or marshes.

Behavior: Secretive; often moves about quietly in bushes or forest understory. When alarmed, flies a few meters before taking cover in grass. Moves quickly on the ground like a mouse.

Distribution: Common migrant. *L. l. lanceolata* breeds in Northeast China and Xinjiang (Altay) and migrates through most of China; *L. l. hendersonii* migrates through eastern China.

Voice: Frequently sings during the breeding season. Song is a fast, repetitive, insectlike trill or reeling *gizi gizi gizi gizi gizi gizi gizi gizi gizi*, very similar to song of Common Grasshopper Warbler but faster, higher

pitched (centered around 6.5 kHz, not 5.0 kHz), and more piercing—hence more insectlike than mechanical. Lower mandible vibrates while singing. (Mandible does not vibrate in Common.) Call is a repeated short *tzhk*, thicker, mellow, and subtly more nasal than the harder click of Common. Alarm call a spluttering two-tone *rzzink-ti-titititititit* lasting 1.2–1.8 sec, very different from the lower-pitched descending purr or chatter of Common.

Locustella luteoventris

棕褐短翅蝗莺 (棕褐短翅莺)

zōng hè duǎn chì huáng yīng (zōng hè duǎn chì yīng)

Brown Bush Warbler

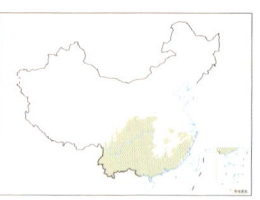

L: 13–14.5 cm

Habitat: In breeding season, inhabits grassy slopes and bushes in alpine coniferous forests and forest edges from 1200 to 3000 m. Moves down to similar habitats below 1200 m in nonbreeding season.

Behavior: Often solitary or in pairs. During breeding season, male perches and sings from top of bushes or grass; also sings in the lower part of grass. May form small flocks in suitable habitats. Timid and secretive; moves quietly and quickly through the lower part of bushes when foraging.

Distribution: Locally common resident. Found in high-altitude mountains in Central, East, South, southeastern, and southwestern China south of the Qinling Mountains; migrates to lower altitudes in winter.

Voice: Song a monotonous, mechanical, chuntering *tk tk tk tk tk tk tk tk tk*, like a fast garden sprinkler. Calls include a hard Common Grasshopper Warbler–like *thak*, often in rapid series, and a Gray-sided Bush Warbler–like *spik*.

Locustella major

巨嘴短翅蝗莺 (巨嘴短翅莺)

jù zuǐ duǎn chì huáng yīng (ju zuǐ duǎn chì ying)

Long-billed Bush Warbler

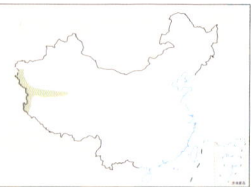

L: 13–15 cm

Habitat: Inhabits hillside bushes and tall grass at open forest edges; also occurs in bushes and weeds on hillsides. In breeding season, inhabits bushes and grassy areas in mountain valleys and above tree line, at 2500–3500 m. Migrates to low mountains and foothills down to 1000 m in winter. Undertakes marked altitudinal migration.

Behavior: Often solitary or in pairs. Timid. Good at hiding; often hides and forages in dense bushes and tall grass. Difficult to see; can be found in breeding season only when male sings for long periods. More often heard than seen.

Distribution: *L. m. major* in SW Xinjiang and SW Tibet; *L. m. innae* in SE Xinjiang. However, no records in China in the last two decades.

Voice: Poorly known call is apparently a calm *tic* or a warning *trrr*. Song is a monotonous metallic *pikha pikha pikha*, with three sounds per second, sometimes lasting for several minutes.

Pleske's Grasshopper Warbler

Relatively grayish-brown upperparts lack brownish yellow

Back lacks conspicuous streaks or spots

Long narrow bill

Off-white throat, breast, and belly lack brownish yellow

Body slightly larger than Middendorff's Grasshopper Warbler, with slightly longer bill, legs, and tail

Lanceolated Warbler

Relatively small body

Relatively dark body, appears olive brown

Dense blackish-brown streaks and spots on body

Middendorff's Grasshopper Warbler

Clear contrast between supercilium and eye stripe

Extremely diffuse dark streaks on back, far less clear than Pallas's Grasshopper Warbler

Distinctive olive-brown body with slightly brownish back, flanks, uppertail coverts, and tail

Distinctive white tips on tail feathers

Pale white fringe on outer web of wing feathers

1st win.

Brown Bush Warbler

Inconspicuous supercilium

All-yellow lower mandible

Brownish overall apart from white throat and belly

Brown undertail coverts lack streaks

Long-billed Bush Warbler

Dusky brown spots on throat and upper breast

Faint white supercilium ends at lore

L. m. major

Relatively slender bill with pinkish-yellow lower mandible

Spotless chin, throat, and upper breast

L. m. innae

Spotless brown undertail coverts

Locustella naevia

黑斑蝗莺 hēi bān huáng yīng

Common Grasshopper Warbler

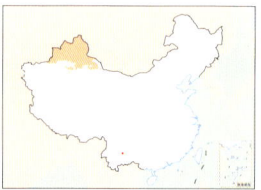

L: 12.5–13.5 cm
Habitat: Inhabits montane forests, forest clearings, and bushes and grass at forest edges, as well as bushes and grass along bodies of water.
Behavior: Often perches on bushtops and sings for long periods before dropping back into the grass. Fast and agile; often wags tail rapidly.
Distribution: *L. n. straminea* is a locally common breeder in the central and western parts of N Tian Shan Mountains in Xinjiang; *L. n. mongolica* is an uncommon breeder in the E and C Altai Mountains in Xinjiang. Vagrant to Yunnan.
Voice: Song is similar to Lanceolated Warbler: a nonstop, fast, tedious *zi zi zi*, like bell of alarm clock. Call shorter, clearer, and more incisive than that of Lanceolated Warbler. Alarm call a chattering rattle about a second long, reminiscent of its own song and very different from that of Lanceolated.

Locustella luscinioides

鸲蝗莺 qú huáng yīng

Savi's Warbler

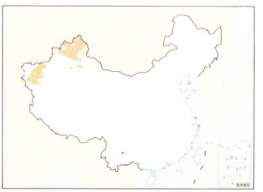

L: 13.5–15 cm
Habitat: Inhabits lakes, reservoirs, marshes, and fishponds on plains, as well as reedbeds at water edges. Also occurs in river-valley bushes with tall grass.
Behavior: Often hides and sings in thick reeds; more often heard than seen.
Distribution: Uncommon breeder in Xinjiang. Vagrant in Yunnan.
Voice: Lengthy, insectlike, Common Grasshopper Warbler–like buzzing. Distinguished from that species by faster pace (more notes per second), lower pitch, and general tone.

Locustella thoracica

斑胸短翅蝗莺 (斑胸短翅莺) bān xiōng duǎn chì huáng yīng (bān xiōng duǎn chì yīng)

Spotted Bush Warbler

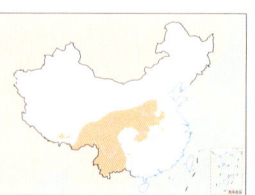

L: 12–14 cm
Habitat: In breeding season, inhabits grassy slopes and bushes at subalpine coniferous forest edges and alpine meadows from 2000 to 4500 m; descends to scrubby environments below 2500 m in nonbreeding season.
Behavior: Similar to Baikal Bush Warbler.
Distribution: Locally common. Breeds from central to southwestern China (Shaanxi, Gansu, Qinghai, Sichuan, Guizhou, Guangxi, Yunnan, and SE Tibet).

Voice: Distinctive song is a far-carrying, hurried, short (0.6–0.7 sec), techno sequence that starts with two short notes followed by three triplet calls and ends with a short buzzy note, *tr-t-zrink-zrink-zrink-bzrrrrt*, of 3.2–7.4 kHz, repeated metronomically with little variation (other than when the bird turns its head). Call very similar to Baikal Bush Warbler, *chut* or *spik*.

Locustella tacsanowskia

中华短翅蝗莺 (中华短翅莺) zhōng huá duǎn chì huáng yīng (zhōng huá duǎn chì yīng)

Chinese Bush Warbler

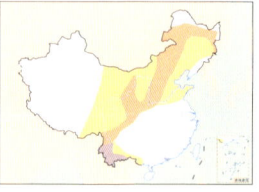

L: 13 cm
Habitat: Inhabits bushes at the edge of larch forests from 2800 to 3600 m in breeding season. The population breeding in Northeast China nests in similar habitats at lower altitudes. In nonbreeding season, moves to grasslands, cultivated lands, and open scrubland at low altitudes.
Behavior: Often solitary or in pairs. In breeding season, male sings at the top and lower parts of bushes or grass; may form small flocks in suitable habitats. Shy and secretive; moves quietly and quickly through the lower part of bushes.
Distribution: Uncommon migrant. Breeds in parts of Northeast, North, and Central China, south to Sichuan and Guangxi. Winters in S and SW Yunnan and SW Guangxi; migrates through North, Central, and southwestern China, with banding records in Hong Kong.
Voice: Song is a lengthy, raspy, insectlike buzzing, *dzzzeep ... dzzzeep.* (See Baikal Bush Warbler.) Calls are a crisp *chir, chir, chir.*

Locustella davidi

北短翅蝗莺 (北短翅莺) bĕi duǎn chì huáng yīng (bĕi duǎn chì yīng)

Baikal Bush Warbler

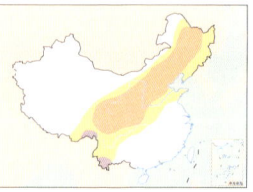

L: 12 cm
Habitat: In breeding season, inhabits subalpine (from 1000 to 1800 m) coniferous forests, bushes near forest edges, and grasslands and bushes near streams and swamps. In nonbreeding season, migrates to grasslands and bushes on low mountains and open plains below 1000 m, occasionally to green spaces.
Behavior: Solitary or in pairs. In breeding season, male sings from top of, or lower down in, bushes or grass; may form small flocks in suitable habitats. Active and secretive; moves quietly and quickly in the lower part of bushes when foraging.
Distribution: Uncommon migrant. Breeds from Northeast and North China to S Gansu, S Shaanxi, C Sichuan, and SW Chongqing. Winters in S Yunnan and SE Tibet; recorded in Northeast, North, Central, South, and southwestern China during migration, with several records in Hong Kong during migration and winter.
Voice: Song starts and ends abruptly and is a uniform buzzing, *dzzzeep ... dzzzeep,* similar to song of Chinese Bush Warbler but with much shorter strophes (0.4–0.5 sec, as opposed to 2.3–2.7 sec in Chinese), with individual notes delivered much faster, and with more intense or "pulsed" quality. Call like Spotted Bush Warbler.

Common Grasshopper Warbler

Discontinuous black spots on head and upperparts

Underparts lack or have a few streaks

Chinese Bush Warbler

Pale brown upperparts and flanks

Pale brown wings

White throat and lower breast; upper breast tinged pale brown with extremely diffuse brown streaks

Diffuse brown bands on off-white undertail coverts

Savi's Warbler

Brownish upperparts lack streaks

Extremely faint bands on undertail coverts

Buff underparts lack streaks

Baikal Bush Warbler

White throat

Blackish-brown patch on upper breast; a few black spots extend to throat and lower breast

Off-white from lower breast to belly

br.

Diffuse blackish-brown bands on off-white undertail coverts

Spotted Bush Warbler

Tinged gray from face and side of neck to lower breast

Black bill

String of black spots on side of neck and breast

White from throat to lower belly

br.

Dark brown bands on off-white undertail coverts

Only a few obscure dark spots on upper breast

non-br.

Obscure pale spots on breast

non-br.

Locustella alishanensis

台湾短翅蝗莺 (台湾短翅莺)

tái wān duǎn chì huáng yīng (tái wān duǎn chì yīng)

Taiwan Bush Warbler　　LC

L: 13 cm
Habitat: In breeding season, inhabits dense grass, bushes, or bamboo thickets at forest edges or in open areas from middle to high altitudes (1200–3000 m); in nonbreeding season, migrates to lower altitudes in mountainous areas, seldom below 1000 m.
Behavior: Often solitary. In breeding season, male sings from top of bushes and grass, or lower down. Shy and secretive; moves quietly and quickly through the lower part of bushes when foraging.
Distribution: Chinese endemic; locally common resident in Taiwan.
Voice: Song is easily recognized—a monotonously repeated, mechanical fricative that starts with a clear, loud, monotonous whistle about 0.3 sec long that is followed by two or three resonant (castanet-like) clicks, *t.t't-jeee*, the whole sequence lasting less than 0.5 sec and repeated metronomically for lengthy periods. Call a harsh, scolding, Lanceolated Warbler–like *tshik*, often repeated.

Locustella mandelli

高山短翅蝗莺 (高山短翅莺)

gāo shān duǎn chì huáng yīng (gāo shān duǎn chì yīng)

Russet Bush Warbler　　LC

L: 13–14 cm
Habitat: In breeding season, inhabits grassy slopes or bushes at edges of broadleaf or mixed forests below 2500 m; breeds down to 200 m in some areas of South China, which is inconsistent with what its Chinese common name implies (高山, which means "alpine" or "high mountain"). In nonbreeding season, migrates to bushes and grass at forest edges on low mountain and plains.
Behavior: Often solitary or in pairs. In breeding season, male sings from top of bushes and grass, or lower down; may form small flocks in suitable habitats. Shy and secretive; moves quietly and quickly through the lower part of bushes when foraging.
Distribution: Locally common. *L. m. mandelli* in S and SE Tibet, NW and W Yunnan, and Sichuan. *L. m. melanorhyncha* in NE Yunnan, Guangxi, Hong Kong, Guangdong, Fujian, SE Hunan, Jiangxi, and S Zhejiang; migrates to lower altitudes during nonbreeding season and wanders to foothill villages, plains, coastal areas, and even offshore islands in winter.
Voice: Song is a long series of buzzy *cre-ut . . . cre-ut . . . cre-ut* notes, repeated metronomically. Call is *tshk*, like many congeners.

Locustella chengi

四川短翅蝗莺 (四川短翅莺)

sì chuān duǎn chì huáng yīng (sì chuān duǎn chì yīng)

Sichuan Bush Warbler　　LC

L: 13 cm
Habitat: In breeding season, inhabits grassy slopes or bushes at edges of broadleaf or mixed forests from 1000 to 2275 m; migrates to similar habitats at lower altitudes in nonbreeding season. Distribution overlaps with Russet Bush Warbler, but usually at lower altitudes when the two species occur in the same region.
Behavior: Similar to Russet Bush Warbler.
Distribution: Uncommon. Known ranges in C and S Shaanxi, N, C, and S Sichuan, N Guizhou, NW Hunan, SW Hubei, and NW Jiangxi. The species has been overlooked, and was formally described and named as a species only very recently, because of its resemblance to Russet Bush Warbler and their overlapping distributions. Separating the pair in the field is extremely difficult and best accomplished using vocalizations.
Voice: Rhythm and melody of song are very similar to song of Russet Bush Warbler, but the pitch is significantly lower and more muffled.

Cincloramphus palustris

沼泽大尾莺 zhǎo zé dà wěi yīng

Striated Grassbird　　LC

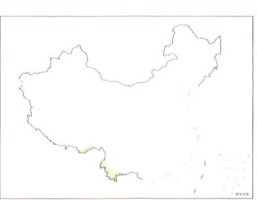

L: 22–28 cm (male larger than female)
Habitat: Inhabits reed swamps, open scrub, and wet grasslands near streams or ponds at altitudes of 350–1800 m. Sometimes occurs near villages and agricultural farmland; commonly seen on field bunds near grass.
Behavior: Solitary or in pairs. Often flicks tail or constantly flaps wings when perched on trees. Usually flies through dense reedbeds or grass; also searches for food while walking on the ground. In breeding season, often perches on top of bushes, bamboo clumps, or reeds; flies up into the air and sings while flying; lands on the ground with wings open after a short flight. Also hides and sings in bushes.
Distribution: Locally common resident in W and C to S Yunnan, S Guizhou, SW Guangxi. Sometimes placed in genus *Megalurus*.
Voice: Song is clear, loud, and sharp; resembles *chot-chot-chot-chot which-u-quieee-chot-trrrrt-kwit-kwit-kwit-cheee-chwot*. Calls a sharp *pwit* and mellow *stup*, both occasionally repeated in lengthy series.

Taiwan Bush Warbler

Black bill with pale base of lower mandible

Olive-brown upperparts

White throat

Diffuse black spots on gray upper breast

Off-white from lower breast to belly

Brown flanks and undertail coverts

Sichuan Bush Warbler

Resembles Russet Bush Warbler, but upperparts tinged olive, not brownish

Relatively long primaries

Relatively short tail

Russet Bush Warbler

Resembles Taiwan Bush Warbler but with dusky brown upperparts

Dark shaft streaks on spotless throat; some individuals have small black patches on throat

Olive brown from flanks to undertail coverts

Striated Grassbird

Pale brown crown and wings

Heavy blackish-brown shaft streaks on back, side of neck, and wings

Pale brown upperparts and flanks

Long strong legs allow straight standing posture

Large strong body, very long tail

扇尾莺科　Cisticolidae

Small songbirds that prefer low vegetation. Plumage similar between sexes. Body usually heavily streaked. Bill long and thin. Wings short and rounded; flight appears clumsy and strenuous. Good at hopping around, with strong legs. Tail long, mostly wedge shaped. Inhabit primarily scrubland, tall grasses, reedbeds, and forest edges. Nest in low vegetation. Feed mostly on insects and other invertebrates. Most species are resident.

　　Cisticolas and allies were previously placed in Sylviidae but now constitute this newly recognized family. Twenty-six genera and 163 species recognized worldwide, distributed mostly in tropical and subtropical regions of the Old World, especially species rich in Africa. Three genera and 12 species recorded in China, found in the Himalayas and eastern and southern China.

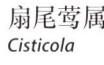

扇尾莺属
Cisticola

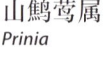

山鷦莺属
Prinia

non-br.

Zitting Cisticola

br.

br.

Plain Prinia

non-br.

缝叶莺属
Orthotomus

Common Tailorbird

Cisticola juncidis

棕扇尾莺 zōng shàn wěi yīng

Zitting Cisticola 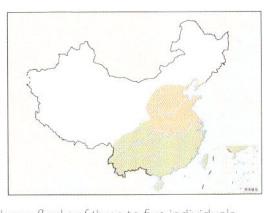 **LC**

L: 10–14 cm
Habitat: Inhabits open fields, grassy areas, bushes, and swamps on plains and hills under 1000 m; prefers low grass environments.
Behavior: Solitary or in pairs in breeding season; highly territorial.
In nonbreeding season, forms loose flocks of three to five individuals. Active; moves about and forages all day. Flies through grass and sometimes perches on bushtops or power lines. Breeding male often performs a characteristic flight display over its territory, taking off straight into sky, soaring, circling, then closing both wings, and diving almost vertically or into grass when approaching the ground. Often fans and flicks tail in flight.
Distribution: Locally common resident or migrant. *C. j. tinnabulans* widespread in North, Central, East, and South China, north to Liaoning, with northern populations migrating south in winter; *C. j. cursitans* in W Yunnan.
Voice: Song is a sharp, repeated *tlip-tlip … tlip-tlip*, given when performing an exaggerated undulating display flight. An alternate is a lengthy penetrating series of very short squeaky-wheel notes, repeated at length, *ri-ri-ri-ri-ri-ri-ri-ri*, and a powerful *pleck*.

Cisticola exilis

金头扇尾莺 jīn tóu shàn wěi yīng

Golden-headed Cisticola  **LC**

L: 9–11.5 cm
Habitat: Inhabits foothills and moist grasslands, scrubland, and paddy fields on plains, often below 1000 m.
Behavior: Similar to Zitting Cisticola but more secretive.
Distribution: Uncommon resident. *C. e. tytleri* in SE Tibet and W Yunnan; *C. e. courtoisi* in South and East China (from SE Yunnan to S Anhui on the east, and south to Guangxi and Guangdong); *C. e. volitans* in Taiwan.
Voice: Song is a peculiar squelch, followed by a blast, *wheeze-plio*, with the second note often given twice, occasionally more often. Call a short nasal mew.

Zitting Cisticola

Off-white or buff supercilium, usually distinct

Dusky brown back; pale yellowish-brown fringe on feathers form dusky streaks on back

White tip and black subterminal band on tail

Feathers of short protruding tail longest in center and gradually shorten toward sides

br.

Fine blackish-brown streaks on crown

Upperparts and flanks more brownish

non-br.

Golden-headed Cisticola

Pale yellowish white from crown to hindneck

Side of neck and flanks tinged buff

White throat and undertail coverts, off-white lower belly

♂ br.

Thick blackish-brown streaks on crown and wing coverts

Distinct brownish-yellow tone from supercilium to side of neck, flanks, wing feathers, and tail

Relatively long tail

♀ non-br.

Prinia striata

山鹪莺 shān jiāo yīng

Striped Prinia 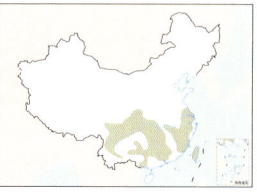 LC

L: 15–16 cm
Habitat: In breeding
season, inhabits bushes
and grassy areas on open
mountains and in foothills
from 1500 to 2000 m; may
come down to 150 m on
China's southeast coast.
In nonbreeding season,
migrates to similar environments at lower altitudes.
Behavior: Solitary or in pairs in breeding season; may form small flocks
of three to five individuals in nonbreeding season. In breeding season,
male often perches high up on grass stems and sings incessantly. Shy
and secretive; often hops about and forages in the lower part of bushes
and grass stems near ground. Flies a short distance to take cover in
bushes when disturbed, dragging its long tail in flight, appearing slow
and clumsy.
Distribution: Uncommon resident. *P. s. catharia* in N, W, and SE Yunnan,
S Shaanxi, SW Sichuan, Hunan, and Guizhou; *P. s. parumstriata* occurs
from Central China to coastal southeastern China; *P. s. striata* in Taiwan.
Voice: Song is a series of swishing sounds, resembling *chitzereet-
chitzereet-chitzereet-chitzereet*. Call is a very high-pitched, thin *ki-ki-ki*.

Prinia crinigera

喜山山鹪莺 xǐ shān shān jiāo yīng

Himalayan Prinia LC

L: 16–18 cm
Habitat: Inhabits forest-
edge bushes, grasslands,
and farmland at middle to
low altitudes.
Behavior: Solitary or in
pairs. Cocks tail when
perched.
Distribution: *P. c. crinigera*
in S Tibet; *P. c. yunnanensis* in W Yunnan; *P. c. bangsi* in SE Yunnan.
Uncommon.
Voice: Song very similar to Striped Prinia but distinguished by usually
longer notes that are typically paired, shorter pauses between phrases,
the presence of thin, rattling, buzzing notes, and a distinctly harsher,
more "squeaky," "nasal" tone.

Prinia atrogularis

黑胸山鹪莺 hēi xiōng shān jiāo yīng

Black-throated Prinia LC

L: 16–20 cm
Habitat: Inhabits
scrubland and grasslands
at middle altitudes.
Behavior: Active among
grass and bushes. Cocks
tail when perched.
Distribution: Uncommon
in S Tibet.
Voice: Song is a rapidly repeated series of bouncy mechanical notes, *zlik
… zlik … zlik* or *ze-szlk … ze-szlk … ze-szlk*. Calls include an abrupt rising
tli-uu, a dry buzz, *kzhrt*.

Prinia rufescens

暗冕山鹪莺 àn miǎn shān jiāo yīng

Rufescent Prinia LC

L: 10.5–12.5 cm
Habitat: Inhabits
scrubland, grasslands, and
secondary forests.
Behavior: Solitary or in
pairs in grass and bushes;
seldom flies.
Distribution: Locally
common in southwestern
and South China.
Voice: Song a rhythmic *t'chew-t'chew-t'chew* or *qipu-qipu*, repeated
rapidly. Calls include a buzzy, peevish *peez-eez-eez-eez*.

Prinia flaviventris

黄腹山鹪莺 huáng fù shān jiāo yīng

Yellow-bellied Prinia 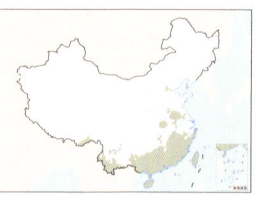 LC

L: 12–14 cm
Habitat: Inhabits forest-
edge bushes, grasslands,
bamboo thickets, and
the edge of cultivated
lands on low mountains,
hills, and plains; especially
prefers dense grass and
swamps near water.
Behavior: Often solitary. Male often sings on the top of manzanita in
breeding season. More secretive in nonbreeding season.
Distribution: Locally common resident. *P. f. delacouri* in SE Tibet and
W to S Yunnan. *P. f. sonitans* from SE Yunnan east to Guangxi,
Guangdong, Hainan, Fujian, Taiwan, and S Zhejiang, and north to
Guizhou, Hunan, Hubei, Jiangxi, Anhui, and S Henan; occasionally in
Shanghai, Jiangsu, and North China, as far north as Beijing. *P. f. sonitans* is
sometimes treated as a separate species, *Prinia sonitans*.
Voice: Pleasant ringing song is a repeated, fast burst of descending
musical trills, *didiliou-didiliou-didiliou*. Call is a shrill catlike *meow-*.

Prinia superciliaris

黑喉山鹪莺 hēi hóu shān jiāo yīng

Hill Prinia  LC

L: 15–20 cm
Habitat: Inhabits grassy
areas and scrubland at
middle altitudes.
Behavior: Solitary or in
pairs in bushes or grass.
Raises tail when perched.
Distribution: Locally
common from
southwestern to South China.
Voice: Song is a loud, clear, and monotonously repetitive *ji-yu … ji-yu …
ji-yu* or, with one syllable missing, a penetrating Little Ringed Plover–like
tiu … tiu … tiu. Calls include a lengthy spluttering.

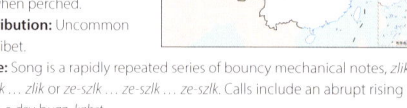

Striped Prinia

Dense streaks on head extend to cheek, side of neck, and throat

non-br.

Dense fine brown or blackish-brown streaks from crown to neck

Diffuse thick brown streaks on back

Brown wings and tail

br.

Rufescent Prinia

Clear supercilium

Rufous-brown upperparts

Long thick bill

Relatively short tail

Yellowish-brown flanks

Yellow-bellied Prinia

Bluish gray from crown to side of neck

Red iris

Olive-brown upperparts and tail

White throat

Buff to pale yellow from breast to belly

P. f. sonitans

Darker gray on head

Brown upperparts and tail

Yellow on belly brighter and more well defined

P. f. delacouri

Himalayan Prinia

Dark streaks from crown to back

Upperparts more grayish black with brownish wash

Relatively large body

Black around eyes

Reddish wing feathers

Pale underparts, whitish throat

Relatively long tail

P. c. crinigera

Black-throated Prinia

Brownish crown, upperparts, wings, and tail

Short white supercilium

White malar stripe

Throat and breast black during breeding season

Hill Prinia

White supercilium

Gray cheek

Extremely long tail

Black streaks on breast

Prinia hodgsonii

灰胸山鹪莺 huī xiōng shān jiāo yīng

Gray-breasted Prinia

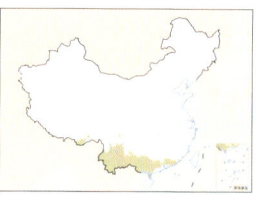

L: 10–12 cm
Habitat: Inhabits grassy areas and scrubland at low altitudes.
Behavior: Solitary or in small flocks among grass and bushes. Often bobs tail up and down when perched.
Distribution: *P. h. rufula* in SE Tibet and W Yunnan; *P. h. confusa* in S Yunnan, S Sichuan, Guizhou, and Guangxi. Locally common.
Voice: Lively, squeaky, scratchy, warbling song, *ti-si-wiwiwiwiwi*, which rises and intensifies in volume but ends abruptly. Calls include dry churrs, *chit-it-trrt*.

Prinia inornata

纯色山鹪莺 chún sè shān jiāo yīng

Plain Prinia

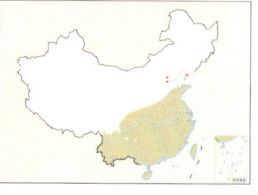

L: 11–15 cm
Habitat: Inhabits agricultural farmland and open grasslands on low mountains, plains, hills, and river valleys below 2000 m; frequents paddy fields, grassy areas, and deserted fields. Also adapted to habitats such as sugarcane fields, bamboo thickets, and urban parks.
Behavior: Solitary or in pairs in breeding season; gathers in small flocks of over 10 individuals in nonbreeding season. Bold, active, and boisterous. Often moves through grass while calling, hopping about and flying deftly among grass stems. Flight appears slow and clumsy because of the long tail; usually flies a short distance before taking cover in bushes.
Distribution: Locally common resident. *P. i. blanfordi* in W and SW Yunnan. *P. i. extensicauda* in E and N Yunnan, W Sichuan, and most of Central, East, and South China, including Hainan; found in recent years expanding north to North China; vagrants as far north as Beijing and Liaoning (Dalian). *P. i. flavirostris* in Taiwan. *P. i. fusca* probably occurs in S Tibet.
Voice: Song similar to Yellow-bellied Prinia, but strophes longer, a hurried bubbling *tidli-idli-lia*, with stress on final, falling syllable. Short, almost cisticola-like calls, *chip-chip-chip*, as well as soft contact calls, *bzzp, bzzp*.

Orthotomus sutorius

长尾缝叶莺 cháng wěi féng yè yīng

Common Tailorbird

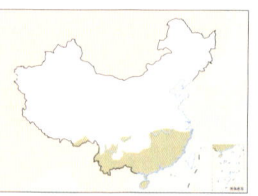

L: 10–14 cm
Habitat: Inhabits low mountains, foothills, and plains below 1000 m; especially perfers woods near human habitation or bushes in artificial forests. Adapted to various wooded habitats.
Behavior: Often solitary. Very active when foraging, hopping constantly among branches. Usually calls from top branches or power lines with tail cocked. Uses the white fibers covering cotton tree seeds to sew a leaf into a nest cup lined with plant fibers, hence the name "tailorbird."
Distribution: Common resident. *O. s. inexpectatus* in SE Tibet, W, SW, and SE Yunnan; *O. s. longicauda* in East and South China, including Hainan, and north to S Zhejiang. In recent years, the species has also been found in Hunan, Jiangxi, and Hubei, potentially a result of population expansion.
Voice: Varied. Song is a loud, repeated, sudden outburst, *chubit, chubit, chubit* or *pwit-pwit-pwit*, varying in intensity and pitch. Calls include loud, rapidly repeated *tiu-tiu-tiu* and repeated *cheep, cheep, cheep, cheep*.

Orthotomus atrogularis

黑喉缝叶莺 hēi hóu féng yè yīng

Dark-necked Tailorbird

L: 11–12 cm
Habitat: Inhabits forests and scrubland at low altitudes.
Behavior: Secretive; moves about in the lower part of bushes. Cocks tail often when perched.
Distribution: Locally common in SW Yunnan and S Guangxi.
Voice: Song is a spluttering, agitated-sounding *piarr-piar-piarr* or a *tiurrrr … tiurrrr*, repeated 2–15 times.

Gray-breasted Prinia

Almost no supercilium

Brown wash on upperparts, brownish wings

br.

Gray breast band

Underparts lack yellowish brown

Short faint supercilium

Indistinct gray breast band

non-br.

Common Tailorbird

Cocks relatively long brown tail

Olive from hindneck and back to uppertail coverts

Rufous chestnut from forehead to crown, gradually fades into olive brown toward nape

Off-white underparts

O. s. inexpectatus

Plain Prinia

Creamy-white supercilium diffuse behind eye

Relatively uniform pale brown crown, upperparts, and tail

Black bill with pale base of lower mandible

Creamy-white underparts with pale yellowish-buff sides of breast, flanks, and undertail coverts

br.

From supercilium, cheek, and throat to underparts tinged brownish yellow

non-br.

Dark-necked Tailorbird

Red crown, nape, forehead, and lores

br.

Distinct black feathers on throat and breast

non-br.

Some black streaks on side of neck

林鹛科 (鹛科)　　Timaliidae

Small to medium-sized babblers, most of which are ground dwellers. Body profile and plumage highly variable. Plumage similar between sexes, mostly brown, rufous, black, and white, while a few species are exceptionally brightly colored. Moderately long bill varies in shape: some bills slightly decurved; some straight, pointed, and wedge shaped; others extremely fine and deeply decurved. Wings short and rounded; usually poor fliers. Legs strong for living on ground. Tail long, squared or wedge shaped. Inhabit primarily edges or bushes of various forest types, as well as parks and plantations. Nest among branches. Feed mostly on insects; also take plant seeds and fruits, sometime even small amphibians and reptiles. Most species are resident; some are nomadic during nonbreeding seasons.

　　Was once a large family that included many babblerlike species, some of which are not closely related phylogenetically. The Chinese name of the family was also revised to avoid confusion with other newly established "babbler" families suggested by recent phylogenetic studies. Nine genera and 52 species recognized worldwide, endemic to Indomalaya. Seven genera and 27 species recorded in China, found along in Himalayas and southern China.

红顶鹛属
Timalia

Chestnut-capped
Babbler

纹胸鹛属
Mixornis

Pin-striped Tit-Babbler

红头穗鹛属
Cyanoderma

Rufous-capped
Babbler

鹪鹛属
Spelaeornis

Bar-winged
Wren-Babbler

大钩嘴鹛属
Erythrogenys

Black-streaked
Scimitar-Babbler

钩嘴鹛属
Pomatorhinus

Streak-breasted
Scimitar-Babbler

穗鹛属
Stachyris

Spot-necked
Babbler

Timalia pileata

红顶鹛 hóng dǐng méi
Chestnut-capped Babbler

L: 15.5–17 cm
Habitat: Inhabits lowland swamps and rivers, as well as grasses, bushes, reedbeds, and bamboo thickets along roads and farmland, at altitudes of 340–880 m.
Behavior: Often in pairs in breeding season and in small flocks of six to eight individuals in nonbreeding season; often mixes with Yellow-eyed Babbler and forages low in vegetation.
Distribution: Common resident in W Yunnan and Guangxi, scarce in Guizhou and Guangdong.
Voice: Extensive vocabulary includes a descending tremulous whistle, *prreeuu-tuu*, repeated every 3 sec or so, and a clear, short, rising whistle, *which-i-tu-ee*. Calls include a scolding chatter, *prr-riti-t-trrr-tr*, a strained, nasal *pwee-te-te-te-t*, and a lengthy series of *spik* notes.

Chestnut-capped Babbler
- White forehead and supercilium
- Black from around eye to base of bill
- Rufous brown from crown to hindneck
- White throat and breast
- Pale ochre-yellow belly

Pin-striped Tit-Babbler
- Pale yellow to olive from supercilium and cheek to underparts, with individual differences
- Yellowish-white iris
- Relatively pale yellow body with some olive or olive brown
- Diffuse fine streaks from throat to breast

Golden Babbler
- Relatively distinct black streaks on head
- Grayish-black lores

C. c. chrysaeum

- Relatively pale black streaks on head
- Grayish lores

C. c. auratum

Mixornis gularis

纹胸鹛 (纹胸巨鹛)
wén xiōng méi (wén xiōng jù méi)
Pin-striped Tit-Babbler

L: 11–14 cm
Habitat: Inhabits scrubland, open evergreen broadleaf and mixed evergreen-deciduous broadleaf forests, bamboo thickets, plantations, gardens, etc., below 1000 m.

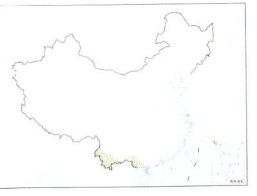

Behavior: Solitary or in pairs during breeding season; in small flocks of 10 or more in nonbreeding season. Often mixes with other small babblers, usually close to the ground; also climbs trees to forage.
Distribution: *M. g. sulphureus* in SW Yunnan; *M. g. lutescens* is common in SE Yunnan and SW Guangxi.
Voice: Simple song is a rhythmically consistent four- or five-note *tit-chut-chut-chut*. Call is a harsh *chrrrt-chrr*.

Cyanoderma chrysaeum

金头穗鹛 jīn tóu suì méi
Golden Babbler

L: 10–12 cm
Habitat: Inhabits primary and secondary evergreen broadleaf forests, secondary dwarf bushes in dense forests, bamboo thickets, and forest edges at altitudes of 450 to 3000 m.

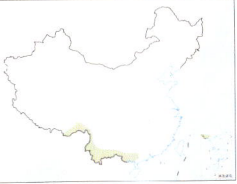

Behavior: Often in pairs in breeding season; mixes with other birds, such as Gray-throated Babbler, outside breeding season (after August). In low bushes or on the ground.
Distribution: *C. c. chrysaeum* in SE Tibet, NW Yunnan to W Yunnan; *C. c. auratum* in S Yunnan, E to SW Guangxi.
Voice: Song is a rapid series of short piping whistles very similar to songs of Black-chinned Babbler, Rufous-capped Babbler, and Buff-chested Babbler. All of these species have scolding churr call notes, *chrrrr-rr-rr*. Golden is often slightly more nasal in tone, especially in the opening few notes, and is usually the least remarkable, with moderately long, evenly paced churrs at medium pitch.

Cyanoderma pyrrhops

黑颏穗鹛 hēi kē suì méi

Black-chinned Babbler `LC`

L: 10 cm
Habitat: Inhabits primary and secondary evergreen broadleaf forests, forest edges, understory bushes, and bamboo thickets at 200–2800 m.
Behavior: Often in flocks outside breeding season; sometimes joins mixed-species flocks. Usually in low bushes or on ground.
Distribution: Resident in SE and S Tibet.
Voice: Song is a rapid series of short, piping, whistled *tu-tu-tu* notes very similar to Golden Babbler, Rufous-capped Babbler, and Buff-chested Babbler but can be in a significantly longer series (up to 15 notes), when delivery is then much faster. The first note of the song is only rarely separated from the remainder. All of these species have scolding churr call notes, *chrrrrr-rr-rr*; that of Black-chinned is usually the longest, fastest, most intense, and lowest pitched. It also slows down very slightly.

Cyanoderma ruficeps

红头穗鹛 hóng tóu suì méi

Rufous-capped Babbler `LC`

L: 12 cm
Habitat: Inhabits evergreen broadleaf forests, bamboo thickets, and dense secondary scrub at 200–2500 m.
Behavior: In pairs during breeding season; otherwise gathers in small flocks or mixes with other species, such as fulvettas. Generally in lower and middle stories.
Distribution: Locally common resident. *C. r. ruficeps* in SE Tibet; *C. r. bhamoense* in W and SW Yunnan; *C. r. davidi* in Central, South, and southeastern China; *C. r. goodsoni* in Hainan; *C. r. praecognitum* in Taiwan.
Voice: Song is a rapid series of short piping whistles, very similar to those of Black-chinned Babbler, Golden Babbler, and Buff-chested Babbler, *tu-tu-tu-tu*. All of these species have scolding churr call notes, *chrrrrr-rr-rr*; that of Rufous-capped is the slowest and driest and the one with the greatest frequency range, often sounding like *trrrtt*.

Cyanoderma ambiguum

黄喉穗鹛 huáng hóu suì méi

Buff-chested Babbler `LC`

L: 12 cm
Habitat: Inhabits primary and secondary evergreen broadleaf forests, forest edges, understory bushes, and bamboo thickets, from 50 to 1200 m.
Behavior: Gathers in small flocks in nonbreeding season; sometimes joins mixed-species flocks. Moves constantly in dense undergrowth or bamboo thickets.
Distribution: *C. a. planicola* in extreme NW Yunnan; *C. a. ambiguum* in SE Tibet.
Voice: Song is a rapid series of short, piping, whistled *tu-tu-tu* notes, very similar to songs of Black-chinned Babbler, Golden Babbler, and Rufous-capped Babbler.

Spelaeornis badeigularis

锈喉鹩鹛 xiù hóu liáo méi

Mishmi Wren-Babbler `VU` `NT`

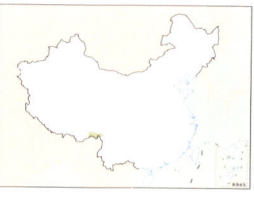

L: 9 cm
Habitat: Inhabits dense undergrowth vegetation in evergreen broadleaf forests, often in moist gullies at 1200–3000 m.
Behavior: Shy. Usually seen alone or in pairs, staying within a meter of the ground; moves among dense herbaceous vegetation.
Distribution: Rare in SE Tibet (Medog).
Voice: Similar to Rufous-throated Wren-Babbler but shorter (0.7–0.8 sec, as opposed to 1.1–1.2 sec) and slightly less rigidly structured.

Spelaeornis caudatus

棕喉鹩鹛 zōng hóu liáo méi

Rufous-throated Wren-Babbler `NT`

L: 9 cm
Habitat: Inhabits moist evergreen broadleaf forests; prefers secluded valleys and steep narrow gulleys, especially in areas with abundant ferns, mossy rocks, and fallen trees.
Behavior: Shy; prefers moving around in secluded, moist bushes near ground. Maintains contact and asserts territory by calling.
Distribution: S Tibet.
Voice: Song is an intense, short outburst, *witchu-witchu-witchu-witchu-witchu*, the first note slightly higher; also a slightly more trisyllabic *whitchitu-* or with the first note slightly drawn out.

Spelaeornis troglodytoides

斑翅鹩鹛 bān chì liáo méi

Bar-winged Wren-Babbler `LC`

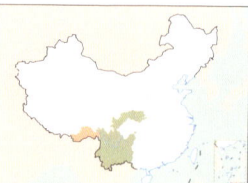

L: 10 cm
Habitat: Lives in dense undergrowth or bamboo thickets in rhododendron forests, broadleaf forests, and mixed broadleaf-coniferous forests at altitudes of 1500–3500 m.
Behavior: In pairs during breeding season and occasionally forms small flocks during nonbreeding season. Often forages on the ground. Also occurs in the middle and upper parts of bamboo thickets, but not in the canopy.
Distribution: *S. t. souliei* in NW Yunnan west of the Nujiang (Salween) River; *S. t. rocki* in NW Yunnan east of the Nujiang River; *S. t. halsueti* in S Gansu, S Shaanxi, N and NE Sichuan; *S. t. nanchuanensis* in C Sichuan, Hubei, and Hunan; *S. t. troglodytoides* in C, W, and SW Sichuan, extreme NE Yunnan, and N Guizhou; *S. t. indiraji* in S Tibet.
Voice: Rapid, rollicking warble of five to eight notes, *chi-wuuuu … t'wi-chi … t'wi-chi … t'wi-chi … twi-chi … twi-chi*, starting with a diphthong, then a longer slightly rising note, before four repeated motifs. The entire song lasts about 1.2 sec at intervals of 2–3 sec.

Black-chinned Babbler

Yellow plumage not as bright as Golden Babbler

Black lores and chin

Mishmi Wren-Babbler

Cheeks concolorous

Rusty throat contrasts strongly with belly

Duller, grayer upperparts

Distinct white breast band

Dense white barring on breast and belly

Rufous-capped Babbler

Buff underparts tinged olive brown

C. r. davidi

Buff underparts tinged pale brown

C. r. ruficeps

Rufous-throated Wren-Babbler

Grayish cheek

Brown throat

Black and white spots on belly

Bar-winged Wren-Babbler

White spots on head, neck, and back

Barred tail

Dense fine bars on wings

Buff-chested Babbler

Relatively white chin

Fine black streaks on pale buff throat

Yellow on underparts relatively pale

Spelaeornis reptatus

长尾鹩鹛 cháng wěi liáo méi

Gray-bellied Wren-Babbler

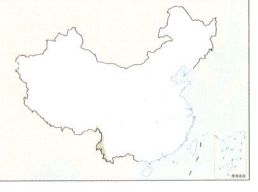

L: 10 cm
Habitat: Inhabits dense undergrowth in evergreen broadleaf forests and moist gullies, at altitudes of 1200–3000 m.
Behavior: Often solitary or in pairs, staying within a meter of the ground. Moves among dense herbaceous vegetation.
Distribution: W and NW Yunnan and Sichuan.
Voice: Song is more varied than that of Pale-throated Wren-Babbler: It is generally shorter (but varies in length from 0.7–1.2 sec), has a rising introductory note, makes several subtle changes in pitch, and usually has fewer (three or four) longer closing notes, which typically rise, *p-w-r-r-r-r-r-tsit-tsit-tsit*. Call is a low, thin *cha-cha*.

Spelaeornis kinneari

淡喉鹩鹛 dàn hóu liáo méi

Pale-throated Wren-Babbler

L: 10 cm
Habitat: Inhabits dense undergrowth of evergreen broadleaf forests and moist gullies, at altitudes of 1200–3000 m.
Behavior: Often solitary or in pairs, staying within a meter of the ground. Moves among dense herbaceous vegetation.
Distribution: SE Yunnan, N Guangxi, and Chongqing.
Voice: Song a loud, strident, sputtering trill about 1.5 sec long that starts with six to eight short, lengthening and rising notes, before slowing and falling away, *p-r-r-r-r-r-r-tiu-tiu-tiu-tiu-tiu*. Frequency range is about 1.2–5.1 kHz. Call is a low, thin *cha-cha*.

Pomatorhinus ferruginosus

红嘴钩嘴鹛 hóng zuǐ gōu zuǐ méi

Coral-billed Scimitar-Babbler

L: 23–24 cm
Habitat: Inhabits evergreen broadleaf forests and bamboo thickets in mountainous areas at 200–2000 m. Prefers habitats with bamboo thickets.
Behavior: Often moves in pairs during breeding season; gathers in small flocks or joins mixed-species flocks in nonbreeding season.
Distribution: *P. f. ferruginosus* in SE Tibet; *P. f. orientalis* in W to S Yunnan. *P. f. orientalis* is sometimes treated, together with other subspecies, as a separate species, Brown-crowned Scimitar-Babbler (*Pomatorhinus phayrei*).
Voice: Extensive vocabulary, with many notes confusingly similar to those of Red-billed Scimitar-Babbler; much more research is needed to distinguish their equally complex vocabularies. Apparent song is a loud, monosyllabic *hoo* or disyllabic *hoo huo*. The broad-spectrum chatter is higher pitched than the equivalent notes of Red-billed.

Pomatorhinus ochraceiceps

棕头钩嘴鹛 zōng tóu gōu zuǐ méi

Red-billed Scimitar-Babbler

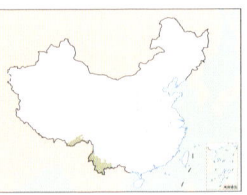

L: 22–23 cm
Habitat: Broadleaf forests and bamboo thickets at 200–2400 m.
Behavior: Often in pairs during breeding season; gathers in small groups or mixes with other birds, such as Coral-billed Scimitar-Babbler and White-hooded Babbler, in nonbreeding season.
Distribution: *P. o. austeni* in W Yunnan and SE Tibet; *P. o. ochraceiceps* in SW Yunnan.
Voice: Wide variety of vocalizations include single and multiple *poip* notes, raucous chattering, female trogonlike spluttering, higher-pitched penetrating whistles, and harsher screams, as well as multiple combinations of the above. Many are similar to Coral-billed Scimitar-Babbler.

Pomatorhinus superciliaris

细嘴钩嘴鹛 (剑嘴鹛)

xì zuǐ gōu zuǐ méi (jiàn zuǐ méi)

Slender-billed Scimitar-Babbler

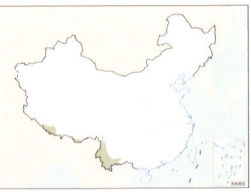

L: 20 cm
Habitat: Bamboo thickets, scrub, dense forest undergrowth, and rhododendron forests, at 900–3500 m. Migrates to lower altitudes in winter.
Behavior: Moves and forages on ground or in bushes; also takes nectar and seeds in trees. Often in pairs during breeding season and in small flocks in winter.
Distribution: *P. s. superciliaris* in SE Tibet; *P. s. forresti* in W to NW Yunnan; *P. s. rothschildi* possibly in SE Yunnan.
Voice: Vocalizations include a series of 7–16 short, slightly rising *put-put-put* notes, given in quick succession, lasting up to 2.5 sec, and with a faint first note. These are occasionally followed by a sequence of intense whistles, *wu-hu-weee-weee-weee*. Other "songs" include a whistled *tu-t-wheet … to-t-whet-tuit*, repeated with an incisive, then rattling *poit-r-r-r-r-r-r*, apparently an alarm call.

Pomatorhinus schisticeps

灰头钩嘴鹛 huī tóu gōu zuǐ méi

White-browed Scimitar-Babbler

L: 19–23 cm
Habitat: Broadleaf forests, bamboo thickets, and scrub on mountain slopes, field margins, and deserted fields, up to 2000 m.
Behavior: Often in pairs in breeding season; sometimes in small flocks or mixed-species flocks during nonbreeding season. Mostly hops on ground; also forages in bushes.
Distribution: Very rare in SE Tibet and S Yunnan.
Voice: Song is an often-intense series of usually three to seven fast, short, resonant, piping notes, *poip-poip-poip*, where the first note is very slightly higher than later ones, and an even faster but gradually slowing *wu-hu-hu-hu-hu*. Calls include lengthy sequences (almost 3 sec) of hoarse, raspy gurgling.

Gray-bellied Wren-Babbler

Scaly pattern on back

Fine flecking and freckling on throat

Gray face

White or pale yellow spots on grayish belly

Red-billed Scimitar-Babbler

No black stripe above white supercilium

Bill more slender than Coral-billed Scimitar-Babbler

Relatively white underparts

Pale-throated Wren-Babbler

Gray face

Cleaner white on throat

Black scaling on breast

Slender-billed Scimitar-Babbler

Extremely slender bill

White supercilium

Black streaks on white throat

Coral-billed Scimitar-Babbler

Brown crown

Black stripe above white supercilium

Buff breast and belly

P. f. orientalis

Almost all-blackish crown

White supercilium

Bill shorter than Red-billed Scimitar-Babbler

Brown breast and belly

P. f. ferruginosus

White-browed Scimitar-Babbler

Gray crown

Body larger than Streak-breasted Scimitar-Babbler

White breast lacks streaks

427

Pomatorhinus ruficollis
棕颈钩嘴鹛 zōng jǐng gōu zuǐ méi
Streak-breasted Scimitar-Babbler

L: 16–19 cm
Habitat: Inhabits
mountain slopes with
dense scrub, undergrowth
in open forests, bamboo
thickets, rhododendron
thickets, brambles, tea
plantations, forest edges,
and hedges, at 200–
3400 m.

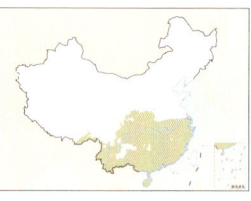

Behavior: Shy. Often in pairs or small groups; also joins mixed-species
flocks. Spends most time foraging in forest undergrowth; also climbs up
moss-covered tree trunks.
Distribution: Common resident. *P. r. godwini* in SW Yunnan; *P. r. similis*
from NW and W Yunnan to SW Sichuan; *P. r. styani* from S Gansu and
NE Sichuan to S Jiangsu and N Zhejiang; *P. r. eidos* in C Sichuan and
adjacent NE Yunnan; *P. r. laurentei* in C Yunnan; *P. r. albipectus* in
SW Yunnan; *P. r. reconditus* in SE Yunnan and SW Guangxi; *P. r. hunanensis*
in SE Sichuan and E Guizhou to W Hunan and N Guangxi; *P. r. stridulus*
in E Hunan, E Guangdong to S Zhejiang, and Fujian; *P. r. nigrostellatus*
in Hainan.
Voice: Extensive vocabulary includes a loud, ringing, very flat *u-hu-hu*,
generally delivered quickly, sometimes slower; a barely trisyllabic *u-wu-
hu*; and various scolding notes.

Pomatorhinus musicus
台湾棕颈钩嘴鹛
tái wān zōng jǐng gōu zuǐ méi
Taiwan Scimitar-Babbler

L: 19–21 cm
Habitat: Inhabits forest-
edge bushes, bamboo
thickets, broadleaf forests,
and secondary forests
on low mountains and
foothill plains. Sometimes
occurs in tea plantations,
orchards, and bushes
near farmland.

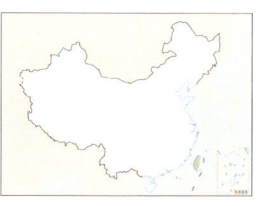

Behavior: Usually gathers in small flocks. Very shy, more frequently
heard than seen. Not good at flying; rarely flaps wings. Often hops about
in dense trees or bushes, sometimes also on the ground.
Distribution: Chinese endemic; uncommon resident in Taiwan.
Voice: Many vocalizations similar to those of Streak-breasted Scimitar-
Babbler but not as flat, with many notes rising slightly from the outset
and others inflected.

Erythrogenys hypoleucos
长嘴钩嘴鹛 cháng zuǐ gōu zuǐ méi
Large Scimitar-Babbler

L: 26–28 cm
Habitat: Inhabits
evergreen broadleaf
and mixed evergreen-
deciduous broadleaf
forests, bamboo thickets,
scrubland, and reedbeds.
Generally distributed
in moist tropical and

subtropical forests below 1200 m, up to 1550 m.
Behavior: Usually in pairs, occasionally in small flocks of five or six
individuals. Hops about clumsily in forest understory. More often on the
ground than other scimitar-babblers.
Distribution: Rare resident. *E. h. hypoleucos* in SW Yunnan (Yingjiang);
E. h. tickelli in S Yunnan and Guangxi; *E. h. hainana* in Hainan.
Voice: Song a variable series of three short, loud, hollow, extremely low-
pitched notes, given in duet, *wu-ip-pu-pyu-pu-pu*.

Erythrogenys erythrogenys
锈脸钩嘴鹛 xiù liǎn gōu zuǐ méi
Rusty-cheeked Scimitar-Babbler

L: 22–26 cm
Habitat: Inhabits
secondary forests,
scrubland, shrubby
hillslopes, field edges,
and deserted fields;
particularly high number
of records from scrub
around villages. Mostly
below 3000 m.

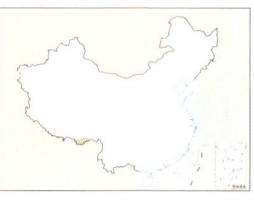

Behavior: Often in pairs, occasionally in flocks. Spends most time on the
ground, hopping around; sometimes forages in trees.
Distribution: S Tibet.
Voice: Many vocalizations are disyllabic, with the female chipping in a
third note antiphonally. There is some variation: in some vocalizations,
the first note is longer; in others, the second dominates. Most notes are
not flat.

Streak-breasted Scimitar-Babbler

Blackish-brown upper mandible with yellow tip

Broad dark chestnut streaks on breast

Brown flanks

P. r. stridulus

Relatively dark chestnut crown

Dark chestnut streaks on breast

P. r. reconditus

Pale iris

Upper mandible half black, half yellow

Brown back

Narrow, sparse, dark chestnut streaks on breast

Dark chestnut-brown belly

P. r. nigrostellatus

Almost all-yellow bill

P. r. similis

Olive-brown streaks on breast

Upper mandible yellow only at tip

Relatively pale olive-brown streaks on breast

P. r. styani

Upper mandible almost all grayish black

Chestnut-brown streaks on breast

P. r. hunanensis

Upper mandible half black, half yellow

Olive-brown streaks on breast

P. r. godwini

Upper mandible yellow only at tip

Relatively heavy olive-brown streaks on breast

P. r. eidos

Almost all-white breast

P. r. albipectus

Pinkish bill

Olive-brown streaks on breast

P. r. laurentei

Taiwan Scimitar-Babbler

All-black upper mandible

Pale iris

Slightly sparse and patchy dark brown streaks on breast

Rusty-cheeked Scimitar-Babbler

Rusty face

Indistinct dusky-gray streaks on whitish breast

Large Scimitar-Babbler

Relatively large body, slightly long bill

Brownish crown

E. h. tickelli

Buff streaks on side of neck, behind chestnut ear coverts and at end of supercilium

Grayish crown

White streaks on grayish-brown flanks

E. h. hainana

No buff streaks behind ear coverts and on side of neck

Chestnut brighter on wing and tail

A few white streaks on blackish-gray flanks

E. h. hypoleucos

Erythrogenys gravivox

斑胸钩嘴鹛 bān xiōng gōu zuǐ méi

Black-streaked Scimitar-Babbler

 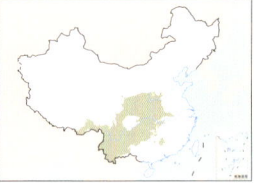

L: 21–25 cm
Habitat: Inhabits open forests, forest edges, scrubland, secondary forests, abandoned agricultural lands, and bamboo thickets. Generally occurs from 200 to 3700 m. Breeds at altitudes up to 3260–3800 m in summer in Tibet.
Behavior: Often in pairs or flocks, searching for food among fallen leaves on the ground.
Distribution: *E. g. gravivox* from S Gansu to S Shaanxi, Sichuan, N Hubei, S Shanxi, and NW Henan; *E. g. cowensae* in SE Sichuan, Chongqing, N Guizhou, SW Hubei, and NW Hunan; *E. g. dedekensi* in E Tibet, W Sichuan, and extreme NW Yunnan; *E. g. decarlei* in NW Yunnan, adjacent SW Sichuan, and extreme SE Tibet; *E. g. odica* in SW Guizhou and most of Yunnan.
Voice: Vocalizations complex and varied; many similar to Gray-sided Scimitar-Babbler and Black-necklaced Scimitar-Babbler. Include a rising, slightly tremulous, nasal *pwerp*, occasionally accompanied by a scolding rattle; a strident, ringing *whi-up*, repeated with the female occasionally interjecting a penetrating, slightly higher-pitched *pweer*; and a slightly descending *pidu-didu-didu*, given in a low series of up to six notes, each with a distinctive tremulous quality.

Erythrogenys swinhoei

华南斑胸钩嘴鹛
huá nán bān xiōng gōu zuǐ méi

Gray-sided Scimitar-Babbler

L: 22–24 cm
Habitat: Inhabits dense moist deciduous broadleaf forests, bush-covered hillslopes, and scrub-covered forest edges; rarely in open woodland.
Behavior: Fond of digging for food on the ground, making noises when turning dead leaves.
Distribution: *E. s. swinhoei* is common in S Anhui, E Jiangxi, S Zhejiang, and Fujian; *E. s. abbreviatus* in NE Guangxi, S Hunan, W Jiangxi, and N Guangdong.
Voice: See Black-streaked Scimitar-Babbler. Antiphonal song of a pair is a ringing *whi-up*, repeated with the female occasionally interjecting a penetrating, slightly higher-pitched *pweer*, similar to Black-streaked but slightly flatter and more muffled. Also a three-note *tu-wic-up* and a *toip*, followed by a scolding rattle.

Erythrogenys erythrocnemis

台湾斑胸钩嘴鹛
tái wān bān xiōng gōu zuǐ méi

Black-necklaced Scimitar-Babbler LC

L: 23–25 cm
Habitat: Inhabits forests, scrubland, bamboo thickets, and low woods on low mountains and hills at middle and low altitudes; sometimes on farmland, villages, or bushes in parks.
Behavior: Often solitary or in pairs. Shy, often heard but not seen. Good at hopping, but not flying; prefers moving under bushes or on the ground.
Distribution: Chinese endemic; uncommon resident in Taiwan.
Voice: Vocalizations very similar to those of Gray-sided Scimitar-Babbler and Black-streaked Scimitar-Babbler.

Stachyris nigriceps

黑头穗鹛 hēi tóu suì méi

Gray-throated Babbler LC

L: 12–15 cm
Habitat: Inhabits evergreen broadleaf forests, secondary scrub and bamboo thickets in dense forests, forest clearings and edges, and undergrowth in artificial forests and plantations. Altitude range: 150–2500 m.
Behavior: Often in small flocks during nonbreeding season, sometimes joins mixed-species flocks. Prefers foraging on the ground or in dense bushes.
Distribution: *S. n. nigriceps* in SE Tibet; *S. n. coltarti* in W Yunnan; *S. n. yunnanensis* occurs in S Yunnan and SW Guangxi.
Voice: Song a very high-pitched, faltering, falling then rising, or simply falling *ti tsuuuuuuueee*. Calls include a buzzy *tzztzztzz*.

Stachyris nonggangensis

弄岗穗鹛 nòng gǎng suì méi

Nonggang Babbler VU

L: 18 cm
Habitat: Inhabits monsoon rain forests of mountains in karst regions.
Behavior: Often in pairs during breeding season, in small flocks in nonbreeding season. Mainly hops about in forest understory; flies short distances only when disturbed or moving. Forages by turning over fallen leaves in rock crevices and on the ground.
Distribution: Uncommon resident; seen only in SW Guangxi.
Voice: Song is a single burry, descending, complex whistle, *rrreeeuw*; about 0.8 sec long, given singly or repeated up to three times in quick succession. Calls include a sputtering *chrrr-rrut* and a dry rattle.

Black-streaked Scimitar-Babbler

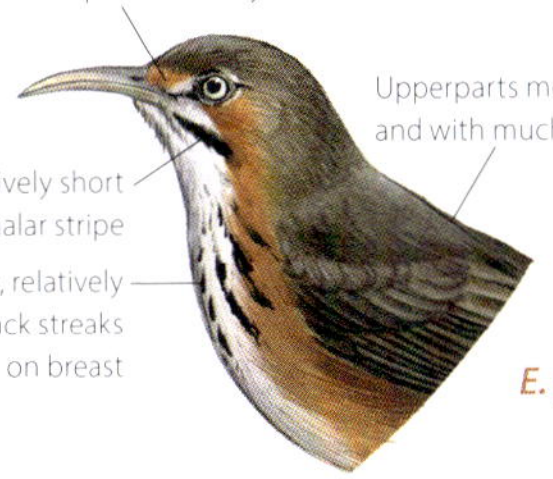

Red supercilium only at lores

Relatively short black malar stripe

Sparse, relatively narrow, black streaks on breast

Upperparts more grayish and with much olive color

E. g. odica

Relatively long black malar stripe

Dense, relatively broad, black streaks on breast

E. g. gravivox

Gray-sided Scimitar-Babbler

Gray flanks

Relatively bright rufous-brown wings

Black-necklaced Scimitar-Babbler

Rusty forehead

Chestnut-brown back contrasts with gray head

Extremely broad, distinct, black streaks on breast

Gray-throated Babbler

Dense fine white streaks on blackish crown

Distinct black supercilium

Grayish-white chin

Nonggang Babbler

Bluish-gray iris

Crescent-shaped white patch behind cheek

A few white feathers on throat

Stachyris humei

黑胸楔嘴穗鹛（黑胸楔嘴鹛鹛）

hēi xiōng xiē zuǐ suì méi (hēi xiōng xiē zuǐ liáo méi)

Sikkim Wedge-billed Babbler　　**NT**

L: 17 cm
Habitat: Inhabits wet areas in primary evergreen broadleaf forests at altitudes of 1000–2500 m.
Behavior: Often in flocks. Prefers moving on the ground or in dense bushes; sometimes also forages on moss-covered tree trunks.
Distribution: SE Tibet.
Voice: Primary song is a variable combination of powerful, rich, fluty whistles, *ter-tui-twee-t-wuu* or *pe-deer-pe'er-witu-widu*, between 0.8–3.8 kHz, lasting between 1.2–1.5 sec, and occasionally preceded by two very short *wit* notes. Alarm call is a monosyllabic *dui* or a dry *zhe, zhe*. Repeats broad-spectrum scolding chatter call, *sprrrrrh*, very similar to that of Cachar Wedge-billed Babbler but slightly lower pitched. The intense, whistled *pe-pee-pee-pee* call is also usually more powerful, doesn't fade, and falls only very slightly in pitch.

Stachyris roberti

楔嘴穗鹛（楔嘴鹛鹛）

xiē zuǐ suì méi (xiē zuǐ liáo méi)

Cachar Wedge-billed Babbler　　**NT**

L: 17 cm
Habitat: Inhabits wet areas in primary evergreen broadleaf forests at altitudes of 1000–2500 m.
Behavior: Often in flocks. Prefers moving on the ground or in dense bushes; sometimes also forages on moss-covered tree trunks.
Distribution: W Yunnan.
Voice: Some song strophes are very similar to those of Sikkim Wedge-billed Babbler but usually include occasional distinctive flatter strophes, *tui-twee-tu-twee* or *dui-dweedu-du*. Agitation calls include a sharp *chick* or *chick-up* and an interrogative rising *pweer*.

Stachyris strialata

斑颈穗鹛 bān jǐng suì méi

Spot-necked Babbler　　**LC**

L: 15.5–16.5 cm
Habitat: Inhabits evergreen broadleaf forests, bamboo thickets, secondary forests, and understory scrub, mostly below 1525 m.
Behavior: Moves about in pairs or small flocks in forest understory or on the ground. Often joins other birds, especially other *Stachyris* babblers, in mixed-species flocks. Shy and secretive.
Distribution: Resident. *S. s. tonkinensis* in S Yunnan and Guangxi; *S. s. swinhoei* in Hainan.

Sikkim Wedge-billed Babbler

Wedge-shaped bluish-black bill

White supercilium

Rufous flecking on dark upperparts

Breast and belly lack distinct scaly pattern

Inconspicuous barring on wings and tail

Cachar Wedge-billed Babbler

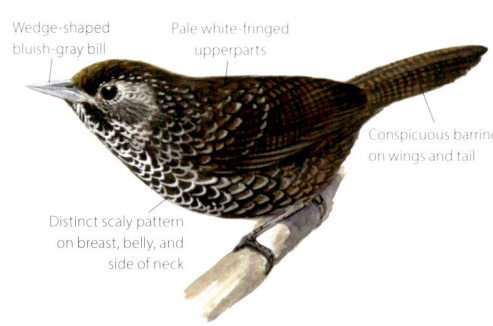

Wedge-shaped bluish-gray bill

Pale white-fringed upperparts

Conspicuous barring on wings and tail

Distinct scaly pattern on breast, belly, and side of neck

Spot-necked Babbler

Gray cheek

White spots on black side of neck

Thick heavy bill

White throat

Voice: Song is a high-pitched, pleasant, slow, two- or three-note whistle, *tuh-tih-tuh*. Calls include a sharp *tip*, *tic-ur*, an indrawn wheeze followed by a scolding chatter, *tirrrirrirr, tchrrrt-tchrrrt*, and a Pin-striped Tit-Babbler–like churr.

幽鹛科 Pellorneidae

Small forest-dwelling babblers. Plumage similar between sexes, mostly brown. Bill thin and weak. Wings short and rounded. Tail of varying length, squared or wedge shaped. Legs moderately strong, adapted for living in trees or on ground. Inhabit primarily forests and scrubland of tropical and subtropical regions. Nest in bushes, grass, or crevices. Feed mostly on insects, and plant seeds and fruits. Most species are resident.

Once included members of *Alcippe* that now form their own family, Alcippeidae. Thirteen genera and 60 species recognized worldwide, distributed mostly in Indomalaya except for Spotted Thrush-Babbler and seven *Illadopsis* species in Africa. Six genera and 17 species recorded in China, found mostly in montane forests of southwestern China; a few also in South China.

大草莺属
Graminicola

Chinese Grassbird

白头鵙鹛属
Gampsorhynchus

White-hooded Babbler

拟雀鹛属
Schoeniparus

Rusty-capped Fulvetta

幽鹛属
Pellorneum

Puff-throated Babbler

鹪鹛属
Gypsophila

Streaked Wren-Babbler

长嘴鹩鹛属
Napothera

Long-billed Wren-Babbler

Graminicola striatus

中华草鹛 （大草莺）

zhōng huá cǎo méi (dà cǎo yīng)

Chinese Grassbird

VU

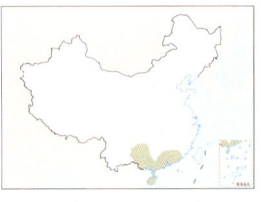

L: 16–18 cm
Habitat: Inhabits humid grasslands, reed swamps, and other wetland vegetation below 900 m. Prefers sedges and grasses.
Behavior: Often solitary, occasionally in pairs or small flocks. Often hidden in vegetation and not easily detected.
Distribution: *G. s. sinicus* has a stable breeding population in Hong Kong; rare in Guangdong, N Guangxi, and S Guizhou. *G. s. striatus* has been recorded in Hainan, but not for more than 100 years.
Voice: Song is a strong down-slurred note followed by a series of clipped, rhythmic warbles, *er-wi-wi-wi-wi-wi-wi-wi*, usually eight or nine syllables long. Calls include rapid chattering and a very strained, teeth-sucking, nasal *chrrrrrr* when agitated.

Gampsorhynchus rufulus

白头鵙鹛 bái tóu jú méi

White-hooded Babbler

LC

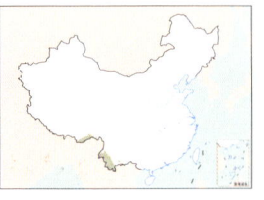

L: 23–24 cm
Habitat: Inhabits evergreen broadleaf forests, secondary forests, forest scrub, and tall grass at forest edges at altitudes of 200–1400 m; especially prefers areas with bamboo thickets.
Behavior: Often in groups. Noisy. Usually seen with other birds, especially Red-billed Scimitar-Babbler, Rufous-headed Parrotbill, and Pale-billed Parrotbill.
Distribution: W Yunnan.
Voice: Song is a twangy, nasal, rapid *pu-pu-kao* that is repeated and occasionally accompanied by strained hissing and squealing.

Gampsorhynchus torquatus

领鵙鹛 lǐng jú méi

Collared Babbler

LC

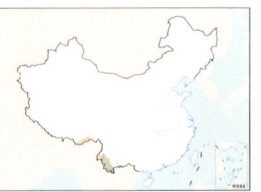

L: 22.5–26 cm
Habitat: Inhabits evergreen broadleaf forests, secondary forests, forest scrub, and tall grass at forest edges at altitudes of 500–1800 m; especially prefers areas with bamboo thickets.
Behavior: Very fond of bamboo thickets. Usually gathers in small flocks. Calls often; noisy. In winter, joins mixed-species flocks.
Distribution: *G. t. torquatus* in S Yunnan; *G. t. luciae* in SE Yunnan.
Voice: Song is poorly known but apparently a loud repeated *vigagagaga* or a slightly shorter *viguiguigui*.

Schoeniparus cinereus

黄喉雀鹛 huáng hóu què méi

Yellow-throated Fulvetta

LC

L: 10–11 cm
Habitat: Inhabits evergreen broadleaf forests, usually occurs in bamboo thickets and dense scrub in the openings, including forest clearings, trails, and stream edges.
Behavior: In nonbreeding season, often in flocks of 5–10 individuals, sometimes as many as 30; also joins mixed-species flocks. Flock moves very quickly. Often found in dense bushes near the ground.
Distribution: Found in SE Tibet, W to NW Yunnan.
Voice: Song is a very fast series of very thin, high *sip* notes that start slowly, descend, and trail off, reminiscent of a coin being dropped on a hard surface. Often accompanied by mellow churrs, *trrrrt*.

Schoeniparus castaneceps

栗头雀鹛 lì tóu què méi

Rufous-winged Fulvetta

LC

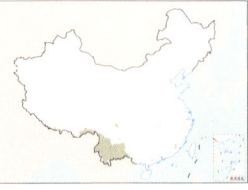

L: 10–11 cm
Habitat: Inhabits moist evergreen broadleaf forests, secondary forests, forest edges, and bamboo thickets, etc.
Behavior: Generally in flocks of 20–40 individuals, sometimes up to 70 individuals in nonbreeding season. Often mixes with nuthatches, Black-throated Parrotbill, and Golden-breasted Fulvetta. Flock moves very quickly. Climbs and forages on moss- and lichen-covered tree trunks.
Distribution: *S. c. castaneceps* in SE Tibet and W to NW Yunnan; *S. c. exul* in S to SE Yunnan and E to W Guangxi.
Voice: Song is a loud, short, high-pitched, descending phrase, *wi-chuw-i-chewi-cheeu*. Calls include various short chatters, chips, and churrs. Flocks often noisy.

Schoeniparus variegaticeps

金额雀鹛 jīn é què méi

Gold-fronted Fulvetta

VU

L: 10–11.5 cm
Habitat: Inhabits evergreen broadleaf forests and mixed larch forests, as well as stream valleys with moss-covered trees and abundant bamboo cover, at altitudes of 650–2000 m.
Behavior: Often in pairs or mixed-species flocks. Climbs on moss-covered tree trunks to forage; also active in bamboo thickets in the understory.
Distribution: Local and uncommon in C to S Sichuan and C to E Guangxi.
Voice: Poorly known. Song is a short high-pitched strophe that lasts about 1.5 sec, *ti-ti-tiia-suuuu*, with the third note falling and the final one flat, repeated about every 2.5 sec. Flocks apparently make a noisy chattering.

Chinese Grassbird

Broad, heavy, blackish-brown streaks from crown, neck, and side of neck to back

Dark red iris

Strong bill and legs

Relatively large body with long tail, like enlarged prinia

White-hooded Babbler

White head contrasts clearly with brown back

White on wing coverts

ad.

Rufous head, distinguished from Rufous-headed Parrotbill by bill shape

juv.

Collared Babbler

Rear crown, neck, and back tinged ochre brown

No white on upperwing coverts

Dark half-collar on side of breast

Yellow-throated Fulvetta

Yellow supercilium

Black lateral crown stripe and eye stripe

Bright yellow throat and breast

Rufous-winged Fulvetta

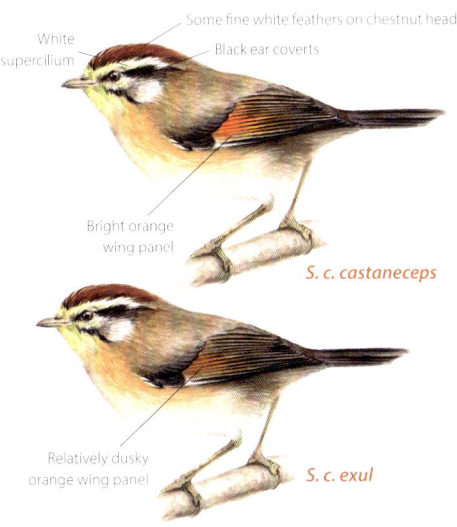

White supercilium

Some fine white feathers on chestnut head

Black ear coverts

Bright orange wing panel

S. c. castaneceps

Relatively dusky orange wing panel

S. c. exul

Gold-fronted Fulvetta

Bright yellow forehead

Chestnut neck

Black patch on grayish-white face

Schoeniparus rufogularis

棕喉雀鹛 zōng hóu què méi

Rufous-throated Fulvetta LC

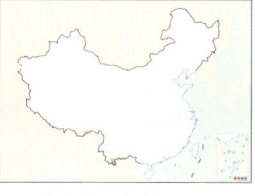

L: 12–13 cm
Habitat: Inhabits evergreen broadleaf forests, as well as understory scrub and bamboo thickets.
Behavior: Solitary or in pairs; sometimes in small flocks of five or six individuals. Moves about on the ground or in dense bushes.
Distribution: Very uncommon in S and SW Yunnan.
Voice: Song a pleasant warbled *tui-twee-tweedle-du*, reminiscent of song of Rusty-capped Fulvetta but longer (with an extra note) and note falling at the end. Calls a powerful, almost explosive, nervous *prrrp* or *wrrreet*.

Schoeniparus dubius

褐胁雀鹛 hè xié què méi

Rusty-capped Fulvetta LC

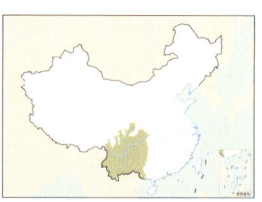

L: 14–15 cm
Habitat: Inhabits broadleaf forests, coniferous forests, mixed broadleaf-coniferous forests, and dense understory of bushes, ferns, brambles, and grass at altitudes of 300–3100 m.
Behavior: In small flocks during nonbreeding season and occasionally joins mixed-species flocks. Often forages on the ground or in bushes near the ground.
Distribution: *S. d. intermedius* occurs west of the Nujiang (Salween) River in W Yunnan; *S. d. genestieri* occurs east of the Nujiang River in Yunnan, E to SW Sichuan, W Hunan, and W Guangxi.
Voice: Song is a repeated, simple, three-note whistle, *di-di-dwei*. Call is a wheezy, descending *bzzz-bz-bz-brt*; alarm call a chirping *chuachua*.

Schoeniparus brunneus

褐顶雀鹛 hè dǐng què méi

Dusky Fulvetta LC

L: 13–13.5 cm
Habitat: Inhabits evergreen broadleaf forests and bushes in understory and at forest edges, at altitudes of 40–1830 m.
Behavior: Often in flocks, turning over fallen leaves or foraging in bushes near the ground.
Distribution: *S. b. olivaceus* in C Sichuan, E Yunnan, E to S Shaanxi, W Hubei, and N Guizhou; *S. b. weigoldi* in W Sichuan; *S. b. superciliaris* in E Guangxi, Hunan, Anhui, SW Zhejiang, Jiangxi, W Fujian, and C Guangdong; *S. b. argutus* in Hainan; *S. b. brunneus* in Taiwan.
Voice: Song is a hurried, pleasant, but nervous warble of four to seven short notes, *wi ti-tiu-wich-itiu-uu* or *tiu wee-witchi-chu*, repeated rapidly. Calls include a tremulous, descending *dii-duu-du*, a low-pitched spluttering *churr*, or a more penetrating *trrrr*.

Pellorneum ruficeps

棕头幽鹛 zōng tóu yōu méi

Puff-throated Babbler LC

L: 15–17 cm
Habitat: Inhabits deciduous or evergreen broadleaf forests, teak forests, secondary forests, bamboo thickets, tea plantations, and bushes, mainly below 1000 m.
Behavior: In pairs or small flocks, searching for food on ground or among dead leaves. Unafraid of people.
Distribution: Resident in Yunnan. *P. r. shanense* in areas west of the Lancang (Mekong) River; *P. r. oreum* between the Lancang River and Honghe (Red) River; *P. r. vividum* east of the Honghe River.
Voice: Penetrating, loud song is a jolly, rapid, descending *tuituitititi-twititi-titiii*. Calls include a monotonously repeated, melancholy *wi-chuu* or *men-yuu*. Wi-ti-chu and harsh churring often associated with other notes, *trrrt-trrrt-wheu*.

Pellorneum albiventre

白腹幽鹛 bái fù yōu méi

Spot-throated Babbler LC

L: 14–15 cm
Habitat: Inhabits scrubland, secondary forests, bamboo thickets, grasslands, clearings, and pine forest undergrowth, at altitudes of 280–2135 m.
Behavior: Often moves about in pairs in dense bushes or on the ground. Very shy and skulky.
Distribution: *P. a. albiventre* in SE Tibet; *P. a. orientalis* in W to S Yunnan and SW Guangxi.
Voice: Loud, complex, lengthy, hwamei-like song, with much repetition and mimicry, clear short whistles, and hard ringing notes, sometimes with subdued guttural buzzy nasal notes and occasionally lasting several minutes. Harsh scolding chatter often accompanied by explosive *tzik* notes.

Pellorneum tickelli

棕胸雅鹛 zōng xiōng yǎ méi

Buff-breasted Babbler LC

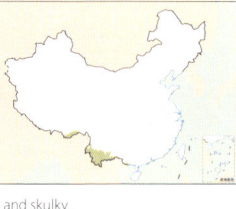

L: 13–15 cm
Habitat: Inhabits evergreen broadleaf forests, secondary forests, bamboo thickets, and dense scrub, at altitudes of 200–1550 m.
Behavior: Often in pairs; sometimes mixes with other small babbler species. Skulks in dense bushes; moves about close to the ground.
Distribution: SW to SE Yunnan and W Guangxi.
Voice: Simple song a repeated, muffled *wi-twee* or *wi-choo*, reminiscent of Common Tailorbird and sometimes repeated incessantly. Calls include a harsh rattling *prrree*, interspersed with higher *pieu* or explosive *whit* notes.

Puff-throated Babbler

Brown crown

White throat

Prominent broad pale
gray-brown streaks from
breast to flanks

Rufous-throated Fulvetta

White supercilium and
black lateral crown stripe

Dark brown ear coverts

Chestnut
throat band

Rusty-capped Fulvetta

Chestnut-tinged
forehead

White supercilium and
black lateral crown stripe

Brownish-gray flanks

Spot-throated Babbler

Gray around eyes

Brown streaks on
throat, sometimes
indistinct

White area on belly

Buff-breasted Babbler

Gray around eyes less distinct
than Spot-throated Babbler

Slightly white
throat lacks
distinct spots

Diffused white on belly,
inconspicuous

Dusky Fulvetta

Lacks distinct supercilium;
only black lateral crown stripe

Gray cheek

Gypsophila brevicaudata
短尾鹩鹛 duǎn wěi jiāo méi
Streaked Wren-Babbler

L: 14–15 cm
Habitat: Inhabits
evergreen broadleaf
forests, often near rocky
areas and limestone
forests at 200–2100 m.
Behavior: Forages in pairs
or small flocks. Often seen
in dense vegetation on

the ground; also around rocks and boulders.
Distribution: *G. b. brevicaudatus* in W Yunnan; *G. b. stevensi* in SE Yunnan
and Guangxi.
Voice: Song variable, loud, clear, melancholy, ringing whistles, repeated
at intervals, *tiuuu, chu-ii*, or *chewee-chui*, and sometimes a single *pweeee*
or *pwee … chi-u-wee*. Lengthy, raspy, scolding notes in alarm.

Gypsophila annamensis
灰岩鹩鹛 huī yán jiāo méi
Annam Limestone Babbler

L: 19 cm
Habitat: Inhabits
limestone forests in rocky
areas.
Behavior: Often in pairs
or small flocks of five or
six individuals; rarely joins
mixed-species flocks.
Usually found on the
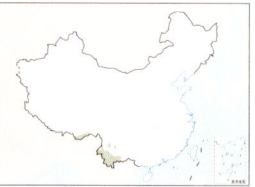
ground or in dense bushes near the ground; also hides in crevices in
rocky cliffs. Shy and secretive.
Distribution: S Yunnan.
Voice: Song is a loud, faltering, or hiccuping series of slowly delivered,
unevenly pitched, harsh notes lasting 4–30 sec and starting abruptly,
chitu-wi-witchu. Alarm call a harsh, scolding rattle, *chrrr*.

Napothera epilepidota
纹胸鹩鹛 wén xiōng jiāo méi
Eyebrowed Wren-Babbler

L: 10–11 cm
Habitat: Inhabits humid
evergreen broadleaf
forests, secondary forests,
and bamboo thickets.
Prefers moss-covered
boulders, epiphyte-laden
fallen trees, old stumps,
etc. Altitude range:
50–2135 m.

Behavior: Often in pairs or small family groups. Forages by turning over
leaves on forest floor, creeping about in low bushes and grasses.
Distribution: *N. e. guttaticollis* in SE Tibet; *N. e. bakeri* SW Yunnan;
N. e. amyae in SE Yunnan; *N. e. delacouri* in Dayao Mountains in Guangxi;
N. e. hainana in Hainan.
Voice: Song is a lengthy, loud, long, melancholy, descending, whistle,
cheeeoo, repeated at intervals of 2–5 sec. Calls are an intense *pwee-pwee-
pwee*, repeated in lengthy series, and a peculiar, slightly nasal, squeaky
chikachik-chikachik, also repeated.

Napothera malacoptila
长嘴鹩鹛 cháng zuǐ liáo méi
Long-billed Wren-Babbler

L: 11–12 cm
Habitat: Inhabits
evergreen broadleaf
forests, bamboo thickets,
and secondary forests at
900–2000 m.
Behavior: Secretive and
often seen in pairs. Usually
found in shady low
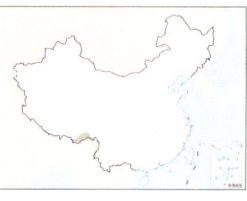
bushes in the understory or on the ground.
Distribution: SE Tibet and Gaoligong Mountains in NW Yunnan .
Voice: Song a penetrating, short, descending, clear whistle, *chiiuuh*,
given every 2–10 sec and often combined with one or more staccato
chip-trru notes. Calls include an abbreviated *trrru-trrru-trrru*.

Napothera naungmungensis
瑙蒙短尾鹛 nǎo méng duǎn wěi méi
Naung Mung Scimitar-Babbler

L: 18–19 cm
Habitat: Inhabits
evergreen broadleaf
forests; prefers
undergrowth of montane
rain forests. Recorded at
about 540 m in Myanmar.
The only acoustic record
from Yunnan (Yingjiang)
was at 950 m.

Behavior: Often solitary or in pairs. Forages by turning over fallen leaves
in the lower layer of forests. Secretive and quiet.
Distribution: W Yunnan (Yingjiang and Baoshan); presumably also in
SE Tibet.
Voice: Poorly known. Song or call a short, nervous, tentative, rising
whistle, repeated at about 2 sec intervals.

Streaked Wren-Babbler

Distinctive white spots on tips of greater coverts and wing feathers

Relatively short tail

Relatively pale streaks on throat

G. b. stevensi

Inconspicuous white spots on tips of greater coverts

G. b. brevicaudatus

Tips of wing feathers lack white spots, except P1

Eyebrowed Wren-Babbler

Pale supercilium

White-tipped wing coverts

Long-billed Wren-Babbler

Relatively long decurved bill

Pale yellow chin and throat

Long, narrow, brown and pale yellow streaks on side of breast and belly

Annam Limestone Babbler

Dark scaly pattern on upperparts

Black streaks on white throat

Tail longer than Streaked Wren-Babbler

Naung Mung Scimitar-Babbler

Relatively long, decurved, grayish-white bill

White from chin and throat to central belly

Body larger than Long-billed Wren-Babbler

雀鹛科　Alcippeidae

Small forest-dwelling babblers. Plumage mostly brown and black, usually with diffuse or conspicuous black or dark brown supercilium. Bill pointed and strong. Wings short and rounded, uniformly brown in color and lacking light-colored fringes. Tail squared. Legs strong, adapted for living in trees. Inhabit vegetation primarily in middle story and understory. Highly vocal. Usually form mixed-species flocks with other babblers. Feed mostly on insects; also take plant seeds, fruits, and shoots. All species are resident.

Alcippe fulvettas were previously placed in Pellorneidae but now constitute this newly recognized family. One genus and 10 species recognized worldwide, endemic to the tropical and subtropical regions of Indomalaya. Six species recorded in China, found in montane forests of southern China, including Hainan and Taiwan.

Brown-cheeked Fulvetta

Gray crown
Black supercilium

Yunnan Fulvetta

Conspicuous supercilium
Buff throat lacks fine streaks
Buff breast

Morrison's Fulvetta

Distinctive white eye ring
Whitish throat
Pale yellow or off-white breast and belly

Huet's Fulvetta

Inconspicuous supercilium
Relatively white throat and breast

David's Fulvetta

Relatively faint supercilium
Fine streaks on buff throat
Whitish breast

Nepal Fulvetta

Chestnut-brownish crown
Conspicuous white eye ring

Alcippe poioicephala
褐脸雀鹛 hè liǎn què méi
Brown-cheeked Fulvetta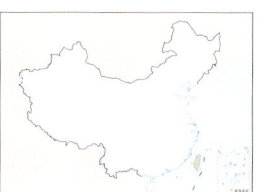

L: 16–16.5 cm
Habitat: Inhabits evergreen broadleaf forests, mixed bamboo forests, and forest-edge bushes from 200 to 1000 m.
Behavior: Often in groups of 6–10 individuals, sometimes up to 20 or more. Prefers middle story and undergrowth of forest; can also move up to the canopy. Usually the nucleus species of mixed-species flocks.
Distribution: *A. p. haringtoniae* in W Yunnan; *A. p. alearis* in S and SW Yunnan.
Voice: Song a pleasant, gentle-sounding, repeated warble of well-spaced notes, the last with a terminal inflection, *twi-wu-uwi-uwee*; harsh, buzzy, spluttering rattles when agitated.

Alcippe morrisonia
台湾雀鹛 tái wān què méi
Morrison's Fulvetta

L: 12.5–14 cm
Habitat: Inhabits montane evergreen broadleaf forests, secondary forests, and bushes in forest understory or at forest edges from 600 to 3000 m (50–2780 m in N Taiwan).
Behavior: Generally gregarious; joins mixed-species flocks.
Distribution: Chinese endemic; restricted to Taiwan.
Voice: Song is a pleasant whistle, *ji-ju ji-ju*, like David's Fulvetta. Calls include an intense, noisy, buzzy chatter.

Alcippe davidi
灰眶雀鹛 huī kuàng què méi
David's Fulvetta

L: 12.5–14 cm
Habitat: Inhabits montane evergreen broadleaf forests, secondary forests, and bushes in forest understory or at forest edges from 200 to 3000 m.
Behavior: Generally gregarious; forms mixed-species flocks with other birds. Often mobs Collared Owlet and other small owls.
Distribution: *A. d. schaefferi* in SE Yunnan, E to S Guizhou, and N Guangxi; *A. d. davidi* in S Gansu, S Shaanxi, S, C, and E Sichuan, SW Hubei, N and E Guizhou, SW Hunan, NE Guangxi, and NW Jiangxi.
Voice: Song is pleasant, clear warble, like song of Yunnan Fulvetta. Calls include a noisy chatter and a rising *ju*.

Alcippe fratercula
云南雀鹛 yún nán què méi
Yunnan Fulvetta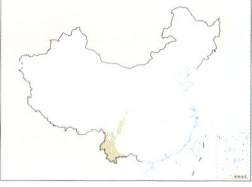

L: 10–11.5 cm
Habitat: Inhabits montane evergreen broadleaf forests, secondary forests, and bushes in forest understory or at forest edges from 200 to 3000 m.
Behavior: Generally gregarious; forms mixed-species flocks with other birds. Often mobs Collared Owlet and other small owls.
Distribution: *A. f. yunnanensis* in NW Yunnan and SW Sichuan; *A. f. fratercula* in W to S Yunnan.
Voice: See David's Fulvetta.

Alcippe hueti
淡眉雀鹛 dàn méi què méi
Huet's Fulvetta

L: 12.5–13 cm
Habitat: Inhabits montane evergreen broadleaf forests, secondary forests, and bushes in forest understory or at forest edges from 200 to 3000 m.
Behavior: Generally gregarious; forms mixed-species flocks with other birds.
Distribution: *A. h. hueti* in SE Anhui, Zhejiang, and S to NE Guangdong; *A. h. rufescentior* in Hainan.
Voice: Sings a *yojiyoyo*, sometimes with an elongated shrill sound, and the alarm call is a noisy chirp.

Alcippe nipalensis
白眶雀鹛 bái kuàng què méi
Nepal Fulvetta

L: 12.5–13 cm
Habitat: Inhabits montane evergreen broadleaf forests, secondary forests, and bushes in forest understory or at forest edges from 200 to 1000 m.
Behavior: Generally gregarious; forms mixed-species flocks with other birds.
Distribution: SE Tibet and W Yunnan.
Voice: Song a loud ringing *chu-chui-chiwi*, lower pitched, slower, and more spaced out than song of Yunnan Fulvetta, and with no buzzy terminal notes. Chattering calls similar to those of Yunnan.

噪鹛科　Leiothrichidae

Brightly colored small to medium-sized babblers that mostly live on the ground, among which Chinese Hwamei is perhaps the best-known member. Plumage highly variable among species; similar or slightly different between sexes. Bill pointed, thick, and strong. Wings short and rounded; usually poor fliers. Legs strong, well adapted to hopping on ground or trees. Tail long, squared or wedge shaped. Inhabit primarily forests, scrubland, forest edges, parks, and plantations on plains and mountains, usually in groups. Nest on branches or in bushes. Feed mostly on insects and other invertebrates; also take plant nectar, seeds, and fruits, sometime even small amphibians and reptiles. Most species are resident or perform altitudinal migration.

　　Laughingthrushes and allies were once regarded as typical members of Timaliidae but now constitute this newly recognized family. Sixteen genera and 135 species recognized worldwide, including laughingthrushes, babaxes, barwings, sibias, liocichlas, and their allies, endemic to the Old World. Twelve genera and 71 species recorded in China, widely distributed except northwestern China.

条纹噪鹛属
Grammatoptila

Striated Laughingthrush

姬鹛属
Cutia

Himalayan Cutia

彩翼噪鹛属
Trochalopteron

Elliot's Laughingthrush

奇鹛属
Heterophasia

Black-headed Sibia

斑翅鹛属
Actinodura

Rusty-fronted Barwing

希鹛属
Minla

Red-tailed Minla

栗背奇鹛属
Leioptila

Rufous-backed Sibia

相思鸟属
Leiothrix

Red-billed Leiothrix

薮鹛属
Liocichla

Scarlet-faced Liocichla

噪鹛属
Garrulax

Chinese Hwamei

蓝噪鹛属
Ianthocincla

Giant Laughingthrush

黑喉噪鹛属
Pterorhinus

Masked Laughingthrush

Grammatoptila striata

条纹噪鹛 tiáo wén zào méi

Striated Laughingthrush

L: 29.5–34 cm
Habitat: Inhabits evergreen broadleaf forests, secondary forests, scrubland, and bamboo thickets and tree groves around villages at altitudes of 600–3060 m, mostly over 1000 m.

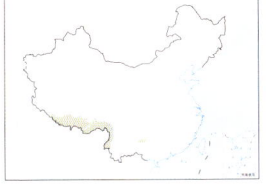

Behavior: More arboreal than most laughingthrushes; forages mainly from low branches and bushes up to tall canopies. Often solitary, in pairs, or in flocks; sometimes in mixed-species flocks with other laughingthrushes and barwings. Usually occurs in flocks in fruit-bearing trees.
Distribution: *G. s. brahmaputra* in SE Tibet; *G. s. cranbrooki* in SE Tibet, W to NW Yunnan, and S Guizhou; *G. s. vibex* in S Tibet; *G. s. striata* in SW Tibet.
Voice: Song a loud, vibrant, bounding *prrrit-you … pre-prrit-prii'u*. When agitated, higher-pitched scolding notes, *pwei-pwei*.

Cutia nipalensis

斑胁姬鹛 bān xié jī méi

Himalayan Cutia

L: 17–19 cm
Habitat: Inhabits primary evergreen broadleaf forests, especially large epiphyte-laden and mossy trees, at altitudes of 1800–2600 m; also recorded up to 3050 m.
Behavior: Prefers

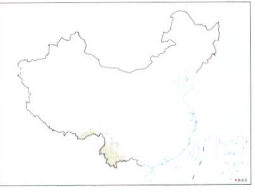

creeping along tree trunks and branches. Searches for food on epiphytes and moss; also sips nectar from flowers such as *Leucosceptrum*. Often in pairs during breeding season, in small flocks during nonbreeding season; joins mixed-species flocks.
Distribution: *C. n. nipalensis* in SE Tibet and extreme NW Yunnan; *C. n. melanchima* in W and S Yunnan.
Voice: Song a rapidly repeated series of 5–20 intense *yip-yip-yip* notes. Calls include excited, intense churring, often accompanied by the occasional *yip* note and a repeated *jert-jert-jert*.

Striated Laughingthrush

- Dark brown overall
- Dense white shaft streaks

Scaly Laughingthrush

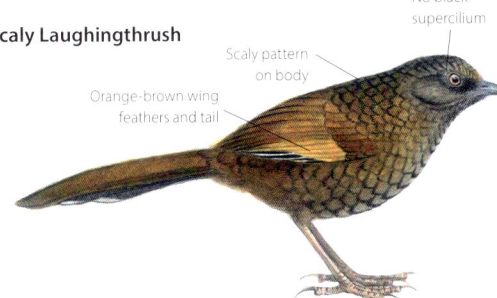

- Scaly pattern on body
- Orange-brown wing feathers and tail

Himalayan Cutia

- Red back lacks spots
- Black bars on flanks
- No black supercilium
- Black spots on pale brown back

Trochalopteron subunicolor

纯色噪鹛 chún sè zào méi

Scaly Laughingthrush

L: 23–25.5 cm
Habitat: Inhabits evergreen broadleaf forests, secondary forests, forests with dense undergrowth, *Rubus* berry bushes, dwarf rhododendron thickets, bamboo thickets, etc., at altitudes of 1500–3960 m.
Behavior: Often in pairs during nonbreeding season; sometimes forms flocks containing up to a dozen individuals. Usually forages on or near the ground. Not very shy.

Distribution: *T. s. subunicolor* in S and SE Tibet; *T. s. griseatum* in W Yunnan; *T. s. fooksi* in S Yunnan.
Voice: Song is a repeated, shrill, rising whistle, *whiuu, whiii-u-u-u-whiiiu*, or, with introductory fricative,

tr'r'r-wiuuu. Calls are an intense scolding, buzzing *thriiii*, interspersed with high-pitched squeaks and a mellower but still rapid, rattling *tr'it'it'it'it'it'*.

Trochalopteron squamatum
蓝翅噪鹛 lán chì zào méi
Blue-winged Laughingthrush LC

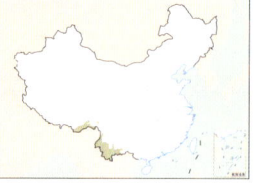

L: 22–25 cm
Habitat: Inhabits evergreen broadleaf forests, dense bushes in the forest understory, and bamboo thickets, often near streams. Altitude range: 900–2440 m.
Behavior: Often in pairs or small family groups, occasionally solitary. Skulking. Not noisy.
Distribution: W and S Yunnan.
Voice: Song thin, high-pitched, rising whistle, *puwiiiii-wit* or *tu-wiiiii-uu-wit*. Call a hesitant stutter that ends with two buzzy notes, *t-t-t-t-t-t-cher-crruuu.*

Trochalopteron lineatum
细纹噪鹛 xì wén zào méi
Streaked Laughingthrush LC

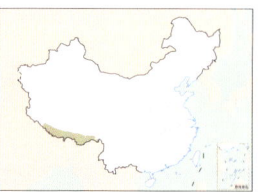

L: 18–20 cm
Habitat: Inhabits open forests and forest edges, bushes and grass on mountain slopes, areas near human habitation, field edges, and gardens at altitudes of 1400–3905 m.
Behavior: In pairs or small flocks of three to six individuals, depending on the season. Often in towns and villages in W Himalayas. Forages in open areas and among bushes.
Distribution: Very local in Tibet (Gyirong and Yadong).
Voice: Loud penetrating song often starts with some spluttered notes, before a lengthy abrupt whistle that falls before rising again, *p-t-t-t-trr-pseeeu*, or where the whistle is shorter and declining, *rrrititit-psee*. The female often responds with high-pitched *twee-tu-tu* notes. An alternate song begins with two to five very short notes, followed by two or three lengthy flat whistles, each lower in pitch than the one before, *t-t-tr-d-pwe-pwee-pwee.* Calls include low, harsh, muffled grumbling and short, high-pitched, scolding notes.

Trochalopteron imbricatum
丽星噪鹛 lì xīng zào méi
Bhutan Laughingthrush LC

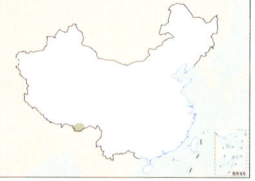

L: 19–20 cm
Habitat: Inhabits scrubland, secondary forests, and open areas at altitudes of 900–2900 m.
Behavior: In pairs or flocks. Moves about and forages mainly on the ground.
Distribution: Very local in SE Tibet (Cona and Medog).
Voice: Songs similar to Streaked Laughingthrush but shorter, lacking the stuttering introductory notes; the first of two long descending whistles often tremulous and buzzy, *bziuud-d-triuu.* Calls perhaps similar.

Trochalopteron variegatum
杂色噪鹛 zá sè zào méi
Variegated Laughingthrush LC

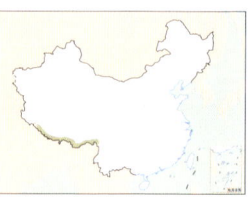

L: 24–26 cm
Habitat: Inhabits open mixed broadleaf-coniferous forests, dense rhododendron thickets, bamboo thickets, and scrubland; also near villages. Occurs in summer at altitudes of 1800–4200 m; some individuals descend to 1000 m in extreme winter weather.
Behavior: In pairs in breeding season; moves about in flocks containing 20 or more individuals in nonbreeding season. Searches for food in bushes and on the ground but often climbs trees. Not shy.
Distribution: S Tibet.
Voice: Song a loud, musical, far-carrying whistle, *weet-we-a-weer*, *que-qu-a-weer*, or *que-weer*, similar in tone to Prince Henry's Laughingthrush. Calls include a nasal raptor- or Eurasian Jay–like mew, *kyear*, and a nasal *qwee-qwee-qwee*, repeated in lengthy series.

Trochalopteron affine
黑顶噪鹛 hēi dǐng zào méi
Black-faced Laughingthrush LC

L: 24–26 cm
Habitat: Inhabits bushes in the undergrowth and edges of montane broadleaf forests, mixed forests, coniferous forests, and bamboo thickets from low to high altitudes; also in alpine scrubland.
Behavior: Moves about in small flocks in forest understory; often mixes with other laughingthrushes and large parrotbills.
Distribution: *T. a. blythii* in S Gansu, Sichuan, and Chongqing; *T. a. affine* and *T. a. bethelae* in S Tibet; *T. a. saturatum* in S Yunnan; *T. a. oustaleti* in SE Tibet and W Yunnan; *T. a. muliense* in SW Sichuan and NW Yunnan.
Voice: Song a loud, shrill, high-pitched *wiee-chiweeoo* or *wi-twi-ti-whuu.* Calls include mellow churrs, a liquid ratcheting splutter, and continuous intense squeals and chattering when agitated.

Trochalopteron morrisonianum
台湾噪鹛 (玉山噪鹛)
tái wān zào méi (yù shān zào méi)
White-whiskered Laughingthrush LC

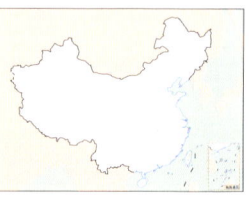

L: 25–28 cm
Habitat: Inhabits the undergrowth and lower story of coniferous forests, mixed forests, secondary forests, or bamboo thickets; also occurs on mountain trails or near cabins. Altitude range: 1800–3300 m.
Behavior: Moves about in low vegetation or on ground.
Distribution: Chinese endemic; restricted to Taiwan.
Voice: Song a clear short series of whistles, *tu-wii-uu*, *tu-wit-wee-uu*, or *chi-wi.* Also a nervous, giddy, descending laugh, vaguely reminiscent of Gray-headed Woodpecker, *hee-hee-hee.* Calls a fine tittering chitter and a loud scolding chatter admixed with short sharp calls when anxious.

Blue-winged Laughingthrush

Black supercilium

Scaly pattern on body

White iris

Blue primaries

Variegated Laughingthrush

Grayish brown overall

Black throat

White lower cheek

Orange primaries with black primary coverts

Blackish tail with grayish subterminal band and white tip

Streaked Laughingthrush

Dark rufous-brown ear coverts

Brown body

Gray rump

Dark streaks and fine, narrow, white shaft streaks on underparts

Black-faced Laughingthrush

Narrow white patch behind eye

Broad white malar stripe forms rounded patch

Olive tail and wings

Bhutan Laughingthrush

No rufous-brown ear coverts

Brown body with less gray

Tail lacks gray tip

White-whiskered Laughingthrush

White supercilium and malar stripe, forming two distinct parallel white lines

Scaly rusty feathers on upperparts

Orange and bluish-gray wings and tail

Trochalopteron henrici
灰腹噪鹛 huī fù zào méi
Prince Henry's Laughingthrush　

L: 24–28 cm

Habitat: Inhabits high-altitude river valleys, hillside scrub, woodlands, and urban parks on the Qinghai-Tibet Plateau.
Behavior: Often moves about in small flocks in bushes and small trees. Forages on ground. Noisy and unafraid of people.
Distribution: Endemic to S and SE Tibet, China. Locally common resident.
Voice: Song is a clear ringing series of whistles, *wi-pu-ti-choo*; male occasionally accompanied by female repeating a mellow slow *tiiiu* note. Calls include scolding mews and a peculiar nasal *zree-zit*.

Trochalopteron elliotii
橙翅噪鹛 chéng chì zào méi
Elliot's Laughingthrush　

L: 22–26 cm

Habitat: Inhabits broadleaf forests, mixed broadleaf-coniferous forests, coniferous forests, forest-edge bushes, and rhododendron forests at middle and high altitudes; also secondary forests and scrub around villages and farmland.
Behavior: Moves about in small flocks in forest understory. Noisy, easy to find, not shy. Forages on the ground.
Distribution: Endemic to the mountains of central and western China. Locally common resident in Hubei, Chongqing, Shaanxi, Shanxi, Guizhou, Sichuan, Qinghai, Gansu, Yunnan, and Tibet.
Voice: Highly vocal and often noisy. Song is variable but always loud, clear, and high pitched, usually consisting of a three- or four-note phrases repeated at intervals of 3–9 sec, the final note often with a slight tremulous quality, *whi-pu-puu* or *whii-tuii-chuu*; sometimes accompanied, as in Prince Henry's Laughingthrush, by an antiphonal *peu*. Calls include subdued and high-pitched chuntering.

Trochalopteron chrysopterum
金翅噪鹛 jīn chì zào méi
Assam Laughingthrush　

L: 23–25 cm

Habitat: Inhabits evergreen broadleaf forests, pine forests, mixed broadleaf-coniferous forests, stunted oak forests, stunted rhododendron thickets, and bamboo thickets at altitudes of 1280–3000 m, mostly above 1500 m.
Behavior: Often solitary, in pairs, or in small flocks. Usually seen on the ground or in bushes near the ground; occasionally forages on branches covered with moss and lichens.
Distribution: *T. c. woodi* in W and NW Yunnan; *T. c. ailaoshanense* in the Ailao Mountains of C Yunnan.
Voice: Not known to differ from vocalizations of Chestnut-crowned Laughingthrush; more research needed.

Trochalopteron melanostigma
银耳噪鹛 yín ěr zào méi
Silver-eared Laughingthrush　

L: 25–26 cm

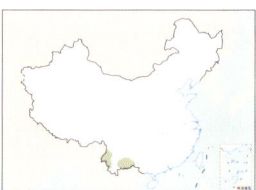

Habitat: Inhabits evergreen broadleaf forests, pine forests, mixed broadleaf-coniferous forests, bushes, grassy areas, bamboo thickets, etc., at altitudes of 1065–2565 m.
Behavior: Often forages solitarily, in pairs, or in small flocks. Generally seen on the ground or in bushes near the ground, occasionally on branches covered with moss and lichens.
Distribution: *T. m. connectans* in SE Yunnan; *T. m. melanostigma* in SW Yunnan.
Voice: Not known to differ from vocalizations of Chestnut-crowned Laughingthrush; more research needed.

Trochalopteron erythrocephalum
红头噪鹛 hóng tóu zào méi
Chestnut-crowned Laughingthrush　

L: 24–26 cm

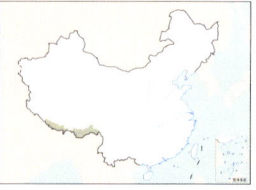

Habitat: Inhabits evergreen broadleaf forests, mixed broadleaf-coniferous forests, understory bushes, and grass and bushes near forests at altitudes of 1100–3500 m, mostly 1800–3400 m; some individuals descend to 600 m in winter.

Behavior: Often in small flocks during nonbreeding season, sometimes in larger groups of more than 30 individuals. Occasionally mixes with other laughingthrushes. Usually digs and searches for food in leaf litter on the ground, sometimes on moderately tall trunks covered with lichens and moss, seldom high in trees. Mostly shy; individuals near human habitation are less shy.
Distribution: S and SE Tibet.
Voice: Repeated melancholy whistle, *wu-wii*; a rising pure *tweeoo*, *pleer-wu*, or *pler-wee-uu*; and a scolding *prrreeer-prrreeer* or *rrreeep-rrrreep*, combined with higher-pitched piping notes when agitated.

Prince Henry's Laughingthrush

Brown mask

Orange-brown bill

Smoky gray body

Whitish submoustachial stripe

Assam Laughingthrush

No scaly pattern on ear coverts

Relatively pale gray throat

Elliot's Laughingthrush

Pure gray-brown head and body

Yellowish-white iris

Gold-olive wing patches

Brick red from lower belly to undertail coverts

Olive-yellow tail with gray subterminal band and white tip

Silver-eared Laughingthrush

Distinct white ear coverts

No distinct spots on neck, back, and breast

Conspicuous black patch on greater coverts

Chestnut-crowned Laughingthrush

Distinct scaly pattern on ear coverts

Relatively heavy black on throat and side of face

Trochalopteron milnei
红尾噪鹛 (赤尾噪鹛)
hóng wěi zào méi (chì wěi zào méi)
Red-tailed Laughingthrush　　LC

L: 26–28 cm
Habitat: Inhabits evergreen broadleaf forests, bamboo thickets, and bushes and grass at forest edges from 610 to 2500 m.
Behavior: In pairs or small flocks in understory or on the ground, hidden in dense forests.
Distribution: *T. m. milnei* in NW Fujian; has not been seen for many years. *T. m. sharpei* in W, S, and SE Yunnan and W Guangxi. *T. m. sinianum* in SE Sichuan, S Chongqing, N Guizhou, C and NE Guangxi, S Hunan, and N Guangdong. The latter two subspecies are locally common residents.
Voice: Song is a clear ringing series of thin whistles, *pwuueer-tu-tu-tuu* or *wiir-pu-ti-choo*, or a simpler *wu-tieer*, *pwu-tii-tii-tee*, or *pwuuuu-eeeerwi*, often gradually rising in pitch. Calls include a throaty, ratcheting splutter.

Trochalopteron formosum
红翅噪鹛 (丽色噪鹛)
hóng chì zào méi (lì sè zào méi)
Red-winged Laughingthrush　　LC

L: 27–28 cm
Habitat: Inhabits evergreen broadleaf forests and bushes and bamboo thickets near forests at altitudes of 900–3500 m.
Behavior: Often in pairs or small flocks. Moves about on the ground or in bushes near the ground; also in thickets up to 4–5 m above the ground. Shy and sneaky.
Distribution: *T. f. formosum* in N and C Sichuan, adjacent area in NE Yunnan, and NW Guizhou; *T. f. greenwayi* in SE Yunnan and SW Guangxi.
Voice: Loud song a clear whistled *chu-weewu*, *pwee-uuu*, or, more rarely, *puu-wee-u-wee* or *pwuu-pwee-chu-wee*. Calls a single, intense, rising whistle, *tweeeeuuu*, and a mellow churring; female often accompanies male, making high-pitched squeals.

Heterophasia auricularis
白耳奇鹛 bái ěr qí méi
White-eared Sibia　　LC

L: 22–24 cm
Habitat: Inhabits evergreen broadleaf forests, mixed forests, coniferous forests, and forest edges at altitudes of 1200–3000 m. Descends to 700 m or even 200 m in winter.
Behavior: Solitary, in pairs, or in small flocks; forages in the upper and middle stories of vegetation, sometimes in understory. Active.
Distribution: Chinese endemic; locally common resident in Taiwan.
Voice: An attractive, loud, resonant, almost wolf-whistling *fee-fee-fee feeyu*, often accompanied by typical sibia *sip* notes.

Heterophasia capistrata
黑顶奇鹛 hēi dǐng qí méi
Rufous Sibia　　LC

L: 21–24 cm
Habitat: Inhabits evergreen broadleaf forests, mixed broadleaf-coniferous forests, and bushes in the understory or at forest edges at altitudes of 1200–3410 m. Sometimes migrates to lower altitudes in winter.
Behavior: In pairs in breeding season, in groups in nonbreeding season; occasionally joins mixed-species flocks. Often forages in the canopy, on mossy tree trunks, and among bushes; occasionally in the understory.
Distribution: Both *H. c. nigriceps* and *H. c. bayleyi* are found in S Tibet.
Voice: Song is a fast, clear, high-pitched, slightly descending, whistled *sweer-ii-ii-ii-lu*, with a rising ending. Calls include sibia's familiar harsh scolding, grating, and rattling notes, as well as a juddering *didididid*.

Heterophasia picaoides
长尾奇鹛 cháng wěi qí méi
Long-tailed Sibia　　LC

L: 28–34.5 cm
Habitat: Inhabits evergreen broadleaf forests, bushes in the understory and forest edges, etc. Altitude range: 100–3000 m.
Behavior: Often in small flocks during breeding season, in larger groups in nonbreeding season; occasionally joins mixed-species flocks. Usually forages in the canopy or on tall understory plants.
Distribution: *H. p. picaoides* in W Yunnan; *H. p. cana* in SE Yunnan and W Guangxi.
Voice: Calls include thin, metallic, high-pitched *tsittsit* and *tsic* notes, interspersed with dry ratcheting rattles, *tsic-tsic-chrrrrrrrt*.

Heterophasia pulchella
丽色奇鹛 lì sè qí méi
Beautiful Sibia　　LC

L: 23 cm
Habitat: Inhabits evergreen broadleaf forests and understory bushes at altitudes of 900–3000 m; sometimes migrates to lower altitudes in winter.
Behavior: In pairs in breeding season; often in flocks during nonbreeding season. Occasionally joins mixed-species flocks. Usually forages in the canopy, among bushes, and on tree trunks covered by mosses and lichen; occasionally comes to bushes in the understory.
Distribution: *H. p. nigroaurita* occurs in SE Tibet; *H. p. pulchella* occurs in W and NW Yunnan.
Voice: Song and calls very similar to those of Rufous Sibia.

Red-tailed Laughingthrush

Brownish yellow from crown to hindneck

White eye ring

Silver-white cheek

Red wings and tail

Red-winged Laughingthrush

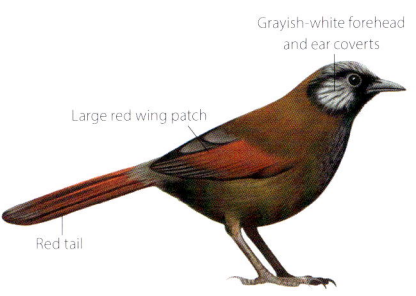

Grayish-white forehead and ear coverts

Large red wing patch

Red tail

White-eared Sibia

White from base of bill, lores, and eye ring to ear coverts, white filamentous feathers behind ears

Brownish yellow on rump and undertail coverts fades to pinkish brown on belly

Rufous Sibia

Grayish back

Black head

Rufous chestnut overall

H. c. bayleyi

Grayish-black wings and tail tips

H. c. nigriceps

Long-tailed Sibia

Mostly gray body

White wing patches

Tail longer than other sibias, with white tip

Beautiful Sibia

Bluish gray from head to back

Black on forehead and around eye

Brown tail with black subterminal band and grayish-white tip

449

Heterophasia gracilis

灰奇鹛 huī qí méi
Gray Sibia

L: 22.5–24.5 cm
Habitat: Inhabits evergreen broadleaf forests and bushes in the understory and forest edges at altitudes of 500–2800 m.
Behavior: In pairs in breeding season; often in flocks during nonbreeding season. Occasionally joins mixed-species flocks. Usually forages in the canopy, among bushes, and on tree trunks covered by mosses and lichen; occasionally comes to bushes in the understory.
Distribution: W Yunnan.
Voice: Song variable but often similar to that of Rufous Sibia but longer, more consistently descending, and lacking the terminal upswing. Calls include a hurried nasal *wit-wit-wit-witarit* and harsh, scolding, Rufous Sibia–like notes, as well as high-pitched *tip* and *sip* and chattering notes more reminiscent of Long-tailed Sibia.

Actinodura morrisoniana

台湾斑翅鹛 tái wān bān chì méi
Taiwan Barwing

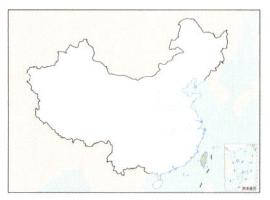

L: 18–19 cm
Habitat: Inhabits mature deciduous broadleaf forests, evergreen broadleaf forests, and mixed forests at altitudes of 1200–3000 m. Migrates to lower altitudes in winter.
Behavior: In pairs or small groups. Forages in canopy or middle story, sometimes in the understory; tends to join White-eared Sibia, Taiwan Yuhina, and Steere's Liocichla in mixed-species flocks. Moves through epiphytes and branches; forages by turning over bark, leaves, moss, and lichens.
Distribution: Chinese endemic; locally common resident in Taiwan.
Voice: Song a clear, loud, rising, nasal, melancholy *pweer*, usually given singly. Call a chattering *chiririririt*, much higher pitched than Hoary-throated Barwing.

Actinodura nipalensis

纹头斑翅鹛 wén tóu bān chì méi
Hoary-throated Barwing

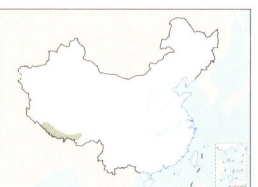

L: 20–21 cm
Habitat: Inhabits broadleaf forests, alpine oak forests, and rhododendron thickets at middle to high altitudes.
Behavior: Often in small flocks in the lower and middle stories of forests and bushes; sometimes mixes with other babbler species.
Distribution: S Tibet.
Voice: Pleasant song is a clear, whistled, slightly melancholy, rising *prrr-wiuu-liu* or *prrt-ti-ti-tiu*. Most common call is a loud, scolding, or spluttering chatter, often accompanied by higher-pitched shrill notes.

Heterophasia desgodinsi

黑头奇鹛 hēi tóu qí méi
Black-headed Sibia

L: 20–24 cm
Habitat: Inhabits montane broadleaf and mixed forests at altitudes of 800–2800 m.
Behavior: Often solitary or in small flocks in the upper and middle stories of forests and canopy; sometimes forages in bushes. Active.
Distribution: Common resident in Sichuan, Chongqing, Yunnan, Guizhou, Hubei, Hunan, and W Guangxi; possibly in S Shaanxi.
Voice: High-pitched whistled song of short notes, *he-wi-wi wi wee*, gradually decreasing in frequency and with a slightly elongated ending. Calls include noisy chatter and high-pitched thin *sip* notes.

Actinodura waldeni

纹胸斑翅鹛 wén xiōng bān chì méi
Streak-throated Barwing

L: 19–22 cm
Habitat: Inhabits broadleaf forests, secondary forests, and alpine scrubland at middle and high altitudes.
Behavior: Moves about in small flocks in the forest understory in nonbreeding season; prefers mature forests.
Distribution: *A. w. daflaensis* in SE Tibet; *A. w. saturatior* in W and NW Yunnan.
Voice: Song a loud strident series of whistles that often start with a slight rattle, *t-t-rrrr-jo-jiewie*. Calls include low nasal grumbling, *grrr-ut grrr-ut*, an intense *grr-grr-grr-grr-grr*, and a *pweer* note very similar to that of Taiwan Barwing but often repeated in short series.

Actinodura souliei

灰头斑翅鹛 huī tóu bān chì méi
Streaked Barwing

L: 22–23 cm
Habitat: Inhabits middle-altitude broadleaf forests, mixed forests, and forest-edge bushes.
Behavior: In pairs or small flocks in the lower and middle stories at forest edges. Noisy and unafraid of people. Rarely mixes with other babbler species.
Distribution: *A. s. souliei* in C and W Sichuan, NW and NE Yunnan; *A. s. griseinucha* in SE Yunnan.
Voice: Song is a soft, nasal *nyut-nyut-nyut*, given in an intensifying, rising sequence. Calls include harsh churrs.

Gray Sibia

Head shades from black forehead to gray nape

Relatively white throat

Grayish-white underparts

Gray tail with black subterminal band

Black-headed Sibia

Black head and wings

White throat

Black tail with gray tip

Taiwan Barwing

Chestnut-brown head

Grayish-brown and white mottling on breast and neck ring

Streak-throated Barwing

Inconspicuous black malar stripe

Earth-brown streaks on pale throat and breast

Hoary-throated Barwing

White eye ring, sometimes inconspicuous

Pale shaft streaks on brown, instead of gray, crown

Black malar stripe

Streaked Barwing

Gray on crown and from behind eye to hindneck

Grayish-white eye ring

Black lores

Dark streaks on upper back

Actinodura cyanouroptera

蓝翅希鹛 lán chì xī méi

Blue-winged Minla

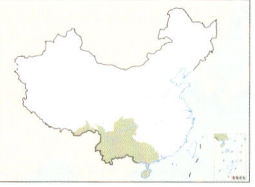

L: 14–15.5 cm

Habitat: Generally occurs in evergreen broadleaf forests; also in pine forests, mixed broadleaf-coniferous forests, and bamboo thickets at altitudes of 250–3000 m.
Behavior: In nonbreeding season, moves about in flocks and tends to join mixed-species flocks. Usually moves among branches to forage.
Distribution: Resident in C and S Sichuan and Yunnan, east to S Hunan, C Guangdong, and Hainan.
Voice: Song is a monotonously repeated, very high-pitched, thin *puiss-si-sweu*. Calls include dry staccato buzzes and short penetrating *whit* notes.

Actinodura strigula

斑喉希鹛 bān hóu xī méi

Chestnut-tailed Minla

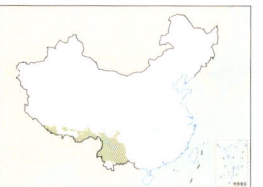

L: 16–18.5 cm

Habitat: Inhabits evergreen broadleaf forests, mixed broadleaf-coniferous forests, rhododendron forests, scrubland, bamboo thickets, etc., at altitudes of 1800–3750 m. May migrate to lower altitudes in winter.
Behavior: Often in trees but also forages in low bushes. In nonbreeding season, often gathers in small flocks and joins mixed-species flocks, especially with Red-tailed Minla and fulvettas.
Distribution: *A. s. strigula* in S Tibet; *A. s. yunnanensis* in SE Tibet, Yunnan, S and SW Sichuan.
Voice: Song is a powerful whistled *tu-wi ti-tu*. Calls include a nasal *yeep*.

Minla ignotincta

红尾希鹛 (火尾希鹛)

hóng wěi xī méi (huǒ wěi xī méi)

Red-tailed Minla

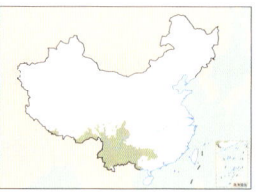

L: 13–14.5 cm

Habitat: Inhabits evergreen broadleaf forests, mixed broadleaf-coniferous forests, rhododendron thickets and other habitats; also found in forests dominated by Fujian Cypress. Altitude range: 300–3750 m.
Behavior: Often in groups; also joins mixed-species flocks. Prefers climbing on tree trunks and scavenging for food in moss and lichens.
Distribution: *M. i. ignotincta* in SE Tibet and W and NW Yunnan; *M. i. mariae* in C and SE Yunnan; *M. i. jerdoni* in C and S Sichuan, Chongqing, N, C, and E Guizhou, S Hunan, and NE Guangxi; *M. i. sini* in Dayaoshan Mountains in E Guangxi.
Voice: Song is a fine, high-pitched whistle, *ti-wi-tu-wutu*. Calls include harsh churrs, short higher-pitched *wit* notes, and scolding calls more reminiscent of those of Chestnut-tailed Minla.

Actinodura ramsayi

白眶斑翅鹛 bái kuàng bān chì méi

Spectacled Barwing

L: 22–25 cm

Habitat: Inhabits broadleaf forests, secondary forests, and woodlands at forest edges at middle and low altitudes.
Behavior: Often solitary or in pairs in understory or forest-edge bushes. Bold and easily spotted.
Distribution: W Guangxi, SE Yunnan, and S Guizhou.
Voice: Song is a slightly hesitant, melancholy, descending, slightly bouncing phrase, *iee-iee-iee-iuu*. Calls include a raptorlike mewing, *kyar-yar*.

Actinodura egertoni

栗额斑翅鹛 (锈额斑翅鹛)

lì é bān chì méi (xiù é bān chì méi)

Rusty-fronted Barwing

L: 21–24 cm

Habitat: Inhabits broadleaf forests, mixed forests, and understory bushes at middle and low altitudes.
Behavior: Seen mostly in pairs or small flocks in the lower and middle stories of forests. Active and noisy.
Distribution: *A. e. egertoni* in SE Tibet; *A. e. ripponi* in W and NW Yunnan.
Voice: Song is a repeated, descending, melancholy sequence of short notes, *ti-ti-ti-wi-wu*. Calls include excited, metallic, high-pitched, swiftlet-like chittering and lower-pitched, harsher, abrupt, even explosive, scolding notes, preceded by a nasal *gursh*.

Leioptila annectens

栗背奇鹛 lì bèi qí méi

Rufous-backed Sibia

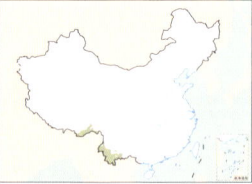

L: 18.5–20 cm

Habitat: Inhabits evergreen broadleaf and sometimes semideciduous forests at altitudes of 200–2650 m.
Behavior: Solitary, in pairs, or in small flocks of three to six individuals; often joins mixed-species flocks. Usually searches for food in the upper and middle stories of forests; prefers large, mossy, and lichen-covered tree trunks or branches. Also feeds on nectar.
Distribution: *L. a. annectens* in W Yunnan and SE Tibet; *L. a. mixta* in SE Yunnan and SW Guangxi.
Voice: Simple song a short, loud, clear, cheerful series of whistles, *pit-wit-weet-wiuuu-twe-twe-twee*. Calls include scolding Rusty-fronted Barwing–like chatter and softer, more nasal, interrogative notes, *pit-wittu*.

Blue-winged Minla

Fine white streaks on blue head, blue crown paler on female than male

Outer parts of wings and tail dark blue, inner parts light blue

Spectacled Barwing

Rusty chestnut from forehead to crown

Distinctive white eye ring

Clean breast and belly lack streaks

Chestnut-tailed Minla

Relatively pale orange-brown crown

More chestnut on central tail feathers

Narrow black bars on throat

A. s. yunnanensis

Wings rich in color

Less chestnut on central tail feathers

A. s. strigula

Rusty-fronted Barwing

Relatively dark orange-brown crown

Rusty-chestnut forehead

Rufous-brown breast and belly

Red-tailed Minla

Black eye stripe

White supercilium

Chocolate-brown back

Pale yellow underparts

Red outer tail feathers

M. i. ignotincta

Olive-brown back

Orange-red outer tail feathers

Olive back

Dark yellow underparts

M. i. mariae

Rufous-backed Sibia

Black head

White throat and breast

Chestnut back

Leiothrix argentauris

银耳相思鸟 yín ěr xiāng sī niǎo

Silver-eared Mesia   LC

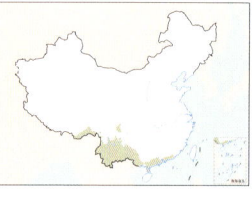

L: 15.5–17 cm

Habitat: Inhabits open evergreen broadleaf forests, mixed broadleaf-coniferous forests, understory bushes, forest edges, abandoned cultivated fields, tea plantations, bamboo thickets, and sometimes grassy areas at altitudes of 175–2100 m.

Behavior: Forms flocks of 5–30 or more individuals in nonbreeding season; often joins mixed-species flocks. Usually moves about in bushes of open forests, sometimes in the canopy.

Distribution: *L. a. argentauris* in SE Tibet, W and NW Yunnan; *L. a. ricketti* in S and SE Yunnan, S Guizhou, SW Guangxi; *L. a. galbana* in SW Yunnan.

Voice: Song is a loud, cheerful, clearly spaced, slowly delivered, descending *che-tchu-tchu che-rit*. Calls include flat piping *pe-pe-pe-pe-pe* and harsh chattering notes similar to those of Red-billed Leiothrix but lower pitched and harsher.

Leiothrix lutea

红嘴相思鸟 hóng zuǐ xiāng sī niǎo

Red-billed Leiothrix   LC

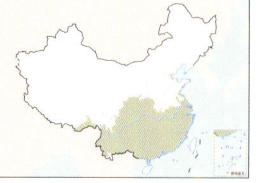

L: 14–15 cm

Habitat: Inhabits evergreen broadleaf forests, mixed forests, forest edges, secondary forests, various scrubland, abandoned cultivated fields, and the understory of tea plantations and bamboo thickets at altitudes of 900–2400 m.

Behavior: In pairs during breeding season, in small groups in nonbreeding season; sometimes joins mixed-species flocks. Prefers scurrying quickly in the understory or climbing tree trunks; also searches for food on the ground.

Distribution: Locally common resident. *L. l. lutea* occurs from C Sichuan, S Gansu, and S Shaanxi to C Anhui, S Jiangsu, Zhejiang, and Fujian, and south to Guizhou and N Guangxi; *L. l. calipyga* occurs in SE Tibet; *L. l. yunnanensis* occurs in W and NW Yunnan; *L. l. kwangtungensis* occurs from SE Yunnan to C Guangxi, S Hunan, and Guangdong. Populations in the wild have declined significantly due to extensive illegal hunting and trade.

Voice: Sings three main types of song: The first is a long complex warble of up to 15 slowly delivered notes that ramble up and down the scale. The second is shorter and has a fixed sequence of syllables. The third is a simple sound produced by males during courtship. Calls include a rapid rattling chatter.

Liocichla bugunorum

黑冠薮鹛 (布坤薮鹛)

hēi guān sǒu méi (bù kūn sǒu méi)

Bugun Liocichla   CR

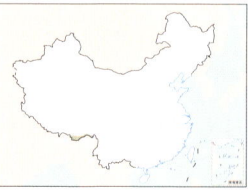

L: 22 cm

Habitat: Inhabits evergreen broadleaf forests and scrubland at middle and low altitudes.

Behavior: Solitary or in pairs in the lower and middle stories of montane forests.

Distribution: Extremely narrow distribution in S Tibet, with a very tiny population.

Voice: Song is loud, simple, penetrating whistle of two to five low flat notes, the first occasionally tremulous, the last two or three decreasing in pitch, and the last often short and descending, *qiu-witty-tii-u*, lasting about 1.5 sec and repeated every 3–5 sec. Similar to Gray-faced Liocichla but more sedate, lower pitched, flatter, and with fewer changes in pitch other than the tremolos.

Liocichla omeiensis

灰胸薮鹛 huī xiōng sǒu méi

Gray-faced Liocichla   VU

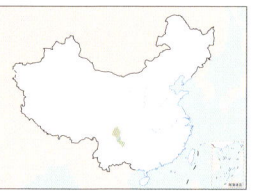

L: 17–20 cm

Habitat: Inhabits middle- to low-altitude montane broadleaf forests, mixed broadleaf-coniferous forests, secondary forests, bamboo thickets, and forest-edge bushes; migrates to lower altitudes in winter.

Behavior: Typical forest-dwelling species, mostly solitary or in pairs.

Distribution: Chinese endemic; found only in C and S Sichuan and NE Yunnan.

Voice: Song is loud penetrating series of lengthy, gently undulating whistles, *chwi-weeiee-eeoo*. Alarm call is a series of agitated, anxious chattering.

Liocichla steerii

黄痣薮鹛 huáng zhì sǒu méi

Steere's Liocichla   LC

L: 17–19 cm

Habitat: Inhabits evergreen broadleaf forests, mixed forests, forest edges, scrubland, and orchards at altitudes of 830–3000 m.

Behavior: Often in low vegetation or on forest floor. Usually solitary, in pairs, or in small flocks; sometimes joins mixed-species flocks.

Distribution: Chinese endemic; locally common resident in Taiwan.

Voice: Song is a high-pitched, penetrating *tsiii tsiuwuu*, with a sharp introductory note and a very slightly longer, strongly inflected second note that is typically repeated every few seconds. Calls include a buzzy, scolding rasp, *djrrrrrr*, that often continues for long periods and a slower, buzzy churr.

Silver-eared Mesia

Black crown

White cheek

♂

L. a. argentauris

Red undertail coverts

Yellow undertail coverts

♀

Vermilion throat and side of neck

L. a. ricketti

Bugun Liocichla

Black spot above eyes

Black crown

Yellow supercilium

Shape of wing patch resembles Gray-faced Liocichla but has olive-yellow base and red tip

Gray-faced Liocichla

Two red wing patches

Red tail tip

♂

Pale gray from lower cheek to breast and belly

Orange-yellow wing patches

♀

Orange-yellow or yellow tail tip

Red-billed Leiothrix

Bright red bill

Yellow and red wing patch

Bright yellow on throat extends to upper breast, becomes orange yellow

Notched tail

L. l. lutea

Yellow wing patch

L. l. yunnanensis

Steere's Liocichla

Bright yellow patch at lores

Gray from crown to back

Gray throat and flanks

Liocichla phoenicea

灰头薮鹛 (赤脸薮鹛)

huī tóu sǒu méi (chì liǎn sǒu méi)

Red-faced Liocichla　

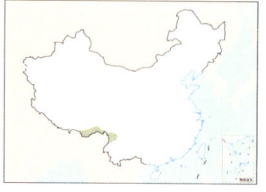

L: 21–24 cm
Habitat: Inhabits
evergreen broadleaf
forests, secondary forests,
and forest-edge bushes at
low and middle altitudes.
Behavior: Mostly in pairs
or in small flocks in the
lower and middle stories.
Distribution: Seen in SE Tibet and extreme NW Yunnan.
Voice: Melodious, clear, cheerful, whistled song of three to nine flat
notes delivered slowly and repeated, *tie-tie-wu-wheeu-wee-ti-wu*, or
shorter *ti-tu-tuwhuu-wheeu*. Calls include strained squeals and rapid
chatters.

Garrulax monileger

小黑领噪鹛 xiǎo hēi lǐng zào méi

Lesser Necklaced Laughingthrush　

L: 24–32 cm
Habitat: Inhabits
evergreen broadleaf
forests, deciduous forests,
secondary forests, and
scrubland below 1650 m.
Behavior: Often in flocks
but does not mix with
other birds in breeding
season. Outside breeding season, joins other laughingthrushes, such as
White-crested, Greater Necklaced, and Black-throated, in mixed-species
flocks. Forages by scraping or turning leaves on the ground.
Distribution: Locally common. *G. m. monileger* in W and SW Yunnan;
G. m. schauenseei in S Yunnan; *G. m. tonkinensis* in SE Yunnan and
Guangxi; *G. m. melli* in S Hunan, S Jiangxi, S Zhejiang, Fujian, and
Guangdong; *G. m. schmackeri* in Hainan.
Voice: Song an intense rich medley of repeated, short, whistled notes,
ti-ti-ti-u … ti-ti-whuu ti-ti-wu wheeuu, lasting 2–5 sec and occasionally
including harsher rattles and short whistles. Various rattles, intense
piping, and squabbling calls from foraging flocks.

Garrulax canorus

画眉 huà méi

Chinese Hwamei　

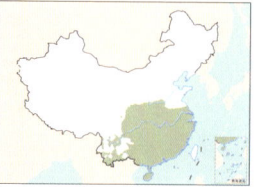

L: 21–24 cm
Habitat: Inhabits
scrubland, open forests,
bamboo thickets,
reedbeds, tall grass, and
gardens below 1800 m.
Behavior: Solitary, in pairs,
or in small flocks; usually
forages on the ground.
Distribution: Common. *G. c. canorus* in southeastern China, including
S Gansu, S Shaanxi, Hebei, SE Henan, S Jiangsu, Guangxi, Guangdong,
and Yunnan (except for NW Yunnan). *G. c. owstoni* in Hainan; sometimes
treated as a separate species: Hainan Hwamei (*G. owstoni*).
Voice: Accomplished song is intense, rich, varied, and often high
pitched and includes regular repetition and some mimicry.

Liocichla ripponi

红翅薮鹛 hóng chì sǒu méi

Scarlet-faced Liocichla　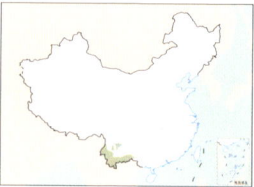

L: 21–24 cm
Habitat: Inhabits
broadleaf forests,
woodlands, and
secondary forests, also
around villages and in
forest-edge bushes, at
middle and low altitudes.
Behavior: Mostly in
pairs or small flocks; sometimes also on the ground. Mixes with other
laughingthrushes.
Distribution: *L. r. ripponi* in W and S Yunnan; *L. r. wellsi* in SE Yunnan.
Voice: Vocalizations very similar to those of Red-faced Liocichla.

Garrulax merulinus

斑胸噪鹛 bān xiōng zào méi

Spot-breasted Laughingthrush　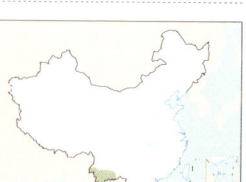

L: 24–26 cm
Habitat: Inhabits
broadleaf forests,
secondary forests, and
forest-edge bushes at
middle and low altitudes.
Behavior: Gathers in
small flocks in forest
understory. Forages on
the ground. Secretive.
Distribution: *G. m. merulinus* in W Yunnan; *G. m. obscurus* in S and
SE Yunnan.
Voice: Song is an often-loud combination of rich, melodious, rambling
notes; harsher strained, scolding notes; and squeaky whistles, similar
to hwamei but harsher and more discordant. Like song of that species,
contains considerable mimicry. Calls include harsh chatter.

Garrulax taewanus

台湾画眉 tái wān huà méi

Taiwan Hwamei　

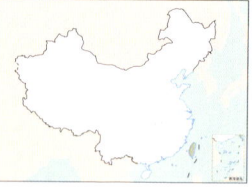

L: 21–24 cm
Habitat: Inhabits
secondary vegetation,
scrubland, and low
thickets in hilly areas
below 1200 m.
Behavior: Forages
solitarily, in pairs, or
in small flocks on the
ground.
Distribution: Chinese endemic; common resident in Taiwan. In recent
years, Taiwan Hwamei is found to hybridize with Chinese Hwamei
introduced from mainland.
Voice: Song is melodious, varied, and loud, similar to Chinese Hwamei
but simpler and more repetitive.

Red-faced Liocichla

Brownish-gray crown

Distinct blue eye ring

Broad black supercilium

Complete red wing patch, differing from Bugun Liocichla and Gray-faced Liocichla

Red color on face extends only to lower cheek

Same color on breast and body

Broad orange-red tip on tail

Scarlet-faced Liocichla

Narrow black line above red supercilium, sometimes inconspicuous

No pale eye ring, an extremely indistinct small triangular patch of pale bare skin behind eye

Complete red wing patch, differing from Bugun Liocichla and Gray-faced Liocichla

Red cheek

Breast tinged olive yellow

Narrow olive-brown tip on tail

Lesser Necklaced Laughingthrush

Well-defined patch of color on cheek, lacks unique black and white mottling of Greater Necklaced Laughingthrush

Pale yellow iris, not black

Olive-brown body

Black lores

Relatively broad black breast band

Pale brown neck and flanks

G. m. monileger

Plumage with more distinctive brown tone than other subspecies

Relatively narrow blackish-brown breast band, often tinged brown

Rufous brown on neck and flanks extends to belly

G. m. melli

Spot-breasted Laughingthrush

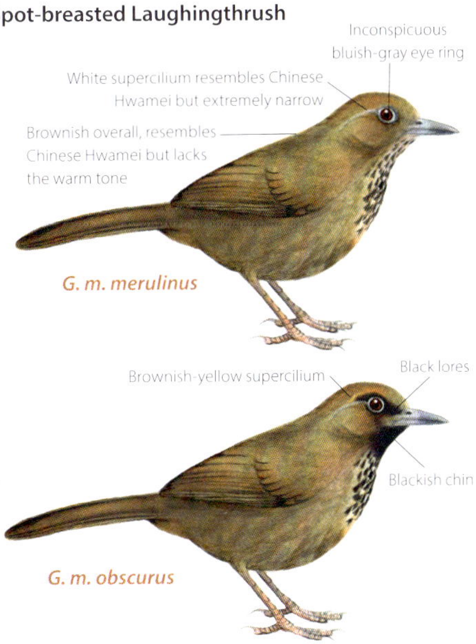

Inconspicuous bluish-gray eye ring

White supercilium resembles Chinese Hwamei but extremely narrow

Brownish overall, resembles Chinese Hwamei but lacks the warm tone

G. m. merulinus

Brownish-yellow supercilium

Black lores

Blackish chin

G. m. obscurus

Chinese Hwamei

Narrow dark streaks on crown, neck, and throat

White around eye reaches behind eye, forming white postocular line

Brown or pale yellow iris

Gray belly

G. c. canorus

Taiwan Hwamei

Bluish gray around eyes, without white postocular line

Gray iris

Many dark brown streaks from crown to back

Garrulax leucolophus
白冠噪鹛 bái guān zào méi
White-crested Laughingthrush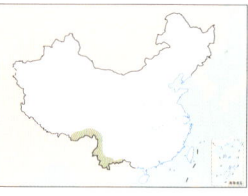

L: 26–31 cm
Habitat: Inhabits evergreen broadleaf forests, mixed deciduous forests, disturbed secondary and regenerated forests, scrubland, bamboo thickets, abandoned cultivated forests, and plantations and gardens near forests.
Behavior: Often in flocks. Noisy. Moves about in the lower and middle stories of forests, foraging mainly on the ground, digging in the soil and turning leaves in search of food.
Distribution: *G. l. leucolophus* occurs in SE Tibet; *G. l. patkaicus* occurs in W and SW Yunnan; *G. l. diardi* occurs in S and SE Yunnan.
Voice: Highly vocal, with flocks chorusing to produce sudden loud outbursts of extended cackling laughter typically involving rapid chatters, croaks, and repeated phrases.

Garrulax strepitans
白颈噪鹛 bái jǐng zào méi
White-necked Laughingthrush

L: 29–31 cm
Habitat: Inhabits low montane rain forests and evergreen broadleaf forests; also occurs in forest-edge bushes and secondary forests.
Behavior: Moves about in small flocks in forest understory. Secretive.
Distribution: Seen in SW Yunnan.
Voice: Song like a group laughing, ending with a short *gur* or a very short, hoarse *ge–ge*.

Garrulax maesi
褐胸噪鹛 hè xiōng zào méi
Gray Laughingthrush

L: 28–30 cm
Habitat: Inhabits evergreen broadleaf forests, secondary forests, and forest-edge bushes at middle and low altitudes.
Behavior: Often in small flocks at montane forest edges. Noisy.
Distribution: Resident in SE Tibet, W and C Sichuan, NE and SE Yunnan, SW Chongqing, Guizhou, Guangxi, and N Guangdong.
Voice: Flocks emit sudden, loud outbursts of White-crested Laughingthrush–like cackling laughter but with more spluttering rapid chatters and bamboo-cracking notes.

Garrulax castanotis
栗颊噪鹛 lì jiá zào méi
Rufous-cheeked Laughingthrush

L: 27–30 cm
Habitat: Inhabits montane evergreen broadleaf forests and rain forests at low altitudes.
Behavior: Often in small flocks in the lower and middle stories of forests and forest-edge bushes. Noisy.
Distribution: *G. c. castanotis* is an endemic subspecies found only in Hainan.
Voice: Rich, melodious, slow song, *tip-poor-wii … poer-wii*, often accompanied by short reed warbler–like *check* notes. Flocks give sudden, loud outbursts of extended cackling, like White-crested Laughingthrush and especially Gray Laughingthrush.

Ianthocincla sukatschewi
黑额山噪鹛 hēi é shān zào méi
Snowy-cheeked Laughingthrush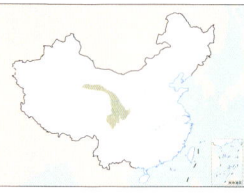

L: 27–31 cm
Habitat: Inhabits mixed forests, coniferous forests, alpine scrubland, and bamboo thickets at middle and high altitudes.
Behavior: Often solitary or in pairs at forest edges or in the lower and middle stories of forests; forages on the ground.
Distribution: Chinese endemic; found in S, SW, and SE Gansu and N Sichuan, possibly also in extreme SE Qinghai.
Voice: Simple song a high-pitched repetition of short shrill notes, *gi-u … gwi-u … gwi-u*; occasionally, a more melodious whistled *puii-pee-u*. Calls include grating, unpleasant chattering.

Ianthocincla cineracea
灰翅噪鹛 huī chì zào méi
Moustached Laughingthrush

L: 21–24 cm
Habitat: Inhabits thickets or brambles at evergreen broadleaf forest edges, mixed broadleaf-coniferous forests, secondary forests, abandoned cultivated lands, and bamboo thickets from 200 to 2750 m; seldom occurs in primary forests.
Behavior: Often in pairs in breeding season and in flocks in nonbreeding season. Forages mainly on the ground. The subspecies in eastern China sometimes flocks with Chinese Hwamei.
Distribution: Uncommon resident. *I. c. strenua* occurs in W and N Yunnan, SE Tibet, and S Sichuan; *I. c. cinereiceps* is more widely distributed, from S Gansu and S Shaanxi in the north, to S Jiangsu, W Shanghai, Fujian, and N Guangdong in the east, and to S and SE Yunnan, Guizhou, and Guangxi in the south. The two subspecies are sometimes treated as two separate species.
Voice: Varied and complex vocabulary. Song a short, sweet, thin whistle, *tip-it-twipi-tuu*. Calls include a repeated, intense, high-pitched, rippling trill, *prree-eue*, vaguely reminiscent of the song of White-throated Kingfisher, a short *pr'r'r'r'ip*, and various low-pitched churring notes.

White-crested Laughingthrush

Distinct white crest

Black mask

White from head to upper breast

Brown belly

G. l. patkaicus

G. l. diardi

White belly slightly tinged brown

Rufous-cheeked Laughingthrush

Black from lores to chin

Chestnut-brown ear coverts

Dark blackish-brown breast

Snowy-cheeked Laughingthrush

Black eye stripe extends to forehead

White cheek, black submoustachial stripe

Ivory-yellow lower mandible

White-necked Laughingthrush

Triangular bluish-gray patch behind eye

White band from hindneck to side of breast

Dark brown from cheek to throat and breast

Moustached Laughingthrush

Dark gray from forehead to crown

Inconspicuous eye stripe

Brownish yellow overall

Blue iris

I. c. cinereiceps

Black malar stripe, black on crown extends to back

Distinct black eye stripe

Grayish overall, distinctively paler than *I. c. cinereiceps* on supercilium and from face to neck

Yellowish-white iris

I. c. strenua

Gray Laughingthrush

Grayish-white ear coverts

Blackish-brown lores

Grayish overall, contrasts slightly with blackish-brown breast

Ianthocincla rufogularis
棕颏噪鹛 zōng kē zào méi
Rufous-chinned Laughingthrush　　LC

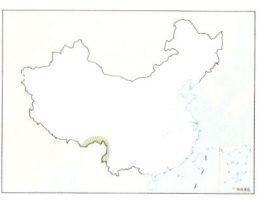

L: 23–25.5 cm
Habitat: Inhabits evergreen broadleaf forests, forest edges, secondary forests, and scrubland at 610–1980 m.
Behavior: Often in pairs or small family groups. Occurs mainly in the lower part of bushes.
Distribution: *I. r. rufogularis* in W Yunnan; *I. r. assamensis* in SE Tibet.
Voice: Song a shrill, slightly husky *whi-whi-whu-whi*. Calls include a low-pitched sputtering and a curious, low, buzzing *jzzzzz*.

Ianthocincla ocellata
眼纹噪鹛 yǎn wén zào méi
Spotted Laughingthrush　　LC

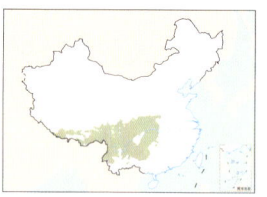

L: 30–34 cm
Habitat: Inhabits broadleaf forests, mixed broadleaf-coniferous forests, alpine scrubland, and bamboo thickets at middle and high altitudes.
Behavior: Often in pairs or small flocks in forest understory. Forages on the ground. Noisy but secretive.
Distribution: Seen from the Himalayas to the mountains of central and western China. *I. o. ocellata* in S and SE Tibet; *I. o. maculipectus* in W Yunnan; *I. o. artemisiae* in W and C Sichuan, S Gansu, W Hubei, NE Yunnan, S Chongqing, Guizhou, and N Guangxi.
Voice: Variable, rich, fluty, low-pitched, but powerful song, *pu-wu-weer-weeer-tweer-twu*. Calls include a repeated *tweeeer*.

Ianthocincla maxima
大噪鹛 dà zào méi
Giant Laughingthrush　　LC

L: 32–36 cm
Habitat: Occurs in bushes, rhododendron thickets, and meadows at montane forest edges at middle and high altitudes.
Behavior: In pairs or small flocks at forest edges and alpine scrubland with few trees. Large and chickenlike.
Distribution: Found in the mountains of southwestern China in S Gansu, S and SE Qinghai, N and W to SW Sichuan, NW Yunnan, and E Tibet.
Voice: Song a variable loud, clear, rich, melodious, almost *Turdus*-like phrase, *chwi-chwi-chwi-wuu* or *fweet-weet-furweet*. Occasionally mimics other species, such as Gray-headed Woodpecker and Sichuan Jay. Calls include a musical *tipit-wheet* and unpleasant, jarring, barked, scolding notes.

Ianthocincla bieti
白点噪鹛 bái diǎn zào méi
Biet's Laughingthrush　　VU

L: 25–28 cm
Habitat: Inhabits the understory bushes of alpine oak forests, mixed forests, and coniferous forests at middle and high altitudes.
Behavior: Often in pairs in the forest understory. Unafraid of people but secretive.
Distribution: Endemic to southwestern China's mountains; found in SW Sichuan and NW Yunnan. Very rare.
Voice: Jolly song is a loud, clear, fluty *chu-whi-wu-wheu … poor-whi-wu-we-weu* or *jiu-hu-huwu*, repeated every few seconds. Calls include a series of nervous, nasal notes mixed with an intense scolding *prrup-rup-rup-queer*.

Ianthocincla lunulata
斑背噪鹛 bān bèi zào méi
Barred Laughingthrush　　LC

L: 23–26 cm
Habitat: Occurs in the understory of broadleaf, mixed broadleaf-coniferous, and coniferous forests; also inhabits bamboo thickets, at middle and high altitudes.
Behavior: Often solitary or in small flocks in the understory or forest-edge bushes; occasionally mixes with other laughingthrushes.
Distribution: *I. l. lunulata* in the mountains of central China in W Hubei, NE Chongqing, S Shaanxi, S Gansu, and N Sichuan; *I. l. liangshanensis* in SW Sichuan.
Voice: Song a loud, clear, sometimes fluty *chu-whi-u-wu-hu-u-wu*, repeated every few seconds, while a second bird adds a tremulous, scolding *tchrr-tchrr*.

Pterorhinus gularis
栗臀噪鹛 (棕臀噪鹛)
lì tún zào méi (zōng tún zào méi)
Rufous-vented Laughingthrush　　LC

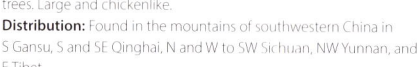

L: 22–27 cm
Habitat: Inhabits evergreen broadleaf forests, secondary forests, scrubland, and bamboo thickets at altitudes of 90–1220 m.
Behavior: Shy. Often in flocks; sometimes mixes with other birds. Moves about in ground-level bushes; occasionally climbs understory trees.
Distribution: Rare in SE Tibet and S Yunnan.
Voice: Presumed song involves simple, descending whistles, repeated seemingly erratically by flocks, *tiuu* or *fweeoou* (similar to those of Blue-crowned Laughingthrush but less intense), occasionally accompanied by quiet scimitar-babbler-like *oip* notes or nasal chattering. Flocks occasionally give explosive White-crested Laughingthrush-like laughter.

Rufous-chinned Laughingthrush

Black crown

Black spots on body

Brown chin and ear coverts

Biet's Laughingthrush

White tip and black subterminal band on feathers on upper back

Inconspicuous fine black spots on underparts with distinct white spots

Spotted Laughingthrush

Brownish-white supercilium

Pale brown ear coverts

Black throat

I. o. maculipectus

Distinct yellowish-brown supercilium

Dark chestnut-brown ear coverts

Blackish-brown throat

Black or blackish-brown spots on breast and belly

I. o. ocellata

No supercilium

Black head

Black throat

I. o. artemisiae

Barred Laughingthrush

Scaly bars on upper back, feathers with brown tip and black subterminal band

Fine black bars on underparts

Rufous-vented Laughingthrush

Long, thick, large bill

Gray crown

Yellow throat

Rufous-brown vent and belly

Giant Laughingthrush

Brown face and throat

White bars on breast

Blackish-brown tail

461

Pterorhinus courtoisi

蓝冠噪鹛 (靛冠噪鹛)

lán guān zào méi (diàn guān zào méi)

Blue-crowned Laughingthrush　**CR**

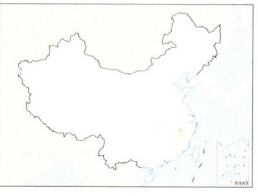

L: 24–25 cm
Habitat: Inhabits evergreen broadleaf forests, mixed evergreen-deciduous broadleaf forests, and forest-edge bushes; also occurs in large trees near human habitation.
Behavior: Often forages in flocks, turning leaves on the ground or foraging for insects or plant seeds on tree trunks.
Distribution: Chinese endemic, with two intermittently distributed subspecies: *P. c. courtoisi* occurs only in the "Fengshui woodland" and montane forests near villages from NE Jiangxi (Wuyuan) to NW Fujian, with a very small population size, and *P. c. simaoensis* occurs only in Pu'er, Yunnan, where it was first collected in 1956 and may now be extinct.
Voice: Apparent song is a resonant *iiddu-tiddu-tiddu*, repeated every few seconds. Calls include an intense, explosive *prhheu*, given in alarm, and descending whistles, *piiuu* or *djuu*, often accompanied by spluttering.

Pterorhinus ruficollis

栗颈噪鹛 lì jǐng zào méi

Rufous-necked Laughingthrush　**LC**

L: 22–27 cm
Habitat: Inhabits broadleaf forests, secondary forests, forest edges, scrub and grass, bamboo thickets, tall grass and reedbeds, margins of cultivated fields, and plantations at altitudes of 120–1645 m.
Behavior: Often in pairs or small flocks in breeding season; forms flocks containing up to 30 or more individuals in nonbreeding season. Usually seen on the ground or in low bushes and grass.
Distribution: W Yunnan and SE Tibet.
Voice: Song consists of fairly quickly repeated, jolly, whistled phrases with a scratchy, husky quality, *wiwi-wi-whu … whi-yi-ha*. Calls include a high-pitched chattering, a shrill *krrrrrrrrttk*, and other notes reminiscent of Striated Bulbul.

Pterorhinus perspicillatus

黑脸噪鹛 hēi liǎn zào méi

Masked Laughingthrush　**LC**

L: 26–32 cm
Habitat: Inhabits green spaces, scrubland, secondary forests, bamboo thickets, and farmland on plains or low-altitude hills, often close to villages.
Behavior: Often in pairs or flocks, hopping and weaving among trees and bushes. Generally does not fly long distances; flight clumsy.
Distribution: Common resident. Occurs in a wide area from S Gansu and S Shaanxi in the north, west to S Yunnan, east to W Shanghai, and south to Guangdong.
Voice: Song a scolding *jhew* or *jhow* note, repeated in series and often accompanied by harsh chattering.

Pterorhinus sannio

白颊噪鹛 bái jiá zào méi

White-browed Laughingthrush　**LC**

L: 22–24 cm
Habitat: Inhabits scrubland, grassy areas, secondary forests, and bamboo thickets at altitudes of 200–1800 m; also occurs in urban parks and green spaces.
Behavior: Often solitary or in pairs; also gathers in small flocks in nonbreeding season. Less shy than other laughingthrushes. Usually forages in understory or on ground by turning over fallen leaves.
Distribution: *P. s. oblectans* occurs in central and southern China, from SE Gansu and S Shaanxi in the north, south to SW Sichuan, and east to W Hubei and N Guizhou; *P. s. sannio* occurs mainly in southeastern China, from E Yunnan and Guizhou to W Zhejiang and Fujian.
Voice: Harsh, shrill, explosive *tcheu*, often repeated; also a harsh, chattering *tcheurrrr*.

Pterorhinus chinensis

黑喉噪鹛 hēi hóu zào méi

Black-throated Laughingthrush　**LC**

L: 23–30 cm
Habitat: Inhabits evergreen broadleaf forests, mixed evergreen-deciduous broadleaf forests, secondary forests, scrubland, and grasslands below 1525 m.
Behavior: Found in pairs or small groups; often mixes with other laughingthrushes. Forages in lower levels of forests and scrub.
Distribution: *P. c. chinensis* in SE Yunnan, Guangxi to C Guangdong; *P. c. lochmius* in SW Yunnan. Locally common.
Voice: Song repetitive, loud, rich, fluty, and thrushlike but with occasional harsher scolding notes and, unusually, a whinny. Call is a low rapid *how*.

Pterorhinus monachus

海南噪鹛 hǎi nán zào méi

Hainan Laughingthrush　**NT**

L: 23–25 cm
Habitat: Inhabits primary forests, secondary forests, scrubland, and grassy fields under 1500 m.
Behavior: Similar to Black-throated Laughingthrush.
Distribution: Chinese endemic; found only in Hainan. Locally common. Sometimes treated as a subspecies of Black-throated Laughingthrush.
Voice: Vocalizations much like Black-throated Laughingthrush.

Blue-crowned Laughingthrush

Bluish gray from crown to hindneck

Bright yellow throat

Yellow lower belly tinged gray

White-browed Laughingthrush

Small crest

White or creamy white from supercilium and lores to cheek

Orange undertail coverts

Rufous-necked Laughingthrush

Rufous chestnut on side of neck

Brown overall

Black face

Black-throated Laughingthrush

Slate-gray crown, some white feathers on forehead

White ear coverts

Slate-gray upper back, neck, breast, flanks, and belly

Pale brown wings

Black around eyes and on forehead, throat, and upper breast

P. c. chinensis

Masked Laughingthrush

Black forehead and mask

Hainan Laughingthrush

Dark brown overall

Black forehead, face, chin, throat, and central breast

Blackish tail tip

Pterorhinus pectoralis

黑领噪鹛 hēi lǐng zào méi

Greater Necklaced Laughingthrush

L: 26.5–34.5 cm
Habitat: Inhabits
evergreen broadleaf
forests, mixed forests,
secondary forests,
scrubland, bamboo
thickets, and artificial
forests below 1830 m.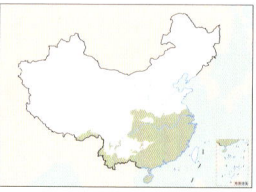
Behavior: Gregarious;
often mixes with other laughingthrushes, Common Green-Magpie, or
large drongos. Usually feeds on insects and fruits on the forest floor;
sometimes moves to middle story of forest.
Distribution: Locally common. *P. p. pectoralis* in W Yunnan. *P. p. robini* in
S Yunnan. *P. p. picticollis* has a wide range, from E Gansu, E Chongqing,
and C and S Shaanxi in the west, to Hubei, S Jiangsu, and NE Zhejiang in
the east, and C Guizhou, Guangxi, and Guangdong in the south.
P. p. semitorquatus in Hainan.
Voice: Huge, complex vocabulary includes unpleasant, combative,
staccato, repeated spluttering, scolding, chattering, and strained
nasal notes; a ringing *ti-didd-tiddid*; and a nervous *weear*. Many notes
confusingly similar to those of Lesser Necklaced Laughingthrush.

Pterorhinus davidi

山噪鹛 shān zào méi

Pere David's Laughingthrush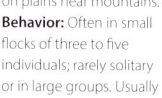

L: 22–27 cm
Habitat: Inhabits
scrubland and thickets in
mountainous areas and
on plains near mountains.
Behavior: Often in small
flocks of three to five
individuals; rarely solitary
or in large groups. Usually
hops around on the ground or in bushes. Active, inquisitive, unafraid
of people.
Distribution: *P. d. davidi* in Northeast and North China, E Gansu, and
P. d. concolor in S Shaanxi, Sichuan, and NE Chongqing.
Voice: Song a loud, varied, usually husky, sometimes melodious, simple,
whistled warble, *wiu-wuyu-weeaow … wuyu-wyu*, and a more complex,
higher-pitched, whistled *twi-wit-tiuu-tiu-wit-wit-tweeit*. Calls a piping
tiu-tiu-tiu-tiu and a scolding, shrike-like *charrr*.

Pterorhinus koslowi

棕草鹛 zōng cǎo méi

Tibetan Babax

L: 28–30 cm
Habitat: Inhabits
high-altitude coniferous
forest edges, scrubland,
farmland, and wooded
areas in gullies.
Behavior: Often moves
about in small flocks in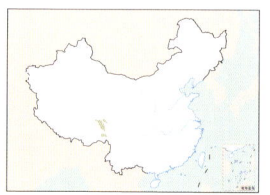
bushes and on ground.
Secretive and not noisy.

Pterorhinus lanceolatus

矛纹草鹛 máo wén cǎo méi

Chinese Babax

L: 25–29 cm
Habitat: Inhabits
broadleaf forests, mixed
broadleaf-coniferous
forests, coniferous forests,
and alpine scrubland in
mountainous areas from
low to high altitudes.
Behavior: Often moves
about in small flocks at forest edges and in understory; also occurs in
villages, orchards, and bushes around farmland.
Distribution: *P. l. bonvaloti* in E Tibet, N Yunnan, and W Sichuan; *P. l. lance-
olatus* in Central and southwestern China; *P. l. latouchei* in southeastern,
South, and eastern southwestern China, except for Taiwan and Hainan.
Voice: Variable song is often a slightly husky, often-melancholy, slowly
delivered series of three or four lengthy, flat whistles, *wee-weeeee-whi-
wooor*, lasting about 2.5 sec and repeated every 8–20 sec, with the first
note rising initially but then flat and the overall strophe descending. Calls
include a repeated series of *pweeu* notes that the female uses to accom-
pany the male's song, and unpleasant, jarring, or scolding notes, *twerrt*,
often in a lengthy, intense sequence or combined with shrill whistles.

Pterorhinus waddelli

大草鹛 dà cǎo méi

Giant Babax

L: 31–34 cm
Habitat: Inhabits
woodlands, scrubland,
and small trees at high
altitudes.
Behavior: Often in small
flocks in bottom layer or
on ground in woodlands
and scrubland; also found
in urban parks and villages.
Distribution: *P. w. jomo* in S Tibet; *P. w. waddelli* in SE Tibet.
Voice: Harsh, loud song a rapid series of galloping notes, *pweer …
ich-pweer … ich-pweer … ich-pweer … ich-pweer*, that lasts 2 sec and is
repeated every 1–2.5 sec in an intensifying sequence. Calls include a
loud, explosive, nasal chattering, *put-pu-dut-purrr-chutttt*, and intense
scolding, *ti-ti-ti-de-urrrr-chch-ch*.

Distribution: Endemic to the eastern edge of the Qinghai-Tibet Plateau;
P. k. yuquensis occurs in SE Tibet; *P. k. koslowi* occurs in S Qinghai and
NE Tibet.
Voice: Song includes falconlike *kyar-kyar-kyar* or *tyut-tyut-tyut*, repeated
rapidly, often 20 or more times, and combined with strained whistles,
chucks, and mimicry of species such as White-backed Thrush. Call a
plaintive, nasal, whining *pwear*, often combined with short chatters and
intense scolding when distressed.

Greater Necklaced Laughingthrush

Unique black and white mottling on cheek

White lores

P. p. picticollis

Dark eye

Chinese Babax

Pale streaks on chestnut-brown upperparts

Dark malar stripe

Lanceolate brown marks on white underparts

Giant Babax

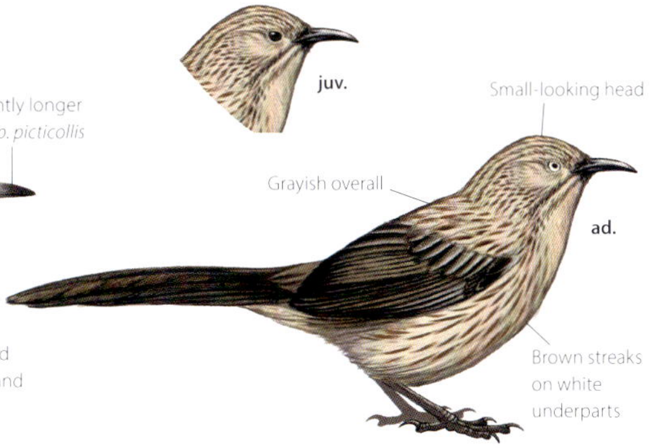

juv.

Small-looking head

Grayish overall

ad.

Brown streaks on white underparts

Bill slightly longer than *P. p. picticollis*

Brownish-yellow neck and back, pale brown flanks

Pale brown body, paler than *P. p. picticollis*

Relatively broad black breast band

P. p. pectoralis

Pere David's Laughingthrush

Pale yellow bill

Well-developed rictal bristles

Tibetan Babax

Inconspicuous pale streaks on back

Brown body

Gray throat contrasts with breast

Chestnut-brown streaks on underparts

465

Pterorhinus albogularis

白喉噪鹛 bái hóu zào méi

White-throated Laughingthrush

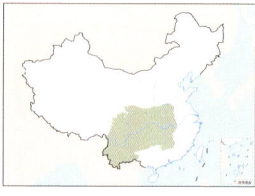

L: 28–30 cm
Habitat: Inhabits evergreen broadleaf forests, deciduous forests, coniferous forests, open secondary forests, and scrubland at altitudes of 300–3800 m; occasionally occurs near farmland. Often breeds above 1200 m; migrates to lower altitudes in bad weather.
Behavior: Often in flocks of 6–15 individuals; also joins mixed-species flocks. Moves about in the lower and middle stories of forests.
Distribution: *P. a. albogularis* in SE Tibet; *P. a. eous* in S Shaanxi, SE Gansu, S Qinghai, Yunnan, N Sichuan, Chongqing, and W Hubei.
Voice: Plaintive, lengthy, descending whistle, *tlliioooo*; thin, shrill, wheezy, asthmatic hisses, *tsuueee*; clicks and intense forced chattering from flocks; *chrrr-chrrr-chrrr* in alarm.

Pterorhinus ruficeps

台湾白喉噪鹛 tái wān bái hóu zào méi

Rufous-crowned Laughingthrush

L: 27–29 cm
Habitat: Inhabits primary forests of oak, fir, and cedar, open secondary forests, and scrubland at altitudes of 850–2300 m; occasionally occurs near farmland.
Behavior: In pairs or flocks; sometimes mixes with Eurasian Jay. Often forages below middle story of forest.
Distribution: Chinese endemic; locally common in Taiwan.
Voice: Very similar to White-throated Laughingthrush.

Pterorhinus berthemyi

棕噪鹛 zōng zào méi

Buffy Laughingthrush

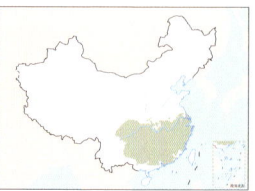

L: 27–29 cm
Habitat: Inhabits evergreen broadleaf forests and bamboo thickets at altitudes of 600–1800 m.
Behavior: Gathers in pairs or small flocks in forest understory; often forages on the ground. Sometimes also moves up to middle story of forest.
Distribution: Chinese endemic; distributed mainly in southeastern China.
Voice: Pleasant, musical, loud song of variable whistled notes, *tu-witi-wittu-titu*. Calls include a sputtering, rising, mellow *ptrrrrrrt*, a repeated *twee-du*, a tremulous *brrrrrrrrrrree*, penetrating short whistles, *pwe-pwe-pwe*, and mellow churring. Sometimes mimics other species, such as Crested Serpent-Eagle.

Pterorhinus poecilorhynchus

台湾棕噪鹛 tái wān zōng zào méi

Rusty Laughingthrush

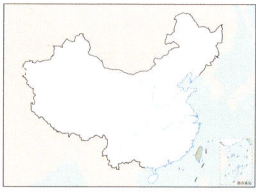

L: 27–29 cm
Habitat: Inhabits evergreen or deciduous broadleaf forests, mixed forests, coniferous forests, and bamboo thickets at altitudes of 340–2100 m.
Behavior: Often moves about in small flocks in the understory or lower vegetation; rarely in the canopy.
Distribution: Chinese endemic; restricted to Taiwan.
Voice: Differences from Buffy Laughingthrush unknown.

Pterorhinus caerulatus

灰胁噪鹛 huī xié zào méi

Gray-sided Laughingthrush

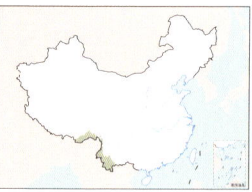

L: 27–29 cm
Habitat: Inhabits bamboo thickets and bushes in evergreen broadleaf forests; sometimes in undergrowth of pine forests. Occasionally seen in mixed forests of trees and rhododendron thickets. Altitude range: 600–2745 m.
Behavior: Often in flocks during nonbreeding season. Forages in low bushes and on the ground, occasionally in tall trees.
Distribution: *P. c. caerulatus* in SE Tibet; *P. c. latifrons* in NW Yunnan; *P. c. kaurensis* in W Yunnan.
Voice: Song a clear, loud, airy, deliberately spaced, highly variable, whistled phrase, such as *tiuu-we-uu-t-tiuoo-weooo*, and occasional mimicry of species such as Crested Serpent-Eagle. Calls include a strained *fweeer-fweeer-fweeer*, often preceded by a scolding oriole-like hiss, and a tremulous, jarring *pwerrrrrt*, often accompanied by descending whistles.

White-throated Laughingthrush

Brownish yellow on forehead and from lower belly to undertail coverts

Pure white throat to breast

P. a. eous

White tip on tail

Brown body, darker than *P. a. eous*

White on throat does not extend to breast

P. a. albogularis

Buffy Laughingthrush

Light blue bare skin around eyes

Brownish-yellow upperparts

Rufous-brown wings and tail feathers

White undertail coverts

Pale gray from lower breast to belly

Rufous-crowned Laughingthrush

Resembles White-throated Laughingthrush but with bright orange from forehead to hindneck

Dark iris

Larger area of white on tail tip

Some individuals have relatively large area of white on belly and distinct dark brown breast band

Rusty Laughingthrush

Rufous-brown head and back

Blue bare skin around eyes

White undertail coverts

Dark gray belly

Gray-sided Laughingthrush

Blue bare skin around eyes

Brown upperparts

White underparts, gray flanks

莺鹛科　Sylviidae

Small forest-dwelling songbirds. Plumage similar between sexes in most species, mostly black, gray, white, and rufous. Bill thin, pointed, and moderately strong. Wings short and rounded. Tail medium to moderately long, squared or wedge shaped. Inhabit primarily scrubland, bamboo forests, or forest edges on plains and mountains. Nest among branches. Feed mostly on insects and insect larvae; also take plant seeds and fruits. Most species are nonmigratory.

Once was a large family that included around 400 warbler-type species, most of which now belong to a handful of newly recognized families. The Chinese name of the family was also revised to indicate a closer relationship between its remaining members and babblers, as suggested by recent phylogenetic studies. Two genera and 34 species recognized worldwide. Two genera and eight species recorded in China, found in the arid and semiarid regions of northwestern China, accidentally in South and southwestern China.

林莺属
Sylvia

Eurasian Blackcap

白喉林莺属
Curruca

Lesser Whitethroat

Sylvia atricapilla

黑顶林莺 hēi dǐng lín yīng

Eurasian Blackcap　LC

L: 13.5–15 cm

Habitat: Inhabits cypress and elderberry thickets in urban parks in winter; occurs on desert scrubland in fall.

Behavior: Often moves about in the upper and middle stories of trees and bushes; rarely comes down to the ground. According to the

observation of wintering birds in Kashgar, Xinjiang, the species inhabits cypress thickets in urban parks and is secretive, feeding on elderberry berries.

Distribution: Only two wintering records from SW Xinjiang (downtown Kashgar) and one fall record from N Xinjiang.

Voice: Unlikely to be heard in China, song is a loud, hurried, pleasant warble. Call is a hard Lesser Whitethroat–like *tek*.

Eurasian Blackcap

Rufous-brown crown

Pale brown underparts

Grayish-brown upper tail, outer feathers lack white

White vent contrasts strongly with gray underparts

Black crown

Slim bill

White lower half of eye ring

Curruca crassirostris

东歌林莺 dōng gē lín yīng

Eastern Orphean Warbler LC

L: 15–17 cm
Habitat: Inhabits open bush-covered woodlands in foothills and low mountains; also occurs in orchards.
Behavior: Often moves among bushes and thickets; occasionally forages on the ground.
Distribution: Recorded only once, in W Tibet (Puran County of Ali Region) in June 2015.
Voice: Unlikely to be heard in China, song is loud and melodic, with many full and low notes similar to those of the Common Nightingale. Call is a *tek*, very similar to that of Eurasian Blackcap.

Curruca nisoria

横斑林莺 héng bān lín yīng

Barred Warbler LC

L: 15.5–17 cm
Habitat: Inhabits artificial forests, farmland shelterbelts, river-valley bushes, urban and rural gardens, and desert oases; most fond of Russian olive forests.
Behavior: Often in pairs. Highly active but extremely vigilant and careful. Moves about and forages mostly in dense canopy; sometimes sings from bushtops or trees.
Distribution: Uncommon to common breeder in Xinjiang and W Gansu; more common in N Xinjiang. Vagrant to Hebei (Tangshan).
Voice: Song a vigorous short warble, like Greater Whitethroat. Call a lengthy throaty rattle that lasts 1–2 sec and slows, *trrrrt-t-t-t*.

Eastern Orphean Warbler

Black crown and ear coverts

Thick stout bill

Pale iris

Black primaries

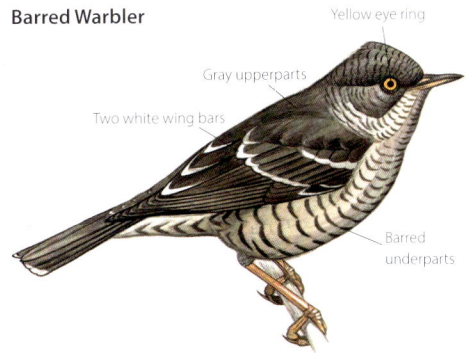

Barred Warbler

Yellow eye ring

Gray upperparts

Two white wing bars

Barred underparts

Curruca minula

漠白喉林莺（沙白喉林莺）
mò bái hóu lín yīng (shā bái hóu lín yīng)

Desert Whitethroat

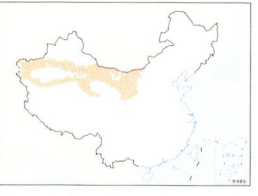

L: 11.5–12.5 cm
Habitat: Adapted to arid and desolate desert environments. Inhabits arid deserts and semidesert areas with scattered bushes and other vegetation; also occurs on premontane desert slopes, lakeside and riverside reeds, dune willows, other stunted low thickets, and in oases and farmland in deserts.
Behavior: More active than other whitethroats. Often twitches its tail and hops among branches and leaves of bushes throughout the day. Also hops and runs on the ground and flies short distances; restless.
Distribution: *C. m. minula* is a locally common breeder in Xinjiang; *C. m. margelanica* is an uncommon breeder in W Inner Mongolia, Ningxia, NW Gansu, and E Qinghai.
Voice: Song is a pleasant, varied trill, lacking the droning rattle of Lesser Whitethroat. Call is a fast scolding trill or churr, *che-che-che-che*.

Curruca curruca

白喉林莺 bái hóu lín yīng

Lesser Whitethroat

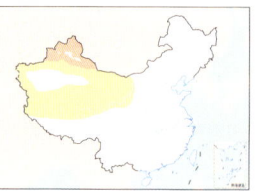

L: 12.5–14 cm
Habitat: Inhabits bushes and grass in plains, lakes, rivers, reed ponds, deserts, and semideserts.
Behavior: Often solitary or in pairs. Active; often hopping among bushes and branches.
Distribution: Seen in Beijing, Tianjin, Hebei, Henan, Shaanxi, Shanxi, Inner Mongolia, Ningxia, Gansu, Xinjiang, Tibet, Qinghai, and Shanghai during migration.
Voice: Song starts with a soft, pleasant trill, develops into an intense rattle, *chikka-chikka-chikka*, then repeats. Call is a hoarse *chek*.

Curruca althaea

休氏白喉林莺 xiū shì bái hóu lín yīng

Hume's Whitethroat

L: 11.5–13.5 cm
Habitat: Inhabits sea buckthorn scrubland in river valleys and premontane areas at altitudes of 2400–3600 m.
Behavior: Often in pairs. Highly mobile; often hops up and down in sea buckthorn bushes, constantly moving and foraging.
Distribution: Common breeder in the mountains of SW Xinjiang.
Voice: Song is very similar to Lesser Whitethroat and usually lacks a final rattle.

Curruca nana

荒漠林莺（亚洲漠地林莺）
huāng mò lín yīng (yà zhōu mò dì lín yīng)

Asian Desert Warbler

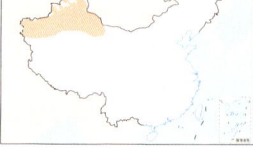

L: 11.5–12.5 cm
Habitat: Inhabits mainly gravelly deserts with low saxaul trees; also found in grassy areas, tamarisk bushes, and saxaul forests in the extremely arid desert hinterland.
Behavior: Often solitary or in pairs. Highly mobile; often hops up and down among the lower branches of bushes, constantly moving and foraging. Weak flier; generally makes short, low flights between bushes.
Distribution: Uncommon breeder throughout Xinjiang and W Inner Mongolia.
Voice: Short song commences with a brief rattle, followed by a rapidly repeated warble that descends at the end and terminates with an up-slurred *krrr-titititititi-teu*. Calls a lengthy churr.

Curruca communis

灰白喉林莺 huī bái hóu lín yīng

Greater Whitethroat

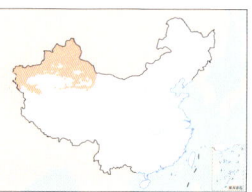

L: 13–15 cm
Habitat: Inhabits bushes in open areas, such as barren slopes, forest edges, streams, and lakeshores, on plains and mountains, even in subalpine areas at altitudes of about 2000 m. Also occurs in shelterbelts, roadside groves, farmland, orchards, and small thickets in parks.
Behavior: Often hops about and forages among the branches and leaves of bushes; also forages on the ground or on the wing. In breeding season, male often sings from top branches of a bush or tree; sometimes dashes into the air before flying back to its original position.
Distribution: *C. c. rubicola* is a locally common breeder in N Xinjiang; also in W Inner Mongolia. *C. c. icterops* is an uncommon breeder in S Xinjiang.
Voice: Short, fast, dry, somewhat scratchy, jerky, or jolting, similar to song of Barred Warbler. Call a dry, nasal *churr*.

Desert Whitethroat

Pale gray ear coverts contrast little with crown

Sandy grayish-brown upperparts

Asian Desert Warbler

Yellow iris

Complete white eye ring

Rufous-brown upperwing coverts

Brick-red rump and uppertail coverts

Lesser Whitethroat

Grayish-black ear coverts contrast with dark gray crown

Grayish-brown back

Greater Whitethroat

Dark gray crown

Rufous-brown iris

Complete white eye ring

Brown upperwing coverts

Hume's Whitethroat

Relatively stout bill

Dark gray ear coverts contrast little with crown

Gray back lacks warm brown

鸦雀科　Paradoxornithidae

Small forest-dwelling songbirds. Plumage similar between sexes in most species. Bill thin and pointed, or thick and strong, resembling bill of parrots. Inhabit primarily forests, understory bushes, and rhododendron thickets; some species heavily reliant on bamboo forests or reedbeds. Feed mostly on insects and insect larvae; also take plant nectar, seeds, and stems. Most species are resident; some are seasonally nomadic.

　　Parrotbills represent a unique group of species. Once placed in Paridae, Timaliidae, or Sylviidae, they now constitute the newly recognized family together with myzornis, hill babblers, and their allies. Sixteen genera and 37 species recognized worldwide. Except for Wrentit of North America, all species belong to the Old World, and mostly to Indomalaya. Fifteen genera and 33 species recorded in China, found mostly in South and southwestern China, with a few species extending to northern China, the east coast, and Taiwan.

绿鹛属
Myzornis

Fire-tailed Myzornis

金胸雀鹛属
Lioparus

Golden-breasted Fulvetta

宝兴鹛雀属
Moupinia

Rufous-tailed Babbler

褐鹛属
Fulvetta

Spectacled Fulvetta

金眼鹛雀属
Chrysomma

Yellow-eyed Babbler

山鹛属
Rhopophilus

Beijing Babbler

红嘴鸦雀属
Conostoma

Great Parrotbill

棕头鸦雀属
Sinosuthora

Vinous-throated Parrotbill

短尾鸦雀属
Neosuthora

Short-tailed Parrotbill

白胸鸦雀属
Psittiparus

Gray-headed Parrotbill

震旦鸦雀属
Calamornis

Reed Parrotbill

褐鸦雀属
Cholornis

Three-toed Parrotbill

金色鸦雀属
Suthora

Golden Parrotbill

黑眉鸦雀属
Chleuasicus

Pale-billed Parrotbill

鸦雀属
Paradoxornis

Spot-breasted Parrotbill

473

Myzornis pyrrhoura

火尾绿鹛 huǒ wěi lǜ méi

Fire-tailed Myzornis LC

L: 11–13.5 cm
Habitat: Inhabits primary forests, bamboo thickets, and rhododendron thickets in alpine and subalpine areas.
Behavior: Solitary or in pairs in bushes and small trees in the lower and middle stories of high-altitude montane forests; also gathers in small flocks in winter. Prefers feeding on nectar.
Distribution: Seen in SE Tibet, NW Yunnan, and W Sichuan.
Voice: Contact call is extremely thin, high-pitched, often rapidly repeated Goldcrest-like notes, *si-si-si*. Alarm call is three to eight low-pitched, strong, repeated trills, *cicici … cicicicici*. Rarely heard complex song is a combination of contact and alarm calls, *si-si-si-ssi-ssi-cicici-cicici-cicicici*.

Lioparus chrysotis

金胸雀鹛 jīn xiōng què méi

Golden-breasted Fulvetta LC

L: 10–11 cm
Habitat: Inhabits broadleaf, mixed broadleaf-coniferous, and coniferous forests at middle and high altitudes; also occurs at forest edges and understory bushes.
Behavior: Moves about in small flocks in nonbreeding season; joins other small babblers, fulvettas, and parrotbills in mixed-species flocks.
Distribution: *L. c. forresti* in NW Yunnan and SE Tibet; *L. c. amoenus* in SE Yunnan; *L. c. swinhoii* in southwestern China, western Central China, and western South China, except Yunnan.
Voice: Flocks can be noisy. Song a rapid, very thin, high-pitched, multinote, buzzy chittering, *si-si-si-see-s-si-sut*. Calls a fast spluttering rattle, *wrrrrrt* or *kwititit*, and a mellow *tut-tut*.

Moupinia poecilotis

宝兴鹛雀 bǎo xīng méi què

Rufous-tailed Babbler LC

L: 13–15 cm
Habitat: Inhabits the middle and lower layers of forest edges in broadleaf, mixed broadleaf-coniferous, and coniferous forests and alpine scrubland, at middle and high altitudes.
Behavior: Often solitary or in pairs in the forest understory; occasionally joins mixed-species flocks.
Distribution: Endemic to the Hengduan Mountains of southwestern China, from NW Yunnan to SW Sichuan.
Voice: Song a high-pitched, plaintive, thin whistle, *tuee-tweeip* or *u-wippy-eeet*, repeated every 3–5 sec and occasionally introduced by a harder, sparrowlike *trrt*. Calls include a rollicking *pu-whippy-t … pu-whippy-t … pu-whippy-t*, with the final note often barely audible, repeated very rapidly, and an intense churring when agitated.

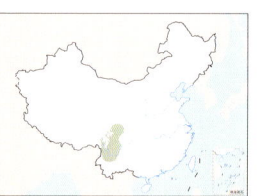

Fulvetta vinipectus

白眉雀鹛 bái méi què méi

White-browed Fulvetta LC

L: 11–12 cm
Habitat: Inhabits broadleaf forests, mixed forests, coniferous forests, alpine rhododendron thickets, and forest-edge bushes at middle and high altitudes.
Behavior: Often in small flocks in the lower and middle stories of forests and understory bushes; also joins other small birds in mixed-species flocks. Active and unafraid of people.
Distribution: *F. v. vinipectus* and *F. v. chumbiensis* in S Tibet; *F. v. perstriata* in NW Yunnan and SE Tibet; *F. v. bieti* in Sichuan and Yunnan.
Voice: Song an easily overlooked, slightly buzzy, tremulous, timid *di-di-di-di-du*. Flocks make soft chattering and high-pitched *seep* or *sip* notes.

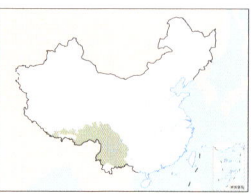

Fulvetta striaticollis

中华雀鹛（高山雀鹛）

zhōng huá què méi (gāo shān què méi)

Chinese Fulvetta LC

L: 12–14 cm
Habitat: Inhabits alpine coniferous forests, rhododendron thickets, and small trees.
Behavior: Often in pairs or small flocks in bushes and rhododendron thickets under alpine coniferous forests. Active.
Distribution: Endemic to midwestern China; found in SE Qinghai, S Gansu, W Sichuan, NW Yunnan, and E Tibet.
Voice: Song a simple high-pitched whistle, *pi-pi-pi tseu*, *tisi-tis-tis-di-deuw*, or *tsi chuu*. Calls include short mellow rattles, *trrrt-t*, and timid, quivering, or tremulous *dwueeer*.

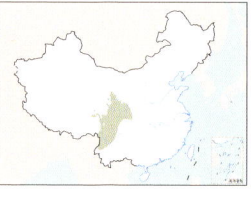

Fulvetta ruficapilla

棕头雀鹛 zōng tóu què méi

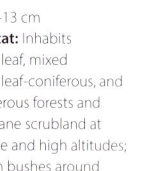

Spectacled Fulvetta LC

L: 10–13 cm
Habitat: Inhabits broadleaf, mixed broadleaf-coniferous, and coniferous forests and montane scrubland at middle and high altitudes; also in bushes around farmland and villages.
Behavior: In small flocks in forest-edge bushes and bamboo thickets.
Distribution: Endemic to midwestern China. *F. r. ruficapilla* in C Gansu, S Shaanxi, Chongqing, Sichuan, and Hubei; *F. r. sordidior* in NW Guizhou and SW to C Yunnan.
Voice: *F. r. ruficapilla* gives a plaintive, two- or three-note whistle, *wuuuu-tiuuu*, each note lasting about 0.3 sec, with the first note flat and the second note descending (2.4–4.6 kHz); a trisyllabic version, *tu-ti-wuuu*; and very intense chattering. The song of *F. r. sordidior* is a very fast, intense, slightly rising, quarrelsome, Eurasian Nuthatch–like trill that lasts just over a second but includes more than 20 notes. Calls include mellow churrs and nasal chattering.

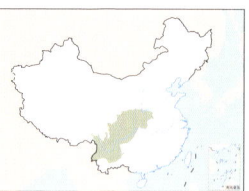

Fire-tailed Myzornis

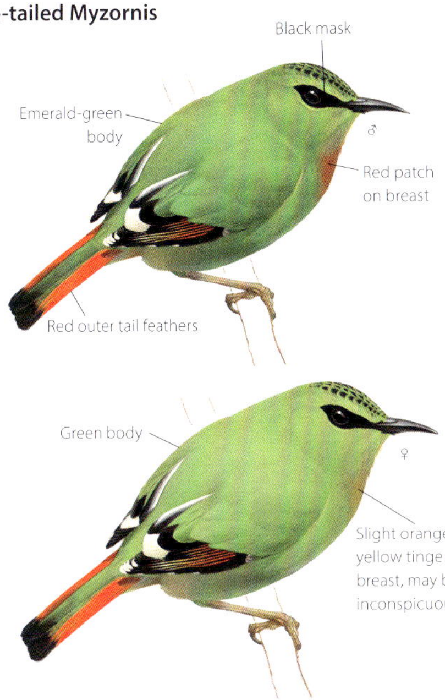

Black mask

Emerald-green body

Red patch on breast

♂

Red outer tail feathers

Green body

♀

Slight orange or yellow tinge on breast, may be inconspicuous

White-browed Fulvetta

Dark brown ear coverts

Black ear coverts

Whitish throat and upper breast

F. v. vinipectus

F. v. bieti

Distinct white supercilium and black lateral crown stripe

Dark brown ear coverts

Dark streaks on throat and upper breast

F. v. chumbiensis

Golden-breasted Fulvetta

Yellow eye ring

L. c. amoenus

Black head, white cheek and crown

L. c. swinhoii

Orange breast, belly, and wings

Chinese Fulvetta

Pale yellow iris

Pale streaks on pure brown head

Bluish-gray outer vanes and black inner vanes on wing feathers

Spectacled Fulvetta

Indistinct supercilium only above eye

Narrow black stripes on side of rufous-brown crown

White eye ring

F. r. ruficapilla

Rufous-tailed Babbler

Short bill with decurved upper mandible

Gray supercilium

Mottled pattern on face

White throat

Rounded body; long tail resembles prinia

Fulvetta danisi

印支雀鹛 yìn zhī què méi

Indochinese Fulvetta

L: 10–13 cm
Habitat: Inhabits evergreen broadleaf and mixed broadleaf-coniferous forests, as well as forest-edge bushes at middle altitudes.
Behavior: In pairs or small flocks in bushes at the forest edge. Secretive but not shy.
Distribution: SE Yunnan and SW Guizhou.
Voice: Song a hoarse, short, variable *cu–ce–e–e–a*.

Fulvetta ludlowi

路氏雀鹛 (路德雀鹛)

lù shì què méi (lù dé què méi)

Brown-throated Fulvetta

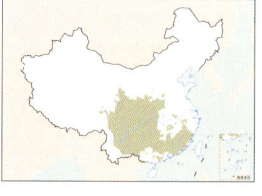

L: 11–12 cm
Habitat: Inhabits rhododendron thickets and bamboo thickets under coniferous forests at middle and high altitudes.
Behavior: Often in small flocks in thickets and the lower and middle stories of forests. Active, bold.
Distribution: SE Tibet.
Voice: Song is a couple of thin, high-pitched, sibilant notes ending with a churr, *see-see-spir-r-r-rt*.

Fulvetta cinereiceps

灰头雀鹛 huī tóu què méi

Gray-hooded Fulvetta

L: 12–14 cm
Habitat: Inhabits understory bushes and bamboo thickets in broadleaf and fir forests at middle and high altitudes.
Behavior: Noisy. Often in pairs or small flocks in the lower and middle stories of forests.
Distribution: Endemic to mountainous areas in southern China. *F. c. fessa* in southern northwestern China and northern southwestern China; *F. c. cinereiceps* in NE Yunnan, Sichuan, Chongqing, W Guizhou, and W Hubei; *F. c. fucata* in N Guizhou, C Hubei, Hunan, Guangxi, and Jiangxi; *F. c. guttaticollis* in mountainous areas in Fujian and Guangdong.
Voice: Song a thin, high-pitched, breathless, two- or three-note series of flat whistles, *tiii wiiurr* or *ti ti wiiuurr*, usually with a slightly burry tremolo toward the end of the descending last note. Calls include sharp, thin, churrs and *pik* notes.

Fulvetta manipurensis

褐头雀鹛 hè tóu què méi

Streak-throated Fulvetta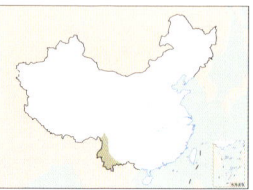

L: 12–14 cm
Habitat: Inhabits the understory bushes, rhododendron thickets, and bamboo thickets in mixed broadleaf-coniferous and coniferous forests at middle and high altitudes.
Behavior: Often in small flocks in the lower and middle stories of forests and on the ground; occasionally joins other small birds in mixed-species flocks.
Distribution: *F. m. manipurensis* in NW and W Yunnan and *F. m. tonkinensis* in C and S Yunnan and Guangxi.
Voice: Song a well-spaced, high-pitched, piping, three-note *ti ti tsew*, similar to the two notes of Gray-hooded Fulvetta but lower pitched and lacking the burry character to the second note. Calls include low churrs, *trrrt*; sharp *pik* notes; and an extended chatter when agitated, *twit'it'it'it-it-it, it, it'it'it*.

Fulvetta formosana

玉山雀鹛 yù shān què méi

Taiwan Fulvetta

L: 12–14 cm
Habitat: Inhabits the edges of broadleaf, mixed broadleaf-coniferous, and coniferous forests and alpine scrubland at middle and high altitudes.
Behavior: Moves about in small flocks in understory bushes; active and easily spotted.
Distribution: Chinese endemic; found in the mountains of C Taiwan.
Voice: Two-note song, *ti tuuu*, very similar to Gray-hooded Fulvetta but higher pitched, with a longer first note, and, when present, the burry tremolo toward the end of the descending second note is more pronounced.

Chrysomma sinense

金眼鹛雀 jīn yǎn méi què

Yellow-eyed Babbler

L: 18–23 cm
Habitat: Inhabits open forest edges, farmland, river valleys, scrubland, and tall grasslands on middle and low mountains.
Behavior: Often in pairs or small flocks in open areas near water; rarely mixes with other birds.
Distribution: Seen in Yunnan, Guizhou, Guangxi, and Guangdong.
Voice: Large vocabulary, with variable song a ringing whistled *wi-chi-chi wu* or *wi-wu-chrieu*, often mixed with intense dry churring, *chrrr-chrrr-chrrr*. Other calls include a spluttering ratchet call and an intense, nasal, Common Tailorbird–like *styyeww*, often combined with a soft *trrrrp*.

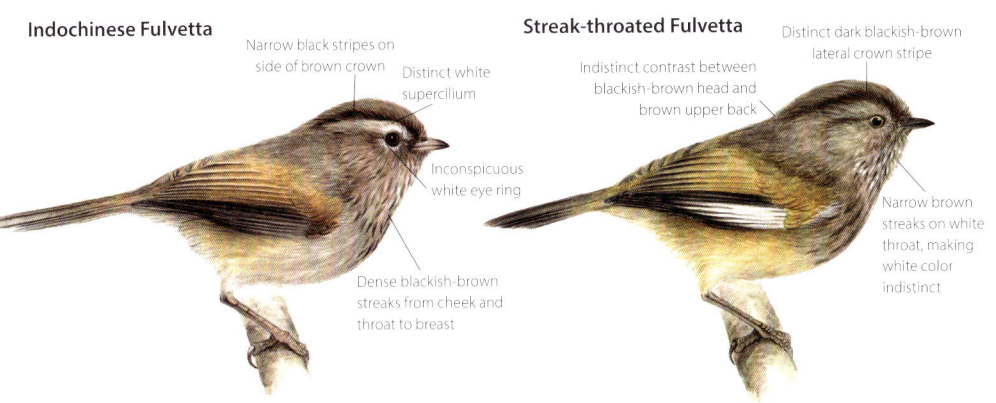

Indochinese Fulvetta

Narrow black stripes on side of brown crown

Distinct white supercilium

Inconspicuous white eye ring

Dense blackish-brown streaks from cheek and throat to breast

Streak-throated Fulvetta

Distinct dark blackish-brown lateral crown stripe

Indistinct contrast between blackish-brown head and brown upper back

Narrow brown streaks on white throat, making white color indistinct

Brown-throated Fulvetta

Chocolate-brown head

Bluish-gray outer vanes and black inner vanes on wing feathers

Broad brown streaks on white throat

Taiwan Fulvetta

Thin dark brown lateral crown stripe

Uniform grayish brown on head and upper back

Distinct white eye ring

Distinct brown streaks on white throat

Gray-hooded Fulvetta

Indistinct pale brown lateral crown stripe extends to forehead

Clear boundary between gray head and chestnut-brown upper back

Narrow brown streaks on white throat

Yellow-eyed Babbler

Yellow iris, orange-red eye ring

Rufous brown from head to upperparts and tail

White lores, cheek, and throat

Rhopophilus pekinensis
山鹛 shān méi
Beijing Babbler　LC

L: 16–18 cm
Habitat: Inhabits scrubland and thickets in mountainous areas and nearby plains.
Behavior: Active. Often gathers in small flocks of two to five individuals. Moves about in bushes and low thickets. Flies short distances across clearings or roads, then quickly hides in vegetation.
Distribution: Found in southern Northeast China, North China, Gansu, S Shaanxi, and E Qinghai.
Voice: Large vocabulary. Complex song is husky, penetrating, and repetitive. Song introduced by a couple of unobtrusive notes; then an extremely rapid series of short notes followed by a variable number of chip-purr combinations, *ti-pi-tu-di-di-di-di-di-di-eh-tip-wurrrrrrr-chip-wurrrrrr … chip-wurrrrrr*, the whole song often lasting for several minutes. Calls include a repeated series of *twik* notes; a *d-dweeerrrr-pweeuuu*, repeated rapidly and at length; and an equally penetrating *dz-ti-tu-weeeeeeeeu*.

Rhopophilus albosuperciliaris
西域山鹛 xī yù shān méi
Tarim Babbler　LC

L: 18 cm
Habitat: Inhabits desert poplar forests, tamarisk thickets, riparian forests, desert highway shelterbelts, reedbeds, and woodlands at oasis edges in plain areas.
Behavior: Often solitary or in pairs. Frequently hops up and down agilely or flies short distances among bushes. Sometimes also runs and hops on the ground.
Distribution: Common to uncommon resident in W Inner Mongolia, NW Gansu, S Xinjiang, and W Qinghai.
Voice: Song and calls mostly very different from those of Beijing Babbler.

Conostoma aemodium
红嘴鸦雀 hóng zuǐ yā què
Great Parrotbill　LC

L: 27–29 cm
Habitat: Inhabits mixed forests, coniferous forests, and rhododendron and bamboo thickets at middle and high altitudes.
Behavior: Often in pairs or small flocks in the lower and middle stories of vegetation; sometimes in mixed-species flocks with laughingthrushes.
Distribution: Found in the mountains from the C Himalayas to southwestern China, including W Hubei, NE Chongqing, S Shaanxi, SW Gansu, N, NE, W, and SW Sichuan, NW Yunnan, and SE Tibet.
Voice: Variable song usually a loud *whip ui-pip-brrahh*, *tuip-pweeoow*, and *tip-tip-bruehh*. Calls include hard *tucks*, growls, nasal wheezes, squeals, churrs, and cackles.

Cholornis paradoxus
三趾鸦雀 sān zhǐ yā què
Three-toed Parrotbill　LC

L: 18–20 cm
Habitat: Inhabits understory bushes and bamboo and rhododendron thickets in montane coniferous forests at middle and high altitudes.
Behavior: Often solitary or in pairs in bushes and bamboo thickets. Bold, inquisitive. Often joins other parrotbills in mixed-species flocks.
Distribution: Endemic to the mountainous areas in central China. *C. p. taipaiensis* in W Hubei and the Qinling Mountains of Shaanxi; *C. p. paradoxus* in N and W Sichuan and S Gansu.
Voice: Songs clear, high, plaintive, and whistled, including *u-plee-wi-tu-woo* and *poo-we-tui-ee*. Most vocalizations lower pitched than those of Brown Parrotbill, but some overlap and many are similar; nevertheless, some, such as Three-toed's rising *pewee-pwee-pwee* and short *tut*, apparently diagnostic. Calls include a plaintive whistled *dwee-dwee-dwee* and raucous, scolding, aggressive chattering. Three-toed's intense growled churrs also less rattling than the equivalent of Brown.

Cholornis unicolor
褐鸦雀 hè yā què
Brown Parrotbill　LC

L: 19–22 cm
Habitat: Inhabits broadleaf, mixed, and coniferous forests and alpine scrubland and bamboo thickets at middle and high altitudes.
Behavior: Often in pairs or small flocks. Active but not agile. Sometimes joins other parrotbills and laughingthrushes in mixed-species flocks.
Distribution: Found from the C Himalayas to southwestern China, including C and W Sichuan, W and NW Yunnan, and SE Tibet.
Voice: See Three-toed Parrotbill.

Neosuthora davidiana
短尾鸦雀 duǎn wěi yā què
Short-tailed Parrotbill　LC

L: 9.5–10 cm
Habitat: Inhabits bamboo thickets, grassy fields, and evergreen broadleaf forest edges from 110 to 1250 m; recorded up to 1830 m in Fujian.
Behavior: Often moves about quickly in flocks in bamboo thickets, foraging for plants and small insect larvae. Sometimes joins Huet's Fulvetta and Rufous-capped Babbler in mixed-species flocks.
Distribution: Uncommon resident. *N. d. davidiana* in S Hunan, N Guangdong, N Fujian, and Zhejiang; *N. d. tonkinensis* in SE Yunnan.
Voice: Song is a very thin, high-pitched series of six to nine rapidly rising notes, *ih-ih-ih-ih-ih-ih*. Flocks chatter and give soft *tip* and *tut* notes.

Beijing Babbler

Rufous-brown tinge
to upper back

Pale iris

Long broad streaks
on relatively dark back

Distinct streaks
on flanks

Three-toed Parrotbill

Distinct white
eye ring

Grayish-brown body

Only three toes

Brown Parrotbill

Narrow black
supercilium

Indistinct pale
eye ring

Brown body

Tarim Babbler

Black iris

Less brown on
upperparts

Sparse chestnut-brown
streaks on flanks

**Short-tailed
Parrotbill**

Rufous brown from head to neck

Gray or grayish-
brown back

Black throat

N. d. davidiana

Grayish white on
breast turns to buff
yellow on belly

Gray on back, breast, and
belly more distinct

N. d. tonkinensis

Great Parrotbill

Grayish-white forehead

Relatively long
bill differs from
parrotlike bill of
other parrotbills

Gray outer vanes
of primaries

Sinosuthora conspicillata
白眶鸦雀 bái kuàng yā què
Spectacled Parrotbill **LC**

L: 12–14 cm
Habitat: Occurs in the understory of montane coniferous forests at middle and high altitudes, also in bamboo thickets and forest-edge bushes on alpine meadows.

Behavior: In pairs or small flocks in bamboo thickets and other habitats; active and rarely mixes with other parrotbills.
Distribution: Endemic to the mountainous areas in central China. *S. c. conspicillata* in S Shaanxi, E and S Chongqing, C, N, and NE Sichuan, S Gansu, and NE Qinghai; *S. c. rocki* in W Hubei.
Voice: Song a plaintive, rising, fine, whistled *tiuu-tiuu-tiuu-tiu*, like a timid Vinous-throated Parrotbill, or a more intense *trriu-trrriu-trrriiu*. Calls include a harsh, slurred *prrrt*, like breaking wind, often combined with the song, *prrrt-zrreu-zree-zreeu*.

Sinosuthora zappeyi
暗色鸦雀 àn sè yā què
Gray-hooded Parrotbill **VU**

L: 12–13 cm
Habitat: Inhabits understory bushes and bamboo thickets in broadleaf, mixed, and coniferous forests at middle and high altitudes.

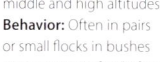
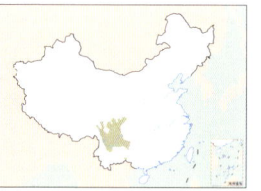

Behavior: Often in pairs or small flocks in bushes near water. Unafraid of people. Relies on bamboo thickets.
Distribution: Endemic to the mountainous areas of southwestern China. *S. z. zappeyi* in C and S Sichuan, NW Guizhou, and NE Yunnan; *S. z. erlangshanica* in SW Sichuan.
Voice: Song a thin, high, piercing, plaintive *si-siuu-siuu-siuu*, reminiscent of Vinous-throated Parrotbill. Calls are harsh, abrupt, scolding, rasping *trr'ik* and *trrrh* notes, similar to the breaking-wind notes of Spectacled Parrotbill.

Sinosuthora webbiana
棕头鸦雀 zōng tóu yā què
Vinous-throated Parrotbill **LC**

L: 11–13 cm
Habitat: Inhabits broadleaf secondary forests, farmland edges, scrubland, reedbeds, and urban parks at middle and low altitudes.
Behavior: Gathers in small to large flocks. Unafraid

of people. Very noisy; often heard before it is seen. Flight low, whirring; often flies short distances.
Distribution: *S. w. mantschurica* in eastern and southern parts of Northeast China; *S. w. fulvicauda* in North China; *S. w. webbiana* in East China; *S. w. elisabethae* in SE Yunnan; *S. w. bulomacha* in Taiwan; *S. w. suffusa* in the Yangtze River Watershed and areas to its south, except for Hainan and Taiwan.
Voice: Song is an aggressive *rit-rit chididi tssu-tssu-tssiu*. Most common call a high piercing *tsiu-tsiu-tsiu-tsiu*. Flocks give subdued rapid chattering, interspersed with occasional thin *tiu-tiu* notes.

Sinosuthora alphonsiana
灰喉鸦雀 huī hóu yā què
Ashy-throated Parrotbill **LC**

L: 11–13 cm
Habitat: Inhabits forest edges, grassy areas, and scrubland at middle and low altitudes; also around farmland, villages, and gardens.

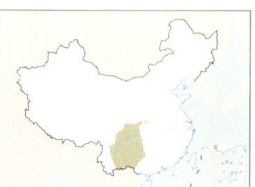
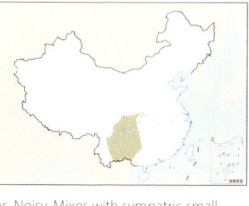

Behavior: Often in small to large flocks in open and scrubby areas; also at forest edges. Noisy. Mixes with sympatric small parrotbills, such as Vinous-throated Parrotbill.
Distribution: *S. a. ganluoensis* in SW Sichuan; *S. a. alphonsiana* in C and S Sichuan and S Chongqing; *S. a. stresemanni* in N and W Guizhou and N Yunnan; *S. a. yunnanensis* in SE and S Yunnan.
Voice: Vocalizations very similar to those of Vinous-throated Parrotbill.

Sinosuthora brunnea
褐翅鸦雀 hè chì yā què
Brown-winged Parrotbill **LC**

L: 11–13 cm
Habitat: Inhabits forest-edge bushes, secondary forests, lianas, grassy areas, reeds, and bamboo thickets at middle and low altitudes.
Behavior: In nonbreeding season, moves about

in small to large flocks in forest understory or open bushes. Noisy and mobile. Rarely mixes with other birds.
Distribution: *S. b. styani* in W, C, and NW Yunnan. *S. b. brunnea* in W Yunnan. *S. b. ricketti*, in S and SW Sichuan and NW Yunnan, sometimes treated as a separate species, *S. ricketti*.
Voice: Song unknown. Calls a *pewee-pwee-pwee*, similar to those of Vinous-throated Parrotbill, and harsh scolding chatter, *tzu–cu–tzu–cu* (*S. b. brunnea*) and *tzu–tzu* (*S. b. ricketti*).

Sinosuthora przewalskii
灰冠鸦雀 huī guān yā què
Rusty-throated Parrotbill **VU**

L: 13–15 cm
Habitat: Inhabits bamboo thickets and alpine scrub under coniferous and mixed broadleaf-coniferous forests at middle and high altitudes.
Behavior: Found in small flocks in alpine bamboo

thickets. Highly mobile. Unafraid of people.
Distribution: Endemic to mountainous areas of central China; found in SE Gansu and N Sichuan.
Voice: Sings a soft insectlike *tzr–cu–tzr–cu–tzr–cu*. Calls include a harsh spluttering *chrr-rr-rr-rr*, a sharp, abrupt, shrill, Meadow Pipit–like *tsip*, and a breaking wind *prrt*, often in series.

Spectacled Parrotbill

Distinct white eye ring

Ashy-throated Parrotbill

Rufous-chestnut crown

Dark gray cheek

Pale iris

Dark gray from breast to lower belly

S. a. alphonsiana

Chestnut-tinged cheek

S. a. ganluoensis

Gray-hooded Parrotbill

Gray head, neck, and breast

Crest often erect

White eye ring

Brown-winged Parrotbill

Rufous-chestnut head

Wings blackish brown, not rufous brown

Yellow bill with dark culmen

Rufous-chestnut streaks on white throat

S. b. brunnea

Vinous-throated Parrotbill

Brown on crown relatively dark

Black iris

Pinkish-red throat

Relatively dark breast

Wings fringed rufous brown

Pale brown crown

Relatively pale breast

S. w. webbiana

Rufous chestnut relatively dark on crown and paler on side of face

Narrow white eye ring

S. b. ricketti

S. w. mantschurica

Rufous-brown crown

Rusty-throated Parrotbill

Gray from crown to hindneck

Long blackish-brown supercilium

Dark rufous-brown lores

Rufous brown around eye and from throat to breast

S. w. suffusa

481

Suthora fulvifrons
黄额鸦雀 huáng é yā què
Fulvous Parrotbill

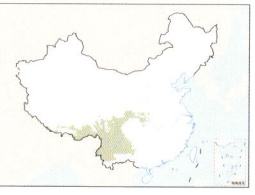

L: 11–12 cm
Habitat: Inhabits understory bushes, bamboo thickets, and rhododendron thickets in broadleaf, mixed, and coniferous forests at middle and high altitudes.
Behavior: In nonbreeding season, moves about in flocks in understory bushes and bamboo thickets. Noisy. Unafraid of people. Seldom mixes with other parrotbills.
Distribution: *S. f. chayulensis* in SE Tibet; *S. f. albifacies* in S Shaanxi, W Hubei, SE Gansu, and W and C Sichuan; *S. f. cyanophrys* in W and NW Yunnan.
Voice: Song a fine, high-pitched, slightly buzzy *tsee-tzeu-tiu, te-ti-ti-dzeu*, and *zee-zee-diu*. Calls include slightly spluttering *trrrip* call notes, often doubled.

Suthora nipalensis
黑喉鸦雀 (橙额鸦雀)
hēi hóu yā què (chéng é yā què)
Black-throated Parrotbill

L: 9–11 cm
Habitat: Inhabits bamboo thickets and understory bushes in broadleaf and mixed forests at middle altitudes.
Behavior: Moves about in small to large flocks in bamboo thickets. Active. Often mixes with small babblers.
Distribution: *S. n. poliotis* in SE Tibet and NW Yunnan; *S. n. beaulieui* in S Yunnan; *S. n. crocotia* in SE Tibet; *S. n. nipalensis* might occur in S Tibet.
Voice: Song is fine, even, wispy, very high-pitched *tu-su-s-su*, with various buzzing and chittering as well as harsh churrs, *trr-ti-trit*, most indistinguishable from Golden Parrotbill.

Suthora verreauxi
金色鸦雀 jīn sè yā què
Golden Parrotbill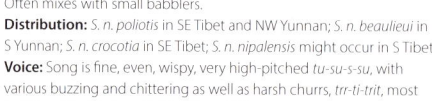

L: 9–11 cm
Habitat: Inhabits bamboo thickets and understory bushes in broadleaf and mixed forests at middle and low altitudes.
Behavior: In winter, moves about in large flocks in forest understory. Noisy and easy to see. Often mixes with small babblers.
Distribution: Resident in montane areas in central and southern China. *S. v. verreauxi* in NE Yunnan, Sichuan, Chongqing, S Shaanxi, and W Hubei; *S. v. craddocki* in S Yunnan, Guangxi, and Hunan; *S. v. pallida* in Guizhou, N Guangdong, NE Jiangxi, and NW Fujian; *S. v. morrisoniana* in Taiwan.
Voice: See Black-throated Parrotbill.

Chleuasicus atrosuperciliaris
黑眉鸦雀 hēi méi yā què
Pale-billed Parrotbill

L: 14–15 cm
Habitat: Inhabits bushes, bamboo thickets, and grassy areas in the understory and edges of broadleaf forests at middle and low altitudes.
Behavior: Often lives in small flocks in open areas at forest edges and in the understory, sometimes in mixed-species flocks with other small birds.
Distribution: SE Tibet and W Yunnan.
Voice: Simple song is a series of intense, abrupt notes, *tzik-tzik-tzik*, often repeated rapidly. Calls include more nasal, plaintive versions of the song.

Psittiparus ruficeps
白胸鸦雀 bái xiōng yā què
White-breasted Parrotbill

L: 16–17 cm
Habitat: Inhabits forest edges, scrubland, bamboo thickets, and tall grasslands at middle and low altitudes in mountainous areas.
Behavior: Moves about in flocks. Noisy and unafraid of people.
Distribution: SE Tibet.
Voice: Song a loud, high, slightly descending series of four to six whistled notes, *kk-twe-twe-twe-ti-tiew* or *wi-wi-twe-we-we-wiu*. Calls include a Rusty-fronted Barwing–like splutter, often introduced by nasal *dyiu-dyiu* notes.

Psittiparus bakeri
红头鸦雀 hóng tóu yā què
Rufous-headed Parrotbill

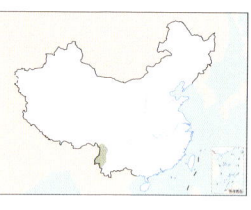

L: 17–19 cm
Habitat: Inhabits open forest edges, reedbeds, scrubland, and bamboo thickets at middle and low altitudes in mountainous areas.
Behavior: Moves about in pairs or small flocks in forest understory; active.
Distribution: W and NW Yunnan.
Voice: Sings a gentle trisyllabic *qiu— qu— gu*.

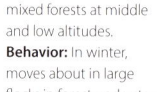

Fulvous Parrotbill

Broad dark bluish-gray lateral crown stripe

Orange-yellow throat and cheek

White half-collar

Pale-billed Parrotbill

Rufous-brown head

Short black supercilium

Creamy underparts

White eye ring

Black-throated Parrotbill

Gray crown

No orange yellow behind eye

Uniform gray on ear coverts

Relatively small area of black on throat

S. n. nipalensis

Orange crown and ear coverts

Gray between ear coverts and nape

Relatively small area of black on throat

S. n. crocotia

Ear coverts dark gray near eyes; gradually fades to lighter gray

Distinct broad black lateral crown stripe bordered below by narrow white supercilium

Orange-yellow crown

Orange-yellow supercilium behind eyes

S. n. poliotis

Gray breast

White-breasted Parrotbill

Rufous-brown head lacks supercilium

White underparts

Sharply defined lower border of ear coverts

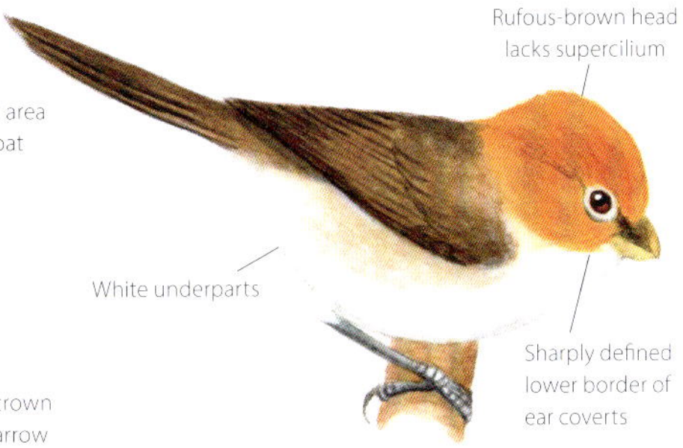

Rufous-headed Parrotbill

Extensive cobalt-blue orbital skin

Rufous-brown head lacks supercilium

Buff-yellow underparts

Ear coverts paler than crown, border with throat slightly diffuse

Resembles White-breasted Parrotbill but with larger body

Golden Parrotbill

Olive-brown crown

Gray ear coverts

S. v. morrisoniana

Orange-yellow crown

Short white supercilium

Yellowish-gray ear coverts

Black throat

S. v. verreauxi

Psittiparus gularis

灰头鸦雀 huī tóu yā què

Gray-headed Parrotbill   **LC**

L: 15.5–18.5 cm

Habitat: Inhabits evergreen broadleaf forests, secondary forests, forest-edge bushes, and bamboo thickets at altitudes of 450–1850 m.

Behavior: Often gathers in small flocks of six to eight individuals, sometimes also in large flocks of 30 individuals or more; usually joins babblers and other passerines in mixed-species flocks. Prefers foraging in treetops more than other parrotbills; sometimes also moves to the understory or the ground.

Distribution: *P. g. transfluvialis* in W and NW Yunnan; *P. g. laotianus* in S Yunnan and SW Guangxi; *P. g. fokiensis* in C, S, and SE Sichuan to N Jiangxi, SE Anhui, and Zhejiang, and south to Guangdong; *P. g. hainanus* in Hainan. Locally common.

Voice: Song a loud, shrill, whistled *eu-chu-chu* and a more plaintive *ti-tiuu*. Calls include harsh, scolding, explosive *jiaw* and short *kip* notes.

Paradoxornis guttaticollis

点胸鸦雀 diǎn xiōng yā què

Spot-breasted Parrotbill   **LC**

L: 18–22 cm

Habitat: Inhabits scrubland, grass, and bamboo thickets in open areas at middle to low altitudes; also occurs in woodlands, secondary forests, orchards, and around farmland.

Behavior: Solitary or in pairs. Noisy and active. Rarely mixes with other parrotbills.

Distribution: Found in the mountains in the Yangtze River Watershed and regions to its south.

Voice: Song a strident, staccato series of 3–10 notes, including *wheet-wheet-wheet-wheet* or the more penetrating *drii-drii-drii-drii* and *jhor-jhor-jhor-jhor.*

Calamornis heudei

震旦鸦雀 zhèn dàn yā què

Reed Parrotbill   **NT**

L: 18–20 cm

Habitat: Relies heavily on reedbeds.

Behavior: Often moves among reedbeds in pairs or in flocks; uses beak to tear open reed stems in search of insects. Does not fly long distances.

Distribution: *C. h. polivanovi* in NE Heilongjiang; *C. h. heudei* in coastal areas from Liaoning to Zhejiang, and in Beijing, Henan, Hubei, Jiangxi, and Anhui.

Voice: Song is a long series of *chut* notes delivered at varying rates but often accelerating, such as *chut-chut-chut-chut-chut'ut'ut'ut'ut'ut* and *chip-chip-brrrrrrr.* Calls include a ringing, vaguely Common Greenshank–like *whew-whew-whew*, often introduced by a sharp *kip*.

Gray-headed Parrotbill

Whitish around eyes

Black supercilium extends to nape

Orange-yellow to orange-red bill

Black from forehead to chin

Spot-breasted Parrotbill

Mottled white cheek

Black ear coverts

Lanceolate spots on breast

Dark supercilium

Large yellow bill

Reed Parrotbill

绣眼鸟科 Zosteropidae

Small forest-dwelling songbirds, some with distinctive green plumage and white eye rings. Plumage similar between sexes. Bill thin and pointed. Some species with crest on head. Wings short and rounded; good at flying. Tail medium in length, mostly squared. Legs moderately strong. Prefer scrubland, broadleaf forests, and mixed forests. Usually live in flocks. Nest on branches. Feed mostly on insects, nectar, and fruits. Most species are nonmigratory.

Some yuhinas were once placed in Timaliidae. Fourteen genera and 141 species recognized worldwide, widely distributed in the Old World and Oceania. Four genera and 14 species recorded in China, including yuhinas and white-eyes, found in the Himalayas and eastern and southern China.

凤鹛属 *Yuhina*	栗耳凤鹛属 *Staphida*	白领凤鹛属 *Parayuhina*	绣眼鸟属 *Zosterops*
Black-chinned Yuhina	Indochinese Yuhina	White-collared Yuhina	Swinhoe's White-eye

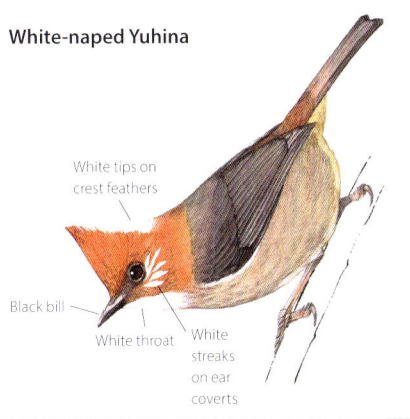

White-naped Yuhina

White tips on crest feathers

Black bill

White throat

White streaks on ear coverts

Whiskered Yuhina

Yellow neck

White around eyes

Yuhina bakeri

白颈凤鹛 (白项凤鹛)

bái jǐng fèng méi (bái xiàng fèng méi)

White-naped Yuhina

L: 13 cm
Habitat: Inhabits evergreen broadleaf forests and bushes in the understory and at forest edges, usually at 350–2200 m, mainly over 900 m.

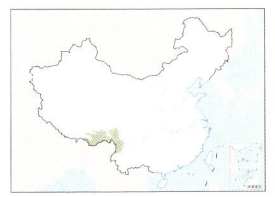

Behavior: Occurs in pairs in breeding season, often in flocks containing up to 20–30 individuals in nonbreeding season. Moves rapidly through forests; noisy. Sometimes joins mixed-species flocks.
Distribution: SE Tibet and NW Yunnan.
Voice: A series of slightly rising, slightly nervous, nasal calls, given singly or in a rapid series of up to five notes, *tsu-tsu-tsu*, similar to Whiskered Yuhina but far less forceful, in shorter series, and rising much less.

Yuhina flavicollis

黄颈凤鹛 huáng jǐng fèng méi

Whiskered Yuhina

L: 12–13.5 cm
Habitat: Inhabits evergreen broadleaf forests and bushes in the understory and at forest edges, often at altitudes of 500–3050 m.

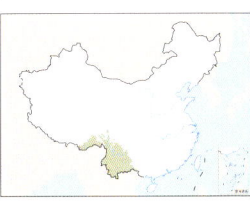

Behavior: Occurs in pairs in breeding season, in flocks in nonbreeding season. Often joins mixed-species flocks.
Distribution: *Y. f. flavicollis* in SE Tibet; *Y. f. rouxi* in Yunnan.
Voice: Vocalizations include a repeated, shrill, high-pitched *tzii-jhu ziddi*; a powerful, rising, squeaky *swee swee-swee*, given singly or in a series of up to 15 notes; and an explosive, buzzy *pzieu*.

Yuhina nigrimenta
黑颏凤鹛 hēi kē fèng méi
Black-chinned Yuhina　LC

L: 9–10 cm
Habitat: Inhabits evergreen broadleaf forests or secondary forests at altitudes of 300–2300 m.
Behavior: In nonbreeding season, often gathers in medium-sized groups of 5–20 individuals and joins mixed-species flocks. Feeds on insects, berries, and seeds in the forest canopy or low bushes.
Distribution: Uncommon resident in Central, South, and southeastern China, from N and NE Sichuan and S Shaanxi to S Zhejiang, south to Yunnan, Guangxi, and Guangdong.
Voice: Song a remarkable, thin, high-pitched sequence of slowly delivered short whistles, similar to a squeaky swing, *uu ii uui ii … uui uu ii … uui … uui uui … uui ii uui*, with no set pattern. Calls include subdued chattering, *tritt-i-trit-it*.

Yuhina brunneiceps
褐头凤鹛 hè tóu fèng méi
Taiwan Yuhina　LC

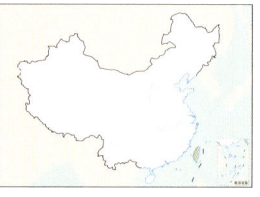

L: 9–10 cm
Habitat: Inhabits evergreen broadleaf, deciduous, and mixed broadleaf-coniferous forests, as well as bushes in forests and at forest edges, orchards, and tea plantations at altitudes of 800–2780 m.
Behavior: Occurs in pairs in breeding season and often moves about in small stable flocks in nonbreeding season. Fast and agile movements.
Distribution: Chinese endemic; restricted to Taiwan.
Voice: Song a powerful, jaunty, mellow *tdt-too, mee, jeeoo* ("to meet you" or "so pleased to meet you"). Short, nasal, bleating *nyeu* calls.

Yuhina gularis
纹喉凤鹛 wén hóu fèng méi
Stripe-throated Yuhina　LC

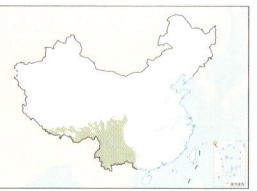

L: 12–16 cm
Habitat: Inhabits evergreen broadleaf, mixed broadleaf-coniferous, and coniferous forests, as well as bushes in understory and at forest edges at altitudes of 1000–3800 m.
Behavior: Occurs in pairs in breeding season; in nonbreeding season, often forms flocks or joins mixed-species flocks. Prefers feeding on berries and nectar.
Distribution: *Y. g. gularis* in S and SE Tibet, W, C, and S Yunnan. *Y. g. omeiensis* in SW, C, and N Sichuan, and south to N and NE Yunnan.
Voice: Presumed song is penetrating, short, descending, nasal *nyerr*, often in long series and sometimes followed by a more intense *spwik*, often in series.

Yuhina occipitalis
棕臂凤鹛 zōng tún fèng méi
Rufous-vented Yuhina　LC

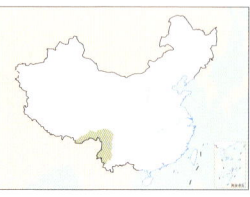

L: 12–14 cm
Habitat: Inhabits evergreen broadleaf forests, bushes in the understory and forest edges, and rhododendron thickets at altitudes of 500–3600 m.
Behavior: Occurs in pairs in breeding season; in nonbreeding season, often forms flocks or joins mixed-species flocks. Prefers berries and nectar; also feeds on sugar-rich sap from tree bark.
Distribution: *Y. o. occipitalis* in S and SE Tibet; *Y. o. obscurior* in W, NW, N, and C Yunnan and SW Sichuan.
Voice: Song a simple, weak, high-pitched *s-swi-swi-swi … s-swi-swi-swi*, often repeated. Calls include variable, short, hard, bouncy, buzzy notes, *trrre, beebee,* and *bzzee, bzzee, bzzee*.

Staphida castaniceps
栗耳凤鹛 lì ěr fèng méi
Striated Yuhina　LC

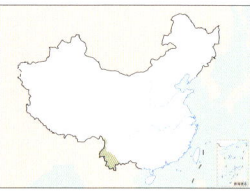

L: 12–15 cm
Habitat: Inhabits upper and middle stories of broadleaf, mixed, and secondary forests at middle and low altitudes.
Behavior: In nonbreeding season, gathers in small to large flocks at forest edges and in bushes; seldom mixes with other birds.
Distribution: W Yunnan.
Voice: Simple song a series of high-pitched, penetrating *tchiu* or *ti-tchiu* notes, with the second note falling strongly, repeated slowly (once every few seconds). Calls include a low spluttering chatter, *prrrt,* and subdued *sip* or *tip* notes.

Staphida torqueola
栗颈凤鹛 lì jǐng fèng méi
Indochinese Yuhina　LC

L: 14–15 cm
Habitat: Inhabits scrubland, the understory of broadleaf forests, low bushes of secondary forests, and low forest canopies at altitudes of 350–2200 m, usually over 900 m.
Behavior: Often in pairs in breeding season; gathers in large flocks of 20–30 individuals in nonbreeding season, usually in single-species flocks. Prefers moving rapidly among vegetation; very noisy. Forages for insects and seeds in moss, lichens, and bark; also feeds on nectar.
Distribution: Locally common resident in Central, East, South, southeastern, and southwestern China; from C and S Sichuan and C Yunnan in the west, to S Shaanxi and Jiangsu in the north, and to S Guangxi and Hong Kong in the south.
Voice: Song a loud ringing *tchu-eet* or *tchu-it*, very different from Striated Yuhina (lower pitched, with a falling first note and a second note that falls and then rises). Calls include a lengthy spluttering chatter, similar to call of Striated.

Black-chinned Yuhina

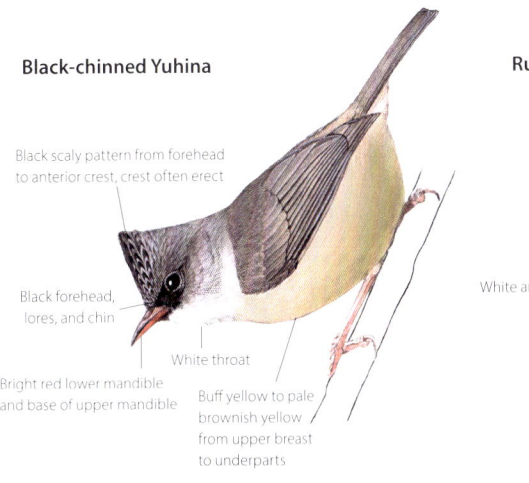

Black scaly pattern from forehead to anterior crest, crest often erect

Black forehead, lores, and chin

Bright red lower mandible and base of upper mandible

White throat

Buff yellow to pale brownish yellow from upper breast to underparts

Rufous-vented Yuhina

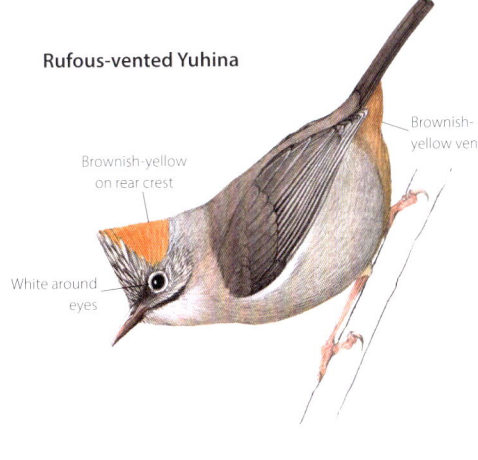

Brownish-yellow on rear crest

Brownish-yellow vent

White around eyes

Taiwan Yuhina

Dark brown forehead and crest

Dark brown stripe on cheek

Striated Yuhina

No chestnut-brown bars on hindneck

Pale gray crest

Chestnut ear spot

Stripe-throated Yuhina

Orange on wings

Narrow black streaks on throat

Indochinese Yuhina

Resembles Striated Yuhina, but chestnut on cheek extends to nape and neck

Gray crest

White flecking on ear coverts, nape, and mantle

487

Parayuhina diademata

白领凤鹛 bái lǐng fèng méi

White-collared Yuhina　LC

L: 14.5–18 cm

Habitat: Inhabits evergreen broadleaf forests, understory bushes, secondary forests, and tea plantations, often at 800–3600 m.

Behavior: Occurs in pairs in breeding season and often in flocks in nonbreeding season; mixes with other species. Prefers feeding on berries and nectar.

Distribution: *P. d. diademata* in S Gansu, S Shaanxi, W Hubei, and south to Sichuan and Guizhou; *P. d. ampelina* in Yunnan, W and NW Guangxi.

Voice: Vocalizations include a spluttering, nasal *wr-wrrri wrrr wrrri*, short *pik* or *pwik* notes, and mellow *trrrt* notes.

Zosterops erythropleurus

红胁绣眼鸟 hóng xié xiù yǎn niǎo

Chestnut-flanked White-eye　LC

L: 10.5–11.5 cm

Habitat: Inhabits deciduous or evergreen broadleaf forests and secondary forests below 2590 m; migrates through coastal areas and oceanic islands.

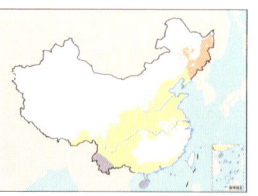

Behavior: Generally moves about in pairs or in small flocks; gathers in large flocks of several hundred individuals during migration. Often mixes with other small passerines.

Distribution: Breeds in Northeast China and migrates through most of North, Central, East, and South China; winters in Sichuan and Central China, with a few wintering records in East and South China.

Voice: Call is a loud, piercing, monosyllabic *tsee* or *psee*, occasionally described as *ping*, often given repeatedly, similar to call of both Warbling White-eye and Swinhoe's White-eye but often slightly shorter, lower pitched, more forceful, sharper, "purer," and less tremulous. Song sounds more confident, with repeated mimicry.

Zosterops simplex

暗绿绣眼鸟 àn lǜ xiù yǎn niǎo

Swinhoe's White-eye　LC

L: 10–11.5 cm

Habitat: Inhabits evergreen and deciduous broadleaf forests, mixed forests, scrubland, open woodlands, secondary forests, cultivated lands, etc.

Behavior: Often moves about gregariously in the upper and middle stories of vegetation. A common species found in mixed-species flocks.

Distribution: Common resident. *Z. s. simplex* in eastern China, west to S Gansu, east to Jiangsu and Taiwan, and south to Sichuan, E Yunnan, Guangxi, Guangdong, and Fujian; *Z. s. hainanus* only in Hainan.

Voice: Similar to Chestnut-flanked White-eye.

Zosterops japonicus

日本绣眼鸟 rì běn xiù yǎn niǎo

Warbling White-eye　LC

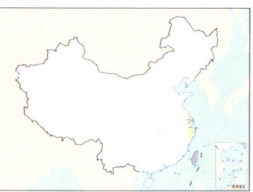

L: 10–11.5 cm

Habitat: Inhabits artificial gardens and green spaces on oceanic islands and coastal areas in winter.

Behavior: Often in pairs or in small flocks in the upper and middle stories of vegetation.

Distribution: Uncommon wintering visitor or passage migrant along the east coast of China; recorded in Taiwan, Shanghai, Zhejiang, and Jiangsu. Easily overlooked because of its resemblance to Swinhoe's White-eye.

Voice: Vocalizations similar to Chestnut-flanked White-eye.

Zosterops palpebrosus

灰腹绣眼鸟 huī fù xiù yǎn niǎo

Indian White-eye　LC

L: 10–11 cm

Habitat: Inhabits evergreen broadleaf forests, deciduous broadleaf forests, swamp forests, forest edges, secondary forests, farms, mangroves, and open woodlands within a wide

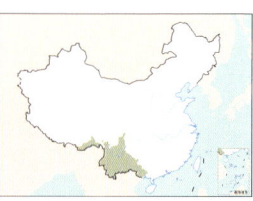

range of altitudes, from plains to mountains at 4000 m.

Behavior: Often gathers in medium-sized to large flocks; forages in the canopy.

Distribution: Common resident in SE Tibet, SW Sichuan, and Yunnan to SW Guangxi.

Voice: Sings a short, sharp, pleasant *qwee-qworr-quwarrr tu-cheeer-tu-cheeer-cheer, tu-cheer-cheeer-tu-cheeeer-cheer* only in breeding season. Call a fast repetitive *cheuw*.

Zosterops meyeni

低地绣眼鸟 dī dì xiù yǎn niǎo

Lowland White-eye　LC

L: 10.2–12 cm

Habitat: Inhabits forests, forest edges, thickets, bamboo thickets, gardens, cultivated lands, or open suburban areas at low altitudes.

Behavior: Forms single- and mixed-species flocks when foraging.

Distribution: Locally common resident on Orchid Island and Green Island in extreme SE Taiwan.

Voice: Males are good at singing. Call is a mumbling *swit* or *swit-tzee*.

White-collared Yuhina

Relatively large area of white on crest and nape

Black forehead and chin

Yellow bill

Warbling White-eye

Resembles Swinhoe's White-eye, but forehead lacks yellow, does not contrast with crown

Bill slightly longer than Swinhoe's White-eye

Diffuse dusky pale ochre from breast to flanks, looks dirty, resembles female Chestnut-flanked White-eye but with larger and lighter color patch on flanks

Chestnut-flanked White-eye

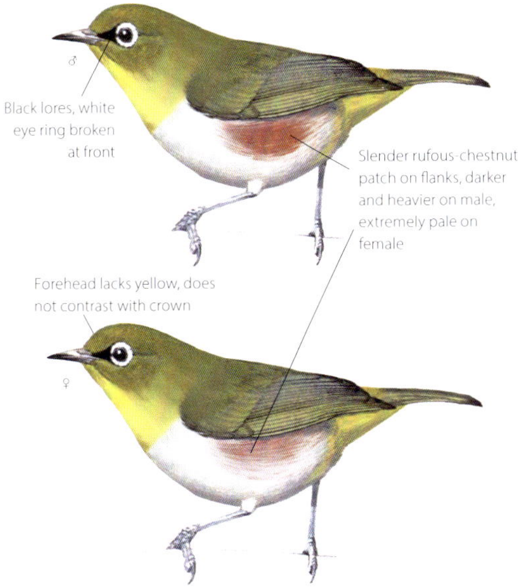

Black lores, white eye ring broken at front

♂

Slender rufous-chestnut patch on flanks, darker and heavier on male, extremely pale on female

Forehead lacks yellow, does not contrast with crown

♀

Indian White-eye

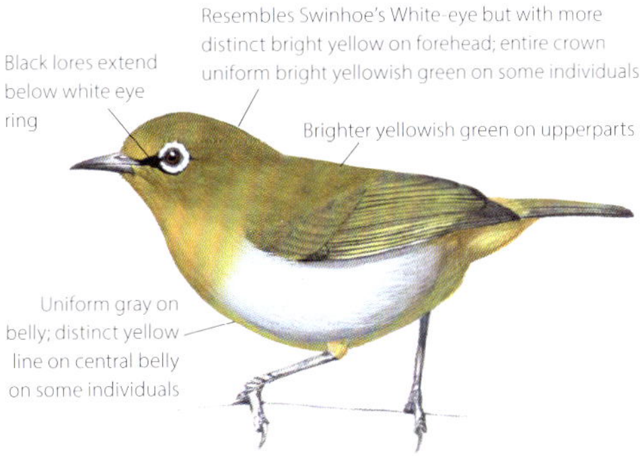

Black lores extend below white eye ring

Resembles Swinhoe's White-eye but with more distinct bright yellow on forehead; entire crown uniform bright yellowish green on some individuals

Brighter yellowish green on upperparts

Uniform gray on belly; distinct yellow line on central belly on some individuals

Lowland White-eye

Relatively large area of yellow on forehead, quite distinct

Resembles Swinhoe's White-eye, but bright yellow on throat only extends to upper breast

Swinhoe's White-eye

From base of upper mandible to forehead tinged yellow, contrasts with yellowish green on crown

White bare skin around eyes

Yellow throat

Grayish white from breast to flanks, sometimes extremely pale pinkish brown

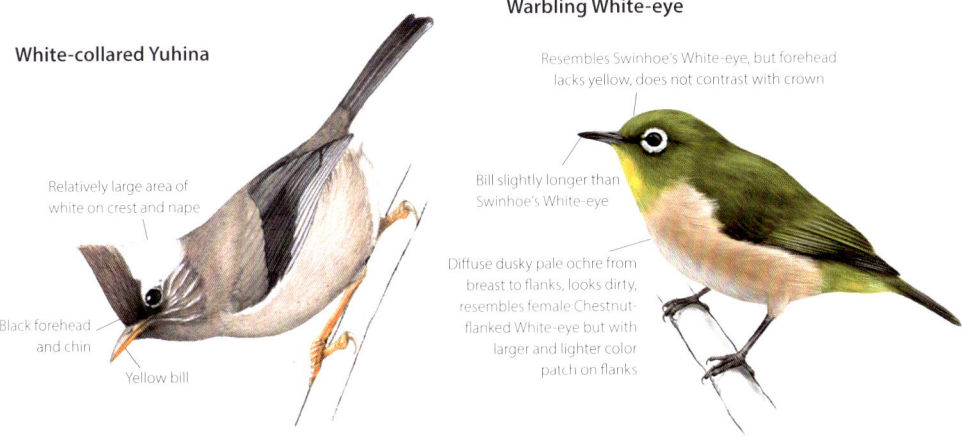

和平鸟科　Irenidae

Small to medium-sized songbirds, endemic to Indomalaya. The family once also included ioras and leafbirds but at present includes only fairy-bluebirds. Body profile similar to orioles. Plumage differs between sexes and is mostly black and blue in both sexes. Bill resembles bill of thrushes but thicker and stronger. Irises red. Wings rounded. Squared tail consists of 12 feathers. Highly reliant on forests. Feed mostly on figs and other fruits; also take insects. Nonmigratory.

Only one genus and two species recognized worldwide. One species recorded in China, found only in SE Tibet and W and S Yunnan.

Asian Fairy-bluebird

Red iris

Bright cobalt-blue crown, hindneck, upper back, rump, and uppertail coverts; remaining body black

♂

Red iris

Bluish plumage, appears greenish at some angles

♀

Blackish-brown wing feathers

Irena puella

和平鸟 hé píng niǎo

Asian Fairy-bluebird LC

L: 21–26 cm

Habitat: Inhabits tall evergreen and semievergreen broadleaf forests in lowlands, as well as similar montane forests, in tropical and subtropical regions often below 1400 m.

Behavior: Solitary or in small flocks; often joins other species in mixed-species flocks. Moves about mainly in the forest canopy or middle forest layer.

Distribution: *I. p. puella* is an uncommon resident in S and SW Yunnan and SE Tibet.

Voice: Song is a loud, liquid, percussive *tu-lip wae-waet-oo*. Calls are similarly powerful whiplash whistle, *wit weet* or *wit-wit-wit*, and quiet *tt*.

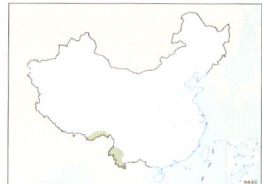

戴菊科　Regulidae

Small forest-dwelling songbirds, widely distributed across the Northern Hemisphere. All species possess warm-colored crown feathers. Resemble leaf warblers and were once put in Sylviidae as a result. Plumage differs slightly between sexes, mostly olive. Irises dark brown, surrounded by conspicuous white eye rings. Nostrils covered by stiff feathers or bristles. Usually with two prominent wing bars and light fringe on tertials. Insectivorous. Prefer coniferous forests. Continental species are migratory.

One genus and six species recognized worldwide. Two species recorded in China, widely distributed except South China and the Qinghai-Tibet Plateau.

Flamecrest

Crest raised

Grayish-white face with broad black eye ring and black moustachial stripe

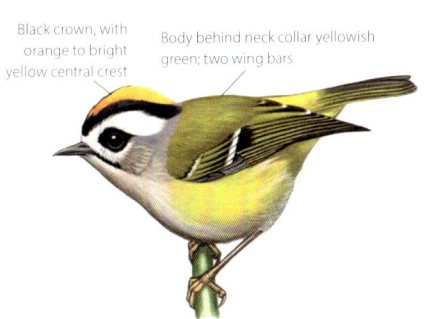

Black crown, with orange to bright yellow central crest

Body behind neck collar yellowish green; two wing bars

Goldcrest

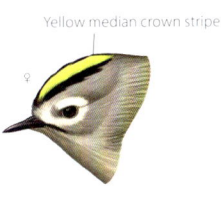

Yellow median crown stripe

Yellow crown with orange stripe

Two white wing bars

White eye ring

Regulus goodfellowi
台湾戴菊 tái wān dài jú
Flamecrest　LC

L: 9 cm
Habitat: Prefers coniferous forests; also occurs in mixed broadleaf-coniferous forests. Often at middle and high altitudes, up to 3700 m.
Behavior: Joins mixed-species flocks in nonbreeding season.
Distribution: Chinese endemic; locally common resident in Taiwan.
Voice: Territorial song is a series of rapidly repeated call notes—an intense, very high-pitched *zree* of constant frequency lacking the terminal trill or flourish of Goldcrest.

Regulus regulus
戴菊 dài jú
Goldcrest　LC

L: 9–10 cm
Habitat: Inhabits subalpine coniferous forests; also occurs on farmland and urban parks on plains during migration or winter.
Behavior: Often solitary. Nests in coniferous trees, laying four to nine eggs per clutch, incubated by the female.
Distribution: *R. r. yunnanensis* is a resident in southwestern China; *R. r. japonensis* breeds in Northeast China and winters in North China and the eastern and central regions south of it; *R. r. tristis* in Xinjiang and Qinghai; *R. r. coatsi* in Xinjiang; *R. r. sikkimensis* on the Qinghai-Tibet Plateau. Relatively common.
Voice: Song comprises a high-pitched main part and a shorter, highly variable terminal flourish, with much geographical variation. Call a high-pitched *seeeh* or *zick*. Song of *R. r. tristis* with a distinctively intensifying or "pulsing" introduction.

丽星鹩鹛科　Elachuridae

A newly recognized monotypic family. Small forest-dwelling songbird resembling Eurasian Wren, with a disjunct distribution in the montane forests of E Himalayas, N Indochina, and southern China. Body length of 10 cm, heavily dotted with white throughout. Once put under *Spelaeornis* with other wren-babblers, but more recent phylogenetic studies suggest that Spotted Elachura is the only extant member of an ancient lineage.

Only one genus and one species recognized, found in South and southwestern China.

鹪鹩科　Troglodytidae

Small brown songbirds. Plumage similar between sexes; most species dark barred. Bill thin and pointed, without rictal bristles. Wings short and rounded. Tail short, often cocked. Legs strong; good at hopping. Ground dwellers; usually solitary, mostly inhabiting the understory of coniferous forests, very few inhabiting forest canopy. Insectivorous. Most species are resident, while some temperate-breeding species are migratory.

Nineteen genera and 88 species recognized worldwide; except for Eurasian Wren, all species distributed in the New World. One genus and one species recorded in China, widely distributed except in South China and on the Qinghai-Tibet Plateau.

Spotted Elachura

Rear body brownish-yellow to rusty red, with blackish-brown bars

Rounded body resembles Eurasian Wren, tail short and stubby

Front body dark brown, dense white spots on side of neck, back, throat, breast, and belly

Eurasian Wren

Short tail often cocked

Slender bill

Dense blackish-brown bars

Elachura formosa

丽星鹩鹛 lì xīng liáo méi

Spotted Elachura

L: 10 cm
Habitat: Inhabits the understory of moist subtropical evergreen broadleaf forests, dense thickets, fern-covered ground, open cliffs with tangles of branches and weeds, and mossy rocks; usually near streams. Altitude range: 1100–2150 m.

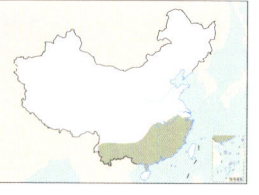

Behavior: Secretive and shy; often in the understory of densely vegetated forests or on the ground. Forages mainly for insects.
Distribution: Locally common resident in southwestern, Central, South, and southeastern China, north to Zhejiang.
Voice: Song a thin, faltering, extremely high-pitched, tinkling *ti-ti-ti-l tit-si-ii ti-ti-ti-i tit-si-ii.* Call a spluttering *put-put-put* trill.

Troglodytes troglodytes

鹪鹩 jiāo liáo

Eurasian Wren

L: 9–11 cm
Habitat: Frequents dense bushes in mountainous areas near river-valley streams or forests.
Behavior: Secretive. Hops quickly among shady bushes with tail cocked.
Distribution: Locally

common with many subspecies: *T. t. idius* in North China and east and central regions south of it; *T. t. dauricus* in Northeast China; *T. t. szetschuanus* in the mountainous areas in southwestern China; *T. t. talifuensis* in Yunnan and Guizhou; *T. t. nipalensis* in SE Tibet and NW Yunnan; *T. t. tianschanicus* in NW Xinjiang; *T. t. taivanus* in Taiwan. Summer breeder or resident in temperate regions and Taiwan; winter visitor in subtropical regions.
Voice: Remarkable song (with considerable geographic variation)—powerful, long, and complex, a series of tinkling trills, one after the other for several seconds. Calls include a sharp *tac* and loud churrs.

鸸科 Sittidae

Small forest-dwelling songbirds, well adapted to climbing on and clinging to trees. Body compact. Plumage differs between sexes in approximately half of the species, but differences are very minor. Plumage mostly white, brown, and bluish gray; some tropical species vivid bluish purple. Bill thin but strong, regularly used to pry open bark or seeds. Well-developed claws and hind toes adapted to clinging onto trunks and branches. Short and squared tail consists of 12 feathers. Exhibit unique and specialized behavior, spiraling up and down tree trunks. Insectivorous; forage from cracks and bark on trees. Nest in cavities. Prefer montane forests, while a few species inhabit rocky cliffs or soil mounds. Nonmigratory.

One genus and 28 species recognized worldwide, mostly in the Palearctic except for four species in North America. One genus and 11 species recorded in China, found in N Xinjiang, as well as the southeastern edge of the Qinghai-Tibet Plateau and regions to the east.

Eurasian Nuthatch

White breast and belly
S. e. asiatica

White throat and cheek
Pale brownish-yellow breast and belly
S. e. amurensis

Grayish-white throat and cheek
Brownish-yellow breast and belly
Rounded white spots on rufous-chestnut undertail coverts
S. e. sinensis

Chestnut-vented Nuthatch

Slightly large white spots on chestnut vent
Rufous-chestnut flanks
Off-white breast and belly

Sitta europaea
普通鸸 pǔ tōng shī
Eurasian Nuthatch

L: 12–14 cm (*S. e. sinensis* group); 17–21 cm (*S. e. asiatica* group)
Habitat: Inhabits montane forests, also forests, orchards, and urban parks on low mountains and plains. Altitude ranges from a few hundred meters to 3000 m.
Behavior: Good at spiraling up tree trunks or descending tree trunks headfirst; searches for insects under tree bark. Very active and agile. Nests in tree cavities made by woodpeckers and other species; sometimes uses mud to plaster cavity opening and smooth out the interior wall.
Distribution: Locally common. *S. e. formosana* in Taiwan; *S. e. seorsa* in Xinjiang; *S. e. asiatica* in northern Northeast China; *S. e. amurensis* in Northeast and northern North China; *S. e. sinensis* is widely distributed in North, East, Central, South, and southwestern China.
Voice: Complex vocabulary. Song is a series of repetitive, high-pitched, monosyllabic whistles, *whip-whip-whip-whip-*, slower than the Snowy-browed Nuthatch and with shorter strophes. Calls include ringing *pwee-u* and nasal *twip-twip*.

Sitta nagaensis
栗臀鸸 lì tún shī
Chestnut-vented Nuthatch

L: 12.5–14 cm
Habitat: Inhabits montane evergreen forests, especially pine, oak, and rhododendron forests. Altitude ranges from 800 m to more than 4500 m.
Behavior: Often solitary or in pairs in breeding season. Occasionally in small flocks of a few individuals in nonbreeding season; sometimes joins mixed-species flocks. Usually forages along tree trunks for insects in bark crevices; also forages on rocks. Occasionally comes to the ground.
Distribution: *S. n. montium* in E Tibet, NE Yunnan, W and SW Sichuan, W and SW Guizhou, Jiangxi, Fujian, and W Guangxi; *S. n. nagaensis* in SE Tibet, W and S Yunnan.
Voice: Song a powerful, usually rapid trill that changes little in pitch and lasts about a second, *trrrrrrrrrre*. Calls include a whining, nasal *tyear* and high-pitched *tsip-sip* notes.

Sitta cinnamoventris
栗腹䴓 lì fù shī
Chestnut-bellied Nuthatch

L: 13–14 cm
Habitat: Inhabits montane evergreen forests, deciduous forests, and pine forests at altitudes of 800–2200 m.
Behavior: Often solitary or in pairs in breeding season, occasionally in small flocks of a few individuals in nonbreeding season; sometimes joins mixed-species flocks. Usually moves along large tree trunks in search of insects in bark crevices; also forages among small branches and twigs.
Distribution: *S. c. neglecta* in SW Yunnan; *S. c. tonkinensis* in S Yunnan and SW Guangxi; *S. c. cinnamoventris* in SE Tibet.
Voice: Song a slurred rapid trill that rises slightly in pitch, *treeeee*, recalling a referee's whistle, similar to, but softer and less intense than, song of Chestnut-vented Nuthatch. Calls include a mellow *tsup* and a thin mouselike *sit* or *sit-sit*.

Sitta himalayensis
白尾䴓 bái wěi shī
White-tailed Nuthatch

L: 12 cm
Habitat: Generally inhabits temperate broadleaf forests and mixed forests; also occurs in mid-altitude oak and rhododendron forests and mixed fir-rhododendron forests at higher altitudes. Prefers forests with moss. Mainly at altitudes of 1500–3500 m. Migrates to lower altitudes in winter.
Behavior: Similar to Chestnut-bellied Nuthatch but prefers foraging on mossy branches; usually occurs in the upper story of trees. Seldom on tree trunks; occasionally in low bushes.
Distribution: Seen in SE and S Tibet and W and S Yunnan.
Voice: Song a rapidly repeated, nasal, ringing *tweep-eep-eep-eep*. Calls include a nasal *nyert* and a hard stone-clicking *tak-chak-chak*.

Sitta przewalskii
白脸䴓 bái liǎn shī
Przevalski's Nuthatch

L: 11–12 cm
Habitat: Inhabits coniferous forests at middle and high altitudes.
Behavior: Arboreal; mostly in the canopy. Also joins mixed-species flocks.
Distribution: Uncommon in S Shaanxi, S Gansu, Sichuan, E Qinghai, N Yunnan, and E and SE Tibet.
Voice: Song is a loud penetrating note, *t-tuui … t-tuui*, repeated every 1–3 sec, or a *ti-tui-tui … ti-tui tui-tui-tui*. Calls include a rapid-fire, nasal *nyit-nyi-nyi-nyi-nyi*.

Sitta villosa
黑头䴓 hēi tóu shī
Snowy-browed Nuthatch

L: 10–11 cm
Habitat: Inhabits montane coniferous and mixed broadleaf-coniferous forests.
Behavior: Agile, spiraling up and down along tree trunks in search of insects under bark. Nests in tree cavities with small opening; generally does not apply soil to the opening.
Distribution: Endemic to northern China. *S. v. bangsi* in N Sichuan, E Qinghai, and Gansu; *S. v. villosa* is widespread across southern Northeast China, northern North China, and eastern northwestern China.
Voice: A prolonged, rapidly repeated, penetrating *weet-weet-weet*, a shriller, more piping *pipipipipipi* or *kikikikikiki*, and an unpleasant, strained, hissing *ksssss* when agitated.

Sitta yunnanensis
滇䴓 diān shī
Yunnan Nuthatch

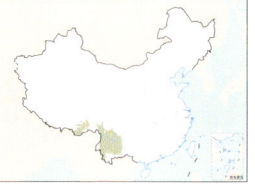

L: 12 cm
Habitat: Prefers large patches of mature pine forest at altitudes of 1200–4000 m.
Behavior: Often solitary or in pairs in breeding season. Occasionally in small flocks of a few individuals in nonbreeding season; sometimes joins mixed-species flocks. Prefers moving along tree trunks, usually in a head-down position.
Distribution: Uncommon in SE Tibet, Yunnan, S and SW Sichuan, and W Guizhou.
Voice: Song is a rapidly repeated, nasal *nyit-nyiy-nyit*. Calls include a scolding, raucous chatter.

Sitta magna
巨䴓 jù shī
Giant Nuthatch

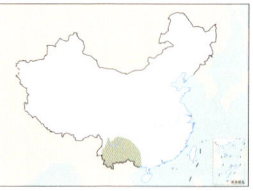

L: 19.5 cm
Habitat: Prefers large patches of mature pine forest; also inhabits evergreen broadleaf forests on hills. Altitude range: 1250–3400 m.
Behavior: Often solitary or in pairs; seldom gathers in groups. Sometimes joins mixed-species flocks. Prefers moving up and down along tree trunks, often in a head-down position.
Distribution: Seen in SW Guangxi, Yunnan, S Sichuan, and SW Guizhou.
Voice: Song is a short, flat, melancholy, toy trumpet–like whistle, *plu* or *nya*, repeated almost metronomically every 1.2 sec and occasionally accompanied by the primary call, a raucous, corvid-like, querulous, growling chatter, *g-drrrrrr*, or a barked Eurasian Jay–like *g-da … g-da-grr*.

Chestnut-bellied Nuthatch

Black and white barring on vent

Broad black eye stripe

White chin and face

♂

Chestnut throat, breast, and belly

Chestnut on underparts slightly pale

♀

Snowy-browed Nuthatch

S. v. villosa

White supercilium

Black crown

Yellowish-brown belly

♂

Gray crown

♀

S. v. bangsi

Off-white belly

♂

White-tailed Nuthatch

White at base of tail feathers

Relatively white chin and throat

Buff-white face

Chestnut pale on breast, belly, and vent, slightly heavier on flanks

Yunnan Nuthatch

Long narrow white supercilium above black eye stripe

Relatively uniform off-white from chin to vent

Przevalski's Nuthatch

Black crown

Brownish-white chin and throat

Brownish-yellow breast and belly

Giant Nuthatch

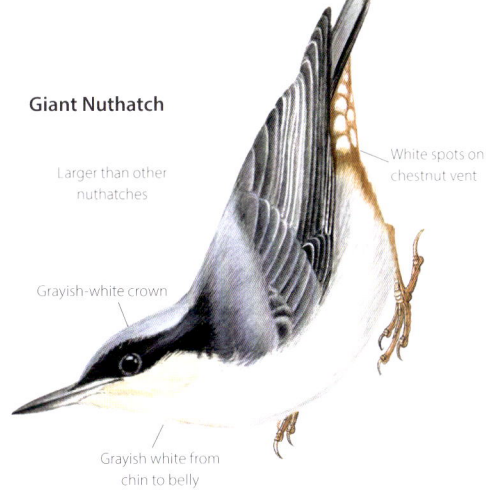

White spots on chestnut vent

Larger than other nuthatches

Grayish-white crown

Grayish white from chin to belly

Sitta frontalis

绒额䴓 róng é shī

Velvet-fronted Nuthatch LC

L: 12–13.5 cm

Habitat: Inhabits tropical rain forests, swamp forests, semievergreen broadleaf forests, mixed forests, and coniferous forests at altitudes of 340–1825 m.

Behavior: In pairs or flocks; often joins mixed-species flocks. Active; climbs tree trunks and branches in the forest canopy in search of food.

Distribution: Locally common resident in southern China, including SE Tibet, W, S, and SE Yunnan, C and S Guizhou, Guangxi, and C and W Guangdong.

Voice: Noisy. Song an intense series of hard, rattling *sit-tit-tit-tit-tit-tit-tit-ti* notes, like a machine gun, lasting 1.5–2 sec. Calls include more timid versions of the song, *trrrt* and *tit-sit-tit*.

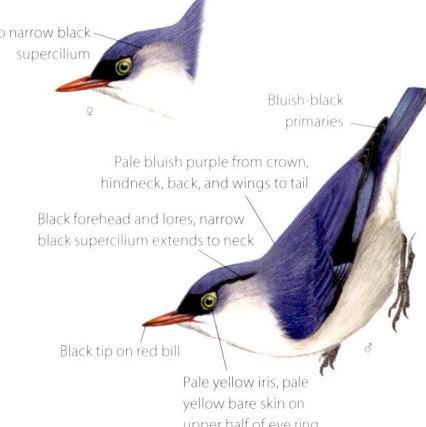

Velvet-fronted Nuthatch

No narrow black supercilium ♀

Bluish-black primaries

Pale bluish purple from crown, hindneck, back, and wings to tail

Black forehead and lores, narrow black supercilium extends to neck

Black tip on red bill

Pale yellow iris, pale yellow bare skin on upper half of eye ring ♂

Sitta solangiae

淡紫䴓 dàn zǐ shī

Yellow-billed Nuthatch NT

L: 12.5–13.5 cm

Habitat: Inhabits montane evergreen broadleaf forests at altitudes of 800–1500 m.

Behavior: Solitary or in small flocks. Often joins other passerines in mixed-species flocks; usually mixes with Sultan Tit in Jianfengling in Hainan.

Distribution: A locally common resident in the mountains of S Hainan.

Voice: Song is a fast *sit-ti-ti-ti-ti-ti-*, lasting 1–2.5 sec, slightly lower pitched, more chattering, more nervous, and less intense than Velvet-fronted Nuthatch. Calls very similar to those of Velvet-fronted.

Yellow-billed Nuthatch

Short broad black supercilium ♀

Pale violet back, wings, and tail feathers, primaries blackish

Black forehead and lores, broad black supercilium extends to neck

Slightly blackish tip on yellow bill

Citrine-yellow iris and eye ring

Pale purple from crown and face to neck ♂

Sitta formosa

丽䴓 lì shī

Beautiful Nuthatch VU

L: 19.5 cm

Habitat: Inhabits large mature subtropical evergreen broadleaf forests at altitudes of 300–2150 m; prefers trees with epiphytes and moss.

Behavior: Often solitary or in pairs; rarely gathers in flocks but sometimes joins mixed-species flocks. Prefers moving up and down tree trunks, usually in a head-down position.

Distribution: Rare in SE Tibet, W, S, and SE Yunnan.

Voice: Song extremely similar to that of Velvet-fronted Nuthatch in speed, tone, and duration but perhaps with a more timid introduction. Calls include short, nervous, nasal *squip* and *quip* notes.

Blue tail

Beautiful Nuthatch

Grayish-white vent

Bluish-black back

White stripes on bluish-black head and neck

Grayish-white chin, throat, and face

Rufous-chestnut breast and belly

White fringe on black wing coverts

旋壁雀科　Tichodromidae

Medium-sized songbird of rocky habitats. Plumage differs inconspicuously between sexes in summer. Bill long and thin. Wings short, rounded, and boldly patterned, each with a prominent large red spot. Tail short, about half the length of wings. Insectivorous. Mostly resident but can be nomadic or perform altitudinal migration.

Wallcreeper was once put in Sittidae but is now in its own monotypic family. Only one genus and one species recognized worldwide, widely distributed across the central part of the Eurasian continent. One genus and one species recorded in China, found in the mountainous areas of central and western China.

Wallcreeper

Black throat

br.

White throat

non-br.

旋木雀科　Certhiidae

Small forest-dwelling songbirds, characterized by unique behavior. Body compactly built. Plumage similar between sexes, mostly brown and finely streaked. Bill long, thin, and decurved, without rictal bristles. Tail constitutes half of total body length, consisting of 12 feathers with stiff and pointed shafts, similar to tails of woodpeckers. Hind claws longer than hind toes, well adapted to climbing and clinging. Exhibit unique behavior, clinging to and spiraling up a tree trunk to the top, before repeating the process after flying to the lower trunk of another tree. Usually join mixed-species flocks but stand out with their behavior. Insectivorous. Nonmigratory; some are nomadic or perform altitudinal migration.

Two genera and 11 species recognized worldwide, including nine treecreepers and two spotted creepers; most species are restricted to Indomalaya except for one species widely distributed in the Palearctic and single species restricted to each of Europe, Africa, and North America. China is the core area of treecreeper distribution, with one genus and seven species recorded, found in the forests of Northeast, North, Central, and Western China.

Eurasian Treecreeper

White breast and belly, gray tinge from lower belly to undertail coverts on some individuals

White supercilium

White streaks on brownish upperparts

Tichodroma muraria

红翅旋壁雀 hóng chì xuán bì què

Wallcreeper　(LC)

L: 13–18 cm
Habitat: Wide altitudinal range, from lowlands to 5000 m. Inhabits mountain cliffs, steep slope faces, or earthen walls.
Behavior: Often solitary; occasionally in small flocks of two or three individuals. Spends almost all time on rock faces, using beak to probe deep into crevices while foraging. Can hop along cliff faces or fly short distances, sometimes spreading wings and clinging to cliff faces. Flight slow and undulating when flying long distances.
Distribution: Resident west of North China, Sichuan, and Yunnan; common at higher altitudes in Western China and occasionally seen in eastern China. Also occurs in East China as a rare winter visitor.
Voice: Quiet. Song is four or five squeezed or strained, rising whistles, *uh-rruuuh-wheiiuu*, occasionally ending with subdued, chattering trills or chirps.

Certhia familiaris

欧亚旋木雀 (旋木雀)

ōu yà xuán mù què (xuán mù què)

Eurasian Treecreeper　(LC)

L: 12–15 cm
Habitat: Inhabits montane broadleaf, mixed, and coniferous forests; also in artificial and secondary forests.
Behavior: Often solitary or in pairs, spiraling around tree trunks in search of food.
Distribution: *C. f. daurica* in Northeast and North China and N Xinjiang; *C. f. bianchii* in S Shaanxi, E and NE Qinghai, Ningxia, and E Gansu; *C. f. tianschanica* in Xinjiang.
Voice: Song a thin, high-pitched but falling, silvery warble, 2–3 sec long, ending with a brief flourish, *tsee-tsee-tsi-tsi-si-si-si-si-sisisisisi-tsee*. Two slightly different calls—a shrill, high-pitched, tremulous, and buzzing but characteristically emphatic, penetrating *srrih*, extremely similar to call of Hodgson's Treecreeper, and a purer, non-tremulous, descending *tiiiih*, possibly not shared by Hodgson's.

Certhia hodgsoni

霍氏旋木雀 huò shì xuán mù què

Hodgson's Treecreeper

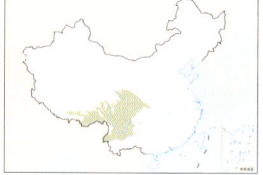

L: 11–13 cm
Habitat: Inhabits broadleaf, mixed, and coniferous forests from low to high altitudes; also in parks and woodlands.
Behavior: Often solitary or in pairs in montane forests; usually joins mixed-species flocks.
Distribution: Seen in W and S Gansu, S Qinghai, S and SE Tibet, Sichuan, Chongqing, W Hubei, and NW Yunnan.
Voice: Song is similar to that of Eurasian Treecreeper but shorter and not quite so "thin," with two, occasionally three, longer, flatter (not quite so tremulous), high-pitched, sibilant *tsree* introductory notes, followed by three to seven rapidly falling notes. The terminal trill is rarely present, and the final note is clearly descending and not flat. See Eurasian Treecreeper for calls.

Certhia himalayana

高山旋木雀 gāo shān xuán mù què

Bar-tailed Treecreeper

L: 13–15 cm
Habitat: Inhabits broadleaf, mixed broadleaf-coniferous, and coniferous forests at middle and high altitudes; migrates to lower altitudes, including urban parks and orchards, in winter.
Behavior: Often solitary or in pairs in forests; joins mixed-species flocks in winter.
Distribution: *C. h. yunnanensis* in the mountains in southwestern and midwestern China; *C. h. taeniura* in the Tian Shan Mountains region of Xinjiang.
Voice: Song a slow, increasingly intense, penetrating rattle, occasionally introduced by a high, fine note, similar to the song of Sikkim Treecreeper but slower and more halting. Calls are lower pitched than both Eurasian Treecreeper and Hodgson's Treecreeper and clearly descending.

Certhia nipalensis

红腹旋木雀 (锈红腹旋木雀)
hóng fù xuán mù què (xiù hóng fù xuán mù què)

Rusty-flanked Treecreeper

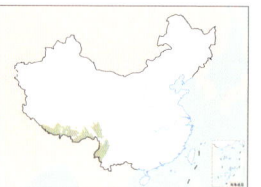

L: 14–16 cm
Habitat: Inhabits mixed broadleaf-coniferous and coniferous forests at middle and high altitudes.
Behavior: Mostly solitary in mature forests. Arboreal; forages on trunks. Often joins other small birds in mixed-species flocks.
Distribution: S and SE Tibet and W and NW Yunnan.
Voice: Distinctive song a short (0.6–1 sec), simple, high-pitched, silvery rattle of 5–10 notes (exceptionally up to usually 30 notes), more or less uniform in pitch and tempo but increasing in volume and introduced by a short thin whistle and ending with two different notes, *twe-sisisisisi-tu-sip*. Calls include a thin *sit* and penetrating *zip*.

Certhia discolor

褐喉旋木雀 hè hóu xuán mù què

Sikkim Treecreeper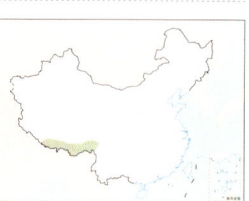

L: 14–16 cm
Habitat: Inhabits broadleaf and mixed broadleaf-coniferous forests at middle and high altitudes.
Behavior: Often solitary on trunks; usually joins other small birds in mixed-species flocks.
Distribution: Seen in S Tibet.
Voice: Song an increasingly intense, simple, fast, accelerating rattle, 1–2 sec long, of simple full *chi* notes given at about 12 notes per second, slightly accelerating and, more obviously, gaining in power and volume toward end. Call an explosive, deliberate *tyip* or a markedly disyllabic *chi-tip*.

Certhia manipurensis

休氏旋木雀 xiū shì xuán mù què

Hume's Treecreeper

L: 14–16 cm
Habitat: Inhabits broadleaf and mixed broadleaf-coniferous forests at middle and high altitudes, especially those in river valleys.
Behavior: Often joins mixed-species flocks. Spirals along tree trunk to forage.
Distribution: Seen only in W Yunnan.
Voice: Song an increasingly intense, simple, fast, accelerating rattle of paired notes, slightly slower than that of Sikkim Treecreeper, but disyllabic units impart stuttering rhythm. Calls like Sikkim.

Certhia tianquanensis

四川旋木雀 sì chuān xuán mù què

Sichuan Treecreeper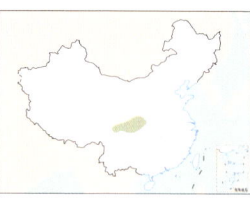

L: 12–14 cm
Habitat: Inhabits deciduous and mixed deciduous-coniferous forests at middle and high altitudes.
Behavior: Mostly solitary or in pairs on trunks in the upper and middle stories of forests; often joins mixed-species flocks in winter.
Distribution: Chinese endemic; found in the mountains of N and W Sichuan, S Gansu, S Shaanxi, and W Hubei.
Voice: Song a rapid high-pitched trill that starts abruptly but rapidly tails off and drops in pitch toward end, usually introduced by a higher, sweeter note that, when present, gives the song a marked two-tone quality, *tsit-lilililililililiuuuuuuuu*. Makes high-pitched Goldcrest-like call notes.

Hodgson's Treecreeper

White breast and belly, from lower belly to undertail coverts tinged brown on some individuals

Grayish-brown flanks and undertail coverts

Brown streaks on chestnut-brown upperparts

Sikkim Treecreeper

Earth-brown underparts

Indistinct supercilium

Bar-tailed Treecreeper

Thin decurved bill

Drab brown breast and belly

White on supercilium disconnected from breast and belly

Barred tail

Hume's Treecreeper

Drab brown from throat to underparts

Indistinct supercilium

Buff eye ring

Rusty-flanked Treecreeper

Yellowish-brown breast

Rusty belly

Sichuan Treecreeper

White chin and throat

Relatively short, slightly decurved bill

Grayish-brown breast and belly

499

椋鸟科 Sturnidae

Small to medium-sized songbirds whose well-proportioned body profile resembles that of thrushes. Plumage similar or slightly different between sexes in most species, mostly gray, black, and white; some are iridescent. Strong bill straight and pointed. Wings pointed. Rounded or squared tail consists of 12 feathers. Legs long and strong, well adapted to grasping and walking. Most species gregarious. Inhabit forests and urban areas. Nest in cavities. Very vocal, with noisy or melodic voices; some species very good at mimicking other sounds. Mostly herbivorous; some also take insects. Northerly breeding species are migratory; some are highly nomadic.

     Thirty-three genera and 123 species recognized worldwide, mostly in the tropics and subtropics, a few more widely spread across Eurasia. Ten genera and 19 species recorded in China, ubiquitously distributed nationwide.

斑翅椋鸟属
Saroglossa

Spot-winged
Starling

树八哥属
Ampeliceps

Golden-crested
Myna

鹩哥属
Gracula

Common
Hill Myna

八哥属
Acridotheres

Crested Myna

丝光椋鸟属
Spodiopsar

Red-billed
Starling

斑椋鸟属
Gracupica

Black-collared
Starling

北椋鸟属
Agropsar

Daurian
Starling

灰背椋鸟属
Sturnia

White-shouldered
Starling

粉红椋鸟属
Pastor

Rosy Starling

椋鸟属
Sturnus

br.

non-br.

European Starling

Saroglossa spilopterus

斑翅椋鸟 bān chì liáng niǎo

Spot-winged Starling 🟢LC

L: 19 cm
Habitat: Inhabits open forests, clearings, and forest edges in hilly regions from lowlands to 1000 m, up to 2000 m in some areas.
Behavior: Forages in forest canopy. Gregarious, especially in nonbreeding season, sometimes in large flocks.
Distribution: Uncommon resident in S Tibet, W and S Yunnan.
Voice: Song is an eclectic mix of dry, harsh, discordant notes and more musical warbling. Calls include a querulous, even aggressive *chek-chek-chek*, a nasal *qwee-e-ek*, and a sharper *chik*.

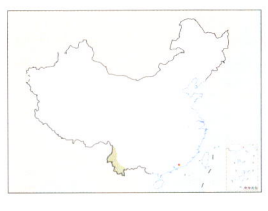

Spot-winged Starling

Grayish-brown upperparts ♀

Black head

Throat and breast washed pale brown

Scaly grayish-white patterns on black neck and back

Dark rufous-chestnut chin and throat ♂

Brown breast and flanks

White wing patch on black wing, sometimes covered when perched

White lower belly and vent

Ampeliceps coronatus

金冠树八哥 jīn guān shù bā gē

Golden-crested Myna 🟢LC

L: 19–21 cm
Habitat: Inhabits lowland evergreen, deciduous, and mixed forests; also in open forests and forest edges with large standing trees; in some areas, found only in primary forests. Occurs from lowlands to about 800 m.
Behavior: Forages in pairs or small flocks; often mixes with Common Hill Mynas in Yingjiang, Yunnan. Found mainly in the canopy.
Distribution: Seen in W and S Yunnan and E Guangdong.
Voice: Poorly known vocalizations include whistles and slightly slurred, bell-like notes, *tik up*, similar to Common Hill Myna.

Golden-crested Myna

Gold area on crown smaller than male

Gold crown and throat ♂

♀

Remaining body mostly black

Yellow wing patch

Gracula religiosa

鹩哥 liáo gē

Common Hill Myna 🟢LC

L: 27–31 cm
Habitat: Inhabits secondary forests, evergreen broadleaf forests, deciduous broadleaf forests, bamboo thickets, and mixed forests in low-altitude hills and plains, especially in forest edges and small open areas in forests.
Behavior: Often forages in pairs or in small flocks for fruits in the canopy; sometimes in flocks with other mynas or starlings in fruiting trees.
Distribution: *G. r. intermedia* is a rare resident in SW and S Yunnan, SW Guangxi, and Hainan. The wild population has declined rapidly due to cage-bird trade, and escaped birds have been recorded in various parts of China and are established in some parts of South China.
Voice: Huge variety of whistles, wails, discordant croaks, and shrieks. Song is loud and varied, and call is harsh. Caged individuals are trained to imitate various human words and other sounds accurately.

Common Hill Myna

Bright yellow lappet extends behind eye and connects on nape

Orange-red bill

White wing patch, visible when perched

Acridotheres grandis
林八哥 lín bā gē
Great Myna LC

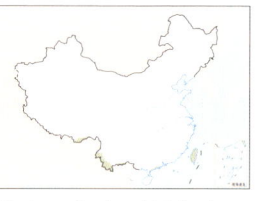

L: 24–27.5 cm
Habitat: Inhabits open areas, such as low-altitude forest edges, farmland, pastures, grasslands, and open fields with scattered trees; also in tall trees near human habitation, such as villages and towns.
Behavior: Often in pairs or small flocks; usually mixes with Collared Myna or other starlings, pecking at earthworms, mole crickets, and other insects disturbed by cattle. Also picks at parasites on cattle.
Distribution: Common resident in W and S Yunnan, W Guangxi, and SE Tibet. Established populations from escaped birds in Taiwan.
Voice: Most vocalizations similar to those of Crested Myna. Song a jumble of repeated discordant chattering and whistled phrases. Contact call a grating *chwur*; flight call a soft *pliu*, often doubled.

Acridotheres albocinctus
白领八哥 bái lǐng bā gē
Collared Myna LC

L: 25 cm
Habitat: Inhabits open grasslands and moist areas, including swamps, cultivated lands, and villages from lowlands to 1200 m.
Behavior: Forages mostly on the ground but also in trees. Often in company of cattle and buffalo. Gregarious; often mixes with Great Myna, Crested Myna, and other starlings.
Distribution: Seen in W Yunnan.
Voice: Discordant, complex combination of pleasant whistles, grating notes, and raucous chatter. Some imitation of other birds.

Acridotheres burmannicus
红嘴椋鸟 hóng zuǐ liáng niǎo
Burmese Myna LC

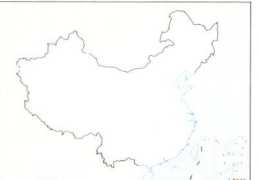

L: 22 cm
Habitat: Inhabits open grasslands, gardens, sports fields, cultivated areas, and scrubland; also in large clearings in forests, mainly in lowlands.
Behavior: Often forages on the ground but also in trees. Generally gregarious; moves about in pairs in breeding season. Often mixes with other starlings.
Distribution: Locally common resident in W Yunnan.
Voice: Scolding, harsh *chew-ii* and various chattering calls.

Acridotheres cristatellus
八哥 bā gē
Crested Myna LC

L: 23–28 cm
Habitat: Inhabits open areas of low-altitude hills and plains; adapted to a variety of habitats, such as cities, villages, farmland, parks, secondary broadleaf forests, bamboo thickets, and open forest edges. Nests in tree cavities, wall cavities, utility poles, pylons, etc.
Behavior: Monogamous. In pairs throughout the year, and pairs may be maintained even in large winter flocks. Forages while walking on the ground, using beak to search for food among grass and picking up food from the ground or from plants; also flies to chase insects. Omnivorous. Often pecks at earthworms, insects, and plant roots in plowed fields or stands on cattle to peck at parasites; in suburban areas, often forages in garbage piles until dusk before flying to roosts. A popular cage bird because it is easy to keep and train to imitate human languages.
Distribution: *A. c. cristatellus* is a common resident in most areas south of the Yellow River; *A. c. brevipennis* in Hainan; and *A. c. formosanus* in Taiwan. Introduced to various parts of China by cage-bird trade, with established populations from escaped individuals in some northern provinces.
Voice: Talented vocalist, especially in the evening, when it is very noisy. Song is loud and clear, and the call is raucous and noisy, often with a short *jaaay, jaaay, jaaay*. Caged individuals are trained to imitate simple human speech and other sounds.

Acridotheres tristis
家八哥 jiā bā gē
Common Myna LC

L: 24–27 cm
Habitat: Inhabits agricultural lands, short grasslands, orchards, towns, and parks in open areas at low altitudes. Highly adaptable; often closely associated with human habitation.
Behavior: In pairs or small flocks; often mixes with other starlings. Bold, unafraid of people. Forages on the ground and often accompanies livestock, sometimes standing on the animals' backs and pecking at parasites.
Distribution: *A. t. tristis* is a locally common resident in W and S Yunnan and Hainan. It has been introduced throughout the country by the cage-bird trade, and escaped individuals have formed large established populations in the southeast coast, W Xinjiang, southwestern China, and Taiwan, affecting the survival of other native mynas. Invasive species in many parts of the world.
Voice: Often highly vocal. A skilled mimic. Particularly noisy at roosts. Call similar to Crested Myna.

Great Myna

Raised crest feathers on forehead

Orange bill

White patch at base of primaries forms distinctive large wing patch

White undertail coverts lack bars

White tip on tail

Crested Myna

Black crest feathers on forehead curved and upturned but shorter and bushier than Great Myna

Ivory-colored bill, base of lower mandible sometimes tinged pale red

White patch at base of primaries forms distinctive wing patch

White wing patch, usually invisible when perched

Dull yellow legs

White bars on black undertail coverts

Collared Myna

Yellow bill

White side of neck

Black and white mottling on vent

Common Myna

Gold bare skin below and behind eye

Yellow bill

White wing patch larger than Great Myna and Crested Myna

Brown back, wing coverts, and belly

Yellow legs

White tip on tail

Burmese Myna

Grayish-white head

Red bill, base of lower mandible black

Dark vinaceous breast and belly

Spodiopsar sericeus

丝光椋鸟 sī guāng liáng niǎo

Red-billed Starling   LC

L: 18–23 cm

Habitat: Inhabits open plains, cultivated lands, and forest edges; also in urban green spaces.

Behavior: Bold. Gregarious; mixes with other starlings. Forages on ground, especially in plowed fields. Perches on trees and power lines to rest.

Distribution: Common resident, widespread in North, Central, South, southeastern, and southwestern China; some populations in North China migrate south in winter; uncommon winter visitor in Taiwan.

Voice: Song a mixture of harsh, scolding notes and some musical ones. Call is a rough *jreee*. Noisy when in flocks.

Spodiopsar cineraceus

灰椋鸟 huī liáng niǎo

White-cheeked Starling   LC

L: 19–23 cm

Habitat: Inhabits open areas close to farmland, short grasslands, and urban parks.

Behavior: Often in flocks. Forages while walking on the ground. Mixes with other birds. Body triangular in flight. Breeds in tree cavities and may excavate on its own.

Distribution: Breeds in Northeast, North, and eastern northwestern China; some populations do not migrate in winter; populations south of the Yellow River are mainly winter visitors plus a few residents; uncommon resident or winter visitor in Taiwan.

Voice: Wide variety of different vocalizations, most coarser and lower pitched than those of Red-billed Starling.

Gracupica nigricollis

黑领椋鸟 hēi lǐng liáng niǎo

Black-collared Starling   LC

L: 27–30.5 cm

Habitat: Inhabits a variety of habitats, such as open short grasslands, cultivated lands, deserted fields near wetlands, and towns.

Behavior: Often in pairs or small flocks; forages on the ground. Adapted to human environments but vigilant.

Distribution: Common resident in most areas south of the Yangtze River and in Hainan; an introduced species in Taiwan (may also involve populations naturally expanding from South China).

Voice: Energetic, distinctive vocalist. Song includes both coarse but melodious notes; a harsh *kraak* and higher-pitched, almost bee-eater-like *prrrp* and *kikilu-tlik* notes are particularly distinctive.

Gracupica contra

斑椋鸟 bān liáng niǎo

Asian Pied Starling   LC

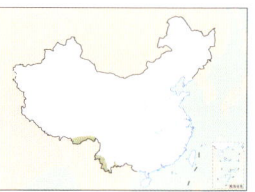

L: 22 cm

Habitat: Inhabits open grasslands, wet swamps, and also among scattered trees, usually near cultivated lands and human habitation. Found mainly in lowlands below 800 m.

Behavior: Forages mostly on the ground but also in trees. Usually in pairs or small flocks; often mixes with other starlings. Often in company of cattle and buffalo.

Distribution: *G. c. superciliaris* in W Yunnan and SE Tibet; *G. c. floweri* in S Yunnan. The latter subspecies is sometimes considered a separate species, Siamese Pied Starling (*G. floweri*).

Voice: Variable song includes shrill notes, churrs, buzzes, and melodic whistled phrases; *queert* calls are distinctive.

Agropsar sturninus

北椋鸟 běi liáng niǎo

Daurian Starling   LC

L: 16–19 cm

Habitat: Inhabits a variety of woodlands, including mixed forests and wooded areas in grasslands, and also occurs in woodlands in villages.

Behavior: Moves about solitarily or in small flocks. Vigilant and not as noisy as other starlings. May breed cooperatively.

Distribution: Breeds in Northeast and North China, westward up to Shaanxi; migrates through most of eastern China, including Taiwan. Uncommon.

Voice: Song includes particularly harsh notes, *spik* and *querp*, while calls include a mellow *quiup* and harsh grating notes.

Agropsar philippensis

紫背椋鸟 zǐ bèi liáng niǎo

Chestnut-cheeked Starling   LC

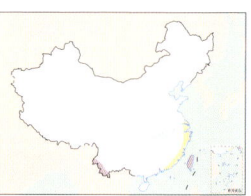

L: 16–19 cm

Habitat: Inhabits sparse woodlands in farmland and open fields; during migration, often occurs in woodlands and parks near oceanic islands or coasts.

Behavior: Often solitary or in flocks during migration; usually mixes with other starlings. Forages for fruit among tree branches; also looks for fallen fruit on the ground or hunts for insects.

Distribution: Rare passage migrant; migrates along island chains; seen on the islands off the east coast (including Taiwan and Hong Kong) and, occasionally, coastal areas on the mainland. Winter visitors recorded in SE and SW Yunnan; also rare.

Voice: Song is an excited, discordant, highly repetitive chattering. Calls include an excited *chair-chair-chair* and intense *tchick*. Other call notes reported to be similar to those of Daurian Starling.

Red-billed Starling

Dark red bill with blackish frontal part

Head white or tinged pale yellow

Distinct white wing patch

Off-white to pale grayish-brown head

Asian Pied Starling

Orange-red bare skin around eye

White face

Black back and wings, white wing patch

Black from throat to nape and neck

White from breast to vent

Lighter-colored wings and bare skin around eye

G. c. floweri

G. c. superciliaris

Daurian Starling

Purplish-black patch on nape; may disappear when feathers are worn

Pale brown head

Dark earth-brown back

White-cheeked Starling

Black or dark gray head

White cheek

Distinct white rump

White undertail coverts

Black-collared Starling

White rump

White tip on tail

Black and white on upper wings, appears mottled

Yellow bare skin around eyes and on cheek

Black from neck to upper breast

Chestnut-cheeked Starling

Distinct yellowish-white rump

White wing patch

Chestnut patch on cheek, enlarges during breeding season; side of neck also chestnut-tinged

Glossy purple on back

Gray flanks

Grayish-brown back

Sturnia sinensis

灰背椋鸟 huī bèi liáng niǎo

White-shouldered Starling　LC

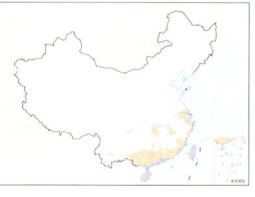

L: 17–20.5 cm
Habitat: Inhabits sparse woodlands in open fields, shelterbelts around wetlands, and woodlands on the outskirts of cities and towns; especially prefers dry fields near woods.
Behavior: Often in small flocks; occasionally solitary during migration. Often mixes with other starlings. Arboreal; rarely forages on the ground. Nests in natural tree cavities, or holes and cracks in walls.
Distribution: Locally common. Breeds in the coastal regions of South and southeastern China, including Hainan and Taiwan. Partly a migrant, with wintering populations in Taiwan and Hainan; often recorded along the east coast and its offshore islands during migration.
Voice: Poorly known. Calls include a soft *prep, prep-prep* and a harsh, explosive *kshaar* or *kt-shar*.

Sturnia pagodarum

黑冠椋鸟 hēi guān liáng niǎo

Brahminy Starling　LC

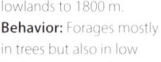

L: 20 cm
Habitat: Inhabits open deciduous forests and scrubland on lowland hills and cultivated areas near human habitation from lowlands to 1800 m.
Behavior: Forages mostly in trees but also in low bushes and occasionally on the ground. In pairs or flocks, sometimes in mixed-species flocks with other starlings.
Distribution: SE Tibet and W Yunnan.
Voice: Song is a very rapid discordant jumble of notes, with some repetition and mimicry. Calls include a series of rapidly repeated harsh shrieks, nasal notes, and various churrs.

Pastor roseus

粉红椋鸟 fěn hóng liáng niǎo

Rosy Starling　LC

L: 19–22 cm
Habitat: Inhabits montane grasslands, hills, deserts, plains, and near rural settlements, as well as on plateaus and mountains. Especially common in open areas with cliffs, water sources, and trees.
Behavior: Colonial breeder; nests in artificial environments, such as rock piles, rock crevices, and brick piles. Forages gregariously on the ground and sometimes in trees or bushes.
Distribution: Breeds in N Xinjiang, W Inner Mongolia, and NW Gansu;

Sturnus vulgaris

紫翅椋鸟 zǐ chì liáng niǎo

European Starling　LC

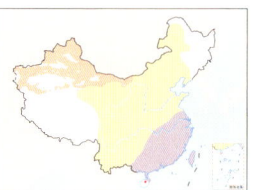

L: 19–22 cm
Habitat: Inhabits open areas, especially forest edges, agricultural areas, vicinity of towns, and desert edges.
Behavior: Gregarious; forms large flocks containing hundreds of individuals during migration. Forages on the ground.
Distribution: *S. v. porphyronotus* is a common breeder in N Xinjiang. *S. v. poltaratskyi* is a common breeder and resident in C and S Xinjiang. Both subspecies migrate south in nonbreeding season, and some wandering individuals have been recorded in various parts of East, South, and southwestern China; overwintering populations in some areas are of considerable size.
Voice: Highly variable complex song is a mixture of twittering whistles, warbles, and considerable mimicry. Calls are similarly variable and include a short metallic *chip* and a soft *prurrp*.

Sturnia malabarica

灰头椋鸟 huī tóu liáng niǎo

Chestnut-tailed Starling　LC

L: 18.5–20.5 cm
Habitat: Inhabits open woodlands and sparsely wooded areas from lowlands to 2000 m; often occurs in plantations, close to human habitation.
Behavior: Forages mostly in trees but also in low bushes and occasionally on the ground. Occurs in pairs or flocks, sometimes in flocks with other starlings.
Distribution: *S. m. malabarica* in SW Tibet; *S. m. nemoricola* in Yunnan, SW Sichuan, SW Guizhou, SW Guangxi, Hong Kong, Macao, and Taiwan.
Voice: Noisy whistles and high-pitched trills; chirpy and noisy when in flocks.

migrates through SW Xinjiang and W Tibet. Vagrant in E Hebei, S Sichuan, Jiangsu, Shanghai, Fujian, Taiwan, Hong Kong, and Macao. Found more commonly in NW Xinjiang.
Voice: Gives *ki–ki–ki* calls, and plaintive *shrr* calls when moving; makes curly-tongued *chik–ik–ik–ik* calls when in flocks. Flight and alarm calls shorter and harsher than those of European Starling. Constant chatter from large flocks includes many different notes and is reminiscent of European.

White-shouldered Starling

♂ ad.

Gray iris

Forehead, face, throat, flanks, and rump tinged pinkish brown; color fades to off-white or buff yellow after breeding season

Bluish-gray bill with dark base

Pure white wing coverts contrast strongly with black wing feathers, white area smaller on female than male

Pale brown body with dark brown wing feathers, gradually turns gray or white as bird ages

Off-white rump

Black on upper tail, pale brown on lower tail

juv.

European Starling

Yellow bill

Bluish-black plumage with metallic sheen

♂ br.

White spots overall

non-br.

Brahminy Starling

Long black crest feathers

Bill with yellow tip and blue base

Grayish brown from back to tail

Brownish yellow from face and throat to belly

Black wing feathers

Chestnut-tailed Starling

Grayish-white head

Bill with yellow tip and blue base

Chestnut from breast and belly to vent

Chestnut outer tail feathers

S. m. malabarica

Rosy Starling

Bluish-black head and neck

Pink body

Whitish underparts

S. m. nemoricola

鸫科　Turdidae

Small to medium-sized songbirds, including various thrushes, cochoas, the charismatic Grandala, and bluebirds, with well-proportioned body profiles. Plumage similar between sexes in more than two-thirds of species. Plumage mostly black, orange, or grayish brown; some with spectacular structural colors. Juveniles usually heavily spotted. Strong bill of medium size, with well-developed rictal bristles. Wings mostly long and pointed. Rounded or squared tail consists of 10–14 feathers. Legs long and strong, well adapted to hopping and walking on ground. Inhabit various habitats, but mostly forests and scrubland. Tree or ground dwellers. Mostly insectivorous, shifting toward herbivory during winter. Voice of many species loud but sweet.

Seventeen genera and 172 species recognized worldwide, with a cosmopolitan distribution. Five genera and 38 species recorded in China, ubiquitously distributed nationwide.

少斑地鸫属
Geokichla

Orange-headed Thrush

地鸫属
Zoothera

White's Thrush

大翅鸫属
Grandala

Grandala

鸫属
Turdus

Chinese Blackbird

宽嘴鸫属
Cochoa

Purple Cochoa

Geokichla citrina

橙头地鸫 chéng tóu dì dōng
Orange-headed Thrush

L: 20–23 cm
Habitat: Inhabits wet evergreen forests, secondary forests, and bamboo thickets; mostly near valley streams, also at forest edges.
Behavior: Solitary or in pairs. Shy and vigilant; more active in the evening. Uses beak to flip over leaves on the ground in search of food and flies to trees and remains stationary when disturbed.
Distribution: *G. c. melli* in Chongqing, Guizhou, Hubei, Hunan, Jiangxi, Guangdong, Hong Kong, Macao, and Guangxi; *G. c. courtoisi* (morphologically very similar to *G. c. melli*) in S Henan, Shaanxi, Anhui, Shandong, Jiangsu, and Zhejiang; *G. c. aurimacula* in Hainan; *G. c. innotata* in SW Yunnan. *G. c. citrina* possibly in China but remains to be confirmed.
Voice: Highly variable, slowly delivered, sweet song, with strophes 2–5 sec long and between 1.2–5 kHz, includes numerous flat whistles, fluty warbles, occasional tremolos, and considerable repetition, *wii-pi … pa-prii-priitilii … wiiuu-piirrpa-wiichii-liit … wiiichii-liit pir-wuu-piirrte.* Calls include a very high-pitched, descending, thin *dzzeet.*

Geokichla sibirica

白眉地鸫 bái méi dì dōng
Siberian Thrush

L: 20–23 cm
Habitat: Breeds in river valley areas; prefers coniferous forests or mixed broadleaf-coniferous forests with rich undergrowth but rarely occurs in broadleaf forests. Occurs in various woodlands and forest edges during migration.
Behavior: Solitary or in pairs. Quiet and secretive; forages on the ground and quickly flies into trees when disturbed.
Distribution: *G. s. sibirica* breeds in N Inner Mongolia and Heilongjiang and migrates through central and eastern China, with overwintering records in South China; *G. s. davisoni* migrates through the east coast, with fewer records.
Voice: Simple, languid song is one or two fluty notes, followed by a brief subdued twittering, with considerable repetition in the introductory notes; overall similar to song of Eyebrowed Thrush but higher pitched, less mournful, with fewer fluty notes, and more emphasis on the terminal twitter. Calls include a very high-pitched thin *tseep.*

Orange-headed Thrush

Two dark brown stripes below eyes and behind ear coverts

Gray back

♂

Distinct white wing bar on greater coverts

Brown back

♀

G. c. melli

Pale yellow lores, chin, and throat

Head lacks brown streaks

G. c. aurimacula

No white wing bar on greater coverts

G. c. innotata

Siberian Thrush

Dull brown wing feathers

♂

White central belly

White bars on undertail coverts

G. s. sibirica

ad.

Almost no white on belly

White bars on undertail coverts

G. s. davisoni

Buff supercilium narrower than male

Indistinct pale wing bar

♀

Long buff submoustachial stripe, dark brown malar stripe

Scaly pattern on breast, belly, and flanks

Buff supercilium and submoustachial stripe

Often with sparse white or pale yellow spots around ear coverts

♂

juv.

Often with white spots on gray-tinged flanks

509

Zoothera mollissima

淡背地鸫（光背地鸫）

dàn bèi dì dōng (guāng bèi dì dōng)

Alpine Thrush LC

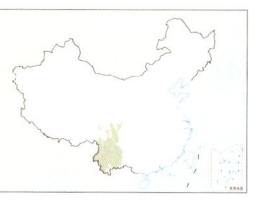

L: 24–27 cm
Habitat: Breeds above tree line, where the ground is usually covered with moss, lichens, grass, low rhododendrons, and sizable rocks; migrates downward to 300–3500 m in nonbreeding season.
Behavior: Prefers moving about in open forest clearings or grassy areas at forest edges and often perches and calls on rocks or other prominent positions.
Distribution: Sichuan and C and NW Yunnan.
Voice: Scratchy, hoarse, unmusical song usually a complex set of short whistles, buzzes, and tremulous notes lasting 2–3 sec and delivered every 5 sec or so. Not especially loud and contains little repetition: *plii-tuu titi zrrrr-triue-plii-up-plup*. Calls include an explosive rattle in alarm and a *chack*. Birds in Yunnan sing more slowly and with a lower voice than in other regions and may be a separate species.

Zoothera griseiceps

四川淡背地鸫（四川光背地鸫）

sì chuān dàn bèi dì dōng (sì chuān guāng bèi dì dōng)

Sichuan Thrush LC

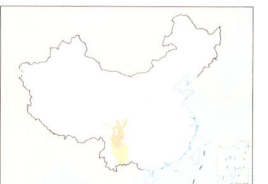

L: 26–27 cm
Habitat: Breeds in secondary broadleaf forests, bamboo thickets, and coniferous forests with rich undergrowth, at altitudes of 2100–3300 m.
Behavior: Secretive; rarely occurs in open areas.
Distribution: Breeds in C Sichuan, possibly also in Shaanxi. Migrates through Yunnan.
Voice: Song is similar to Himalayan Thrush but richer and lower pitched, with longer, more musical, fluty notes and slower overall speed.

Zoothera salimalii

喜山淡背地鸫（喜山光背地鸫）

xǐ shān dàn bèi dì dōng (xǐ shān guāng bèi dì dōng)

Himalayan Thrush LC

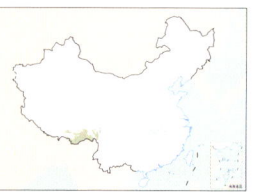

L: 25–27 cm
Habitat: In breeding season, inhabits coniferous forests with rich undergrowth and mixed with broadleaf trees or rhododendrons at altitudes of 3200–3800 m; prefers denser broadleaf forests at lower altitudes in nonbreeding season.
Behavior: Often secretive but sometimes perches and sings from tree branches in breeding season.
Distribution: Breeds in S Tibet and NW Yunnan; possibly also in S Sichuan and SE Yunnan (Lvchun).
Voice: Jaunty, pleasant, rich song includes very few harsh scratchy notes, compared to Sichuan Thrush.

Zoothera dixoni

长尾地鸫 cháng wěi dì dōng

Long-tailed Thrush LC

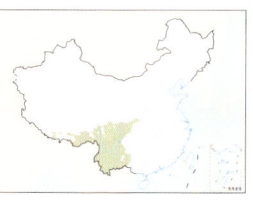

L: 27–28 cm
Habitat: Inhabits coniferous forests or alpine scrubland near tree line in breeding season and migrates to lower altitudes in winter, where it may appear near forest edges, streams, and cultivated land.
Behavior: Solitary or in pairs. Secretive; flies into bushes and remains stationary when disturbed.
Distribution: S Gansu, E Qinghai, Sichuan, Guizhou, Yunnan, S and SE Tibet, and W Guangxi.
Voice: Slow strained song is a series of phrases of variable length (1.8–5.2 sec), consisting mostly of short, low-pitched (1.3–3.9 kHz), fluty chortles, dry trills, tremulous twitters, and longer descending rasps, *tddt-wut-zweer-brzt chut-plwrrrrr-zwer plit*. Call unknown.

Zoothera aurea

虎斑地鸫（怀氏虎鸫）

hǔ bān dì dōng (huái shì hǔ dōng)

White's Thrush LC

L: 26–30 cm
Habitat: Breeds in coniferous, broadleaf, or mixed forests in northern China, particularly near streams; inhabits a variety of scrubby habitats during migration, including urban parks and forest edges.
Behavior: Often solitary. Vigilant. Walks on the ground to forage; quickly flies into nearby trees and stays stationary when disturbed. Rarely appears in the canopy.
Distribution: Seen throughout China. Breeds in northern Northeast China, winters in East, South, and southwestern China, and migrates through other regions. Fairly common.
Voice: Song a drawn-out, slow, weak-sounding, yet far-carrying series of thin, melancholy, flat whistles lasting just over one second and repeated monotonously every 4–5 sec, *weeee … weeee*; pairs occasionally duet. Rarely calls.

Zoothera dauma

小虎斑地鸫（虎斑地鸫）

xiǎo hǔ bān dì dōng (hǔ bān dì dōng)

Scaly Thrush LC

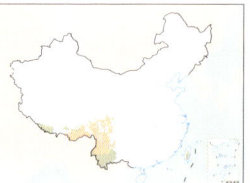

L: 23–27 cm
Habitat: Often breeds in broadleaf forests above 2000 m.
Behavior: Mostly solitary. When foraging, sometimes stops suddenly after running quickly, resembling plover. When resting, sometimes bobs body while head remains stationary.
Distribution: Resident or breeder in SE Tibet, W Sichuan, W and S Yunnan, Guangxi, and Taiwan; relatively rare.
Voice: Slow song a series of unmusical, seemingly disconnected, abrupt, simple, mostly tremulous, grating, low-pitched (1.6–4.3 kHz) notes lasting about 0.5 sec and given every 2–3 sec, *plueeei-chit-tweee-pleee-triuuu*, like White's Thrush. Rarely calls.

Alpine thrush

Black spots behind
ear coverts

No strong contrast
between head and back

Pale pink or pale
yellow base of
lower mandible

Breast slightly
tinged brown

Black claws

Long-tailed Thrush

Black spots at
ear coverts

Two wing bars

Scaly pattern on
breast and belly

White fringe on
outer tail feathers

Sichuan Thrush

No black patch from
behind eye to nape

From forehead and crown to
neck distinctly tinged gray,
contrasting with back

Black base
of lower
mandible

White's Thrush

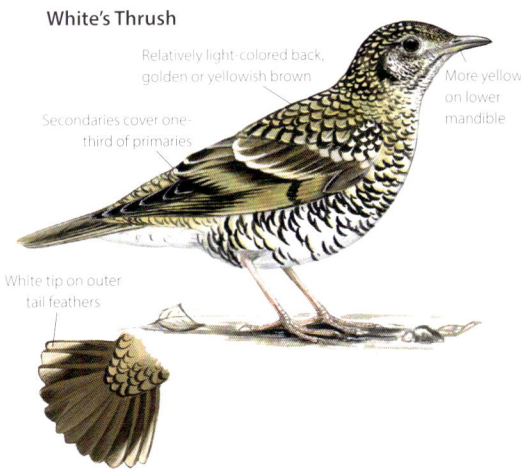

Relatively light-colored back,
golden or yellowish brown

More yellow
on lower
mandible

Secondaries cover one-
third of primaries

White tip on outer
tail feathers

Himalayan Thrush

Black at lores and
below eyes

Black base of lower mandible

Relatively broad
black malar stripe

Short tail

Pale claws

Scaly Thrush

Almost entirely black
lower mandible, only
base dull yellow

Rufous-brown back,
relatively dark overall

Secondaries cover
half of primaries

No white tips on
outer tail feathers

Zoothera monticola

大长嘴地鸫 dà cháng zuǐ dì dōng

Long-billed Thrush　LC

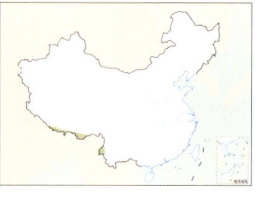

L: 26–28 cm
Habitat: Inhabits
undisturbed, moist,
shady fir, bamboo, and
rhododendron forests;
also occurs in evergreen
broadleaf forests, often
near streams at altitudes
of 900–3800 m.
Behavior: Forages in moist leaf litter and on open muddy ground, digs
with bill in search of invertebrates, often flicking wings.
Distribution: S and SE Tibet and W Yunnan.
Voice: Simple song is a flat, low-pitched (2 kHz), slowly delivered, two-
and occasionally three-note whistle, *tuuu-tweeee*, lasting about 1.5 sec
and given every 3–10 sec.

Zoothera marginata

长嘴地鸫 cháng zuǐ dì dōng

Dark-sided Thrush　LC

L: 24–25 cm
Habitat: Inhabits
evergreen broadleaf
forests and forest edges
and bamboo thickets,
mainly in damp areas
near rocky streams; also in
dense reeds inside forests
at altitudes of 750–2100 m.
Migrates to lower altitudes in winter.
Behavior: Forages in moist leaf litter and on open muddy ground, digs
with bill in search of invertebrates.
Distribution: SE Tibet and W Yunnan.
Voice: Thin, reedy, three-note, flat (around 3.9 kHz) whistle, *tweee-de-
dee*, lasting about 3.4 sec, higher pitched and much tinnier than White's
Thrush. Calls include a hard, short *kuk*.

Grandala coelicolor

蓝大翅鸲 lán dà chì qú

Grandala　LC

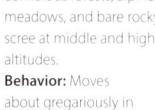

L: 19–23 cm
Habitat: Inhabits alpine
coniferous forests, alpine
meadows, and bare rocky
scree at middle and high
altitudes.
Behavior: Moves
about gregariously in
high-altitude treeless
environments outside breeding season; also descends to coniferous
forests in winter.
Distribution: Seen along the Himalayas to E Qinghai-Tibet Plateau,
including NW and SW Gansu, NE and S Qinghai, SE and S Tibet,
NW Yunnan, Sichuan, and Chongqing.
Voice: Song and call not well distinguished. Calls include a finchlike,
husky, scolding, slightly tremulous, descending *tsheuu*, repeated every
second or so.

Turdus hortulorum

灰背鸫 huī bèi dōng

Gray-backed Thrush　LC

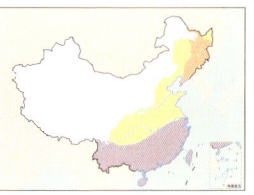

L: 18–23 cm
Habitat: Breeds in forests
on hills below 1500 m,
prefers areas near water;
in open bamboo thickets
and woodlands in winter.
Behavior: Solitary or in
pairs in breeding season,
often hides in canopy.
Secretive and vigilant during migration. Gathers in small, loose flocks
in winter.
Distribution: Breeds in Northeast China; migrates along east coast.
More commonly seen south of the Yangtze River in winter.
Voice: Powerful song a series of simple, unhurried, evenly spaced
phrases of three to seven short notes. Each strophe usually contains just
one or two different notes, with some sounding paired (with the second
note invariably higher and usually shorter than the first), generally in
a narrow frequency range and involving considerable repetition and
mimicry. Occasional strophes are finished with weaker chuckling. Call is
a thin, insectlike *zizi*.

Turdus unicolor

蒂氏鸫（梯氏鸫）dì shì dōng (ti shi dōng)

Tickell's Thrush　LC

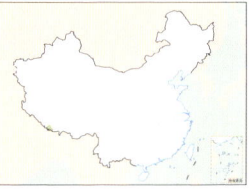

L: 20–25 cm
Habitat: Inhabits open
broadleaf or mixed forests
in breeding season; also
near villages.
Behavior: Solitary or in
pairs. Often sings from
treetops before sunrise
in breeding season. Shy
and vigilant. Forages in forest clearings, by digging in soil or lifting leaves.
Quickly flies to trees when disturbed.
Distribution: All records in China from Tibet, mostly around Zhangmu.
Voice: Song is a relaxed slow series of three to six short, wheezy, slurred,
tremulous, scolding, staccato notes, as well as some more musical ones,
some repetition, *dzzwee-wu … tuu-chup-chup*. Calls include a loud,
Fieldfare-like splutter and chuckling, *juk-juk-juk*.

Turdus dissimilis

黑胸鸫 hēi xiōng dōng

Black-breasted Thrush　LC

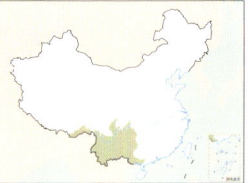

L: 22–24 cm
Habitat: In moist broadleaf
and mixed forests with
rich undergrowth on low
mountains and hills; also in
villages and parks in winter.
Behavior: Vigilant in
breeding season. Bold in
nonbreeding season; even
on roadsides. Forages
under trees or bushes; attracted to fruiting trees.
Distribution: Found in Yunnan, Guizhou, SW Sichuan, Guangxi, and
Tibet. Records in Guangdong are probably escaped cage birds.
Voice: Attractive, slow, powerful song is highly varied, with numerous
types of notes, occasional repetition of the same note, as well as mimicry.
Calls a rich staccato *tuc-tuc-tuc* or *chup-chup-chup* and a thin *see*.

Long-billed Thrush

Relatively thick, strong, slightly long, and decurved bill

Grayish-blue upperparts

White throat

Black breast

White belly, black spots on upper belly

Gray-backed Thrush

Slightly dark lores

♂

More brownish on greater coverts and wing feathers

Arrow-shaped black marks on breast

♀

Dark-sided Thrush

Crescent-shaped black patch on mottled off-white face

Slightly slender bill

White throat

White spots on blackish-brown breast and belly

Tickell's Thrush

Slightly dark lores

Pale gray fringe on wing feathers

Grayish-white chin and throat

♂

Relatively pale lower belly

Fine pale streaks around ear coverts

Black spots on pale brown breast band

Brown-tinged flanks

♀

Grandala

Black lores

Black wings and tail

♂

Bright bluish-purple body

Grayish brown overall; with short fine brown streaks except for wings

♀

Black-breasted Thrush

Black head and breast

♂

Orange belly

♀

Rounded spots on breast become arrow shaped lower down

Turdus cardis
乌灰鸫 wū huī dōng
Japanese Thrush · LC

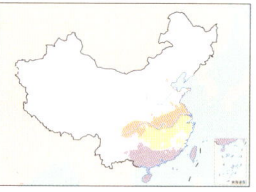

L: 18–23 cm
Habitat: Breeds in densely wooded valleys or near streams; occurs in woodlands, scrubland, even gardens during migration. Altitude range: 400–1500 m.
Behavior: Mostly solitary or in pairs. Vigilant; gathers in small flocks in winter.
Distribution: "Gray-backed" individuals (mainland breeding population) breed in central China, migrate through most of eastern China, and winter in South and southwestern China. "Black-backed" individuals (Japanese breeding population) move through coastal areas south from S Jiangsu during migration and winter in Fujian, Guangdong, and Hainan.
Voice: Song lasts 3–4 sec, often starts with three to six whistled, flat notes, most paired, and includes higher-pitched, subdued twittering. Calls apparently include a thin *tsweee* or *tsuuu* and a hollow *chuk*.

Turdus albocinctus
白颈鸫 bái jǐng dōng
White-collared Blackbird · LC

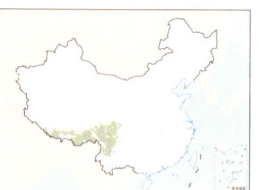

L: 26–28 cm
Habitat: Breeds in broadleaf, coniferous, and mixed broadleaf-coniferous forests with well-developed undergrowth at altitudes of 2270–4300 m, sometimes above tree line.
Behavior: Vigilant. Often solitary or in pairs, perching on treetops and sometimes foraging on open grasslands. Gathers in small flocks in nonbreeding season.
Distribution: S Tibet, NW Yunnan, and W Sichuan.
Voice: Simple, slightly melancholy, three- to eight-note song involves repetition of one or two short whistled notes in a lengthy, slowly delivered series notable for its narrow frequency range and repetition. Calls include a throaty Chinese Blackbird–like *tuck-tuck* or *chack-chack*.

Turdus boulboul
灰翅鸫 huī chì dōng
Gray-winged Blackbird · LC

L: 28–29 cm
Habitat: Breeds in coniferous forests, oak forests, and rhododendron thickets; winters in woodlands, scrubland, and near villages.
Behavior: Vigilant; flies to the middle and upper parts of trees to hide when disturbed. Solitary or in pairs in breeding season. Aggressive. Gathers in small flocks in nonbreeding season.
Distribution: Found in southwestern and Central China; occasionally in South China.
Voice: Powerful, fluty, variable song, with short strophes (2.5–5 sec) and beginning with paired introductory whistles that vary in pitch (but no tremolos) and often ending with a harsher twitter. Includes mimicry. Calls reportedly include a low *chuck* and mellow, descending, repeated *tiu*.

Turdus merula
欧亚乌鸫 (欧乌鸫) ōu yà wū dōng (ōu wū dōng)
Eurasian Blackbird · LC

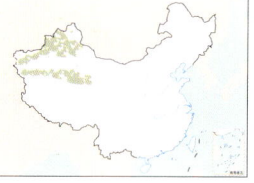

L: 24–27 cm
Habitat: Adapted to a wide range of habitats, from woodlands to urban green spaces.
Behavior: Small. Very well adapted to human environments. Often forages on the ground, nests in tall trees, and rests with slightly drooping wings.
Distribution: Common in Xinjiang and W Qinghai.
Voice: Both call and song are varied. See Chinese Blackbird.

Turdus mandarinus
乌鸫 wū dōng
Chinese Blackbird · LC

L: 28–29 cm
Habitat: Adapted to a variety of habitats, from woodlands to urban green spaces; may even nest and breed on balconies.
Behavior: A common black thrush in eastern China. Large; adapted to a variety of habitats, including urban parks and residential areas. Often moves about on the ground. Bold; sometimes in small flocks, resting with slightly drooping wings.
Distribution: Widespread in eastern and central China. Seen from Beijing and E Inner Mongolia in the north, to Gansu, Sichuan, and Yunnan in the west, and to Hainan in the south. Very common.
Voice: Song is often faster and more vigorous than song of Eurasian Blackbird, with more pronounced and longer, lengthy, descending, wheezy, sliding, Oriental Magpie-Robin–like notes and fewer tremolos, and usually with much more repetition. Most phrases lack a high-pitched ending. The *chack* call is thinner than call of Eurasian. Also makes a much more mellow *stup-tup-tup* that has no real equivalent in Eurasian.

Turdus maximus
藏乌鸫 zàng wū dōng
Tibetan Blackbird · LC

L: 23–28 cm
Habitat: Inhabits subalpine scrubland, rhododendron forests, and meadows at altitudes of 2000–4500 m; moves to lower altitudes in winter.
Behavior: Similar to Chinese Blackbird but more vigilant and seldom calls.
Distribution: Common in E and S Tibet.
Voice: Song is lower pitched, flatter, far less ambitious, and more repetitive than the songs of Eurasian and Chinese Blackbirds, and it lacks their warbling phrases. Calls include a low White-backed Thrush–like *chut-ut-ut* and a harder *chak-chak-chak-chak*.

Japanese Thrush

Japanese breeding population

Blackish back

♂

Black flanks

Gray back

♂

Gray flanks

Mainland breeding population

Black spots on chestnut-tinged flanks

White-collared Blackbird

Distinct yellow eye ring

White neck

♂

Some individuals have pure white throat

White streaks on undertail coverts

♂

Some individuals have grayish throat

♀

Gray-winged Blackbird

Yellow or grayish-white eye ring

♂

Fine white streaks on flanks

Brown wings

♀

Eurasian Blackbird

Relatively large bill often with orange hue

♂

♀

Whitish chin and central throat

Chinese Blackbird

Relatively narrow yellow eye ring often with teardrop effect at rear

Yellow bill

Rusty-tinged body

♂

Proportionately longer tailed than Eurasian Blackbird

♀

Streaked chin, throat, and upper breast

Brown crown

Off-white throat and breast, sparse black spots on breast

juv.

Tibetan Blackbird

Narrow eye ring, usually none

Pure black body lacks rusty color or sheen

♂

Feathering extends slightly farther down upper mandible

Narrow eye ring, usually none

Brown overall

♀

Turdus poliocephalus
岛鸫 dǎo dōng
Island Thrush

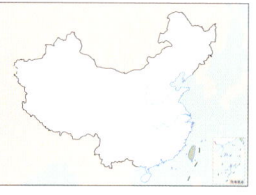

L: 17–25 cm
Habitat: Broadleaf forests at 1000–3000 m.
Behavior: Shy and secretive. Sings from treetops in breeding season; often forages in small flocks in the middle and upper stories in nonbreeding season, sometimes mixes with other thrushes. Favors fruits of *Idesia polycarpa* trees.
Distribution: Widespread across islands in the Pacific Ocean, with marked morphological variation. *T. p. niveiceps* is endemic to Taiwan and relatively rare. Sometimes treated as Taiwan Thrush (*T. niveiceps*).
Voice: Variable and occasionally complex song is a short series of melodious whistles, repeated lower-pitched flutelike notes, and broad-spectrum chattering and fast warbling, at times vaguely reminiscent of European Starling. Call is *kekekeke* and *tzee tzee*.

Turdus feae
褐头鸫 hè tóu dōng
Gray-sided Thrush

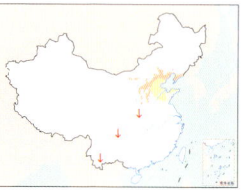

L: 22–23.5 cm
Habitat: Breeds in coniferous forests at 1200–2000 m; nests in *Larix principis-rupprechtii* or *Abelia biflora*.
Behavior: Secretive. Often calls before dawn; mostly silent after sunrise, hiding in canopy and difficult to find. Relatively quiet during migration.
Distribution: Breeds in the mountains of northern North China; passage migrants recorded in Henan, Hubei, and Sichuan, may occur throughout southwestern China.
Voice: Simple, monotonous song is a series of alternating, repeated, simple, short phrases similar to those of Eyebrowed Thrush but often with terminal twittering (short trills and richer melodious burry notes) and then more reminiscent of Siberian Thrush. Calls include a thin *zeeee* and an explosive rattling chatter, also very similar to Eyebrowed.

Turdus kessleri
棕背黑头鸫 zōng bèi hēi tóu dōng
White-backed Thrush

L: 25–28 cm
Habitat: Inhabits rhododendron thickets, scrubland, or open rocky areas at 3600–4500 m in breeding season; moves to low altitudes in winter.
Behavior: Forages on ground. Flies low; often combining flaps and glides.
Distribution: Relatively common in E Qinghai-Tibet Plateau.
Voice: Simple unmusical song is a slow-paced repetition of two, harsh, tremulous, coarse, almost squawking notes every 3–10 sec. Calls include short, rapid, nasal rattles and chuckles, some similar to Tibetan Blackbird.

Turdus obscurus
白眉鸫 bái méi dōng
Eyebrowed Thrush

L: 20–24 cm
Habitat: Breeds in taiga or mixed forests; often nests near streams. Occurs at forest edges, orchards, and parks during migration and winter.
Behavior: Solitary or in pairs. Mixes with Dusky and Red-throated Thrushes in winter. Vigilant; when disturbed, flies into trees and stands still.
Distribution: Breeds in N Heilongjiang, migrates through most of the country, and winters in South and southwestern China.
Voice: Simple song of short repetitive strophes, very similar to Siberian Thrush and especially Gray-sided Thrush but usually delivered more timidly than Gray-sided, with the final note of each strophe lower pitched and normally lacking the terminal twittering. Calls also similar and include a thin *tseep* and crackling rattle.

Turdus rubrocanus
灰头鸫 huī tóu dōng
Chestnut Thrush

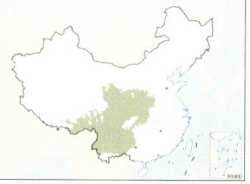

L: 25–28 cm
Habitat: Inhabits mixed forests with dense undergrowth at 2300–3300 m. In orchards and green spaces in winter.
Behavior: Vigilant. Often solitary or in pairs; gathers in small flocks in winter and sometimes mixes with other thrushes, such as Gray-winged and White-collared Blackbirds. Often sings from canopy in early breeding season at dawn and dusk.
Distribution: *T. r. rubrocanus* in S Tibet, N and W Sichuan, Shanxi, and Henan. *T. r. gouldii* widespread in central and southwestern China; vagrant in Shandong (Jinan), Jiangxi, and Guangxi.
Voice: Simple song a measured series of repeated, pleasant, well-spaced (2.5–5 sec apart), loud phrases of three to eight clear, short notes (about 0.2 sec in length and separated by about 0.3 sec). Most strophes are just one note repeated; others are two notes. Calls include a dry staccato rattle and a harsh, deep *chuck-chuck-chuck*.

Turdus pallidus
白腹鸫 bái fù dōng
Pale Thrush

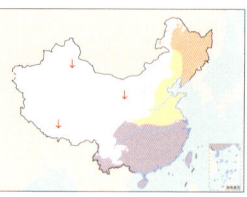

L: 22–23 cm
Habitat: Breeds in dense mixed forests; in orchards, parks, and fields during migration and winter.
Behavior: Vigilant; solitary or in pairs. Forages on ground in winter; sometimes mixes with other thrushes, such as Eyebrowed Thrush.
Distribution: Breeds in Northeast China, migrates through North China, winters in East, South, and southwestern China; more commonly seen in the eastern wintering areas. Occasionally in Western China.
Voice: Simple, stereotyped song is a series of loud, cheerful, but short, slow, monotonous strophes, *trrer-treer-trrer tuudle*, occasionally ending with terminal twittering. Calls include a harsh *chuck-chuck* and a bubbling alarm.

Island Thrush

All-white head

Dark spots on ear coverts

Rufous-brown belly

T. p. niveiceps

Gray-sided Thrush

Dark brown lores

Relatively pale throat

White supercilium

Black lores

White subocular line

Slate-gray flanks differ from Eyebrowed Thrush

White wing bar

1st win.

White-backed Thrush

Fine pale streaks on ear coverts

Color on head and neck distinctly darker than back

Eyebrowed Thrush

Slightly dark lores

Pale yellow fringe on wing feathers

Grayish-white submoustachial stripe extends to side of neck

Chestnut Thrush

Grayish-white head

Gray head

Grayish-white neck

T. r. rubrocanus

Vent and central belly very pale, nearly white

Smoky-gray head

Relatively heavy chestnut on body

Brownish head

T. r. gouldii

Pale Thrush

Grayish-brown head

Brown or olive-brown forehead

Chestnut back

Turdus atrogularis
黑颈鸫 hēi jǐng dōng
Black-throated Thrush　　

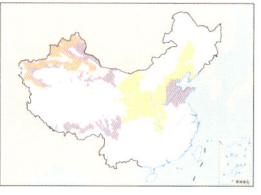

L: 20–24 cm
Habitat: Breeds in coniferous or mixed forest edges; winters on farmland, forest edges, and other open environments. Birds wintering in eastern China prefer forests in mountains.
Behavior: Typical thrush but bolder. Prefers perching in the middle and upper parts of trees. Often gathers in flocks; also mixes with other thrushes, such as Red-throated Thrush and Naumann's Thrush, and sometimes in winter with smaller birds, such as Brambling. Forages on the ground and also prefers feeding on cypress seeds.
Distribution: Breeds in W and N Xinjiang and winters in northwestern and North China. Migrates through southwestern China. More often seen in the west than in the east.
Voice: Song a very simple series of slow-paced, unpleasant, short phrases separated by 3–5 sec. Some individuals open each phrase with a harsh, broad-spectrum, complex chatter and follow that with two or three short warbled notes, while other individuals make more ambitious efforts, *tlip-tlip-tlip … che-che-che … cherr-viu*. Calls similar to Red-throated Thrush.

Turdus naumanni
红尾斑鸫 (红尾鸫)
hóng wěi bān dōng (hóng wěi dōng)
Naumann's Thrush

L: 20–24 cm
Habitat: Breeds in open woodlands; winters in open woodlands, farmland edges, urban green spaces, etc.
Behavior: Moves about in small to large flocks during migration and winter; sometimes mixes with other thrushes. Forages while walking and hopping on the ground or searches for seeds on cypress and other plants. Bolder than other thrushes.
Distribution: Occurs as passage migrant or winter visitor in most parts of China except Tibet and Hainan; common in eastern China and occasional or rare in Western China.
Voice: Noisier than the many other thrushes. Calls include a subdued *tut-tut-tut*, a more staccato *chuck*, and *spirr-spirr* in flight.

Turdus eunomus
斑鸫 bān dōng
Dusky Thrush

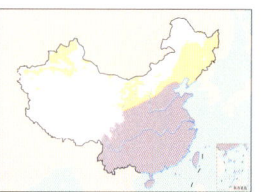

L: 19–24 cm
Habitat: Breeds in open forests and sometimes mountainous areas; wintering environment similar to that of Naumann's Thrush.
Behavior: Similar to Naumann's Thrush.
Distribution: Seen widely across China as passage migrant or winter visitor.
Voice: Calls like Naumann's Thrush.

Turdus ruficollis
赤颈鸫 chì jǐng dōng
Red-throated Thrush

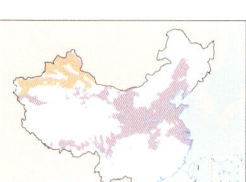

L: 22–25 cm
Habitat: Breeds in open coniferous forests; migrates and winters in open environments, including farmland, parks, forest edges, etc.
Behavior: Similar to Black-throated Thrush.
Distribution: Breeds in W and N Xinjiang, migrates through most of Northeast and northwestern China, and winters in southern Northeast, North, and southwestern China, with a few records in East and Central China. More common in the central and eastern parts of the wintering range.
Voice: Vocalizations like Black-throated Thrush.

Turdus pilaris
田鸫 tián dōng
Fieldfare

L: 22–27 cm
Habitat: Inhabits a variety of woodlands and scrubland; also occurs in open grasslands and rocky areas in winter.
Behavior: Bold; often stands on treetops. Gathers in high density in breeding season and in flocks of hundreds of individuals in winter.
Distribution: Breeds in N Xinjiang and winters in S Xinjiang and W Qinghai, with occasional records in W and NE Inner Mongolia and W Gansu in winter. Vagrant in Beijing and Inner Mongolia (Chifeng).
Voice: Unpleasant harsh song is a tuneless, scratchy, chattering medley of chuckles, wheezes, and trills, with some call notes admixed. Call a loud two- or three-note *shack-shack-shak*.

Turdus iliacus
白眉歌鸫 bái méi gē dōng
Redwing　　

L: 19–23 cm
Habitat: Inhabits Russian olive forests, orchards, urban parks, and scrubland near water in winter.
Behavior: Joins other thrushes in large mixed-species flocks in winter. Feeds on fruit, such as crabapples and Russian olives, in trees or on ground; also catches insects near water.
Distribution: *T. i. iliacus* is a rare winter visitor in N and S Xinjiang and vagrant in Beijing.
Voice: Flight call a lengthy, high-pitched *shriiii*, occasionally combined with a harsh *grek-grek-tek*.

Black-throated Thrush

♂ Heavy black on cheek, throat, and breast

♀ Pale throat

Scaly breast

Dark gray at base of tail feathers

Pale throat

Dark streaks on flanks

1st win.

Red-throated Thrush

Red supercilium

Gray back

Dark wing feathers

Red outer tail feathers

♂ Rufous-brown chin, throat, and breast

♀ Throat color usually relatively pale

Black spots on malar stripe and breast, color of breast feathers slightly bright

Hybrids of Red-throated Thrush and Black-throated Thrush, and of Dusky Thrush and Naumann's Thrush, are common, especially in winter. Hybrids of Red-throated Thrush and Naumann's Thrush have also been reported.

♂ Chocolate-colored breast, looks dirty

Relatively pale supercilium

♀ Pale throat

Black spots on chocolate-colored breast

Black-throated Thrush x Red-throated Thrush

Naumann's Thrush

Rufous-brown supercilium

Red spots on wing coverts

♂ Red throat

A few black spots on malar stripe

Less brown on wings

♀ Relatively pale throat

Relatively dark tail

Pale supercilium

Pale throat

Distinct malar stripe

1st win.

Fieldfare

Relatively long tail

Large triangular black spots on flanks

Redwing

Prominent white supercilium

Dense black streaks on upper breast and flanks

Rusty flanks and under wing

Dusky Thrush

Black crown

Black back contrasts strongly with brown wings

♂

Brownish back

♀

Scaly black pattern on flanks

Hybrids of Dusky Thrush and Naumann's Thrush are common. Hybrids usually join flocks of Dusky Thrush or Naumann's Thrush. Their behavior and profile are similar to the two parental species.

Strong contrast between black and white on head, resembles Dusky Thrush

Scaly rufous-brown pattern on flanks, resembles Naumann's Thrush

White throat

Dusky Thrush x Naumann's Thrush

Scaly dark brown on flanks

Turdus chrysolaus
赤胸鸫 chì xiōng dōng
Brown-headed Thrush   LC

L: 23–24 cm
Habitat: Inhabits mixed forests and subalpine coniferous forests in breeding season; occurs in woodlands and parks in winter.
Behavior: Vigilant; mostly hides in the undergrowth, moving under weeds or bushes. Sometimes mixes with other birds, such as Eyebrowed Thrush.
Distribution: During migration, found in coastal areas south from Shandong; winters in Guangdong, Hong Kong, Taiwan, and Hainan. Vagrant in Beijing, Tianjin, and Hebei.
Voice: Song is a simple repetition of one or two different notes. Calls include a harsh Pale Thrush–like *chuck-chuck*.

Turdus philomelos
欧歌鸫 ōu gē dōng
Song Thrush   LC

L: 20–22 cm
Habitat: Inhabits edges of coniferous, mixed, and broadleaf forests; particularly mixed broadleaf-coniferous forests.
Behavior: In pairs in breeding season; forages in small flocks on the forest floor or in bushes in nonbreeding season.
Distribution: Uncommon breeder in the Altai Mountains in northernmost Xinjiang.
Voice: Highly variable song is a loud, lively, sustained series of short, fairly rapidly delivered phrases, each generally repeated two or three times. Calls include a thin *sip*.

Turdus mupinensis
宝兴歌鸫 bǎo xīng gē dōng
Chinese Thrush   LC

L: 20–24 cm
Habitat: Breeds in montane broadleaf forests, mixed forests, or a variety of other forests with rich understory bushes; occurs in a variety of wooded environments during migration.
Behavior: Solitary or in small flocks. Relatively rounded profile. Often forages on the ground, suddenly trotting for a while, then standing and keeping silent.
Distribution: Resident in central and southwestern China, as well as S Shaanxi, Ningxia, and S Gansu; summer breeder in the Taihang Mountains to Beijing and N Hebei.
Voice: Song is an often-lengthy series of 2–15 (usually 4–6) mostly pleasant, hugely varied, mostly short, whistled notes, some rising, others falling, some tremulous or even slightly burry, with often only very short pauses between strophes; contains limited repetition and often mimicry of species such as Large Hawk-Cuckoo. Rarely calls.

Turdus viscivorus
槲鸫 hú dōng
Mistle Thrush   LC

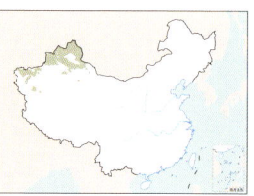

L: 26–29 cm
Habitat: Often inhabits mixed forests, coniferous forests, forest clearings, forest-edge bushes, and river-valley woodlands in mountainous areas; also occurs in shelterbelts, orchards, and urban parks.
Behavior: Solitary or in pairs during breeding season; gathers in flocks in other seasons. Often forages in the forest undergrowth or on ground. Immediately flies into trees when disturbed and makes harsh chattering notes.
Distribution: Resident throughout Xinjiang, more common in N Xinjiang than S Xinjiang. Partially migratory.
Voice: Song is a mournful, melancholy, descending, fluty warble, similar to the song of Eurasian Blackbird. Call is a dry rattle, *zerrrrrr*.

Cochoa purpurea
紫宽嘴鸫 zǐ kuān zuǐ dōng
Purple Cochoa   LC

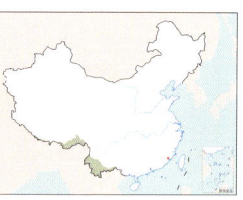

L: 26–28 cm
Habitat: Inhabits evergreen broadleaf forests at middle altitudes.
Behavior: Solitary or in pairs in the canopy, often silent.
Distribution: Rare in SE Tibet, W Yunnan, and W Guizhou, with a few records from C Sichuan and Hong Kong.
Voice: Song a rich, mellow, low, pure whistle, *fwhiiiiiiiit*, very similar to song of Green Cochoa but slightly shorter (about 1.3 sec, as opposed to 1.5–1.8 sec). Song also initially rises slightly (Green is even pitched throughout) and is slightly lower pitched (2.5 kHz, while Green is about 3.1 kHz).

Cochoa viridis
绿宽嘴鸫 lǜ kuān zuǐ dōng
Green Cochoa   LC

L: 27–29 cm
Habitat: Inhabits evergreen broadleaf forests at middle and low altitudes.
Behavior: Solitary or in pairs. Secretive; perches quietly in the canopy.
Distribution: Rare in SE Tibet, W and S Yunnan, and Fujian.
Voice: Song similar to song of Purple Cochoa.

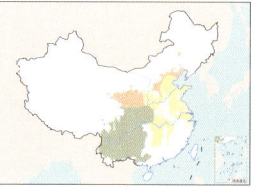

Brown-headed Thrush

Brown forehead

♂

Heavy black on cheek, throat, and breast

Orange flanks

Dark spots on white undertail coverts

♀

White throat

Mistle Thrush

Pale ear coverts

White tip on wing coverts

Rounded spots on underparts

White tip on tail

Purple Cochoa

♂

Purplish brown overall except for head and tail

♀

Brownish overall except for head and tail

Song Thrush

Dark ear coverts

Buff fringe on wing coverts

Lanceolate spots on underparts

Green Cochoa

Greenish overall except for head and tail

♂

Large, complex, distinct wing patch

♀

Large yellowish-brown wing patch

Chinese Thrush

Crescent-shaped black patch behind ear

Two wing bars

Rounded spots on breast

鶲科　Muscicapidae

The most species-rich family currently recognized among Old World birds, consisting of a large collection of small to medium-sized songbirds, including robins, chats, redstarts, Old World flycatchers, forktails, shortwings, whistling-thrushes, and rock-thrushes. Body profile well proportioned. Plumage differs between sexes in roughly half of species. Plumage mostly black, blue, or grayish brown, while juveniles usually spotted. Bill in most species flat and wide, with well-developed rictal bristles. Wings mostly long and pointed; good at flying. Long tail consisting of 12 feathers. Legs long and strong in most species except flycatchers. Most species inhabit forests and scrubland, some on scree slopes and rocky cliffs. Many species wag or vibrate tails when standing. Mostly insectivorous. Northerly breeding species are migratory.

　　Fifty-one genera and 331 species recognized worldwide, widespread in the Old World. Twenty-four genera and 107 species recorded in China, ubiquitously distributed nationwide.

棕薮鸲属
Cercotrichas

Rufous-tailed Scrub-Robin

鹊鸲属
Copsychus

Oriental Magpie-Robin

鶲属
Muscicapa

Asian Brown Flycatcher

白喉姬鶲属
Anthipes

White-gorgeted Flycatcher

蓝白鶲属
Cyanoptila

Zappey's Flycatcher

铜蓝鶲属
Eumyias

Verditer Flycatcher

蓝仙鶲属
Cyornis

Chinese Blue Flycatcher

仙鶲属
Niltava

Large Niltava

短翅鸫属
Brachypteryx

Chinese Shortwing

欧亚鸲属
Erithacus

European Robin

栗背短翅鸫属
Heteroxenicus

Gould's Shortwing

蓝歌鸲属
Larvivora

Siberian
Blue Robin

歌鸲属
Luscinia

Bluethroat

红喉歌鸲属
Calliope

Siberian
Rubythroat

地鸲属
Myiomela

White-tailed Robin

林鸲属
Tarsiger

Red-flanked Bluetail

燕尾属
Enicurus

White-crowned
Forktail

啸鸫属
Myophonus

Blue
Whistling-Thrush

蓝额地鸲属
Cinclidium

Blue-fronted
Robin

姬鹟属
Ficedula

Yellow-rumped
Flycatcher

红尾鸲属
Phoenicurus

Daurian
Redstart

矶鸫属
Monticola

Blue Rock-Thrush

石䳭属
Saxicola

Siberian
Stonechat

䳭属
Oenanthe

Northern
Wheatear

523

Cercotrichas galactotes

棕薮鸲 zōng sǒu qú

Rufous-tailed Scrub-Robin

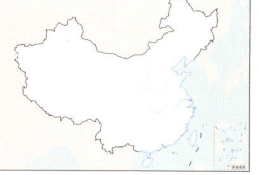

L: 15–17 cm
Habitat: Inhabits deserts with fixed and semifixed sand dunes, and saxaul forests interspersed with dense tamarisk thickets.
Behavior: Moves about in pairs on the ground or in thickets. On ground or dead trees, flicks tail up and down or cocks tail over back, exposing the black and white tips of its rufous-brown tail. Prefers perching and singing on top of thickets.
Distribution: Rare breeder restricted to a small area on the southern edge of the Junggar Basin in N Xinjiang.
Voice: Song a high-speed, scratchy, unmusical warbling, 2–4 sec long, with slight repetition and some mimicry of species such as Desert Whitethroat. Most strophes end abruptly. Calls include hard *tek tek* or *chak chak.*

Copsychus saularis

鹊鸲 què qú

Oriental Magpie-Robin

L: 19–22 cm
Habitat: Inhabits forest-edge bushes, bamboo thickets, and secondary forests at altitudes below 2000 m; especially prefers villages, scrub, orchards, and parks.
Behavior: Often solitary or in pairs. Active, bold; often fights for mates in breeding season. Often cocks tail when resting.
Distribution: Common resident in southern China.
Voice: Song is complex, varied, but mostly short strophes of thin clear warbles and whistles. Calls include an indrawn hiss.

Copsychus malabaricus

白腰鹊鸲 bái yāo què qú

White-rumped Shama

L: 20–28 cm
Habitat: Inhabits dense forests below 1500 m; especially prefers forest edges, secondary forests, and bamboo thickets.
Behavior: Shy, often hides solitarily in the undergrowth. Good at singing; often holds tail upright when singing.
Distribution: Uncommon resident in Hainan, S and W Yunnan, and W Guangxi. Opinions vary regarding subspecies delineation; some also suggest that this species should be placed in genus *Kittacincla.*
Voice: Pleasant, fluty song is arguably one of the finest in the region, a series of rich, melodious, thrushlike phrases that start slowly but build, have frequent changes in pitch and tempo, and include considerable mimicry. Calls include a breathless *ksssh* and a harsh *tshak.*

Anthipes monileger

白喉姬鹟 bái hóu jī wēng

White-gorgeted Flycatcher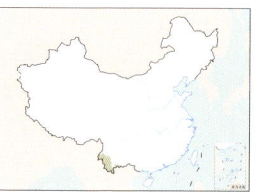

L: 11–14 cm
Habitat: Inhabits evergreen broadleaf forests, secondary forests, and forest edges at altitudes below 2000 m; particularly prefers open woodlands and forest edges.
Behavior: Often solitary. Shy, afraid of people and other flycatchers; usually hides in the undergrowth. Fans and bobs tail up and down when resting. When foraging, flies to catch aerial insects from perch and then falls back to the original perch.
Distribution: Uncommon resident found only in S and W Yunnan.
Voice: Song a lengthy unmusical series of high-pitched, thin, strained whistles, scratchy notes, and occasional churrs, *ji ji zhui zhui zhui zhui.* Calls include a short penetrating whistle and a metallic *tlik-cik.*

Cyanoptila cyanomelana

白腹蓝鹟 bái fù lán wēng

Blue-and-white Flycatcher

L: 14–17 cm
Habitat: Inhabits montane broadleaf forests, mixed forests, and forest edges; also occurs in secondary forests and tall scrubland during migration.
Behavior: Solitary or in pairs. Moves about in the middle story of forests and rarely visits ground; often perches on tree branches to rest or search for food. Male often stands in thickets and sings for a long time in summer.
Distribution: Seen mainly in eastern China. Summer breeder in Northeast China; common passage migrant in most of eastern China. *C. c. cyanomelana* in Heilongjiang and Hebei; *C. c. intermedia* in E Heilongjiang, Jilin, Liaoning, Hebei, Shandong, Guizhou, Jiangsu, Hong Kong, and Taiwan.
Voice: Song is rich, powerful, and melodious, with strophes lasting 2.5–3.5 sec separated by pauses of similar length. Considerable repetition of notes within strophes, but each strophe very different from its neighbor, *hi-hwi-pipipipi … ti-ti-we-wee-wwi-dzwee-dzwee-dzwee.* Calls like Zappey's Flycatcher.

Cyanoptila cumatilis

白腹暗蓝鹟 (琉璃蓝鹟)

bái fù àn lán wēng (liú li lán wēng)

Zappey's Flycatcher

L: 14–17 cm
Habitat: Often inhabits broadleaf forests, mixed forests, and forest edges on hills.
Behavior: Similar to Blue-and-white Flycatcher.
Distribution: Found in central and southwestern China; uncommon breeder in northern China and passage migrant in other places.
Voice. Song similar to Blue-and-white Flycatcher but at lower frequency and less variable in pitch.

Rufous-tailed Scrub-Robin

Distinct white supercilium

Sandy-brown
upperparts

Black and white
spots on tail tip

Rufous-brown
rump and tail

Oriental Magpie-Robin

Black head and back

♂

Grayish head and back

♀

White-rumped Shama

Grayish-brown
head and back

Metallic black
head and back

♂

♀

Rusty breast

Orange breast

White-gorgeted Flycatcher

Pale brown back

Blue-and-white Flycatcher

Bluish-purple back

Blackish throat

♂

♂ imm.

C. c. cyanomelana

Cobalt-blue back

♂

Dark blue throat

Brownish upperparts

♀

C. c. intermedia

Zappey's Flycatcher

Turquoise-blue back

♂

Bluish-black throat

Grayish-brown
upperparts

♀

Muscicapa striata
斑鹟 bān wēng
Spotted Flycatcher　LC

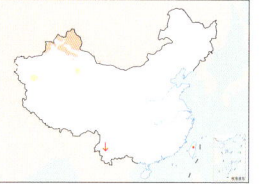

L: 13.5–15 cm
Habitat: Inhabits forest edges of montane coniferous and mixed forests, broadleaf forests in river valleys, shelterbelts in agricultural areas, and urban parks.
Behavior: Solitary in nonbreeding season. Quiet and tame, unafraid of people. Often sits upright on horizontal branches, flicking tail up and down; immediately dashes out to capture passing insects and returns to the original perch.
Distribution: Common breeder in NW Xinjiang and Altai Mountains in N Xinjiang; occasionally seen in S Xinjiang and C Yunnan during migration. Vagrant to Taiwan.
Voice: Most vocalizations quiet and easily overlooked—song a slow disjointed sequence of high-pitched squeaky and scratchy notes. Calls include a thin *tseet* and mellower *chup, chup*.

Muscicapa sibirica
乌鹟 wū wēng
Dark-sided Flycatcher　LC

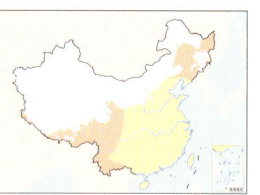

L: 12–14 cm
Habitat: Inhabits mixed, coniferous, and secondary forests and forest edges at middle and low altitudes; prefers open woodlands and forest edges during migration. *M. s. rothschildi* found in forests over 4000 m in summer.
Behavior: Often solitary. Moves about mostly in the middle story or canopy of forests, rarely on the ground. Usually perches on horizontal branches to rest or search for food; flies out to chase insects once spotted and returns to the original perch.
Distribution: Found in most regions in central and eastern China. *M. s sibirica* breeds in Heilongjiang, Jilin, and Liaoning and migrates through Beijing, Shandong, Shanghai, Zhejiang, Fujian, Guangdong, Taiwan, Hubei, and Jiangxi; *M. s. rothschildi* breeds in S Gansu, SE Tibet, S Qinghai, W and S Yunnan, Sichuan, and Guizhou; *M. s. cacabata* breeds in S Tibet.
Voice: Song a prolonged, rapid, rambling series of weak, high-pitched, thin, tinkling trills and whistles, with occasional slightly more melodious phrases, trills, and whistles. More continuous and even higher pitched and not as harsh as Asian Brown Flycatcher and includes more mimicry. Calls include a short, metallic, tinkling *chip* or *tsip*.

Muscicapa griseisticta
灰纹鹟 huī wén wēng
Gray-streaked Flycatcher　LC

L: 13–15 cm
Habitat: Similar to Dark-sided Flycatcher.
Behavior: Similar to Dark-sided Flycatcher.
Distribution: Found in eastern China mostly as uncommon passage migrant; some breed in Northeast China.
Voice: Song unobtrusive, high-pitched twittering, very similar to Dark-sided Flycatcher and Asian Brown Flycatcher. Calls include a descending *tsee*.

Muscicapa dauurica
北灰鹟 běi huī wēng
Asian Brown Flycatcher　LC

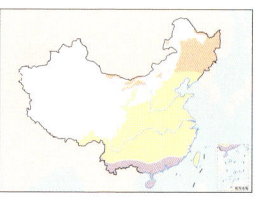

L: 12–14 cm
Habitat: Inhabits broadleaf forests, mixed forests, and coniferous forests at middle and low altitudes; also occurs in secondary forests, forest edges, or urban parks during migration.
Behavior: Similar to Dark-sided Flycatcher.
Distribution: *M. d. dauurica* is found in most of the eastern and central parts of China; breeds mostly in Northeast China, seen in eastern China during migration, and overwinters in southeastern China. *M. d. siamensis* in Yunnan.
Voice: See Dark-sided Flycatcher.

Muscicapa muttui
褐胸鹟 hè xiōng wēng
Brown-breasted Flycatcher　LC

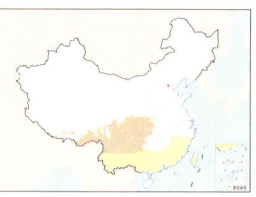

L: 12–14 cm
Habitat: Inhabits montane broadleaf forests, bamboo thickets, and forest edges at middle and low altitudes; also occurs in secondary forests, forest edges, or urban parks during migration.
Behavior: Often solitary. Timid; usually sits still for long periods on branches in the middle story of dense forests. Hawks passing insects and returns to its original perch; rarely moves to the ground.
Distribution: Found mainly in southwestern, North, and East China; mostly as uncommon summer breeder or passage migrant. Vagrant in Beijing.
Voice: High-pitched, thin, strained song similar to songs of Dark-sided Flycatcher, Asian Brown Flycatcher, and Gray-streaked Flycatcher. Calls include a thin descending *tzii*, similar to Gray-streaked.

Muscicapa ferruginea
棕尾褐鹟 zōng wěi hè wēng
Ferruginous Flycatcher　LC

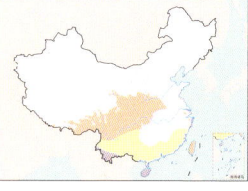

L: 11–13 cm
Habitat: Inhabits broadleaf forests, mixed forests, and open woodlands at forest edges at middle altitudes; also occurs in secondary forests, bamboo thickets, and urban parks during migration.
Behavior: Often solitary. Usually sits still for long periods on open branches or power lines watching for prey; hawks passing insects and returns to its original perch; rarely moves to the ground.
Distribution: Found mainly in southern China; mostly as uncommon summer breeder.
Voice: Call a very intense, high-pitched *tzii* or *tziit*. Song usually lasts about 2 sec and starts with a call, followed by a rapid series of similarly intense notes that accelerate before finishing with three buzzier notes, *tzi-tititu-trrrrr-dzrrr-dzrrr-dzrrr*.

Spotted Flycatcher

Streaked crown

Diffuse streaks on breast

Asian Brown Flycatcher

Pale lores

Relatively large bill, distinct yellow at base of lower mandible

Few streaks on pale gray breast and belly

Tips of primaries do not reach halfway down closed tail

Dark-sided Flycatcher

Relatively small bill, indistinct yellow at base of lower mandible

Diffuse dark brown streaks on breast and belly, extending to flanks

Tips of primaries reach about halfway down closed tail

Brown-breasted Flycatcher

Brown upperparts

Relatively large bill, distinct yellow at base of lower mandible

Grayish-brown breast

Gray-streaked Flycatcher

Dark lores

Clear dark gray streaks on breast and belly

Tips of primaries reach beyond halfway down closed tail

Ferruginous Flycatcher

Rufous-brown rump and uppertail coverts

Rufous-brown breast and belly

Eumyias thalassinus
铜蓝鹟 tóng lán wēng
Verditer Flycatcher LC

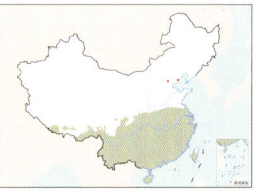

L: 13–16 cm
Habitat: Inhabits mixed and evergreen broadleaf forests and forest edges at 1000–3500 m; also in secondary forests, artificial forests, forest-edge bushes, and orchards in winter.
Behavior: Often solitary or in pairs. Mainly in forest canopy, sometimes in middle story or bushes, rarely on ground. Bold and unafraid of people. Usually captures flying insects on the wing.
Distribution: Seen in most areas in southern China; mostly as uncommon summer breeder or resident. Vagrant in Hebei and Beijing.
Voice: Pleasant song is a prolonged (3 sec), very fast, rambling series of short whistled notes, with a slight burry quality and a falling final syllable, *pwe-pe-p-pwe-p-p-pwe-pe-tititi-wu-pititi-weu*. Calls include a short, plaintive, white-eye-like *pseeut*.

Niltava davidi
棕腹大仙鹟 zōng fù dà xiān wēng
Fujian Niltava LC

L: 16–19 cm
Habitat: Montane evergreen broadleaf and deciduous forests at middle and low altitudes; sometimes also in secondary forests or forest-edge bushes.
Behavior: Often solitary or in pairs. Mainly in middle story and bushes, rarely in canopy or on ground. Usually perches on branches and sallies to capture passing insects.
Distribution: Uncommon summer breeder and resident in Central, southwestern, and South China, including Hainan and Taiwan.
Voice: Song is a thin, very high-pitched (7.1–9.4 kHz), almost descending, whistled *ssssew* lasting about 0.5 sec and repeated almost every 3 sec, reminiscent of Rufous-bellied Niltava (but higher pitched and longer) and also Slaty Bunting. Other calls included sharp metallic *tit tit tit*.

Niltava sundara
棕腹仙鹟 zōng fù xiān wēng
Rufous-bellied Niltava LC

L: 13–16 cm
Habitat: Similar to Fujian Niltava; also inhabits coniferous and mixed forests, sometimes in orchards.
Behavior: Similar to Fujian Niltava. Quiet.
Distribution: Uncommon breeder and resident. *N. s. sundara* in W Yunnan and S Tibet; *N. s. denotata* in S Shaanxi, SE Gansu, Yunnan, Sichuan, Hubei, Hunan, Guangdong, Guangxi, and Taiwan.
Voice: Rarely heard song a high-pitched, flat call note followed by either an intense, strained, scratchy jumble of notes or a churring call note, such as *seee-eh tre-tri-trr-tii*. Calls include a thin *tsiii* or *seee*, very similar to the song of Fujian Niltava but flat and not descending, slightly lower pitched (centered on about 7.5 kHz), and shorter (0.3 sec). Other calls include a rattling churr, *trrrrt*, often in lengthy series or with *siiii*.

Niltava vivida
棕腹蓝仙鹟 zōng fù lán xiān wēng
Vivid Niltava LC

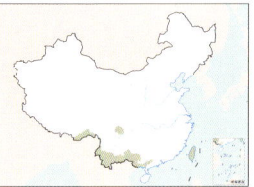

L: 16–18 cm
Habitat: Montane mixed and evergreen broadleaf forests at middle and low altitudes; sometimes also in secondary forests, forest-edge bushes, and orchards.
Behavior: Often solitary or in pairs. Bold, unafraid of people. Prefers middle story and canopy; also on power lines and in forest understory. Catches insects on the wing and on ground opportunistically.
Distribution: Uncommon resident or summber breeder in southern China. *N. v. oatesi* in E and SE Tibet, Yunnan, and Sichuan; *N. v. vivida* in Taiwan. *N. v. vivida* is sometimes treated as a separate species (Taiwan Vivid Niltava), very similar to *N. v. oatesi* but smaller in size.
Voice: Song a slow series of four mellow whistles of similar length (all about 0.2–0.3 sec): the first descending and increasing in volume, the second and third flat, and the last initially rising but then falling in pitch. Pauses between whistles are about 0.1 sec long but almost twice that between the two middle notes. Song of nominate *N. v. vivida* differs slightly by being slightly longer, having much longer individual notes and a shorter pause between the two pairs, and with more changes in pitch.

Niltava grandis
大仙鹟 dà xiān wēng
Large Niltava LC

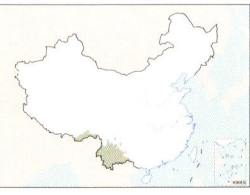

L: 20–22 cm
Habitat: Montane mixed and evergreen broadleaf forests at middle altitudes; also in secondary forests and forest-edge bushes. Sometimes descends to foothills in winter.
Behavior: Similar to Vivid Niltava but prefers middle story of forests and understory bushes; rarely to the canopy.
Distribution: Common summer breeder and resident. *N. g. grandis* in S Tibet and W Yunnan; *N. g. griseiventris* in S Yunnan and W Guangxi.
Voice: Simple song a stepped sequence of three or four melancholy whistles rising up the scale and given every 3–4 sec, *do-ray-me*. An uncommon alternate version is *weet-di-duu-du-wee-di-du-weet*. Calls include a penetrating, whistled *diu* and a harsh *tuck*, often combined.

Niltava macgrigoriae
小仙鹟 xiǎo xiān wēng
Small Niltava LC

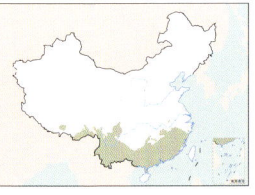

L: 11–14 cm
Habitat: Similar to Large Niltava.
Behavior: Often solitary or in pairs. Active; hopping about or flying in forests. Prefers middle story and undergrowth; occasionally to canopy. Catches insects on the wing and on ground.
Distribution: Southwestern and southeastern China.
Voice: Song is a very thin, high-pitched *tu-weee dwee*, where the first note falls. Calls include a high-pitched churr, *trrrrrrt*.

Verditer Flycatcher

Verditer-blue body

Black lores

♂

Dark gray lores

♀

Body slightly paler than male

Vivid Niltava

♂

Triangular orange patch extends to center of blue throat

No blue patch on side of neck

♀

Yellowish-brown undertail coverts

Fujian Niltava

Relatively small area of bright blue on crown

Inconspicuous bright blue on lesser coverts

♂

Brown-tinged fringe on bright blue wing feathers

Orange underparts paler than breast

♀

Blue patch on side of neck

White throat-patch

Nearly white undertail coverts

Large Niltava

Bright blue crown

Bright blue patch on side of neck

♂

Blue patch on side of neck

♀

Whitish throat

Rufous-bellied Niltava

Relatively large area of bright blue on crown

Distinct bright blue on lesser coverts

♂

Bright blue wing feathers lack brown-tinged fringe

Relatively dark orange on both breast and belly

♀

Blue patch on side of neck

White throat patch

Grayish-brown undertail coverts

Small Niltava

Bright blue crown

Bright blue patch on side of neck

♂

Whitish belly

♀

Blue patch on side of neck

Cyornis unicolor
纯蓝仙鹟 chún lán xiān wēng
Pale Blue Flycatcher

L: 13–17 cm
Habitat: Inhabits evergreen broadleaf forests, mixed forests, and bamboo thickets in mountainous areas at low and middle altitudes.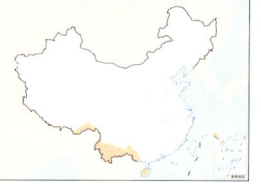
Behavior: Solitary. Hops about and forages mainly in the middle story of forest; sometimes stays motionless in trees for a long time and then darts out for passing insects in the air.
Distribution: Uncommon summer breeder or resident. *C. u. unicolor* in SE Tibet, SW Yunnan, and Guangxi; *C. u. diaoluoensis* in Hainan.
Voice: Song is a low-pitched (1.7–3.4 kHz), fairly lengthy (2.2 sec), descending series of short unhurried whistles, *trii-tru-ti-ti-ti-tu-d-did-di-di*, occasionally ending with a buzzy *chizz*. Call a hard *ticc*.

Cyornis poliogenys
灰颊仙鹟 huī jiá xiān wēng
Pale-chinned Blue Flycatcher

L: 13–15 cm
Habitat: Inhabits evergreen broadleaf forests and secondary forests on low mountains and foothill plains below 1500 m; sometimes also occurs in bamboo thickets and forest-edge bushes.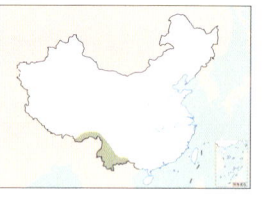
Behavior: Solitary or in pairs. Mainly in the middle story of forests or the canopy layer; rarely comes to the ground. When foraging, spends a long time in trees scanning for insects and darts out once insects pass by.
Distribution: Uncommon summer breeder or resident. *C. p. laurentei* in southwestern China; *C. p. cachariensis* in SE Tibet and NW Yunnan.
Voice: Song is a strained, rising and falling series of up to 11 short hurried notes, with no obvious pattern, *du-dae-chi-cha-du-srrr-do-weent*, given every 3 sec or so. Calls include a soft *tuc-tuc-tuc*.

Cyornis whitei
山蓝仙鹟 shān lán xiān wēng
Hill Blue Flycatcher

L: 14–15.5 cm
Habitat: Inhabits broadleaf forests, bamboo thickets, and secondary forests from 400 to 2500 m; sometimes also occurs in forest-edge bushes. Prefers dense broadleaf forests.
Behavior: Often solitary or in pairs. Moves about mainly in the middle story of forests, rarely in the canopy or on the ground. Hides in dense branches and leaves and flies out to capture passing insects.
Distribution: Common breeder or resident in Yunnan, Sichuan, Guizhou, and Hunan.
Voice: Simple song is a fast medley of mostly falling short whistles between 2–7.5 kHz and lasting almost 2 sec, often with a slightly juddering ending, very different from the much longer song of Chinese Blue Flycatcher. Calls include a soft *tac*, similar to Pale-chinned Blue Flycatcher but slightly harder.

Cyornis rubeculoides
蓝喉仙鹟 lán hóu xiān wēng
Blue-throated Flycatcher

L: 14–15 cm
Habitat: Inhabits evergreen broadleaf forests, mixed forests, bamboo thickets, and secondary forests in mountainous areas below 2000 m; also occurs in scrubland at forest edges.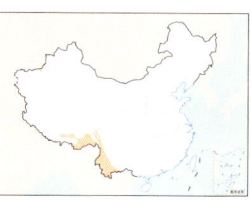
Especially prefers forest understory near streams.
Behavior: Solitary or in pairs. Moves about in the lower and middle stories of forests, rarely in the canopy; often scans for insects from branches and leaves and makes aerial sallies after passing insects.
Distribution: Summer breeder. *C. r. rubeculoides* in SE Tibet; *C. r. dialilaemus* in W Yunnan.
Voice: Song is a *si si yi yu yu yu yu*, with a heavy trill at the end.

Cyornis glaucicomans
中华仙鹟 zhōng huá xiān wēng
Chinese Blue Flycatcher

L: 14–15 cm
Habitat: Inhabits evergreen broadleaf forests, mixed forests, secondary forests, and bamboo thickets at altitudes of 800–2000 m; sometimes also occurs in scrubland at forest edges.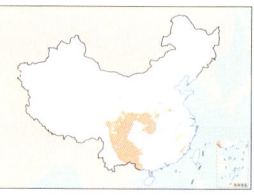
Behavior: Similar to Blue-throated Flycatcher.
Distribution: Common summer breeder in S Shaanxi, SE and W Yunnan, Sichuan, Chongqing, Guizhou, Hubei, Hunan, Jiangxi, Guangdong, and Guangxi. Recorded in South China during nonbreeding season.
Voice: Bustling, jolly, rich song, longer and lower pitched than songs of both Blue-throated and Hill Blue Flycatchers, with fewer harsh, discordant notes.

Cyornis concretus
白尾蓝仙鹟 bái wěi lán xiān wēng
White-tailed Flycatcher

L: 15–18 cm
Habitat: Inhabits low-mountain evergreen broadleaf forests below 1500 m and occasionally in bamboo thickets and secondary forests.
Behavior: Solitary or in pairs. Shy, afraid of people.
Moves about mainly in the middle story of forests; seldom moves to the canopy or ground. Often perches quietly on tree branches and scans for prey, making aerial sallies after passing insects. The white on outer tail feathers is easier to see when the bird rests and spreads its tail.
Distribution: Rare resident in W and S Yunnan.
Voice: Short highly variable song is a series of mostly falling, muffled whistles, *tzze-di-ii-duu-lu-lu-lu*, occasionally including some mimicry. Calls include a soft, long, flat whistle, *pweee*.

Pale Blue Flycatcher

Black lores, less distinct than Verditer Flycatcher

Pale blue body

♂ Bill longer than Verditer Flycatcher

♀

Grayish-brown underparts

Blue-throated Flycatcher

♂ Blue throat

♀

Pale-chinned Blue Flycatcher

Gray head

♂

White-tinged throat

Orange breast

♀

Chinese Blue Flycatcher

♂ Triangular orange area on throat

♀

Hill Blue Flycatcher

♂ Large area of heavy orange on chin and throat

Distinct orange flanks

♀

White-tailed Flycatcher

Bluish-purple body

♂

Whitish underparts

White outer tail feathers

♀

Relatively broad white ring on throat

Cyornis hainanus

海南蓝仙鹟 hǎi nán lán xiān wēng

Hainan Blue Flycatcher

L: 13–15 cm
Habitat: Inhabits evergreen broadleaf forests, mixed forests, secondary forests, and forest-edge bushes in low mountains.
Behavior: Often solitary or in pairs. Moves about mainly in tall bushes and middle story of forests, hopping about and flying short distances; occasionally moves to the ground or canopy level.
Distribution: Uncommon summer breeder or resident in southwestern and southeastern China. Vagrant recorded in Taiwan.
Voice: Short, eclectic, ill-structured song of multiple notes, with no obvious pattern and very similar to Hill Blue Flycatcher. Transcriptions have included "hello mummy." Wide variety of calls include a repeated, sharp *tic* and mellower *tuc* and *trrt-trrt* notes.

Cyornis brunneatus

白喉林鹟 bái hóu lín wēng

Brown-chested Jungle Flycatcher

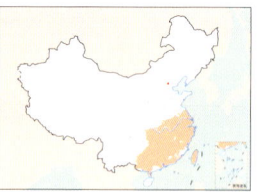

L: 14–16 cm
Habitat: Inhabits evergreen broadleaf forests and bamboo thickets in low mountains below 1000 m; sometimes also occurs in secondary forests or forest-edge bushes.
Behavior: Solitary or in pairs. Shy; seldom hops about in forests. Often hides in dense woods or bamboo thickets and remains still; rarely moves to the ground.
Distribution: Uncommon summer breeder in East, Central, and South China; vagrant to North China.
Voice: Distinctive, loud song starts with an intense, indrawn, high-pitched *teeze* or similar sound, followed by a rapid, piping, two-pitched trill, *pset tu-tu-tu-ri-ri-ri-ri.* Calls include harsh churring.

Erithacus rubecula

欧亚鸲 ōu yà qú

European Robin

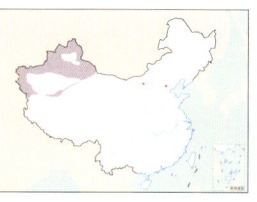

L: 13–15 cm
Habitat: Inhabits woodlands at forest edges, orchards, and city parks on foothill plains or at low altitudes. Sometimes also occurs in forest-edge bushes on low mountains and hills.
Behavior: Solitary or in pairs. Active; good at hopping and running on the ground. Prefers moving about on the ground under bushes; occasionally hops into bushes.
Distribution: Uncommon winter visitor and passage migrant in Xinjiang. A few records in North China during migration and winter.
Voice: Pleasant song (unlikely to be heard in China) is a complex, lengthy, varied, melancholy warbling, with frequent changes of speed. Call is a sharp *tic*, often repeated.

Heteroxenicus stellatus

栗背短翅鸫 lì bèi duǎn chì dōng

Gould's Shortwing

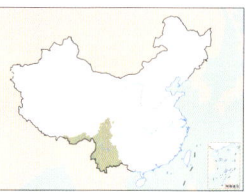

L: 12–14 cm
Habitat: Inhabits middle- to high-altitude montane forests from 2000 to 4000 m. Prefers bamboo thickets, forest-edge bushes, and especially moist river valleys and streams; also occurs on grassy rocky slopes and meadows near mountaintops. Makes seasonal altitudinal movements.
Behavior: Solitary. Timid; prefers moving about on ground under bamboo thickets or bushes.
Distribution: Uncommon winter visitor or resident in Yunnan, Tibet, and Sichuan.
Voice: Song a series of breathlessly high, strained notes, gradually increasing in volume and speed in a descending 10 sec series, *tsiu-t-tssiu, tssiu, tssiu, tssiu-tssiu-tsitsitssiutssiutssiu.* Calls include a descending *zwee* and a hard *tik* when alarmed.

Brachypteryx hyperythra

锈腹短翅鸫 xiù fù duǎn chì dōng

Rusty-bellied Shortwing

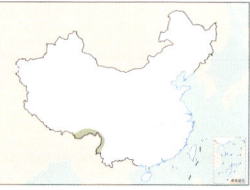

L: 12–14 cm
Habitat: Inhabits evergreen broadleaf forests, bamboo thickets, and bushes below 3000 m; especially prefers oak-dominated evergreen broadleaf forests.
Behavior: Often solitary. Moves about on the forest floor, foraging or resting mainly in bushes and bamboo thickets.
Distribution: Rare resident in Yunnan and Tibet.
Voice: High-pitched, hurried, rising, warbled song, like Lesser Shortwing but longer.

Brachypteryx leucophris

白喉短翅鸫 bái hóu duǎn chì dōng

Lesser Shortwing

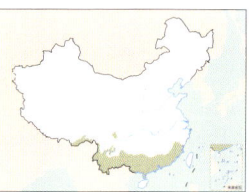

L: 12–13 cm
Habitat: Inhabits evergreen broadleaf forests with well-developed undergrowth at altitudes of 1000–2500 m; sometimes also comes to bamboo thickets, woodlands at forest edges, and abandoned farmland.
Behavior: Solitary or in pairs. Timid; often moves about on ground beneath understory bushes.
Distribution: Uncommon resident in southeastern and southwestern China. *B. l. nipalensis* in SE Tibet, W Yunnan, and W Sichuan; this subspecies also has a blue morph, in which males have a bluish-gray upper body, recorded in South Asia. *B. l. carolinae* in W and S Yunnan, Guangxi, Guangdong, and NW Fujian.
Voice: Loud, fast song is similar to Rusty-bellied Shortwing but shorter .

Hainan Blue Flycatcher

Dark blue head, breast, and back

♂

Pale gray from lower breast to belly

♀

Buff from throat to breast

Gould's Shortwing

Rufous-chestnut upperparts

Dark gray underparts; fine, small, distinct, white spots on belly

Rusty-bellied Shortwing

Short white supercilium

Dark blue upperparts

♂

Rufous-orange underparts

No white supercilium

♀

Rufous orange on underparts paler than male

Brown-chested Jungle Flycatcher

Relatively large bill, yellow lower mandible

White throat

European Robin

Rufous orange from forehead to breast

Lesser Shortwing

Brown upperparts

Short, narrow, white supercilium

Pale throat

Short white supercilium

Bluish-gray upperparts

♂

White underparts with bluish-gray tinge

Blue morph

Brachypteryx cruralis

喜山短翅鸫 xǐ shān duǎn chì dōng

Himalayan Shortwing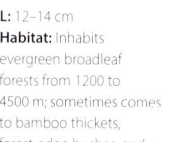

L: 12–14 cm
Habitat: Inhabits evergreen broadleaf forests from 1200 to 4500 m; sometimes comes to bamboo thickets, forest-edge bushes, and grasslands.
Behavior: Solitary, timid. Often in the undergrowth of dense forests; takes cover under bushes when people are around. Good at running or hopping on the ground.
Distribution: Uncommon resident in SE Tibet, Yunnan, and C and SW Sichuan.
Voice: Song a loud, rambling, formless warble of thin high-pitched notes of different pitches, introduced by a series of thin, wheezing, tremulous whistles. Call a hard *tak*.

Brachypteryx sinensis

中华短翅鸫 zhōng huá duǎn chì dōng

Chinese Shortwing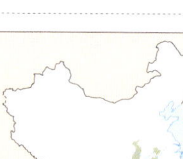

L: 12–13 cm
Habitat: Inhabits evergreen broadleaf forests at middle and high altitudes; sometimes also occurs in bamboo thickets, forest-edge bushes, and grasslands.
Behavior: Often solitary. Timid; quickly takes cover under bushes when people are around. Ground dweller. Often in the undergrowth; good at running on the ground.
Distribution: Uncommon resident in N Guizhou, Hubei, Hunan, Jiangxi, Fujian, Guangdong, and Guangxi.
Voice: Typical song a short phrase with few notes, starting with a fairly short note on a flat pitch and ending abruptly. Differs from song of Himalayan Shortwing by having shorter phrases at lower pitch, shorter pauses between strophes, and much repetition. Calls an abrupt *tack* and a hard rattle, given in alarm.

Brachypteryx goodfellowi

台湾短翅鸫 tái wān duǎn chì dōng

Taiwan Shortwing

L: 12–13 cm
Habitat: Inhabits evergreen broadleaf forests at middle altitudes; sometimes also occurs in bamboo thickets and forest-edge bushes.
Behavior: Solitary or in pairs. Timid; quickly takes cover under dense bushes when people are around. Ground dweller; prefers moving about in the undergrowth near ground.
Distribution: Chinese endemic; uncommon resident in Taiwan.
Voice: Typical song is a brief verse starting with a fairly short note on a flat pitch, similar to Chinese Shortwing. Song is a loud metallic *yi ju ju ju*.

Larvivora brunnea

栗腹歌鸲 lì fù gē qú

Indian Blue Robin

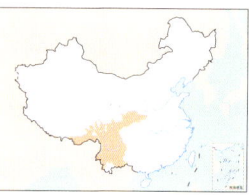

L: 13–15 cm
Habitat: Inhabits evergreen broadleaf forests and forest-edge bushes in mountainous areas at altitudes of 1000–3500 m; sometimes comes to bamboo thickets, secondary forests, and mixed forests.
Behavior: Solitary or in pairs. Timid; often moves around in the undergrowth, good at running fast on the ground. Tends to flick tail up and down or spread tail feathers.
Distribution: Uncommon summer breeder and passage migrant in Yunnan, Tibet, Sichuan, Gansu, and Shaanxi.
Voice: Song begins with a series of often six increasingly intense short whistles before a rapid explosive jumble of trilling notes, *tiu … tiu … tiu … tiu … tiu … tiu … tiu …chichichichi*. Calls include hard guttural *tuk*, often in series, and a penetrating, whistled, Daurian Redstart–like *tsee*.

Larvivora cyane

蓝歌鸲 lán gē qú

Siberian Blue Robin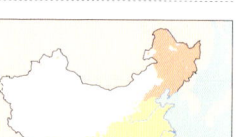

L: 12–14 cm
Habitat: Inhabits mainly mixed forests, coniferous forests, and forest edges in mountainous areas; especially prefers forested areas along roads in breeding season. Also occurs in urban parks, coastal windbreak forests, bamboo thickets, secondary forests, and scrubland during migration.
Behavior: Often solitary or in pairs. Ground dweller; usually moves about among understory bushes on the ground. Good at running fast on the ground; seldom in trees. Timid; often hides in bushes. Stands and walks in a "crouched" posture; flicks tail up and down when walking.
Distribution: *L. c. cyane* in Northeast, North, Central, East, South, southeastern, and southwestern China; summer breeder and passage migrant in Northeast and North China, winter visitor in southwestern China, and mainly as passage migrant in other areas. *L. c. bochaiensis* breeds in northern Northeast China.
Voice: Usually introduced by a brief series of thin, often barely audible *tsit* notes, song is a loud, rapid, explosive trill, *tjuree-tjuree-tjuree-ree*. Call a very mellow, throaty *tuk*, often in series.

Himalayan Shortwing

Slender white supercilium

Black lores

Dark blue body

Blue on wings contrasts
little with upperparts

♂

Brownish overall

♀

Indian Blue Robin

Distinct white supercilium

Bluish-gray upperparts

♂

Rufous-orange
underparts

Pale brown upperparts

♀

Whitish underparts

Chinese Shortwing

No dark streak
at lores

♂

Wing color darker than
Himalayan Shortwing

Blue on underparts paler
than upperparts

Brown upperparts paler
than Himalayan Shortwing

♀

Whitish belly

Siberian Blue Robin

Blue upperparts

♂

White underparts

Brown upperparts

♀

Slightly bluish rump
and uppertail coverts

Yellowish-brown
underparts,
indistinct scaly
pattern on
breast

Taiwan Shortwing

Distinct white supercilium

Indistinct white supercilium

Dark brown
body

♂

Brown overall, paler
than male

♀

Larvivora sibilans

红尾歌鸲 hóng wěi gē qú

Rufous-tailed Robin  LC

L: 13–15 cm
Habitat: In breeding season, inhabits mainly montane mixed forests and coniferous forests; especially prefers woodland bushes at forest edges. During migration, occurs at foothill forest edges, urban parks, orchards, and bamboo thickets and bushes in coastal windbreak forests.
Behavior: Solitary or in pairs. Ground dweller; often moves around among undergrowth on the ground. Good at running fast on the ground, flicking tail from time to time. Timid, secretive; often hides in bushes.
Distribution: Found in Northeast, North, Central, East, South, and southwestern China. A summer breeder in Northeast China, a winter visitor in southwestern China, and mainly a passage migrant in most other regions.
Voice: Song is a constant speedy trill almost 3 sec long, *tiuuuuuuuuuuuuuuuuw*, that starts quietly but rapidly increases in vigor. Calls include an unobtrusive, mellow *tuc*.

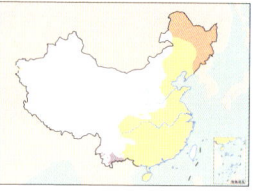

Larvivora ruficeps

棕头歌鸲 zōng tóu gē qú

Rufous-headed Robin 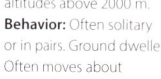 EN

L: 13–15 cm
Habitat: Inhabits montane fir and birch forests, as well as woodland scrub at altitudes above 2000 m.
Behavior: Often solitary or in pairs. Ground dweller. Often moves about among forest understory near ground; good at running fast on the ground. Timid, secretive; often hides swiftly under dense bushes when people are spotted.
Distribution: Rare summer breeder in Shaanxi and Sichuan.
Voice: Powerful, rich, and yet very simple song is a short melodious phrase preceded by a short introductory note, *tiiii lu-lu-lu-lu*.

Larvivora komadori

琉球歌鸲 liú qiú gē qú

Ryukyu Robin NT

L: 14–15 cm
Habitat: Inhabits dense woods and woodland scrub along valley streams and on low mountains and hills; especially prefers the bottom layer of moist forests.
Behavior: Solitary or in pairs. Often forages or rests in low bushes or in forest understory near ground.
Distribution: Occasional winter visitor or passage migrant; recorded only in Taiwan.
Voice: Song reminiscent of song of Rufous-headed Robin; unlikely to be heard in China. Calls include a high penetrating *tsiii* and trilled *trrrrrrrrrr*.

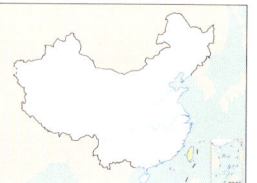

Larvivora akahige

日本歌鸲 rì běn gē qú

Japanese Robin LC

L: 13–15 cm
Habitat: Inhabits mixed forests, broadleaf forests, secondary forests, and woodland scrub in low mountains, urban parks, or coastal areas.
Behavior: Solitary or in pairs. Ground dweller; moves about mostly beneath understory bushes. Good at running fast on the ground, seldom in trees. Secretive; often hides under dense bushes to rest and watch the surroundings.
Distribution: Seen along the east coast. Uncommon winter visitor in southeastern China; uncommon passage migrant or vagrant elsewhere.
Voice: Song, occasionally heard in winter, is a series of simple, well-spaced, quavering, or trilled phrases (likened to a ringing telephone), reminiscent of song of Rufous-headed Robin. Calls include a thin metallic *tsip*.

Luscinia svecica

蓝喉歌鸲 lán hóu gē qú

Bluethroat LC

L: 14–16 cm
Habitat: Highly adaptable to various habitats. Breeds in a wide range of habitats in summer, from alpine scrubland and woodland scrub to low-altitude reedy wetlands. Also occurs in bushes, bamboo thickets, and reedbeds in urban parks, villages, and coastal windbreak forests during migration.
Behavior: Often solitary or in pairs; sometimes in small scattered flocks during migration. Shy; active in the morning and evening, hiding under dense bushes or in reedbeds most of the time. Usually shakes or fans tail.
Distribution: *L. s. svecica* is widely distributed throughout China as uncommon summer breeder in Northeast China, uncommon winter visitor in southern China, and passage migrant in most areas; *L. s. saturatior* breeds in N Xinjiang; *L. s. kobdensis* breeds in N and W Xinjiang; *L. s. abbotti* breeds in W Tibet; *L. s. przevalskii* breeds in eastern northwestern China and winters in SW Yunnan.
Voice: Song starts slowly, with a metallic *zruu*, followed by a series of powerful, varied, melodious, squeaky phrases often incorporating clicks, buzzes, and mimicry. Calls include a hard, metallic, clipped *track*, often doubled.

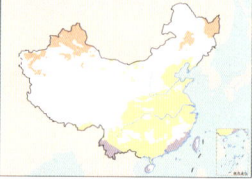

Rufous-tailed Robin

Pale brown upperparts

Rufous-brown rump and tail feathers

Scaly pattern on pale brown underparts

Rufous-headed Robin

Rufous-brown crown

♂

White throat

Brown upperparts

Yellowish-brown underparts, relatively distinct scaly pattern on throat and breast

♀

Ryukyu Robin

Rufous-brown upperparts

♂

Black throat and breast

Black patch on white flanks

Rufous-brown upperparts

♀

Whitish underparts, scaly dark brown pattern on breast and belly

Japanese Robin

Orange face and throat

♂

Blackish breast band

Gray breast and flanks

Orange red on face paler than male

♀

Grayish-brown flanks

Bluethroat

Distinct white supercilium

♂

Brown spots on blue throat

Relatively broad brown and black bands on breast

L. s. svecica

♀

White throat and black malar stripe

♂

White band on throat

L. s. abbotti

♂

Relatively narrow black band

L. s. kobdensis

537

Luscinia megarhynchos

新疆歌鸲 xīn jiāng gē qú

Common Nightingale

L: 15–16.5 cm
Habitat: Inhabits
vegetated plains, desert
scrubland, river valley
broadleaf forests, and
urban and rural gardens.
Behavior: Moves about
solitarily or in pairs among
bushes; also perches on

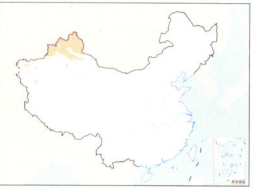

low branches of small trees or bushes when resting. Secretive, often hard
to see. During breeding season, hides almost all day in bushes and sings
constantly; easier to locate by its songs.
Distribution: Common breeder in N Xinjiang but rare in S Xinjiang.
Voice: Song, often given at night, a powerful, rich, varied, intense warble
of phrases 2–4 sec long, broken by short pauses, with clear whistles,
churrs, bubbling, and considerable mimicry. Calls include a grating *krrrr*,
hard *tek*, and whistled *uiit*.

Luscinia phaenicuroides

白腹短翅鸲 bái fù duǎn chì qú

White-bellied Redstart

L: 16–18 cm
Habitat: Inhabits
montane forests and
forest-edge bushes at
altitudes of 1500–
4000 m; especially
prefers woodland scrub.
Sometimes also moves
down to mixed forests,

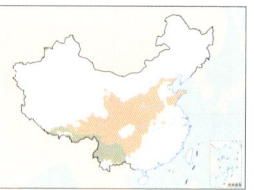

bamboo thickets, and secondary forests in fall and winter.
Behavior: Solitary. Shy; often hides under bushes and seldom moves.
Highly territorial, chasing out individuals that enter its territory. Prefers
moving on the ground or in bushes, rarely up in trees. Once landed in a
new location, often cocks tail over back and fans tail.
Distribution: Seen from southwestern to North China; uncommon
resident in southwestern China and uncommon summer breeder in
North China. *L. p. ichangensis* in Beijing, Hebei, Shandong, Shaanxi,
Ningxia, Qinghai, Yunnan, and Sichuan; *L. p. phoenicuroides* in SE Tibet.
Voice: Loud, slightly melancholy song is three slightly burry whistles,
then two buzzier notes and an abrupt end, *dze-peeee-pee-pzee-zrrre-pt*.
Calls include a mellow *tuk* and *tsie tsie -tsie-tek tek*.

Calliope pectoralis

黑胸歌鸲 hēi xiōng gē qú

Himalayan Rubythroat

L: 13–16 cm
Habitat: Inhabits
Juniperus pseudosabina
scrub and low thickets
in subalpine regions and
above tree line.
Behavior: Timid and
vigilant; mostly hides in
bushes. Good at walking
quickly or running on the ground. Male often sings from bushtops or
rocks during breeding season.
Distribution: *C. p. ballioni* is an uncommon breeder in C and
SW Xinjiang; *C. p. confusa* is a rare breeder in S Tibet.
Voice: Song similar to Siberian Rubythroat and especially Chinese
Rubythroat—a series of variable, loud, warbling phrases typically lasting
3–4 sec, sometimes much longer, and rich in trills. Calls include a short,
thin, descending whistle, *fyeww*, and harsher *etch*.

Calliope tschebaiewi

白须黑胸歌鸲 bái xū hēi xiōng gē qú

Chinese Rubythroat

L: 13–16 cm
Habitat: Often inhabits
alpine scrubland at
altitudes of 3000–4500 m;
also occurs in subalpine
coniferous forests
and bamboo thickets.
Sometimes descends to
foothills of mountains in
winter.

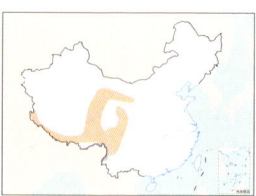

Behavior: Often solitary or in pairs. Shy; mostly hides under bushes.
Good at running on the ground; cocks tail high and slightly fans tail
when running stops. Male often sings on bushtops or rocks during
courtship.
Distribution: Uncommon summer breeder in Tibet, Yunnan, Sichuan,
Qinghai, and Gansu.
Voice: Song similar to the other two rubythroats but usually with shorter
strophes than Himalayan Rubythroat.

Calliope calliope

红喉歌鸲 hóng hóu gē qú

Siberian Rubythroat

L: 14–16 cm
Habitat: Often inhabits
secondary forests and
mixed forests on low
mountains, hills, and
foothill plains; also occurs
in bamboo thickets and
reedbeds.

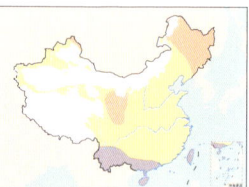

Behavior: Often solitary
or in pairs; sometimes in small flocks during migration. Shy; mostly hides
under bushes. Ground dweller. Good at running and foraging on the
ground; sometimes moves to low branches of bushes. Male often sings
on bushtops or power lines during courtship.
Distribution: Found mostly in eastern China. Summer breeder in
Northeast China, Gansu, and Qinghai; winter visitor in southern China;
passage migrant in other regions.
Voice: Vocalizations similar to other rubythroats.

Common Nightingale

Indistinct supercilium

Pale brown upperparts

Rufous-brown tail feathers

White underparts

White-bellied Redstart

Bluish gray overall

Orange base of outer tail feathers

Two white spots on wings

♂

Brown overall

♀

Black breast

imm.

Himalayan Rubythroat

White supercilium

Lacks white submoustachial stripe

Deep red throat

♂

Broad black breast band

Resembles Chinese Rubythroat but with indistinct supercilium

♀

Chinese Rubythroat

Grayish-black cheek with white submoustachial stripe

♂

Orange-red throat

Black breast

White supercilium

♀

White throat

Siberian Rubythroat

Narrow white supercilium

♂

Grayish-brown cheek with white submoustachial stripe

Red throat

White supercilium

♀

White throat, with scattered red in some individuals

Calliope pectardens

金胸歌鸲 jīn xiōng gē qú

Firethroat

 NT

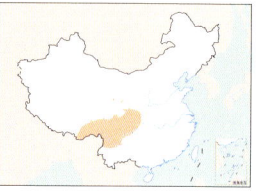

L: 13–15 cm

Habitat: Inhabits dense bushes and bamboo thickets under evergreen broadleaf forests and mixed forests at middle and high altitudes; also in bamboo thickets and secondary forests on low mountains and foothill plains during migration.

Behavior: Often solitary or in pairs. Shy; often hides under bushes. Ground dweller; mostly on the ground and in bushes. Male often sings on bushtops during courtship.

Distribution: Occasional summer breeder from SE Tibet to S Gansu, Sichuan, and N Yunnan.

Voice: Intense song a series of cheerful, rich phrases that contain considerable mimicry, indistinguishable from song of Blackthroat. Calls include a quiet *tup*.

Calliope obscura

黑喉歌鸲 hēi hóu gē qú

Blackthroat

 VU

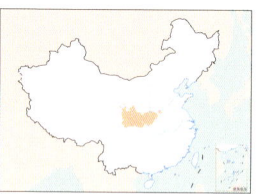

L: 12–14 cm

Habitat: Inhabits mixed forests, woodlands at forest edges, and bamboo thickets in mountainous areas, and sometimes migrates to secondary forests, bamboo thickets, and scrub on foothill plains.

Behavior: Solitary or in pairs. Shy; mostly hides under bushes and immediately takes cover in shady understory when disturbed. Ground dweller; moves about mainly on the ground or among bushes.

Distribution: Rare summer breeder in Sichuan, Shaanxi, and Gansu.

Voice: Vocalizations like Firethroat.

Myiomela leucura

白尾蓝地鸲 bái wěi lán dì qú

White-tailed Robin

 LC

L: 15–18 cm

Habitat: Inhabits montane evergreen broadleaf and mixed forests below 3000 m; prefers shady, humid forest, also in secondary forests, bamboo thickets, and scrub on foothill plains.

Behavior: Often solitary or in pairs. Shy; immediately takes cover in shady understory when disturbed. Ground dweller; mainly on the ground or in undergrowth. Fans tail often, exposing the white patches on outer tail feathers. Feeds mainly on insects; quickly sallies for insects on the ground or in the air.

Distribution: Uncommon resident. *M. l. leucura* in Shaanxi, Ningxia, Gansu, Sichuan, Yunnan, Tibet, Hubei, Hunan, Guangxi, and Hainan; *M. l. montium* in Taiwan.

Voice: Pleasant though hurried warbled song is a short ditty of up to eight notes that vaguely recall Lesser Shortwing, *eeeh-tu-pr-liuuuu*. Calls heard far less often but include a quiet *tuc*.

Tarsiger indicus

白眉林鸲 bái méi lín qú

White-browed Bush-Robin

 LC

L: 13–15 cm

Habitat: Inhabits alpine and subalpine coniferous forests, mixed broadleaf-coniferous forests, and forest-edge bushes at altitudes of 2000–4000 m; especially prefers gully forests and rhododendron thickets and sometimes occurs in woodland bushes and rhododendron thickets in rocky alpine areas.

Behavior: Often solitary or in pairs. Shy; often hides in shady undergrowth. Ground dweller; mainly on the ground or in bushes.

Distribution: Occasional resident. *T. i. yunnanensis* in S Gansu, NW Yunnan, and Sichuan; *T. i. indicus* in SE Tibet; *T. i. formosanus* in Taiwan.

Voice: Mellow song has an undulating quality, a series of rapidly repeated almost identical notes, *tu-whi-wich-u-wi-rr*. Calls include a penetrating, rising, Daurian Redstart–like *tyii*, often repeated rapidly, and a mellow *tuk-tyuk*.

Tarsiger hyperythrus

棕腹林鸲 zōng fù lín qú

Rufous-breasted Bush-Robin

 LC

L: 12–14 cm

Habitat: Inhabits montane forests and forest-edge bushes at altitudes of 1500–4000 m, especially along forest streams, sparse grasslands, and alpine rhododendron thickets. Altitudinal migrant; descends to evergreen forests and forest-edge bushes on low mountains in winter.

Behavior: Often solitary or in pairs. Ground dweller; mainly on the ground or in undergrowth. Good at running and feeding on the ground.

Distribution: Uncommon resident in Tibet and Yunnan.

Voice: Short (1.2 sec), simple, six-note song is an intense frenzied warble with a burry tone and often one or two buzzy notes repeated with little variation, *t-t-zeewe'zzwee'tititi-twediu*, where the introductory notes are barely audible. Calls include low Golden Bush-Robin–like *truk-tuk* notes.

Tarsiger johnstoniae

台湾林鸲 tái wān lín qú

Collared Bush-Robin

 LC

L: 12–13 cm

Habitat: Inhabits montane forests and forest-edge bushes at altitudes of 2000–3500 m; especially prefers woodland scrub and roadside scrub.

Behavior: Often solitary. Not shy. Sometimes occurs on roadsides; perches in bushes to rest. Ground dweller; moves about mainly in grasslands or undergrowth. Good at running and foraging on the ground.

Distribution: Chinese endemic; common resident in Taiwan.

Voice: Song similar to that of White-browed Bush-Robin but shorter, less hurried, higher pitched, and with a slightly burry quality. Calls include an intense *tip-tip-tip*, repeated rapidly and often in a long series, and a scolding *churk-urk-kk*.

Firethroat

White stripe on side of neck

♂

Orange red from throat to breast

Brownish belly

♀

White-browed Bush-Robin

White supercilium

Dark blue upperparts

♂

Pale brownish-yellow underparts

White supercilium

Yellowish brown overall

♀

Yellowish-brown undertail coverts

Blackthroat

♂

Black throat and breast

Off-white belly

♀

Pale brown belly

Rufous-breasted Bush-Robin

Bright blue supercilium

Bright blue upperparts

♂

Rufous underparts

Blue rump and tail

♀

White-tailed Robin

Bright blue forehead

Bright blue patch on wing coverts

♂

White base of outer tail feathers

♀

White base of outer tail feathers

Collared Bush-Robin

White supercilium

Orange-red neck collar extends to back

♂

White supercilium

White undertail coverts, differing from female White-browed Bush-Robin

♀

541

Tarsiger cyanurus

红胁蓝尾鸲 hóng xié lán wěi qú

Red-flanked Bluetail　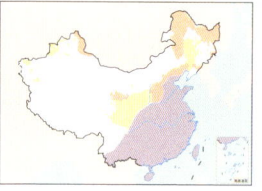

L: 12–14 cm
Habitat: Inhabits coniferous forests, mixed broadleaf-coniferous forests, and forest-edge bushes above 1000 m in breeding season. Also moves to secondary forests, bamboo thickets, woodland scrub, and roadside scrub on low mountains, hills, and foothill plains during migration and winter.
Behavior: Often solitary or in pairs; sometimes in small flocks during migration. Not very shy. Sometimes occurs on roadsides. Ground dweller; moves about mainly on the ground or in undergrowth. Good at running and feeding on the ground.
Distribution: Breeds in North China, Northeast China, Qinghai, and Gansu; winters in areas south of the Yangtze River; widespread during migration.
Voice: Melancholy, slightly desolate song, repeated with little variation, *itri-chirr-tre-tre-tru-tru-rr*. Calls include a soft, up-slurred, mournful *heed* and a low throaty *guk-guk*.

Tarsiger rufilatus

蓝眉林鸲 lán méi lín qú

Himalayan Bluetail　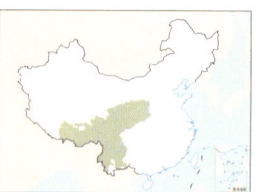

L: 12–14 cm
Habitat: Breeds in montane evergreen broadleaf forests, mixed broadleaf-coniferous forests, and forest-edge bushes above 1500 m. Descends to secondary forests, woodland scrub, and side of trails on low mountains, hills, and foothill plains during migration and winter.
Behavior: Often solitary or in pairs; similar to Red-flanked Bluetail.
Distribution: Widespread throughout southwestern China, north to Shaanxi and Ningxia. Mostly as resident but migrates altitudinally.
Voice: Song similar to Red-flanked Bluetail but shorter and far simpler, *ti-du-du-i ... ti-du-du-di*. Call basically identical to Red-flanked.

Tarsiger chrysaeus

金色林鸲 jīn sè lín qú

Golden Bush-Robin　

L: 12–14 cm
Habitat: Inhabits montane coniferous forests, bamboo thickets, and scrubland at altitudes of 2000–4500 m; especially prefers low scrub and bamboo thickets. In fall and winter, often migrates down to evergreen broadleaf forests, mixed forest, and forest-edge bushes at about 2000 m.
Behavior: Often solitary or in pairs. Timid; usually hides in dense bushes. Ground dweller; mainly on the ground or in undergrowth. Good at running on the ground; often cocks tail over back after a brief rapid walk. Rarely flies long distances.
Distribution: Summer breeder or resident in southwestern China.
Voice: Song a hurried, buzzy, short trill, *tze-tu-tze-tze-tze-tu-tize-dze*. Calls include a distinctive, soft, nasal, rattling *trrr'rk*.

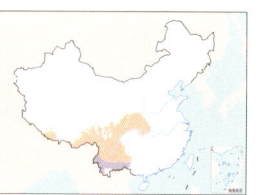

Myophonus caeruleus

紫啸鸫 zǐ xiào dōng

Blue Whistling-Thrush　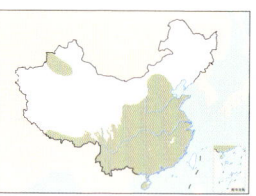

L: 29–35 cm
Habitat: Breeds in evergreen broadleaf forests, mixed forests, and gorges at altitudes of 1000–4000 m; may migrate to lower altitudes in winter.
Behavior: Turns over leaves on wet ground in search of food. Often moves along gravelly banks or digs in soft ground.
Distribution: Locally common resident. *M. c. caeruleus* is widely distributed, from SE Gansu and E Sichuan, to N Hebei and Zhejiang in the east, and to Guangdong and Guangxi in the south, with some individuals migrating short distances south in winter, occasionally to islands off the east coast; *M. c. temminckii* in Xinjiang, Tibet, W Yunnan, and Guizhou, east to W Sichuan; *M. c. eugenei* in C and S Yunnan and SW Guangxi.
Voice: Lengthy, rambling, even haphazard song is a disjointed string of lazy-sounding, straight, high-pitched whistles of varying pitches and lengths. Calls include a far-carrying, strident, shrill *skreee* that recalls Spotted Forktail and White-crowned Forktail (but is shorter and more markedly descending) and a multisyllabic *tzeet tze-tze-tzeet*.

Myophonus insularis

台湾紫啸鸫 tái wān zǐ xiào dōng

Taiwan Whistling-Thrush　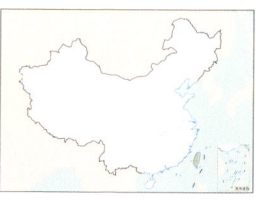

L: 28–30 cm
Habitat: Often inhabits evergreen broadleaf forests and forest-edge woodland and scrubland in mountainous areas at middle and low altitudes, especially in moist forest understories and near streams.
Behavior: Often solitary. Vigilant. Seldom occurs in open areas; often moves about in shady and wet areas at forest edges. Mainly on the ground or in understory. Good at running on the ground; often fans tail and "squats" when resting.
Distribution: Chinese endemic; resident in Taiwan.
Voice: Unhurried, harsh, unmusical song is a mix of rising and falling whistles and strained notes. Calls include a loud screeching *zi* or *sui yi*.

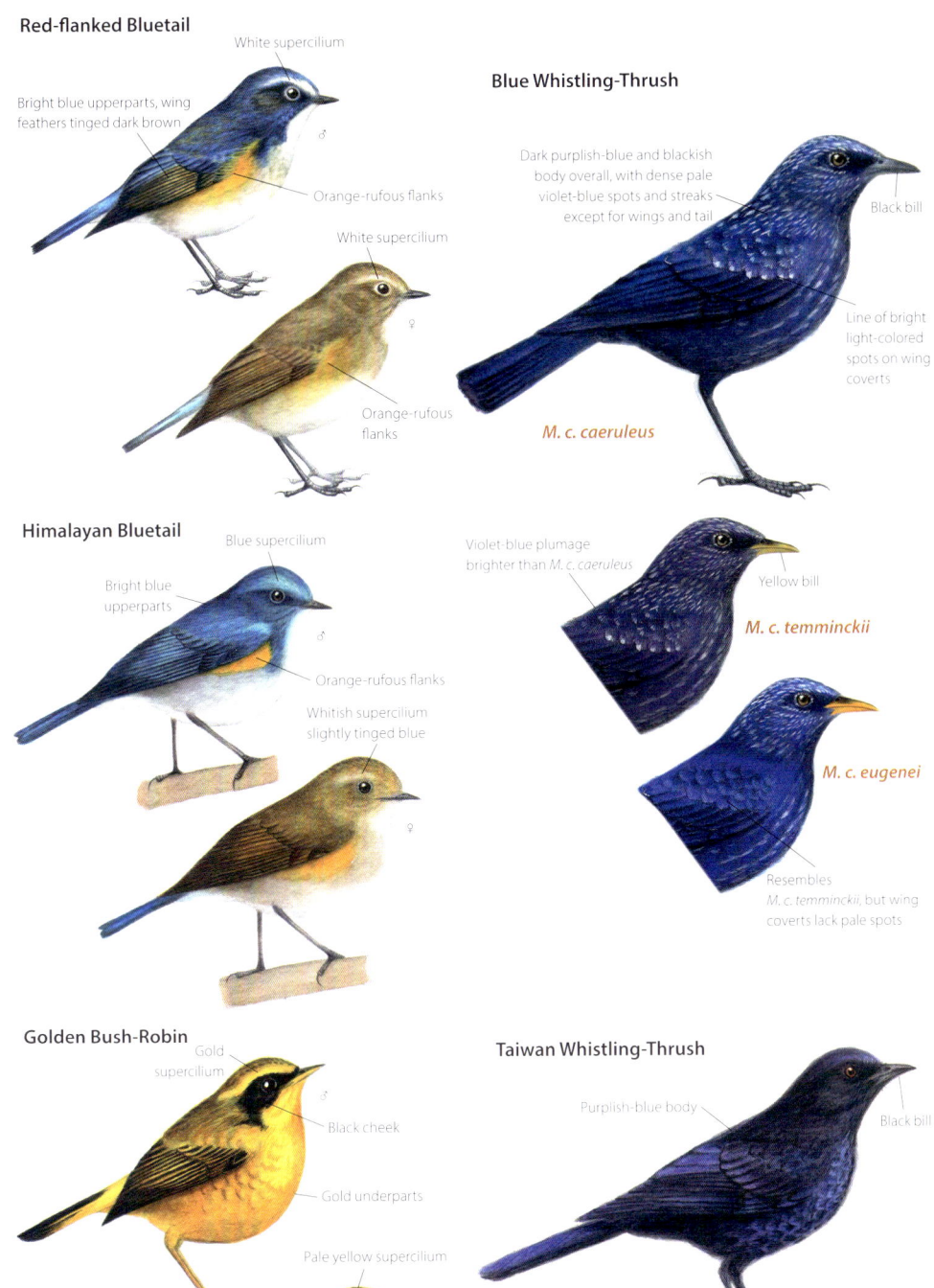

Red-flanked Bluetail

White supercilium

Bright blue upperparts, wing feathers tinged dark brown

♂

Orange-rufous flanks

White supercilium

♀

Orange-rufous flanks

Blue Whistling-Thrush

Dark purplish-blue and blackish body overall, with dense pale violet-blue spots and streaks except for wings and tail

Black bill

Line of bright light-colored spots on wing coverts

M. c. caeruleus

Himalayan Bluetail

Blue supercilium

Bright blue upperparts

♂

Orange-rufous flanks

Whitish supercilium slightly tinged blue

♀

Violet-blue plumage brighter than *M. c. caeruleus*

Yellow bill

M. c. temminckii

M. c. eugenei

Resembles *M. c. temminckii*, but wing coverts lack pale spots

Golden Bush-Robin

Gold supercilium

♂

Black cheek

Gold underparts

Pale yellow supercilium

♀

Pale yellow underparts

Taiwan Whistling-Thrush

Purplish-blue body

Black bill

Enicurus scouleri

小燕尾 xiǎo yàn wěi

Little Forktail `LC`

L: 12–14 cm
Habitat: Prefer small to medium-sized mountain streams at altitudes of 1100–2500 m.
Behavior: Hops among rocks in streams to search for food; also flies short distances to catch prey.

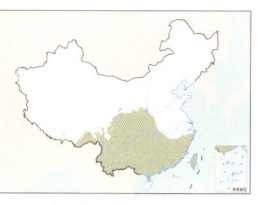

Distribution: Locally common resident; found in mountain streams south of S Gansu and N Zhejiang.
Voice: Silent; song and calls easily overlooked. Song is a fast, torrent-adapted, high-pitched, scratchy warbling, vaguely reminiscent of White-gorgeted Flycatcher, but phrases very short. Calls include a flat or slightly rising, short *ee* or *weee*.

Enicurus schistaceus

灰背燕尾 huī bèi yàn wěi

Slaty-backed Forktail `LC`

L: 22–25 cm
Habitat: Inhabits wide rivers in canyons and often found in open countryside at altitudes of 400–1800 m.
Behavior: Moves mostly among rocks in rivers.
Distribution: Common

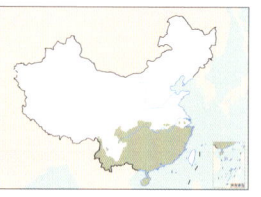

resident in southwestern, Central, southeastern, and South China.
Voice: Call a short, high, thin, sharp, penetrating whistle, *seet*, very similar to that of Black-backed Forktail but higher pitched (centered at about 5.5 kHz, as opposed to 3.5 kHz).

Enicurus maculatus

斑背燕尾 bān bèi yàn wěi

Spotted Forktail `LC`

L: 25–26 cm
Habitat: Inhabits small streams hidden in dense forests. Breeds at altitudes of 600–3000 m and migrates to areas below 2300 m in winter.
Behavior: Moves about in the upper reaches of rivers

or along rocky shores around riverbeds; sometimes forages in shallow waters or rock crevices.
Distribution: Locally common or uncommon resident. *E. m. maculatus* in SE Tibet; *E. m. guttatus* in SW Sichuan, C and W Yunnan; *E. m. bacatus* in SE Yunnan, Guangxi, Guangdong, Fujian, and S Zhejiang.
Voice: Call an intense loud *tseek*, like both Blue Whistling-Thrush and White-crowned Forktail.

Enicurus immaculatus

黑背燕尾 hēi bèi yàn wěi

Black-backed Forktail `LC`

L: 20–25 cm
Habitat: Inhabits rocky mountain streams, rocky banks along rapids, and waterfalls in moist dense forests below 1450 m.
Behavior: Forages for insects on rocks in streams or in shallows.

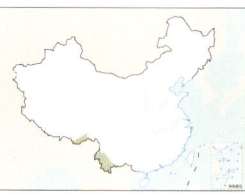

Distribution: Locally common resident in SW Yunnan.
Voice: Calls similar to Slaty-backed Forktail.

Enicurus leschenaulti

白额燕尾 (白冠燕尾)

bái é yàn wěi (bái guān yàn wěi)

White-crowned Forktail `LC`

L: 25–28 cm
Habitat: Inhabits rocky rapids, streams, and rivers in evergreen broadleaf forests below 1200 m; also occurs in ditches and village drainage channels.
Behavior: Forages among rocks at water edge or riverbeds.

Distribution: Locally common resident. *E. l. indicus* in Yunnan; *E. l. sinensis* more widespread, from extreme E Qinghai and S Gansu, to Zhejiang in the east, and to Guangdong and Hainan in the south.
Voice: Unmusical song is strained and scratchy. Calls include a strident *skreee* that recalls Blue Whistling-Thrush and especially Spotted Forktail but is husky and flatter (not rising) than the forktail. Also makes a plaintive, high-pitched, monosyllabic whistle, *zeeee*, very similar to Slaty-backed Forktail but even higher pitched (centered at about 6.3 kHz, as opposed to 5.5 kHz) and much longer.

Cinclidium frontale

蓝额地鸲 (蓝额长脚地鸲)

lán é dì qú (lán é cháng jiǎo dì qú)

Blue-fronted Robin `LC`

L: 18–20 cm
Habitat: Inhabits mid-altitude montane evergreen broadleaf forests, bamboo thickets, and scrubland; especially prefers moist forest understory.
Behavior: Often solitary.

Vigilant; afraid of people. Often moves about in the forest shade. Ground dweller; mainly on the ground or in undergrowth; good at running.
Distribution: Rare summer breeder or resident; recorded only in Yunnan and Sichuan.
Voice: Enchanting, powerful, yet melancholy song a series of simple, clear, short (0.8 sec), melodious, three- or four-note phrases repeated every 6 sec or so, *tliee-ti-tueh … tuuii-t-te-tue … pluo-tlee-tu*, only modest variations in pitch between each phrase. Calls include a mellow *tuk*, similar to Siberian Blue Robin.

Little Forktail

Large area of white on forehead

Short tail, white outer tail feathers and base of inner tail feathers

Small compact body

Extremely pale pink legs and toes

Black-backed Forktail

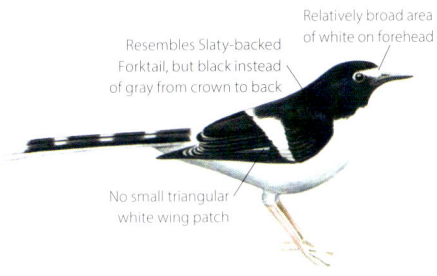

Relatively broad area of white on forehead

Resembles Slaty-backed Forktail, but black instead of gray from crown to back

No small triangular white wing patch

Slaty-backed Forktail

Narrow white line from upper eye ring and lores to forehead

Slate-gray crown, ear coverts, hindneck, upper back, and scapular

White tip on long, deeply forked, black tail

Black wings; unique triangular white patch when primaries closed

Extremely pale legs and toes

White-crowned Forktail

Long white wing patch

White rump

White tip on long deeply forked tail

Extremely pale pink legs and toes

Spotted Forktail

Relatively large white patch on forehead

Dense scaly white spots on hindneck become sparse on back

E. m. guttatus

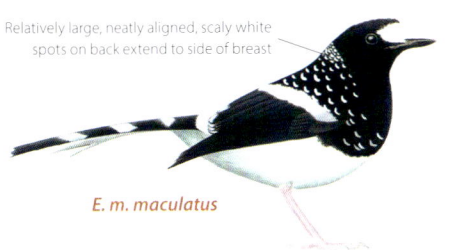

Relatively large, neatly aligned, scaly white spots on back extend to side of breast

E. m. maculatus

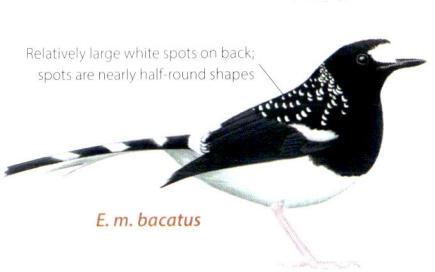

Relatively large white spots on back; spots are nearly half-round shapes

E. m. bacatus

Blue-fronted Robin

Bright blue forehead and wing coverts

Dark purplish-blue body

♂

Brown body

Resembles female White-tailed Robin, but tail lacks white spots

♀

545

Ficedula ruficauda

栗尾姬鹟 lì wěi jī wēng

Rusty-tailed Flycatcher LC

L: 13–15 cm

Habitat: Breeds in open montane forests at 1500–3000 m; migrates to lower altitudes at forest edges and clearings in winter.

Behavior: Often solitary. Arboreal; moves about mainly in the middle story of forests. Makes aerial sallies after passing insects; sometimes also forages on the ground.

Distribution: Rare vagrant recorded in Chengdu, Sichuan, in April 2016.

Voice: Song unlikely to be heard in China. Calls include a mellow rattling purr and a more intense *peep* or *weet*, both very similar to calls of Little Pied Flycatcher, but the rattle is shorter and slower.

Ficedula hypoleuca

斑姬鹟 bān jī wēng

European Pied Flycatcher LC

L: 12–14 cm

Habitat: Breeds in montane broadleaf and mixed forests at middle altitudes. Also in lowland forests, sparse grasslands, and courtyard gardens in winter.

Behavior: Solitary or in pairs. Forages for insects in various parts of forests, especially among leaves or in the air; sometimes on ground.

Distribution: Rare vagrant; only a few records in China. First recorded in October 2008 in Xinjiang (Yutian), then in Sichuan (Chengdu), Zhejiang (Shaoxing), and Jiangsu (Rudong).

Voice: Calls include a quiet *tek*.

Ficedula zanthopygia

白眉姬鹟 bái méi jī wēng

Yellow-rumped Flycatcher LC

L: 12–14 cm

Habitat: Inhabits broadleaf and mixed forests on low mountains, hills, and foothills below 1200 m. Also in secondary forests, forest edges, and urban parks.

Behavior: Often solitary or in pairs. Hops and forages among branches in middle story or canopy of forests; sallies to catch flying insects and returns to different and often higher perches. Sometimes to small trees or bushes.

Distribution: Found in eastern China. An uncommon summer breeder and passage migrant in Northeast, North, and Central China; passage migrant in other regions.

Voice: Song a series of low, melodious, thrushlike, fluty whistles, very similar to song of Green-backed Flycatcher but separated, with difficulty, by slower speed, shorter strophes, and occasional presence of tremulous notes. Calls include a mellow, dry, rattle, *trrrrt*, and a penetrating *weet*, again both very similar to Green-backed, although the rattles are often shorter and higher pitched, with fewer notes, and the *weet* is slightly lower and flatter.

Ficedula narcissina

黄眉姬鹟 huáng méi jī wēng

Narcissus Flycatcher LC

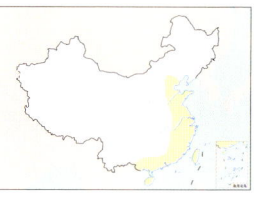

L: 13–13.5 cm

Habitat: Inhabits open woodlands and parks during migration.

Behavior: Usually found in the upper and middle stories of vegetation; tends to return to the original perch after flycatching.

Distribution: Migrates through coastal areas of North, East, and South China, including Hainan and Taiwan, as a common or uncommon passage migrant; potentially wintering in Hainan.

Voice: Song is variable, with a typical series of fast trills, *pipipiityu–ito–foi*. Calls like Yellow-rumped Flycatcher and Green-backed Flycatcher, but bullfinch-like whistle is falling, longer, and lower than both, while the dry rattle is long and low, like Green-backed, but often also slows.

Ficedula owstoni

琉球姬鹟 liú qiú jī wēng

Ryukyu Flycatcher LC

L: 12–13.5 cm

Habitat: Occurs mostly on oceanic islands or coastal woodlands during migration.

Behavior: Solitary or in pairs. Usually in understory bushes or the lower and middle stories of vegetation. Catches insects in aerial pursuits.

Distribution: Rare in coastal areas in eastern and southeastern China, including Hainan and Taiwan, during migration.

Voice: Vocalizations poorly known but probably very similar to those of Narcissus Flycatcher.

Ficedula elisae

绿背姬鹟 lù bèi jī wēng

Green-backed Flycatcher LC

L: 12–14 cm

Habitat: Inhabits broadleaf forests and mixed broadleaf-coniferous forests in mountainous areas at altitudes of 1200–2500 m. Also found in secondary forests, urban parks, and orchards.

Behavior: Often solitary or in pairs. Forages in the middle story of forests; sallies in pursuit of aerial insects. Sometimes also comes to small trees or thickets.

Distribution: Uncommon summer breeder and passage migrant in North China; seen throughout eastern China during migration.

Voice: Vocalizations very similar to those of Yellow-rumped Flycatcher.

Rusty-tailed Flycatcher

Rufous-brown rump and tail feathers

Yellow lower mandible

Narcissus Flycatcher

Yellow supercilium

Long white wing patch

♂

Color on upper belly gradually turns orange on throat; throat color more saturated during breeding season, appears orange red

Yellow on belly turns to white on undertail coverts

Brownish gray overall, slight yellowish-green wash on back

Rufous-brown uppertail coverts and outer tail feathers

♀

European Pied Flycatcher

White forehead

♂

Black upperparts

Distinct white wing patch

Grayish-brown upperparts

♀

Distinct white wing patch

Ryukyu Flycatcher

Resembles male Narcissus Flycatcher, but with olive from crown to back

♂

Resembles female Narcissus Flycatcher with olive plumage, difficult to distinguish in field

♀

Yellow-rumped Flycatcher

White supercilium

Yellow rump

♂

White wing patch

Yellow rump

♀

White wing patch

Green-backed Flycatcher

Relatively short yellowish-green supercilium

Yellowish-green rump

♂

White wing patch

Greenish upperparts

♀

Yellowish-green underparts

547

Ficedula mugimaki

鸲姬鹟 qú jī wēng

Mugimaki Flycatcher LC

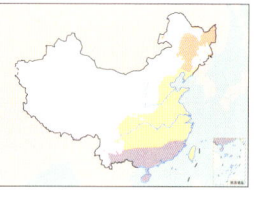

L: 12–14 cm
Habitat: Broadleaf and mixed forests on mountains and plains below 1000 m. Also in secondary forests, artificial forests, forest edges, urban parks, and orchards on foothill and plains.
Behavior: Solitary or in pairs; sometimes in small flocks during migration. Active; mainly in forest canopy, hopping among branches or flying short distances; sometimes in understory or on ground.
Distribution: Uncommon summer breeder in Northeast China; winter visitor in southeastern China; seen throughout eastern China during migration.
Voice: Song is a loud, stuttering, 3 sec rattle that starts slowly, with two or three short notes, and accelerates to a fast staccato trill, almost tripping over itself, *ti-di-ti-di-tidititi'wichitiwichiti'tidzeu*. Rattle call is long, like that of Green-backed Flycatcher and Narcissus Flycatcher but faster and higher pitched than both.

Ficedula erithacus

锈胸蓝姬鹟 xiù xiōng lán jī wēng

Slaty-backed Flycatcher LC

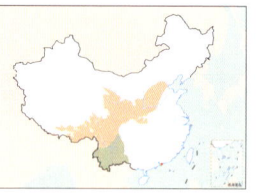

L: 12–14 cm
Habitat: Montane evergreen broadleaf and mixed forests, bamboo thickets, and forest-edge bushes at 2000–4000 m. Often winters down to secondary forests, bamboo thickets, and understory bushes on hills and foothills.
Behavior: Solitary or in pairs. Active; prefers undergrowth and bamboo thickets, hopping among branches. When foraging, watches from perch for passing insects and flies to the air or ground to catch them.
Distribution: Found in southwestern to North China, as uncommon summer breeder or resident. Vagrant in Hong Kong.
Voice: Pleasant musical song starts gradually, like song of Mugimaki Flycatcher, and accelerates and rises in pitch before falling at the end; lacks the staccato, hurried climax of Mugimaki. Calls include a descending *che*, often combined with a shorter flat note, *che-di* or *che-di … chliu*, and a deep Mugimaki-like rattle.

Ficedula strophiata

橙胸姬鹟 chéng xiōng jī wēng

Rufous-gorgeted Flycatcher LC

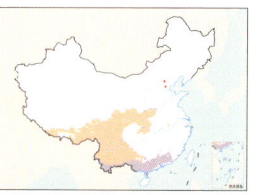

L: 13–15 cm
Habitat: Montane evergreen broadleaf and mixed forests, woodlands, and scrub below 3000 m; winters down to secondary forests and understory bushes in low mountains.
Behavior: Similar to Slaty-backed Flycatcher but prefers the middle story of forests or bushes.
Distribution: Uncommon summer breeder or resident in southwestern, Central, and South China. Vagrant in Hebei and Beijing.
Voice: Very simple song a thin *twee- ree-zick*, repeated, with little variation, every 1.5–2 sec. Calls include a mellow chatlike *tuk* or *tchuk-tchuk-tuk*.

Ficedula albicilla

红喉姬鹟 hóng hóu jī wēng

Taiga Flycatcher LC

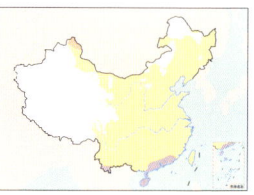

L: 12–14 cm
Habitat: Breeds in broadleaf and mixed forests in low mountains and hills below 1800 m; often in secondary forests, artificial forests, forest-edge woodlands, and understory bushes in low mountains and plains during migration and winter.
Behavior: Often solitary or in pairs; in small flocks during migration and winter. Active; usually moves about in the middle story of forests or bushes, hopping or flying among branches. Returns to the original perch after catching aerial insects in flight; sometimes also in forest undergrowth or on the ground. Often flicks tail up and down.
Distribution: Migrates throughout the country; a common passage migrant in most areas.
Voice: Dry rattle, *trrrrt*, similar to that of Red-breasted Flycatcher but twice as fast (so that individual notes can't be distinguished).

Ficedula parva

红胸姬鹟 hóng xiōng jī wēng

Red-breasted Flycatcher LC

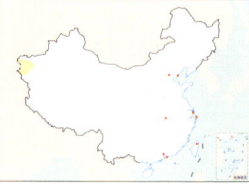

L: 11–13 cm
Habitat: Breeds in dense forest undergrowth near water sources; often occurs in secondary forests, artificial forests, forest-edge woodlands, and understory bushes on low mountains or plains during migration and in winter.
Behavior: Similar to Taiga Flycatcher.
Distribution: Records as vagrant in the east coast and Xinjiang, usually in fall and winter.
Voice: Slow dry rattle, *trrrt*, and occasionally a high flat *tiu*.

Ficedula hyperythra

棕胸蓝姬鹟 zōng xiōng lán jī wēng

Snowy-browed Flycatcher LC

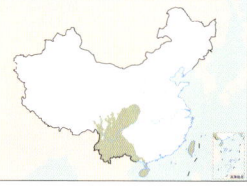

L: 10–12 cm
Habitat: Evergreen broadleaf, mixed, and secondary forests, bamboo thickets, forest-edge woodlands, and understory bushes below 1500 m; descends to secondary forests and understory bushes on low mountains, hills, and foothill plains in winter.
Behavior: Similar to Taiga Flycatcher but more timid, hiding in bushes most of the time. Often fans tail or bobs tail up and down.
Distribution: *F. h. hyperythra* in Shaanxi, Sichuan, Chongqing, SE Qinghai, Guizhou, W and S Yunnan, Guangxi, and Hainan; *F. h. innexa* in Taiwan. Uncommon summer breeder and resident.
Voice: Song a quiet, high-pitched, tinkling, sometimes wheezy, two- or occasionally three-part series of notes, *tsii-tsidi … tstsee-tsi … chiziwee-chiziwee*, where the second note is lower than the first. Calls include a thin *tsee* and an intense, high-pitched, penetrating *dzik*.

Mugimaki Flycatcher

Relatively short white supercilium behind eye

White wing patch

♂

Rufous-orange throat and breast

Two pale wing bars

♀

Pale red throat and breast

Taiga Flycatcher

Black bill

♂

Orange-red throat

Black uppertail coverts

White base of outer tail feathers

♀

Grayish-white breast

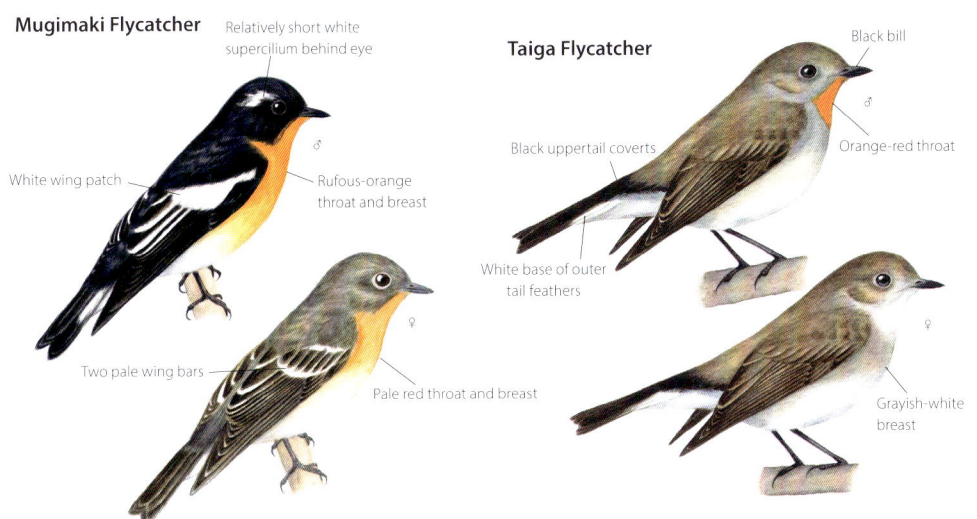

Slaty-backed Flycatcher

Bluish-gray upperparts

♂

Orange-red throat and breast

White base of outer tail feathers

Brown overall

♀

Red-breasted Flycatcher

Grayish cheek

Brownish-black uppertail coverts

♂

Pinkish base of lower mandible

Orange-red throat and upper breast

Pinkish base of lower mandible

White base of outer tail feathers

♀

Whitish breast tinged rufous brown

White base of outer tail feathers

Rufous-gorgeted Flycatcher

Short white supercilium

♂

Orange patch on upper breast

Diffuse white supercilium

♀

Indistinct orange patch on upper breast

White base of outer tail feathers

Snowy-browed Flycatcher

Thick short white supercilium

Dark purplish-blue upperparts

♂

Brown overall

♀

Short tail

Brownish-yellow underparts

Ficedula westermanni

小斑姬鶲 xiǎo bān jī wēng

Little Pied Flycatcher

L: 10–12 cm
Habitat: Inhabits
evergreen broadleaf
forests, mixed broadleaf-
coniferous forests,
bamboo thickets,
secondary forests, and
scrubland at altitudes of
1000–3000 m. In winter,
sometimes descends to secondary forests, orchards, and gardens on low
mountains, hills, and foothill plains.
Behavior: Often solitary or in pairs. Bold; usually forages in parks,
villages, and courtyards. Prefers middle story of forests; often hops or
flies among branches and leaves. Makes aerial sallies from perch after
passing insects.
Distribution: Uncommon resident in southwestern China. *F. w. collini* in
S Tibet; *F. w. australorientis* in SE Tibet, W and S Yunnan, S Guizhou, and
Guangxi.
Voice: Simple song is a single *tiu* followed either by a rattle of call
notes or by a series of thin notes, *tiu-e-e-e-e-e-e* or *tiu-titititit*. Calls
include a single low *chur* and a mellow *weet*, both similar to Rusty-tailed
Flycatcher.

Ficedula superciliaris

白眉蓝姬鶲 bái méi lán jī wēng

Ultramarine Flycatcher

L: 10–12 cm
Habitat: Inhabits
montane evergreen
broadleaf forests, mixed
broadleaf-coniferous
forests, coniferous forests,
and bamboo thickets at
altitudes of 1800–2500 m;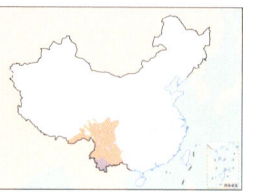
prefers moist forests.
Sometimes migrates down to secondary forests, orchards, and farmland
on low mountains, hills, and foothill plains in fall and winter.
Behavior: Often solitary or in pairs. Prefers moving about in the middle
story and canopy of coniferous forests; occasionally comes to the
ground. Active; often hops or flies constantly among branches and
leaves. Makes aerial sallies from perch after passing insects.
Distribution: Uncommon summer breeder in SE Tibet, W Yunnan, and
Sichuan.
Voice: Song is the rattling call note preceded by a short *pip*. Calls include
a decelerating rattle, extremely similar to Pygmy Flycatcher and slower
and deeper than Little Pied Flycatcher and not rising.

Ficedula hodgsoni

侏蓝姬鶲 (侏蓝仙鶲)

zhū lán jī wēng (zhū lán xiān wēng)

Pygmy Flycatcher

L: 8–10 cm
Habitat: Often inhabits evergreen broadleaf and mixed forests at
altitudes of 1000–2500 m; especially prefers streams and forest edges.
Sometimes descends to foothills in winter.

Ficedula tricolor

灰蓝姬鶲 huī lán jī wēng

Slaty-blue Flycatcher

L: 10–13 cm
Habitat: Inhabits
evergreen broadleaf
forests, mixed broadleaf-
coniferous forests,
bamboo thickets, and
scrubland in mountains at
altitudes of 1500–3000 m;
sometimes migrates down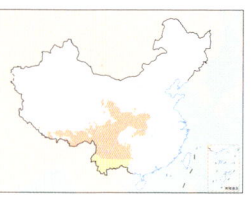
to secondary forests, scrub, and grassy patches on low mountains, hills,
and foothill plains in winter.
Behavior: Often solitary or in pairs; sometimes in small flocks during
nonbreeding season. Prefers moving about in understory and on the
ground; occasionally comes to the middle story of forests. Often cocks
tail high over back when resting. Makes aerial sallies from perch after
passing insects.
Distribution: *F. t. minuta* in SE Tibet; *F. t. diversa* in S Shaanxi, S Gansu,
Yunnan, Sichuan, Guizhou, Hunan, and Guangxi.
Voice: Simple, short song of three or four high-pitched whistles, the
first drawn out, the second slightly higher and louder and occasionally
becoming a brief trill, and the last two differently toned, *chee-tt-titit* or
chee-dzz-whit-it. Hard *tic* call note, like European Robin, often repeated
in lengthy series.

Ficedula sapphira

玉头姬鶲 yù tóu jī wēng

Sapphire Flycatcher

L: 11–12 cm
Habitat: Inhabits
evergreen broadleaf
forests and mixed forests
below 2000 m.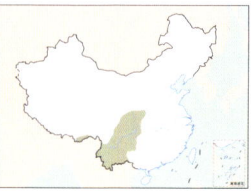
Behavior: Often solitary
or in pairs. Typical arboreal
species; prefers moving
around on treetops and
occasionally in the canopy and understory bushes. Usually hops or flies
among branches and leaves and frequently catches flying insects.
Distribution: *F. s. sapphira* in W and S Yunnan and W Sichuan;
F. s. laotianna in W and NW Yunnan; *F. s. tienchuanensis* in the Qinling
Mountains of Shaanxi, Shennongjia in Hubei, and C and W Sichuan.
F. s. tienchuanensis is very similar to Ultramarine Flycatcher and may be
misidentified or overlooked.
Voice: Song essentially unknown. Dry rattle calls very similar to
Ultramarine Flycatcher.

Behavior: Often solitary
or in pairs. Active; hops
frequently among
branches and leaves
of treetops. Prefers
moving around in the
undergrowth; sometimes
also moves to the air and
ground to capture insects.
Distribution: Uncommon resident found only in W Yunnan.
Voice: Short weak song a high-pitched, stuttering trill, *tzi-te-te-te-te-te-
heee*. Calls very similar to Ultramarine Flycatcher.

Little Pied Flycatcher

Distinct white supercilium

White wing patch

♂

Grayish overall

Rufous-brown tail

♀

Slaty-blue Flycatcher

Bluish-gray upperparts

♂

White base of outer tail feathers

Grayish brown overall

♀

Ultramarine Flycatcher

Diffuse white or pale blue supercilium

Dusky grayish-blue upperparts and side of breast not as glossy as subspecies *F. s. tienchuanensis* of Sapphire Flycatcher

♂

White underparts

Grayish overall

♀

Sapphire Flycatcher

Bright purplish-blue upperparts and side of breast

♂

F. s. sapphira

Orange-red throat and breast

♀

Cobalt-blue crown with sheen

No supercilium

F. s. tienchuanensis

♂

Pale brown head

F. s. laotianna

Orange throat

♂

Pygmy Flycatcher

Glossy blue forehead

Purplish-blue upperparts

♂

Brownish-yellow underparts

Grayish-brown upperparts

♀

Pale brownish-yellow underparts

Phoenicurus alaschanicus

贺兰山红尾鸲 hè lán shān hóng wěi qú

Ala Shan Redstart NT

L: 18–20 cm
Habitat: Breeds in mountainous areas with dense bushes; also occurs in forest clearings in coniferous forests. In winter, inhabits valleys with streams, mostly in sea buckthorn bushes, and sometimes in urban parks.
Behavior: Occurs in bushes; often perches on rocks or the upper part of bushes, with tail flicking slightly.
Distribution: Breeds in C Gansu to E Qinghai; winters in Inner Mongolia (Alashan/Alxa), Ningxia, Gansu to N Shanxi and Beijing; may be a resident in the Helan Mountains. Rare.
Voice: Lengthy song a discordant scratchy series of erratic notes, with many elements reminiscent of Eurasian Siskin. Call a repeated mellow *tuc*, very similar to Rufous-backed Redstart.

Phoenicurus ochruros

赭红尾鸲 zhě hóng wěi qú

Black Redstart 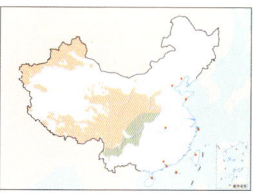 LC

L: 14–15 cm
Habitat: Inhabits rocky, scrubby grasslands and river flats.
Behavior: Often appears in open environments; also perches on buildings, earthen slopes, and other conspicuous places. Stance horizontal; bobs head and flicks tail.
Distribution: *P. o. phoenicuroides* in Xinjiang and W Tibet. *P. o. rufiventris* in northwestern and southwestern China except for Xinjiang; common; recorded as vagrant in Beijing, Hebei, Shandong, Zhejiang, Hubei, Guangdong, Hainan, Hong Kong, and Taiwan. *P. o. gibraltariensis* was first recorded in Xinjiang (Korla) in December 2013 and only in Xinjiang so far.
Voice: Song is a trill, followed by a peculiar dry crackle and musical flourish (but order may change). Call a mellow *tuc*, often in a series, *weet*, or *vit-vit-t-t-t-t*.

Phoenicurus erythronotus

红背红尾鸲 hóng bèi hóng wěi qú

Rufous-backed Redstart LC

L: 14–17 cm
Habitat: Breeds in subalpine coniferous forests and montane scrubland; especially common along mountain streams near dense understory and in forest-edge bushes. The

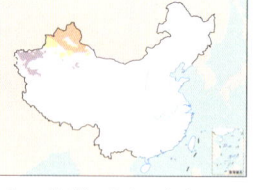

northern population migrates southward in fall and winters in river-valley thickets and woodlands on the plains.
Behavior: Male noisy and lively in spring; calls clearly and frequently wags tail up and down but not sideways. Often solitary; moves about and forages on the ground and on top of bushes.
Distribution: Breeds in the Altai Mountains, W Junggar Mountains, and Tian Shan Mountains in Xinjiang. Northern breeding populations migrate south to S Xinjiang in winter.
Voice: Short song a mellow, slowly delivered, thrushlike warble. Calls include a mellow, low, croaking *trukk-uk* or similar.

Phoenicurus phoenicurus

欧亚红尾鸲 ōu yà hóng wěi qú

Common Redstart LC

L: 13–14.5 cm
Habitat: In breeding season, inhabits various wooded areas ranging from foothill scrubland to subalpine coniferous forests. Sometimes occurs in urban and rural gardens in river valleys.

Behavior: In pairs in breeding season and in small flocks during migration. Often perches on low tree branches or shrubs and frequently flies to the ground to catch prey.
Distribution: Breeds in the Tarbagatai Mountains and Altai Mountains in NW Xinjiang; locally common to uncommon. Occasional in C and SW Xinjiang during migration.
Voice: Two-part song starts with a short indrawn whistle, followed by a very rapid repetition of three or four notes and a variable flourish that contains skilled mimicry. Calls include plaintive *hwiiit*, often combined with a harder *tic*.

Phoenicurus coeruleocephala

蓝头红尾鸲 lán tóu hóng wěi qú

Blue-capped Redstart LC

L: 13–15 cm
Habitat: Breeds in subalpine coniferous forests, especially in steep rocky mountainside thickets and woodlands with dense undergrowth close to tree line, but also in mixed broadleaf-coniferous forests. Found in river-valley forests and urban parks in hilly areas near mountains in winter.
Behavior: Moves about and forages mainly on the ground and in bushes; sometimes also perches on rocks, bushes, and trees and swoops down immediately when food is detected on the ground. Frequently

flies up and down between the ground and branches to hunt; also flies between bushes and tree branches in search of food. When resting, often wags tail from side to side. Song is very pleasant during breeding season.

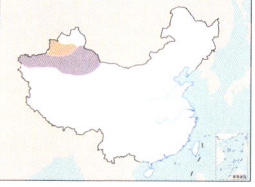

Distribution: Common breeder in the Tian Shan Mountains and Alatau Mountains in W and C Xinjiang; winters on the south slopes of Tian Shan Mountains and in Turpan region in Xinjiang.
Voice: Song similar to *tik-tik* sound of robins. Alarm sound near the nest is a piping *tit, tit, tit*. Song is loud and clear, like a silver bell, and fast and high pitched, similar to Godlewski's Bunting.

Ala Shan Redstart

Bluish-gray head

Distinct white eye ring

Slender white wing bar

♂

Narrow eye ring

Double wing bars, wing panel, and white fringe on tertials

♀

Grayish-brown belly

Rufous-backed Redstart

Bluish-gray head

Black patch on cheek

Long white wing patch

Chestnut throat and breast

♂

Wing like female Ala Shan Redstart

Underparts whiter than female Ala Shan Redstart

♀

Blue-capped Redstart

Bluish gray from crown to neck

Black upperparts and tail feathers

Large white wing patch

♂

Relatively dark body

♀

Brown rump and uppertail coverts

Narrow brown fringe on blackish-brown tail feathers

Black Redstart

White supercilium

Gray-tinged crown and upper back

♂

Grayish overall

♀

P. o. phoenicuroides

Black back

♂

Orange-rufous belly

P. o. rufiventris

Brownish overall

♀

White wing patch

Dark gray belly lacks orange

P. o. gibraltariensis

Common Redstart

White supercilium and forehead

♂

Black on throat does not extend to breast

Pale upperparts

♀

Buff underparts

553

Phoenicurus hodgsoni

黑喉红尾鸲 hēi hóu hóng wěi qú

Hodgson's Redstart　LC

L: 13–16 cm

Habitat: Inhabits scrubland, river valleys, forest edges, and meadows at middle and high altitudes; also occurs near villages or in urban green spaces in nonbreeding season.

Behavior: Often solitary, in pairs, or in small flocks. Generally stays in bushes or on rocks. Habitually flicks its tail and can catch insects in flight.

Distribution: Seen across the region from S Qinling Mountains to NW Yunnan, east to Hubei and Hunan.

Voice: Hurried song a series of short (1.2–1.8 sec) phrases, repeated with only minor changes, each containing a series of mostly short buzzes, trills, and musical notes. Most start quietly and build, *hwe-ti'ti-ti'ti-he'he-zwe-he-he-tii-tii-heh-hii*. Calls include a very fast rattle, a shrill Daurian Redstart–like *peep*, and a Hawfinch-like *pik*.

Phoenicurus schisticeps

白喉红尾鸲 bái hóu hóng wěi qú

White-throated Redstart　LC

L: 14–16 cm

Habitat: Inhabits coniferous forests, mixed forests, or forest edges in subalpine zones, mostly near streams.

Behavior: Solitary or in pairs. Active, mobile, and not shy.

Distribution: Seen in the region south of Qinling Mountains, from E Qinghai-Tibet Plateau to W Hubei; multiple winter records in Beijing.

Voice: Short, jolty, scratchy song that alternates between high- and low-pitched notes, often paired or tripled, in a rapid series, slower and more varied than the song of Hodgson's Redstart. Calls include a flat *tsee* whistle (longer than the equivalent of Hodgson's) and a sharp *tik*.

Phoenicurus auroreus

北红尾鸲 běi hóng wěi qú

Daurian Redstart　LC

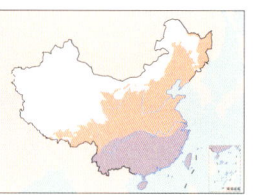

L: 13–15 cm

Habitat: Inhabits mountainous areas, forest edges, river valleys, and vicinity of villages in breeding season; nests in rock crevices and wall cavities. In winter, occurs in various broadleaf woodland and scrub environments; also in cities.

Phoenicurus erythrogastrus

红腹红尾鸲 hóng fù hóng wěi qú

White-winged Redstart　LC

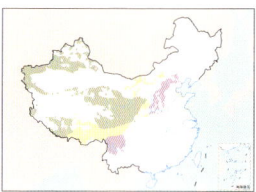

L: 15–17 cm

Habitat: Inhabits open areas at middle and high altitudes, such as bare rock zones, meadows, and scrubland.

Behavior: Solitary or in small flocks; often perches on prominent rocks or bush branches but constantly changes perching location.

Distribution: Breeds in Western China except for C, S, and W Tibet; winters in Beijing, Hebei, Sichuan, and N Yunnan. Relatively common.

Voice: Brief hurried song is a series of low-pitched, melancholy, whistled phrases mixed with a variety of quiet chirps, clicks, and twitters. Calls include an easily overlooked *tuk* and a short rattle, *drrrrt*.

Phoenicurus frontalis

蓝额红尾鸲 lán é hóng wěi qú

Blue-fronted Redstart　LC

L: 15–16 cm

Habitat: In breeding season, inhabits dwarf juniper, rhododendron, birch thickets, and open grasslands in subalpine zones at altitudes of 3000–5200 m; in winter, migrates down to open woodlands, river-valley thickets, cultivated lands, tea plantations, and orchards at altitudes of 1000–2700 m.

Behavior: Often takes off from perch and flies short distances to catch insects. Typical tail-flicking behavior of redstarts.

Distribution: Locally common resident in central and southern China, including S Shaanxi, Ningxia, Gansu, W Inner Mongolia, S and E Qinghai, Tibet, Yunnan, Guizhou, Sichuan, Chongqing, and Hubei, with scattered records in East and South China.

Voice: Song is less varied, short, and melodic, ending with a sharp trill. Alarm call is a low repetitive *ee–tit–tit*.

Behavior: Solitary or in pairs. Constantly bobs head and flicks tail from prominent perch; often returns to the original perch after catching insects.

Distribution: *P. a. leucopterus* in northwestern China except for Xinjiang; winters in southwestern China. *P. a. auroreus* is widely distributed throughout the country except for Xinjiang, W Tibet, and W Qinghai; winter visitor in areas south of the Yangtze River; passage migrant, summer breeder, or resident in areas north of the Yangtze River; common throughout the range.

Voice: Hurried song is a combination of scratchy and attractively whistled notes that typically have a stereotyped beginning, a scratchy central portion, and some mimicry in the closing few notes. Typical calls include a sharp, high-pitched *heep*, similar to call of Red-flanked Bluetail, and a mellow *tek* or *tek tek*.

Hodgson's Redstart

Gray-tinged back

Small wing patch with diffuse boundary

♂

Relatively pale grayish belly

♀

White-winged Redstart

Distinct white on head

♂

Almost square white wing patch

♀

White-throated Redstart

Bluish-gray head

Long narrow white wing patch

♂

Bluish-black tail

White patch on throat

Long narrow white wing patch

White patch on throat

♀

Blue-fronted Redstart

Glossy blue forehead

Blue head, back, throat, and wing coverts

♂ br.

Rufous orange from breast, flanks, and belly to undertail coverts

Brownish-yellow base of uppertail coverts and outer tail feathers outline a black inverted T on tail

Grayish-brown upperparts slightly tinged olive

♀

Buff underparts

Tail like male

Pale brown fringe of feathers on upperparts, making blue plumage duskier

non-br.

Daurian Redstart

Slightly grayish-white crown

Black back

Well-defined triangular white wing patch

♂

Triangular white wing patch

♀

555

Phoenicurus fuliginosus
红尾水鸲 hóng wěi shuǐ qú
Plumbeous Redstart

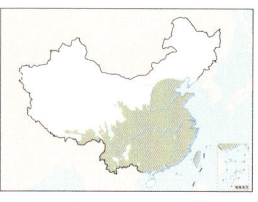

L: 12–13 cm
Habitat: Inhabits montane streams and river valleys, especially along rocky streams in forests or forest edges; also occurs in river valleys and streams on plains. Occasionally seen on banks of lakes, reservoirs, and ponds. Breeds from lowlands up to 3600 m.
Behavior: Solitary or in pairs. Perches on rocks by or in water, rocky walls by highways, or power lines, and sometimes also on roofs in villages; when perched, wags tail and sometimes spreads feathers like a fan. Flies quickly to catch bugs detected on water or the ground and then flies back to the original perch. Sometimes also runs quickly on the ground to peck at insects. When disturbed, flies along rivers above the water's surface and calls while flying.
Distribution: *P. f. fuliginosus* is common in most areas in southern and eastern China, *P. f. affinis* in Taiwan.
Voice: Song a lengthy, high-pitched (torrent-adapted), shrill, metallic, unmusical jingling, with only very short pauses between phrases and many notes repeated twice. Also a series of three to five shrill *striiii-trii-trii-trí* notes that slow as the notes become shorter. Calls include a sharp, strident *zwet*, sometimes repeated.

Phoenicurus leucocephalus
白顶溪鸲 bái dǐng xī qú
White-capped Redstart

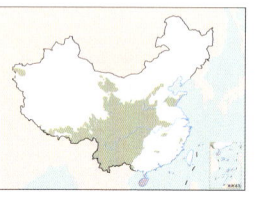

L: 18–19 cm
Habitat: Inhabits montane river valleys, rocks along mountain streams, riverbanks, and huge rocks exposed in rivers; sometimes in valleys or on dry riverbeds. Altitudinal migrant, inhabiting high-altitude mountains in summer and migrating down to lower altitudes in fall and winter, generally between 1800 and 4800 m.
Behavior: Often perches on prominent rocks in or close to water; bobs head and flicks tail upon landing. Captures small invertebrates mainly from water's surface. Display includes a peculiar head-bobbing behavior.
Distribution: Common in North, northwestern, Central, and southwestern China; a few records in the mountainous areas of Anhui, Zhejiang, and N Jiangxi, probably as summer breeder; occasionally wanders to the islands off China's east coast.
Voice: Song is an increasingly tremulous, flat, and weak *zreeeerrrrrr* lasting about a second and repeated every 3–12 sec. Calls include a shorter but still fairly lengthy (0.4 sec), shrill, rising *tseee*.

Monticola saxatilis
白背矶鸫 bái bèi jī dōng
Rufous-tailed Rock-Thrush

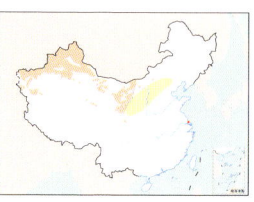

L: 17–20 cm
Habitat: Inhabits rocky barren slopes, scrubland, and grasslands with sparse vegetation from low hills to subalpine and plateau areas, especially on floodplain meadows with scattered bushes and open, treeless, rocky grassy slopes.
Behavior: Often found in pairs in breeding season; forms small loose flocks during migration. Ground dweller; often perches on prominent rocks or bare treetops. In display flight, male flaps wings with tail feathers spread, and then glides down with both wings and tail outstretched.
Distribution: Breeder or passage migrant in Beijing, N Hebei, Shanxi, Shaanxi, C Inner Mongolia, Ningxia, Gansu, Xinjiang, and Qinghai; common in N Xinjiang in breeding season; vagrant in Jiangsu.
Voice: Song is fast, melodic, fluty notes mixed with trills and occasional grating notes and mimicry, very similar to Blue Rock-Thrush. Calls include a mellow *tak*, often repeated and combined with a peculiar squeaky *viii*.

Monticola solitarius
蓝矶鸫 lán jī dōng
Blue Rock-Thrush

L: 20–23 cm
Habitat: Breeds on steep cliffs, rocks, stone buildings, or quarries below 1830 m; *M. s. philippensis* prefers coastal zones more than other subspecies.
Behavior: While hunting, often perches low and scans the ground before striking. Can also hop on ground in search of food and occasionally catches prey directly in flight.
Distribution: Common resident. *M. s. pandoo* in most of central and eastern China; *M. s. philippensis* is resident in most of eastern and southern China and a summer breeder in coastal areas of Northeast and North China; *M. s. longirostris* in SW Tibet.
Voice: Rich, melodious song includes whistles, churrs, trills, some repetition, and occasional mimicry.

Plumbeous Redstart

Compact rounded body,
dark slate blue overall

Bluish-black wings

Ochre red from
rump to tail

♂

Slate-gray upperparts

Two white wing bars

Blackish-brown tail

♀

Dense scaly pattern
on underparts

Pure white undertail coverts,
outer tail feathers, and rump

Rufous-tailed Rock-Thrush

Bluish-gray head, neck,
and upper back

White back

Chestnut-
brown tail

Rusty-brown
underparts

♂

Grayish-brown
upperparts

Chestnut-brown tail

Scaly buff underparts

♀

White-capped Redstart

White crown

Black-tipped tail

Rufous chestnut from breast
to belly, rump, and tail;
remaining body black

Blue Rock-Thrush

Glossy blue overall,
wings bluish black

Dark lores

♂

M. s. pandoo

Dull bluish-gray upperparts,
wings grayish black

♀

Scaly blackish-brown pattern
from cheek, breast, and belly
to undertail coverts

Blue head, breast, back, and
tail; wings bluish black

♂

Rufous-chestnut lower breast,
belly, flanks, and undertail coverts

M. s. philippensis

Monticola rufiventris

栗腹矶鸫 lì fù jī dōng

Chestnut-bellied Rock-Thrush

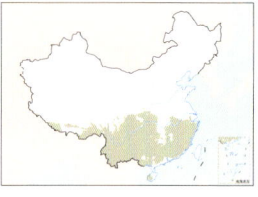

L: 21–23 cm
Habitat: In breeding season, inhabits open and moist primary forests, oak forests, stone oak forests, rhododendron thickets on cliffs, boulders, riverbeds, and forest-edge bushes at altitudes of 1200–2440 m.
Behavior: Often forages on the ground but also perches on top of dead trees to make aerial sallies.
Distribution: Common resident in Central, southeastern, southwestern, and South China.
Voice: Repeated simple short song begins with three short notes that rise in pitch, followed by a longer, buzzier, slightly descending note and then a warbled note, *du-du-et-zeeee-trui-it*. Calls include a piping *diuu* and a scolding Eurasian Jay–like *kssssss*.

Monticola cinclorhyncha

蓝头矶鸫 lán tóu jī dōng

Blue-capped Rock-Thrush

L: 17–19 cm
Habitat: Inhabits low-mountain evergreen broadleaf forests and mixed forests, especially near streams and rocky areas.
Behavior: Solitary or in pairs. Active; often calls from top of rocks. Prefers moving about on forest floor or among understory bushes.
Distribution: Rare summer breeder found only in S Tibet.
Voice: Gives a monotonous and metallic *yu-yu-yu*.

Monticola gularis

白喉矶鸫 bái hóu jī dōng

White-throated Rock-Thrush

L: 17–19 cm
Habitat: Inhabits low-mountain broadleaf forests and mixed forests; especially prefers wet and rocky areas. Occurs in coastal windbreak forests and secondary forests during migration.
Behavior: Solitary or in pairs. Active; often stands on top of rocks and calls. Prefers moving about on the forest floor.
Distribution: Found in eastern China; summer breeder and passage migrant in North and Northeast China; passage migrant in most of eastern China.
Voice: Slow melancholy song is a series of drawn-out rising whistles followed by stuttering shorter whistles and slightly more melodious or burry notes, *te-wee-tueee-wee-we-t-ttueee-tut-zree-t-t-t-t-zweeuw*; each phrase is different and can last anywhere from 2.5 sec to 6 sec. Calls include a thin *tsip* and harder *tack*.

Saxicola insignis

白喉石䳭 bái hóu shí jí

White-throated Bushchat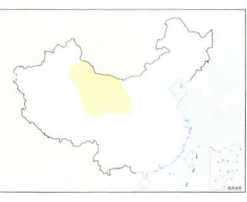

L: 14–15 cm
Habitat: Inhabits bushes on plateaus and alpine bare rocks above 2000 m.
Behavior: Solitary or in pairs. Often perches on the top of rocks, bushes, and fences on plateaus. Prefers moving about on the ground.
Distribution: Rare passage migrant, with scattered records in northwestern China.
Voice: Song poorly known but similar to Siberian Stonechat and Amur Stonechat but apparently with even more trills than Siberian and occasionally some mimicry. Calls include a *whit* that is lower pitched and flatter than those of Siberian and Amur and a *chack* that is apparently lower, huskier, and intermediate in length.

Saxicola maurus

黑喉石䳭 hēi hóu shí jí

Siberian Stonechat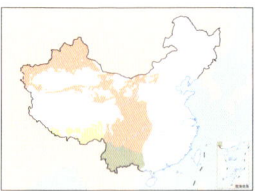

L: 12–15 cm
Habitat: Inhabits low mountains, hills, plains, marshes, and open fields; especially prefers scrubby environments on plains and open fields; sometimes on plateaus above 4000 m.
Behavior: Often solitary or in pairs. Prefers open areas of scrubland and grasslands. Usually stands still on bushtops; when insects are detected, makes sallies after insects in the air or on the ground and then flies back to the original perch.
Distribution: *S. m. presvalskii* is a summer breeder in N Gansu, E Xinjiang, NE Qinghai, and a resident in S Gansu, Ningxia, Sichuan, Hubei, Yunnan, and Tibet; *S. m. maurus* breeds in N and NW Xinjiang and migrates through Shaanxi and Tibet.
Voice: See Amur Stonechat.

Saxicola stejnegeri

东亚石䳭 dōng yà shí jí

Amur Stonechat

L: 12–14 cm
Habitat: Similar to Siberian Stonechat but not on plateaus.
Behavior: Often solitary or in pairs; in small flocks during migration. Other behaviors resemble those of Siberian Stonechat.
Distribution: Widespread in eastern China; breeds in Northeast China; winters in southeastern China; common throughout the range during migration.
Voice: Song is similar to, but shorter than, song of Siberian Stonechat, with a narrower frequency range, and trills are rarely included, making the song sound more melodious. Calls include a descending *whit* and a harsh *chack*, both similar to those of Siberian. The *whit* calls are very similar (but again with a narrower frequency range in Amur), while the *chack* is slightly longer.

Chestnut-bellied Rock-Thrush

Bright blue from forehead to crown

Bluish black at lores and ear coverts extends to scapulars

Blue upperparts and wings

♂

Blue from throat to upper breast contrasts well from chestnut on breast and belly

Distinguished from *M. s. philippensis* of Blue Rock-Thrush by black mask and bright glossy blue forehead

Prominent crescent-shaped white patch behind ear coverts

Brown upperparts

♀

Dense scaly blackish-brown pattern on buff underparts

Blue-capped Rock-Thrush

Blue crown

♂

Blue throat

White wing patch

Grayish-brown upperparts relatively uniform

♀

White breast and belly with scaly grayish-brown pattern

White-throated Rock-Thrush

Blue crown

♂

White wing patch

White throat

Scaly blackish-brown pattern on grayish-brown upperparts

♀

Scaly blackish-brown pattern on whitish breast and belly

White-throated Bushchat

♂

White throat

♀

Whitish rump

Siberian Stonechat

♂

Relatively heavy brown on underparts and flanks

♀

Amur Stonechat

♂

Relatively pale brown on underparts and flanks

♀

Saxicola caprata

白斑黑石䳭 bái bān hēi shí jí

Pied Bushchat

L: 13–15 cm

Habitat: Often inhabits open areas, such as low mountains, hills, foothill plains, farmland, and open fields; especially prefers grasslands, farmland, and deserted fields with sparse scrub.

Behavior: Often solitary or in pairs, and in small flocks during migration. Usually rests on tops of bushes or rocks in open areas, and sometimes on small trees or power lines; flicks tail up and down frequently. Prefers running on the ground or chasing insects on the wing.

Distribution: Uncommon resident in S Tibet, Yunnan, and Sichuan.

Voice: Short hurried song begins hesitantly but soon accelerates and includes whistles, warbles, trills, and some repetition, *hiu-hiu-t-t'hiuu'we'chizu'we'we'ww'dz-rr-dziu*. Calls include a falling *piu* and a markedly disyllabic *chu-it*, harsh *check*, and buzzier *dreet*.

Saxicola jerdoni

黑白林䳭 hēi bái lín jí

Jerdon's Bushchat

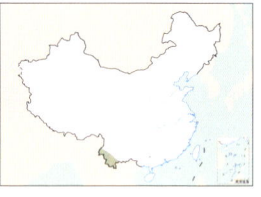

L: 14–16 cm

Habitat: Inhabits scrubland and scrubby grasslands on low mountains, hills, and foothill plains; especially prefers bushes and tall grass near water, such as along streambanks, river valleys, and reed swamps.

Behavior: Often solitary or in pairs. Active; often flies constantly among bushes and tall grass, frequently flicking or fanning tail. Scans for insects when perched; makes sallies after insects on the ground or in the air.

Distribution: Rare resident in W and S Yunnan.

Voice: Simple short song is a repeated series of sweet, clear, thin, mellow notes, *weeee-TI-TI-dee'rrrrr*. Calls include loud, high, nasal whistle, *hweee*, often combined with a rasping, buzzing churr, *dzrrrr*.

Saxicola ferreus

灰林䳭 huī lín jí

Gray Bushchat

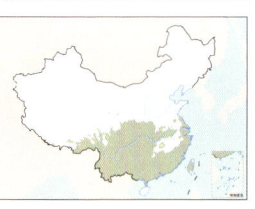

L: 14–15 cm

Habitat: Inhabits open scrubland, densely vegetated mountain slopes and hills, forest edges of coniferous or evergreen broadleaf forests, cultivated lands, farms, and gardens at altitudes below 3300 m.

Behavior: Often perches on vegetation. Prefers flying from perch to the ground to catch prey; also forages on the wing.

Distribution: Locally common resident. *S. f. ferreus* in SE Tibet, W and S Yunnan; *S. f. haringtoni* in southwestern, Central, South, and East China, and vagrant in Beijing and C Inner Mongolia.

Voice: Rapid but variable song includes phrases where the same note is repeated at different pitches three or four times, followed by an abrupt end on an often-buzzy descending note, such as *t-wee-wii-wee-wii-wee-wii-wee-t-driuu*. Calls include a nasal, buzzy, rising *djuee* and a falling Pied Bushchat–like *piu* and a *dzzuee*.

Oenanthe oenanthe

穗䳭 suì jí

Northern Wheatear

L: 14–16.5 cm

Habitat: Inhabits montane meadows, plateaus, and rocky grasslands at middle and high altitudes in breeding season; also occurs in premontane arid grasslands, deserts, and semidesert areas. Especially prefers open, gravelly grasslands with sparse vegetation.

Behavior: Ground dweller; moves about and forages mostly on grass. Perches often on protruding rocks or bushes; flicks tail up and down slowly. Immediately flies to the ground or in the air to hunt insects and then returns to the same perch. The song is variable and can imitate other birds' songs.

Distribution: Breeds in N Hebei, Shanxi, Shaanxi, Inner Mongolia, Ningxia, and Xinjiang, more common in N Xinjiang; vagrant in Zhejiang and Taiwan.

Voice: Short vigorous song includes trills, whistles, squeaks, and often mimicry. Calls include a mellow *chack*, often repeated, and a softer *tuc*.

Pied Bushchat

White wing patch

♂

Pale streaks on grayish-brown upperparts

♀

Gray Bushchat

Grayish-black mask contrasts strongly with narrow white supercilium and white throat

♂

Pale gray underparts

White throat contrasts strongly with brown cheek

♀

Chestnut-brown rump and uppertail coverts

Jerdon's Bushchat

Bluish-black upperparts

♂

White underparts

Brown upperparts

♀

Whitish throat

Northern Wheatear

Black ear coverts and lores

Gray head and upper back

♂

Black wings

Black projects into central tail relatively far

Black sides no more than half length of tail

Blackish-brown lores

Grayish-brown head and upper back

♀

Dark brown wings

561

Oenanthe isabellina

沙鹛 shā jí

Isabelline Wheatear LC

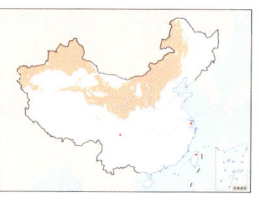

L: 15–16.5 cm
Habitat: Inhabits arid plains, deserts, and semideserts with sparse vegetation in premontane hilly areas; sometimes in subalpine desert grasslands and saline meadows.
Behavior: Often in pairs; highly territorial. Stance upright; flicks tail up and down constantly and runs swiftly on the ground. During display, male leaps into the air, hovers with tail open, and then glides to land. Good singer; often imitates other birds and animals.
Distribution: Breeds in N Hebei, Shanxi, Shaanxi, Inner Mongolia, Ningxia, Gansu, Xinjiang, N Tibet, and Qinghai; vagrant in Sichuan, Shanghai, and Taiwan.
Voice: Lengthy song (3–10 sec or longer) is an ill-formed, rambling series of variable phrases that contain an eclectic mix of whistles, chacks, twills, and often-skilled mimicry. Variety of calls include a buzzy *dzzzk* and harsh *chack*.

Oenanthe deserti

漠鹛 mò jí

Desert Wheatear LC

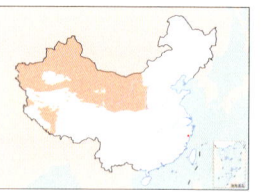

L: 14.5–15.5 cm
Habitat: Inhabits mainly desert and semidesert areas, such as arid desert plains and sand dunes. Also inhabits bare mountain rocks, rocky scrub and grasslands, and even desert and semidesert areas at 4000–5000 m. Prefers rocky deserts and deserted fields.
Behavior: Solitary or in pairs. Ground dweller; mostly runs fast on the ground to forage or hops on both feet. Sometimes also stands on rocks or bushes to scan the surroundings; sallies after prey when detected on the ground or in the air. When standing, often flicks tail up and down and sometimes shakes body along with it. Often perches on low vegetation. Shy; often flies to hide behind rocks. Male makes brief wing-fluttering display flight near nest.
Distribution: *O. d. atrogularis* breeds in N Shaanxi, C Inner Mongolia, Ningxia, N Gansu, and N and E Xinjiang. *O. d. oreophila* breeds in W and S Xinjiang, W and SE Tibet, and Qinghai; vagrant in Sichuan and Taiwan.
Voice: Song a distinctive series of two to four mournful, short, descending whistles, often broken or ended by a grating churr, such as *siie-drool-e-drruue-trrrrt*. Calls include a hard *tak*.

Oenanthe pleschanka

白顶鹛 bái dǐng jí

Pied Wheatear LC

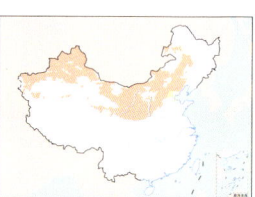

L: 14–16.5 cm
Habitat: Inhabits various habitats, such as arid grasslands, semideserts, barren mountains, gullies, forest-edge bushes, and rocky barren slopes; especially common in sparsely vegetated or barren and gravelly deserts and semideserts, as well as grassy areas on plains, fields, and even near urban parks and human settlements.
Behavior: Ground dweller; mostly runs after food on the ground but also perches often on rocks or bushes, taking off suddenly to capture prey once detected.
Distribution: Seen in W Liaoning, Beijing, Tianjin, Hebei, Henan, Shaanxi, Shanxi, Inner Mongolia, Ningxia, Gansu, Xinjiang, Qinghai, and S Sichuan.
Voice: Song is a unpleasant, short, explosive, buzzy, but varied trill, *tri-tri-trrriii*, that occasionally includes some mimicry. Calls include a harsh *chack*.

Oenanthe picata

东方斑鹛 dōng fāng bān jí

Variable Wheatear LC

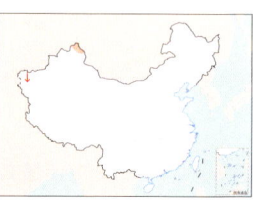

L: 17–18 cm
Habitat: Inhabits mainly low-mountain bare rocks, desert, and semidesert areas below 2500 m; especially prefers rocky barren mountains and ravines with sparse plants, as well as overhanging rocks in river valleys and rocky scrubland.
Behavior: Ground dweller; perches often on buildings or large stones on the ground, rarely in trees. Forages mostly on the ground but also waits on fixed perches for prey to approach and makes aerial sallies once insects are detected; sometimes also hawks insects.
Distribution: Breeding recorded in N Xinjiang; occasionally seen in W and S Xinjiang during migration.
Voice: Song apparently similar to that of Isabelline Wheatear. Calls include an indrawn *huid* or *huid-it* and a hard *tuk*.

Isabelline Wheatear

Black lores

Sandy-brown head and upperparts

Black on wings restricted to alula

Black sides no more than half length of tail

Black projection into central tail relatively short

Pied Wheatear

White on head and hindneck

White throat

Black on face, neck and back connects with wings

"*vittata*" morph

Black projects into central tail extremely far

Black sides no more than one-third length of tail

Black throat

♂

Desert Wheatear

Sandy-brown head and upper back

Black face, neck, and throat

Black on wings joins black throat

♂

No black projection

Black sides more than two-thirds length of tail

Dark brown head and upper back

Dull brown on wings does not extend to throat

♀

Streaks on grayish-brown cheek, throat, and upper breast

Sandy-brown head and upper back

Pale lesser coverts and scapulars

Brownish cheek, neck, and throat

♀

Variable Wheatear

All-black head, back, upper breast, and wings

♂

"*picata*" morph

Black projects into central tail, thick black band at tip

Black sides no more than one-third length of tail

Sooty-brown head, back, upper breast, and wings

♀

河乌科　Cinclidae

Small to medium-sized songbirds highly specialized to live and forage in streams. Plumage similar between sexes, mostly brown, black, and white, while juveniles usually scaly patterned. Bill thin and straight, without rictal bristles. Nostrils covered by flaps. Wings short and rounded; usually fly just above water. Short tail about half the length of wings, consisting of 12 feathers. Powerful legs well adapted to swimming and diving. Inhabit primarily mountain streams and creeks. Usually solitary or in pairs. Feed mostly on aquatic insects, mollusks, and small fish and shrimp. Highly territorial, nonmigratory.

Only one genus and five species recognized worldwide: two in Eurasia, one in North America, and two in South America. One genus and two species recorded in China, widely distributed except in the grasslands and deserts in northern China.

White-throated Dipper

Brownish head and mottled body

Dusky brown throat and breast

Dusky brown belly

ad.

Dark morph

juv.

All-white underparts

C. c. leucogaster

ad.

Pale morph

White throat and breast

Dusky brown belly

C. c. przewalskii

ad.

Pale morph

Pale morph

Pale brown underparts

Dark morph

ad.

ad.

C. c. cashmeriensis

Brown Dipper

Inconspicuous white eye ring

Longer tail

ad.

Relatively slender, blackish-brown body

juv.

White streaks on belly

C. p. pallasii

juv.

Relatively rounded, paler body

ad.

C. p. dorjei

Cinclus cinclus

河乌 hé wū

White-throated Dipper　🟢 LC

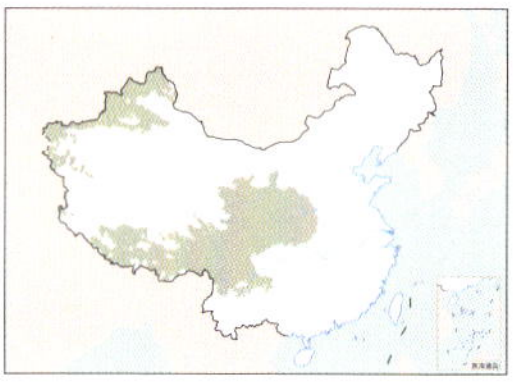

L: 16–20 cm

Habitat: Inhabits clear fast-flowing streams and rivers in valleys at altitudes of 800–4000 m.

Behavior: Often flies rapidly above water and forages in river valleys. Strong diver; capable of

moving on the bottom of river. Usually bobs head and rapidly flicks tail when perched on stones.

Distribution: *C. c. leucogaster* in Xinjiang; *C. c. przewalskii* in the mountains of southwestern China; *C. c. cashmeriensis* in S and SE Tibet and NW Yunnan. Not hard to find in suitable habitats.

Voice: Torrent-adapted song is a scratchy, high-pitched, piercing, loud warbling or trilling. Call a loud *zit*, similar to Arctic Warbler.

Cinclus pallasii

褐河乌 hè hé wū

Brown Dipper　🟢 LC

L: 18–24 cm

Habitat: Inhabits streams or fast-flowing rivers in mountain valleys.

Behavior: Prefers standing on bare rocks in rivers; often flicks tail. Strong diver. Often flies rapidly for a short distance above water.

Distribution: *C. p. tenuirostris* in NW Xinjiang and S Tibet; *C. p. dorjei* in NW Yunnan; *C. p. pallasii* in central, eastern, and southern China. Not hard to find in suitable habitats.

Voice: Song is a scratchy, uncomfortable warbling that includes dry buzzes, clipped trills, and rattles. Call is a buzzy *zit*, very similar to that of White-throated Dipper but fractionally lower pitched, longer, and coarser.

叶鹎科　Chloropseidae

Small forest-dwelling tropical songbirds whose body profile resembles bulbuls. Plumage differs between sexes. Face and throat of most species covered with a large black or bluish-purple patch. Bill thin and pointed, slightly decurved, and with less-developed rictal bristles. Wings short and rounded. Tail rounded, consisting of 12 feathers. Strong legs relatively short. Insectivorous; also take plant fruits and pollen. Voices clear and sweet. Inhabit the middle and upper stories of broadleaf forests in the tropics and subtropics. All species are resident.

Leafbirds were formerly put in Irenidae but now constitute this recently recognized family. Only one genus and 11 species recognized worldwide, endemic to Indomalaya. One genus and 3 species recorded in China, found in South and southwestern China.

Chloropsis aurifrons

金额叶鹎 jīn é yè bēi
Golden-fronted Leafbird　LC

L: 18–19 cm
Habitat: Inhabits broadleaf forests at middle and low altitudes.
Behavior: Gathers in small flocks in the canopy. Active; often hangs and forages on thin branches or flowers.

Distribution: Uncommon in SE Tibet and S and W Yunnan.
Voice: Highly variable, often-powerful song includes raucous chattering, musical whistles, and some repetition, as well as considerable mimicry of birds such as drongos and accipiters.

Chloropsis hardwickii

橙腹叶鹎 chéng fù yè bēi
Orange-bellied Leafbird　LC

L: 15–19 cm
Habitat: Inhabits the canopy or edges of evergreen and deciduous broadleaf forests and secondary forests at 500–2000 m; also in parks and gardens.
Behavior: Solitary or in pairs; often joins mixed-species flocks.

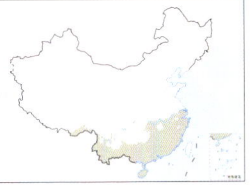

Distribution: Locally common resident. *C. h. hardwickii* in SE Tibet, Yunnan, and southwestern China; *C. h. melliana* in most regions of southeastern and South China; *C. h. lazulina* in Hainan. The latter two subspecies are sometimes treated as one separate species.
Voice: Variable and melodious songs, with many multisyllabic *tshiwatshishi-watshishi-watshishi* or *cheat chewee-chewee-cheweei* calls. Loud, rapid, and hoarse alarm calls *ti-ti-tsyi-tsyi-tsyi-tsyi-tsyi* or *fweeew-whew*.

Chloropsis cochinchinensis

蓝翅叶鹎 lán chì yè bēi
Blue-winged Leafbird　LC

L: 16–18 cm
Habitat: Inhabits various types of woodlands and scrubland at low altitudes.
Behavior: Often gathers in small flocks; also joins mixed-species flocks.
Distribution: Locally common in S and W Yunnan and SW Guangxi.
Voice: Makes *qi-wiwi* sounds.

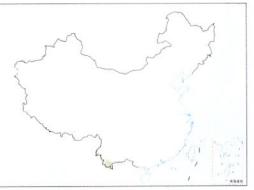

Blue-winged Leafbird

Yellow forehead
Blue outer wing feathers
Black throat with blue flash at jawline
Green underparts
Green forehead and throat
Blue outer wing feathers, differing from female Orange-bellied Leafbird (*C. h. melliana*)

Golden-fronted Leafbird

Gold forehead and crown
Green upperparts, wings, breast, and belly
Black throat bordered yellow

Orange-bellied Leafbird

Yellowish green from crown to neck
Yellowish-green upperparts
Black from face and throat to breast
Yellowish-green body
Violet-blue flash at jawline
C. h. hardwickii
Orange yellow from lower breast and belly to undertail coverts

Grayish green from crown to neck, gray in some individuals
Green upperparts
Bluish-black face, throat, and breast
Green body with bluish-green head
C. h. melliana
Pale violet flash at jawline

啄花鸟科 Dicaeidae

Small, brightly colored, forest-dwelling songbirds with a compact body profile. Plumage differs between sexes in most species. Bill short but strong. Wings short and rounded. Tail short, mostly squared. Inhabit primarily broadleaf and mixed forests but also in orchards and plantations; usually active in canopy. Nest on tree branches. Feed mostly on insects, nectar, and plant fruits; well adapted to process fruits of mistletoe rapidly. Most species are resident, while a few are migratory.

Two genera and 49 species recognized worldwide, distributed in Indomalaya and Oceania. One genus and six species recorded in China, found in the forests of southern China.

Dicaeum agile
厚嘴啄花鸟 hòu zuǐ zhuó huā niǎo
Thick-billed Flowerpecker LC

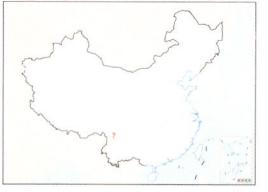

L: 9–10.4 cm
Habitat: Active from lowlands to altitudes up to 3000 m. Occurs in evergreen broadleaf forests, secondary forests, forest edges, plantations, orchards, and gardens, especially in flowering and fruiting trees and shrubs; prefers mistletoe and small fig trees.
Behavior: Primarily in canopy and middle story of trees; sometimes comes down to bushes in the understory. Solitary, in pairs, or in small flocks; also joins mixed-species flocks.
Distribution: Rare in S and W Yunnan and Sichuan.
Voice: Intense *zip* call becomes the song when given in an accelerating, then slowing, series, *tzip-ip-ip-ip-ip-ip-ip*.

Dicaeum ignipectus
红胸啄花鸟 hóng xiōng zhuó huā niǎo
Fire-breasted Flowerpecker LC

L: 7–9 cm
Habitat: Inhabits montane forests, subtropical evergreen forests and deciduous forests, forest edges, secondary forests, orchards, and flowering and fruiting trees at altitudes of 900–3950 m; migrates down to 300 m in winter.
Behavior: Usually solitary or in pairs in the canopy or middle story; gathers in small flocks or joins mixed-species flocks during nonbreeding season.
Distribution: Locally common resident. *D. i. ignipectus* in South and southeastern China, from S Shaanxi, S Henan to SE Tibet, Yunnan, Guangdong, Fujian, S Zhejiang, and Hainan; *D. i. formosum* in Taiwan.
Voice: Song a strident, high-pitched, usually bisyllabic trill (with a note repeated extremely rapidly five to eight times), *t'zee-t'zee-t'zee-t-zee-t'zeee … chi'zu-chi'zu-chi'zu-chi'zu … see-bit, see-bit, see-bit … sit-sit-sit-sit-sit-sit*, and a shrill *titty-titty-titty*. Calls include a very intense, very short, low-pitched *tup*, often repeated in lengthy series.

Dicaeum melanozanthum
黄腹啄花鸟 huáng fù zhuó huā niǎo
Yellow-bellied Flowerpecker LC

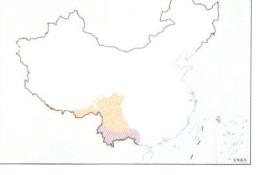

L: 11.5–13 cm
Habitat: Inhabits evergreen broadleaf forests, pine forests, and clearings and forest edges of tropical rain forests, especially in flowering and fruiting trees and shrubs; prefers mistletoe and small fig trees. At altitudes of 1400–3915 m in summer and 775–1550 m (highest record up to 2450 m) in winter.
Behavior: Usually slower than other flowerpeckers. Often perches on bare branches for a long time. Solitary, in pairs, or gathers in small flocks.
Distribution: Locally common in SE Tibet, Yunnan, S and W Sichuan, and SW Guangxi.
Voice: Agitated *tzeet* (3.4–9.8 kHz and lasting almost 0.1 sec—a long time for a flowerpecker!), repeated every second or so.

Dicaeum cruentatum
朱背啄花鸟 zhū bèi zhuó huā niǎo
Scarlet-backed Flowerpecker LC

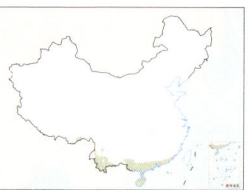

L: 7–9 cm
Habitat: Active in various types of forests, forest edges, coastal bushes, orchards, and gardens at altitudes below 1200 m.
Behavior: Often solitary or in pairs. Seldom gathers in flocks, except for the family groups formed in late breeding season. Forest dweller; especially prefers foraging in flowering trees or trees with parasitic plants, usually in the canopy or upper story of trees. Occasionally comes down to bushes.
Distribution: *D. c. cruentatum* is locally common resident in SE Tibet, S and E Yunnan, Guangxi, Guangdong, Fujian, and Hainan.
Voice: Song a rising *see-bi … see-bi … see-bi*, with the emphasis on the first syllable (almost a Yellow-browed Warbler–like *tsooest*). Call a very short, hard *tik*, often repeated in lengthy series and very similar to Plain Flowerpecker.

Dicaeum chrysorrheum

黄臀啄花鸟 huáng tún zhuó huā niǎo

Yellow-vented Flowerpecker

L: 9–10 cm
Habitat: Inhabits open woodlands, secondary forests, forest edges, plantations, orchards, and gardens from lowlands to 2000 m; especially in flowering and fruiting trees and shrubs; prefers mistletoe and small fig trees.

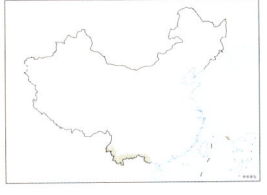

Behavior: Forages at all levels of vegetation. Other behaviors resemble those of Thick-billed Flowerpecker.
Distribution: S and W Yunnan and SW Guangxi.
Voice: High-pitched, rising *tzee* that, when given in an evenly spaced sequence, is also the species' song.

Dicaeum minullum

纯色啄花鸟 chún sè zhuó huā niǎo

Plain Flowerpecker

L: 7.5–9 cm
Habitat: Inhabits various types of vegetation, such as open woodlands, secondary forests, forest edges, plantations, orchards, and gardens from sea level to 3660 m; especially prefers flowering and fruiting trees and shrubs.

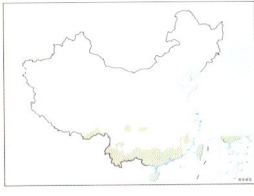

Behavior: Similar to Yellow-bellied Flowerpecker.
Distribution: *D. m. olivaceum* in Yunnan, C and E Sichuan, Chongqing, Guizhou, S Hunan, NE Jiangxi, Fujian, W Guangdong, Hong Kong, and Guangxi; *D. m. uchidai* in Taiwan; *D. m. minullum* in Hainan.
Voice: Song a high-pitched, accelerating, then slowing, slightly descending *tsit-tsit-tsit-si-si-si-si*. Calls include a *tik*, often repeated, very similar to Scarlet-backed Flowerpecker.

Yellow-vented Flowerpecker

Short white stripe at lores

Yellow vent

Black streaks on underparts

Thick-billed Flowerpecker

Dark red iris

Slightly thick, strong bill

White vent

Gray streaks from breast to flanks

Fire-breasted Flowerpecker

Metallic greenish blue black on crown and back

Bluish black from cheek to side of breast

Scarlet patch on breast does not extend to throat

Olive flanks

Thick black stripe on central belly

D. i. ignipectus

Dusky olive upperparts with gray-tinged face

Yellowish-green uppertail coverts

Yellowish brown from throat to underparts, sides of breast olive

Scarlet patch on breast extends to lower chin

D. i. formosum

Plain Flowerpecker

Olive upperparts

Relatively thin bill

Pale underparts lack streaks

Yellow-bellied Flowerpecker

Black head and back

Narrow white patch from throat to breast

Yellow belly and vent

Mostly gray from head to breast

Relatively light yellow from belly to vent

Scarlet-backed Flowerpecker

Bright scarlet from head and back to rump

Black cheek

Metallic bluish-black wings

Gray flanks

Brownish-gray upperparts, head with slight olive wash

Scarlet rump and uppertail coverts

Buff yellow from throat to underparts

Flanks tinged grayish brown

Dark brown tail and wings

花蜜鸟科（太阳鸟科）　Nectariniidae

Small, colorful, forest-dwelling songbirds with a slender body profile. Plumage different between sexes. Bill long and pointed, straight or slightly decurved. Wings short and rounded. Most species with relatively long tail. Inhabit primarily forests, scrubland, parks, and plantations; attracted by plants with abundant flowers. Nest on tree branches. Feed mostly on nectar; also take insects and other arthropods. Most species are resident, while a few are migratory.

　　Sixteen genera and 145 species recognized worldwide, most distributed in Africa, others in Indomalaya and Oceania. Six genera and 13 species recorded in China, found in the forests of southern China.

直嘴太阳鸟属
Chalcoparia

Ruby-cheeked Sunbird

食蜜鸟属
Anthreptes

Brown-throated Sunbird

蓝枕花蜜鸟属
Kurochkinegramma

Purple-naped Spiderhunter

花蜜鸟属
Cinnyris

Olive-backed Sunbird

太阳鸟属
Aethopyga

Fork-tailed Sunbird

捕蛛鸟属
Arachnothera

Streaked Spiderhunter

Chalcoparia singalensis
紫颊太阳鸟 (紫颊直嘴太阳鸟)
zǐ jiá tài yáng niǎo (zǐ jiá zhí zuǐ tài yáng niǎo)
Ruby-cheeked Sunbird LC

L: 10–11 cm
Habitat: Occurs in various habitats up to 1000 m, including open woodlands, secondary forests, forest edges, plantations, orchards, and gardens. Especially prefers flowering trees and shrubs.

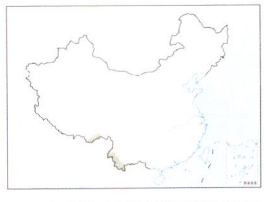

Behavior: Occurs primarily in upper and middle stories of trees; also comes down to bushes. Solitary, in pairs, or gathers in small flocks. Often forages for spiders, probes flowers for nectar, and gleans insects from leaves.
Distribution: SE Tibet and S and W Yunnan.
Voice: Two-part song starts with two *fwisp* notes, then a decelerating trill, followed by several scratchy, high-pitched, warbled notes, *fwisp-fwisp-isis-r·r·r·r·r·r·r·r—tr'tr'r-si-si-si-si.* Call is a weak *twissu.*

Anthreptes malacensis
褐喉食蜜鸟 hè hóu shí mì niǎo
Brown-throated Sunbird LC

L: 12.1–13.5 cm
Habitat: Occurs in various habitats up to 1200 m, including open woodlands, secondary forests, forest edges, plantations, orchards, and gardens. Especially prefers flowering trees and shrubs.

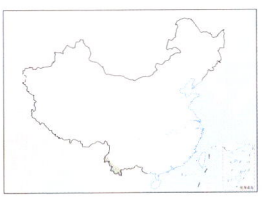

Behavior: Primarily in the upper and middle stories of trees; also comes down to bushes. Solitary or in pairs. Probes flowers for nectar and gleans insects from leaves; also feeds on mistletoe.
Distribution: Local and uncommon resident in S and W Yunnan.
Voice: Distinctive seesawing song of alternately rising, then falling notes almost 0.5 sec apart, *swit-tuuu* or *wee-chew.* Calls include a piercing, frequently repeated, double note and a loud cheerful *kelichap* and a harder *swit* or *twit-twit.*

Kurochkinegramma hypogrammicum
蓝枕花蜜鸟 lán zhěn huā mì niǎo
Purple-naped Spiderhunter LC

L: 12.7–15 cm
Habitat: Similar to Brown-throated Sunbird.
Behavior: Forages for insects, spiders, nectar, fruits, and seeds. Usually forages below 5 m in trees, occasionally above 5 m. Raids spiderwebs for trapped prey.

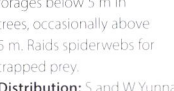

Distribution: S and W Yunnan.
Voice: Song is an erratic, slowly delivered combination of a strident Asian Fairy-bluebird-like *shwep* and fine, high-pitched, Yellow-bellied Fairy-Fantail-like *tsiu* and *sip* notes. Call is a high-pitched, descending, Common Tailorbird-like *tiuu-tiuu-tiuu* or *sweet-sweet-sweet*, repeated at a rate of about four per second, often for long periods.

Ruby-cheeked Sunbird

Purplish-red cheek
Metallic green from head to back
Orange-red throat and breast
Light yellow belly and vent
♂

Olive from head to back
Orange red from throat to breast
♀

Brown-throated Sunbird

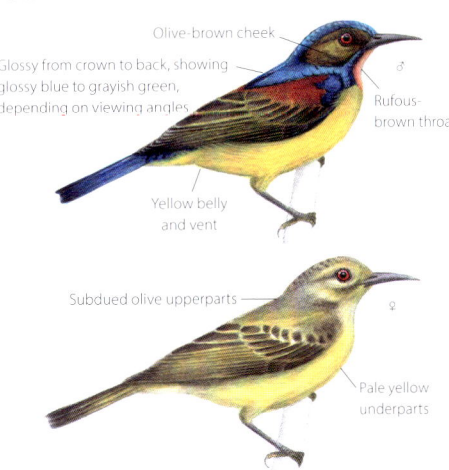

Olive-brown cheek
Glossy from crown to back, showing glossy blue to grayish green, depending on viewing angles
Rufous-brown throat
♂
Yellow belly and vent

Subdued olive upperparts
Pale yellow underparts
♀

Purple-naped Spiderhunter

Metallic blue sheen on nape and rump
♂
Multiple black streaks from throat to belly

Olive upperparts
Black streaks from throat to belly
♀

Cinnyris asiaticus

紫花蜜鸟 (紫色花蜜鸟)
zǐ huā mì niǎo (zǐ sè huā mì niǎo)

Purple Sunbird

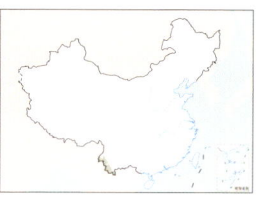

L: 10–11 cm
Habitat: Inhabits various types of habitats up to 1600 m, including open woodlands, secondary forests, forest edges, plantations, orchards, and gardens. Especially prefers flowering trees and shrubs.
Behavior: Solitary, in pairs, or gathers and forages in small and medium flocks; forms larger flocks during nonbreeding season.
Distribution: Common resident in W Yunnan.
Voice: Song is an exuberant, descending trill, swi-swi-swi-a-col-a-oli, often introduced by a pair of chweit-chwit calls and warbles. Calls include a powerful chwees.

Aethopyga gouldiae

蓝喉太阳鸟 lán hóu tài yáng niǎo

Mrs. Gould's Sunbird

L: 14–15 cm
Habitat: Inhabits various types of habitats, including open woodlands, secondary forests, and forest edges; sometimes in plantations, orchards, and gardens. Especially prefers flowering trees and shrubs, at altitudes of 1200–4270 m in breeding season and 330–700 m in winter.
Behavior: Solitary or in pairs; individuals gather at places with abundant food.
Distribution: A. g. gouldiae in SE Tibet; A. g. dabryii in Yunnan, Sichuan, Chongqing, Guizhou, W Hubei, W Hunan, Guangxi, Guangdong, Hong Kong, Henan, S Shaanxi, and SE Gansu. Common resident.
Voice: Calls include a sharp tzit-tzit or fwisp, often repeated very rapidly in a lengthy series, and a more rapid tshi-stshi-ti-ti-ti in alarm.

Cinnyris jugularis

黄腹花蜜鸟 huáng fù huā mì niǎo

Olive-backed Sunbird

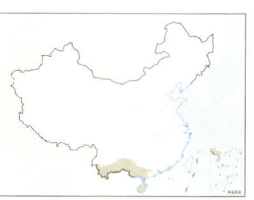

L: 10–11.4 cm
Habitat: Inhabits various types of habitats, including open woodlands, secondary forests, forest edges, plantations, orchards, and gardens. Especially prefers flowering trees and shrubs. Usually on lowland but recorded up to 1700 m.
Behavior: Solitary, in pairs, or gathers and forages in small and medium flocks; forms larger flocks during nonbreeding season.
Distribution: SE Yunnan, Guangxi, Guizhou, Hong Kong, and Hainan.
Voice: Song starts slowly, with a couple of repeated twees notes, before a rapid, intense trill of discordant notes and then switches to repeating some rising nasal notes, twees-twees-zree'ree'ree'ree'ree'ree'dwee'dwee'd wee'dwee'dwee'dwe. Calls are a short zip and an inflected Yellow-browed Warbler–like tweess or toeess.

Aethopyga nipalensis

绿喉太阳鸟 lǜ hóu tài yáng niǎo

Green-tailed Sunbird

L: 14–15 cm
Habitat: Inhabits various types of habitats, including open woodlands, secondary forests, and forest edges; sometimes in plantations, orchards, and gardens. Especially prefers flowering trees and shrubs. At altitudes of 300–3665 m, lower in winter.
Behavior: Solitary or in pairs; individuals gather at places with abundant food.
Distribution: S and SE Tibet, Yunnan, and W Sichuan.
Voice: Two apparent song types are a high-pitched tinkling trill, rising and falling slightly in pitch and lasting up to 1.5 sec, and a more intense, lower-pitched repetition of rising fwisp notes, similar to but lower pitched than Mrs. Gould's Sunbird and given in an accelerating, descending sequence.

Purple Sunbird

Dark metallic bluish purple from head to upperparts

Brown wings

Sometimes a small red breast band

Dark black belly

Olive upperparts

Relatively pale yellow on underparts

Olive-backed Sunbird

Yellowish-brown upperparts

Metallic black from throat to breast

Glossy bluish-green malar stripe

Pale yellow supercilium

Olive upperparts

Pale yellow underparts

Mrs. Gould's Sunbird

Glossy blue feathers on crown and side of neck

Bright red back and breast

Relatively long blue tail

Yellow belly and vent

Metallic blue throat, appears black in dim light

A. g. dabryii

Olive back

Gray head

Pale yellow underparts

A. g. gouldiae

Yellow breast, sometimes with red streaks

Green-tailed Sunbird

Metallic bluish-green head

Dark red side of neck and back

Blackish-green tail

Yellow underparts with rufous-tinged lower breast and upper belly

Gray head

Olive back

Pale yellow underparts

Aethopyga christinae

叉尾太阳鸟 chā wěi tài yáng niǎo
Fork-tailed Sunbird

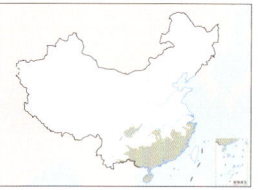

L: 9–11 cm
Habitat: Inhabits forests, forest edges, villages, and urban parks below 1400 m; occurs at the edges of dense primary or secondary broadleaf forests.
Behavior: Often occurs on treetops and in bushes or flowering vegetation; especially prefers mistletoe. Often hovers above flowers and uses its slightly downcurved bill and long tubular tongue to feed on nectar.
Distribution: Locally common resident. *A. c. latouchii* in S Yunnan, Guizhou, Sichuan, Chongqing, S Hunan, Jiangxi, Zhejiang, Fujian, Guangdong, and Guangxi. *A. c. christinae* in Hainan; sometimes treated as a separate species, "Hainan Sunbird" (海南太阳鸟).
Voice: Song is a fast trill of 10–15 high-pitched notes, similar to song of Green-tailed Sunbird but slightly lower pitched and given in a rising, not rising-and-falling, series. Calls include a powerful, sharply rising *twisp*, given singly or repeated up to five times. Another vocalization is a very lengthy trill (up to 7 sec), extremely similar to White-browed Piculet.

Aethopyga saturata

黑胸太阳鸟 hēi xiōng tài yáng niǎo
Black-throated Sunbird

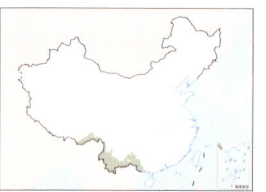

L: 14–15 cm
Habitat: Inhabits various types of habitats, including open woodlands, secondary forests, and forest edges; sometimes in plantations, orchards, and gardens. Especially prefers flowering trees and shrubs. At altitudes of 820–2200 m, lower in winter.
Behavior: Solitary or in pairs; individuals gather at places with abundant food.
Distribution: *A. s. saturata* in SE Tibet; *A. s. assamensis* in S Tibet, S and W Yunnan; *A. s. petersi* in E Yunnan, C and S Guizhou, and SW Guangxi.
Voice: Song is a twittering series of up to eight sharp, high-pitched, but descending and slowing *swi'swi'swi* notes that vaguely recall Gray-headed Canary-Flycatcher. Calls include a *twis*.

Aethopyga ignicauda

火尾太阳鸟 huǒ wěi tài yáng niǎo
Fire-tailed Sunbird

L: 15–20 cm (♂)
　　8.5–11 cm (♀)
Habitat: Inhabits coniferous, oak, and especially rhododendron forests at altitudes of 3000–4880 m in summer and 610–2900 m in winter.
Behavior: Solitary or in pairs; individuals gather at places with abundant food.
Distribution: Yunnan and S and SE Tibet.
Voice: Song starts with a series of three to five descending, accelerating, well-separated *tzip* call notes, before these degenerate into an evenly pitched, very fast trill. Other calls include one with an introductory consonant *t-tzip* and a multisyllabic *tchizziwik*.

Aethopyga siparaja

黄腰太阳鸟 huáng yāo tài yáng niǎo
Crimson Sunbird

L: 11.7–15 cm
Habitat: Inhabits various types of habitats up to 2000 m, including open woodlands, secondary forests, and forest edges; sometimes in plantations, orchards, and gardens. Especially prefers flowering trees and shrubs.
Behavior: Solitary or in pairs; individuals gather at places with abundant food.
Distribution: Locally common resident. *A. s. labecula* in S to SW Yunnan; *A. s. tonkinensis* in SE Yunnan, S Guangxi to W Guangdong; *A. s. owstoni* once recorded in Guangdong (the Leizhou Peninsula and Naozhou Island); *A. s. seheriae* is possibly in SW Tibet.
Voice: Song a high-pitched, fine *pissi'pissi'pit*.

Arachnothera longirostra

长嘴捕蛛鸟 cháng zuǐ bǔ zhū niǎo
Little Spiderhunter

L: 13.3–16 cm
Habitat: Inhabits various types of habitats up to 2100 m, including open woodlands, secondary forests, and forest edges; sometimes in plantations, orchards, and gardens. Especially prefers flowering trees and shrubs.
Behavior: Solitary or in pairs; individuals gather at places with abundant food.
Distribution: Locally common resident. *A. l. longirostra* in W and SW Yunnan; *A. l. sordida* in S and SE Yunnan and W and SW Guangxi.
Voice: Calls include a nasal, broad-spectrum, abrasive *tyich* or *chyit*. A very similar note is used as the song, when it is repeated monotonously for lengthy periods at a rate of about four per second.

Arachnothera magna

纹背捕蛛鸟 wén bèi bǔ zhū niǎo
Streaked Spiderhunter

L: 17–20.5 cm
Habitat: Similar to Little Spiderhunter. From lowland to 2150 m.
Behavior: Solitary or in pairs.
Distribution: SE Tibet, S and W Yunnan, SW Guizhou, and W Guangxi.
Voice: Song is a loud continuous chatter of *chiddick* or *chiriririk* calls.

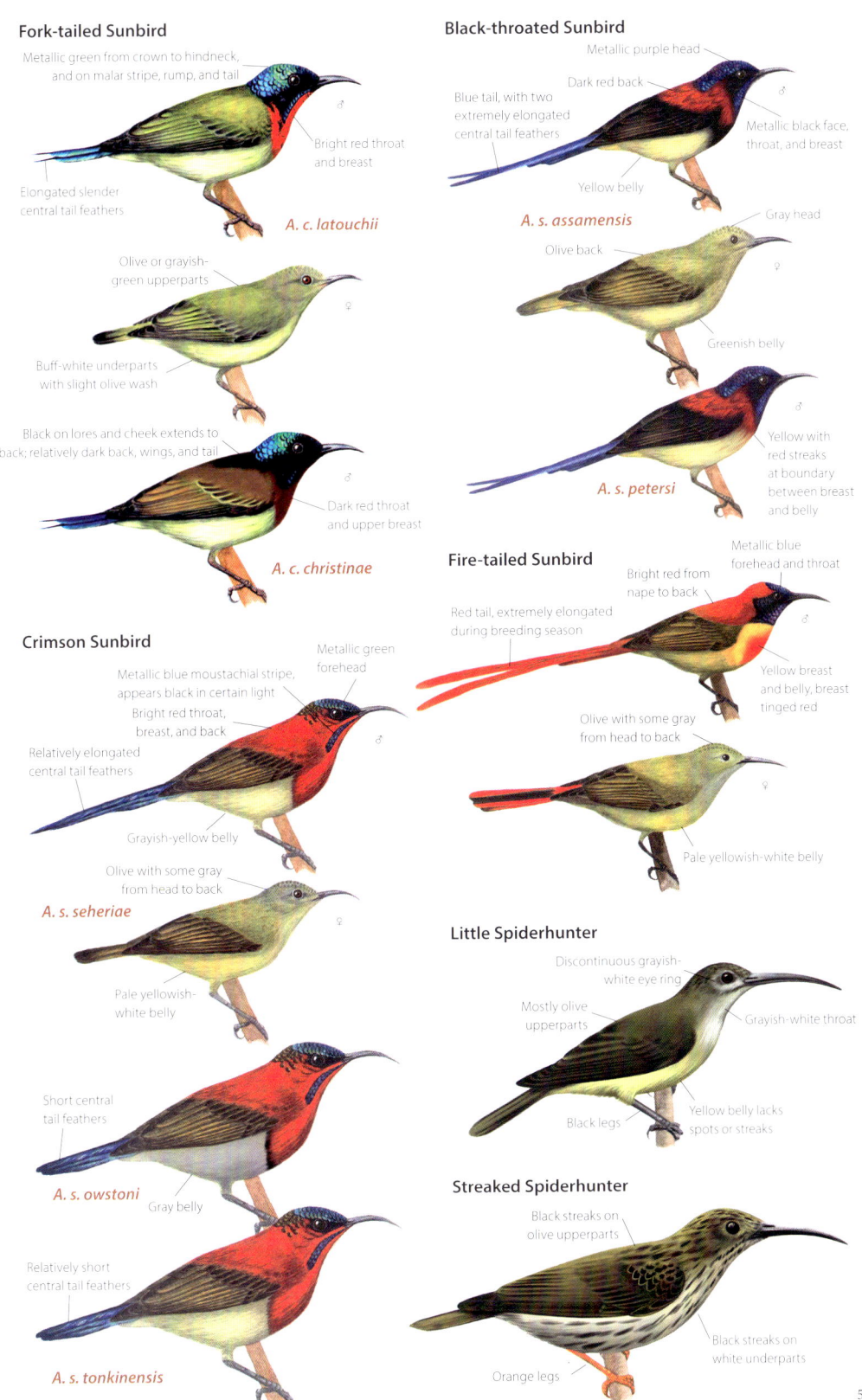

Fork-tailed Sunbird

Metallic green from crown to hindneck, and on malar stripe, rump, and tail

Bright red throat and breast

♂

Elongated slender central tail feathers

A. c. latouchii

Olive or grayish-green upperparts

♀

Buff-white underparts with slight olive wash

Black on lores and cheek extends to back; relatively dark back, wings, and tail

♂

Dark red throat and upper breast

A. c. christinae

Black-throated Sunbird

Metallic purple head

Dark red back

♂

Blue tail, with two extremely elongated central tail feathers

Metallic black face, throat, and breast

Yellow belly

A. s. assamensis

Gray head

♀

Olive back

Greenish belly

♂

Yellow with red streaks at boundary between breast and belly

A. s. petersi

Crimson Sunbird

Metallic green forehead

Metallic blue moustachial stripe, appears black in certain light

Bright red throat, breast, and back

Relatively elongated central tail feathers

♂

Grayish-yellow belly

A. s. seheriae

Olive with some gray from head to back

♀

Pale yellowish-white belly

Short central tail feathers

♂

Gray belly

A. s. owstoni

Relatively short central tail feathers

♂

A. s. tonkinensis

Fire-tailed Sunbird

Metallic blue forehead and throat

Bright red from nape to back

♂

Red tail, extremely elongated during breeding season

Yellow breast and belly, breast tinged red

Olive with some gray from head to back

♀

Pale yellowish-white belly

Little Spiderhunter

Discontinuous grayish-white eye ring

Mostly olive upperparts

Grayish-white throat

Black legs

Yellow belly lacks spots or streaks

Streaked Spiderhunter

Black streaks on olive upperparts

Black streaks on white underparts

Orange legs

573

雀科 Passeridae

Small ground-dwelling and tree-dwelling songbirds. Plumage similar between sexes in most species; mostly brown, rufous, black, and white, usually streaked or spotted. Wings moderately long and pointed; good at flying. Tail medium in length, squared or slighted emarginated. Inhabit primarily deserts, grasslands, forest-edge bushes, and bare rock zones at higher latitudes. Some seen mostly in urban or rural areas. Nest in crevices, under roofs, and in various natural or artificial cavities. Feed mostly on plant seeds and fruits; also take insects. Most species are resident, but some species form nomadic flocks during nonbreeding seasons.

Once was a large family that also included weavers, waxbills, and their allies, but these species now constitute the newly recognized Ploceidae and Estrildidae. Eight genera and 43 species recognized worldwide, endemic to the Old World. Five genera and 13 species recorded in China, ubiquitously distributed nationwide.

麻雀属
Passer

Eurasian Tree Sparrow

石雀属
Petronia

Rock Sparrow

雪雀属
Montifringilla

White-winged Snowfinch

白腰雪雀属
Onychostruthus
(merged with
Montifringilla in eBird/
Clements Checklist)

White-rumped
Snowfinch

黑喉雪雀属
Pyrgilauda
(merged with
Montifringilla in eBird/
Clements Checklist)

Pere David's Snowfinch

Passer montanus
麻雀 má què
Eurasian Tree Sparrow 🟢LC

L: 12–15 cm
Habitat: Inhabits cities and countryside; adapted to human habitations.
Behavior: Loud and noisy. Often gathers in flocks; large flocks contain over a hundred individuals in winter. Omnivorous;

forages on ground or in bushes. Nests in wall crevices and tree cavities.
Distribution: Very abundant resident. *P. m. montanus* in Northeast China; *P. m. dilutus* in northwestern China; *P. m. kansuensis* in W Gansu and E Qinghai; *P. m. tibetanus* at the eastern edge of the Qinghai-Tibet Plateau; *P. m. malaccensis* in Yunnan and Hainan; *P. m. hepaticus* in SE Tibet; *P. m. saturatus* widely across China except for Xinjiang and Tibet.
Voice: Most vocalizations, including chattering song, distinctive hard *tek*, and *tsuwit*, higher pitched than House Sparrow.

Passer domesticus
家麻雀 jiā má què
House Sparrow 🟢LC

L: 14–16 cm
Habitat: Dwells in or close to human habitation on plains, foothills, and plateaus, including villages, towns, farmland, and nearby river valleys, forests, bushes, deserts, and meadows.

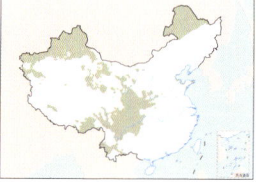

Behavior: Prefers living in flocks; often mixes with Eurasian Tree Sparrow. Occurs and forages in residential areas and farmland. The subspecies in Tibet, *P. d. parkini*, shows seasonal altitudinal migration.
Distribution: *P. d. domesticus* in Heilongjiang, NE Inner Mongolia, and NW Xinjiang; *P. d. bactrianus* in Shaanxi, C and S Xinjiang, Qinghai, and Sichuan; *P. d. parkini* in W Tibet, Yunnan, and Guangxi. Expanding eastward in recent years.
Voice: Gives simple *chirrup* or *chirp*. Song a simple unmusical chattering of call-type notes.

Passer cinnamomeus
山麻雀 shān má què
Russet Sparrow 🟢LC

L: 11–14 cm
Habitat: Inhabits forest-edge bushes and cultivated fields on montane hills from middle to low altitudes.
Behavior: Similar to Eurasian Tree Sparrow. Often gathers in small flocks in nonbreeding season. After breeding season, migrates to montane areas at lower altitudes or migrates southward in short distances.

Distribution: *P. c. rutilans* in eastern and central China except for Northeast China; *P. c. cinnamomeus* in S and SE Tibet; *P. c. intensior* in southwestern China. Common resident or summer breeder.
Voice: Primary call a monosyllabic, descending

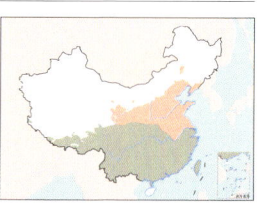

tsleep, less burry than call of Eurasian Tree Sparrow, *cheep* or *chleep*. Short, not particularly musical song is faster and more muffled than Eurasian Tree, such as *cheep chrrup cheweep*.

Passer hispaniolensis

黑胸麻雀 hēi xiōng má què

Spanish Sparrow

L: 14–16 cm

Habitat: Inhabits farmland with trees, towns, orchards, and sparse woodlands or scrubland; also occurs in semideserts and reedbeds. Especially prefers foraging in sparse woodlands, orchards, and young planted tress in towns and farmland.

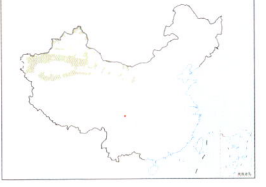

Behavior: Gathers in flocks; mixes often with other sparrows and gathers in large flocks particularly in nonbreeding season.

Distribution: N and S Xinjiang, W Gansu, and W Inner Mongolia. More common in S Xinjiang; vagrant to Sichuan (Balang Mountain).

Voice: Songs extremely similar to House Sparrow but slightly higher pitched and more rhythmic. Calls are also similar to House Sparrow but, again, fractionally higher pitched.

Passer ammodendri

黑顶麻雀 hēi dǐng má què

Saxaul Sparrow

L: 14–16 cm

Habitat: Inhabits desert poplar forests, tamarisk bushes, and saxaul forests in deserts and semideserts; also in vegetated areas, river valleys, and farmland in deserts.

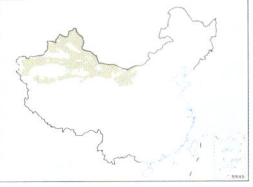

Behavior: Often gathers in small flocks in trees or bushes of desert poplar, tamarisk, and saxaul; frequently hops about in trees.

Distribution: *P. a. nigricans* in N Xinjiang; *P. a. stoliczkae* in S Xinjiang, NW Gansu, W Inner Mongolia, and Ningxia.

Voice: Various calls include *cheerp, dzerp,* a nasal *tlerp,* and a buzzier, rising *dzeep,* often softer, simpler, and more melodious than Eurasian Tree Sparrow. Song typically a subdued, slow, simpler version of song of Eurasian Tree Sparrow, *tlip-chip-chip-tlip,* a sparrowlike repetition of chirps and sweeter notes.

Eurasian Tree Sparrow

Black spot on cheek

Black chin and throat

White neck collar

Russet Sparrow

Rufous chestnut from head to back

Short black eye stripe

♂

Black throat

Buff-yellow supercilium

♀

Brown eye stripe

Spanish Sparrow

Rufous-brown crown

White supercilium

Dense black streaks on back

Dense black streaks on breast and flanks

Pale buff-yellow supercilium

♂

White wing patch

♀

Pale streaks on flanks

House Sparrow

No supercilium

Gray crown

♂

Buff-yellow supercilium

No streaks on flanks

♀

Saxaul Sparrow

Black from crown to hindneck

Chestnut supercilium and side of neck

White wing patch

♂

Ochre-red supercilium

Dark patch on throat

White wing patch

No streaks on flanks

♀

Montifringilla nivalis

白斑翅雪雀 bái bān chì xuě què

White-winged Snowfinch　　LC

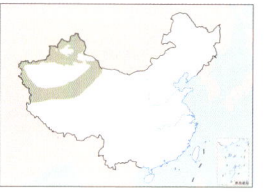

L: 16–18 cm
Habitat: In breeding season, inhabits alpine meadows, grasslands, and plateaus at altitudes of 2500–4500 m; especially prefers rocky mountains. Also often occurs in alpine valleys, near houses and residential areas on plateaus. In winters with heavy snow, migrates to lower altitudes. *M. n. groumgrzimaili* migrates down to deserts at altitudes of 1000 m in winter.
Behavior: Often in pairs or gathers in small flocks on cliffs or bare rocks in breeding season; gathers in large flocks in winter. Usually on ground. Strong runner; hops and forages among rocks. Short-distance flight rapid and low.
Distribution: *M. n. kwenlunensis* in S Xinjiang and Tibet; *M. n. groumgrzimaili* in E Xinjiang; *M. n. tianshanica* in C and W Xinjiang.
Voice: Song is a complex, stuttering mix of varied buzzy notes and sparrowlike chirps, with considerable repetition. Calls include a twangy, goldfinch-like *jie'jie'jie*, *prrrrit*, or *chet-et-et* and an intense *spzee*.

Montifringilla henrici

藏雪雀 zàng xuě què

Tibetan Snowfinch　　LC

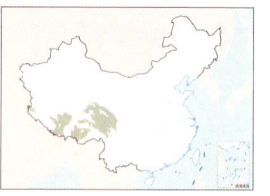

L: 16–18 cm
Habitat: In breeding season, inhabits alpine meadows, grasslands, and plateaus at altitudes of 3500–4500 m; especially prefers rocky mountains. Also often occurs in alpine valleys, near houses and residential areas on plateaus. Typically does not undertake seasonal altitudinal migration; moves to lower altitudes only when snow is heavy.
Behavior: Similar to White-winged Snowfinch.
Distribution: E Tibet and NE and S Qinghai.
Voice: Song is a repeated, ringing, seesawing, vaguely White Wagtail–like *too-z-leet … tuuzi-leet, chiz-eet … chiz-eet*, or *chi-e-leet-dzzzs*, occasionally paired or repeated every 1–3 sec. Call is a Eurasian Tree Sparrow–like nasal chattering and includes a purring *prrret*.

Montifringilla davidiana

黑喉雪雀 hēi hóu xuě què

Pere David's Snowfinch　　LC

L: 12–13 cm
Habitat: Inhabits rocky mountain slopes, plains, valleys, streams, and deserts and semideserts with sparse vegetation, usually near water; also near river valleys, farmland, and human habitation.
Behavior: Often in pairs in breeding season. Usually gathers in small flocks and wanders about in nonbreeding season; often joins sparrows and Horned Larks in mixed-species flocks. Prefers running on ground.

Montifringilla adamsi

褐翅雪雀 hè chì xuě què

Black-winged Snowfinch　　LC

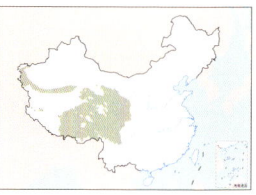

L: 14–18 cm
Habitat: Inhabits primarily highlands and alpine bare rocky areas at altitudes of 3000–4500 m, sometimes at altitudes up to 5000 m in summer; usually occurs in rocky alpine meadows, grasslands, deserts, and semideserts with sparse vegetation. In winter, mostly in valleys, sometimes close to human habitation.
Behavior: Similar to White-winged Snowfinch. Often in pairs or gathers in small flocks; gathers in larger flocks in fall and winter, sometimes containing more than a hundred individuals.
Distribution: *M. a. xerophila* in SE Xinjiang and N and W Qinghai; *M. a. adamsi* in S Gansu, Tibet, SE Qinghai, and W Sichuan.
Voice: Song a bouncing, musical *bzz-er-wichy-tuu*, repeated with little pause. Calls include sparrowlike notes, such as *chip-chip-dzzr* and apparently a strident *pink pink*.

Montifringilla taczanowskii

白腰雪雀 bái yāo xuě què

White-rumped Snowfinch　　LC

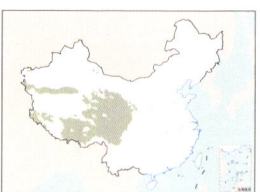

L: 14–18 cm
Habitat: Inhabits alpine meadows, grasslands, and deserts and semideserts with sparse vegetation at altitudes of 3000–4500 m. A hardy species in highlands and alpine meadows and deserts.
Behavior: In pairs or gathers in small flocks. Stable distribution range, especially during breeding season; wanders in a small range or migrates altitudinally in winter. Sometimes occurs and forages in livestock sheds near human habitation. Strong ground runner and hopper.
Distribution: Resident in S Gansu, S Xinjiang, Tibet, S Qinghai, and NW Sichuan.
Voice: Variable, unobtrusive song is an unpleasant, scratchy combination of very short clicks, whistles, purring notes, and buzzes, *kik-k-weer'kee'k* or *kle'r-plzzzz'k'iuh'pleer'k'ik'it*, given every few seconds or in a lengthy sequence. Calls include sparrowlike chatter, a rising *kluu* or *klu-it*, or a sand-plover-like *krrt*.

Good flier; usually flies low above ground. Bold and quiet, unafraid of people.
Distribution: W and C Inner Mongolia, Ningxia, N Gansu, NE Xinjiang, and E Qinghai.
Voice: Song a complex, short (0.5 sec), Horned Lark–like chatter of multiple, short, high-pitched notes, *tseeliu*, repeated irregularly, sometimes in rapidly combination. Calls include a nasal *weer*, given in flight, a quiet *chit*, and *te-liu*.

White-winged Snowfinch
Grayish-brown head
Large white wing patch
White flanks

Black-winged Snowfinch
Grayish-brown head
Relatively small white wing patch

Tibetan Snowfinch
Brown head
Large white wing patch
Brown flanks

White-rumped Snowfinch
Black lores
Black bill
Pale overall
White rump
br.

Pere David's Snowfinch
Black forehead connects with black lores, forming black band
Black on throat extends to upper breast

Yellow bill
non-br.

Montifringilla ruficollis
棕颈雪雀 zōng jǐng xuě què
Rufous-necked Snowfinch

L: 14–16 cm

Habitat: Often on alpine bare rocks, meadows, grasslands, and plateaus at altitudes of 3800–5000 m; also seen in deserts, semideserts, farmland, and near houses.

Behavior: Often in pairs in breeding season, active in a small range. Gathers in small flocks or mixes with other snowfinches in other seasons and moves over a larger range. Also makes short-distance altitudinal migration. Not shy. Often on ground and visits pika tunnels. Strong runner and agile. Short-distance flight strong and fast.

Distribution: *M. r. isabellina* in S Xinjiang and N Qinghai; *M. r. ruficollis* in S Gansu, Tibet, S Qinghai, and W Sichuan.

Voice: Extensive vocabulary includes a poorly structured, stuttering flight song, an eclectic mix of sparrowlike notes, short buzzes, whistles, trills, and some mimicry, similar to song of Pere David's Snowfinch but not as hurried. Gives a much simpler, more repetitive version from the ground, *dut-ple-dut-t-plee*. Calls include a short Horned Lark–like *tzee*; a nasal *duuid*, given in flight; a sharp sparrowlike *chit*, *chi-it-it*, or very sharp *kit*, frequently in a sequence of two or three; and a harsh, scolding European Starling–like note, *tshk*.

Montifringilla blanfordi
棕背雪雀 zōng bèi xuě què
Blanford's Snowfinch

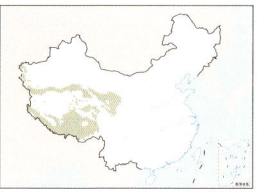

L: 13–15 cm

Habitat: Inhabits dry, rocky alpine grasslands and plateaus with short grass at altitudes of 4200–5000 m; also inhabits alpine deserts and semideserts with sparse vegetation. Especially prefers grasslands with abundant pika tunnels. In winters with heavy snow, often moves to valleys, livestock sheds, and human habitation in low and sheltered places.

Behavior: Similar to Rufous-necked Snowfinch; gathers in large flocks with other snowfinches in fall and winter.

Distribution: *M. b. blanfordi* in SW Xinjiang, S and W Tibet, and S Qinghai; *M. b. ventorum* in SE Xinjiang and NW Qinghai; *M. b. barbata* in S and E Qinghai.

Voice: Rapid twittering song, given both from ground and in display flight. Calls include a sparrowlike *chiddick* or *chut-ut*, a nasal *truuu*, and a rapid twittering *kut … chitrrrr*.

Rufous-necked Snowfinch

Brown on side of neck and side of breast

Black eye stripe

Black malar stripe

Blanford's Snowfinch

Black on lores extends above eyes, forming characteristic short "horns"

Black streak on central forehead

Back and wings plainer than Rufous-necked Snowfinch

Pale fringe on wings less contrasting

Thick bill with yellow lower mandible

Yellow patch on breast

Dusky brown streaks on flanks

White tail tip

Rock Sparrow

Barred undertail coverts

Petronia petronia
石雀 shí què
Rock Sparrow

L: 15–17 cm

Habitat: Often inhabits deserts and semideserts with sparse bushes on plateaus, barren mountains, and hills; also in artificial environments, such as sheep pens and residential areas.

Behavior: Often in pairs or small flocks in breeding season, on bare rocks. Strong runner and hopper on ground. Good flier. Gathers in large flocks in winter and migrates to low mountains and foothill plains; also usually mixes with other passerines, such as Horned Lark.

Distribution: *P. p. intermedia* in Xinjiang; *P. p. brevirostris* in Beijing, W Inner Mongolia, Ningxia, S and E Gansu, Tibet, Qinghai, and NW Sichuan.

Voice: Often noisy. Wide variety of nasal calls, such as *tyeee-uu-it*, *pee-yeer*, or *jee-wee*. Song is a simple combination of a large number of calls.

织雀科　Ploceidae

Small tree-dwelling songbirds with unique breeding behavior. Plumage similar between sexes in most species, while males of some species differ in plumage between breeding and nonbreeding seasons. Thick bill pointed and cone shaped, very wide at base. Wings long and pointed; good at flying. Inhabit primarily plains, marshes, scrubland, grasslands, savanna, reedbeds, and river valleys that are close to water. Most species are colonial and weave complex and delicate nests, hence their common names. Feed mostly on grains, grass seeds, and other plant seeds; also take insects. Resident and nonmigratory, while some species are nomadic during nonbreeding seasons.

Weavers and their allies were once placed in Passeridae, but recent phylogenetic studies suggest they are only distantly related to sparrows and buntings; thus they are currently in their own family. Fifteen genera and 117 species recognized worldwide; except for five species in Indomalaya and three on the islands of the Indian Ocean, all are in Africa. One genus and two species recorded in China, found in S to W Yunnan.

Streaked Weaver

Yellow crown

♂

Black face, throat, and neck

Black streaks on breast and flanks

♀ Pale gray face, yellow-tinged supercilium and neck

Pale yellowish white from throat to belly

Some streaks on breast and flanks

Baya Weaver

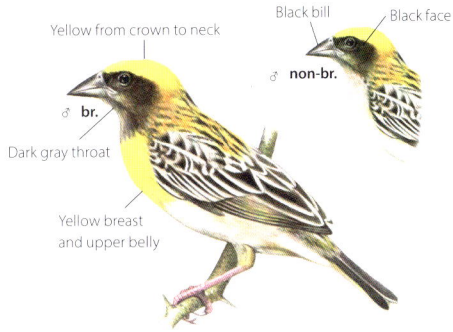

Black bill

Black face

Yellow from crown to neck

♂ non-br.

♂ br.

Dark gray throat

Yellow breast and upper belly

Dense fine streaks on pale yellow head

Yellow bill ♀

Whitish throat

Pale yellowish-white underparts lack streaks

Ploceus manyar
纹胸织雀 wén xiōng zhī què
Streaked Weaver

L: 15 cm
Habitat: Inhabits wetlands, reedbeds, and paddy fields; prefers marshes with reedbeds and cattails.
Behavior: Gathers in flocks. Weaves nest in high grass, such as reeds, in breeding season; usually breeds in colonies.
Distribution: *P. m. williamsoni* in W Yunnan; *P. m. peguensis* in NW Yunnan.
Voice: Calls are essentially sparrowlike notes, often with a muffled, buzzy quality. Song is a combination of chattering, excited call notes, buzzes, and wheezes.

Ploceus philippinus
黄胸织雀 huáng xiōng zhī què
Baya Weaver

L: 15 cm
Habitat: Inhabits grasslands, bushes, clearings with scattered trees, and cultivated fields. Less restricted to marshy habitats than Streaked Weaver. Seen primarily at middle and low altitudes.
Behavior: Gathers in flocks. Often nests colonially in trees.
Distribution: S and W Yunnan and Hong Kong.
Voice: Song similar to that of Streaked Weaver, a discordant combination of sparrowlike chattering, buzzes, and wheezes. Calls include a harsh *chit*.

梅花雀科　Estrildidae

Small granivorous songbirds. Plumage similar or slightly different between sexes; highly variable among species, mostly brown, chestnut, green, and red. Strong bill cone shaped. Wings long and pointed; good at flying. Tail squared or rounded, medium in length. Inhabit primarily marshes, scrubland, grasses, reedbeds, and forest edges in groups. Nest on trees. Feed mostly on grains, grass seeds, and other plant seeds and fruits; also take insects.

　　Though now usually placed in their own family, waxbills and allies were formerly placed in Passeridae. Thirty-one genera and 141 species recognized worldwide, distributed mainly in Oceania and tropical regions of the Old World. Three genera and five species recorded in China, found in regions south of the Yellow River Watershed.

红梅花雀属
Amandava

Red Avadavat

鹦雀属
Erythrura

Pin-tailed
Parrotfinch

文鸟属
Lonchura

ad.　　　　　juv.

White-rumped Munia

Amandava amandava

红梅花雀 hóng méi huā què

Red Avadavat

LC

L: 9–10 cm
Habitat: Inhabits low mountains, hills, plains, villages, and orchards at altitudes below 1500 m.
Behavior: Social; often gathers in flocks and forages in bushes, farmland, and reedbeds. Flight rapid and agile.

Distribution: *A. a. flavidiventris* is occasionally seen in S and SW Yunnan and S Guizhou. *A. a. punicea* rare in Hainan, Guangdong, and Hong Kong, and populations have become established from escaped cage birds.
Voice: Song is a series of thin, high-pitched, rising, then descending whistles, followed by a melodious, rapid twitter, *tleeeo-weet … weet-weet … tleoo-weet … tui'tui't't'tui'pl't'pl'eee'ti'tid'qrr.* Calls are a high-pitched, falling *tsi*.

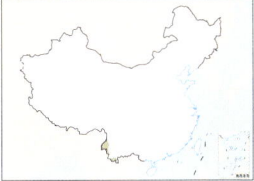

L: 13–15 cm (♂)
     12–13 cm (♀)
Habitat: Inhabits forest-edge bushes, bamboo thickets, and farmland on low mountains and plains.
Behavior: Often in pairs or small flocks. Feeds primarily on grass seeds and crops, such as rice; sometimes also searches for food among fallen leaves.

Distribution: First recorded in 2013 in Yunnan (Xishuangbanna); rare but regularly recorded in S and W Yunnan in recent years.
Voice: Gives a series of high-pitched chittering and buzzy notes, often in quick succession. Calls include buntinglike *tic* and *dzzt*.

Red Avadavat

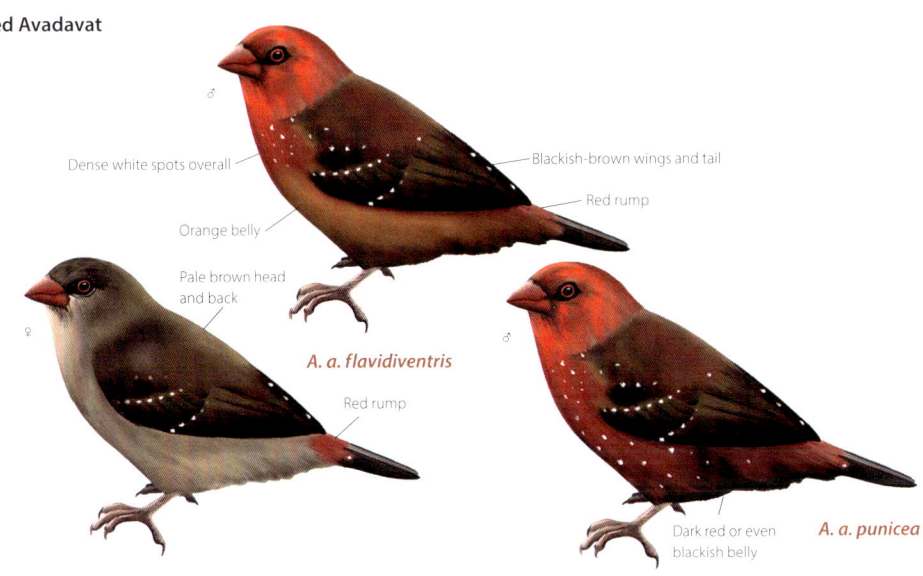

Dense white spots overall

Orange belly

Blackish-brown wings and tail

Red rump

Pale brown head and back

Red rump

A. a. flavidiventris

Dark red or even blackish belly

A. a. punicea

Pin-tailed Parrotfinch

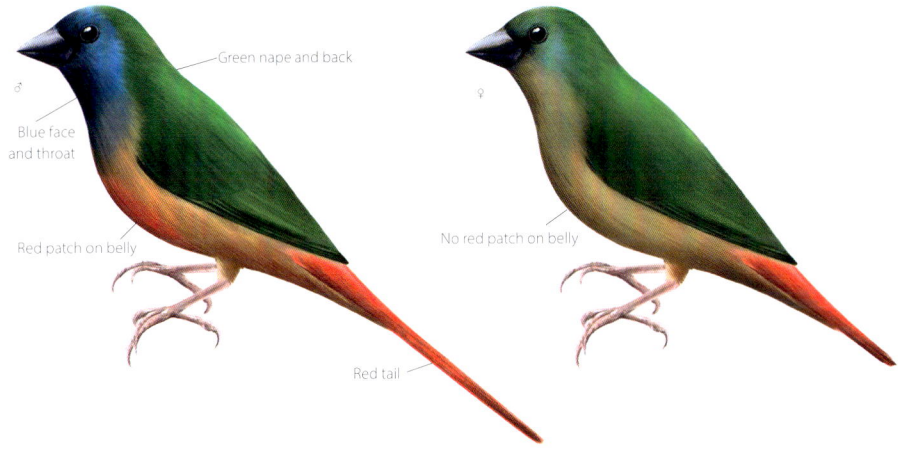

Green nape and back

Blue face and throat

Red patch on belly

Red tail

No red patch on belly

581

Lonchura striata

白腰文鸟 bái yāo wén niǎo

White-rumped Munia LC

L: 10–12 cm

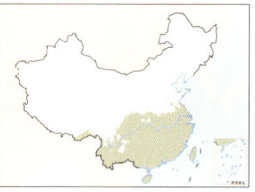

Habitat: Adaptable.
Inhabits forest edges,
shrublands, nurseries,
gardens, paddy fields,
and around villages
from plains to highlands,
usually at altitudes lower
than 2000 m; descends to
middle and low altitudes in winter.

Behavior: In pairs during breeding season. In nonbreeding season,
forages and roosts in flocks containing over a hundred individuals.
Occasionally mixes with Scaly-breasted Munia. Prefers using bill to take
grass seeds off the stem; also often takes filamentous green algae from
streams. Does not make long-distance migration but makes short-
distance migrations according to changes in food availability.

Distribution: Common resident. *L. s. swinhoei* in most regions south of
the Yellow River, including Hainan and Taiwan; *L. s. subsquamicollis* in
SW Yunnan and SE Tibet.

Voice: Song a rapid sparrowlike chattering. Calls include a loud *plee* and
a penetrating *dzee*.

Lonchura atricapilla

栗腹文鸟 lì fù wén niǎo

Chestnut Munia LC

L: 11–12 cm

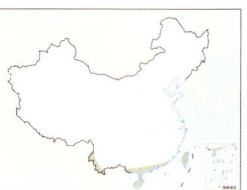

Habitat: Inhabits deserted
fields, grasslands, and
forests on hills and plains
at low altitudes, usually
lower than 200 m.

Behavior: Gregarious;
joins Scaly-breasted Munia
and sparrows in mixed-
species flocks and feeds on grass seeds. Does not make long-distance
migration but makes short-distance migrations according to changes in
food availability.

Distribution: *L. a. atricapilla* is locally common resident in SW Yunnan.
L. a. deignani is a rare resident in Guangdong, S Guangxi, Macao, and
Hainan. *L. a. formosana* is a native subspecies in Taiwan; uncommon
resident. Introduced populations also seen in Taiwan, including
L. a. sinensis (native to Malaysia) and *L. a. jagori* (native to the Philippines).

Voice: Gentle fluty *wee, wee* calls, similar to White-rumped Munia. Calls
are slightly different among subspecies.

Lonchura punctulata

斑文鸟 bān wén niǎo

Scaly-breasted Munia LC

L: 10–12 cm

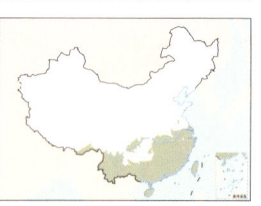

Habitat: Inhabits open
areas, such as low
moutains, rural areas,
plains, and hills. Forages
on bushy grasslands in
farmland and deserted
fields; prefers silvergrass.
Usually at altitudes lower
than 500 m.

Behavior: Gregarious; forages in flocks containing several dozen
individuals; sometimes mixes with White-rumped Munia. Diet similar
to White-rumped Munia. Does not make long-distance migration
but makes short-distance migrations according to changes in food
availability.

Distribution: Common resident. *L. p. topela* in most regions south of the
Yangtze River, including Hainan and Taiwan. *L. p. yunnanensis* in SE Tibet
and southwestern China. *L. p. subundulata* is also recorded in SE Tibet.

Voice: Calls include a sharp *chit-tit* and a sparrowlike *tseep*, sometimes
repeated in staccato.

White-rumped Munia

Darker back

Dark brown or nearly blackish breast

ad.

L. s. subsquamicollis

Relatively pale body

Buff-yellow to pale brown rump

juv.

Off-white from lower breast to belly, contrasts strongly with dark brown upper breast

ad.

White rump

Dark brown tail

L. s. swinhoei

Scaly-breasted Munia

Pale brown rump

Dark and densely scalloped breast and belly

ad.

juv.

Plain belly lacks streaks or patches, paler than adults

Chestnut Munia

Dark brown from head and face to upper breast

Grayish brown from nape to hindneck

Remaining body chestnut brown

Chestnut breast band separates blackish-brown belly and upper breast

ad.

L. a. formosana

Uniform black from head to breast

Remaining body chestnut

Pale yellow or orange-brown uppertail coverts

Chestnut breast band

ad.

Blackish-brown belly

L. a. atricapilla

Resembles juvenile Scaly-breasted Munia, but with browner body and chestnut-tinged wings

juv.

Resembles *L. a. atricapilla*, but with chestnut-brown uppertail coverts

ad.

L. a. deignani

Black head, face, and upper breast; only nape appears dark brown

ad.

L. a. jagori

Blackish-brown belly almost connects with upper breast

Uniform black head

ad.

No black patch on belly

L. a. sinensis

岩鹨科   Prunellidae

Small ground-dwelling songbirds living in cold and high-altitude habitats. Plumage similar or slightly different between sexes; mostly gray, rufous, brown, and black. Pointed bill resembles bill of pipits. Wings moderately wide; good at flying. Tail medium in length, mostly squared. Inhabit primarily scrubland, grasslands, deserts, and bare rock zones from middle to high altitudes; some also at forest edges. Nest among bushes or in crevices. Usually solitary or in small groups. Feed on ground or in mid-level and lower vegetation, mostly on insects, insect larvae, other small invertebrates, and plant seeds and fruits. Most species undertake altitudinal or short-distance migrations.

Accentors' plumage appears similar to plumage of pipits or sparrows, but they are phylogenetically more closely related to thrushes or warblers. One genus and 13 species recognized worldwide, endemic to Eurasia. One genus and nine species recorded in China, found in Western, North, and Northeast China.

Prunella collaris

领岩鹨 lǐng yán liù

Alpine Accentor   LC

L: 15–18 cm
Habitat: Inhabits mountainous areas at middle and high altitudes. Forages in areas with bare rocks or grasslands; breeds in rock crevices or bushes.
Behavior: Solitary or gathers in small flocks. Active; often perches on rocks in upright stance, often slightly flaps wings and tail. Undulating flight rapid and agile.
Distribution: *P. c. rufilata* in Xinjiang; *P. c. whymperi* in W Tibet; *P. c. nipalensis* in S Shaanxi, Gansu, Tibet, NW Gansu, and Sichuan; *P. c. tibetana* in NW Gansu and S and E Qinghai; *P. c. erythropygia* in Northeast, North, and Central China; *P. c. fennelli* in Taiwan.
Voice: Song is a loud, musical, low-pitched, highly variable warble. Calls include a dry, rippling, larklike *trrrt … truiririp*, occasionally more buzzy or purring.

Prunella himalayana

高原岩鹨 gāo yuán yán liù

Altai Accentor   LC

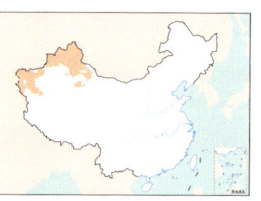

L: 15–17.5 cm
Habitat: Inhabits bare alpine rocks, moraine gravel piles, and stony alpine meadows at altitudes of 3000–5000 m; particularly common on rocky grasslands with sparse bushes.
Behavior: In pairs during breeding season; often gathers in large flocks containing hundreds of individuals when arriving or leaving the breeding ground. Ground dweller; prefers running rapidly and foraging on rocky grasslands.
Distribution: Breeders in the high mountains in W and N Xinjiang; locally common.
Voice: Song is a loud warble, with trills and fluted whistles, similar to song of Alpine Accentor but discordant, rambling, and less repetitive, with higher-pitched, scratchier, far less musical phrases. Calls include soft *plew* and *tiew* notes in flight.

Prunella rubeculoides

鸲岩鹨 qú yán liù

Robin Accentor   LC

L: 14–17 cm
Habitat: Inhabits alpine willow scrub, pastures, and river valleys at altitudes of 3000–5000 m.
Behavior: Often gathers in small flocks on open ground.
Distribution: Seen in S Xinjiang, Qinghai, S and W Gansu, N and W Sichuan, and Tibet.
Voice: Most common call an extremely fast, lengthy, metallic trill. Also a Ground Tit–like *ti-di*.

Prunella strophiata

棕胸岩鹨 zōng xiōng yán liù

Rufous-breasted Accentor   LC

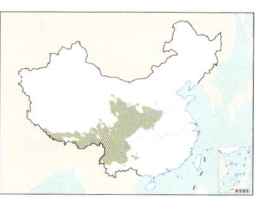

L: 13–16 cm
Habitat: Inhabits alpine scrubland and grasslands in river valleys at altitudes of 1800–4500 m. Migrates to mountains at lower altitudes in winter.
Behavior: Gathers in small flocks and forages on ground or in bushes.
Distribution: Seen in Qinghai, S and W Gansu, S Shaanxi, N and W Sichuan, W Yunnan, and Tibet.
Voice: Song is melodious warble and trill, similar to song of Brown Accentor in its repetition of individual notes two to five times. Generally lower pitched than that species, with the same notes repeated in subsequent phrases but often in a different order! Call a penetrating trill *tr-r-r-r-r-* or *trr-r-rit* that slows slightly, similar to Pink-rumped Rosefinch, and faster, lower pitched, and not quite as "tinny" as Brown Accentor.

Alpine Accentor

Yellow base of lower mandible

Grayish throat differs from Altai Accentor

Broad white streaks on flanks

Two white wing bars, inconspicuous in autumn and winter

P. c. erythropygia

Relatively pale body, more sandy grayish

No white streaks on flanks

P. c. rufilata

Larger area of red on scapulars

Lacks or has a few white streaks on flanks

P. c. fennelli

Lacks or has a few white streaks on flanks

P. c. nipalensis

Altai Accentor

Reddish-brown iris

White patch on side of neck

White throat

Black breast band

Brown streaks on flanks

Robin Accentor

Grayish-brown head and neck

Black streaks on back

Broad brownish-red breast band

Rufous-breasted Accentor

Thick chestnut-yellow supercilium behind eye

Blackish-brown cheek

Black streaks on flanks and belly

Prunella montanella

棕眉山岩鹨 zōng méi shān yán liù
Siberian Accentor

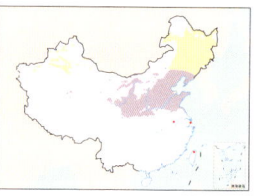

L: 15–16 cm
Habitat: Inhabits bushes and broadleaf forests on mountains and plains near mountains.
Behavior: Solitary or in small flocks; sometimes mixes with buntings. Forages on ground.
Distribution: Passage migrant in Northeast and northwestern China; winters from North China to E Qinghai. Not rare. Vagrant to the Yangtze River Watershed and Taiwan.
Voice: Song melodious and powerful, very similar to song of Black-throated Accentor. Contact call a fine, trisyllabic *dididi*, also very similar to that of Black-throated.

Prunella fulvescens

褐岩鹨 hè yán liù
Brown Accentor

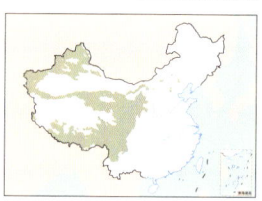

L: 13–16 cm
Habitat: Inhabits plateau grasslands, deserted fields, farmland, and pastures at middle and high altitudes; sometimes comes to human habitation. Also occurs in deserts, semideserts, and alpine grasslands with bare rocks. Especially prefers rocky hills and gravelly areas with sparse bushes. Wanders to mountain valleys, river valleys, and lakeshores at low altitudes in winter.
Behavior: Often in pairs during breeding season; flocks in nonbreeding season. Ground dweller; forages on ground, on rocks, or in bushes.
Distribution: *P. f. fulvescens* in C and W Xinjiang and W Tibet; *P. f. dresseri* in SE Xinjiang and N Tibet; *P. f. dahurica* in Heilongjiang, Beijing, Inner Mongolia, NW Gansu, and N Xinjiang; *P. f. nanschanica* in Ningxia, Gansu, Tibet, Qinghai, Yunnan, and W Sichuan; *P. f. khamensis* in W Yunnan and W Sichuan.
Voice: Song is a fairly loud, very rapid (and short), but simple warble that repeats four or five short notes two to five times in a slightly descending sequence, *tuk-tuk-tuk-tuk-t-ti-ti-ti-uh-uh-tr-tr-tr-tr-tr—uhr-uhr-uhr*. Subsequent strophes generally repeat different notes (unlike Rufous-breasted Accentor). Call higher pitched and thinner than call of Mongolian Accentor.

Prunella atrogularis

黑喉岩鹨 hēi hóu yán liù
Black-throated Accentor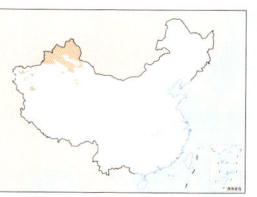

L: 13–14.5 cm
Habitat: In breeding season, inhabits primarily montane coniferous forests; also occurs in broadleaf forests, secondary forests, and bushes near coniferous forests. In winter, occurs in river-valley forests, orchards, urban parks, and woods and bushes in vegetated farmland.
Behavior: Often solitary or in pairs during breeding season; gathers in small flocks in other seasons. Often sings from top of coniferous forests or bushes; also active in understory and on ground. Active and prefers hiding; usually hides in trees when in danger.
Distribution: *P. a. huttoni* is a breeder or a resident in Xinjiang and Tibet.
Voice: Stereotyped song is an extremely hurried complex of short buzzes, whistles, and piping notes, similar to song of Siberian Accentor, but many notes are higher pitched and with fewer buzzes. Contact call is a soft trill, very similar to Brown Accentor and Siberian Accentor.

Prunella koslowi

贺兰山岩鹨 hè lán shān yán liù
Mongolian Accentor

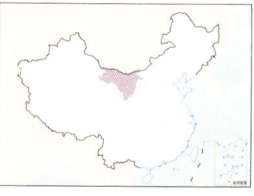

L: 14–15 cm
Habitat: Inhabits dry mountains and open scrubland in semideserts.
Behavior: Vigilant. Often gathers in small flocks under bushes or on ground; sometimes mixes with Siberian Accentor and Saxaul Sparrow. Flight undulating when flying a long distance.
Distribution: All records are in N Ningxia and Inner Mongolia (Alashan Zuo Qi/Alxa Left Banner).
Voice: Song is short, like song of Brown Accentor but generally much fuller and richer, with longer individual notes and much less repetition. Call like Siberian Accentor and Black-throated Accentor but flatter and slightly slower.

Prunella immaculata

栗背岩鹨 lì bèi yán liù
Maroon-backed Accentor

L: 13–16 cm
Habitat: Inhabits alpine coniferous forests and woodland bushes at altitudes of 3000–4500 m; migrates to mountainous areas in river valleys at lower altitudes in winter.
Behavior: Often moves about and forages on ground and in bushes; usually gathers in small flocks.
Distribution: Seen in Tibet, Yunnan, N and W Sichuan, S Gansu, S Qinghai, S Shaanxi, and W Hubei.

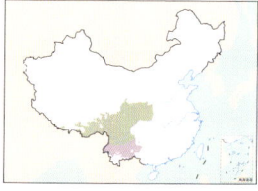

Voice: Distinctive, piercing, high-pitched (7.7 kHz), three-part song is a lengthy (1.5 sec) whistle. The first two-thirds are tremulous; then there's a sudden drop in pitch (to 6.2 kHz) before an equally sudden jump back to the original level as the song fades, *tre'e'e'e'e'e'e'e'du'u'u'u'de'e'e'e'e'e*. Occasionally, the order and relative lengths of the three parts are altered, and the central portion is tremulous and highest pitched, *ti'i'i'i'i't re'e'e'e'tri'i'i'i*. Rarely heard calls are also high pitched, *ti-si-si*.

Siberian Accentor

Brownish-yellow superciliumBrownish-yellow throat

Rufous hue to upperparts

Black-throated Accentor

Black throat

Brown Accentor

Long broad white supercilium

Uniform buff yellow on throat and underparts

Mongolian Accentor

Slender black bill

Streaked back

Maroon-backed Accentor

Creamy-white iris, red outer ring

Dark gray head, neck, and breast

Chestnut-brown back, wings, and lower belly

鹡鸰科 Motacillidae

Small ground-dwelling songbirds with slender profile. Plumage similar or slightly different between sexes; mostly black, white, and grayish brown. Bill thin and pointed. Wings long and pointed. Flight usually deeply undulating. Tail medium to long, usually with white outer tail feathers, constantly wagged sideways or up and down. Legs long and strong; good at walking. Inhabit mostly plains, marshes, rivers, lakes, open grasslands, and forest edges from coastal area to alpine meadows; a few live in forests. Nest among bushes or in crevices. Feed mostly on insects and other invertebrates, occasionally plant seeds and fruits. Most species are migratory.

Five genera and 69 species recognized worldwide, including wagtails, pipits, and allies, with a cosmopolitan distribution. Three genera and 20 species recorded in China, ubiquitously distributed nationwide.

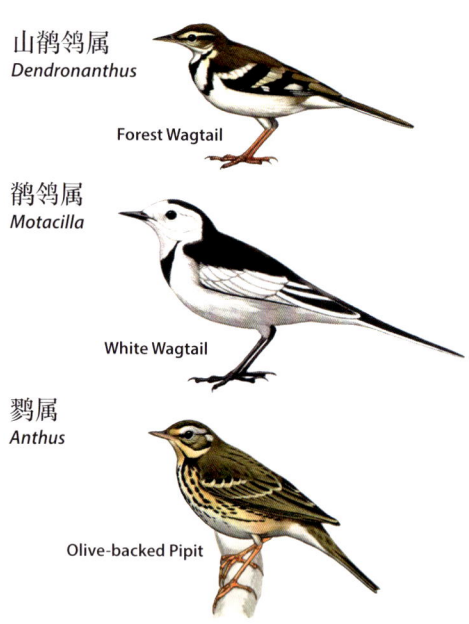

山鹡鸰属
Dendronanthus

Forest Wagtail

鹡鸰属
Motacilla

White Wagtail

鹨属
Anthus

Olive-backed Pipit

Dendronanthus indicus
山鹡鸰 shān jí líng
Forest Wagtail

L: 16–18 cm
Habitat: Adapted to various types of deciduous forests and secondary forests; also occurs in parks and plantations during migration or winter.
Behavior: Forest dweller.

Usually solitary; walks effortlessly along branches. Mostly near trees and bushes when foraging on ground; flies to perch on trees when disturbed. Often wags tail laterally, unlike other wagtails that wag tail up and down.
Distribution: Breeds in Northeast, North, Central, and East China and northern southwestern China; winters in southwestern and South China. Rare passage migrant or winter visitor in Taiwan.
Voice: Song a disyllabic, seesawing, Japanese Tit–like *fee-see*, repeated four or five times. Call is a metallic *tlip*.

Motacilla flava
西黄鹡鸰 xī huáng jí líng
Western Yellow Wagtail

L: 15–16 cm
Habitat: Inhabits reservoirs, lakesides, farmland, rivers, and swamps on plains.
Behavior: Often in pairs or gathers in small flocks of three to five individuals; also in large flocks

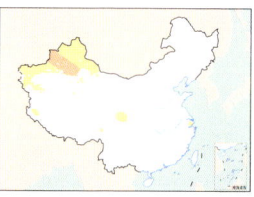

containing several dozen individuals during migration. Prefers walking along watersides; usually wags tail up and down. Calls during undulating flight. Often forages on ground; sometimes makes short flights to catch insects on the wing. During migration, often follows livestock and takes advantage of insects disturbed by the animals.
Distribution: *M. f. leucocephala* is an uncommon passage migrant to N Xinjiang, occasionally seen in Shanghai; *M. f. feldegg* breeds in W and C Tian Shan Mountains in N Xinjiang; *M. f. beema* seen in N Xinjiang, S Gansu, E and SW Tibet, Qinghai, and Sichuan during migration.
Voice: Gives piercing and melodious *tsweep* when flying in flocks; pitch rises at the end. Songs are repeated calls, mixed with some trills.

Motacilla tschutschensis
黄鹡鸰 huáng jí líng
Eastern Yellow Wagtail

L: 16–18 cm
Habitat: In breeding season, prefers wet habitats with short grass, such as wet meadows and lakeside grasslands. Almost anywhere near water during migration; also occurs in cropland.

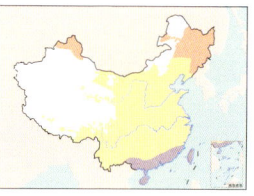

Behavior: Wags tail but to smaller extent than White Wagtail. Undulating flight also smoother.
Distribution: Eastern Yellow Wagtail (*M. tschutschensis*) and Western Yellow Wagtail (*M. flava*) were once a single species, and subspecies allocation is not well resolved. This book follows the IOC list, which holds that three subspecies of Western Yellow Wagtail (see above) and four subspecies of Eastern Yellow Wagtail occur in China: *M. t. macronyx, M. t. taivana, M. t. plexa*, and *M. t. tschutschensis*. The subspecies *M. t. angarensis, M. t. similima*, and *M. t. zaissanensis*, regarded as Western Yellow Wagtail in some other checklists, are not considered valid subspecies but populations of *M. t. tschutschensis*. *M. t. macronyx* is widespread except for northwestern China and C Qinghai-Tibet Plateau and breeds in Northeast China and winters in South China. *M. t. taivana* migrates through Northeast, North, East, and southwestern China and winters in South China, including Hainan and Taiwan. *M. t. plexa* migrates through Heilongjiang, NE Inner Mongolia, Sichuan, and Hubei. *M. t. tschutschensis* breeds in N Xinjiang and migrates through northwestern, Northeast, North, East, Central, and South China; some individuals winter in Taiwan.
Voice: Often gives repeating *psie* calls in flight; songs monosyllabic.

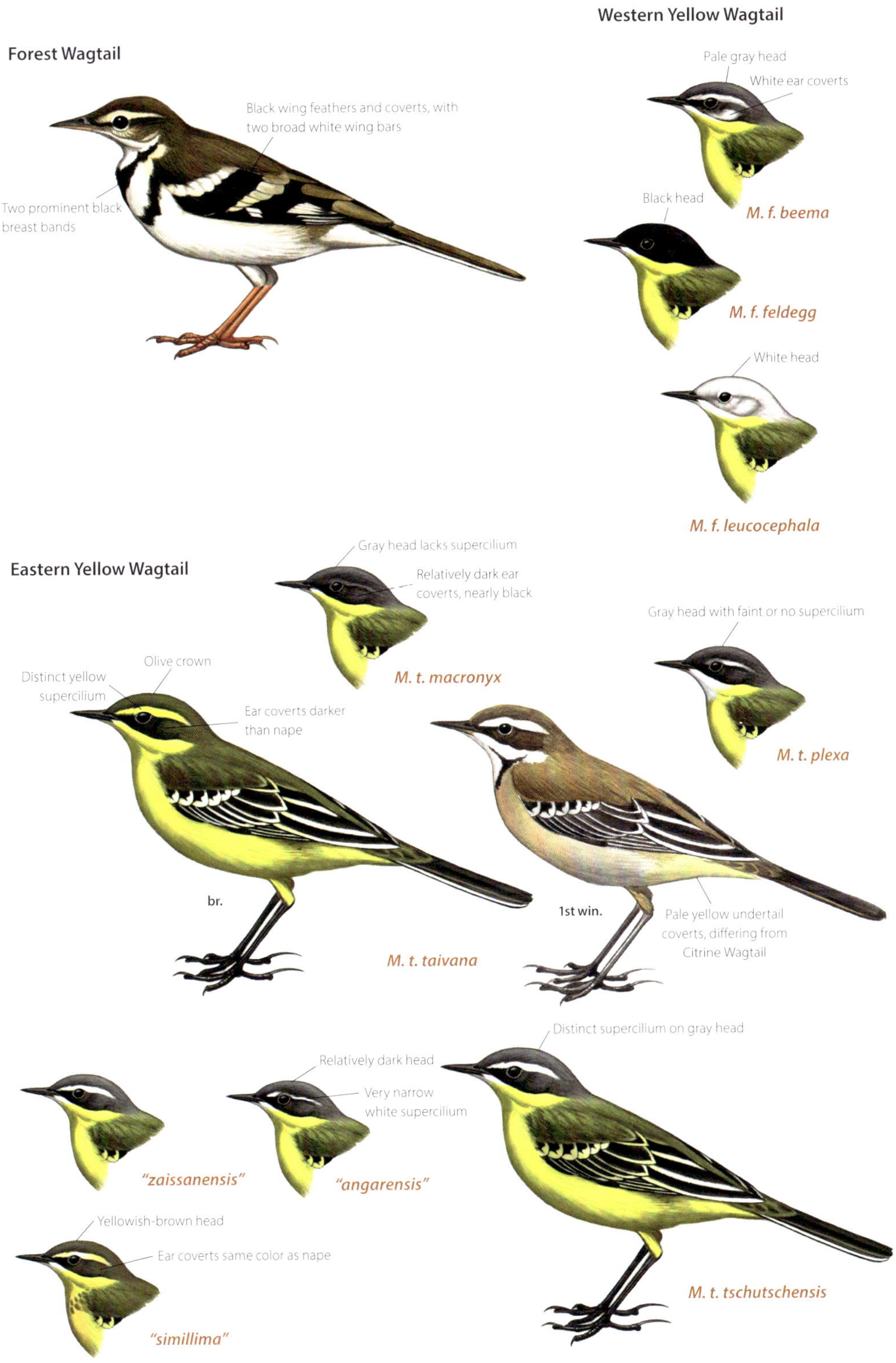

Forest Wagtail

Two prominent black breast bands

Black wing feathers and coverts, with two broad white wing bars

Western Yellow Wagtail

Pale gray head

White ear coverts

M. f. beema

Black head

M. f. feldegg

White head

M. f. leucocephala

Eastern Yellow Wagtail

Gray head lacks supercilium

Relatively dark ear coverts, nearly black

M. t. macronyx

Distinct yellow supercilium

Olive crown

Ear coverts darker than nape

Gray head with faint or no supercilium

M. t. plexa

br.

M. t. taivana

1st win.

Pale yellow undertail coverts, differing from Citrine Wagtail

Distinct supercilium on gray head

Relatively dark head

Very narrow white supercilium

"zaissanensis"

"angarensis"

Yellowish-brown head

Ear coverts same color as nape

"simillima"

M. t. tschutschensis

589

Motacilla alba

白鹡鸰 bái jí líng

White Wagtail (LC)

L: 17–20 cm
Habitat: Adapted to various types of habitats but prefers water edges.
Behavior: Often wags tail up and down when in standing posture; when walking, moves head back and forth, as if "nodding."
Flight undulating. Calls in flight when disturbed; seldom flies high in the air. Gathers in large flocks containing over a hundred individuals during migration or winter; perches in trees at night during migration. Many subspecies recognized, with extensive overlaps in their distribution.
Distribution: Common. For detailed distribution of different subspecies, see the diagram below.
Voice: Gives loud and piercing *eehi*.

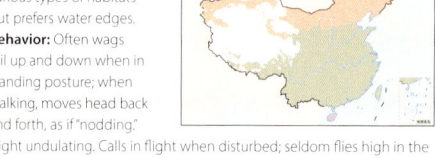

2 With or without eye stripe

M. a. lugens
Characteristics: White chin, black throat; black on hindneck separate from black on breast
Distribution: Eastern parts of China

♂ ad.win.

Black (white in winter)

3 Chin color

M. a. alboides
Characteristics: Black throat; black on hindneck connects with black on breast
Distribution: Northern and western parts of China and South China

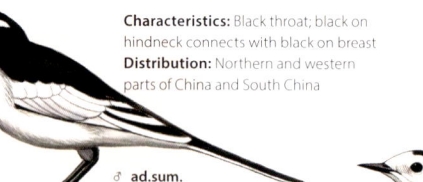

♂ ad.sum.

♂ ad.sum.

juv.

♀

M. a. leucopsis
Characteristics: White throat, black on hindneck separate from black on breast
Distribution: Nationwide

♂ ad.win.

---

1 Back color

Black back   Gray back

2 With or without eye stripe

Without eye stripe   With eye stripe

3 Chin color

Black chin (white in winter)   White chin

4 Whether black on hindneck connects with black on breast
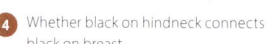
Connected   Disconnected

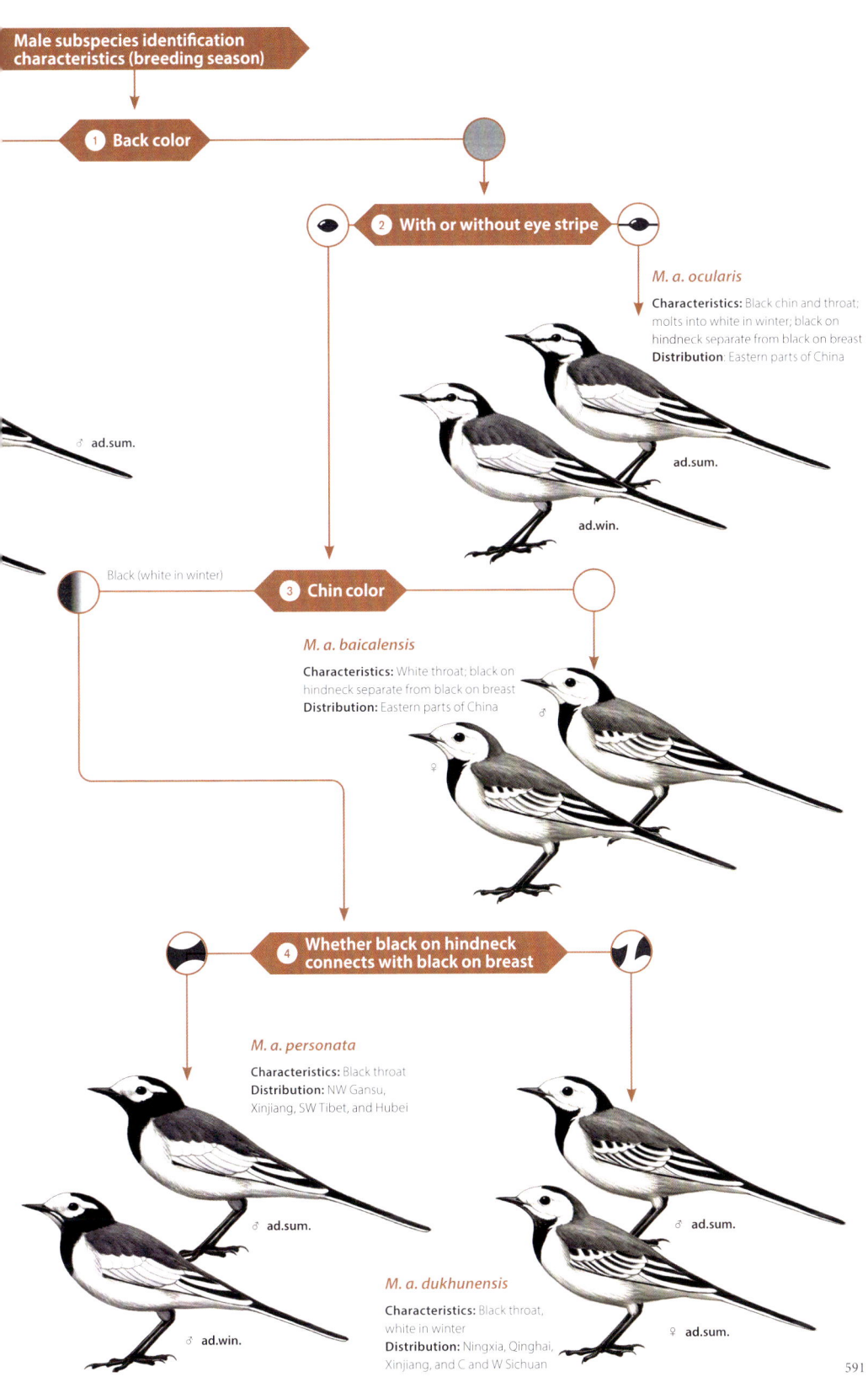

Male subspecies identification characteristics (breeding season)

1 Back color

2 With or without eye stripe

M. a. ocularis

Characteristics: Black chin and throat; molts into white in winter; black on hindneck separate from black on breast
Distribution: Eastern parts of China

♂ ad.sum.

ad.sum.

ad.win.

Black (white in winter)

3 Chin color

M. a. baicalensis

Characteristics: White throat; black on hindneck separate from black on breast
Distribution: Eastern parts of China

♂

♀

4 Whether black on hindneck connects with black on breast

M. a. personata

Characteristics: Black throat
Distribution: NW Gansu, Xinjiang, SW Tibet, and Hubei

♂ ad.sum.

♂ ad.win.

♂ ad.sum.

M. a. dukhunensis

Characteristics: Black throat, white in winter
Distribution: Ningxia, Qinghai, Xinjiang, and C and W Sichuan

♀ ad.sum.

Motacilla citreola

黄头鹡鸰 huáng tóu jí líng

Citrine Wagtail   LC

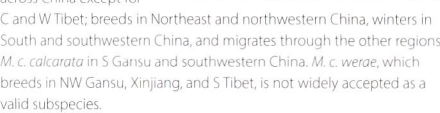

L: 16–20 cm
Habitat: Similar to Eastern Yellow Wagtail but more often associated with water margins.
Behavior: Similar to Eastern Yellow Wagtail.
Distribution: *M. c. citreola* across China except for C and W Tibet; breeds in Northeast and northwestern China, winters in South and southwestern China, and migrates through the other regions. *M. c. calcarata* in S Gansu and southwestern China. *M. c. werae*, which breeds in NW Gansu, Xinjiang, and S Tibet, is not widely accepted as a valid subspecies.
Voice: Calls are a short and dry *trzz-trzz*. Songs are simple and mostly monosyllabic, occasionally with high-pitched notes.

Motacilla cinerea

灰鹡鸰 huī jí líng

Gray Wagtail   LC

L: 16–18 cm
Habitat: In breeding season, prefers watersides surrounded by forests, especially near streams. In winter or during migration, occurs in various wetland habitats.
Behavior: Usually solitary or in pairs; also gathers in small loose flocks. Standing posture flatter than other wagtails. Usually near water, and more often flies into trees than other wagtails except for Forest Wagtail. Frequently wags tail up and down. Sometimes wades in shallow water or hovers in the air for food. Undulating flight distinct.
Distribution: Widespread except for W Tibet. Breeds in Northeast, northwestern, North, and East China; winters in East, Central, South, and southwestern China; migrates through other regions.
Voice: Song is a short mechanical repetition of several *zliss-liss-liss* and similar notes. Calls are a short *zizi*, shorter and higher pitched than call of White Wagtail.

Anthus richardi

田鹨 (理氏鹨) tián liù (lǐ shì liù)

Richard's Pipit   LC

L: 17–18 cm
Habitat: Inhabits open low grasslands, farmland, or deserted fields; prefers meadows near water during breeding season.
Behavior: Solitary or in pairs. Upright standing posture; undulating flight ponderous. Hovers and calls in the air during breeding season.
Distribution: Widespread but absent from the Qinghai-Tibet Plateau. Breeds north of the Qinling-Huaihe Line, winters in southeastern and South China, and migrates through most regions in China. Common during migration.
Voice: Simple song, usually given in undulating display flight, includes 3–6 and occasionally as many as 12 buzzy *chewee* notes. Call a loud, harsh, explosive, sparrowlike *schreep*, given primarily in flight.

Motacilla grandis

日本鹡鸰 rì běn jí líng

Japanese Wagtail   LC

L: 21–23 cm
Habitat: Occurs more frequently near water than White Wagtail, usually at riverbanks, lakesides, and dikes.
Behavior: Similar to White Wagtail but usually in pairs.
Distribution: Several vagrant records in coastal Taiwan. According to *Birds of Hopei*, one record in Hebei (Dongling) on January 29, 1903. Recorded in Zhejiang (Anji) in February 2022.
Voice: Distinctive harsh *dzzr* call given at rest and in flight, as reminiscent of Olive-backed Pipit as it is of White Wagtail.

Anthus rufulus

东方田鹨 (田鹨) dōng fāng tián liù (tián liù)

Paddyfield Pipit   LC

L: 15–16 cm
Habitat: Inhabits low grasslands, paddy fields, plantations, airports, roadsides, wetland edges, and savanna. Usually at altitudes up to 1000 m.
Behavior: Forages on ground. Solitary or in pairs; occasionally gathers in small flocks.
Distribution: Seen in Yunnan, Sichuan, N Guangdong, Guangxi, and Guizhou.
Voice: Simple, repetitive flight song is several lengthening series of two to six complex *tseep* notes; series gradually lengthens to a climax of perhaps as many as 24 notes, immediately followed by far fewer *clink* notes, as the bird descends. Call a mellow *chup*.

Anthus godlewskii

布氏鹨 bù shì liù

Blyth's Pipit   LC

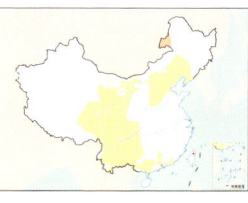

L: 15–17 cm
Habitat: Breeds in mountainous areas; occurs in montane grasslands, mountain valleys, and forests. Also occurs in deserted fields on plains and grasslands during migration; prefers drier habitats than Richard's Pipit.
Behavior: Similar to Richard's Pipit but with low and flat standing posture. Sometimes wanders among tall trees like Tree Pipit or Olive-backed Pipit; Richard's seldom occurs in forests.
Distribution: Breeds in western Northeast China, northern North China, C Inner Mongolia, Qinghai, Ningxia, Gansu, W Sichuan, and E Tibet and winters in southwestern China. Rare. Vagrant to Taiwan.
Voice: Song far more complex and very different from song of Richard's Pipit: a series of rasping *zet* notes, rising whistles, complex buzzes, and occasional call notes in an increasingly rapid and intensifying series that usually ends with a fast repetition of harsh notes and can last more than 20 sec. Calls include a clipped *chup*, very similar to Paddyfield Pipit, and a species-specific, longer, explosive *psheeu* or *spzeuw*.

Citrine Wagtail

Black back

Gray back

♂ ad.sum.

♂ ad.sum.

M. c. citreola

Gray ear coverts, disconnected from nape

Gray ear coverts, disconnected from nape

M. c. calcarata

Pale brownish body

♀ ad.sum.

Whitish undertail coverts

juv.

Whitish undertail coverts

1st win.

Gray Wagtail

Black throat

Dark gray back

Tail longer than other wagtails

♂ ad.sum.

White throat

win.

♀

Japanese Wagtail

Broad white supercilium

White chin

Paddyfield Pipit

Black lores

Black streaks on breast

White outer tail feathers

No streaks on flanks

Hind claws and hind toes similar in length

Richard's Pipit

Relatively long, thick, strong bill

Relatively diffuse streaks on back

Long, pointed, black spots on median coverts

Upright posture

Relatively long legs

Relatively long tail

Very long, relatively flat, straight hind claw

Blyth's Pipit

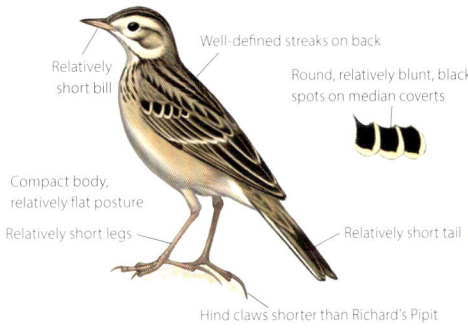

Well-defined streaks on back

Relatively short bill

Round, relatively blunt, black spots on median coverts

Compact body, relatively flat posture

Relatively short legs

Relatively short tail

Hind claws shorter than Richard's Pipit

Anthus campestris

平原鹨 píng yuán liù

Tawny Pipit

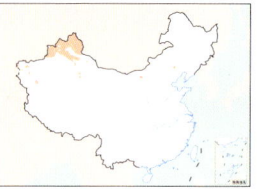

L: 15.5–18 cm

Habitat: Inhabits open low mountains, hills, and foothills, especially dry steppes and semideserts; sometimes also occurs at forest edges, meadows in forests, and streams; seldom comes to croplands and regions with dense and tall grass.

Behavior: Often solitary or in pairs. Usually runs and forages on ground or among gravel; flies to bushes or rocks when disturbed.

Distribution: Locally common to uncommon breeder in Xinjiang and E, C, and W Inner Mongolia.

Voice: Flight song is a monotonous, slow repetition of *tzirilee*, repeated at regular 1–3 sec intervals for up to a minute or more. Calls include an explosive *chilip*.

Anthus trivialis

林鹨 lín liù

Tree Pipit

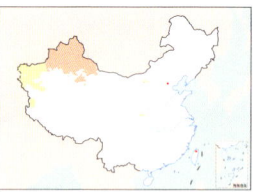

L: 14–16 cm

Habitat: Inhabits primarily edges of montane coniferous forests, broadleaf forests, or mixed forests; especially prefers sparsely wooded areas, such as forest gaps or forest meadows. Also occurs on grasslands with bushes, meadows, and even alpine meadows.

Behavior: In pairs during breeding season; gathers in small flocks in nonbreeding season. Often perches and sings on treetops or power lines. Forages on ground; flies immediately to trees when disturbed.

Distribution: *A. t. trivialis* is a common breeder in the Altai Mountains in N Xinjiang; occasionally seen in S Shaanxi, C Inner Mongolia, Ningxia, and Tibet during migration; winter visitor recorded in Guangxi; vagrant to Beijing and Taiwan. *A. t. haringtoni* is a common breeder in the Tian Shan Mountains in C and W Xinjiang.

Voice: Song is very similar to that of Olive-backed Pipit—a confident series of trills and repeated notes, generally with a far-carrying *seee-er, seee-er, seee-er* toward the end. Call a *tseep* or *teez*, very similar to Olive-backed Pipit.

Anthus pratensis

草地鹨 cǎo dì liù

Meadow Pipit

L: 14–15.5 cm

Habitat: Inhabits desert grasslands, farmland, and fields during migration; occurs in unfrozen rivers, ponds, and nearby grasslands during winter.

Behavior: Often solitary or in loose flocks during winter; also gathers in flocks during migration. Often on meadows or near water. Seldom flies.

Distribution: Rare in Xinjiang, Beijing, and Gansu (Lanzhou) in winter; occasionally seen in Xinjiang, Gansu, E Inner Mongolia, and Liaoning during migration. Vagrant in Ningxia and Henan.

Voice: Call an often-repeated, thin, high-pitched *pssip* or *isst*.

Anthus hodgsoni

树鹨 shù liù

Olive-backed Pipit

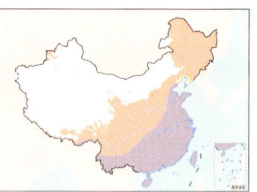

L: 15–17 cm

Habitat: Inhabits various types of forests during breeding season. Occurs at forest edges, grasslands, cultivated fields, or parks in plains, hills, or mountainous regions during nonbreeding season. Favors forests more than other pipits.

Behavior: Moves about in small flocks. Walks on ground and flicks tail; takes cover in nearby trees when disturbed.

Distribution: Common migrant. *A. h. yunnanensis* breeds from Northeast China to North China, migrates southward, and winters in most regions in Central, East, and South China; *A. h. hodgsoni* breeds in S and E Tibet, Qinghai, Inner Mongolia, S Shanxi, Ningxia, Gansu, Sichuan, Hubei, Guizhou, Yunnan, and Northeast China and migrates south in winter.

Voice: Song very similar to that of Tree Pipit. Calls include a loud *teaze*, very similar to Tree (slightly higher pitched, more descending, sounding "thinner," subtly less burry, and inviting comparison with Red-throated Pipit), and a short, weak, almost Goldcrest-like *tsi* or *sip*, also given in flight, higher pitched than the similar but inflected note of Tree.

Anthus sylvanus

山鹨 shān liù

Upland Pipit

L: 17–18 cm

Habitat: In breeding season, occurs in scrubby grasslands and steep rocky regions; in winter, occurs in similar habitats at low altitudes. The population in Hong Kong breeds on low hills at altitudes of about 500 m; individuals in other regions mostly in montane areas at altitudes of 1200–3000 m.

Behavior: Often solitary. Forages for various small invertebrates on ground; stance upright. Flicks tail when disturbed; flies fast and low in a straight line when flushed. Makes display flights similar to grasshopper warblers during breeding season.

Distribution: Uncommon resident in southern China; some populations make altitudinal migrations in winter.

Voice: Powerful, far-carrying, melancholy song is a simple two-note squeaky-gate whistle, repeated twice each second and three to five times in each strophe, *t-lee … t-lee … te-lee* or *see-yu … see-yu*, where the notes are at different pitches and with a buzzy quality reminiscent of Japanese Tit.

Tawny Pipit

Black lores

Clear back, usually lacks streaks

Few or no streaks on side of breast

Meadow Pipit

Almost no supercilium

Yellow lower mandible

Thick streaks on back, no streaks on rump

Streaks of equal width on flanks

Tree Pipit

No black spots on ear coverts

Well-defined streaks on back

Black streaks on flanks become narrower from top to bottom

Olive-backed Pipit

Broad white or buff-yellow supercilium

Inconspicuous blackish-brown streaks on olive back

Well-defined thick black streaks on white belly

A. h. yunnanensis

More well-defined streaks on back, with more uniform olive

More and denser streaks on belly

A. h. hodgsoni

Upland Pipit

Orderly thick blackish-brown streaks from crown and nape to back

A few fine blackish-brown streaks from crown and side of neck to flanks

Thick bill with pinkish-brown lower mandible

Strong pinkish-brown legs

Large body, easily distinguished from other pipits

Anthus roseatus

粉红胸鹨 fěn hóng xiōng liù

Rosy Pipit LC

L: 15–16.5 cm
Habitat: Breeds on alpine meadows and rocky or grassy plateaus at altitudes of 2700–4400 m; migrates to low altitudes in winter on paddy fields or grasslands.
Behavior: Often occurs near streams and forages for insects and seeds on ground.
Distribution: Locally common. Breeds from S and W Xinjiang, the Qinghai-Tibet Plateau, to North China, south to Sichuan and Hubei, migrates southward, and winters in SE Tibet and Yunnan; vagrant to Hainan. Isolated resident populations in Jiangxi (Wuyi Mountains).
Voice: Two-part flight song involves twittering during ascent, followed by a long series of pleasant *tsuli tsuli tsuli* notes during parachute descent; lower pitched, slower, and usually with shorter strophes than song of Water Pipit. Flight call very similar to Meadow Pipit and less shrill than call of Water Pipit.

Anthus rubescens

黄腹鹨 huáng fù liù

American Pipit LC

L: 14–17 cm
Habitat: Breeds on gravelly alpine or subalpine tundra at altitudes up to 2400 m; in nonbreeding season, occurs on open wetlands, swamps, riverbanks, and cultivated fields on plains and near coasts.
Behavior: Solitary or in small flocks. Walks rapidly on ground when foraging; wags tail slightly up and down, like wagtails.
Distribution: *A. r. japonicus* breeds in Northeast China and winters in most regions south of the Yangtze River (including Taiwan); common.
Voice: Call in flight a high *tssip*, sometimes a disyllabic *tsipit*, or a rapidly repeated *si-si-si-si-si*, lacking the power and shrillness of Water Pipit; the shorter notes are similar to one of the calls of Meadow Pipit.

Anthus gustavi

北鹨 běi liù

Pechora Pipit LC

L: 14–15 cm
Habitat: Occurs in open wet grassy areas and coastal scrubby woods during migration.
Behavior: Mostly solitary, secretive; often descends to nearby small trees or hides in dense grass when disturbed.
Distribution: Two subspecies are seen in China, uncommon passage migrant or breeder. *A. g. gustavi* and *A. g. menzbieri* migrate through Northeast China and coastal regions in eastern China, including Taiwan.

Anthus spinoletta

水鹨 shuǐ liù

Water Pipit LC

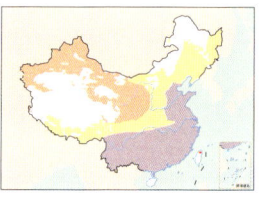

L: 15–17.5 cm
Habitat: Breeds on alpine gravelly or grassy slopes; in nonbreeding season, occurs on wetlands near lakes, rivers, and other waterbodies, especially along streams.
Behavior: Often solitary. Walks rapidly on ground when foraging, wags tail slightly up and down, like wagtails. Perching posture horizontal.
Distribution: *A. s. blakistoni* breeds in N and W Xinjiang, Qinghai, and Gansu; migrates through and winters in most regions in North China and central, eastern, and southern China. Common migrant; vagrant to Taiwan.
Voice: Song a rapid series of notes often arranged in four to five phrases, similar to the song of Rosy Pipit. Call a short, sharp, explosive *wisst* or *chui*.

Anthus cervinus

红喉鹨 hóng hóu liù

Red-throated Pipit LC

L: 14–15 cm
Habitat: Occurs on wetlands, ponds, rivers, paddy fields, and moist grasslands on plains.
Behavior: Solitary or in loose small flocks; often forages in the same area with Richard's Pipit and Eastern Yellow Wagtail. Adopts horizontal posture when foraging.
Distribution: Common passage migrant in northern, East, and Central China; winters south of the Yangtze River, including Hainan and Taiwan.
Voice: Call a high-pitched, piercing *teeeze*, longer and higher pitched than call of Olive-backed Pipit.

The migration time of *A. g. menzbieri* (peaks during mid- to late April and early September) is usually earlier than that of *A. g. gustavi* (peaks during mid- to late May and early October). Breeding records of *A. g. menzbieri* in E Heilongjiang; vagrant to Xinjiang.
Voice: Song an unobtrusive but lengthy and exuberant sequence of short, sharp, mechanical notes, buzzes, and intensifying trills, often incorporating calls. Less likely to call when flushed than other pipits—a bold, clearly enunciated *pwit*, reminiscent of both Zitting Cisticola and Gray Wagtail.

Rosy Pipit

Dark lores

Pink supercilium

Pale pink breast and upper belly (extent of pink varies)

br.

Thick, clear, creamy-white supercilium

Broad clear blackish-brown streaks on back, differing from Olive-backed Pipit and American Pipit

Relatively heavy olive on upperparts, differing from pale brown upperparts of Red-throated Pipit in nonbreeding plumage

Lacks pink

Dense thick black spots or streaks on breast and flanks

non-br.

Olive fringe on flight feathers

Water Pipit

Lores not black

Pale brownish-gray head

Thick short creamy-white supercilium, relatively broad behind eyes

Black bill

Inconspicuous dark shaft streaks on pale brown back

Clean pinkish-buff throat and underparts, almost without streaks

br.

Blackish-brown legs

Triangular dusky black patch on side of neck

Pale yellowish-brown back

Sparse, relatively narrow, pale black streaks on breast and flanks

non-br.

Blackish-brown to brownish-pink legs, differing among individuals

American Pipit

Lores not black

Short, thick, buff-yellow supercilium

Pinkish-gray head and back

Triangular dusky black patch on side of neck

A few blackish-brown streaks from breast to flanks

br.

Brownish-pink legs

Distinct triangular black patch on side of neck

Inconspicuous streaks on grayish-brown back

Dense black streaks from breast and flanks to upper belly

non-br.

Red-throated Pipit

Like male, but red color relatively pale

♀ br.

Rufous chestnut from supercilium and face to breast

♂ br.

Yellowish-brown lower mandible

Buff-yellow streaks on back

Buff-yellow throat

Brownish-pink legs

1st win.

Pechora Pipit

Body more compact than Red-throated Pipit, thus head and eyes appear large

White streaks on back

Pinkish brown on base of lower mandible

Wing patch and wing coverts contrast more starkly than first winter Red-throated Pipit

Pure white underparts

A. g. gustavi

Thicker and heavier black streaks on back

Whiter and narrower fringe on wings; usually with buff-yellow streaks on underparts and back

A. g. menzbieri

597

朱鹀科　Urocynchramidae

A monotypic family. Plumage mostly pink and buffy. Bill cone shaped, resembling bills of buntings. Wings short and rounded. Tail long. Inhabit the scrubby areas of meadows, forest edges, and river valleys at high altitudes; solitary or in pairs. Nest among bushes. Highly vocal during breeding season. Feed mostly on plant seeds and fruits; also take insects. Resident.

Przevalski's Pinktail was once placed in Emberizidae or Fringillidae owing to bill and tail morphology but is now in its own family, closely related to other rosefinches. Only one genus and one species recognized worldwide. Chinese endemic, found only on the eastern edge of the Qinghai-Tibet Plateau.

Przevalski's Pinktail

Rosy lores and throat, pinkish on remainder of underparts

Slender bill

Slender bill

Relatively long tail with pinkish-red outer tail feathers

Relatively long tail with orangish outer tail feathers

Urocynchramus pylzowi

朱鹀 zhū wú

Przevalski's Pinktail   **LC**

L: 15–17 cm
Habitat: Inhabits bushes at middle to high altitudes.
Behavior: Solitary or in pairs; also gathers in small flocks. Often perches on bushtops and calls.
Distribution: Endemic to the eastern edge of the Qinghai-Tibet Plateau; uncommon in NW Gansu, Qinghai, and N Sichuan. Vagrant to Chongqing.
Voice: Song is a short, hurried, poorly structured, chattering *chitri-chitri-chitri-chitri* that can include mimicry of species such as White-browed Tit, Ground Tit, and Twite, reminiscent of song of Long-tailed Rosefinch but longer and more complex. Rarely heard call is *kvuit kvuit*.

燕雀科　Fringillidae

Small ground-dwelling or tree-dwelling songbirds. Plumage differs between sexes in most species; highly variable among species, mostly brown, red, yellow, and black. Bill short, thick, and pointed. Wings relatively long and narrow; good at flying. Tail emarginated and medium in length. Inhabit a variety of habitats, including forests, scree, scrubland, grasslands, deserts, and agricultural fields. Nest among branches; some species also on ground or among bushes. Feed mostly on grains, grass seeds, or other plant seeds, shoots, and fruits; also take insects and other arthropods.

Fifty genera and 228 species recognized worldwide, widely distributed except for Oceania. Twenty-two genera and 63 species recorded in China, ubiquitously distributed nationwide.

燕雀属
Fringilla

Brambling

拟蜡嘴雀属
Mycerobas

White-winged Grosbeak

锡嘴雀属
Coccothraustes

Hawfinch

蜡嘴雀属
Eophona

松雀属
Pinicola

Yellow-billed Grosbeak

Pine Grosbeak

红翅沙雀属
Rhodopechys

灰雀属
Pyrrhula

蒙古沙雀属
Bucanetes

Crimson-winged Finch

Gray-headed Bullfinch

Mongolian Finch

赤朱雀属
Agraphospiza

红眉金翅雀属
Callacanthis

黑雀属
Pyrrhoplectes

Blanford's Rosefinch

Spectacled Finch

Gold-naped Finch

暗胸朱雀属
Procarduelis

岭雀属
Leucosticte

金翅雀属
Chloris

Dark-breasted Rosefinch

Plain Mountain Finch

Oriental Greenfinch

朱雀属
Carpodacus

沙雀属
Rhodospiza

朱顶雀属
Linaria

Common Rosefinch

Desert Finch

Twite

红额金翅雀属
Carduelis

白腰朱顶雀属
Acanthis

黄雀属
Spinus

European Goldfinch

Common Redpoll

交嘴雀属
Loxia

Eurasian Siskin

Red Crossbill

599

Fringilla coelebs

苍头燕雀 cāng tóu yàn què

Common Chaffinch

L: 14–16 cm

Habitat: Inhabits various
types of forests, such
as broadleaf, mixed
broadleaf-coniferous,
coniferous, and secondary
forests in river valleys; also
occurs in woodlands and
bushes at forest edges
and woods along riverbanks. During migration and winter, in orchards,
urban parks, and bushes and trees near farmland; sometimes also near
villages and residential areas.

Behavior: In pairs during breeding season. Great singer. Joins Brambling
in mixed-species flocks in fall and winter; gathers in large flocks
containing several dozen individuals. Often perches in trees and bushes.
Moves about and forages on ground. Bold and not shy; easy to approach.

Distribution: F. c. coelebs is a common breeder in NW Xinjiang, seen in
Heilongjiang, Jilin, Liaoning, Beijing, Tianjin, Hebei, Shanxi, N Shaanxi,
C Inner Mongolia, and S Xinjiang during migration and winter; vagrant
to Yunnan and Sichuan.

Voice: Cheerful, loud song is an accelerating, descending, musical rattle,
followed by an accelerating final flourish, chip chip chip … tet tet tet-erry-
erry-erry-tissi cheweeeoo. Calls include a sharp, distinctive chink, often
doubled, a loud husky hweet, and a loud hooeed as a so-called rain call.

Mycerobas carnipes

白斑翅拟蜡嘴雀 bái bān chì nǐ là zuǐ què

White-winged Grosbeak

L: 21–24 cm

Habitat: Inhabits
coniferous forests, mixed
forests, and bushes at
middle and high altitudes.
Behavior: Solitary or in
small flocks. Quiet.
Distribution:
M. c. merzbacheri in
northwestern China; M. c. carnipes in central and southwestern China.
Uncommon.

Voice: Rarely heard song is a mellow, slow whistle, plieu or plieeeu …
tip-u, repeated about every 1.5 sec and often combined with calls but
easily overlooked. Calls include a nasal "in your face" chattering, nyit-yit-
yit, or a scolding chet-et-et-et.

Mycerobas melanozanthos

白点翅拟蜡嘴雀 bái diǎn chì nǐ là zuǐ què

Spot-winged Grosbeak

L: 21–23 cm

Habitat: Inhabits various
types of forests at middle
altitudes.

Behavior: Often gathers
in flocks. Active and noisy.
Distribution: Uncommon
in W Gansu, W Sichuan,
Yunnan, and S Tibet.

Voice: Musical song a lengthy melancholy series of plaintive, very low-
pitched, flat whistles, wuuu-weeeee-di-di-di-deee … wuee-ti-di-. Calls are
similarly low-pitched, almost bullfinch-like, short phu or pu notes, often
given in excited, rapid series.

Fringilla montifringilla

燕雀 yàn què

Brambling

L: 13–16 cm

Habitat: Inhabits
coniferous, mixed, and
broadleaf forests at middle
and low altitudes; also
occurs in secondary forests,
farmland, and parks.
Behavior: Gathers in large
flocks during nonbreeding
season. Perches or forages in trees; often flies onto the ground in flocks
and then returns to trees.

Distribution: Widely distributed across China, except for Tibet and
Hainan. Seasonally common.

Voice: Highly distinctive song is an unpleasant wheezy buzz, zweeee,
repeated every 6 sec or so. Calls include a rasping, nasal chweee or tswee-
ik and a mellow chup.

Mycerobas affinis

黄颈拟蜡嘴雀 huáng jǐng nǐ là zuǐ què

Collared Grosbeak

L: 22–24 cm

Habitat: Inhabits
broadleaf forests, mixed
forests, coniferous forests,
and rhododendron
thickets at middle and
high altitudes.
Behavior: Solitary or in
pairs; gathers in small

flocks in winter. Often moves about in trees; occasionally comes to the
ground.

Distribution: Uncommon in S Shaanxi, S and W Gansu, Sichuan, Yunnan,
and S Tibet.

Voice: Variable song is a loud, clear, ringing, simple series of 5–12 notes,
ki-ki-du … ki-ki-du or plee-ti-tu-tt-ti-plee-ti-tu-tee-tu-clid.

Coccothraustes coccothraustes

锡嘴雀 xī zuǐ què

Hawfinch

L: 16–18 cm

Habitat: Inhabits various
forests at middle and low
altitudes.
Behavior: Often solitary
or in pairs. Quiet.
Distribution: C. c.
coccothraustes across
China except for Tibet,

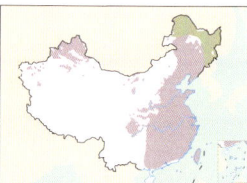

Yunnan, and Hainan, but often in the regions north of the Yangtze River;
seasonally common. C. c. japonicus occasionally seen in Fujian.

Voice: Call a piercing, buntinglike tik or pix.

Common Chaffinch

Black forehead

Bluish-gray crown and hindneck

Chestnut-brown upper back

Reddish-brown face and breast

♂

Green crown and neck

Grayish-green upper back

♀

Pale brown face and breast

Brambling

♂

♀

Pale brown breast and scapulars

White rump

White-winged Grosbeak

♂

Grayish-black head, neck, back, and breast

Large white patch on primaries, yellow patch on rest of wings

Large area of black extends to entire breast

Yellow rump

♀

Large white patch on primaries

Collared Grosbeak

♂

Black head

Orange-yellow hindneck

Black wings and tail

Bright yellow on remaining upperparts, breast, and belly

Dark gray from head to upper breast

♀

Grayish-green head, back, and wing coverts

Yellowish-green underparts

Black primaries

Black tail

Spot-winged Grosbeak

♂

Relatively small area of black extends only to upper breast

Yellow supercilium

Whitish wing patches

Rump not yellow

Black streaks on crown and upper back

♀

Broad black streaks on yellow underparts

Hawfinch

Large thick bill

non-br.

♂ br.

Uniquely shaped inner primaries

♀ br.

Yellowish-brown overall except for wing feathers

Eophona migratoria

黑尾蜡嘴雀 hēi wěi là zuǐ què

Yellow-billed Grosbeak　

L: 15–18 cm
Habitat: Inhabits various types of forests at low altitudes.
Behavior: Gathers in large flocks during nonbreeding season. Moves about in or among trees. Wingbeats audible.
Distribution: *E. m. migratoria* throughout China except for northwestern China and Hainan; *E. m. sowerbyi* in southwestern, Central, and South China, including Hainan. Common.
Voice: Pleasant, slow song is a series of loud fluty whistles, *kuu-ki-che-ku-ch … kuu-ki-che-ku-ch*, repeated at length and with little variation, very similar to the song of Japanese Grosbeak but slightly higher pitched. Call a penetrating *tek*, similar to that of Japanese but fractionally higher pitched, cleaner, and sharper.

Eophona personata

黑头蜡嘴雀 hēi tóu là zuǐ què

Japanese Grosbeak　

L: 20–24 cm
Habitat: Inhabits various forests at middle and low altitudes.
Behavior: Seldom gathers in large flocks; often solitary or in small flocks. Usually hides in the upper trees.
Distribution: *E. p. magnirostris* in central and eastern China; *E. p. personata* in Japan and Taiwan. Uncommon.
Voice: Song a series of four or five short, fluty, rising and falling whistles, *tsuki-hi-hoshi*, with the last note usually protracted. Call similar to Yellow-billed Grosbeak.

Pyrrhula erythrocephala

红头灰雀 hóng tóu huī què

Red-headed Bullfinch　

L: 16–17 cm
Habitat: Inhabits coniferous forests, mixed forests, and bushes at middle and high altitudes.
Behavior: Solitary or in pairs during breeding season; gathers in small flocks during nonbreeding season. Bold.
Distribution: Uncommon in S Tibet.
Voice: Song a low mellow *terp-terp-tee* or *heer-t-yeer, heer-t-yeer, yeer-phew*. Calls include a flat whistled *puu* or *puu-tii*, very similar to call of Eurasian Bullfinch.

Pinicola enucleator

松雀 sōng què

Pine Grosbeak　

L: 19–22 cm
Habitat: In winter, inhabits coniferous and mixed forests; also in orchards and parks.
Behavior: Gathers in small flocks in trees during winter; also in bushes or forages on ground. Irruptive migrations in some years.
Distribution: *P. e. pacata* in N Heilongjiang, NE Inner Mongolia, and N Xinjiang; *P. e. kamtschatkensis* in S and E Heilongjiang, Jilin, and Liaoning. Uncommon.
Voice: Varied song is a short, fast, loud, melancholy, musical warble of clear fluty whistles and trills, often with considerable repetition of individual notes and occasionally including some mimicry. Calls include a fluty *wheet-idd … wheet-idd-it*.

Pyrrhula nipalensis

褐灰雀 hè huī què

Brown Bullfinch　

L: 16–17 cm
Habitat: Inhabits broadleaf and mixed broadleaf-coniferous forests at middle altitudes.
Behavior: Often gathers in small flocks. Bold.
Distribution: *P. n. nipalensis* in SE Tibet and NW Yunnan; *P. n. ricketti* in Shandong, Shaanxi, Hubei, Hunan, Jiangxi, Fujian, Guangdong, Guangxi, and Yunnan; *P. n. uchidai* in Taiwan. Uncommon.
Voice: Song is an easily overlooked, mellow *per-whitty-pu-w* or *ppi-puee*. Calls include a soft *terrilip*.

Pyrrhula pyrrhula

红腹灰雀 hóng fù huī què

Eurasian Bullfinch　

L: 15.5–17.5 cm
Habitat: Inhabits coniferous and mixed forests; also occurs in artificial forests, orchards, and parks.
Behavior: Often gathers in small flocks in trees; also forages on ground.
Distribution: Various subspecies are seen in Northeast China, northern North China, and Xinjiang; occasionally seen in Shanghai, Jiangsu, and Henan. *P. p. griseiventris* and *P. p. cineracea* were once treated as Gray-bellied Bullfinch (灰腹灰雀). Uncommon.
Voice: Most common call a slow, soft, melancholy, descending *peu*. Nervous-sounding song mixes call notes with strained, higher-pitched, scratchy notes and wheezes.

Yellow-billed Grosbeak

Black head with relatively large hood

Brownish upperparts

♂

♀

White wing tips

Orange flanks

Japanese Grosbeak

Very small black hood

juv.

ad.

Predominantly gray

No white on wing tips

Red-headed Bullfinch

♂

♀

Orange-red crown and nape

Orange-red breast

♂

Grayish-yellow crown and nape

Salt licking

Pine Grosbeak

Thick strong bill

White wing patch and white outer fringe of inner wing feathers

♂

♀

♀

juv.

♀

Largest finch in region

Brown Bullfinch

Relatively uniform grayish-brown body lacks brownish-yellow tinge

More spots on crown

Relatively pale scaly spots on crown

Darker on upper back

Black lores on male; both female and male lack white forehead

P. n. ricketti

P. n. nipalensis

juv.

White on shafts of central tail feathers

Whiter belly

P. n. uchidai

Eurasian Bullfinch

Gray on tip of greater coverts

♀

Dark pink on cheek, throat, breast, and belly

♂

♀

P. p. pyrrhula

Pink cheek and throat

♀

Gray or pinkish breast and belly

♂

P. p. griseiventris

♂

♀

No pink on body

P. p. cineracea

Pale gray upperparts, occasionally tinged pink

Pure white wing patch

♂

Resembles *P. p. pyrrhula*, but lighter on cheek, throat, breast, and belly

P. p. cassinii

603

Pyrrhula erythaca

灰头灰雀 huī tóu huī què

Gray-headed Bullfinch LC

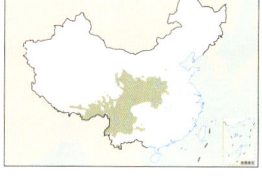

L: 15–16 cm
Habitat: Inhabits coniferous forests, mixed forests, and bamboo and rhododendrons thickets at middle and high altitudes.
Behavior: Gathers in small flocks and forages in trees during nonbreeding season; active.
Distribution: Locally common in North, Central, and southwestern China.
Voice: Song is basically a series of two to six short, slightly descending call notes that differ slightly in pitch, *pu-ti-ti-ti-ti-teu* or *tliue-pi-ti … tliue.*

Pyrrhula owstoni

台湾灰头灰雀 tái wān huī tóu huī què

Taiwan Bullfinch NR

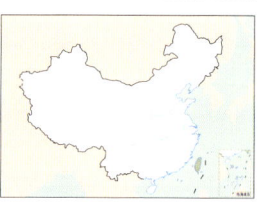

L: 15–17 cm
Habitat: Inhabits coniferous and mixed forests at middle and high altitudes.
Behavior: In pairs or small flocks in trees or on ground. Flight straight, slightly undulating. Perches on treetops.
Distribution: Chinese endemic; locally common in mountainous areas at middle and high altitudes in Taiwan.
Voice: Not known to differ from Gray-headed Bullfinch.

Rhodopechys sanguineus

红翅沙雀 hóng chì shā què

Crimson-winged Finch LC

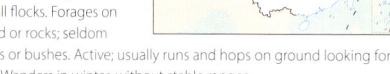

L: 13–15 cm
Habitat: Inhabits sparsely vegetated rocky slopes, scrubby grasslands, river ditches, and steep cliffs in low mountains and hills.
Behavior: Often gathers in small flocks. Forages on ground or rocks; seldom in trees or bushes. Active; usually runs and hops on ground looking for seeds. Wanders in winter, without stable ranges.
Distribution: Very rare resident in SW and NW Xinjiang.
Voice: Song a scratchy, disjointed, grating twitter, *pleit … tchwilit … dzik-tchup … tchup … spzzz'bz'bz … tchwilichip.* Calls include a strained, vaguely sparrowlike *chlivv* and a multisyllabic *werr-ttlr-tt'eee.*

Bucanetes mongolicus

蒙古沙雀 měng gǔ shā què

Mongolian Finch LC

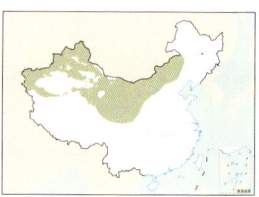

L: 11.5–13 cm
Habitat: Inhabits open areas in deserts and semideserts on hills and mountains; especially prefers bare rocky slopes, cliffs, scrubby dry grasslands, and broad gravelly river valleys and lowlands. Avoids habitats with dense vegetation.
Behavior: Usually on ground in flocks. Tame and not shy. Drinks at fixed places. Wanders in fall and winter.
Distribution: Resident in N Hebei, Inner Mongolia, Ningxia, Gansu, Xinjiang, Tibet, Qinghai, and Sichuan; vagrant in C Heilongjiang.
Voice: Spluttering, scratchy song includes Rock Sparrow–like notes, nasal wheezes, and multiple Eurasian Linnet–like notes in an erratic combination, *tu-wit-chup-tuu whit-tu-tu-chrrrh … weit … veer.* Common calls include a rising, sometimes buzzy *weep* or *tu-weep* and several other calls like Eurasian Linnet or vaguely like Common Chaffinch, *chink.*

Agraphospiza rubescens

赤朱雀 chì zhū què

Blanford's Rosefinch LC

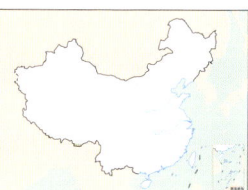

L: 14–15 cm
Habitat: Inhabits coniferous forests, scrub, and grasslands at middle and high altitudes.
Behavior: At high altitudes in summer; migrates down to forests and bushes at middle altitudes in winter. Also forages on ground.
Distribution: Rare in S and E Tibet, Yunnan, Sichuan, S Shaanxi, and S Gansu.
Voice: Simple, hurried song is a rapidly repeated staccato sequence of about six vaguely sparrowlike disyllabic notes, *tu'wit-tu'wit-tu'wit-tu'whit-tu'whit,* lasting about 1.5 sec and repeated every 3–5 sec. Call is essentially the first note of the song.

Callacanthis burtoni

红眉金翅雀 hóng méi jīn chì què

Spectacled Finch LC

L: 17–18 cm
Habitat: Inhabits coniferous forests and grassy slopes at middle and high altitudes.
Behavior: Gathers in small flocks. Quiet. Forages on ground or in grassy bushes.
Distribution: Very rare in S Tibet.
Voice: Song is a rapid series of 3–17 falling, then rising, short chittering notes lasting 0.5–3 sec and repeated with little variation, *chit-chit-chit-chit-chittrr-chitrr-chitrr.* Most phrases become louder toward the end; some also slow down. Calls include loud, clear, slightly rising, whistled *tweee-tweee, pewee, pweeu,* or *pu-weee.*

Gray-headed Bullfinch

Grayer back, with slightly short wings ♂

No scaly pattern on crown

Black forehead and lores bordered with white ♀

Some males have yellow breast and belly

Orange-red breast and belly

Taiwan Bullfinch

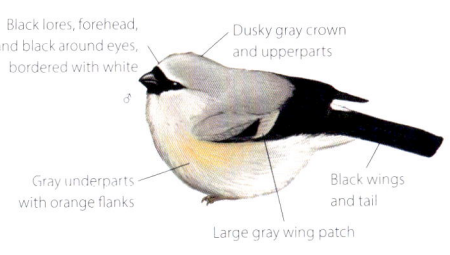

Black lores, forehead, and black around eyes, bordered with white ♂

Dusky gray crown and upperparts

Gray underparts with orange flanks

Black wings and tail

Large gray wing patch

Mongolian Finch

Relatively small pale yellow bill

Often pinkish red on breast

Two white wing bars

Crimson-winged Finch

Pure black crown

Thick yellow bill

Crimson around eyes ♂

Crimson on wings

Grayish-black crown

No red around eyes ♀

Primary projection relatively long

Pink on wings

Blanford's Rosefinch

Relatively long straight bill ♂

Resembles Common Rosefinch, but with relatively dark body

Slightly thick pale bill ♀

Inconspicuous wing patch

Red-tinged upper back and rump, almost without streaks, resembles Dark-breasted Rosefinch

Spectacled Finch

Black head with red forehead and mask

Black wings with white patch ♂

Yellow forehead and mask ♀

605

Pyrrhoplectes epauletta
金枕黑雀 jīn zhěn hēi què
Gold-naped Finch　　LC

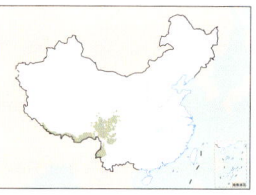

L: 13–15 cm
Habitat: Inhabits the dense undergrowth and edges of oak and rhododendron forests on mountains; also in rhododendron thickets and bushes in mixed broadleaf-coniferous forests, low bamboo thickets, and nettle fields. At altitudes of 1400–4000 m. Descends to low altitudes in winter.
Behavior: Forages on ground, in low vegetation and dense undergrowth, occasionally venturing into open areas near forest or bush edges. Often perches motionless in bushes. Solitary or in pairs; sometimes in small flocks. Often in mixed-species flocks with other seed-eating birds in nonbreeding season.
Distribution: Seen in S Gansu, S and SE Tibet, W and NW Yunnan, and W Sichuan.
Voice: Song a very high-pitched, fine *teeeeeeeee* lasting about 0.6 sec and repeated every 2.5 sec or so.

Procarduelis nipalensis
暗胸朱雀 àn xiōng zhū què
Dark-breasted Rosefinch　　LC

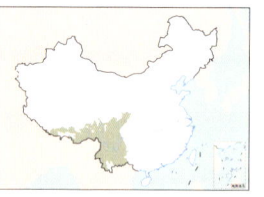

L: 14–16 cm
Habitat: Inhabits the undergrowth and bamboo and rhododendron thickets of coniferous and mixed broadleaf-coniferous forests at middle and high altitudes; also moves up to alpine meadows and scree.
Behavior: In pairs or small flocks at high altitudes with forests or open habitats. Secretive but not shy; seldom joins mixed-species flocks.
Distribution: Found in southwestern China: E, N, W, and SW Sichuan, N and W Yunnan, and S Tibet.
Voice: Song poorly known; apparently a monotonously repeated chipping. Calls include a harsh, nasal, sparrowlike *chiarr*.

Leucosticte nemoricola
林岭雀 lín lǐng què
Plain Mountain Finch　　LC

L: 14–17 cm
Habitat: Inhabits alpine meadows, scrub, and scree at middle and high altitudes; also seen on farmland, deserted fields, and grasslands.
Behavior: Prefers gathering in large flocks in open habitats. Forages on ground. Flies to trees when disturbed or flies fast in turning and wheeling flock.
Distribution: *L. n. altaica* in Xinjiang (except for the Tarim Basin) and W Tibet; *L. n. nemoricola* in S Tibet, Qinghai, C Inner Mongolia, Gansu, Shaanxi, Sichuan, and Yunnan.
Voice: Song is a monotonously repeated, unmusical, sparrowlike *tssliu-tssliu-tssliu-*. Calls include various hard, mostly short, twittering notes, *spziu*, *tick-it*, *tup-tup-tup*, given singly, in a series, or combined with other notes.

Leucosticte brandti
高山岭雀 gāo shān lǐng què
Black-headed Mountain Finch　　LC

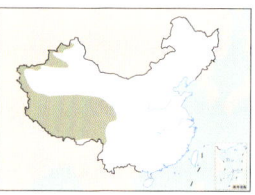

L: 15–17 cm
Habitat: Inhabits rocky areas with sparse vegetation at high altitudes.
Behavior: Gathers in flocks; also joins mixed-species flocks. Active on ground.
Distribution: Locally common at high altitudes in Western China. *L. b. margaritacea* in northwestern China; *L. b. brandti* in NW Xinjiang; *L. b. pamirensis* in W Xinjiang; *L. b. haematopygia* in S and W Tibet and S Qinghai; *L. b. pallidior* in NW Gansu, S Xinjiang, and N Qinghai; *L. b. intermedia* in Gansu and Qinghai; *L. b. walteri* in E Tibet, N Yunnan, and Sichuan.
Voice: Rarely heard song a fine, weak-sounding, very high-pitched, strained whistle about 0.2 sec long, repeated after a pause of also about 0.2 sec and often followed by sparrowlike chattering. Calls include a buzzy *tziu*, repeated occasionally rapidly; a loud, chattering, sparrowlike *twitt-twitt*, *twee-ti-ti*, or dryer *chut*, often doubled; and a *chrup*.

Leucosticte arctoa
粉红腹岭雀 fěn hóng fù lǐng què
Asian Rosy-Finch　　LC

L: 14–18 cm
Habitat: Inhabits alpine meadows, tundra, and bare rocks.
Behavior: Gathers in medium-sized to large flocks. Forages on ground. Some individuals wander in valleys in winter.
Distribution: *L. a. arctoa* in N Xinjiang. *L. a. brunneonucha* breeds in northern Northeast China and winters in Northeast and northern North China. Occasional.
Voice: Song a measured, slow series of *clew* or *tiew* notes. Calls include a buzzy, nasal *chzew* and a spluttering, sparrowlike *clip-tlip-tip* or *tup-tup-tupp* when taking flight.

Carpodacus erythrinus
普通朱雀 pǔ tōng zhū què
Common Rosefinch　　LC

L: 13–15 cm
Habitat: Breeds in montane forests at middle and high altitudes; migrates down to broadleaf forests, secondary forests, and bushes on plains in winter.
Behavior: Solitary, in pairs, or in small flocks.
Distribution: *C. e. grebnitskii* in Northeast, North, Central, and East China; *C. e. roseatus* in northwestern and southwestern China. Common.
Voice: Simple song is a slow and monotonously repeated whistle, *weedy-wu-weeeja-wu*, often rendered as "Pleased to meet you." Call, given at rest and in flight, is a distinctive rising *djoooee*.

Gold-naped Finch

Gold nape

♂
Mostly black body

White inner edge of tertials

Some yellow on side of breast

Gray face and forehead

♀
Pale yellow nape and side of neck

Gray back

Brownish yellow from throat to underparts and rump

Brownish-yellow wing coverts, black wing feathers, white inner edge of tertials

Dark-breasted Rosefinch

Rosy forehead connects with supercilium

♂
Blackish-red eye stripe and breast band

Straight pointed conical bill, differing from other rosefinches

♀
All blackish-brown body

Two grayish-brown wing bars

Plain Mountain Finch

Brown head

juv.

Buff-white supercilium

Distinct white wing patch

ad.

Black-headed Mountain Finch

Resembles Plain Mountain Finch, but lacks streaks on crown and has narrower streaks on upperparts and larger pale wing patches

Pale cheek contrasts with brown crown and nape

L. b. margaritacea

Pale head and black forehead

Pink on wing coverts

Subspecies with darkest body; more extensive black on head

L. b. brandti

L. b. walteri

Asian Rosy-Finch

L. a. brunneonucha

Brownish yellow on side of neck

Pinkish-red fringe on wing feathers and coverts

Wing feathers and coverts lack pinkish red

L. a. arctoa

Darker body, nearly brown

Common Rosefinch

♂
Dark rosy head and throat

C. e. roseatus

Red on underparts extends to belly

♂
Bright red head and throat

Upperparts dark red with some grayish brown

Red on underparts extends to flanks

C. e. grebnitskii

607

Scarlet Finch

Yellow bill

♂

Bright red body

Yellowish-brown head and back, fine small black spots on back

♀

Black wing and tail feathers

Citrine rump

Gray underparts

Black wings and tail feathers

Blyth's Rosefinch

Rear part of supercilium whitish

♂

Relatively large area of silvery spots on side of face

Primary projection relatively long

White supercilium

♀

Slightly thin bill

Relatively pale body

Larger than Red-mantled Rosefinch

Streaked Rosefinch

Back darker than Great Rosefinch with more streaks; smaller white spots on head and breast

♂

Extremely heavy streaks on dark body, almost without supercilium

♀

Dark legs

Red-mantled Rosefinch

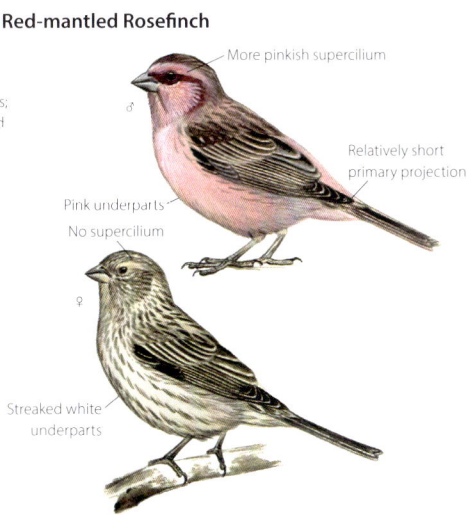

More pinkish supercilium

♂

Relatively short primary projection

Pink underparts

No supercilium

♀

Streaked white underparts

Great Rosefinch

Pink ear coverts

♂

Pinkish-brown upper back, almost no streaks

♀

Inconspicuous narrow streaks on upper- and underparts

Pink-browed Rosefinch

Dark red crown lacks dark streaks

Broad rosy forehead and supercilium

♂

Pinkish-brown upperparts

Bright pink underparts and rump

Buff-yellow supercilium

♀

Relatively thick streaks on olive-brown body

Buff-yellow underparts

Carpodacus pulcherrimus
红眉朱雀（喜山红眉朱雀）
hóng méi zhū què (xǐ shān hóng méi zhū què)

Himalayan Beautiful Rosefinch LC

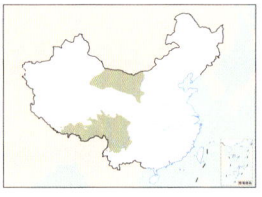

L: 14–15 cm

Habitat: Inhabits dwarf forests, scrubland, alpine meadows, and scree at high altitudes.

Behavior: Forages in pairs or small flocks on ground; often joins other rosefinches in mixed-species flocks.

Distribution: *C. p. pulcherrimus* seen in S and E Tibet; *C. p. argyrophrys* seen in NW Yunnan, C, SW, N, and W Sichuan, Shaanxi, Qinghai, NW and C Gansu, Ningxia, and W Inner Mongolia.

Voice: Song unknown. Call a *chillip* or *tip*, both very similar to calls of Chinese Beautiful Rosefinch.

Carpodacus davidianus
中华朱雀（红眉朱雀）
zhōng huá zhū què (hóng méi zhū què)

Chinese Beautiful Rosefinch NR

L: 14–15 cm

Habitat: Inhabits edges of oak and pine forests at middle and high altitudes; also occurs in river-valley bushes, alpine meadows, and scree.

Behavior: In pairs or small flocks. Active, not shy, and easy to locate; forages on the ground and in the understory.

Distribution: Endemic to mountains in North China; seen in W Beijing, Shanxi, N Shaanxi, N Hebei, and SE Inner Mongolia.

Voice: Song a rapidly repeated *wissiwissiwissiwi*, repeated every 3–4 sec and often accompanied by call notes, which are more metallic and subtly less dry than call of Himalayan Beautiful Rosefinch. Calls include a shrill *tsink*, repeated once or twice, and also *tsink-it*, with second syllable higher and softer.

Carpodacus waltoni
曙红朱雀 shǔ hóng zhū què

Pink-rumped Rosefinch LC

L: 12–15 cm

Habitat: Inhabits mixed and coniferous forest edges at high altitudes; also in alpine bushes, meadows, and shrubs in farmland.

Behavior: Often gathers in small to large flocks in open habitats. Forages on ground. Often joins other rosefinches in mixed-species flocks.

Distribution: E Qinghai-Tibet Plateau and nearby mountains. *C. w. waltoni* in S and SE Tibet; *C. w. eos* seen in NW Yunnan, C, N, and W Sichuan, and S and SE Qinghai.

Voice: Song a coarse, buzzy *dzzzzeet*, repeated with little variation about once a second. Calls include a Rufous-breasted Accentor–like rattle, *tvitt-itt-itt*.

Carpodacus edwardsii
棕朱雀 zōng zhū què
Dark-rumped Rosefinch LC

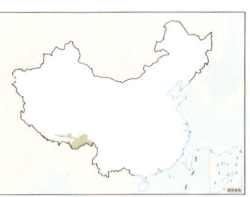

L: 14–17 cm

Habitat: Inhabits broadleaf, mixed, and coniferous forests at middle and high altitudes.

Behavior: Often solitary or in pairs in understory bushes; also seen at forest edges. Primarily a forest dweller; secretive.

Distribution: *C. e. rubicundus* in S and SE Tibet; *C. e. edwardsii* in W and NW Yunnan, N, C, and S Sichuan, and S Gansu.

Voice: Song unknown. Calls include a short, explosive, rising *twink*.

Carpodacus rodopeplus
点翅朱雀（喜山点翅朱雀）
diǎn chì zhū què (xǐ shān diǎn chì zhū què)

Spot-winged Rosefinch LC

L: 13–14 cm

Habitat: Inhabits forest edges, alpine meadows, scrubland, and bamboo thickets at middle and high altitudes.

Behavior: In pairs or small flocks. Forages on the ground. Also joins other rosefinches, accentors, and mountain finches in mixed-species flocks.

Distribution: Seen in the C Himalayas in S Tibet.

Voice: Calls include an unusual rising *pler-wee*, reminiscent of Pink-browed Rosefinch but thinner, buzzier, less nasal, and markedly two-tone.

Carpodacus verreauxii
淡腹点翅朱雀（点翅朱雀）
dàn fù diǎn chì zhū què (diǎn chì zhū què)

Sharpe's Rosefinch LC

L: 13–15 cm

Habitat: Inhabits forest-edge bushes of broadleaf, mixed, and coniferous forests, sparse woodlands, farmland, and orchards at middle and high altitudes.

Behavior: Often in pairs or small flocks in open areas at forest edges; not shy.

Distribution: Seen in southwestern China, including N, C, and SW Sichuan, W and NW Yunnan, and E Tibet.

Voice: Call is a short sharp *spink*.

Himalayan Beautiful Rosefinch

Pink underparts

Relatively narrow streaks on underparts, a few or inconspicuous streaks on flanks

Prominent streaks on brown upperparts

Relatively narrow streaks on underparts, a few or inconspicuous streaks on flanks

Dark-rumped Rosefinch

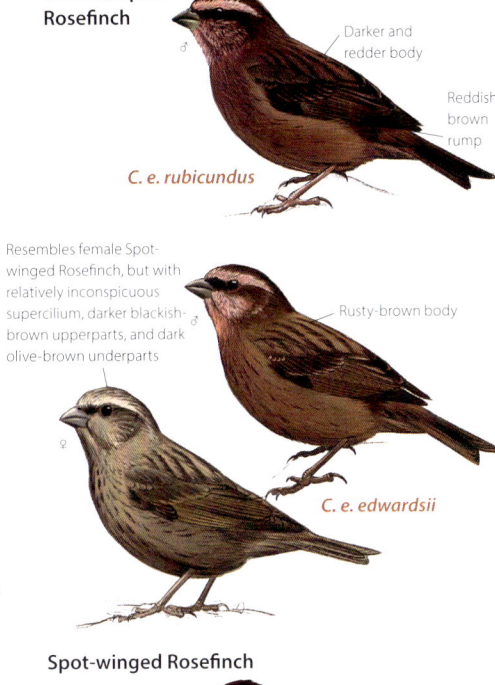

Darker and redder body

Reddish-brown rump

C. e. rubicundus

Resembles female Spot-winged Rosefinch, but with relatively inconspicuous supercilium, darker blackish-brown upperparts, and dark olive-brown underparts

Rusty-brown body

C. e. edwardsii

Chinese Beautiful Rosefinch

Thicker stronger bill

Distinctively thick streaks on brown upperparts

Pink underparts

More grayish-white and less brown upperparts

Relatively thick streaks on underparts, a few or inconspicuous streaks on flanks

Spot-winged Rosefinch

Rosy to pinkish-white streaks on rufous-brown back

Relatively small area of pink on chin and throat

Spotty pinkish-white wing patch

Distinct buff-yellow wing patch

Pink-rumped Rosefinch

Brown on upperparts slightly paler than Himalayan Beautiful Rosefinch, with more distinct streaks

Pale rosy underparts with relatively narrow streaks, densely distributed on flanks

Relatively narrow streaks on underparts, densely distributed on flanks

Sharpe's Rosefinch

Brownish back, without pink, with rosy to pinkish-white streaks

Larger area of pink on chin and throat

Pinkish-brown wing patch, less distinct than Spot-winged Rosefinch

Pale pink on breast and belly

Larger area of pale red on rump

Prominent supercilium

Grayish-brown wing patch, less distinct than Spot-winged Rosefinch

Carpodacus vinaceus

酒红朱雀 jiǔ hóng zhū què

Vinaceous Rosefinch

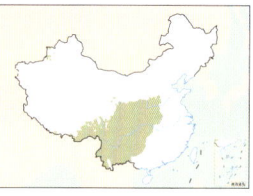

L: 13–15 cm

Habitat: Inhabits understory bushes of broadleaf, mixed, and coniferous forests, edges of secondary forests, scrubland, and bamboo thickets at low and high altitudes; also seen in villages, grasslands, and trees near farmland.

Behavior: Often solitary or in pairs. Forages in open habitats in the understory and middle story of vegetation.

Distribution: Seen in western North China, western Central China, and southwestern China.

Voice: Calls include a short, sharp, high-pitched, buntinglike *tink* or *zik*.

Carpodacus formosanus

台湾酒红朱雀 tái wān jiǔ hóng zhū què

Taiwan Rosefinch

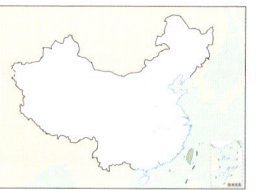

L: 14–16 cm

Habitat: Inhabits broadleaf forests, mixed broadleaf-coniferous forests, coniferous forests, alpine scrubland, and bamboo thickets at middle and high altitudes.

Behavior: Often solitary or in pairs in understory bushes and open habitats. Not active; seldom joins mixed-species flocks.

Distribution: Chinese endemic; seen in mountains in C Taiwan.

Voice: Vocalizations like Vinaceous Rosefinch, although very brief *zip* call is slightly longer and higher pitched.

Carpodacus stoliczkae

沙色朱雀 shā sè zhū què

Pale Rosefinch

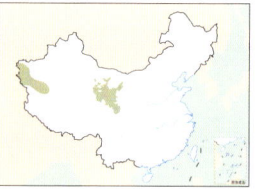

L: 14–16 cm

Habitat: Inhabits deserts and scree at middle and high altitudes; also in farmland and rocky habitats near villages.

Behavior: Often in pairs or small flocks in open dry habitats with few trees; forages on ground.

Distribution: *C. s. stoliczkae* in W Xinjiang; *C. s. beicki* in W and E Qinghai and C Gansu.

Voice: Song is a short, simple, understated trill, *zwee-zwee-wee*, becoming louder with each note and vaguely reminiscent of the introductory notes of the song of Richard's Pipit but lacking the power. Calls include a short rising *trizp*, often doubled, and a slightly mellow *tip*.

Carpodacus roborowskii

藏雀 zàng què

Tibetan Rosefinch

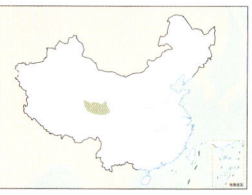

L: 17–18 cm

Habitat: Inhabits grasslands, deserts, meadows, and scree at high altitudes.

Behavior: Often solitary or in pairs. Forages on ground. Often joins other plateau birds, such as snowfinches and rosefinches, in mixed-species flocks.

Distribution: Chinese endemic; seen in SW Qinghai and NE Tibet.

Voice: Quiet but gives subdued, mellow, sparrowlike *tup* and a more penetrating, slightly ringing *tlip-tlip-tlip* note.

Carpodacus sillemi

褐头朱雀（褐头岭雀）

hè tóu zhū què (hè tóu lǐng què)

Sillem's Rosefinch

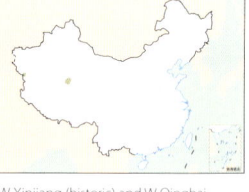

L: 18 cm

Habitat: Inhabits alpine meadows, deserts, and scree at high altitudes.

Behavior: Similar to Tibetan Rosefinch.

Distribution: Endemic to the Qinghai-Tibet Plateau. Once placed within *Leucosticte.* Very local and rare in SW Xinjiang (historic) and W Qinghai.

Voice: Poorly known. One call includes a simple, fine, descending *tiuu* or *ziuu* and a slightly shorter, burrier *zhie*.

Carpodacus sibiricus

长尾雀 cháng wěi què

Long-tailed Rosefinch

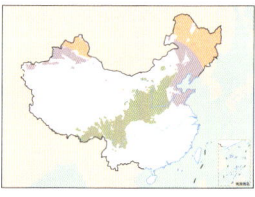

L: 14–18 cm

Habitat: Inhabits edges of woodlands, mixed forests, and broadleaf forests in temperate to polar regions; also in bushes, parks, and farmland.

Behavior: Solitary or in small flocks in understory and middle story; forages in bushes and small trees.

Distribution: *C. s. sibiricus* breeds in N Heilongjiang and NE Inner Mongolia and winters in Shanxi and Xinjiang; *C. s. ussuriensis* breeds in S and E Heilongjiang, Jilin, Liaoning and migrates through or winters in Beijing, N Hebei, Henan, Shandong, and C Inner Mongolia; *C. s. henrici* is resident in E Tibet, NW Yunnan, Sichuan, and Chongqing; *C. s. lepidus* is resident in W Hebei, W Beijing, Henan, SW Shanxi, Shaanxi, S Gansu, E Tibet, and E Qinghai. Some consider *C. s. henrici* and *C. s. lepidus* a separate species.

Voice: Rapid song is a short musical series of rippling high-pitched trills, *churu chiru fee fee fee*, lasting almost a second and repeated about every 2 sec. Calls may vary racially (more research needed) and include a very short, rising *wit*, given singly, in pairs, or in a lengthy series by both *C. s. ussuriensis* and *C. s. sibiricus*, and a slightly longer, also rising *whet*, *wit-it*, or *whit-t-leeoot*, given by *C. s. lepidus* and *C. s. henrici*. At least *C. s. ussuriensis* and *C. s. sibiricus* also give a complex sparrowlike *tlooet* when agitated.

Vinaceous Rosefinch

Pinkish-white supercilium

Wine-red body

♂

Pale fringe of tertials

♀

Tibetan Rosefinch

Dark red head, appears blackish

Brown to yellow bill

♂

Fine white streaks on chin and throat

Bright yellow bill

Primaries reach about halfway down closed tail

♀

Grayish-brown body with fine streaks, not brown

Sillem's Rosefinch

Bill resembles Tibetan Rosefinch but weaker, smaller and brown ♂

Resembles Tibetan Rosefinch, but streaks paler brown

Yellow bill

♀

White from lower belly to undertail, lacks streaks

Taiwan Rosefinch

Pinkish-white supercilium

Wine-red body, more purplish than Vinaceous Rosefinch

♂

Pale Rosefinch

White supercilium extends backward and curls on hindneck

Sandy-brown upperparts

♂

Dark crimson face

Long-tailed Rosefinch

Rosy forehead, white crown mixes with fine rosy color

♂

Broad white wing patch

Prominent thick streaks on dark blackish-brown body

Long tail

♀

C. s. sibiricus/C. s. ussuriensis

White forehead mixes with rosy color

♂

White cheek

Bill shorter than other rosefinches

♀

Second wing bar broader

C. s. henrici/C. s. lepidus

Pale grayish-brown body; fine streaks on breast and upperparts, inconspicuous streaks on underparts

Carpodacus roseus
北朱雀 běi zhū què
Pallas's Rosefinch

L: 15–17 cm
Habitat: Inhabits temperate sparse woodlands, coniferous forests, rhododendron thickets, and scrubland; prefers open woodlands and scrubland with fewer trees.
Behavior: Gathers in small flocks and forages in low bushes and on ground; often mixes with other small birds, such as rosefinches and accentors.
Distribution: Common winter visitor. Seen in Northeast, North, East, and Central China; also seen in N Xinjiang during migration.
Voice: Call a short, intense, penetrating, Hawfinch–like *zrri*, sometimes given in a short series. Song is essentially unknown: reputedly fluty and reminiscent of Pine Grosbeak or Great Rosefinch.

Carpodacus trifasciatus
斑翅朱雀 bān chì zhū què
Three-banded Rosefinch

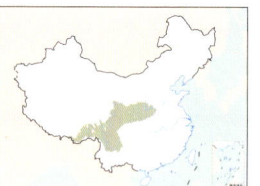

L: 17–19 cm
Habitat: Inhabits edges of broadleaf forests, mixed forests, and coniferous forests at middle and high altitudes; also occurs in bushes and small trees around gardens and farmland.
Behavior: Often solitary or gathers in small flocks in open bushes; often joins other small birds, such as rosefinches and mountain finches, in mixed-species flocks.
Distribution: Endemic to mountains in southwestern China; seen in SW Gansu, S Shaanxi, N, C, W, and SW Sichuan, NW and W Yunnan, and SE Tibet.
Voice: Calls include a high-pitched Sichuan Leaf Warbler–like *fwisp*, often given in series of four or five notes.

Carpodacus thura
喜山白眉朱雀 xǐ shān bái méi zhū què
Himalayan White-browed Rosefinch

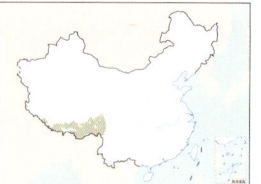

L: 16–18 cm
Habitat: Inhabits coniferous forests, sparse woodlands, alpine meadows, forest-edge bushes, and alpine scree at high altitudes.
Behavior: Gathers in small flocks and moves about in understory of vegetation and on ground; active and not shy.
Distribution: S Tibet.
Voice: Rarely heard song is apparently a short series of loud whistles followed by three or four short warbled notes and then several longer whistles. Calls vaguely reminiscent of calls of Chinese White-browed Rosefinch.

Carpodacus puniceus
红胸朱雀 hóng xiōng zhū què
Red-fronted Rosefinch

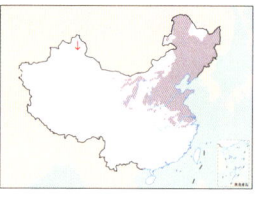

L: 19–22 cm
Habitat: Inhabits alpine meadows, bushes, and scree on plateaus and mountains at high altitudes.
Behavior: Often in pairs or small flocks. Forages on ground. Bill stronger than other rosefinches and better at digging.
Distribution: *C. p. kilianensis* in S and W Xinjiang; *C. p. puniceus* in S and E Tibet and NW Sichuan; *C. p. sikangensis* in NW Yunnan and W and SW Sichuan; *C. p. longirostris* in N Sichuan and SE Gansu.
Voice: Loud song is a far-carrying, slightly falling series of six or seven short whistles, *twi-di-di-di-di—di-diu*, or "Are you quite ready?" Calls include a buzzy *dzzzr*, given singly or in intensifying, accelerating series.

Carpodacus subhimachalus
红眉松雀 hóng méi sōng què
Crimson-browed Finch

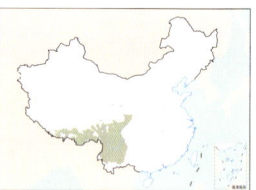

L: 16–21 cm
Habitat: Inhabits mixed broadleaf-coniferous forests, coniferous forests, forest-edge bushes, rhododendron thickets, bamboo thickets, and grassy patches at middle and high altitudes.
Behavior: Often in pairs; also gathers in small flocks in the understory and middle story of vegetation. Quiet but not shy.
Distribution: Seen in southwestern China, including SE and S Tibet, NW and W Yunnan, and C, W, and SW Sichuan.
Voice: Song an often-loud, rapid, scratchy series of 5–15 shrill warbled notes, repeated with only modest changes every 5–15 sec. Rarely heard calls include a mellow *stup*, similar to the call of Sulphur-bellied Warbler.

Carpodacus dubius
白眉朱雀 bái méi zhū què
Chinese White-browed Rosefinch

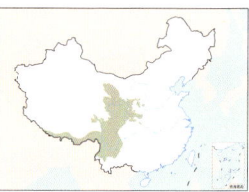

L: 16–18 cm
Habitat: Inhabits mixed forests, coniferous forests, forest-edge bushes, and alpine meadows at middle and high altitudes.
Behavior: In pairs or small flocks at forests edges. Forages in open bushes and on ground. Sings from high places; calls are unique.
Distribution: Endemic to plateaus and mountains in midwestern China. *C. d. femininus* in NW Yunnan, S and SE Tibet, SE Qinghai, and W Sichuan; *C. d. deserticolor* in S and E Qinghai and W Inner Mongolia; *C. d. dubius* in W Gansu, Ningxia, NE Qinghai, and W Sichuan.
Voice: Rarely heard song is a ringing *pleeu-tit* by *C. d. dubius* and a *pleuw* or *pleuw-pleuw* by *C. d. femininus*. Calls are a nasal, sheeplike bleating, similar to those of Himalayan White-browed Rosefinch but distinctly flatter, lower pitched, and delivered much faster (with an impossible-to-count 16 notes in a one-second phrase—twice as many as Himalayan).

Pallas's Rosefinch

Scaly white pattern on forehead and throat

Red breast and flanks, white belly

Pinkish-tinged forehead, lower cheek, throat, and breast

Pink rump

Red-fronted Rosefinch

Bright red supercilium and forehead

Bill length is more than double the width of bill base

Bright red chin, throat, and upper breast

Pink to bright red rump

Olive-yellow rump

Three-banded Rosefinch

Characteristic three "wing" bars, including one formed by white outer scapulars

White speckles on forehead, cheek, and throat

Dark rosy body

Olive-yellow breast and rump

Crimson-browed Finch

Bright red supercilium

Fine white spots on red throat and breast

Pure dark gray lower breast and belly

Olive-yellow supercilium

Gray face

Olive-yellow breast

Himalayan White-browed Rosefinch

Rosy color covers almost entire forehead

Broad supercilium tinged white

Brown patch between supercilium and pink cheek

Distinct olive yellow on front part of supercilium

Brown patch between supercilium and cheek

Distinct pale yellowish brown from throat to breast

Chinese White-browed Rosefinch

Broad white supercilium from forehead to nape; some rosy color at mid-part of white supercilium and on forehead

Brown patch between white supercilium and pink cheek does not extend to eye; pink behind eye

No distinct olive yellow on front part of supercilium

Brown patch between white supercilium and cheek does not extend to eye; streaky behind eye

Dark brown streaks on grayish-white breast and belly; no other colors

Chloris chloris

欧金翅雀 ōu jīn chì què

European Greenfinch　LC

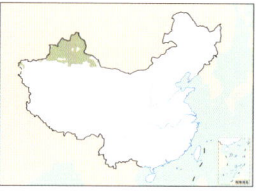

L: 14–16 cm

Habitat: Inhabits various types of forests, including montane coniferous, broadleaf, and mixed forests, as well as river-valley forests, urban parks, countryside woodlands, and natural forests in deserts and semideserts on plains.

Behavior: In pairs during breeding season; gathers in flocks of several to dozens of individuals in nonbreeding season. Often mixes with European Goldfinch. Calls loudly in spring; easy to be found. Forages on ground; usually perches on trees or power lines. Adaptable to various habitats, from deserts to alpine woodlands.

Distribution: *C. c. turkestanica* is locally common to uncommon resident in N and W Xinjiang; some populations in north migrate southward during fall. Rapid range expansion after being first recorded in Xinjiang (Ili) in the early 1990s; now widespread in N Xinjiang and the northwestern edge of the Tarim Basin in S Xinjiang.

Voice: Rapid canarylike song begins with a dry nasal trill and a slowly rising, accelerating series of *teu-teu-teu-teu* notes and incorporates twittering and whistles mixed with nasal buzzes or wheezing *tzzzwee*. Wide variety of calls include a mellow *chut-chut-chut*, harsher *chit*, and rising *teu-teu*.

Chloris sinica

金翅雀 jīn chì què

Oriental Greenfinch　LC

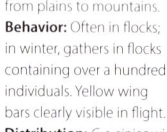

L: 12–14 cm

Habitat: Inhabits bushes, forests, and urban parks from plains to mountains.

Behavior: Often in flocks; in winter, gathers in flocks containing over a hundred individuals. Yellow wing bars clearly visible in flight.

Distribution: *C. s. sinica* widely across China except for Northeast China, Xinjiang, Tibet, and Hainan; common. *C. s. ussuriensis* in Northeast China. *C. s. kawarahiba* in Taiwan.

Voice: Song similar to that of European Greenfinch—a mix of chattering notes, wheezes, and calls but generally faster, with fewer pauses, and more repetition. Species-specific notes include a distinctive *tiu-tiu-tiu-tiu*, while the buzzes are shorter and do not descend.

Chloris spinoides

高山金翅雀 gāo shān jīn chì què

Yellow-breasted Greenfinch　LC

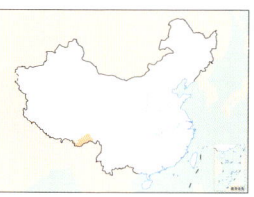

L: 13–14 cm

Habitat: Inhabits alpine coniferous forests or forest edges at altitudes of 2000–4000 m; in winter, migrates to mountains, river valleys, and the vicinity of farmland and villages at lower altitudes.

Behavior: Often gathers in small flocks in the canopy of trees or bushes in farmland. Flight rapid, undulating.

Distribution: Local and occasional in S Tibet.

Voice: Vocalizations are reminiscent of the other greenfinches, especially Black-headed Greenfinch, but distinctive elements include a nasal wheeze that rises initially, then descends and falls away; rising *djoo-ee* notes that are more similar to European Goldfinch; and chittering notes given in a series of often four *tick'y't'tu* notes that don't descend and where the second note is stressed. A species-specific call, *quieoo*, is more reminiscent of Scarlet Finch than any other greenfinch.

Chloris ambigua

黑头金翅雀 hēi tóu jīn chì què

Black-headed Greenfinch　LC

L: 12–14 cm

Habitat: Inhabits subalpine coniferous forests and forest edges at altitudes above 1500 m; also near farmland and villages. In winter, migrates to mountains at lower altitudes.

Behavior: Often gathers in flocks containing over a hundred individuals. Forages in the canopy of tall trees or near farmland; rapidly flies to tall trees when disturbed.

Distribution: *C. a. taylori* in S and E Tibet; *C. a. ambigua* in Yunnan, W Sichuan, NW Guizhou and NE Qinghai.

Voice: Vocalizations are reminiscent of the other greenfinches, particularly Yellow-breasted Greenfinch, but distinctive elements include a wheeze that is almost as long as that of European Greenfinch but is thin, flat, and lower pitched. The series of twittering notes are generally higher pitched and given in a descending series of often five to eight notes, while a rising *dweeoo* is apparently species specific.

Rhodospiza obsoleta

巨嘴沙雀 jù zuǐ shā què

Desert Finch　LC

L: 13–14 cm

Habitat: Prefers dry saxaul forests, desert poplar forests, urban parks, street trees, and countryside gardens.

Behavior: Often in pairs in breeding season; otherwise in flocks. Forages on ground; perches in dense bushes or trees when resting. Often wanders in a large range during winter.

Distribution: N Shaanxi, W Inner Mongolia, Ningxia, Gansu, Xinjiang, and Qinghai.

Voice: Song a series of rambling, disconnected buzzes and harsher trills. Flight call is a melodious *prrrrp'whikku-prp-chink-miuw-zank-uuuhrrr*, with elements reminiscent of both Eurasian Linnet and European Greenfinch. Calls include a quiet, soft, purring *prrrr-r-r* or *prrrt*, a sharper *pink*, a *shreep*, and a rolling *kittrr*.

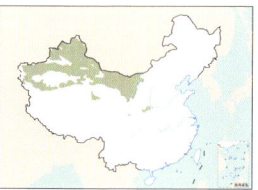

European Greenfinch

Green head and upper back lack streaks

Yellowish-green underparts

♂

More yellow on outer tail feathers

♀

More brown on head and upper back, with diffuse streaks

Dark green underparts

Less yellow on outer tail feathers

Yellow-breasted Greenfinch

Yellow supercilium

Black patch on ear coverts

Bright yellow throat, breast, and belly

♂

♀

Pale brown supercilium, breast, and belly

juv.

Small yellow wing patch

Oriental Greenfinch

Black patch at lores

Slate-gray head

♂

Large yellow wing patch

Gray head

♀

Pale grayish-brown streaks on breast

Black-headed Greenfinch

Black head

Relatively small yellow wing patch

Olive-brown belly

Desert Finch

Black lores

Large black bill

No black lores

Black bill

♂

♀

Linaria flavirostris

黄嘴朱顶雀 huáng zuǐ zhū dǐng què

Twite LC

L: 12.5–14 cm

Habitat: In breeding season, inhabits primarily small trees, bushes, meadows, and rocky slopes on high mountains and plateaus at altitudes above 3000 m; also inhabits deserts and semideserts with sparse vegetation.

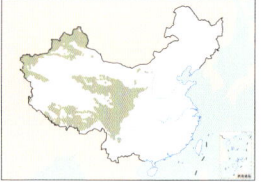

Behavior: Prefers gathering in flocks. In pairs during breeding season; in other seasons, often gathers in large flocks with several or several dozen individuals. Active in sparse trees, bushes, bare rocks, and meadows near mountains, river valleys, streams, or lakes. In winter, wanders at low altitudes in farmland. Moves about and forages on ground; rests on bushes or protruding rocks.

Distribution: *L. f. korejevi* in N Xinjiang; *L. f. montanella* in S Xinjiang; *L. f. rufostrigata* in Tibet; *L. f. miniakensis* in W Inner Mongolia, Ningxia, NW Gansu, E Xinjiang, Qinghai, and W Sichuan.

Voice: Song consists of fast trills, buzzes, low rattles, *trrrrrr*, and some call notes. Calls include a drawn-out, slightly rising *tweeet* and redpoll-like *tup-tup-up* notes.

Linaria cannabina

赤胸朱顶雀 chì xiōng zhū dǐng què

Eurasian Linnet LC

L: 12.5–14 cm

Habitat: Inhabits primarily forest edges, logged forests, barren mountains, and sparse woodlands from foothills to middle altitudes, as well as river valleys, rocky slopes, farmland, roadsides, and

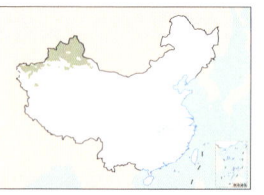

similar open regions with sparse trees or bushes; also occurs in orchards and gardens. Avoids forests and grasslands.

Behavior: In pairs during breeding season; often gathers in small flocks containing several to a dozen individuals in other seasons. Wanders and forages on ground or in bushes; perches on small trees or bushes when resting.

Distribution: Locally common to uncommon resident in N and W Xinjiang.

Voice: Song is a soft, fast twitter, with whistles and trills interspersed with call notes. Calls include a rapid trilled *chi-chi … chi-chi-chi-it … chi-chit*.

Acanthis flammea

白腰朱顶雀 bái yāo zhū dǐng què

Common Redpoll LC

L: 11.5–14 cm

Habitat: Inhabits broadleaf, mixed, and artificial forests during migration and winter.

Behavior: Often forages in flocks in trees, bushes, or on ground.

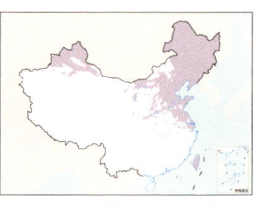

Distribution: Locally common in Northeast, North, and northwestern China. Occasionally seen in the Yangtze Plain and even south to Taiwan.

Voice: Distinctive flight call is a rapid series of *chet-chet-chet-chet* notes. Song a dry buzzy trill, *seerrrrrrr*.

Acanthis hornemanni

极北朱顶雀 jí běi zhū dǐng què

Hoary Redpoll NR

L: 12–14 cm

Habitat: Inhabits broadleaf forests and scrubland in winter.

Behavior: Gathers in flocks in trees; also forages on ground.

Distribution: Uncommon in Northeast and northwestern China. Sometimes treated as a subspecies of Common Redpoll.

Voice: Very similar to Common Redpoll.

Carduelis carduelis

红额金翅雀 hóng é jīn chì què

European Goldfinch LC

L: 12–13.5 cm

Habitat: Inhabits montane coniferous forests, broadleaf forests in river valleys, bushes, farmland, and vegetated areas.

Behavior: Gathers in large flocks containing several dozen to over a hundred individuals outside breeding season.

Distribution: *C. c. subulata* is common resident in the Altai Mountains in N Xinjiang; *C. c. paropanisi* is resident in W and C Xinjiang and NW Gansu; *C. c. caniceps* is resident in W Tibet; *C. c. frigoris* is an uncommon winter visitor in N Xinjiang and occasionally seen in W Inner Mongolia.

Voice: Varied, cheerful, tinkling, or twittering calls include a *pee-uu*, somewhat reminiscent of Eurasian Siskin, while the rapidly delivered song is essentially an erratic combination of call notes.

Twite

Tiny gray bill

Streaked brown head

Streaked underparts

Yellow bill

br.

Pink rump

non-br.

Common Redpoll

Red forehead

♂

Relatively heavy streaks on upperparts, also streaks on rump

Streaks on flanks

♀

Streaks on rump

Eurasian Linnet

Red forehead

♂

Chestnut-brown back

Red upper breast

Relatively large gray bill

♀

Black streaks on white throat

Hoary Redpoll

♂

Resembles Common Redpoll, with paler plumage and no streaks on rump or undertail coverts

♀

No streaks

European Goldfinch

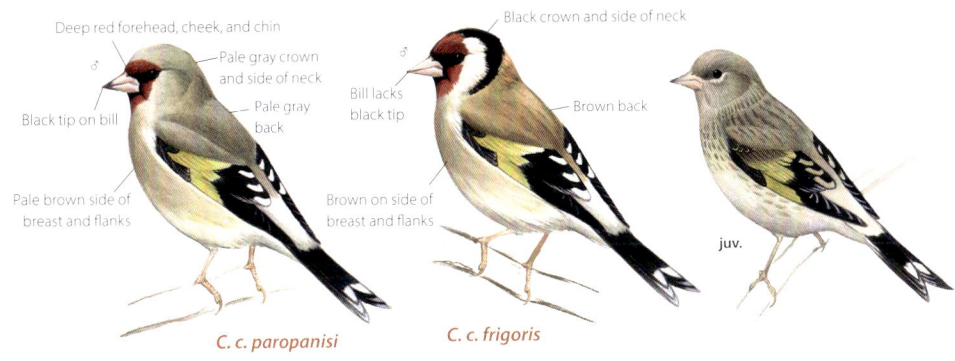

Deep red forehead, cheek, and chin

♂

Pale gray crown and side of neck

Black tip on bill

Pale gray back

Pale brown side of breast and flanks

Black crown and side of neck

♂

Bill lacks black tip

Brown back

Brown on side of breast and flanks

juv.

C. c. paropanisi

C. c. frigoris

619

Loxia curvirostra
红交嘴雀 hóng jiāo zuǐ què
Red Crossbill **LC**

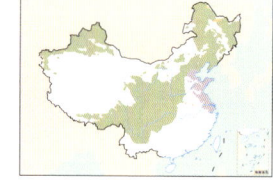

L: 15–17 cm

Habitat: In summer, inhabits coniferous forests at middle and high altitudes; in winter, descends to coniferous forests or mixed forests at low altitudes.

Behavior: Solitary or in pairs; also forages on ground. Always calls in flight.

Distribution: *L. c. curvirostra* in Qinghai; *L. c. tianschanica* in Xinjiang; *L. c. himalayensis* in S Tibet, Qinghai, NW Yunnan, Sichuan, Gansu, Chongqing, and W Hubei; *L. c. japonica* in Northeast, North, Central, and East China. Locally common.

Voice: Most frequent call is a hard high-pitched *chip* or *chip chip*, but there is considerable variation, some of it no doubt geographic.

Loxia leucoptera
白翅交嘴雀 bái chì jiāo zuǐ què
White-winged Crossbill **LC**

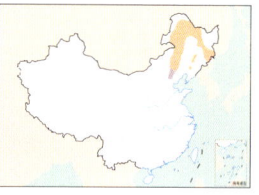

L: 14.5–16 cm

Habitat: Often inhabits coniferous forests.

Behavior: Solitary or in pairs. Often wanders. Noisy.

Distribution: Uncommon in Northeast China, occasionally in North China.

Voice: "Chip" call is weaker and higher pitched than call of Red Crossbill; also gives a distinctive, nasal, trumpeting *pleep*, often in a slow sequence or combined with calls.

Serinus pusillus
金额丝雀 jīn é sī què
Fire-fronted Serin **LC**

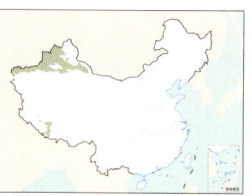

L: 11–13 cm

Habitat: Inhabits coniferous forests, montane grasslands, broadleaf forests, and river-valley bushes; especially prefers habitats near residential areas of herders.

Behavior: Gathers in small flocks during breeding season, making noisy calls; in fall and winter, wanders in large flocks containing several dozen to over a hundred individuals. Bold and unafraid of people.

Distribution: Locally common to uncommon resident in C and SW Xinjiang and SW Tibet.

Voice: European Goldfinch–like song is a fast series of rippling trills combined with softer twittering. Calls include a rippling trill, *trrrrr*, and a mellow *dueet*.

Spinus thibetanus
藏黄雀 zàng huáng què
Tibetan Serin **LC**

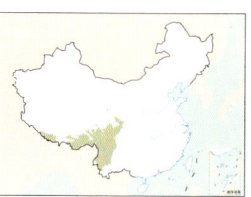

L: 10–12 cm

Habitat: Inhabits alpine coniferous forests, mixed broadleaf-coniferous forests, and surrounding bushes in breeding season; in nonbreeding season, descends to similar habitats at low altitudes.

Behavior: Forages in bushes near ground or on treetops. Active when foraging; often moves between trees. Usually in pairs or small flocks but also gathers in flocks containing hundreds of individuals in winter.

Distribution: Seen in S and SE Tibet, W and NW Yunnan, Xinjiang, and W Sichuan.

Voice: Hurried, even frenzied, twittering song phrases, basically an extended version of various repeated call notes. Songs can last up to 10 sec and include nasal buzzes, wheezes, and redpoll-like chittering, with most including a buzzing *choo-loot-chweeeeze*, repeated two or three times, similar to Eurasian Siskin but containing far less mimicry.

Spinus spinus
黄雀 huáng què
Eurasian Siskin **LC**

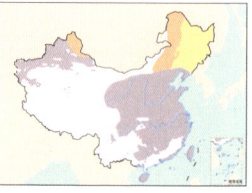

L: 11–12 cm

Habitat: Inhabits various types of forests at middle and low altitudes.

Behavior: Migratory; gathers in flocks in nonbreeding season. Active. Primarily arboreal.

Distribution: Widespread and seasonally common across China except for Ningxia and Tibet.

Voice: Song a jumbled mix of thin or high-pitched twitters, trills, wheezes, and often-considerable mimicry. Call a distinctive, high-pitched, ringing *toolee* or *tsuu-ee*, often given in flight.

Red Crossbill

Upper and lower mandible crossed

No wing bar

♂

Orangish body

♀

L. c. tianschanica

Red body

♂

♀

L. c. himalayensis

White-winged Crossbill

Upper and lower mandible crossed

Two prominent white wing bars

juv.

♂

♀

♂

♀

Fire-fronted Serin

Bright red forehead

Black head

Dense black streaks on breast

ad.

juv.

Tibetan Serin

More extensive yellow on head and body

♂

Black wing feathers

♀

More heavily black-streaked belly and back

Eurasian Siskin

Black crown and chin

♂

Yellowish-green upperparts

Yellow wing bar on black wings

Yellow rump

Yellow breast and white belly

♀

Dusky streaks on body

Black streaks on flanks and undertail coverts

铁爪鹀科 Calcariidae

Small ground-dwelling songbirds. Plumage similar or slightly different between sexes. Bill short and pointed, resembling bills of buntings. Inhabit primarily open habitats, including tundra, grasslands, seashores, alpine meadows, and agricultural fields. Nest on ground or among bushes. Feed mostly on plant seeds and fruits; also take insects and insect larvae.

Longspurs and snow buntings were once placed in Emberizidae but are now in their own family. Three genera and six species recognized worldwide, distributed in the higher latitudes of the Northern Hemisphere. Two genera and two species recorded in China, both as wintering birds in Northeast and northwestern China and occasionally south to the Yangtze River Watershed.

Lapland Longspur

Snow Bunting

Calcarius lapponicus

铁爪鹀 tiě zhǎo wú

Lapland Longspur (LC)

L: 14–18 cm
Habitat: In breeding season, inhabits open tundra and meadows at high latitudes; in winter, often inhabits open plains, grass fields, and desert steppes at low altitudes.
Behavior: Often gathers

in large flocks containing dozens to over a hundred individuals in winter; forages for grass seeds or grains on ground. Flies in tight formations at high speed.
Distribution: Winters in Northeast, North, East, and Central China and Sichuan; vagrant to Fujian (Ningde) and Taiwan.
Voice: Calls include a falling *tiew* and a dry rattle, *tr'r'r'r'r*, both given in flight and often combined. Both recall Snow Bunting.

Plectrophenax nivalis

雪鹀 xuě wú

Snow Bunting (LC)

L: 16–18 cm
Habitat: Inhabits open tundra and wet grasslands in breeding season; migrates through, and winters in, scrub and farmland on plains or low mountains.
Behavior: Often gathers

in flocks on ground in winter and forages for grass seeds. Flies a distance when disturbed and then falls back to ground and continues feeding.
Distribution: Winters in Northeast China and W Xinjiang; occasionally seen in northern North China, Jiangsu, and Taiwan.
Voice: Call include a descending *chwee* and rattling *tr'r'r'r'r*, both similar to Lapland Longspur.

鹀科　Emberizidae

Small ground-dwelling songbirds. Plumage differs slightly between sexes; mostly brown, black, white, rufous, yellow, and slate. Upper body usually heavily streaked. Bill short, pointed, and cone shaped. Wings moderately long and pointed; very good at flying. Tail relatively long, with white outer tail feathers. Inhabit primarily scrubland, tundra, grasslands, plains, marshes, and forest understory. Nest on ground or among bushes. Feed mostly on plant seeds and fruits; also take insects. Most species are migratory.

One genus and 44 species recognized worldwide, endemic to the Old World. Twenty-nine species recorded in China, ubiquitously distributed nationwide.

鹀属
Emberiza

Meadow Bunting

Slaty Bunting

Dark bluish-gray body

White underparts contrast strongly with upperparts

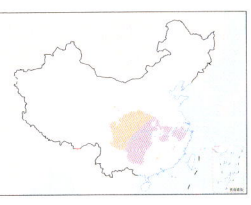

Almost entirely uniform brown body, differing from other buntings

Crested Bunting

Long upturned crest

Black and glossy body

Chestnut wings

Chestnut tail with black tip

Crest shorter than male but still prominent

Black streaks on grayish-brown back

Chestnut wings

Emberiza lathami

凤头鹀 fèng tóu wú
Crested Bunting

LC

L: 16–18 cm
Habitat: Breeds on open, dry, rocky slopes; prefers foraging on cultivated fields in foothills.
Behavior: Solitary or in pairs. Active and vigilant. Often perches on power lines.
Distribution: Uncommon resident in southwestern, South, Central, and East China and coastal southeastern China; some populations make altitudinal or short-distance migrations in winter.
Voice: Simple song is a repeated, buzzy, descending, five- or six-note phrase, *tizz … tizz-tiz-tssii-tsuu* or *tiss- … tiss-tiss-ts-tisseeu*. Typical call a mellow *tup*, reminiscent of Sulphur-bellied Warbler.

Emberiza siemsseni

蓝鹀 lán wú
Slaty Bunting

LC

L: 12–14 cm
Habitat: Occurs in the understory of mixed broadleaf-coniferous forests and broadleaf forests at middle and low altitudes; in winter, migrates down to low altitudes.
Behavior: Often gathers in small flocks in the understory; prefers forests more than other buntings.
Distribution: Chinese endemic; found to the east of W Sichuan, to the south of Qinling Mountains, to the north of northern South China, and east to mountainous areas in NW Zhejiang.
Voice: Distinctive, very high-pitched song is individually variable, but the lengthy, call-like, introductory note (rising, flat, or even descending) is usually repeated two to four times and reminiscent of one of the elements in the song of Olive-backed Pipit, *seeeeuu-seeeeuu-seeeeuu-seeeeuu … seeeeuu-seeeeuu-seeeeuu'tst'st'st'st-seeeeu*, or with a slightly rising introductory note, *tleeee-tleeee-tleeee-tleeee-tlt … tleeee-tleeee-tleeee-tlt'tlt*. Calls include a very high-pitched, lengthy (0.7 sec), slightly descending Fujian Niltava–like *seeeooo* that becomes slightly tremulous or buzzy toward the end.

Emberiza calandra
黍鹀 shǔ wú
Corn Bunting LC

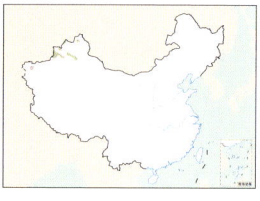

L: 18–19 cm
Habitat: Inhabits open low mountains and hills; especially prefers bushes and meadows on plains and piedmont and foothills without or with sparse trees; also occurs at forest edges, orchards, fields, and farmland.
Behavior: Often solitary or in pairs during breeding season; gathers in small flocks in nonbreeding season. Seldom comes to the ground except for foraging. Often active and rests in grass, or on top of bushes and small trees; sometimes also perches on power lines or fences.
Distribution: Locally common to uncommon breeders or resident in N Xinjiang; winters in SW Xinjiang.
Voice: Short song is a buzzy phrase, hesitant initially but accelerating and occasionally likened to the sound made by a jangling bunch of keys, *tik tik zreeississ*, repeated every 5 sec or so. Calls include a similarly distinctive, quiet, unobtrusive, mellow *plt* and a harsh, descending, buzzy *zweer*.

Emberiza citrinella
黄鹀 huáng wú
Yellowhammer LC

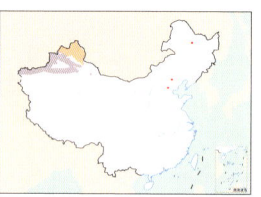

L: 15.5–17 cm
Habitat: Inhabits river-valley forests on sparsely wooded mountains and plains during breeding season; especially prefers forest edges and clearings. In winter, gathers in flocks in river-valley forests, scrubland, orchards, and farmland shelterbelts on hills and piedmont.
Behavior: In pairs during breeding season; often forms large flocks with Pine Bunting in winter. Active and bold, unafraid of people. Forages on ground or in bushes and grasslands; flies to trees when disturbed.
Distribution: Uncommon breeder in NW Xinjiang. Winters in S and C Xinjiang; occasional vagrant to Heilongjiang, Beijing, and Hebei. More common in winter.
Voice: Simple stereotyped song is a rapid series of 5–12 lengthening, accelerating, and intensifying similar notes, usually ending in a drawn-out wheeze, such as *zi-zi-zi-zi-zi-zi-zi-ziiiiiiii*, often transcribed as "a little bit of bread and no cheese!" Most common call a loud, buzzy, metallic *tszik*, very similar to that of Pine Bunting.

Emberiza leucocephalos
白头鹀 bái tóu wú
Pine Bunting LC

L: 16–17.5 cm
Habitat: In breeding season, inhabits primarily clearings, river-valley bushes, sparsely wooded grassy slopes, and farmland shelterbelts in montane areas and foothill plains. Also occurs in farmland, deserted fields, and orchards on plains in winter.
Behavior: Often in pairs during breeding season; usually gathers in small flocks containing several dozen individuals during nonbreeding season. Moves about in bushes or trees; also forages on the ground and grasslands.
Distribution: *E. l. leucocephala* breeds in Xinjiang and Northeast China; winters sporadically in North, Central, and East China; vagrant to Taiwan. *E. l. fronto* is resident in Gansu and Qinghai.
Voice: Song very similar to and perhaps a trifle shorter, slower, and higher pitched than Yellowhammer, with fewer, more evenly spaced notes that don't lengthen or give the impression of accelerating, *zi-zi-zi-zi-zi-zi-ziiiiiiii*. Call notes include a single *trp* or more complex *perett*, or a thin *dzee* in alarm.

Emberiza jankowskii
栗斑腹鹀 lì bān fù wú
Rufous-backed Bunting EN

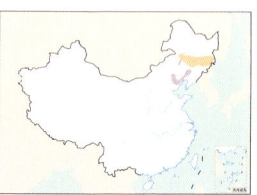

L: 15–16 cm
Habitat: Breeds on needle grass steppes interspersed with Siberian apricot (*Prunus sibirica*) and large-fruited elm (*Ulmus macrocarpa*); wintering habitats unclear, possibly in bushes on open fields.
Behavior: Often solitary. Perches and calls from treetops and power lines; takes cover in bushes when disturbed.
Distribution: Bred historically from S Heilongjiang and E Jilin to North Korea and S Primorsky Krai of Russia, but breeding range is now hugely contracted; recorded only in NE Inner Mongolia and W Jilin (also recorded recently in Mongolia). Wintering range unclear; multiple records in Beijing in recent years.
Voice: Short simple song, *chi-ut-ut-zweeer*. Calls include a sharp *tsit*, often doubled and very similar to Ochre-rumped Bunting.

Corn Bunting

Large bill with pale yellow lower mandible

Grayish-brown upperparts

Black streaks on side of breast and flanks

Buff-white belly

Pine Bunting

White crown

Large white patch on ear coverts

Rufous-chestnut throat

White belly

♂

Black streaks on crown, with a touch of white

Inconspicuous white patch on ear coverts

Black streaks on white throat

White belly

♀

Yellowhammer

Bright yellow head

Bright yellow throat

Yellow belly

♂ br.

Grayish-brown head tinged yellow

Inconspicuous pale yellow patch on ear coverts

Pale yellow throat

Pale yellow belly

♀ non-br.

Rufous-backed Bunting

Dark brown or black submoustachial stripe

Grayish ear coverts

♂

Central chestnut patch on gray belly

Streaks on breast

Small chestnut patch on belly

♀

Emberiza cia

淡灰眉岩鹀 (灰眉岩鹀)
dàn huī méi yán wú (huī méi yán wú)

Rock Bunting

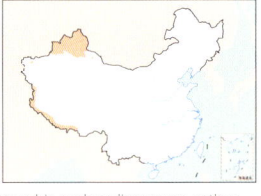

L: 15–17 cm
Habitat: Inhabits coniferous forests and forest-edge bushes on rocky mountain slopes at altitudes of 1000–2500 m (*E. c. par*) or above 3000 m (*E. c. stracheyi*).
Behavior: Often forages for insects and grass seeds on ground; in nonbreeding season, gathers in small loose flocks. In breeding season, male stands on prominent branches and bushtops, singing continuously.
Distribution: *E. c. stracheyi* in SW Tibet; *E. c. par* in N and W Xinjiang. Locally common.
Voice: Calls and song like Godlewski's Bunting.

Emberiza godlewskii

灰眉岩鹀 (戈氏岩鹀)
huī méi yán wú (gē shì yán wú)

Godlewski's Bunting

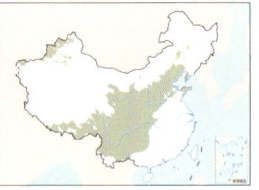

L: 15–18 cm
Habitat: Inhabits forest-edge bushes in montane areas at altitudes of 1000–3000 m; especially prefers habitats with abundant cliffs and rocks.
Behavior: Male strongly territorial in breeding season; often perches on branches or high places and sings. Omnivorous; forages on ground or in bushes. Nests in rock crevices on cliffs or branches in low bushes.
Distribution: All subspecies are common. *E. g. omissa* in Northeast and North China; *E. g. godlewskii* in northwestern China and the Qinghai-Tibet Plateau; *E. g. decolorata* in N and W Xinjiang; *E. g. yunnensis* in E Tibet and southwestern China; *E. g. khamensis* in S and E Tibet, NW Yunnan, S Qinghai, and W Sichuan. Some studies suggest that there is distinct genetic differentiation between the southern and northern populations, possibly as two separate species.
Voice: Song is variable series of repeated musical twittering notes, very similar to both Rock Bunting and Meadow Bunting, with differences still essentially unresolved. Calls include a *tzii*, similar to Rock Bunting but slightly higher pitched.

Emberiza cioides

三道眉草鹀 sān dào méi cǎo wú

Meadow Bunting

L: 15–18 cm
Habitat: Inhabits forest edges and bushes on plains and hills.
Behavior: Typical scrub-dwelling bird. In breeding season, male often perches on branches and sings continuously. Feeds on insects while rearing chicks; otherwise feeds primarily on grass seeds. Nests at base of bushes or in grass.
Distribution: Widespread but absent from Tibet and Hainan; common. *E. c. tarbagataica* in N Xinjiang; *E. c. cioides* in northern Northeast China and N Xinjiang; *E. c. weigoldi* in Northeast China and northern North China; *E. c. castaneiceps* across China except for Xinjiang, Tibet, Northeast China, and Hainan.
Voice: Short simple song extremely similar to that of Godlewski's Bunting. Call short, thin, multisyllabic, *tsip it tsip*.

Emberiza stewarti

白顶鹀 bái dǐng wú

White-capped Bunting

L: 17 cm
Habitat: Inhabits rocky slopes with grass and sparse bushes on mountains and piedmont hills.
Behavior: Nests on ground or in low bushes.
Distribution: Rare in China; vagrant to Taiwan and northern foothills of the Kunlun Mountains in SW Xinjiang.
Voice: Song a simple rattling phrase consisting of a buzzy note repeated six to nine times with no change in the rate of about four or five notes every second (half the speed of Yellowhammer), *tzee-tzee-tzee*, recalling Yellowhammer and Pine Bunting. Calls include a very high-pitched, descending *tsit* and a multisyllabic *tchirit* in flight.

Rock Bunting

White or grayish-white supercilium

Black lateral crown stripe and eye stripe

Meadow Bunting

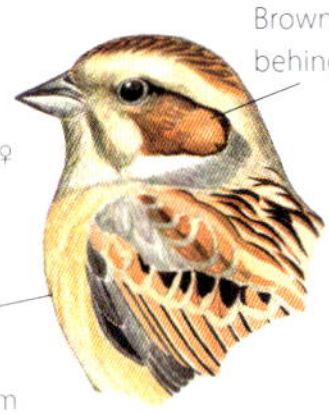

Brown patch behind eye

♀

Pale brown breast

White supercilium

Rufous-chestnut crown

Large rufous-chestnut patch behind eye

♂

Relatively heavy rufous chestnut on breast

Godlewski's Bunting

Brownish-red lateral crown stripe

Gray supercilium

Black moustachial stripe and loral stripe

Brown belly

E. g. yunnensis

Browner and less reddish brown on upperparts

Pale belly

E. g. godlewskii

White-capped Bunting

Streaks on crown

♀

Grayish-white throat, with black malar stripe

Pale grayish-white head

♂

Black eye stripe

Black throat

Brownish-chestnut back

Brownish-chestnut breast and flanks

Emberiza buchanani

灰颈鹀 huī jǐng wú

Gray-necked Bunting　LC

L: 14–15.5 cm
Habitat: In breeding season, inhabits bare hills, as well as dry grassy slopes and rocky ground with sparse bushes on low mountains and foothills.
Behavior: Often forages on ground in pairs in breeding season. Active and vigilant but shy; immediately hides in grass or bushes, or flies away, when people are around.
Distribution: Locally common breeder in N and W Xinjiang; vagrant in Guangdong (Zhongshan).
Voice: Remarkably stereotyped song is lower pitched, huskier, and not as musical or pleasant as that of Ortolan Bunting: four or five short, intensifying, flat notes, followed by three paired notes, one slightly buzzy, *sru-sru-sru-sru-sru-du'zrrrr-du'zrrr-du'zrrrr*, repeated monotonously every 6–10 sec or so. Typical call is a soft short *tup*, *tlup*, or *tsip*.

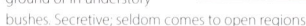

Emberiza tristrami

白眉鹀 bái méi wú

Tristram's Bunting　LC

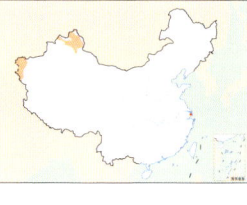

L: 13–16 cm
Habitat: Inhabits mixed forests, broadleaf forests, and forest-edge bushes on low mountains at altitudes of 500–1200 m.
Behavior: Forages for grass seeds and insects on ground or in understory bushes. Secretive; seldom comes to open regions.
Distribution: Breeds in Heilongjiang, NE Inner Mongolia, and N Jilin; migrates through most regions in eastern China; and winters in regions south of the Yangtze River.
Voice: Short unremarkable song begins with one or two flat introductory notes, followed by a more rapid series of often-paired notes, *seee swe'te-swe'te swe'tsirririri*, all within a very modest frequency range. Call is a short, high-pitched, explosive *tik*, very similar to call of Yellow-browed Bunting.

Emberiza hortulana

圃鹀 pǔ wú

Ortolan Bunting　LC

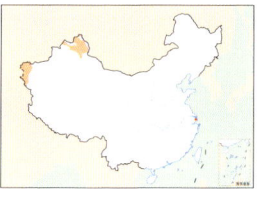

L: 15–16.5 cm
Habitat: In breeding season, inhabits primarily open hillside fields, steep grassy slopes, and river-valley bushes with sparse trees; also occurs in trees or bushes near farmland, and in subalpine scrubland.
Behavior: In pairs during breeding season; usually gathers in flocks in nonbreeding season. Forages on ground or in bushes and grass. Active and bold, unafraid of people.
Distribution: Locally common breeder in NW Xinjiang and far N Xinjiang. Uncommonly seen during migration in S Xinjiang; vagrant in Shanghai (Nanhui Dongtan).
Voice: Song a short phrase of five or six notes, with a ringing quality in the first, dropping markedly in pitch with the final two, *swee-swee-swee, dee-dee*, repeated every 5–7 sec or so. Wide variety of calls include a short sharp *plik* (principal flight call), a mellower *pluk*, a descending *tew*, and a sometimes more disyllabic *tsleee*.

Emberiza fucata

栗耳鹀 lì ěr wú

Chestnut-eared Bunting　LC

L: 14–16 cm
Habitat: Inhabits forest-edge bushes, grassy areas in farmland, and the vicinity of villages on low mountains and plains; especially prefers open meadows near wetlands.
Behavior: Solitary or in small flocks. Forages for grass seeds or insects on ground. Sometimes joins other buntings in mixed-species flocks.
Distribution: *E. f. fucata* breeds in most regions in northern China except for Xinjiang and Qinghai, migrates through North, East, and southwestern China, and winters in southern China. *E. f. arcuata* breeds in S Shaanxi, S Ningxia, and SE Tibet; resident in southwestern China. *E. f. kuatunensis* is a resident or breeder that migrates altitudinally in the Wuyi Mountains in Jiangxi and Fujian, a winter visitor in Guangdong, and a passage migrant in Taiwan.
Voice: Short, rapid, twittering song is faster than most other regional buntings and individually variable but otherwise heavily stereotyped; often begins with a stressed *zwee*. Strophes often last about one second, are repeated every 2.5–5 sec, and are vaguely reminiscent of Horned Lark. Call a mellow *plup*.

Gray-necked Bunting

Gray head

White eye ring

Yellowish-white throat and submoustachial stripe

Chestnut hue to scapulars

Thin faint black streaks on back

No dark breast band

Ortolan Bunting

Olive head

Pale yellow eye ring

Yellow throat and submoustachial stripe

Broad olive breast band

Distinct black streaks on back

Tristram's Bunting

White median crown stripe, supercilium, and submoustachial stripe

Black lateral crown stripe, cheek, and throat

White spot behind ear

♂

Chestnut-brown breast

Buff-yellow median crown stripe, supercilium, and submoustachial stripe

♀

Short black streaks on brown throat

Chestnut-eared Bunting

Rufous-chestnut cheek

♂

Black spots form black breast band

Narrow rufous-chestnut breast band

E. f. fucata

Brown crown

♀

No rufous-chestnut breast band

♂

Broad chestnut breast band

E. f. arcuata

Emberiza pusilla

小鹀 xiǎo wú

Little Bunting

L: 11–14 cm

Habitat: Inhabits forests, bushes, grasslands, and farmland on plains and mountains.

Behavior: Prefers understory of trees or open areas, such as farmland and grasslands.

Forages for grass seeds, grains, and insects. Often migrates in flocks; spreads out or is solitary in winter.

Distribution: Widespread in China; common in fall and winter. Migrates through northern China and the Qinghai-Tibet Plateau; winters in southern China.

Voice: Variable song is short, usually accelerates, and often includes rapid trills and finchlike notes. Call a sharp *tik*.

Emberiza chrysophrys

黄眉鹀 huáng méi wú

Yellow-browed Bunting

L: 13–17 cm

Habitat: In breeding season, occurs near bushes and wetlands in taiga forests; during migration and winter, inhabits broadleaf forests and bushes near wetlands in lowland plains. Also

seen in sparse woodlands, scrubland, and farmland.

Behavior: In winter, often gathers in small flocks on ground or in bushes and grass, feeding on grass seeds. Secretive; sometimes mixes with other buntings.

Distribution: Passage migrant or winter visitor in China. Migrates through regions in Northeast and East China; winters in South and East China.

Voice: Simple short song often begins with a single rising note, followed by a short chatter, *hweee-chit'it'it'it'it*. Call is a sharp *zik*, similar to Tristram's Bunting and higher pitched and sharper than Little Bunting.

Emberiza rustica

田鹀 tián wú

Rustic Bunting

L: 13–15 cm

Habitat: During migration and winter, occurs in open weedy areas, cultivated fields, mixed forests, and nurseries from plains to piedmonts.

Behavior: Often gathers in small flocks and forages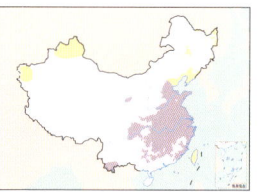

on ground. Not shy. Often erects crest when perched.

Distribution: Locally common. Migrates through Xinjiang and Northeast and North China; winters in Central, East, and South China (including Taiwan), and occasionally in Yunnan. Population clearly declining in recent years.

Voice: Mellow, medium-paced, melodious song lacks sharp notes of many eastern buntings, *duuu-dele-duudo-deluu … delu* or *dudeleu-dewee-deweea-weeu*. Calls include a sharp *tzik*.

Emberiza elegans

黄喉鹀 huáng hóu wú

Yellow-throated Bunting

L: 15–16 cm

Habitat: Breeds in open and dry deciduous forests, mixed forests, and shrublands; moves about at forest edges, open areas, and grassy slopes, usually near streams. Migrates through and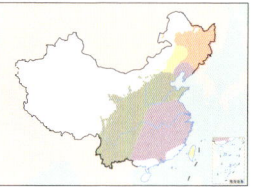

winters in forests, orchards, and deserted fields on plains and hills at low altitudes.

Behavior: Solitary or in pairs during breeding season; often gathers in small flocks in nonbreeding season. Vigilant; takes cover in bushes when disturbed. Forages on ground, primarily for plant seeds; in breeding season, also takes insects and other small invertebrates.

Distribution: Common. *E. e. elegans* breeds in Northeast and North China and winters in eastern and southeastern China (including Taiwan); *E. e. elegantula* breeds in Central and southwestern China and in winter migrates southward or altitudinally.

Voice: Variable but always extremely hurried twittering song with two to four different notes repeated three to five times and often including short trills, such as *tswit tsu ri tu tswee witt tsuri weee dee tswit tsuri tu*, usually repeated, with only slight changes every 2–4 sec, or *u'tu'lu'tu'zre'u'zre'u'zre'u't't't't'zrrrr*. Call a sharp *tzik*, similar to Little Bunting and Rustic Bunting.

Emberiza aureola

黄胸鹀 huáng xiōng wú

Yellow-breasted Bunting

L: 14–16 cm

Habitat: Inhabits tall grass, cultivated fields, paddy fields, or reedbeds on plains.

Behavior: Migrates in flocks; often forages with other buntings in mixed-species flocks.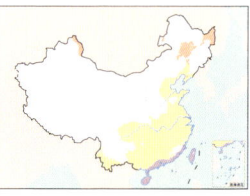

Distribution: *E. a. ornata* breeds in Northeast China; *E. a. aureola* breeds in Xinjiang (Altay). The two subspecies migrate through most regions of China (*E. a. ornata* primarily in the east and *E. a. aureola* primarily in the west). A few winter in coastal southeastern and South China. Very common in the past, but the population has declined rapidly due to long-term, extensive hunting in South China, where the species is regarded as a natural invigorant with medicinal value, commonly known as 禾花雀 ("rice flower bird" or "rice bird").

Voice: Song variable but typically a slow series of four to eight mostly paired units that often increase in pitch and volume usually before a short terminal flourish, such as *djuu-djiuu weee-weee ziii-zi* or *dwee-di-dee wuuu cha-cha zreee*. Call a mellow *tsup*.

Little Bunting

- Brownish-black lateral crown stripe
- Buff-yellow supercilium
- Inconspicuous white spots on ear coverts
- Brownish-red face

Yellow-browed Bunting

- Broad black lateral crown stripe and black cheek
- Yellow front half and white rear half of supercilium
- White spots on ear coverts
- White submoustachial stripe
- ♂

- ♀
- Brown crown and cheek

Rustic Bunting

- Pale brown crest and ear coverts
- Buff-yellow supercilium
- Short black crest
- Chestnut-brown hindneck
- ♂
- Rufous-chestnut streaks on breast and flanks
- Scaly rufous-chestnut pattern on rump

Yellow-throated Bunting

- Relatively distinct crest, brown in front and pale yellow behind
- Pale brown ear coverts
- ♀
- No black spots on breast
- Prominent crest, can be lowered
- Black, white, and yellow on head and breast contrast strongly; in nonbreeding season, yellow and black plumage becomes duskier
- ♂
- Brown streaks on flanks

E. e. elegans

- Yellow and black on head darker than *E. e. elegans*
- Relatively heavy black streaks on back
- Gray on hindneck and rump relatively dark
- ♂
- Streaks on flanks relatively dark (sometimes dark brown)

E. e. elegantula

Yellow-breasted Bunting

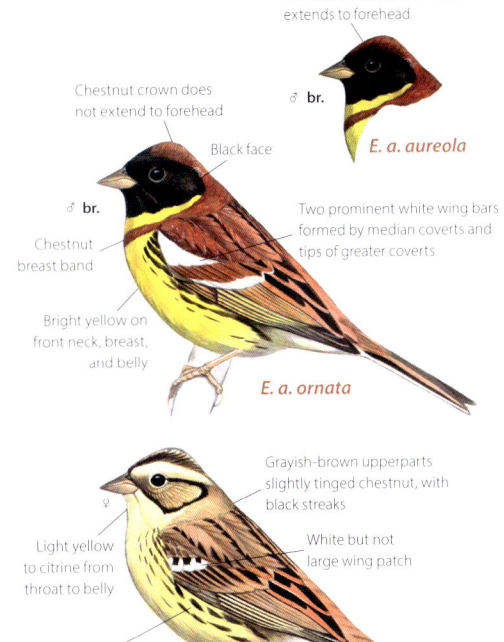

- Closely resembles *E. a. ornata*, but chestnut on crown often extends to forehead
- ♂ br.

E. a. aureola

- Chestnut crown does not extend to forehead
- Black face
- ♂ br.
- Chestnut breast band
- Two prominent white wing bars formed by median coverts and tips of greater coverts
- Bright yellow on front neck, breast, and belly

E. a. ornata

- Grayish-brown upperparts slightly tinged chestnut, with black streaks
- White but not large wing patch
- ♀
- Light yellow to citrine from throat to belly
- Blackish-brown streaks on flanks

Emberiza rutila
栗鹀 lì wú
Chestnut Bunting LC

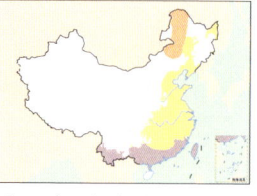

L: 14–15 cm
Habitat: Breeds in open coniferous forests or mixed forests, nesting on mountain slopes with dense bushes; migrates through and winters at piedmont forest edges, woodlands in farmland, cultivated fields, paddy fields, bushes, and grasslands at low altitudes.
Behavior: Gathers in small flocks.
Distribution: Breeds in Northeast China, migrates through North and Central China and most regions in eastern China (including Taiwan), and winters in coastal South China. Usually uncommon; clear population decline due to involvement in the illegal hunting and trade of the similar-looking Yellow-breasted Bunting.
Voice: Simple, fairly low-pitched song of three or four different notes starts with four slow notes (reminiscent of part of the song of Olive-backed Pipit), followed by a characteristic buzzy trill, *wee-wee-wee-zrru-zrru-zr'ee-zr'ee-ti'it*. Calls include a sharp *zik*, similar to Yellow-browed Bunting and Tristram's Bunting.

Emberiza koslowi
藏鹀 zàng wú
Tibetan Bunting NT

L: 17–19 cm
Habitat: Inhabits sun-facing, bush-covered, grassy slopes at high altitudes.
Behavior: Often solitary or in pairs; also gathers in small flocks. Occurs in grass or bushes. Makes only short-distance movements.
Distribution: Chinese endemic, uncommon in Tibet, Qinghai, S Gansu, and NW Sichuan.
Voice: Song an uncomplicated, stereotyped twittering, *tisp tsi tsi chiriree teetew*, similar to Godlewski's Bunting, *tslip-tsle-tsle-tzede-tzede … tslip-tsle-tsle-tzede-tzede-tz-ti'di'di*, or *tup … dzee-dzee-cher-dzr-dzr-wheer*. Strophes from the same bird don't vary much, but they do vary between individuals; some sing particularly slow songs, others make an accelerated ending. Songs occasionally include call notes or end with a rising buzzy note. Strophes last about 1.5 sec and are repeated every 2–5 sec. Calls include a mellow sparrowlike *chup* and a Godlewski's Bunting–like *tsip*.

Emberiza bruniceps
褐头鹀 hè tóu wú
Red-headed Bunting LC

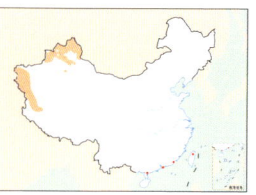

L: 15–16.5 cm
Habitat: Occurs in various scrubs and herbaceous vegetation on low mountains, hills, and open plains; especially prefers dry plains, streams, and farmland with sparse trees and bushes on hills. Also inhabits small trees and scrubland in semideserts and deserts.
Behavior: Often in pairs or solitary in breeding season; moves about and forages in bushes and along grass stems. Male prefers singing from bushtops for a long time.
Distribution: Locally common to uncommon breeder in Xinjiang; more common in N Xinjiang. Vagrant to Guangdong, Hong Kong, and Taiwan.
Voice: Low-pitched song is an accelerating and descending 2.5 sec verse that usually starts with a call note, followed by two repeated notes and a rapid jumble, *zit … zrit-zrit-too'lueet'toolu'treet*, repeated with minor modifications every 6–16 sec. Calls varied but mostly sparrowlike, such as *chup* or *chlip*, indistinguishable from Black-headed Bunting.

Emberiza sulphurata
硫黄鹀 (硫磺鹀) liú huáng wú (liú huáng wǔ)
Yellow Bunting VU

L: 13–14 cm
Habitat: During migration, inhabits cultivated fields, bushes, wet grasslands on plains and low mountains; commonly seen on oceanic islands and in coastal regions.
Behavior: Often solitary; gathers in small flocks in suitable habitats during migration. Feeds on plant seeds on ground; also forages for insects. Often perches on bushtops; takes cover in dense grass when disturbed.
Distribution: Uncommon passage migrant; migrates through coastal South and East China; and small numbers of individuals winter in coastal South China (including Taiwan).
Voice: Simple, often-spluttering song usually involves about five different notes; introductory notes are repeated two to six times before an often-simple ending, *chip-chip-chip-see-see-tr'r'r'r'r-piu-tiuu*. The most common call is slightly shorter, burrier, and more descending than call of Black-faced Bunting.

Emberiza melanocephala
黑头鹀 hēi tóu wú
Black-headed Bunting LC

L: 15.5–17.5 cm
Habitat: Inhabits open forests, scrubland, farmland, and orchards.
Behavior: Solitary or in pairs. Perches on treetops or power lines; also forages on ground. Active.

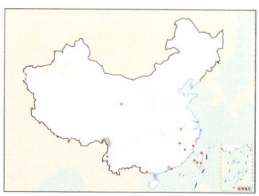

Distribution: Rare winter visitor or vagrant in the regions south of the Yangtze River and southwestern and northwestern China; recorded in Xinjiang, Tibet, Qinghai, Yunnan, Anhui, Zhejiang, Jiangxi, Fujian, Guangdong, Guangxi, Hong Kong, and Taiwan.
Voice: Vocalizations very similar to Red-headed Bunting.

Chestnut Bunting

♂ **br.**

Chestnut upperparts and breast

Bright yellow belly

♀

Pale brown wing feathers

Light yellow underparts

Rufous-chestnut rump

♂ **non-br.**

Chestnut on head and back, and yellow on belly appear slightly faint

Tibetan Bunting

Chestnut brown from lores to chin

Black crown

♂

White supercilium

White throat

Rufous-chestnut upper back

Black upper breast

Streaks on grayish-brown head

Streaks on chestnut back

♀

Red-headed Bunting

Rufous-chestnut head and upper breast

♂

Yellowish-green rump

Gold underparts

♀

Grayish-brown head and upperparts

Buff-yellow underparts

Some have diagnostic greenish-tinged rump

Black-headed Bunting

Black head

♂ **br.**

Rufous-chestnut upper back

Streaks on crown

No streaks on yellow underparts

♀ **non-br.**

Outer tail feathers not white

Brownish rump, differing from Red-headed Bunting

Yellow Bunting

No black on lores and chin

Body more grayish than males

♀

Yellowish-green and slightly gray-tinged head and face

Black lores and chin, dusky during nonbreeding season

Prominent white eye-ring

♂ **br.**

Grayish-green back, with blackish-brown streaks

Light yellow on throat to belly

Pale brown edges on wing feathers

Gray-tinged flanks, a few blackish-brown streaks on rear part of flanks

Yellowish-white undertail coverts

♂ **juv.**

Brown on back and fringe of wing feathers relatively heavy

Brown-tinged from head to breast

633

Emberiza spodocephala
灰头鹀 huī tóu wú
Black-faced Bunting　LC

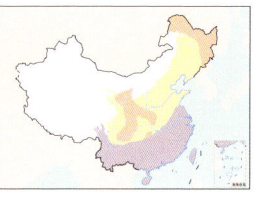

L: 13.5–16 cm
Habitat: Breeds in bushes in the understory of broadleaf, coniferous, and mixed forests at altitudes above 600 m; active in open areas at forest edges. In nonbreeding season, inhabits farmland, deserted fields, nurseries, gardens, and riverbanks on plains and hills at low altitudes, often hidden in bushes. *E. s. personata* usually occurs on oceanic islands and near coasts.
Behavior: In nonbreeding season, often solitary or gathers in small loose flocks. Hops and forages on ground. Vigilant; flies immediately into bushes when disturbed.
Distribution: Common migrant. *E. s. spodocephala* breeds in Northeast China and winters in most regions of South and East China, including Hainan and Taiwan. *E. s. sordida* breeds in central and eastern China and winters in southwestern, South, and East China (including Taiwan). *E. s. personata* is rare in China, with occasional records during migration in coastal eastern China; rare passage migrant or winter visitor in Taiwan; vagrant in Beijing. *E. s. personata* sometimes treated as a separate species, Masked Bunting (*E. personata*).
Voice: Highly variable, usually short, but complex, scratchy song includes clear whistles, trills, and some repetition, *chi chi chu chirri chu chi zeee chu chi chi*. Typical call a thin *tzii*, subtly more lisping, longer, and weaker than Little Bunting and Rustic Bunting. Songs differ slightly among the three subspecies; differences also occur among populations within one subspecies in different breeding grounds. *E. s. spodocephala* sings with the fastest rhythm; *E. s. personata* sings with discontinuous segments and the slowest rhythm; *E. s. sordida*, which breeds at the highest altitudes, sings with complete but short segments, with a gentle rhythm.

Emberiza variabilis
灰鹀 huī wú
Gray Bunting　LC

L: 14–17 cm
Habitat: Inhabits parks, nurseries, lowland bushes, forest edges, and cultivated fields on oceanic islands or coasts.
Behavior: Often solitary or in pairs; gathers in small flocks containing about five individuals during migration. Not shy. Often feeds on grass seeds on ground; sometimes joins Black-faced Bunting in mixed-species flocks.
Distribution: Rare vagrant. Recorded occasionally in Taiwan during migration; also sporadically in coastal East China.
Voice: Simple, two- to five-note song usually starts with soft drawn-out note and is slow, with three to five distinct notes, *swee hwee-we'er-we'er … hwee-tis-uu … hwee-chi'it … hwee-chiu*. Call is a short sharp *tzi*, even higher pitched than Yellow-browed Bunting and Tristram's Bunting and, unlike those species, often doubled, *tzi-dit*.

Emberiza pallasi
苇鹀 wěi wú
Pallas's Bunting　LC

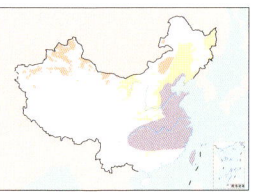

L: 13–15 cm
Habitat: Inhabits primarily reedbeds and bushes near marshes and streams on plains.
Behavior: Often perches on reed stems and scans the surroundings in horizontal stance.
Distribution: *E. p. polaris* is a passage migrant or winter visitor in eastern China; *E. p. pallasi* is a breeder or passage migrant in northwestern China; *E. p. lydiae* breeds in Inner Mongolia. Common.
Voice: Calls include a complex Eurasian Tree Sparrow–like *tsleu*, a clearly rising, subtly disyllabic *chlip* or *tsilip*, and a more Reed Bunting–like *tsweep*. Song of *E. p. pallasi* and *E. p. polaris* is a simple monotonous repetition of short raspy notes, such as *chi-chi-chi-chi* or *srri-srri-srri-srri*, reminiscent of the song of White-capped Bunting. The song of *E. p. lydiae* is a series of four to six much buzzier, more evenly pitched notes, *dzzz-dzzz-dzzz-dzzz-du*, with the later notes occasionally fading.

Emberiza yessoensis
红颈苇鹀 hóng jǐng wěi wú
Ochre-rumped Bunting　NT

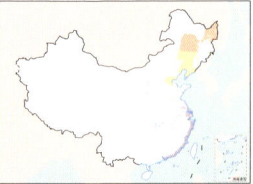

L: 13–15 cm
Habitat: Inhabits lowland bushes near wetlands, wet meadows, and reed marshes.
Behavior: Gathers in small flocks on ground and in bushes.
Distribution: Breeds in Heilongjiang and Jilin, migrates through North and East China, and winters in the Yangtze Plain; occasionally seen in South China.
Voice: Simple, stereotyped, short song (1–1.5 sec) often starts with a repeated note or notes and has a buzzier finish, *tsi'ui … tsi'ui-tsi'i-dzeu … tsi'ui … tsi'ui-tsi'i-dzeu-ee-it* or *tse … cup'tse-cup'ts-tsew-zree-sew-wee*. Call is an intense, short *tic*, unlike other "reed" buntings.

Emberiza schoeniclus
芦鹀 lú wú
Reed Bunting　LC

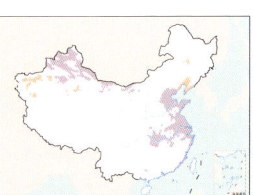

L: 15–17 cm
Habitat: Inhabits open areas near wetlands on low mountains and plains, such as bushes, reedbeds, and grasslands. In winter, also on farmland and pastures.
Behavior: Often moves about in small flocks on ground or in bushes.
Distribution: *E. s. minor* breeds in Northeast China and winters in South China; *E. s. pyrrhuloides* breeds in W Xinjiang; *E. s. zaidamensis* is resident in Qinghai. Multiple subspecies migrate through North and East China; several winter in Xinjiang, Qinghai, Inner Mongolia, and Gansu.
Voice: Slow, short, simple, and reluctant or hesitant song is usually a series of repeated notes, often followed by a different note, often a buzz or trill, such as *tweep-tweep-tweep-zrrrrreeu* or *triu-triu-triu-pli'uu*. Calls include a diagnostic falling *siuu* and a husky *bziu*.

Black-faced Bunting

♂ br.
Black lores and chin

Disorderly blackish-brown streaks on pale brown back

Black on lores and chin more conspicuous

Blackish-brown streaks on flanks

Largest body among three subspecies

Buff-yellow supercilium and submoustachial stripe

Dusky gray neck

E. s. spodocephala

Dark dense streaks on brown body

Black malar stripe

Whitish to bright yellow belly (population that breeds in W Siberia has palest belly; the closer to Northeast Asia, the yellower the belly)

Resembles *E. s. spodocephala*, but with clearer boundary between gray breast and yellow belly

E. s. sordida

Pale yellow or nearly white belly

Diffuse light yellow eye stripe

Citrine submoustachial stripe and from throat to belly; gray-tinged breast

Smallest body among three subspecies

Heavily streaked flanks

Yellow supercilium, throat, and belly, distinctly yellower than female of other two subspecies

E. s. personata

Gray Bunting

Pink on most of lower mandible and base of upper mandible

Black streaks on upper back

♂

Dark slate-gray body

No white on outer tail feathers

♀

Bill coloration similar to male

Brownish-gray body, prominent streaks on back, breast, and belly

Brown rump (more distinct in flight)

Resembles Black-faced Bunting but with brown tail and no white on outer tail feathers

Pallas's Bunting

White neck collar

White submoustachial stripe

Gray lesser coverts

♂ br.

Different colors on upper and lower mandibles

Short blackish-brown malar stripe

Gray lesser coverts

♂ non-br. / ♀

Grayish-brown crown, mixed with some black

♂ Molting

Ochre-rumped Bunting

Black head

♂ br.

Triangular gray patch on lesser coverts

Brick-red rump

Different colors on upper and lower mandibles

Chestnut-tinged nape and hindneck

Reddish-brown tinge on back

♀ / ♂ non-br.

Blackish-brown cheek

Reed Bunting

Inconspicuous buff-yellow neck collar

White neck collar

♀

White submoustachial stripe

♂

White submoustachial stripe

Brownish-red lesser coverts

Grayish-brown rump

雀鹀科　Passerellidae

Small buntinglike songbirds that prefer scrubby habitats. Plumage similar or slightly different between sexes; mostly gray, white, and brown. Bill short, pointed, and cone shaped. Wings moderately long and pointed; very good at flying. Tail relatively long, usually emarginated. Inhabit primarily scrubland, grasslands, tundra, and seashores. Feed mostly on insects and plant seeds and fruits. Most species are migratory and gather in flocks during winter.

New World sparrows, closely related to Old World buntings, were once placed in Emberizidae and are now in their own family. Twenty-eight genera and 136 species recognized worldwide, endemic to the New World. Two genera and two species recorded in China, both as vagrants, in Qinghai, Taiwan, and Inner Mongolia.

Savannah Sparrow

Slight yellow tinge on front part of supercilium

Black upper mandible and pinkish lower mandible

Dense thick streaks on breast and flanks

White-crowned Sparrow

White median crown stripe

Broad white supercilium

Black lateral crown stripe and eye stripe

♂

Passerculus sandwichensis

稀树草鹀 xī shù cǎo wú

Savannah Sparrow　LC

L: 11–15 cm
Habitat: Occurs in open grasslands with sparse bushes; also in farmland.
Behavior: Gathers in small loose flocks. Usually forages on ground; occasionally searches for food in bushes or small trees.
Distribution: Vagrant in Qinghai and Taiwan.
Voice: Calls include a thin, high-pitched *tsip*.

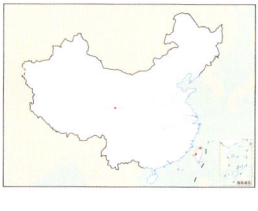

Zonotrichia leucophrys

白冠带鹀 bái guān dài wú

White-crowned Sparrow　LC

L: 14–17 cm
Habitat: In breeding season, inhabits forest-edge bushes and subalpine meadows. In winter, occurs in open scrubland and grasslands.
Behavior: Often forages for grass seeds or small arthropods on ground. Often mixes and migrates with other buntings.
Distribution: Vagrant. First recorded in Inner Mongolia (Hulunbuir) in the winter of 2012; scattered records afterward.
Voice: Call a hard *pik* or *pink*.

Recorded Species That May Have Been Ship-Assisted

白鞘嘴鸥 Snowy Sheathbill
Chionis albus

霍氏金鹃
Horsfield's Bronze Cuckoo
Chrysococcyx basalis

黑白扇尾鹟 Willie Wagtail
Rhipidura leucophrys

褐头牛鹂 Brown-headed Cowbird
Molothrus ater

Species That Potentially Occur in China (Those potentially extinct or without recent/definite records)

冠麻鸭 Crested Shelduck
Tadorna cristata

普通潜鸟 Common Loon
Gavia immer

花斑鹱 Cape Petrel
Daption capense

克岛圆尾鹱 Kermadec Petrel
Pterodroma neglecta

黑腹蛇鹈 Oriental Darter
Anhinga melanogaster

马来鸻 Malaysian Plover
Charadrius peronii

黑腰滨鹬 Baird's Sandpiper
Calidris bairdii

姬滨鹬 Least Sandpiper
Calidris minutilla

银鸥 Herring Gull
Larus argentatus

大贼鸥 Brown Skua
Stercorarius antarcticus

乌雕鸮 Dusky Eagle-Owl
Ketupa coromandus

灰腰金丝燕 Gray-rumped Swiftlet
Collocalia marginata

大金丝燕 Black-nest Swiftlet
Aerodramus maximus

长尾鹦鹉 Long-tailed Parakeet
Psittacula longicauda

白胸燕鵙 White-breasted Woodswallow
Artamus leucorynchus

盘尾树鹊 Racket-tailed Treepie
Crypsirina temia

里海地鸦 Turkestan Ground-Jay
Podoces panderi

西伯利亚山雀
Gray-headed Chickadee
Poecile cinctus

白眼河燕 White-eyed River Martin
Pseudochelidon sirintarae

布氏柳莺 Brooks's Leaf Warbler
Phylloscopus subviridis

大嘴苇莺 Large-billed Reed Warbler
Acrocephalus orinus

岩䴓 Eastern Rock Nuthatch
Sitta tephronota

灰背岸八哥 Bank Myna
Acridotheres ginginianus

欧歌鸲 Thrush Nightingale
Luscinia luscinia

大斑鹡鸰 White-browed Wagtail
Motacilla maderaspatensis

Species Presumably Occurring in South Tibet and on Islands of South China Sea

South Tibet

沼泽鹧鸪 Swamp Francolin
Ortygornis gularis

丛林鹑 Jungle Bush-Quail
Perdicula asiatica

印度鸬鹚 Indian Cormorant
Phalacrocorax fuscicollis

印度兀鹫 Indian Vulture
Gyps indicus

南亚鸨 Bengal Florican
Houbaropsis bengalensis

棕翅歌百灵
Bengal Bushlark
Mirafra assamica

须草莺 Bristled Grassbird
Schoenicola striatus

南亚大草莺 Indian Grassbird
Graminicola bengalensis

细嘴鹛鹛
Slender-billed Babbler
Argya longirostris

丛林鹛鹛 Jungle Babbler
Argya striata

阿氏雅鹛 Abbott's Babbler
Malacocincla abbotti

奥氏穗鹛 Snowy-throated Babbler
Stachyris oglei

斑胸鸦雀 Black-breasted Parrotbill
Paradoxornis flavirostris

大蓝仙鹟 Large Blue Flycatcher
Cyornis magnirostris

黑喉织雀 (黑胸织雀)
Black-breasted Weaver
Ploceus benghalensis

Islands of South China Sea

澳洲石鸻 Beach Thick-knee
Esacus magnirostris

银鸽 Silvery Wood-Pigeon
Columba argentina

爪哇斑鸠 Sunda Collared-Dove
Streptopelia bitorquata

尼柯巴鸠 Nicobar Pigeon
Caloenas nicobarica

马来皇鸠 Gray Imperial-Pigeon
Ducula pickeringii

斑皇鸠 Pied Imperial-Pigeon
Ducula bicolor

红颈绿鸠 Pink-necked Green-Pigeon
Treron vernans

小绿鸠 Little Green-Pigeon
Treron olax

棕头绿鸠
Cinnamon-headed Green-Pigeon
Treron fulvicollis

灰腹杜鹃 Gray-bellied Cuckoo
Cacomantis passerinus

白腹金丝燕 Glossy Swiftlet
Collocalia esculenta

Introduced Species

黑天鹅 Black Swan
Cygnus atratus

珠鸡 Helmeted Guineafowl
Numida meleagris

蓝孔雀 Indian Peafowl
Pavo cristatus

圣鹮 African Sacred Ibis
Threskiornis aethiopicus

斑姬地鸠 Zebra Dove
Geopelia striata

戈氏凤头鹦鹉 Tanimbar Corella
Cacatua goffiniana

葵花鹦鹉 Sulphur-crested Cockatoo
Cacatua galerita

小葵花鹦鹉 Yellow-crested Cockatoo
Cacatua sulphurea

白凤头鹦鹉 White Cockatoo
Cacatua alba

红鹦鹉 Red Lory
Eos bornea

椰果彩虹鹦鹉 Coconut Lorikeet
Trichoglossus haematodus

虎皮鹦鹉 Budgerigar
Melopsittacus undulatus

亚洲辉椋鸟 Asian Glossy Starling
Aplonis panayensis

爪哇八哥 Javan Myna
Acridotheres javanicus

东非金织雀
African Golden-Weaver
Ploceus subaureus

黑头织雀 Village Weaver
Ploceus cucullatus

橙颊梅花雀 Orange-cheeked Waxbill
Estrilda melpoda

梅花雀 Common Waxbill
Estrilda astrild

白喉文鸟 Indian Silverbill
Euodice malabarica

白头文鸟 White-headed Munia
Lonchura maja

禾雀 Java Sparrow
Padda oryzivora

针尾维达雀 Pin-tailed Whydah
Vidua macroura

黄额丝雀 Yellow-fronted Canary
Crithagra mozambica

Funding Agencies

We thank the following organizations for their financial support:

The Beijing Entrepreneur Environmental Protection Foundation (SEE Foundation) was founded in 2008 by the Society of Entrepreneurs and Ecology. The SEE Foundation aims to support and fund the growth of local nongovernmental environmental organizations in China by providing a pan-society platform for entrepreneurs, environmental organizations, and the public to work together to promote ecological conservation and sustainable development.

The SEE Foundation became a public foundation in 2014. The main area of the foundation's work is facilitating the development of environmental organizations, focusing on three main themes: desertification prevention, climate and sustainable business, and nature conservation and nature education. To date, the SEE Foundation has supported more than 700 individuals and local nongovernmental environmental organizations either directly or indirectly, invested more than RMB 800 million, and facilitated public participation in environmental activities totaling more than 600 million times. The SEE Foundation has gradually grown into one of the most influential and well-regarded environmental organizations in China.

The Mangrove Conservation Foundation (MCF), China's first publicly founded public environmental foundation, aims to conserve wetlands and biodiversity through public participation. Three main strategic areas of work are protecting Shenzhen Bay, saving the Spoon-billed Sandpiper, and bringing back mangrove forests.

The MCF was founded by the Society of Entrepreneurs and Ecology, philanthropic entrepreneurs, and relevant governmental agencies of Shenzhen in July 2012. WANG Shi and MA Weihua served as founding presidents; ZHANG Bigong (former president of Shenzhen University) and AI Luming (president of the Society of Entrepreneurs and Ecology) served as honorary committee chairs; and Prof. LEI Guangchun (School of Ecology and Nature Conservation, Beijing Forestry University) served as the committee chair.

The Changjiang Conservation Foundation (CCF) was jointly founded by the Society of Entrepreneurs and Ecology and 38 entrepreneurs of the Yangtze River Center of the Society of Entrepreneurs and Ecology, the first local publicly founded private environmental foundation in Hubei Province. The CCF aims to promote a harmonious relationship between people and wildlife (symbolized by the Yangtze Finless Porpoise) by protecting the ecology of the Yangtze River, focusing on the flagship species the Yangtze Finless Porpoise and Baer's Pochard; by facilitating and supporting nongovernmental environmental organizations; and by providing a platform that unifies the power of entrepreneurs, universities, research institutes, nongovernmental organizations, and the general public.

Sustainable development is a common theme that human society needs to address collaboratively to ensure a shared prosperous future. Huatai Securities deeply recognized its social responsibility and proactively launched the Yixin Huatai: One Yangtze River Project as part of the national strategy to protect the Yangtze River. The environmental-protection project focuses on biodiversity conservation in the Yangtze River Watershed in collaboration with environmental organizations, including the Shan Shui Conservation Center, World Wildlife Fund, and Green River Environmental Protection Promotion Association. Huatai Securities also works through the Yixin Huatai: Citizen Scientists Program to promote the mainstreaming of biodiversity conservation, to guide the public to nature's beauty, to provide valuable data to species monitoring and research, and to nurture a generation of young conservation minds.

The Tencent Foundation was founded by Tencent Holdings Ltd. in June 2007. Registered in the Ministry of Civil Affairs as a national private foundation, the Tencent Foundation is the first charitable foundation founded by an internet company in China. The Tencent Foundation focuses on promoting public welfare through innovative digital technology and cultural approaches. In 2020, building upon its experience fighting COVID-19 empowered by advanced technology, the Tencent Foundation initiated the Tech for Good Project, exploring innovative and collaborative ways to address societal problems using digital technologies, together with volunteers and organizations with needed expertise. The Tech for Good Project will address a suite of sustainable development topics, including biodiversity conservation, environmental protection, and accessible technology.

Naturewin promotes a harmonious relationship between people and nature in its own practice, delivering outstanding natural experiences and cultural values to the public through nature education, ecotourism, ecological design, ecological and conservation consulting, and online tools and courses.

Acknowledgments

We thank the following experts, researchers, and organizations for valuable support and comments:

Vocalizations:
Zhejiang Museum of Natural History, FAN Zhongyong, SUN Qingsong, YONG Ding Li, LIAO Xiaodong, GOU Jun, WANG Ruiqing, HUANG Qin, HUANG Hanchen, YANG Xiaojing, WANG Aizhen, ZHAO Chao, LIU Aitao, ZHOU You, TAN Chen, LIANG Dan, XING Xiaoying, ZHANG Yanyun, DONG Lu, WANG Nan, and WANG Ning

Distribution maps and data:
China Birdwatching Association, Beijing Jinglang Ecology and Techniques Co. Ltd., HUANG Hanchen, CHENG Yuan, LIU Haobing, GU Yiyun, TONG Hao, GAO Xiangyu, TIAN Dongyu, HAN Shouqing, ZHANG Binghua, SHI Rui-De, YU Changhao, and LI Fei

References

Alström, P., and K. Mild. 2003. *Pipits and Wagtails of Europe, Asia and North America: Identification and Systematics*. London: Christopher Helm.

Alström, P., P. C. Rasmussen, G. Sangster, et al. 2019. "Multiple species within the Striated Prinia *Prinia crinigera*-Brown Prinia *P. polychroa* Complex Revealed by Integrative Taxonomy." *Ibis* 162 (3): 936–967.

Alström, P., P. C. Rasmussen, C. Zhao, et al. 2016. "Integrative Taxonomy of the Plain-backed Thrush (*Zoothera mollissima*) Complex (Aves, Turdidae) Reveals Cryptic Species, Including a New Species." *Avian Research* 7 (1).

Ayé, R., M. Schweizer, and T. Roth. 2013. *Birds of Central Asia: Kazakhstan, Turkmenistan, Uzbekistan, Kyrgyzstan, Tajikistan, and Afghanistan*. Princeton: Princeton University Press.

Beaman, M., and S. Madge. 2010. *The Handbook of Bird Identification for Europe and the Western Palearctic*. London: Christopher Helm Press.

Bhushan, B., K. Sonobe, and S. Usui. 1993. *A Field Guide to the Waterbirds of Asia*. Tokyo: Kodansha International.

BirdLife International. 2020. IUCN Red List for Birds. http://www.birdlife.org.

Bird Report: http://www.birdreport.cn/ [中国观鸟记录中心 : http://www.birdreport.cn/].

Birds of the World–Cornell Lab of Ornithology: https://www.birdsoftheworld.org/.

Birds of Xinjiang, China: http://xinjiang.birds.watch/ [新疆鸟类 : http://xinjiang.birds.watch/].

Bo, S. Q., X. Yuan, and W. P. Lu. 2013. "*Puffinus pacificus* (Wedge-tailed Shearwater) and Six Other New Records of Birds in Shanghai." *Journal of Fudan University (Natural Science)* 52 (4): 150–53 [薄顺奇、袁晓、陆万鹏、2013. 楔尾鹱等 7 种上海市鸟类新记录. 复旦学报 (自然科学版)、(4): 150–53].

Brazil, M. 2009. *Birds of East Asia: China, Taiwan, Korea, Japan, and Russia*. London: Christopher Helm Press.

Cai, Q. K. 1987. *The Birds of Beijing*. Beijing: Beijing Press [蔡其侃、1987. 北京鸟类志 . 北京 : 北京出版社].

Chen, F. G., and S. Y. Luo. 1998. *Fauna Sinica. Aves*. Vol. 9. Beijing: Science Press [陈服官、罗时有、1998. 中国动物志 鸟纲、第九卷 . 北京 : 科学出版社].

Cheng, T. H. 1964. *The Keys to the Birds of China*. Beijing: Science Press [郑作新,1964. 中国鸟类系统检索. 北京:科学出版社].

Cheng, T. H. 1979. *Fauna Sinica. Aves*. Vol. 2. Beijing: Science Press [郑作新、1979. 中国动物志 · 鸟纲、第二卷 . 北京 : 科学出版社].

Cheng, T. H. 1983. *The Avifauna of Xizang*. Beijing: Science Press [郑作新、1983. 西藏鸟类志 . 北京 : 科学出版社].

Cheng, T. H. 1994. *A Complete Checklist of Species and Subspecies of the Chinese Birds*. Beijing: Science Press [郑作新、1994. 中国鸟类种和亚种分类名录大全 . 北京 : 科学出版社].

Cheng, T. H. 2002. *The Keys to the Birds of China*, 3rd ed. Beijing: Science Press [郑作新、2002. 中国鸟类系统检索 (第三版). 北京 : 科学出版社].

Cheng, T. H., Z. Y. Long, and T. C. Lu. 1995. *Fauna Sinica. Aves*. Vol. 10. Beijing: Science Press [郑作新、龙泽虞、卢汰春、1995. 中国动物志 · 鸟纲、第十卷 . 北京 : 科学出版社].

Cheng, T. H., Y. H. Xi, and G. X. Guan. 1991. *Fauna Sinica. Aves*. Vol. 6. Beijing: Science Press [郑作新、冼耀华、关贯勋、1991. 中国动物志 · 鸟纲、第六卷 . 北京 : 科学出版社].

Cheng , T. S. 1987. *A Synopsis of the Avifauna of China*. Beijing: Science Press.

China Bird Report. 2020. *The CBR Checklist of Birds of China*. Version 8.0 [中国观鸟年报编辑、2020. 中国观鸟年报 - 中国鸟类名录 8.0 版].

China Ornithological Society. 2004. *China Bird Report 2003*. Beijing: China Ornithological Society [中国动物学会鸟类学分会、2004. 中国观鸟年报 2003. 北京 : 中国动物学会鸟类学分会].

China Ornithological Society. 2005. *China Bird Report 2004*. Beijing: China Ornithological Society [中国动物学会鸟类学分会、2005. 中国观鸟年报 2004. 北京 : 中国动物学会鸟类学分会].

China Ornithological Society. 2006. *China Bird Report 2005*. Beijing: China Ornithological Society [中国动物学会鸟类学分会、2006. 中国观鸟年报 2005. 北京 : 中国动物学会鸟类学分会].

China Ornithological Society. 2007. *China Bird Report 2006*. Beijing: China Ornithological Society [中国动物学会鸟类学分会、2007. 中国观鸟年报 2006. 北京 : 中国动物学会鸟类学分会].

China Ornithological Society. 2008. *China Bird Report 2007*. Beijing: China Ornithological Society [中国动物学会鸟类学分会、2008. 中国观鸟年报 2007. 北京 : 中国动物学会鸟类学分会].

Clement, P., and R. Hathway. 2000. *Thrushes*. London: Christopher Helm.

Clements, J. F., T. S. Schulenberg, M. J. Iliff, T. A. Fredericks, J. A. Gerbracht, D. Lepage, S. M. Billerman, B. L. Sullivan, and C. L. Wood. 2022. *The eBird/Clements Checklist of Birds of the World*: v2022. https://www.birds.cornell.edu/clementschecklist/download/.

Collar, N. J. 2005. *Handbook of the Birds of the World*, Vol. 10, Cuckoo-shrikes to Thrushes. Barcelona: Lynx Edicions.

Couzens, D. 2019. *Identifying Birds by Behaviour*. Translated by X. He and Y. X. Cheng. Changsha: Hunan Science and Technology Press [多米尼克·卡曾斯, 2019. 鸟类行为图鉴. 何鑫、程翊欣, 译. 长沙：湖南科学技术出版社].

Craig, R. 2008. *A Field Guide to the Birds of South-East Asia*. 2nd ed. London: New Holland Publishers.

Del Hoyo, J., A. Elliott, and J. Sargatal. 1992. *Handbook of the Birds of the World*, Vol. 1, *Ostrich to Ducks*. Barcelona: Lynx Edicions.

Devvratsinh, M., and S. S. Kasam. 2017. "Breeding of White-tailed Lapwing *Vanellus leucurus* in Nal Sarovar Bird Sanctuary, Gujarat, India." *Indian BIRDS* 13 (1): 24–25.

Ding, P., Z. W. Zhang, W. Liang, and X. T. Li. 2019. *The Forest Birds of China*. Changsha: Hunan Science and Technology Press [丁平、张正旺、梁伟、等, 2019. 中国森林鸟类. 长沙：湖南科学技术出版社].

Ding, Z. F., H. F. Cao, M. Ma, B. Z. Li, and H. J. Hu. 2017. "Distribution of the Eurasian Griffon (*Gyps fulvus*) in China." *Chinese Journal of Zoology* 52 (1): 129–32 [丁志锋、曹宏芬、马鸣、等, 2017. 兀鹫在中国的分布. 动物学杂志, (52):129].

Eaton, J. A. 2018. "Identification of *Hierococcyx* Hawk Cuckoos, and the First Record of Northern Hawk Cuckoo *Hierococcyx hyperythrus* for Continental South-East Asia." *Birding Asia* 30: 68–73.

Eaton, J. A., B. van Balen, N. W. Brickle, et al. 2016. *Birds of the Indonesian Archipelago: Greater Sundas and Wallacea*. Barcelona: Lynx Edicions.

Eck, S., and J. Martens. 2006. "Systematic Notes on Asian Birds 49: A Preliminary Review of the Aegithalidae, Remizidae and Paridae." *Zoologische Mededelingen* 80 (5): 1–63.

Ganbold, D., and C. Smith. 2019. *A Field Guide to Birds of Mongolia*. Oxford: John Beaufoy Publishing.

Gao, W. 2014. *Handbook of Common Wild Birds in Beijing*. Beijing: China Machine Press [高武, 2014. 北京地区常见野鸟图鉴. 北京：机械工业出版社].

Garner, M. 2014. "Velvet, White-winged and Stejneger's Scoters Photo Guide." *Birdwatch* 260: 45–52.

Gavrilov, E. I., and A. E. Gavrilov. 2005. *The Birds of Kazakhstan*. Almaty: Tethys.

Gill, F., D. Donsker, and P. Rasmussen. 2020. IOC World Bird List (v10.2. Harrap, S., and D. Quinn. 1996. *Tits, Nuthatches and Treecreepers*. London: Christopher Helm.

Ho, G. W. 2010. "Orange-breasted Green Pigeon *Treron bicinctus* on Po Tai Island: The First Hong Kong Record and the First in China for 30 Years," In *Hong Kong Bird Report 2005–06*, ed. G. J. Carey. Hong Kong: Hong Kong Bird Watching Society.

Hsiao, M. C. 2014. *A Field Guide of Birds of Taiwan*. Taipei: Forestry Bureau, Council of Agriculture, Executive Yuan [萧木吉、2014. 台湾野鸟手绘图鉴. 台北：台湾农业委员会林务局].

Kamp, J., M. A. Koshkin, and R. D. Sheldon. 2010. "Historic Breeding of Sociable Lapwing (*Vanellus gregarius*) in Xinjiang." *Chinese Birds* 1 (1): 70–73.

Kang, Z. J., M. S. Liu, D. D. Yang, X. J. Deng, and W. F. Yin. 2014. "Six New Records of Passerine Birds in Hunan Province." *Chinese Journal of Zoology* 49 (1): 116–20 [康祖杰、刘美斯、杨道德、等, 2014. 湖南省雀形目鸟类新纪录 6 种. 动物学杂志 49 (1): 116–20].

Kennedy, R. S., P. C. Gonzales, E. C. Dickinson, et al. 2000. *A Guide to the Birds of the Philippines*. London: Oxford University Press.

König, C., F. Weick, and J. H. Becking. 2008. *Owls of the World*. 2nd ed. London: Christopher Helm.

La Touche, J. D. D. 1925–1930. *A Handbook of Birds of Eastern China*. Vol. 1. London: Taylor & Francis.

Leader, P. J. 2011. "Taxonomy of the Pacific Swift *Apus pacificus* Latham, 1802, Complex." *Bulletin of the British Ornithologists' Club* 131 (2): 81–93.

Leader, P. J., G. J. Carey, and P. I. Holt. 2013. "Species Limits within *Rhopophilus pekinensis*." *Forktail* 29: 31–36.

Lei, J. Y., L. Y. Zhang, S. Y. Zhang, and X. M. Zhu. 2012. "Latest Number of Bird Species of Hubei Province." *Sichuan Journal of Zoology* 31 (6): 145–49 [雷进宇、张立影、张叔勇、等, 2012. 湖北鸟类种数的新统计. 四川动物, (6):145–49].

Li, G. Y., B. L. Zheng, and G. Z. Liu. 1982. *Fauna Sinica. Aves*. Vol. 13. Beijing: Science Press [李桂垣、郑宝赉、刘光佐、1982. 中国动物志·鸟纲, 第十三卷. 北京：科学出版社].

Li, X. J., D. W. Ying, D. J. Gong, and T. Ma. 2019. "Mistle Thrush (*Turdus viscivorus*) Found in Jiuquan and Yuzhang, Gansu Province." *Chinese Journal of Zoology* 54 (4): 611–12 [李晓军、殷大文、龚大洁、等, 2019. 甘肃酒泉及榆中县发现槲鸫. 动物学杂志, 54 (4): 611–12].

Liang, H., X. Luo, L. Ma, and G. Chen. 2020. "The First Record of *Turdus boulboul* in Fujian Province." *Sichuan Journal of Zoology* 39 (2): 213 [梁晖、罗萧、马良、等, 2020. 福建省鸟类新纪录 — 灰翅鸫. 四川动物, 39 (2): 213].

Lin, D. L. 2014. "*Zoothera dauma*'s Taxonomy Revisited." *Nature Conservation Quarterly* 86: 62–71 [林大利, 2014. 台湾的小虎鸫：一种陌生的留鸟. 自然保育季刊, (86): 62–71].

Liu, S. M., Y. Liu, E. Jelen, et al. 2020. "Regional Drivers of Diversification in the Late Quaternary in a Widely distributed Generalist Species, the Common Pheasant *Phasianus colchicus*." *Journal of Biogeography* 47 (12): 2714–2727.

Liu, S. M., and B. Smith. 2012. "Great Stone-curlew *Esacus recurvirostris* at Mai Po NR: The First Hong Kong Record." In *Hong Kong Bird Report 2009–10*, ed. J. A. Allcock, G. J. Carey, et al. Hong Kong: Hong Kong Bird Watching Society.

Liu, Z. Y., T. C. Wang, and S. G. Liu. 2012. "The First Record of *Turdus pilaris* in Heilongjiang Province." *Sichuan Journal of Zoology* 31 (4): 667 [刘志远、王天成、刘曙光，2012. 黑龙江省鸟类新纪录 — 田鸫 . 四川动物, 31 (4): 667].

Lu, X. 2005. "Reproductive Ecology of Blackbirds (*Turdus merula maximus*) in a High-altitude Location, Tibet." *Journal of Ornithology* 146 (1): 72–78.

Lu, X. 2018. *The Birds of Tibetan Plateau of China*. Changsha: Hunan Science and Technology Press [卢欣、2018. 中国青藏高原鸟类 . 长沙 : 湖南科学技术出版社].

Ma, M. 2011. *A Checklist on the Distribution of the Birds in Xinjiang*. Beijing: Science Press [马鸣、2011. 新疆鸟类分布名录 . 北京 : 科学出版社].

Ma, Z. J., and S. H. Chen. 2018. *The Birds in the Sea and Wetlands of China*. Changsha: Hunan Science and Technology Press [马志军、陈水华、2018. 中国海洋与湿地鸟类 . 长沙 : 湖南科学技术出版社].

MacKinnon, J. R., and K. Phillipps. 2000. *A Field Guide to the Birds of China*, English ed. Oxford: Oxford University Press.

MacKinnon, J. R., K. Phillipps, and F. Q. He. 2000. *A Field Guide to the Birds of China*, Chinese ed. Changsha: Hunan Education Press [约翰·马敬能、卡伦·菲利普斯、2000. 中国鸟类野外手册 . 何芬奇，译 . 长沙 : 湖南教育出版社].

Madge, S., and H. Burn. 2013. *Crows and Jays*. London: Christopher Helm.

Mi, X. Q., K. J. Guo, J. W. Xiong, D. Dan, J. P. Wu, and X. J. Deng. 2016. "*Terpsiphone paradisi* Found in Zhada County, Tibet." *Sichuan Journal of Zoology* 35 (1): 123 [米小其、郭克疾、熊嘉武、等、2016. 西藏札达发现印度寿带 . 四川动物, 35 (1): 123].

Miao, X. L., D. J. Sai, and C. D. Liu. 2017. "New Distribution of *Zoothera citrina* in Shandong Province." *Sichuan Journal of Zoology* 36 (6): 685 [苗秀莲、赛道建、刘传栋、2017. 橙头地鸫在山东省的新分布 . 四川动物, 36 (6): 685].

Moores, N. 2002. "Probable Hybrid Red-throated Thrush *Turdus ruficollis ruficollis* and Naumann's Thrush *Turdus naumanni naumanni*." http://www.birdskorea.org/Birds/Identification/ID_Notes/BK-ID-Thrush.shtml.

Moores, N. 2007. "Selected Records from Socheong Island, South Korea." *Forktail* 23: 102–124.

Nylander, J. A. A., O. Urban, A. Per, et al. 2008. "Accounting for Phylogenetic Uncertainty in Biogeography: A Bayesian Approach to Dispersal-Vicariance Analysis of the Thrushes (Aves: *Turdus*)." *Systematic Biology* 57 (2): 257–268.

Olsen, K. M., and H. Larsson. 2003. *Gulls of Europe, Asia and North America*. London: Christopher Helm.

Oriental Bird Club Image Database. http://orientalbirdimages.org/.

Pilgrim, J. D., P. Bijlmakers, T. De Bruyn, et al. 2009. "Updates to the Distribution and Status of Birds in Vietnam." *Forktail* 25: 130–136.

Qiu, Y., B. J. Cui, F. C. Ao, T. X. Zhang, H. T. Li, Y. Liu, and Z. W. Zhang. 2009. "Nest Site Selection and Home Range of the Grey-sided Thrush." *Sichuan Journal of Zoology* 28 (4): 572–74 [邱阳、崔本杰、敖飞成、等、2009. 褐头鸫巢址选择和活动区的初步研究 . 四川动物, 28 (4): 572–74].

Rasmussen, P. C., and J. C. Anderton. 2005. *Birds of South Asia: The Ripley Guide*. Barcelona: Lynx Edicions.

Reeber, S. 2015. *Wildfowl of Europe, Asia and North America*. London: Christopher Helm.

Salzburger, W., J. Martens, A. A. Nazarenko, et al. 2002. "Phylogeography of the Eurasian Willow Tit (*Parus montanus*) Based on DNA Sequences of the Mitochondrial Cytochrome b Gene." *Molecular Phylogenetics and Evolution* 24 (1): 26–34.

Schweizer, M., C. Etzbauer, H. Shirihai, T. Töpfer, and G. M. Kirwan. 2020. "A Molecular Analysis of the Mysterious Vaurie's Nightjar *Caprimulgus centralasicus* Yields Fresh Insight into Its Taxonomic Status." *Journal of Ornithology* 161: 635–50. https://doi.org/10.1007/s10336-020-01767-8.

Schweizer, M., Y. Liu, U. Olsson, et al. 2018. "Contrasting Patterns of Diversification in Two Sister Species of Martins (Aves: Hirundinidae): The Sand Martin, *Riparia riparia*, and the Pale Martin, *R. diluta*." *Molecular Phylogenetics and Evolution* 125: 116–126.

Severinghaus, L. L., T. S. Ding, W. H. Fang, W. H. Lin, M. C. Tsai, and C. W. Yen. 2012. *The Avifauna of Taiwan*, 2nd ed. Taipei: Forest Bureau, Council of Agriculture [刘小如、丁宗苏、方伟宏、等、2012. 台湾鸟类志 (上、中、下) 第二版 . 台中 : 台湾农业委员会林务局].

Sibley, D. A. 2000. *The Sibley Guide to Birds*. New York: Alfred A. Knopf.

Simpson, K., and N. Day, eds. 2010. *Birds of Australia*, 8th ed. Princeton: Princeton University Press.

Song, Y., and C. Wen. 2016. *A Photographic Guide of the Birds of China*, Raptor Version. Fuzhou: The Straits Publishing and Distributing Group [宋晔、闻丞、2016. 中国鸟类图鉴 : 猛禽版 . 福州 : 海峡书局].

Svensson, L., K. Mullarney, and D. Zetterström. 2022. *Collins Bird Guide: The Most Complete Field Guide to the Birds of Britain and Europe*. 3rd ed. London: HarperCollins Publishers.

Ujihara, O., and M. Ujihara. 2015. *An Identification Guide to the Ducks of Japan*. Tokyo: Seibundo Shinkosha Publishing Co. Ltd. [氏原巨雄、氏原道昭、2015. 決定版日本のカモ識別図鑑 . 東京 : 誠文堂新光社].

Ujihara, O., and M. Ujihara. 2019. *An Identification Guide to the Gulls of Japan*. Tokyo: Seibundo Shinkosha Publishing Co. Ltd. [氏原巨她、氏原道昭、2019. 決定版日本のカモメ識別図鑑 . 东京 : 誠文堂新光社].

Vaurie, C. 1960. "Systematic Notes on Palearctic Birds, No. 39: Caprimulgidae: A New Species of *Caprimulgus*." *American Museum Novitates* 1985: 1–10.

Wang, C. H., S. H. Wu, H. Y. Yang, et al. 1991. *A Field Guide to the Wild Birds of Taiwan*. Taiwan Wild Bird Information Centre, Taipei, and Wild Bird Society of Japan, Tokyo [王嘉雄、吴森雄、黄光瀛、等、1991. 台湾野鸟图鉴 . 台北 : 台湾野鸟资讯社].

Wang, H. T., Y. L. Jiang, and W. Gao. 2010. "Jankowski's Bunting (*Emberiza jankowskii*): Current Status and Conservation." *Chinese Birds* 1 (4): 251–58.

Wang, H. Y., B. X. Wu, J. J. Xiang, C. L. Liao, Z. Q. Zhang, and D. D. Yang. 2016. "Long-tailed Thrush (*Zoothera dixoni*) Found in Badagongshan, Hunan." *Chinese Journal of Zoology* 51 (6): 1105 [王海燕、吴炳贤、向建军、等、2016. 湖南八大公山发现长尾地鸫 . 动物学杂志、51 (6): 1105].

Wang, Q. S., M. Ma, and Y. R. Gao. 2006. *Fauna Sinica. Aves*. Vol. 5. Beijing: Science Press [王岐山、马鸣、高育仁、2006. 中国动物志 · 鸟纲、第五卷 . 北京 : 科学出版社].

Welch, G. 2015. "Whistling Green Pigeon *Treron formosae* of Ryukyu Islands ssp. *Permagnus medioximus* on Po Tai Island: The First Hong Kong Record." In *Hong Kong Bird Report 2013*, ed. G. Welch and G. Chow. Hong Kong: Hong Kong Bird Watching Society.

Wildlife Division, Nature Conservation Bureau, Ministry of the Environment of Japan. 2016. *Japanese Night Heron Protection Plan*. [2016-06-08]. https://www.env.go.jp/content/900506754.pdf [日本環境省自然環境局野生生物課 . ミゾゴイ保護の進め方 . (2016-06-08). http://www.env.go.jp/press/files/jp/103126.pdf].

xeno-canto: Sharing Wildlife Sounds from Around the World. https://www.xeno-canto.org/.

Xia, C. G., and L. Luo. 2019. "The First Record of *Turdus cardis* in Shanxi Province." *Sichuan Journal of Zoology* 38 (5): 552 [夏川广、罗磊、2019. 陕西省鸟类新记录 — 乌灰鸫 . 四川动物、38 (5): 552].

Xia, C.W., W. Liang, G. J. Carey, et al. 2016. "Song Characteristics of Oriental Cuckoo *Cuculus optatus* and Himalayan Cuckoo *Cuculus saturatus* and Implications for Distribution and Taxonomy." *Zoological Studies* 55: 1–9.

Xing, L. L., G. S. Yang, and M. Ma. 2020. *The Birds in the Grasslands and Deserts of China*. Changsha: Hunan Science and Technology Press [邢莲莲、杨贵生、马鸣、2020. 中国草原与荒漠鸟类 . 长沙 : 湖南科学技术出版社].

Yang, L., and X. J. Yang. 1994. *The Avifauna of Yunnan China*. Vol. 1, *Non-Passeriformes*. Kunming: Yunnan Science and Technology Press [杨岚、杨晓君、1994. 云南鸟类志 (上卷 : 雀形目). 云南 : 云南科技出版社].

Yang, L., and X. J. Yang. 2004. *The Avifauna of Yunnan China*. Vol. 2, *Passeriformes*. Kunming: Yunnan Science and Technology Press [杨岚、杨晓君、2004. 云南鸟类志 (下卷 : 雀形目). 昆明 : 云南科技出版社].

Yu, L. J., A. W. Jiang, and F. Zhou. 2015. "Update on the Distribution Range of the White-browed Crake (*Porzana cinerea*): A New Record from Mainland China." *Zoological Research* 36 (1): 59–61.

Zhang, L., and M. Zhang. 2018. *A Photographic Guide of the Birds of China*. Vol. 2, *Shorebirds*. Fuzhou: The Straits Publishing and Distributing Group [章麟、张明、2018. 中国鸟类图鉴 : 鸻鹬版 . 福州 : 海峡书局].

Zhang, L. X., X. L. Zeng, Y. L. Du, J. Wang, and Q. Zhang. 2019. "White-winged Duck *Asarcornis scutulata* Found in Yingjiang, Yunnan." *Chinese Journal of Zoology* 54 (6): 902 [张利祥、曾祥乐、杜银磊、等、2019. 云南盈江发现白翅栖鸭 . 动物学杂志、54 (6): 902].

Zhang, R. 1999. *Zoogeography of China*. Beijing: Science Press [张荣祖、1999. 中国动物地理 . 北京 : 科学出版社].

Zhao, X. R. 2018. *A Photographic Guide to the Birds of China*. Beijing: The Commercial Press [赵欣如、2018. 中国鸟类图鉴 . 北京 : 商务印书馆].

Zhao, Z. J. 2001. *A Handbook of the Birds of China (Non-Passerine)*. Changchun: Jilin Science and Technology Publishing House [赵正阶、2001. 中国鸟类志 (上卷 : 非雀形目). 长春 : 吉林科学技术出版社].

Zhao, Z. J. 2001. *A Handbook of the Birds of China (Passerine)*. Changchun: Jilin Science and Technology Publishing House [赵正阶、2001. 中国鸟类志 (下卷 : 雀形目). 长春 : 吉林科学技术出版社].

Zheng, B. L. 1985. *Fauna Sinica. Aves*. Vol. 8. Beijing: Science Press [郑宝赉、1985. 中国动物志 · 鸟纲 — 第八卷 . 北京 : 科学出版社].

Zheng, G. M. 2018. *A Checklist on the Classification and Distribution of the Birds of China*, 3rd ed. Beijing: Science Press [郑光美、2018. 中国鸟类分类与分布名录 (第三版). 北京 : 科学出版社].

Zhu, L., W. G. Zhao, K. Yang, W. P. Lei, and Y. H. Sun. 2014. "The Spring and Summer Distribution of Pine Grosbeak (*Pinicola enucleator*) in China." *Chinese Journal of Zoology* 49 (1): 121–25 [朱磊、赵文阁、杨琨、等、2014. 松雀在中国的春夏季分布 . 动物学杂志、49 (1): 121–25].

Index of Common Names

659

Index of Scientific Names

p. 30	p. 36	p. 38	p. 52	p. 54	p. 62	p. 68	p. 78
p. 80	p. 83	p. 84	p. 84	p. 88	p. 90	p. 91	p. 92
p. 96	p. 98	p. 98	p. 100	p. 110	p. 110	p. 112	p. 114
p. 115	p. 118	p. 124	p. 132	p. 136	p. 138	p. 146	p. 152
p. 152	p. 154	p. 154	p. 156	p. 162	p. 164	p. 176	p. 186
p. 188	p. 196	p. 202	p. 204	p. 206	p. 208	p. 212	p. 220
p. 222	p. 228	p. 229	p. 240	p. 244	p. 250	p. 252	p. 253
p. 256	p. 258	p. 260	p. 261	p. 262	p. 264	p. 266	p. 278

p. 284	p. 288	p. 289	p. 292	p. 293	p. 294	p. 295	p. 300
p. 306	p. 308	p. 312	p. 316	p. 317	p. 326	p. 334	p. 335
p. 338	p. 346	p. 347	p. 348	p. 355	p. 365	p. 370	p. 373
p. 380	p. 384	p. 402	p. 408	p. 417	p. 423	p. 434	p. 441
p. 443	p. 454	p. 469	p. 474	p. 485	p. 490	p. 491	p. 492
p. 492	p. 493	p. 497	p. 497	p. 501	p. 509	p. 526	p. 544
p. 552	p. 558	p. 564	p. 565	p. 566	p. 569	p. 574	p. 579
p. 581	p. 584	p. 588	p. 592	p. 598	p. 600	p. 604	p. 623